J. Yamada · T. Sugimoto (Eds.)

TTF Chemistry

Kodansha

Springer

Berlin
Heidelberg
New York
Hong Kong
London
Milan
Paris
Tokyo

Jun-ichi Yamada, Toyonari Sugimoto (Eds.)

TTF Chemistry

Fundamentals and Applications of Tetrathiafulvalene

With 262 Figures, 196 Schemes, 75 Charts and 47 Tables

Kodansha

Springer

Professor Jun-ichi Yamada
Department of Material Science,
Graduate School of Science,
Himeji Institute of Technology
Hyogo 678-1297
Japan
E-mail: yamada@sci.himeji-tech.ac.jp

Professor Toyonari Sugimoto
Research Institute for
Advanced Science and Technology,
Osaka Prefecture University
Osaka 599-8570
Japan
E-mail: toyonari@riast.osakafu-u.ac.jp

ISBN 4-06-211164-0 Kodansha Ltd., Tokyo
ISBN 3-540-21004-0 Springer-Verlag Berlin Heidelberg New York

Library of Congress Cataloging-in-Publication Data
TTF Chemistry / Jun-ichi Yamada, Toyonari Sugimoto (eds.)
ISBN 3540210040 (hc : alk. paper) – ISBN 4062111640
1. TTF Chemistry. I Yamada, Jun-ichi, 1957-   II Sugimoto, Toyonari, 1945-   III Series
QD716.P45 P44 2002   541.3'95-dc21   2002072789

Springer-Verlag is a part of Springer Science+Business Media

http://www.springeronline.com

Cover: Künkel + Lopka, Heidelberg/Design and Production, Heidelberg
Printed on acid-free paper   2/3020mh   5 4 3 2 1 0

List of Contributors

Numbers in parentheses refer to the chapters.

Becher, Jan (15) Department of Chemistry, University of Southern Denmark, DK-5230 Odense M, Denmark

Bryce, Martin R. (17) Department of Chemistry, University of Durham, Durham, DH1 3LE, UK

Fabre, Jean-Marc (1) Hétérochimie et Matériaux Organiques, ENSCM/ESA-5076, France

Garín, Javier (16) Departamento de Química Orgánica, ICMA, Universidad de Zaragoza-CSIC, E-50009 Zaragoza, Spain

Imakubo, Tatsuro (3) RIKEN, Saitama 351-0198, Japan

Iyoda, Masahiko (8) Department of Chemistry, Graduate School of Science, Tokyo Metropolitan University, Tokyo 192-0397, Japan

Izuoka, Akira (6) Department of Materials and Biological Science, Faculty of Science, Ibaraki University, Ibaraki 310-8512, Japan

Jeppesen, Jan O. (15) Department of Chemistry, University of Southern Denmark, DK-5230 Odense M, Denmark

Kikuchi, Koichi (11) Department of Chemistry, Graduate School of Science, Tokyo Metropolitan University, Tokyo 192-0397, Japan

Martín, Nazario (16) Departamento de Química Orgánica, Facultad de Química, Universidad Complutense, E-28040 Madrid, Spain

Misaki, Yohji (10, 14) Department of Molecular Engineering, Graduate School of Engineering, Kyoto University, Kyoto 615-8510, Japan

Nakazaki, Jotaro (6) Department of Chemistry, Graduate School of Arts and Sciences, The University of Tokyo, Tokyo 153-8902, Japan

Nishikawa, Hiroyuki (11) Department of Chemistry, Graduate School of Science, Tokyo Metropolitan University, Tokyo 192-0397, Japan

Otsubo, Tetsuo (5, 9) Department of Applied Chemistry, Graduate School of Engineering,

Hiroshima University, Higashi-Hiroshima 739-8527, Japan

Papavassiliou, George C. (2) Theoretical and Physical Chemistry Institute, National Hellenic Research Foundation, Athens 116/35, Greece

Sugimoto, Toyonari (7, 14) Research Institute for Advanced Science and Technology, Osaka Prefecture University, Osaka 599-8570, Japan

Takahashi, Kazuko (13) Center for Interdisciplinary Research, Tohoku University, Sendai 980-8578, Japan

Takimiya, Kazuo (5, 9) Department of Applied Chemistry, Graduate School of Engineering, Hiroshima University, Higashi-Hiroshima 739-8527, Japan

Yamada, Jun-ichi (11, 14) Department of Material Science, Graduate School of Science, Himeji Institute of Technology, Hyogo 678-1297, Japan

Yamashita, Yoshiro (12) Department of Electronic Chemistry, Interdisciplinary Graduate School of Science and Engineering, Tokyo Institute of Technology, Yokohama 226-8502, Japan

Yamochi, Hideki (4) Division of Chemistry, Graduate School of Science, Kyoto University, Kyoto 606-8502, Japan

Foreword

Large streams from little fountains flow
Tall oaks from little acorns grow—David Everett

In this volume, the very patient and gifted editors, Jun-ichi Yamada and Toyonari Sugimoto, have achieved a remarkable collection of fundamental aspects of "TTF Science". This book is also a testament to the enormous growth of TTF chemistry alone from 1973 until 2002. There are 17 chapters dedicated, *inter alia*, to the synthesis of symmetric, unsymmetric, heteroatom-substituted, halogenated, stable radical-substituted, dimeric (simple, fused or cyclophane-type), oligomeric (linear, macrocyclic and dendritic) and TTF-acceptor linked molecules for special applications. With the exception of Chapter 16 (rectifiers, photovoltaics, nonlinear optics), the emphasis of the book is strictly on chemistry in general and synthesis in particular. To badly quote the David Rueben book (and Woody Allen's movie), this remarkable TTF Chemistry book you are holding in your hands contains "everything you always wanted to know about TTF chemistry but [were] afraid to ask". This makes this monograph unique in the field because in the past, compendia dedicated to TTF science have always had a large physics component.

An organic metal? An organic superconductor? An organic solar cell? Any scientist or engineer living in the early part of the 20th century, up to the 1960's would have scoffed at the idea. But, starting in the 1970's, a special, small sulfur heterocycle was the gateway to organic metals, superconductors and semiconductors and will very likely have a role to play in organic ferromagnets (see Chapter 6 by J. Nakazaki and A. Izuoka) and organic solar cells (see Chapter 16 by N. Martín and J. Garín).

The radical cation TTF$^{(+\cdot)}$ was first published under its official IUPAC name, bis-1,3-dithiolium "BDT$^{(+\cdot)}$" in 1970 but derivatives of its neutral form had already been prepared by Prinzbach, Takamizawa and Hünig. In the meantime, Cowan had been exploring the properties of fulvalene-iron complexes and their mixed valence properties so that when the Johns Hopkins group published their first paper on the temperature dependence of the conductivity of the TCNQ salt of bis-1,3-dithiolium radical ion, they labeled it (TTF)(TCNQ) [rather than (BDT)(TCNQ)] and, except for Chapter 13 by K. Takahashi, where the old name needed to be resurrected, the less "legal" name TTF stuck for perpetuity. Hünig and Coffen had also made the heterocycle but did not give it an unusual name. The fact that TTFCl$_n$ was discovered to be a highly conducting solid led to the formation of the first organic metal in 1973 as well as a, never substantiated, claim for superconducting fluctuations above 57 K in this solid in the same year. The last two publications were the catalyst which initiated the growth of interest in TTF in the chemistry and physics communities.

One could trace the study of all modern organic conductors, including conducting polymers, back to the principles learnt from TTF and its salts. Many of today's "players" in conducting polymers were either trained with TTF-based systems or developed TTF-based semiconductors, metals and superconductors before extending their work to conducting polymers.

It is clear that the chemistry of TTF is a vibrant and active part of the organic chemistry scene. It has sparked the imagination of that rare breed of organic chemists, heterocyclic chemists and sulfur chemists who are interested not only in the synthesis of a molecule but also in its applications. In practically every issue of the most important chemistry journals today one can find a paper containing some aspect of TTF-based science. One of the most recently fashionable trends in organic materials science research is that of "molecular machines" and molecular-based computers. Here TTF is also playing an important role [see C. P. Collier, J. O. Jeppesen, Y. Luo, J. Perkins, E. W. Wong, J. R. Heath and J. F. Stoddart, *J. Am. Chem. Soc.*, **123**, 12632 (2001)].

Fred Wudl

Department of Chemistry and Biochemistry,
University of California,
Los Angeles, CA 90095-1569, USA

Preface

The synthesis of tetrathiafulvalene (TTF) independently reported by the three research groups of Wudl, Coffen and Hünig during the period 1970–1971 inaugurated the TTF era, and TTF chemistry has been undergoing remarkable development. The first great milestone was marked in 1973 by the discovery of metallic behavior in the charge-transfer (CT) complex composed of TTF and tetracyanoquinodimetane (TCNQ). This discovery stimulated work on new donor compounds related to TTF, and, simultaneously, considerable attention was focused on the field of organic conductors, which has its origin in the discovery of the conducting perylene-bromine complex by Akamatsu, Inokuchi and Matsunaga in 1954. The next two striking milestones occurred with the discovery of superconductivity in the CT salt of tetramethyltetraselenafulvalene (TMTSF) in 1980, followed by that in the CT salt of bis(ethylenedithio)tetrathiafulvalene (BEDT-TTF) in 1983. Needless to say, the TMTSF and BEDT-TTF donors belong to the TTF family. Since then, great momentum has been sustained by organic chemists, inorganic chemists, physical chemists, solid-state physicists, theoretical physicists and materials scientists. A large number of metallic and superconducting organic materials are now known, and new TTFs, TTF-type compounds and their CT materials continue to be reported week by week. Recently, attention has also been directed to the novel magnetic and optical properties that CT materials based on TTFs and related compounds can display.

The study of design and synthesis of new TTF-based systems to provide organic materials with more interesting and exciting solid-state properties is an obvious first step. However, no single volume has so far compiled review articles pertaining to various aspects of TTF chemistry authored by prominent organic chemists, although some collections of data and a number of individual review articles have been published. The aim of this volume is to highlight the chemistry of functionalized TTFs (Part I), dimeric TTFs (Part II) and 1,3-dithiol-2-ylidene donors (Part III) and the applications of TTFs from the molecular level to the supra- and macromolecular levels (Part IV), which will be very useful to specialists and nonspecialists, with the aim of interpreting current advances in TTF chemistry. We hope that this volume will serve not only as a reference work for a broad audience, but also as a review citing much of the extensive literature related to this field.

We are grateful to all the invited contributors who readily consented to write chapters of this book, and are convinced that every chapter will make a contribution as a guideline to further progress in this interdisciplinary and important field of modern materials chemistry. We sincerely thank Professor Fred Wudl who, as a pioneer of TTF chemistry, kindly contributed the foreword to this book. Thanks are due to Mr. Ippei Ohta of Kodansha Scientific Ltd. for very good cooperation in the editing this book.

We are most thankful to all those who have contributed their time and effort to the work cited in this book. Finally, generous financial support by Grant-in-Aid for Publication of Scientific Research Results from Japan Society for the Promotion of Science is gratefully acknowledged.

September 2003

Jun-ichi Yamada
Toyonari Sugimoto

Contents

II Dimeric TTFs

III   1,3-Dithiol-2-ylidene Donors

IV   Applications of TTFs

I Functionalized TTFs

1

Synthetic Approaches to Unsymmetrical, Dimeric and Oligomeric TTFs

1.1 Introduction

Interest in synthetic chemistry of tetrathiafulvalene (TTF, Fig. 1.1)[1,2] began with the discovery of the first organic metal TTF-TCNQ (TCNQ = tetracyanoquinodimethane) in 1973,[3] and led to the production of superconducting salts based on symmetrical and unsymmetrical TTF-type donors containing the selenium (TMTSF), selenium/sulfur (DMET) and sulfur atoms (BEDT-TTF and MDT-TTF).[4–7] Since then, the subsequent discovery in the physical properties (the electrical, optical and magnetic properties) of charge-transfer (CT) materials incorporating tetrachalcogenafulvalenes (TCFs) has stimulated an intense activity in the search for a novel series of TCF donors.[8–11] During the last decade, further activity intensified by the use of TTF derivatives as building blocks for macromolecular and supramolecular architectures[8,12,13] has resulted in the synthesis of more complicated TTF systems, such as functionalized TTFs,[11,14,15] oligomeric TTFs,[16,17] TTF phanes[18] and dendritic TTFs.[19]

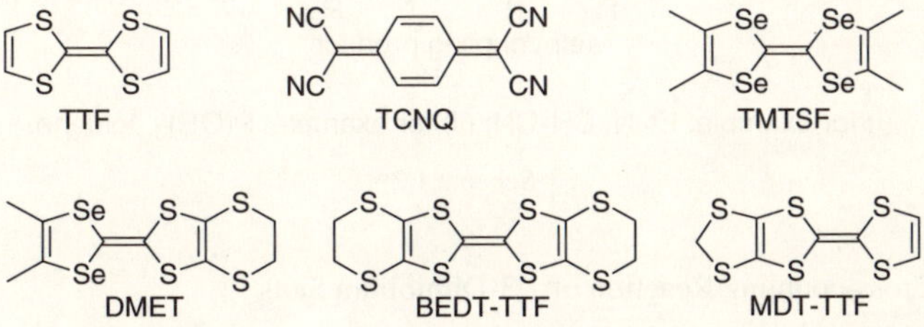

Fig. 1.1

    In this chapter, the main strategies reported to date for constructing unsymmetrical TTFs, dimeric and oligomeric TTFs, particularly including functionality for chemical transformation, are reviewed.

1.2 Synthesis of Functionalized Unsymmetrical TTFs *via* Condensation of Two 1,3-Dichalcogenole Rings

1.2.1 Nonselective Strategy: Cross-coupling Reactions

This strategy has been described in many review articles[1,2,6,10] and is still one of the most useful

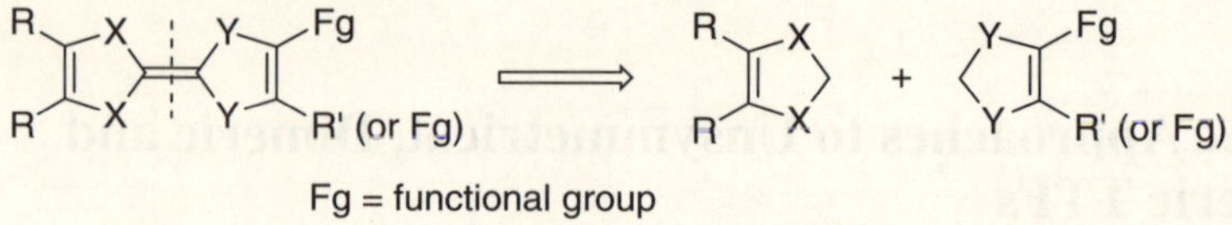

Fg = functional group

Scheme 1.1

strategies for constructing unsymmetrical TTFs, which are the most numerous compounds in the TTF family so far reported. As shown in Scheme 1.2, this strategy consists of mainly two different methods: the cross-coupling reaction of two different 1,3-dithiolium salts in the presence of a basic medium, *e.g.*, a tertiary amine in CH_3CN (Method A), and the trivalent phosphorous mediated cross-coupling of two different 1,3-dichalcogenole-2-chalcogenones (Method B).[1,2] The main drawback of this strategy is the concomitant formation of two symmetrical byproducts, so that the chromatographic separation of the desired unsymmetrical TTF donor from the resulting mixture is essential. However, the separation sometimes meets with difficulties.

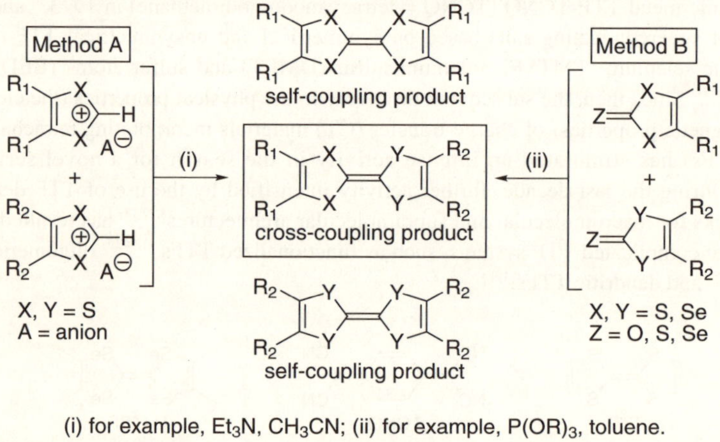

(i) for example, Et_3N, CH_3CN; (ii) for example, $P(OR)_3$, toluene.

Scheme 1.2

A. Method A: Cross-coupling Reaction of 1,3-Dithiolium Salts

Although this method is not applicable to the air-sensitive 1,3-diselenolium salts which easily decompose in a basic medium to generate elemental selenium,[2] it has been occasionally used for the synthesis of TTFs from the beginning of the 1970's.[20,21] When this method involves the condensation of two different dithiolium salts, a mixture of two symmetrical and one unsymmetrical TTFs is generally obtained. It is appreciated that this condensation reaction proceeds *via* the generation of a dithiolium carbene by deprotonation of a dithiolium salt followed by dimerization with another carbene or followed by reaction with the remaining dithiolium salt and subsequent deprotonation of the intermediate adduct (Scheme 1.3). When the polarities of the three TTFs obtained as a mixture are too close, it is difficult to separate the desired unsymmetrical TTF donor from the two symmetrical byproducts by column chromatography. For example, the low yields were obtained in the synthesis of EDT-TTF (20%) and its dimethyl analogue EDT-DMTTF (18%).[22] In addition, this difficulty becomes more serious when the reactivities of the two dithiolium salts used are quite different from each other. As shown in Scheme 1.4, the functionalized

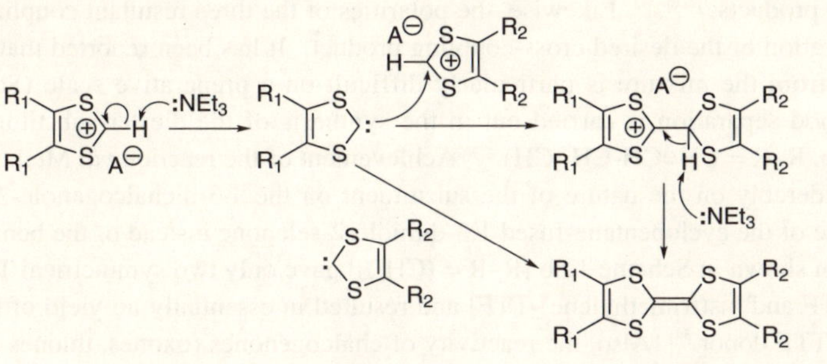

Scheme 1.3

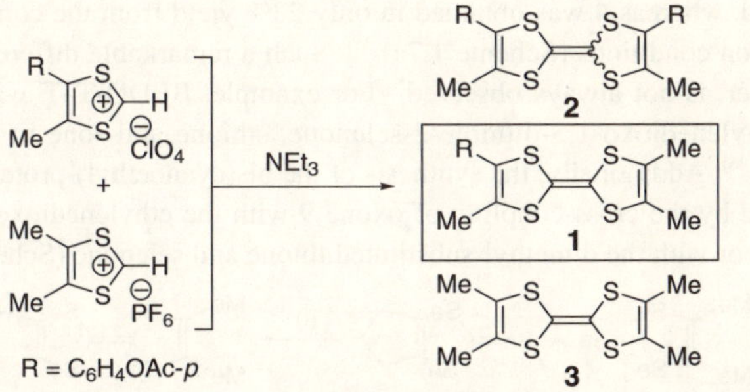

R = C₆H₄OAc-*p*

Scheme 1.4

unsymmetrical TTF derivative **1** was obtained in only 15% yield together with symmetrical TTFs **2** (12%) and **3** (31%).[23]   In the synthesis of **1**, TMTTF **3**, which is one of the self-coupling byproducts, was mainly formed then isolated as a major product, indicating the superior reactivity of the corresponding dimethyl-substituted dithiolium salt.   Because of these difficulties, the utilization of Method A has grown less and less, except for the synthesis of a few new functionalized symmetrical and unsymmetrical TTFs by this method (Schemes 1.5a,b).[24,25]

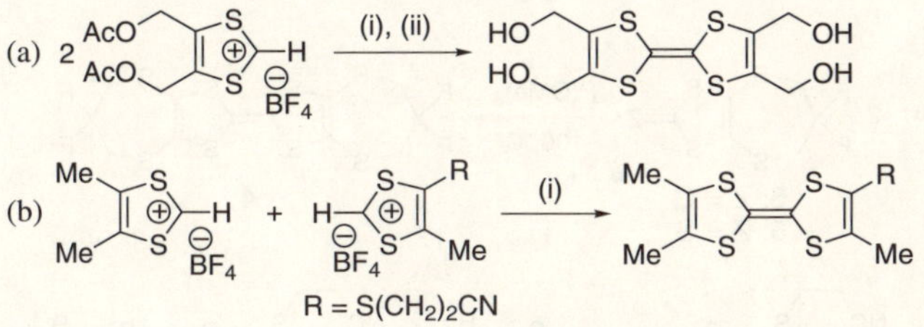

(i) NEt₃, CH₃CN; (ii) KOH, EtOH, reflux.

Scheme 1.5

B.   Method B: Cross-coupling Reaction of 1,3-Dichalcogenole-2-chalcogenones

The phosphine- or phosphite-mediated coupling of 1,3-dichalcogenole-2-thiones, -selenones or -ones (Scheme 1.2) has been the most widely used for the synthesis of unsymmetrical TTFs,[10] but has also encountered the same disadvantage as Method A, *i.e.*, the formation of two symmetrical self-coupling products.[1,2,26–31] Likewise, the polarities of the three resultant coupling products affect the separation of the desired cross-coupling product. It has been reported that the separation of DMtTSF from the mixture is particularly difficult on a preparative scale (Scheme 1.6a),[32] whereas a good separation is carried out in the synthesis of the diester-substituted benzo-TTF (Scheme 1.6b, R–R = CH=CH-CH=CH).[33] Achievement of the reaction *via* Method B appears to depend considerably on the nature of the substituent on the 1,3-dichalcogenole-2-chalcogenone used. The use of the cyclopentane-fused 1,3-dithiole-2-selenone instead of the benzene-fused one in the reaction shown in Scheme 1.6b [R–R = $(CH_2)_3$] gave only two symmetrical TTF derivatives [tetraester-TTF and bis(trimethylene)-TTF] and resulted in essentially no yield of the desired unsymmetrical TTF donor.[34] Also, the reactivity of chalcogenones (oxones, thiones and selenones) sometimes has a great influence on the yield of the self- or cross-coupling product. As illustrated by a recent synthesis of the fluorinated symmetrical TTF donor **4**, the self-coupling of oxone **5** gave **4** in 91% yield, whereas **4** was obtained in only 23% yield from the corresponding thione **6** under similar reaction conditions (Scheme 1.7a).[35] Such a remarkable difference in the yields of the product, however, is not always observed. For example, BEDO-TTF was obtained by self-coupling of 4,5-ethylenedioxo-1,3-dithiole-2-selenone, -thione and -one in 35%, 15% and 5% yields, respectively.[36] Additionally, the synthesis of the bis(cyanoethyl)-protected TTF derivative **7** or **8** was achieved by the cross-coupling of oxone **9** with the ethylenedioxo-substituted thione, selenone and oxone or with the dimethyl-substituted thione and selenone (Scheme 1.7b).[37]

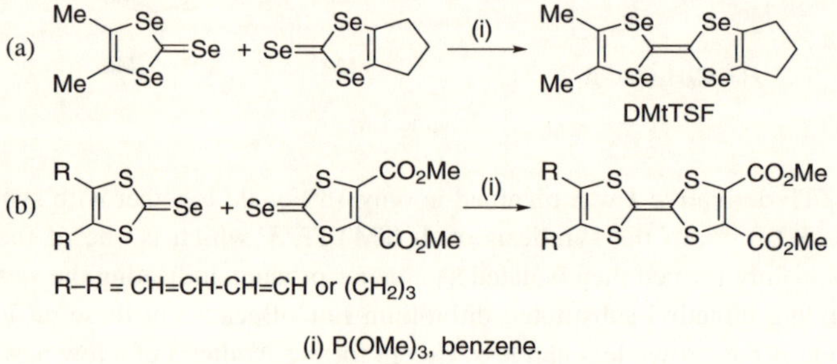

R–R = CH=CH-CH=CH or $(CH_2)_3$

(i) $P(OMe)_3$, benzene.

Scheme 1.6

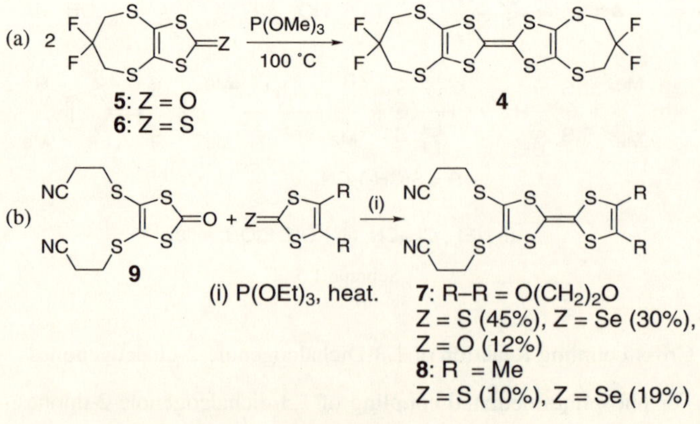

5: Z = O
6: Z = S

4

9

(i) $P(OEt)_3$, heat.   **7**: R–R = $O(CH_2)_2O$
Z = S (45%), Z = Se (30%),
Z = O (12%)
8: R = Me
Z = S (10%), Z = Se (19%)

Scheme 1.7

The synthetic examples shown in Schemes 1.7a,b clearly indicate that, in this type of coupling reaction, there is no guideline as to what kind of chalcogenone should be used to obtain the desired TTF donor in the highest yield. However, it has been demonstrated that the yield of an unsymmetrical TTF donor can be significantly enhanced when a derivative of oxone is coupled with a different derivative of thione rather than the corresponding oxone.[38] This result may be attributed to the fact that the thione, compared to the oxone, is less active to the nucleophilic attack on the carbon atom of the C=S bond and more favorable for the formation of a reactive ylid (for example, see ylid **A** in Scheme 1.8). Nevertheless, it is still very difficult to predict what combination of chalcogenones (oxone/oxone, oxone/thione, oxone/selenone, *etc.*) will give the best yield of the desired product.

Although the mechanistic details of this coupling reaction have not yet been studied, several postulated mechanisms have been proposed in specific cases,[39–41] all of which invoke the initial chalcogenophilic attack of the trialkyl phosphite reagent [P(OR)$_3$] at the C=X (X = chalcogen) bond to give an ylid. More recently, a mechanistic basis has been given on the formation of the diester-substituted unsymmetrical TTF donor **10** by the P(OEt)$_3$-mediated cross-coupling between two different thiones (Scheme 1.8).[42] This study showed that the temperature, the concentration and the nature of a phosphorous reagent play a crucial role in achievement of the reaction, and also that P(OEt)$_3$ more easily reacts with thione **11** having two electron-withdrawing ester functions than thione **12** with two appended electron-donating methylthio groups. As illustrated in Scheme 1.8, the ylid intermediate **A** derived from thione **11** is generated as an equilibrium mix-

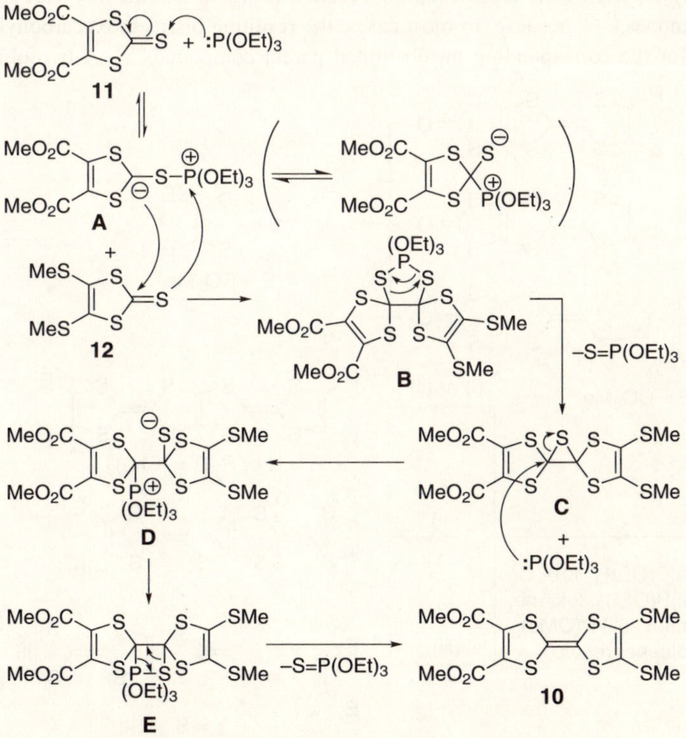

Scheme 1.8

ture, which then reacts with thione **12** to allow the successive formation of the adducts **B** and **C** *via* the carbophilic attack of the carbanion of **A** on the C=S bond of thione **12** followed by loss of S=P(OEt)$_3$. Subsequent nucleophilic attack of a second P(OEt)$_3$ to **C** leads to the intermediates **D** and **E**, the latter of which gives rise to the unsymmetrical TTF donor **10** after the second loss of S=P(OEt)$_3$.

1.2.2 Recent Applications of Method B: Synthesis of Functionalized Unsymmetrical TTFs, TSFs and Bis-fused TTF-type Donors

During the last decade, the P(OR)$_3$-mediated cross-coupling of two different chalcogenones has been intensively carried out to obtain unsymmetrical TTFs and tetraselenafulvalenes (TSFs) with functional groups, such as CO$_2$R, CN, halogen and carbonyl- and hydroxy-protecting groups (see also Schemes 1.6b and 1.7b), which are generally easy to separate from the concomitant self-coupling products by column chromatography due to their different polarities from those of the self-coupling products. Linkage patterns of these functional groups are classified as follows: (i) direct linkage to the TTF core (TTF-Fg, Type I), (ii) linkage through an alkylene chain (or a cyclic alkylene spacer) [TTF-(CH$_2$)$_n$-Fg, Type II] and (iii) linkage through an alkylenechalcogeno chain (or a cyclic alkylenechalcogeno spacer) [TTF-X-(CH$_2$)$_n$-Fg; X = chalcogen, Type III].

A. Synthesis of TTF-CO$_2$Me (Type I)

There are many reports on the cross-coupling reaction using 4,5-bis(methoxycarbonyl)-1,3-dithi-ole-2-chalcogenones,[43–47] because, in most cases, the resulting bis(methoxycarbonyl)-TTFs serve as precursors for the corresponding unsubstituted parent compounds.[45] A twofold decarboxyla

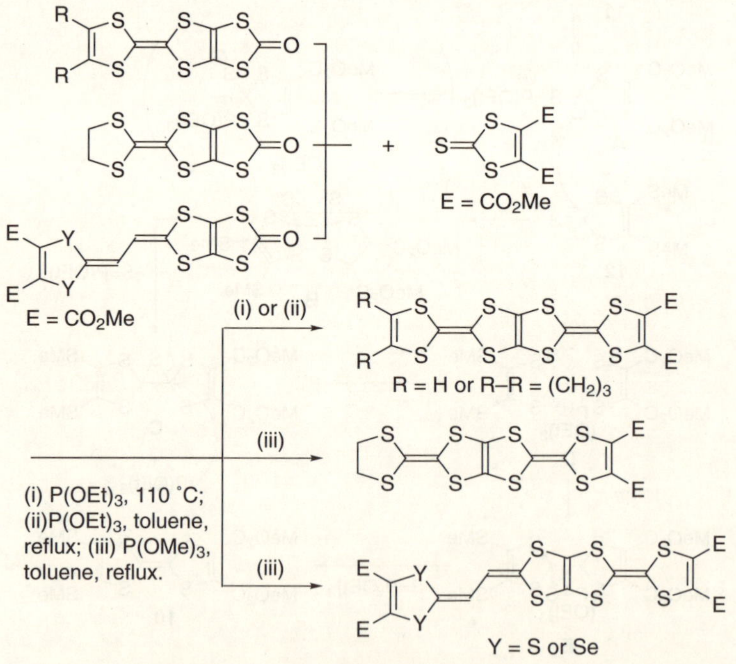

Scheme 1.9

tion is generally carried out by heating the diester-substituted TTFs at around 110–150 °C in the presence of LiBr·H_2O in HMPA (or DMF).[22,43,48] When the heating temperature is lower (70–80 °C), the monoester-substituted TTFs are isolated.[44,48] This cross-coupling reaction was also applied to the synthesis of bis-fused TTF-type donors containing one tetrathiapentalene (TTP) unit and two or four ester groups (Scheme 1.9).[49–52] These compounds then underwent decarbonylation to give the corresponding unsubstituted bis-fused TTF-type donors, which have been found to be efficient donor components in the formation of organic metals.

B. Synthesis of Types I–III

As exemplified in Schemes 1.10a,b, the diester-attached TTFs also serve not only as key compounds leading to the bis-functionalized TTFs carrying, in particular, the hydroxy- and halogenomethyl groups, but also as a source of the mono-functionalized TTFs containing, for example, carboxylic acid, acid chloride and amide.[43,44,46,47] Other functionalized unsymmetrical TTFs with a variety of substituents, such as alcohol (Schemes 1.11a,b),[15,53–56] formyl,[57] ester (which can be

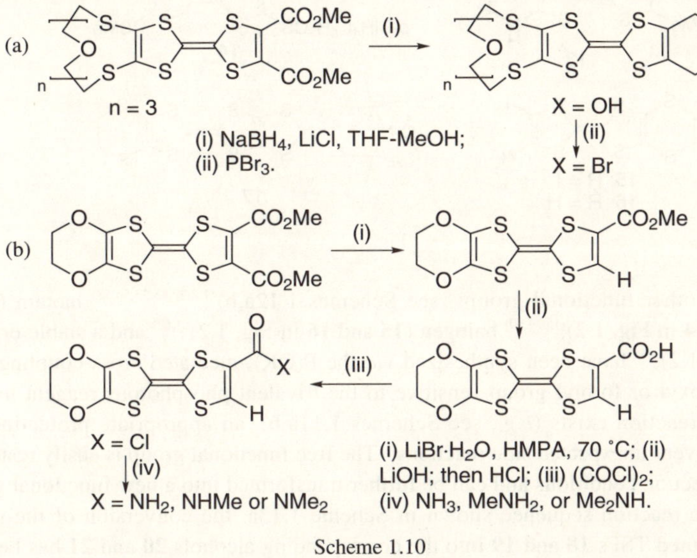
(a)
CO2Me
CO2Me
n
n = 3
(i)
X = OH
(ii)
X = Br
(i) NaBH4, LiCl, THF-MeOH;
(ii) PBr3.
(b)
CO2Me
CO2Me
(i)
CO2Me
H
(ii)
CO2H
(iii)
X = Cl
(iv)
X = NH2, NHMe or NMe2
(i) LiBr·H2O, HMPA, 70 ˚C; (ii)
LiOH; then HCl; (iii) (COCl)2;
(iv) NH3, MeNH2, or Me2NH.

Scheme 1.10

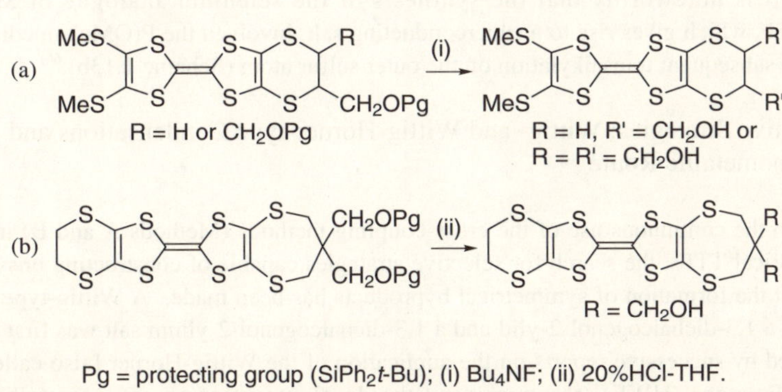
(a)
MeS
MeS
R
CH2OPg
R = H or CH2OPg
(i)
R
R'
R = H, R' = CH2OH or
R = R' = CH2OH
(b)
CH2OPg
CH2OPg
(ii)
R
R
R = CH2OH
Pg = protecting group (SiPh2t-Bu); (i) Bu4NF; (ii) 20%HCl-THF.

Scheme 1.11

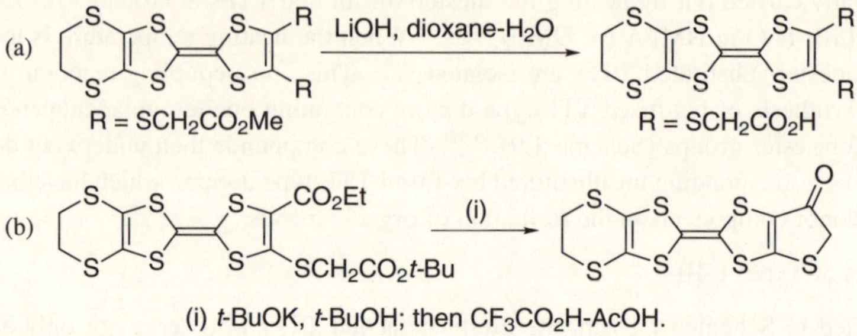

(i) t-BuOK, t-BuOH; then CF_3CO_2H-AcOH.

Scheme 1.12

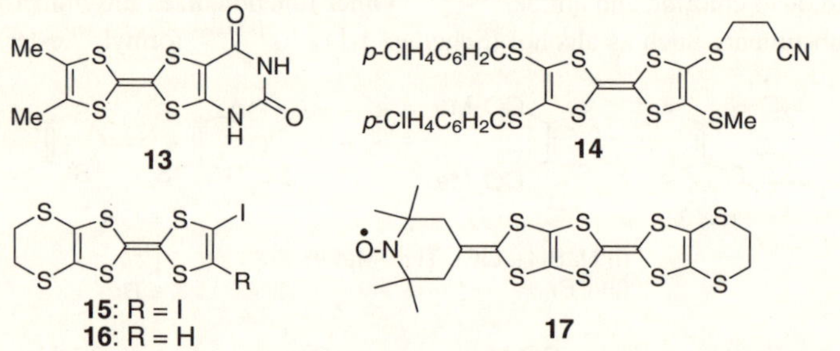

Fig. 1.2

converted into other functional groups, see Schemes 1.12a,b),[1,2,22,43,53,56,58] lactam (**13** in Fig. 1.2),[59] nitrile (**14** in Fig. 1.2),[15,37,60] halogen (**15** and **16** in Fig. 1.2)[61,62] and a stable organic radical (**17** in Fig. 1.2),[63] have been synthesized *via* the P(OR)$_3$-mediated cross-coupling reactions. When the hydroxyl or formyl group sensitive to the trivalent phosphorous reagent used for the cross-coupling reaction exists (*e.g.*, see Schemes 1.11a,b), an appropriate protecting group is necessary to prevent unexpected side reactions. The free functional group is easily restored under ordinary deprotection conditions and can be further transformed into a new functional group. For example, by the reaction sequence shown in Scheme 1.13a, the conversion of the mono- and di-acetal-substituted TSFs **18** and **19** into the corresponding alcohols **20** and **21** has been accomplished.[64] It is noteworthy that the synthesis of the selenium analogue of MDT-TTF, *i.e.*, MDT-TSF, which gives rise to a superconducting salt, involved the P(OMe)$_3$-mediated cross-coupling and subsequent transalkylation on the outer sulfur atom (Scheme 1.13b).[65]

1.2.3 Selective Strategies: Wittig- and Wittig-Horner-type Condensations and Organometallic Route

In addition to the continuous use of the cross-coupling methods (Methods A and B) in synthesis of unsymmetrical TTFs, the search for selective strategies capable of constructing unsymmetrical TTFs without the formation of symmetrical byproducts has been made. A Wittig-type condensation between a 1,3-dichalcogenol-2-ylid and a 1,3-dichalcogenol-2-ylium salt was first investigated,[34] followed by successive reports on the application of the Wittig-Horner [also called Horner-Wadsworth-Emmons (HWE)] reaction[66] and the development of an organometallic route by Yamada *et al.*[67]

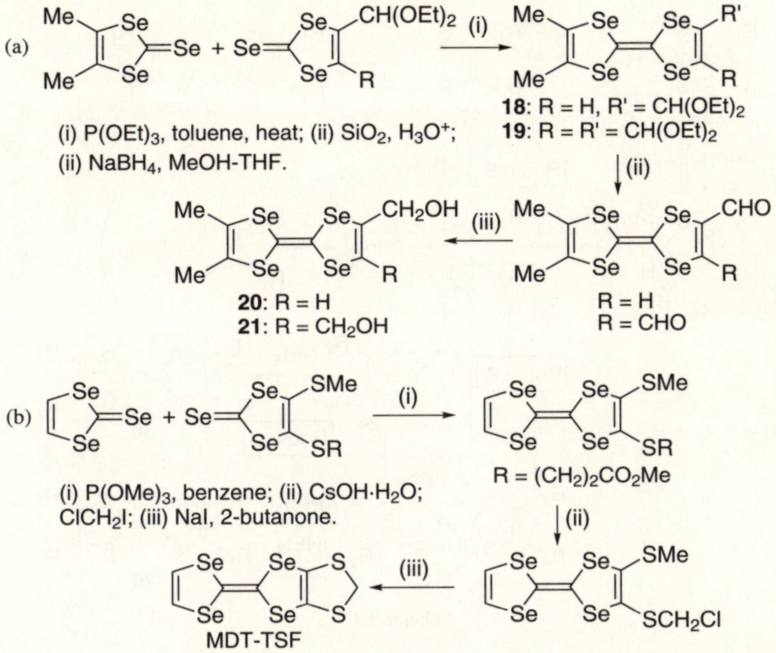

(a)

18: R = H, R' = CH(OEt)$_2$
19: R = R' = CH(OEt)$_2$

(i) P(OEt)$_3$, toluene, heat; (ii) SiO$_2$, H$_3$O$^+$;
(ii) NaBH$_4$, MeOH-THF.

20: R = H
21: R = CH$_2$OH

R = H
R = CHO

(b)

R = (CH$_2$)$_2$CO$_2$Me

(i) P(OMe)$_3$, benzene; (ii) CsOH·H$_2$O;
ClCH$_2$I; (iii) NaI, 2-butanone.

MDT-TSF

Scheme 1.13

A. Method C: Wittig-type Condensation

This type of condensation was first described by Cava *et al.* in the synthesis of benzo-TTF.[34]
Later, it was pointed out that the condensation is not completely selective.[68,69] For example, the
formation of undesirable symmetrical TTFs was observed in the synthesis of dimethyl(tetramethylene)-TTF by two possible condensation reactions (Scheme 1.14).[68] Subsequent investigations
advanced a general mechanism to explain the formation of two symmetrical TTFs.[69,70] As illustrated in Scheme 1.15, the phosphonium ylid **22** undergoes deprotonation by *n*-BuLi to give the
corresponding ylid **23**, which reacts with the dithiolium salt **24** to lead to the expected adduct **25**
(Route A). Deprotonation of **25** by NEt$_3$ followed by loss of PPh$_3$ produces the desired unsym-

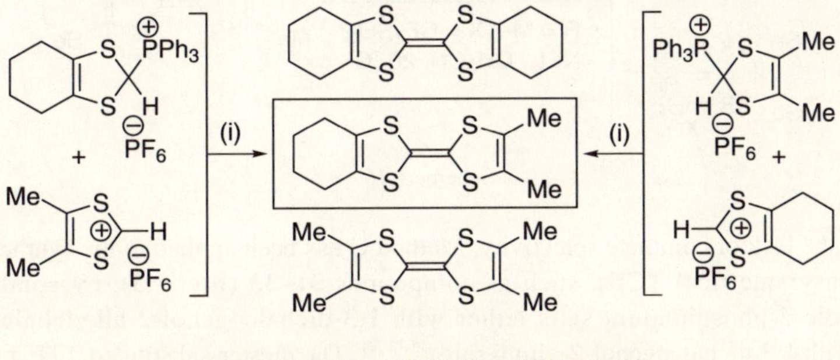

(i) *n*-BuLi, THF, −78 ˚C; then NEt$_3$, 20 ˚C.

Scheme 1.14

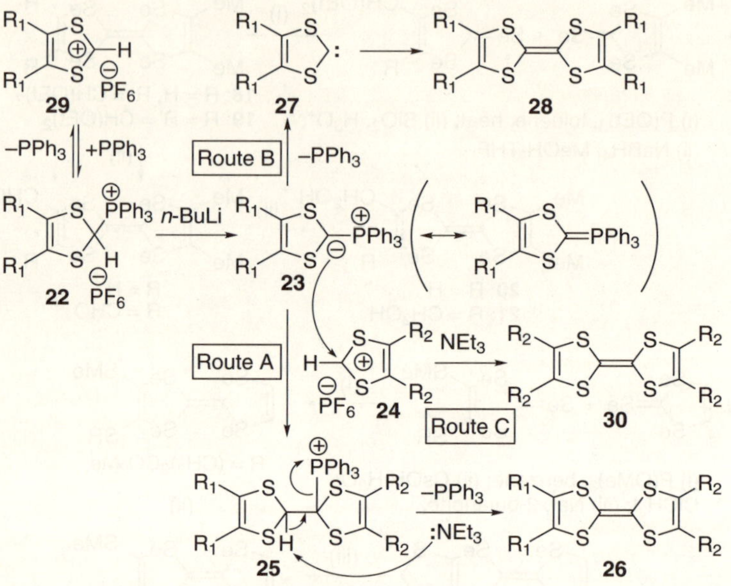

Scheme 1.15

metrical TTF derivative **26** as a main product. On the other hand, as suggested by Sudmale and coworkers,[70] the intrinsically unstable ylid **23** is assumed to be decomposed into the carbene **27** (Route B), which then undergoes dimerization to give the undesirable symmetrical TTF derivative **28**. Similar to the two reaction pathways shown in Scheme 1.3, it is thought that reaction of **27** with the dithiolium salt **29** generated from **22** by a PPh$_3$ loss gives **28** and that the NEt$_3$-induced self-coupling of the dithiolium salt **24** is responsible for the formation of another symmetrical TTF derivative **30** (Route C). Route C can be suppressed by using the 1,3-dichalcogenol-2-alkylchalcogenolium salt instead of the dithiolium salt (Scheme 1.16).[53,69]

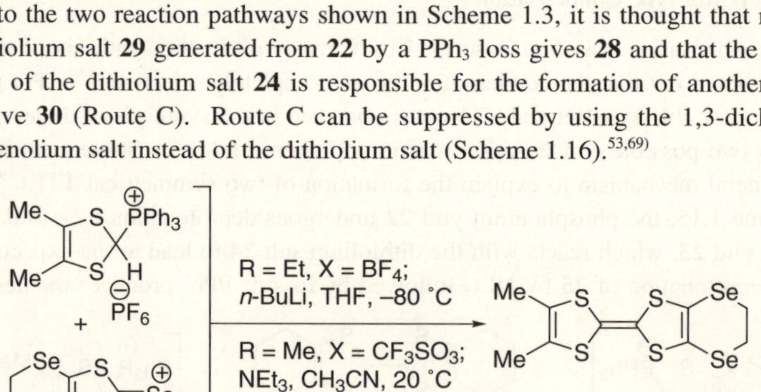

Scheme 1.16

Despite the lack of complete selectivity, Method C has been applied to the synthesis of functionalized unsymmetrical TCFs, such as compounds **31–33** (Fig. 1.3), by combining 1,3-dichalcogenole-2-phosphonium salts either with 1,3-dichalcogenol-2-alkylchalcogenolium salts[71,72] or with 1,3-dichalcogenol-2-ylium salts.[15,72,73] The diester-substituted TTF and diselenadithiafulvalene (DSDTF) derivatives **31a,b** were demethoxycarbonylated by heating with LiBr·H$_2$O in HMPA at 85–150 °C, and the bis-protected TTF and DSDTF derivatives **32a,b** and

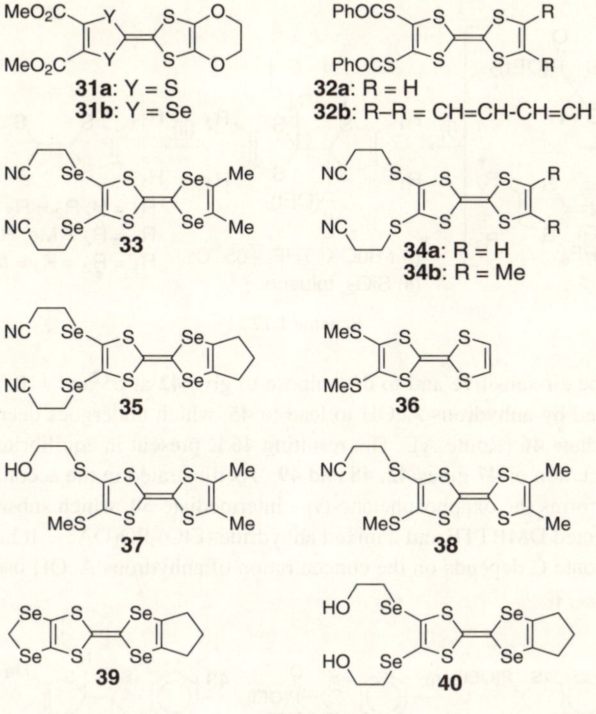

31a: Y = S
31b: Y = Se

32a: R = H
32b: R–R = CH=CH-CH=CH

33

34a: R = H
34b: R = Me

35

36

37

38

39

40

Fig. 1.3

33 were subjected to deprotection by EtONa in EtOH followed by alkylation.[71,72] Another bis(cyanoethyl)-protected TTF and DSDTF derivatives **34a,b** and **35** (Fig. 1.3) were also synthesized *via* Method C, and further transformations of (i) **34a** into **36**, (ii) **34b** into **37** *via* **38** and (iii) **35** into both **39** and **40** (Fig. 1.3) were attained by applying stepwise deprotection/alkylation protocols developed by Becher and coworkers.[74] Furthermore, the use of Method C permitted the synthesis of the functionalized single-bridged bis-TTFs.[70]

B. Method D: Wittig-Horner-type Condensation

In order to suppress the formation of the self-coupling products in Method C, a useful modification has been developed. Lerstrup *et al.* applied Wittig-Horner (or HWE) olefinations using 1,3-dichalcogenole-2-phosphonates to the synthesis of TTFs (Scheme 1.17).[75] The reaction was reported to fail to proceed when an unsubstituted imminium salt ($R_2 = R_3 = H$ in Scheme 1.17) was used.[75] However, this result was later contradicted through the successful formation of a series of new TTFs,[76] in which no significant side reaction took place. Since then, this selective method has been employed in the synthesis of a large number of unsymmetrical TTFs,[48,77,78] DSDTFs[79] and their functionalized derivatives.[23,43]

Since low yields of the products were obtained in several cases,[76,79,80] a study on the mechanism of a Wittig-Horner reaction leading to DMBTTF was conducted to clarify the cause.[81] The reaction is thought to proceed *via* the following two steps (Scheme 1.18). First, the phosphonate carbanion **41**, generated from benzo-1,3-dithiole-2-phosphonate **42** by proton abstraction, reacts with the piperidium salt **43** to allow the quantitative formation of the adduct **44**, which, however,

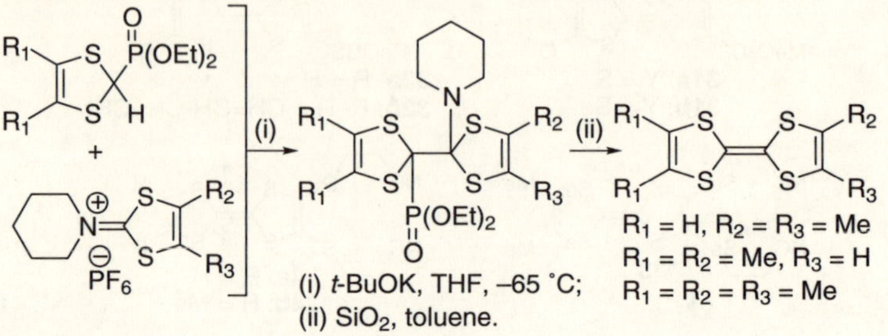
R1 S P(OEt)2 O
R1 S H
+
N S R2
PF6 S R3
(i)
R1 S N S R2
R1 S S R3
P(OEt)2 O
(ii)
R1 S S R2
R1 S S R3
R1 = H, R2 = R3 = Me
R1 = R2 = Me, R3 = H
R1 = R2 = R3 = Me
(i) t-BuOK, THF, –65 °C;
(ii) SiO2, toluene.

Scheme 1.17

has been found to be air-sensitive and to decompose to give **42** at 65 and 110 °C. In the second step, **44** is protonated by anhydrous AcOH to lead to **45**, which undergoes deamination to afford the acetate intermediate **46** (Route A). The resulting **46** is present in equilibrium with **47** (Route B), and the fragmentation of **47** gives **42**, **48** and **49**. As illustrated in the accompanying equation (Route C), **46** also forms the oxaphosphetane-type intermediate **50**, which subsequently collapses to produce the expected DMBTTF and a mixed anhydride $(EtO)_2PO(OAc)$. It has been found that the preference of Route C depends on the concentration of anhydrous AcOH used in the reaction.

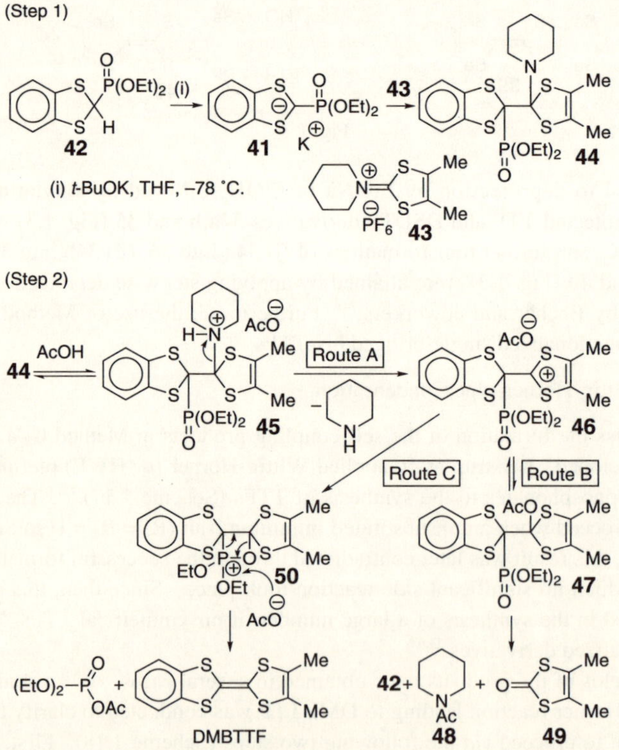
(Step 1)
(i)
42
41
43
44
(i) t-BuOK, THF, –78 °C.
43
(Step 2)
AcOH
44
45
Route A
46
Route C
Route B
50
47
DMBTTF
48
49

Scheme 1.18

On the basis of these results, an attempt to improve the yield of DMBTTF was successful with the use of a one-pot procedure without isolation of **44**. In addition, the application of this one-pot procedure to the synthesis of other unsymmetrical TTFs resulted in enhanced overall yields of the products.[82] Nevertheless, low yields were observed when the unsubstituted and monosubstituted 1,3-dithiole-2-phosphonates were used, because the corresponding phosphonate carbanions favor the ring-opened forms, which easily undergo dimerization.[83]

On the other hand, even if the imminium salt bearing the ester or nitrile group was used, this type of olefination proceeded smoothly without serious competing reactions (Scheme 1.19).[43] Subsequent transformation of respective ester and nitrile functions into the corresponding carboxylic acid and amine followed by a peptide coupling of the resulting two TTFs gave the single-bridged bis-TTF derivative (Scheme 1.19). By a similar reaction sequence, the tris-TTF derivative linked by two peptide bonds can be obtained.[43] Another type of single-bridged bis-TTFs was also obtained by Method D using bis(imminium) salts prepared by the reaction of a mesoion with alkylenedihalogenides (Scheme 1.20).[84] It has recently been reported that a successful repetition of a twofold Wittig-Horner olefination leads to the macrocyclic compound incorporating two π-extended TTF units in each of which the two 1,3-dithiole rings are separated by a π-conjugated

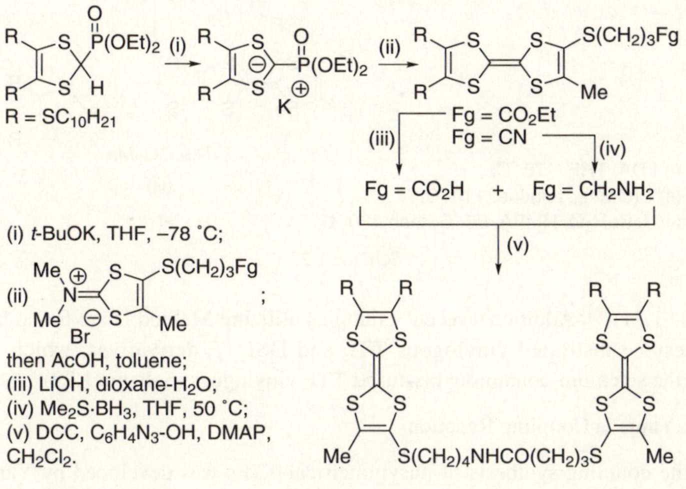

Scheme 1.19

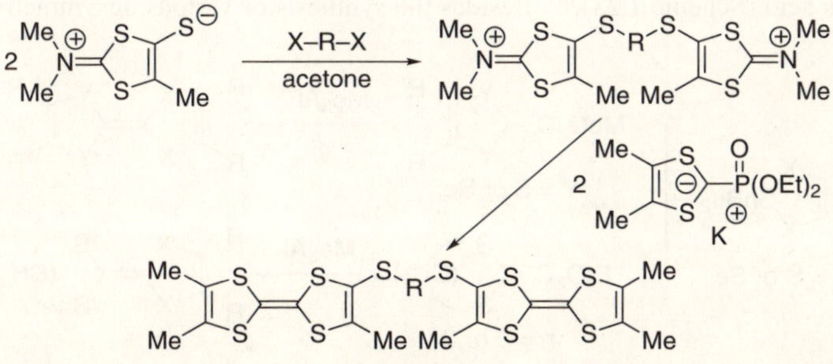

Scheme 1.20

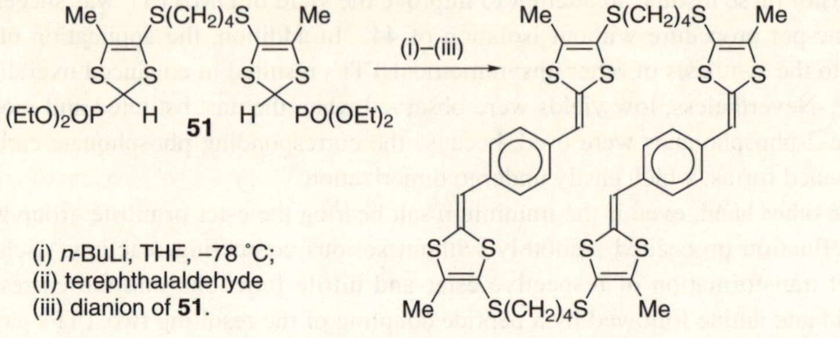
Me
S(CH2)4S
Me
(EtO)2OP
H
51
H
PO(OEt)2
(i)–(iii)
S(CH2)4S
(i) n-BuLi, THF, –78 ˚C;
(ii) terephthalaldehyde
(iii) dianion of 51.

Scheme 1.21

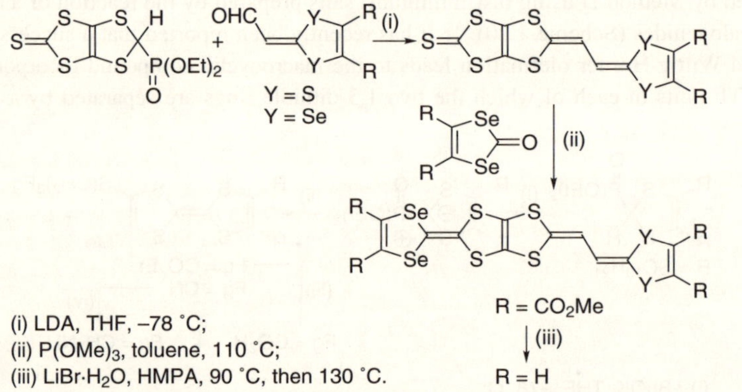
P(OEt)2
OHC
Y = S
Y = Se
(i)
(ii)
R = CO2Me
(iii)
R = H
(i) LDA, THF, –78 ˚C;
(ii) P(OMe)3, toluene, 110 ˚C;
(iii) LiBr·H2O, HMPA, 90 ˚C, then 130 ˚C.

Scheme 1.22

spacer (Scheme 1.21).[85] Additional recent examples utilizing Method D are found in the preparation of the diester-substituted vinylogous TTF and DSDTF derivatives, which were used as precursors for the selenium-containing bis-fused TTF vinylogues (Scheme 1.22).[86]

C. Method E: Yamada Coupling Reaction

A non-phosphite coupling synthesis of unsymmetrical TTFs was developed by Yamada *et al.* in 1995.[67,87] This predominant synthesis was accomplished by the Lewis acid-promoted reaction of organotin compounds with esters, and the best yield of the product was obtained when Me_3Al was used as a Lewis acid (Scheme 1.23).[67] Besides the synthesis of various unsymmetrical TTFs and

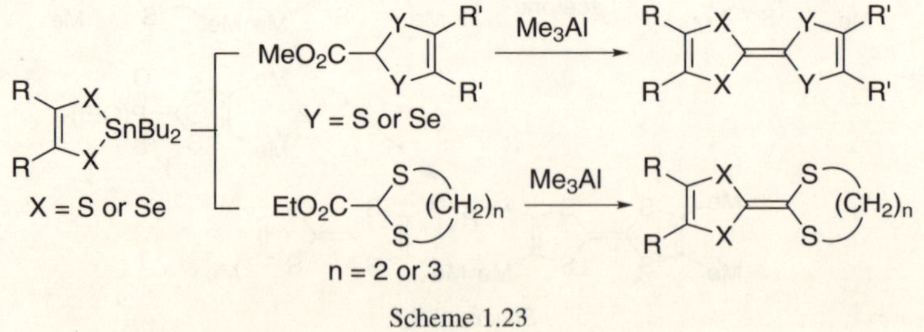
MeO2C
Me3Al
Y = S or Se
SnBu2
X = S or Se
EtO2C
(CH2)n
Me3Al
n = 2 or 3

Scheme 1.23

DSDTFs, this method was extended to the synthesis of dihydro-TTFs and their selenium and dithiane analogues. It has been found that the Me_3Al-promoted coupling reaction of tin diseleno-lates is relatively slow in comparison with that of tin dithiolates. The organotin dithiolates and diselenolates essential to the reaction are easily accessible from the corresponding 1,3-dithiol- and 1,3-diselenol-2-ones *via* reaction with MeMgBr in THF followed by trapping with Cl_2SnBu_2 (Scheme 1.24). As for the desired ester parts, they can be obtained by transmetalation of the cor-responding organotin compounds with 2 equiv. of *n*-BuLi and subsequent treatment with methyl dichloroacetate (Scheme 1.24). The usefulness of this method has also been established in the synthesis of new TTF-fused and TTP donors, which give rise to organic metals and superconduc-tors.[88-91] For example, Scheme 1.25 shows the synthetic route to the TTF-DSDTF fused donor **52**. Basic cleavage of thiapendione **53** followed by trapping with Cl_2SnBu_2 gave tin dithiolate **54**, which reacted with ester **55** in the presence of Me_3Al to lead to DSDTF derivative **56** with a fused 1,3-dithiol-2-one ring. The final product **52** was obtained by using the $P(OMe)_3$-mediated cross-coupling reaction of **56** with the corresponding thione. The yields of the products in these Me_3Al-promoted and cross-coupling reactions have been reported to depend on the type of substrate.[67]

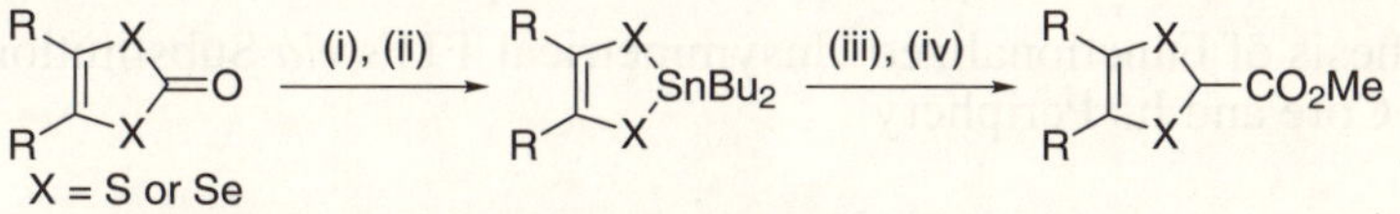

(i) MeMgBr, THF; (ii) Cl_2SnBu_2, THF, −78 °C; (iii) *n*-BuLi (2 equiv.),
THF, −78 °C; (iv) Cl_2CHCO_2Me, THF, −78 °C.

Scheme 1.24

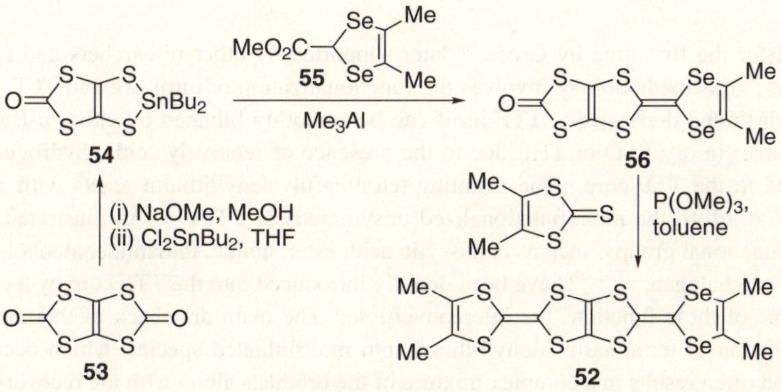

Scheme 1.25

Although the mechanistic aspect of this Me_3Al-promoted reaction has not yet been described in detail, the reaction pathway, if an opportunity of transmetalation between tin dichalcogenolate and Me_3Al is given, can be written as in Scheme 1.26. A successive Sn/Al metal exchange reaction leads to bis(dimethylaluminum) dichalcogenolate **57**, which subsequently reacts with es-ter to give chalcogenoester **58**. The intramolecular nucleophilic attack of the remaining aluminum chalcogenolate on the carbonyl function followed by loss of Me_2AlOH produces the unsymmetri-cal TCF derivative **59**.

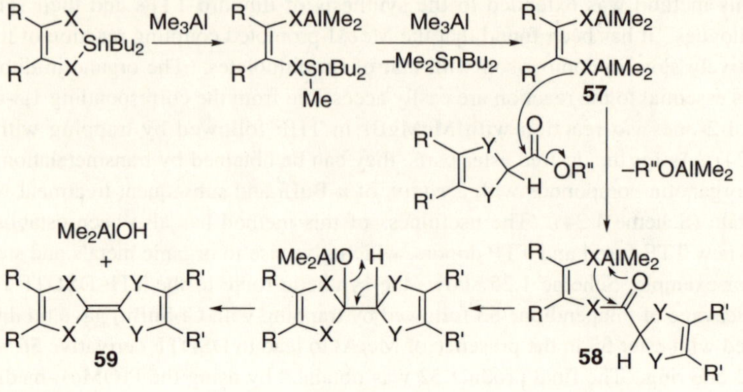

Scheme 1.26

1.3 Synthesis of Functionalized Unsymmetrical TTFs *via* Substitution on the TTF Core and Its Periphery

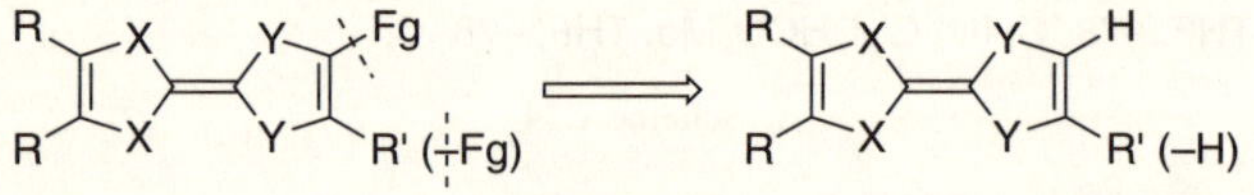

Scheme 1.27

As proposed for the first time by Green,[92] later supported by other researchers and reviewed by Garín in 1995,[9] this methodology involves the functionalization of unsubstituted TCF and its preliminarily substituted derivatives. TTF itself can be smoothly lithiated by either *n*-BuLi or LDA (more favorable) in dry Et_2O or THF due to the presence of relatively acidic hydrogen atoms directly bonded to the TTF core. The resulting tetrathiafulvalenyllithium reacts with a variety of electrophiles to afford the monofunctionalized unsymmetrical TTFs.[92] As illustrated in Scheme 1.28, many functional groups, such as carboxylic acid, ester, amide, thioamide, alcohol, amine, ketone, formyl and halogen,[9,77,92–99] have been directly introduced into the TTF core by a one-pot procedure. Some of these functions are interconvertible. The main drawback of this method is the disproportionation of tetrathiafulvalenyllithium into multilithiated species, which occurs even at $-70\,^{\circ}C$[9,92] and often results in a complex mixture of the products along with the recovery of TTF.

1.3.1 Conversion of Functional Groups

Reduction of carbonyl compounds, including esters, to alcohols,[53,93] oxidation of alcohols to aldehydes,[100] transformation of a carboxylic acid into an amide *via* the corresponding acidic chloride[44,96] or fluoride,[101] *etc.* are the frequently encountered reactions in TTF chemistry. Since these reactions published up to the beginning of 1994 have been surveyed in an earlier review,[9] this section covers particular or recent examples dealing with such conversions. TTF-COOR and TTF-CHO have been utilized as starting materials in the multi-step synthesis of functionalized TTFs as described below.

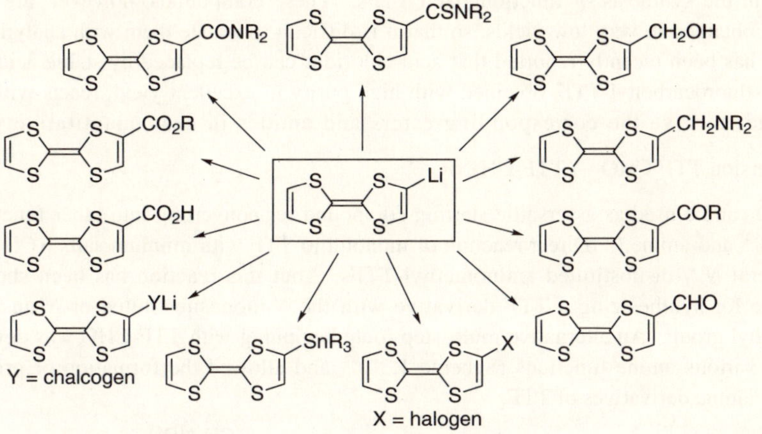

Scheme 1.28

A. Conversion TTF-CO$_2$R → TTF-CH$_2$OH and TTF-CONR$_2$

Reduction of the di- and mono-ester-containing TTFs by appropriate reducing reagents produces the corresponding diols and alcohols, respectively.[94,100] In addition to the example shown in Scheme 1.10a, this conversion has been applied to the synthesis of vinylogous TTF systems (Scheme 1.29).[101] Diester **60** was reduced by NaBH$_4$/ZnCl$_2$ to afford diol **61** in 82% yield, whereas the reduction of monoester **62** (a mixture of *E* and *Z* isomers) required DIBAH as a reducing reagent at a controlled temperature region in CH$_2$Cl$_2$ to lead to alcohol **63** in 72% yield.

While a *N*-monosubstituted amide derivative of trimethyl-TTF can be obtained from the corresponding monolithio-TTF species by a one-pot procedure,[77] the synthesis of EDO-TTF with the *N*-unsubstituted amide group necessitated stepwise procedures as previously shown in Scheme 1.10b. Like the construction of EDO-TTF-CONH$_2$, its sulfur analogue EDT-TTF-CONH$_2$ was synthesized from EDT-TTF-CO$_2$Me *via* a successive formation of EDT-TTF-CO$_2$H and EDT-TTF-COCl.[96] In this way, TTF derivatives with acid chloride functionality serve as useful inter-

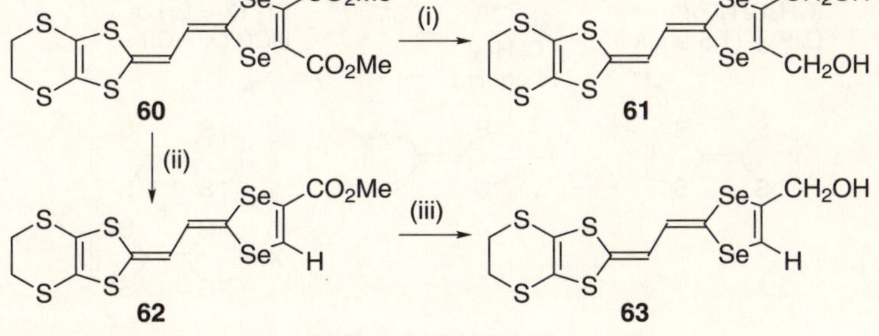

(i) NaBH$_4$, ZnCl$_2$; (ii) LiBr·H$_2$O, HMPA, 60–70 ˚C; (iii) DIBAH, CH$_2$Cl$_2$, −60 → > 40 ˚C; then 6N HCl, MeOH, −60 → > 20 ˚C.

Scheme 1.29

mediates in the synthesis of functionalized TTFs. These compounds, however, are generally labile and obtained in very low yields, so that it is difficult to isolate them with analytically high purity. It has been recently reported that acid chlorides can be replaced by stable acid fluorides because 4-fluorocarbonyl-TTF, obtained with high purity in excellent yield, reacts with alcohols and amines to give the corresponding esters and amides in near quantitative yields.[102]

B. Conversion TTF-CHO → TTF-CH$_2$NR$_2$

TTF-CHO can be used as a versatile starting compound for converting into other functions such as alcohol[94] and amine.[97] Direct reaction of monolithio-TTF with iminium salts [(CH$_2$=NRR')I] gave several *N,N*-disubstituted aminomethyl-TTFs,[97] but this reaction has been shown to be ineffective for synthesizing a TTF derivative with the *N*-monosubstituted or *N*-unsubstituted aminomethyl group. An alternative multi-step route beginning with TTF-CHO was developed to introduce various amine functions (Scheme 1.30)[97] and allowed the formation of primary and secondary amine derivatives of TTF.

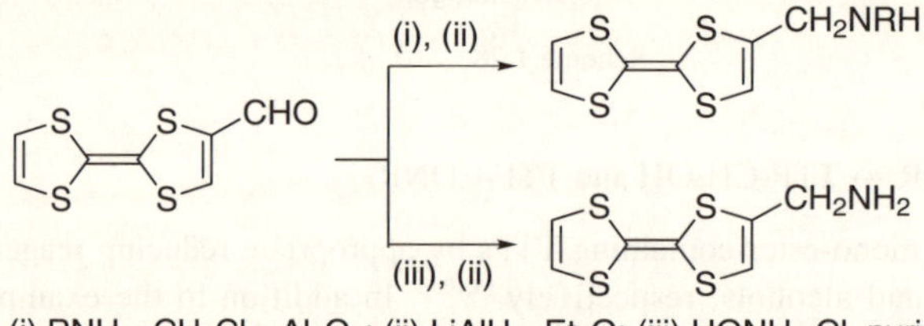

(i) RNH$_2$, CH$_2$Cl$_2$, Al$_2$O$_3$; (ii) LiAlH$_4$, Et$_2$O; (iii) HONH$_3$·Cl, pyridine-EtOH.

Scheme 1.30

C. Conversion into TTF-X (X = halogen), TTF-C$_{60}$ and TTF-SnR$_3$

Halogenated TTFs belong to a new class of useful intermediates in TTF synthesis and have been mainly prepared from lithiated TTFs by reaction with various halogenating reagents (Scheme 1.31a). When TTF was used as a starting material, it was found that the number of halogenated positions on the TTF molecule depends both on the amount of base used (1.5 or 4 equiv. of

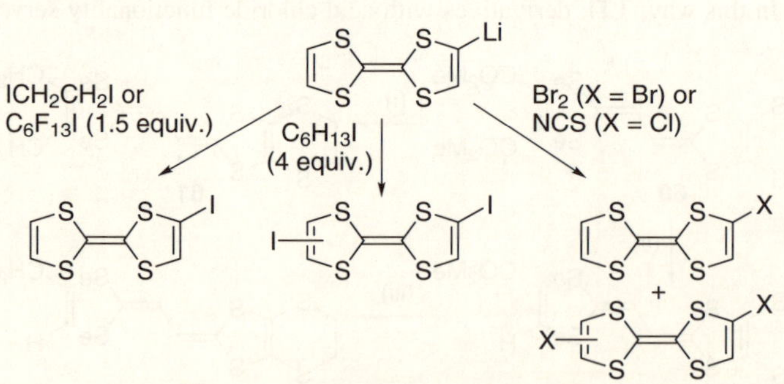

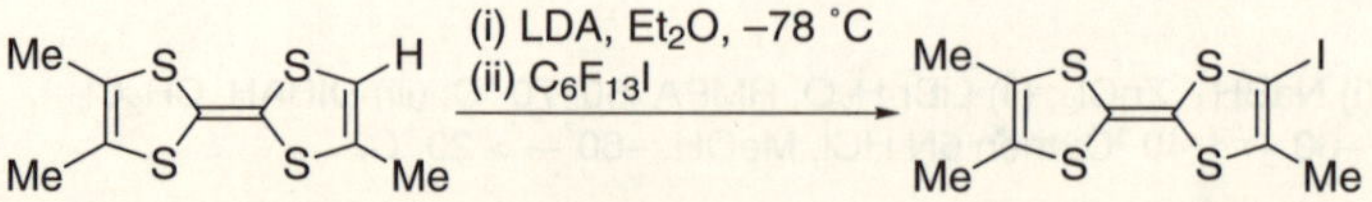

Scheme 1.31

LDA)[98] and on the nature of the halogenating reagent ($C_6F_{13}I$, Br_2, NCS, *etc.*).[97,98] In addition to monohalogenation,[103] the coexistence of mono- and di-halogenations has been observed.[98] In the case of halogenation of the trisubstituted TTF derivative, only monohalogenation occurs naturally (Scheme 1.31b).[99]

Another recent interesting example is the conversion of the bis(hydroxymethyl)-attached TTF derivative **64** into the corresponding bis(bromomethyl)-TTF derivative **65** by using PBr_3 (Scheme 1.32).[104] Subsequently, the attachment of C_{60} acting as an electron acceptor to **64** and **65** was examined to construct photo-active and donor-acceptor systems, respectively. A twofold esterification of **64** with C_{60} derivative **66** bearing the benzoic acid moiety led to the C_{60}-TTF-C_{60} triad,[105] and conversion of **65** into the corresponding cisoid diene **67** followed by Diels-Alder cycloaddition with C_{60} gave the double-bridged TTF-C_{60} system.[106] Besides these compounds, a wide range of TTF-containing fullerenes have been reported.[8,107]

Recently, it has been reported that lithiated TTFs can be transformed into the corresponding stannyl TTFs, which undergo cross-coupling reactions in the presence of Pd catalysts to permit introduction of specific functional groups into the TTF moiety.[108–110] Scheme 1.33 shows two typical examples. With the aid of constructing organic ferro- and ferri-magnets, TTFs linking the nitronyl nitroxide group (which is one of the stable organic radicals) through benzene were derived from the corresponding aldehydes obtained by the Pd-catalyzed cross-coupling of trimethyl-stannyl TTFs with 4-halogenobenzaldehyde.[108] A similar Pd-catalyzed coupling reaction with 4-bromopyridine hydrochloride led to the pyridine-attached TTF derivative, which reacted with methyl iodide to give the *N*-methylpyridinium-substituted TTF derivative as a donor-acceptor system exhibiting intermolecular CT.[110]

The functionalized TTFs shown in Scheme 1.28 and their analogues can also serve as building blocks for highly substituted architectures containing the bis- and oligo-TTF units.[11,12,17,111]

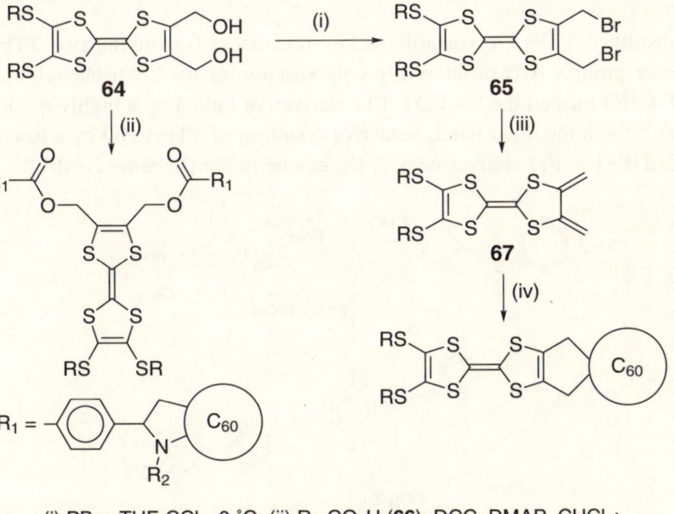

(i) PBr_3, THF-CCl_4, 0 °C; (ii) R_1-CO_2H (**66**), DCC, DMAP, $CHCl_3$;
(iii) KI, 18-crown-6, toluene, 75 °C; (iv) C_{60}.

Scheme 1.32

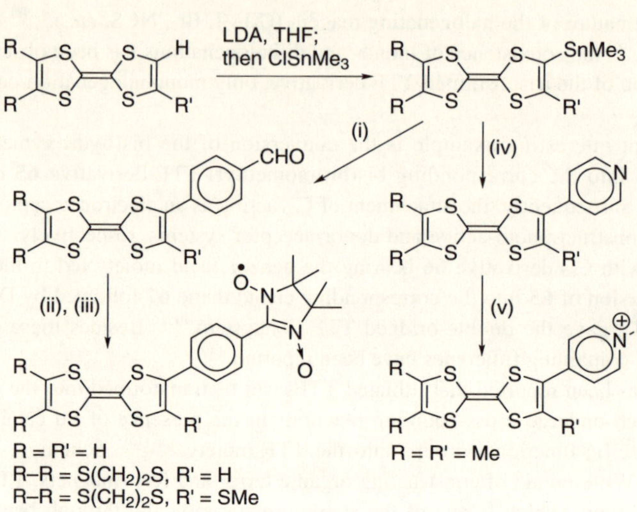

(i) XC$_6$H$_4$CHO (X = Br or I), cat. Pd(PPh$_3$)$_2$Cl$_2$ or Pd(PPh$_3$)$_4$, toluene,
reflux; (ii) [C(CH$_3$)$_2$(NHOH)]$_2$, cat. [C(CH$_3$)$_2$(NHOH)]$_2$·H$_2$SO$_4$,
CH$_2$Cl$_2$-hexane, reflux; (iii) PbO$_2$, K$_2$CO$_3$, THF; (iv) 4-BrC$_5$H$_4$N·HCl,
NaHCO$_3$, toluene, cat. Pd(Ph$_3$)$_4$, reflux; (v) MeI, acetone, reflux.

Scheme 1.33

1.3.2 Synthesis of Bis- and Bi-TTFs

The formyl-substituted TTFs are versatile as key precursors for linking two TTFs through a π-conjugated spacer group. A twofold Wittig condensation of the bis(triphenylphosphonium) salt with EDT-TTF-CHO formed the bis-EDT-TTF derivative linked by a highly π-conjugated spacer (Scheme 1.34a).[112] On the other hand, reductive coupling of TTF-CHO by a low-valent titanium reagent produced the bis-TTF derivative with the ethene linker (Scheme 1.34b).[113]

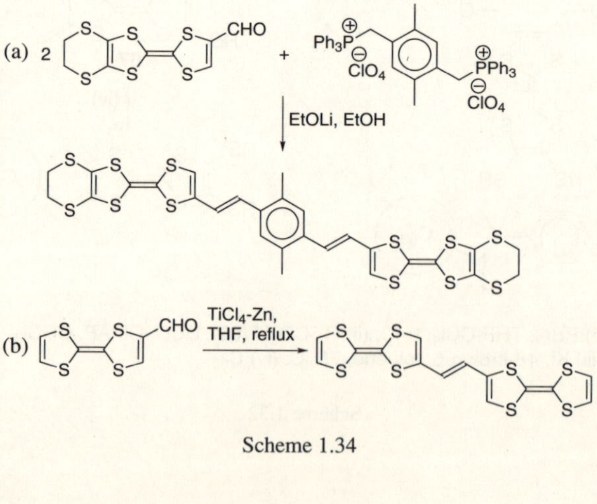

Scheme 1.34

Also, TTFs having the hydroxy group capable of leading to ester functionality serve as building blocks for the bis- and oligo-TTFs.[111] For example, treatment of alcohols **68** and **69** (Schemes 1.35a,b) with acid chlorides gave the bis-TTFs **70–72** linked by ester-containing spacer groups. Compound **72** was used as a stopperless axle suitable for a [2]pseudorotaxane molecular shuttle.[111]

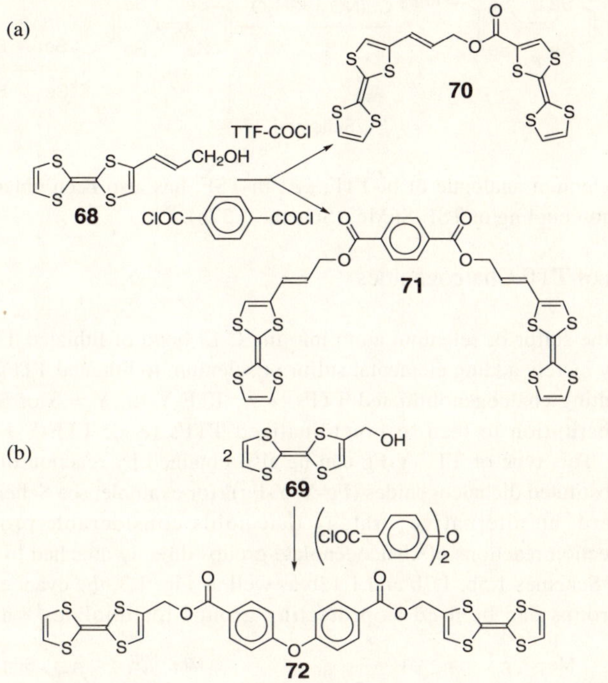

Scheme 1.35

Bi-TTFs, in which two TTF units are directly connected, can be derived from iodo-TTFs by treating with copper in refluxing DMF (Ullmann coupling) or with copper(I) thiophene-2-carboxylate (CuTC) in *N*-methylpyrrolidinone (NMP).[114] For example, bi(trimethyltetrathiafulvalenyl) and its methylthio-analogue were obtained by the CuTC-mediated coupling of the corresponding iodo-TTF precursors in 75% and 72% yields, respectively (Scheme 1.36).[114]

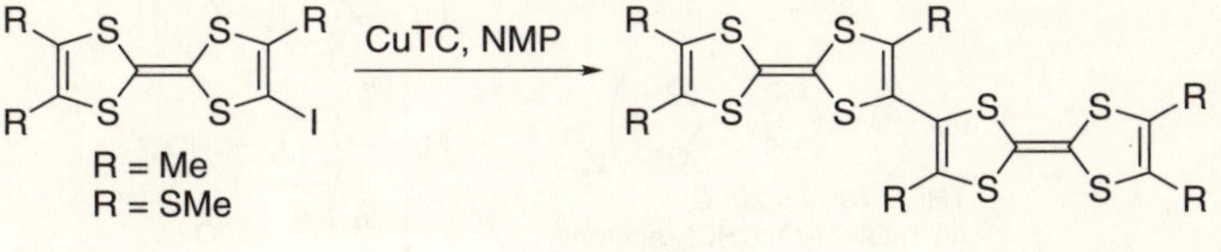

Scheme 1.36

It has been demonstrated that trialkylstannyl-TTF is subjected to the Pd-catalyzed homo- and cross-coupling reactions to give bi-TTF and aryl-substituted TTFs, respectively.[115] Similarly, TTF-SnMe$_3$ reacted with various aryl dihalides in the presence of a catalytic amount of Pd(PPh$_3$)$_4$ to produce bis-TTFs with aryl linkers, such as benzene, thiophene, pyridine and azulene (Scheme

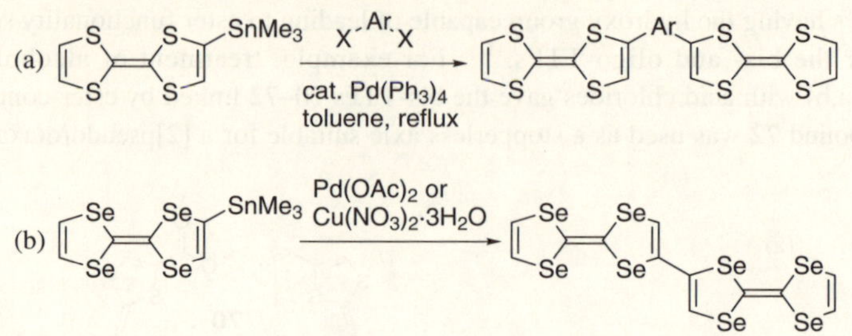

Scheme 1.37

1.37a).[116] The selenium analogue of bi-TTF, *i.e.*, bi-TSF, has also been obtained by the Pd- or Cu-mediated homo-coupling of TSF-SnMe$_3$ (Scheme 1.37b).[109]

1.3.3 Synthesis of TTF Chalcogenides

The insertion of the sulfur or selenium atom into the C-Li bond of lithiated TTFs (*e.g.*, TTF-Li) can be attained by merely adding elemental sulfur or selenium to lithiated TTFs at a low temperature.[77] The resulting chalcogenolithiated TTFs (*e.g.*, TTF-Y-Li, Y = S or Se) easily undergo electrophilic substitution to lead to functionalized TTFs (*e.g.*, TTF-Y-Fg) as outlined in Scheme 1.38a.[77] This type of TTF-Y-Fg can be also obtained by reaction of the lithiated TTF species with disubstituted dichalcogenides (Fg-Y-Y-Fg) (for example, see Scheme 1.38b).[108]

In this regard, an alternative strategy that holds considerable promise is to apply protection/deprotection reactions of chalcogenolate groups directly attached to the TTF core. As already shown in Schemes 1.5b, 1.7b and 1.13b as well as Fig. 1.3, the cyanoethyl and methoxycarbonylethyl groups can be used as protecting groups for thiolates[15] and selenolates.[64]

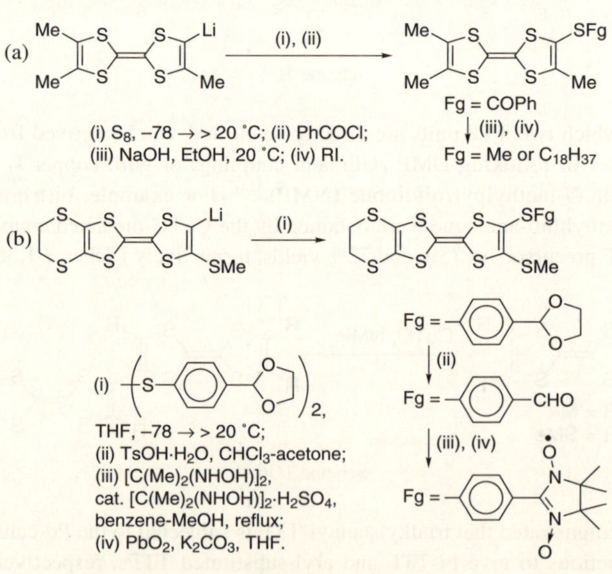

Scheme 1.38

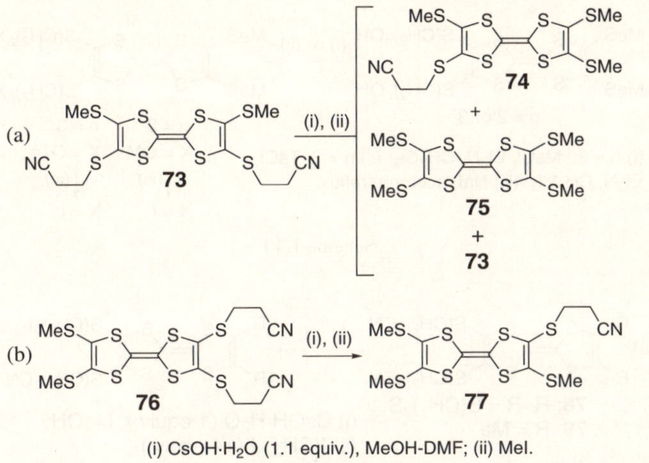

(i) CsOH·H₂O (1.1 equiv.), MeOH-DMF; (ii) MeI.

Scheme 1.39

Deprotection is generally carried out *in situ* under basic conditions, such as NaOEt in EtOH, KO*t*-Bu in *t*-BuOH or CsOH·H₂O in DMF. However, when equimolecular amounts of base were used, the selectivity in monodeprotection of the bis(cyanoethyl)-protected TTFs has been found to depend considerably on the relative positions of the two protected chalgogenolates. Deprotection/methylation of the bis-protected TTF derivative **73**, in which one cyanoethythio group exists on each 1,3-dithiole ring (hereafter designated as Type I), gave a mixture of the mono- and bis-deprotected products **74** (67%) and **75** (19%) along with the recovered **73** (11%) (Scheme 1.39a).[18] On the contrary, the same reaction of **76** bearing two cyanoethylthio groups on one 1,3-dithiole ring (designated as Type II) provided only the monodeprotected product **77** in 94% yield (Scheme 1.39b).[18] On the other hand, the use of more than two equiv. of base permitted a facile deprotection/alkylation sequence for various bis(cyanoethylchalcogeno)-substituted TTFs irrespective of Types I and II, making this reaction sequence very versatile. The cyanoalkyl chain can be not only transformed into other alkyl substituents equipped with various functions, such as hydroxy,[15,37,73,119] halogeno,[117,118] amino[43,119] and phosphino groups,[120] but also used as a key function leading to a variety of single- and double-bridged bis-TTFs and oligo-TTFs.[117,121–124] These examples are shown in the following sections.

1.3.4 Conversion of Protected TTF Chalcogenolates into Other Functionalized TTFs

Deprotection of the bis(cyanoethylchalcogeno)-substituted TTFs of Type II by NaOEt (4 equiv.) in EtOH followed by reaction with an excess of chloroethanol allowed the formal substitution of each cyano function with the hydoxy group.[15,73] The hydroxyalkyl chain can be further converted into the corresponding iodoalkyl one *via* mesylation or tosylation as illustrated in Scheme 1.40.[117,118] Additionally, selective monodeprotection of the bis-protected TTFs **78** and **79** followed by alkylation with diiodopropane gave the corresponding monoiodinated TTFs (Scheme 1.41a).[117] Also, TTF derivative **80** of Type I successively underwent deprotection/alkylation, mesylation and bromination to furnish the dibrominated TTF derivative **81** (Scheme 1.41b), which was subsequently used for synthesis of the aza-crown-annelated TTF systems capable of trapping various cations.[118]

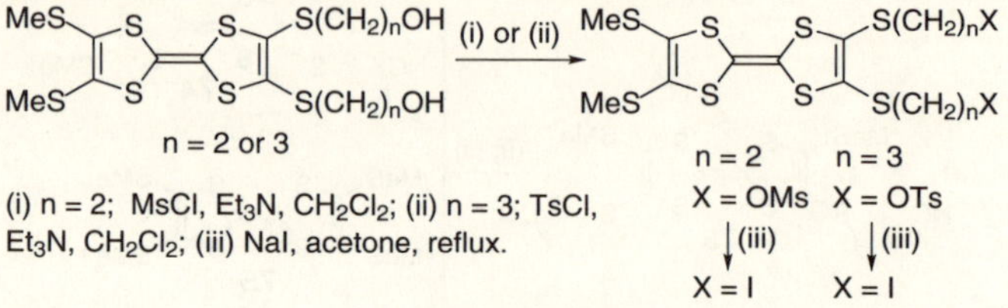
MeS
MeS
S(CH2)nOH
S(CH2)nOH
(i) or (ii)
S(CH2)nX
S(CH2)nX
n = 2 or 3
(i) n = 2; MsCl, Et3N, CH2Cl2; (ii) n = 3; TsCl,
Et3N, CH2Cl2; (iii) NaI, acetone, reflux.
n = 2 n = 3
X = OMs X = OTs
(iii)
X = I X = I

Scheme 1.40

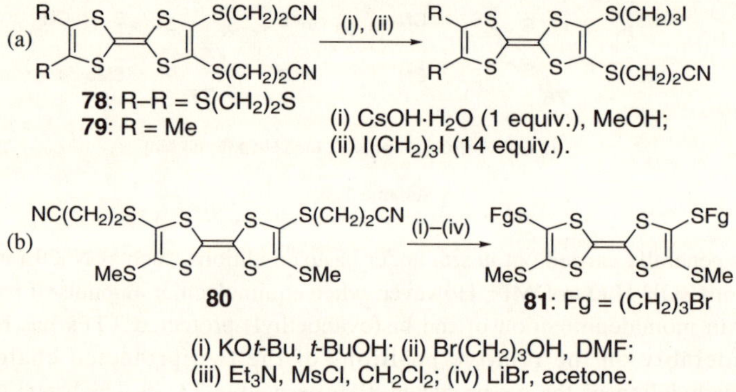
(a)
S(CH2)2CN
(i), (ii)
S(CH2)3I
78: R–R = S(CH2)2S
79: R = Me
(i) CsOH·H2O (1 equiv.), MeOH;
(ii) I(CH2)3I (14 equiv.).
(b)
NC(CH2)2S
S(CH2)2CN
(i)–(iv)
FgS
SFg
MeS
SMe
80
81: Fg = (CH2)3Br
(i) KOt-Bu, t-BuOH; (ii) Br(CH2)3OH, DMF;
(iii) Et3N, MsCl, CH2Cl2; (iv) LiBr, acetone.

Scheme 1.41

As for additional substitution on the side chain of the TTF molecule, transformation of the cyano or hydroxy group into amine or phosphine functionality has been reported (Scheme 1.42).[119,120] In order to obtain TTFs with two primary alkyl amines, two synthetic strategies were employed: one is the reduction of the cyano and azido groups, and the other is the application of Gabriel synthesis. A one-pot reduction of the two cyano groups in TTF derivative **82** with $Me_2S{\cdot}BH_3$ led to the bis(aminopropylthio)-substituted TTF derivative **83** (Scheme 1.42a),[43,119] but this method was not applicable to the conversion of $–SCH_2CN$ into $-S(CH_2)_2NH_2$ due to a solubility problem. Therefore, other routes starting from the bis(hydroxyethylthio)-attached TTF derivative **84** were examined (Scheme 1.42b).[119] Substitution of the hydroxy groups in **84** with the azido groups (Route A in Scheme 1.42b) and subsequent reduction with $LiAlH_4$ gave the bis(aminoethylthio)-containing TTF derivative **85**. However, since this route could not be reproduced, the multi-step route (Gabriel synthesis, see Route B in Scheme 1.42b) was used. As a result, **85** was obtained in good yield. It is noteworthy that the diphenylphosphino group can be introduced into the side chain of TTF derivative **86** *via* deprotection/alkylation reactions (Scheme 1.42c).[120]

1.3.5 Conversion of Protected TTF Chalcogenolates into Dimeric and Oligomeric TTFs

Synthesis of the single- and double-bridged bis-TTFs from the bis(cyanoethyl)-protected TTFs of Type II has been reported. As shown in Scheme 1.43, the linkage of two same TTFs, which leads to a symmetrical bis-TTF derivative, can be carried out by monodeprotection of TTF derivative **87** and subsequent reaction with 0.5 equiv. of dihalide in a high dilute solution (Route A).[12)]

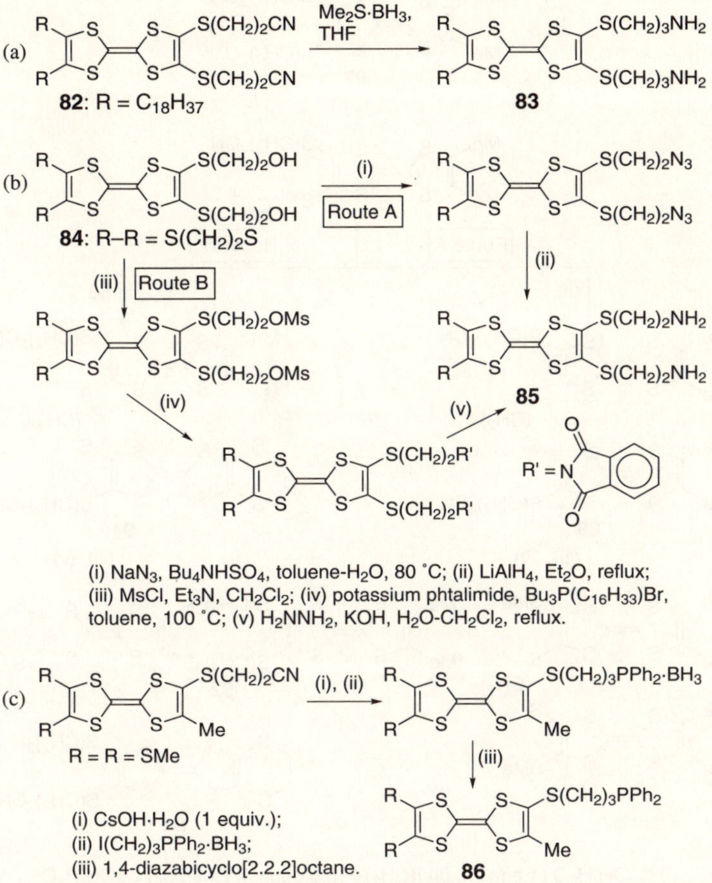

(i) NaN$_3$, Bu$_4$NHSO$_4$, toluene-H$_2$O, 80 °C; (ii) LiAlH$_4$, Et$_2$O, reflux;
(iii) MsCl, Et$_3$N, CH$_2$Cl$_2$; (iv) potassium phtalimide, Bu$_3$P(C$_{16}$H$_{33}$)Br,
toluene, 100 °C; (v) H$_2$NNH$_2$, KOH, H$_2$O-CH$_2$Cl$_2$, reflux.

(i) CsOH·H$_2$O (1 equiv.);
(ii) I(CH$_2$)$_3$PPh$_2$·BH$_3$;
(iii) 1,4-diazabicyclo[2.2.2]octane.

Scheme 1.42

However, this method was found to be ineffective for the linkage of two different TTFs, which leads to an unsymmetrical bis-TTF derivative.[117] A general reaction sequence for constructing unsymmetrical as well as symmetrical bis-TTFs involved monodeprotection of the bis-protected TTF derivative followed by reaction with the mono-halogenoalkylated TTF derivative (Route B).[117,122] The resulting symmetrical bis-TTF derivative **89** underwent deprotection/methylation to give a new single-bridged bis-TTF derivative **90**, and intramolecular cyclization of the unsymmetrical bis-TTF derivative **91** *via* deprotection followed by reaction with dihalide under high dilution conditions led to the double-bridged bis-TTF macrocycle **92**.[121,122] A reaction sequence similar to Route B has been applied to the synthesis of oligo-TTFs containing more than two TTF units.[121] For example, tris-TTF macrocycles **93** and **94** (Scheme 1.44), as well as tetrakis- and pentakis-TTF macrocycles,[122] can be obtained.

Bis-protected TTFs of Type I have also been employed for synthesizing bis-TTF macrocycles.[122,125] As shown in Scheme 1.45, a rigid bis-TTF macrocycle **95** comprising two TTF units and two alkynyl linkers can be obtained *via* deprotection/alkylation of TTF derivative **96**.[122] In this case, because the length of the alkynyl linker is considerably shorter than the distance

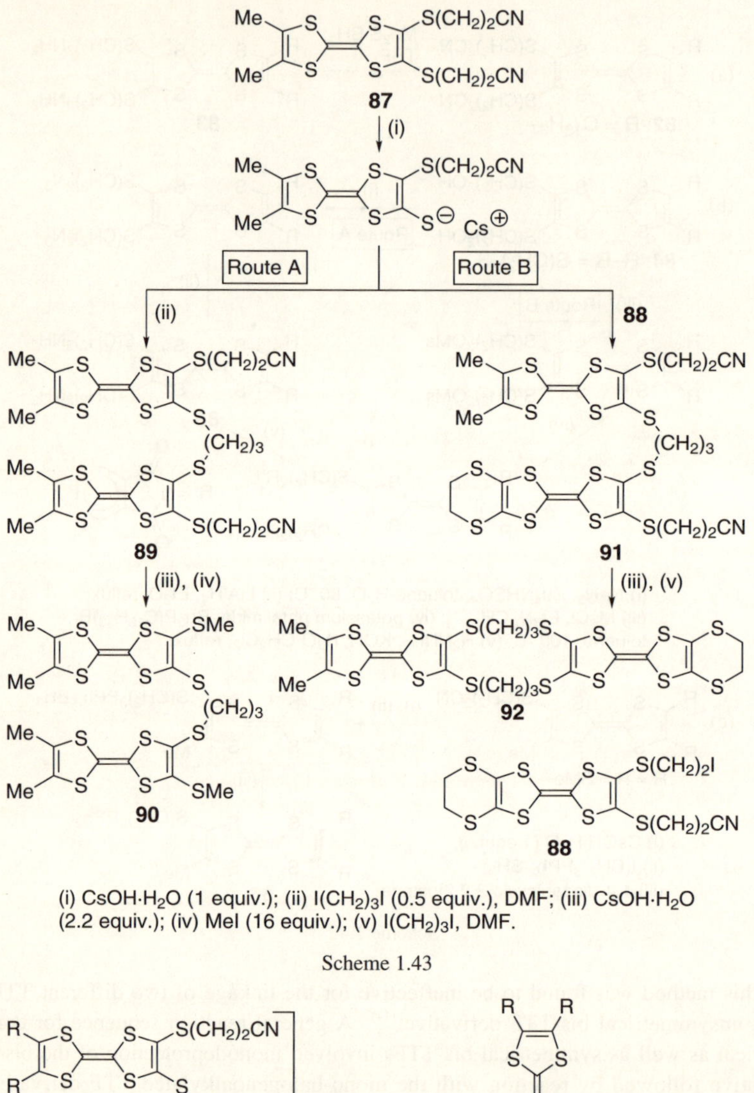
87
(i)
Route A
Route B
(ii)
88
89
91
(iii), (iv)
(iii), (v)
90
92
88

(i) $CsOH{\cdot}H_2O$ (1 equiv.); (ii) $I(CH_2)_3I$ (0.5 equiv.), DMF; (iii) $CsOH{\cdot}H_2O$ (2.2 equiv.); (iv) MeI (16 equiv.); (v) $I(CH_2)_3I$, DMF.

Scheme 1.43

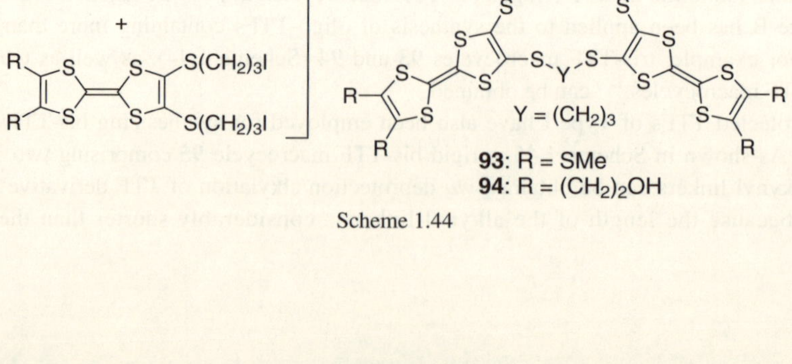
CsOH·H2O
(2.2 equiv.),
DMF
Y = (CH2)3
93: R = SMe
94: R = (CH2)2OH

Scheme 1.44

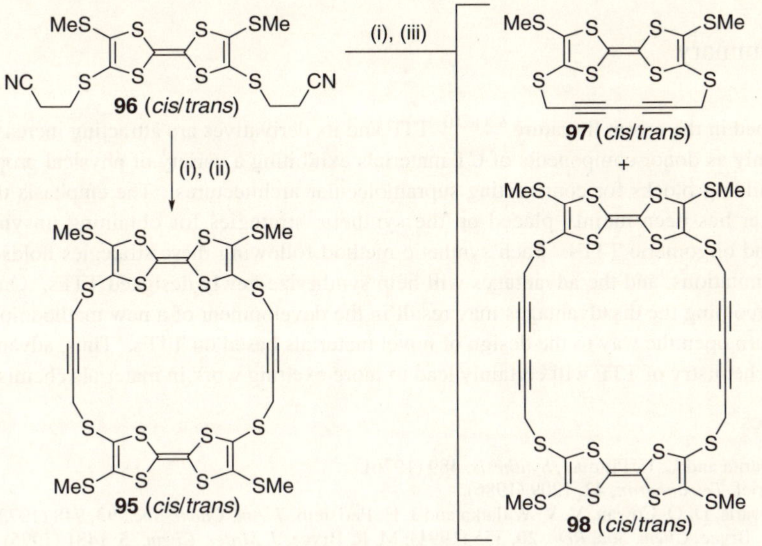

96 (*cis/trans*) (i), (iii) **97** (*cis/trans*)

+

95 (*cis/trans*) **98** (*cis/trans*)

(i) CsOH·H₂O (2.2 equiv.), MeOH-DMF; (ii) ClCH₂-C≡C-CH₂Cl;
(iii) TsOCH₂-C≡C-C≡C-CH₂OTs.

Scheme 1.45

between two thiolates in **96**, **95** was isolated as a sole product. A similar tendency was observed in the reaction of the dithiolate of **96** with bis[4-(bromomethyl)phenyl]methane.[125] On the other hand, deprotection of **96** followed by reaction with hexa-2,4-diyne-1,6-diol ditosylate gave a mixture of TTF phane **97** (major) and bis-TTF macrocycle **98** (minor).[122] These results indicate that the formation of macrocyclic bis-TTFs is dependent on the length of the linkers arising from the alkylating reagents used. In order to avoid formation of the TTF phane, a stepwise reaction sequence involving monodeprotection/alkylation (using 0.5 equiv. of dihalide) and deprotection/alkylation was used.[125] Moreover, it has been reported that a twofold deprotection of **96** followed by reaction with TTF derivative **99** (Scheme 1.46) bearing two iodinated side chains on one 1,3-dithiole ring gives another type of bis-TTF macrocycle.[123]

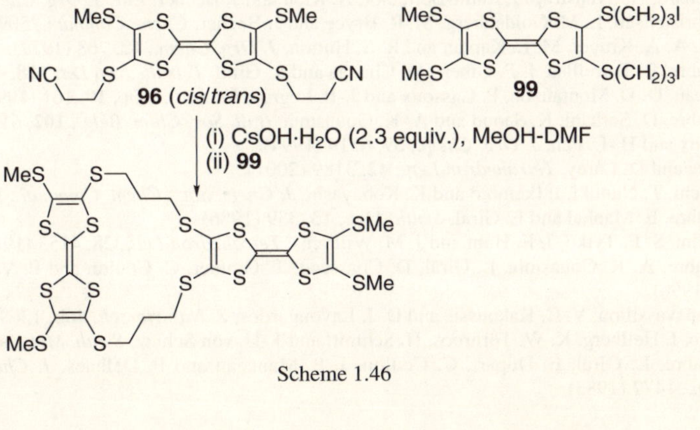

96 (*cis/trans*) **99**

(i) CsOH·H₂O (2.3 equiv.), MeOH-DMF
(ii) **99**

Scheme 1.46

1.4   Summary

As described in the recent literature,[6–8,16,21] TTF and its derivatives are attracting increasing attention not only as donor components of CT materials exhibiting a variety of physical properties but also as building blocks for constructing supramolecular architectures. The emphasis throughout this chapter has been mainly placed on the synthetic strategies for obtaining unsymmetrical, dimeric and oligomeric TTFs. Each synthetic method following these strategies holds possibilities and limitations, and the advantages will help synthesize newly designed TTFs. On the other hand, overcoming the disadvantages may result in the development of a new methodology, which could in turn open the way to the design of novel materials based on TTFs. Thus, advances in the synthetic chemistry of TTF will certainly lead to more exciting work in materials chemistry.

References

1. M. Narita and C. U. Pittman, *Synthesis*, 489 (1976).
2. A. Krief, *Tetrahedron*, **42**, 1209 (1986).
3. J. Ferraris, D. O. Cowan, V. V. Walatka and J. H. Perlstein, *J. Am. Chem. Soc.*, **93**, 948 (1973).
4. M. R. Bryce, *Chem. Soc. Rev.*, **20**, 355 (1991); M. R. Bryce, *J. Mater. Chem.*, **5**, 1481 (1995).
5. T. Ishiguro, K. Yamaji and G. Saito, *Organic Superconductors*, 2nd ed., Springer Ser. Solid-State Sci., Vol. 88, Springer, Berlin (1998).
6. J. M. Williams, J. R. Ferraro, R. J. Thorn, K. D. Carlson, U. Geiser, H. H. Wang, A. M. Kini and M.-H. Whangbo, *Organic Superconductors (Including Fullerenes)*, Prentice Hall, Englewood Cliffs, NJ (1992).
7. J.-M. Fabre, *J. Phys. IV France*, **10**, 3 (2000).
8. J. L. Segura and N. Martín, *Angew. Chem. Int. Ed.*, **40**, 1372 (2001).
9. J. Garín, *Adv. Heterocycl. Chem.*, **62**, 249 (1995).
10. G. Schukat and E. Fanghänel, *Sulfur Rep.*, **18**, 1 (1996).
11. K. B. Simonsen, N. Svenstrup, J. Lau, O. Simonsen, P. Mork, G. J. Kristensen and J. Becher, *Synthesis*, 407 (1996).
12. K. B. Simonsen and J. Becher, *Synlett*, 1211 (1997); M. B. Nielsen, C. Lomholt and J. Becher, *Chem. Soc. Rev.*, **29**, 153 (2000).
13. J. Roncali, *J. Mater. Chem.*, **7**, 2307 (1997).
14. M. R. Bryce, *J. Mater. Chem.*, **10**, 589 (2000).
15. J. P. Legros, F. Dahan, L. Binet, C. Carcel and J.-M. Fabre, *J. Mater. Chem.*, **10**, 2685 (2000); J. P. Legros, F. Dahan, L. Binet and J.-M. Fabre, *Synth. Met.*, **102**, 1632 (1999).
16. M. Adam and K. Müllen, *Adv. Mater.*, **6**, 439 (1994).
17. T. Otsubo, Y. Aso and K. Takimiya, *Adv. Mater.*, **8**, 203, (1996).
18. P. Blanchard, N. Svenstrup, J. Rault-Berthelot, A. Riou and J. Becher, *Eur. J. Org. Chem.*, 1743 (1998).
19. C. A. Christensen, L. M. Goldenberg, M. R. Bryce and J. Becher, *Chem. Commun.*, 509 (1998).
20. F. Wudl, A. A. Kruger, M. L. Kaplan and R. S. Hutton, *J. Org. Chem.*, **42**, 768 (1977).
21. J.-M. Fabre, E. Torreilles, J.-P. Gibert, M. Chanaa and L. Giral, *Tetrahedron Lett.*, **18**, 4033 (1977).
22. B. Garreau, D. D. Montauzon, P. Cassoux and J.-P. Legros, *New J. Chem.*, **19**, 161 (1995).
23. J.-M. Fabre, D. Serhani, K. Saoud and A.-K. Gouasmia, *Bull. Soc. Chim. Belg.*, **102**, 615 (1993).
24. M. A. Fox and H.-L. Pan, *J. Org. Chem.*, **59**, 6519 (1994).
25. N. Bellec and D. Lorcy, *Tetrahedron Lett.*, **42**, 3189 (2001).
26. K. Kikuchi, T. Namiki, I. Ikemoto and K. Kobayashi, *J. Chem. Soc., Chem. Commun.*, 1472, (1986).
27. J.-M. Fabre, E. Manhal and L. Giral, *Synth. Met.*, **13**, 339 (1986).
28. A. M. Kini, S. F. Tytko, J. E. Hunt and J. M. Williams, *Tetrahedron Lett.*, **28**, 4153 (1987).
29. J.-M. Fabre, A. K. Gouasmia, L. Giral, D. Chasseau, T. Granier, C. Coulon and P. Vaca, *Synth. Met.*, **35**, 57 (1990).
30. G. C. Papavassiliou, V. C. Kakoussis and D. J. Lavougardos, *Z. Naturforsch*, **46b**, 1269 (1991).
31. M. Moge, J. Hellberg, K. W. Törnroos, H. Schmitt and J.-U. von Schütz, *Synth. Met.*, **86**, 1877 (1997).
32. J.-M. Fabre, L. Giral, E. Dupart, C. Coulon, J.-P. Manceau and P. Delhaes, *J. Chem. Soc., Chem. Commun.*, 1477 (1983).

33. H. K. Spencer, M. P. Cava and A. F. Garito, *J. Chem. Soc., Chem. Commun.*, 966 (1976).
34. C. Gonnella and M. P. Cava, *J. Org. Chem.*, **43**, 369 (1978).
35. O. J. Dautel and M. Fourmigué, *J. Org. Chem.*, **65**, 6479 (2000).
36. M. Fettouhi, L. Ouahab, D. Serhani, J.-M. Fabre, L. Ducasse, J. Amiell, R. Canet and P. Delhaes, *J. Mater. Chem.*, **3**, 1101 (1993).
37. L. Binet, J.-M. Fabre, C. Montginoul, K. B. Simonsen and J. Becher, *J. Chem. Soc., Perkin Trans. 1*, 783 (1996).
38. T. Konoike, K. Namba, T. Shinada, K. Sakaguchi, G. C. Papavassilou, K. Murata and Y. Ohfune, *Synlett*, **9**, 1476 (2001).
39. G. Scherowski and J. Weiland, *Chem. Ber.*, **109**, 3155 (1974).
40. M. G. Miles, J. S. Wager, J. D. Wilson and A. R. Siedle, *J. Org. Chem.*, **40**, 2575 (1975).
41. S. Yoneda, T. Kawase, M. Inaba and Z. Yoshida, *J. Org. Chem.*, **43**, 595 (1978).
42. R. D. McCullough, M. A. Petruska and J. A. Belot, *Tetrahedron*, **55**, 9979 (1999).
43. R. P. Parg, J. D. Kilburn, M. C. Petty, C. Pearson and J. G. Rayan, *Synthesis*, 613 (1994); R. P. Parg, J. D. Kilburn, M. C. Petty, C. Pearson and J. G. Rayan, *J. Mater. Chem.*, **5**, 1609 (1995).
44. C. Mézière, M. Fourmigué and J.-M. Fabre, *C. R. Acad. Sci., Ser. IIc*, **3**, 387 (2000).
45. E. Ribera, J. Veciana, E. Molins, I. Mata, K. Wurst and C. Rovira, *Eur. J. Org. Chem.*, 2867 (2000).
46. S.-G. Liu and L. Etchegoyen, *Eur. J. Org. Chem.*, 1157 (2000); S.-G. Liu, H. Liu, K. Bandyopadhyay, Z. Gao and L. Etchegoyen, *J. Org. Chem.*, **65**, 3292 (2000).
47. P. Hudhomme, S. L. Moustarder, C. Durand, N. Gallego-Planas, N. Mercier, P. Blanchard, E. Levillain, M. Allain, A. Gorgues and A. Riou, *Chem. Eur. J.*, **7**, 5070 (2001).
48. J.-M. Fabre, D. Serhani, K. Saoud, S. Chakroune and M. Hoch, *Synth. Met.*, **60**, 295 (1993).
49. Y. Misaki, H. Fujiwara, T. Yamabe, T. Mori, H. Mori and S. Tanaka, *Chem. Lett.*, 1653 (1994).
50. Y. Misaki, K. Kawakami, H. Fujiwara, T. Miura, T. Kochi, M. Taniguchi, T. Yamabe, T. Mori, H. Mori and S. Tanaka, *Mol. Cryst. Liq. Cryst.*, **296**, 77 (1997).
51. J. Yamada, S. Mishima, H. Anzai, M. Tamura, Y. Nishio, K. Kajita, T. Sato, H. Nishikawa, I. Ikemoto and K. Kikuchi, *Chem. Commun.*, 2517 (1996); J. Yamada, N. Akashi, H. Anzai, M. Tamura, Y. Nishio and K. Kajita, *Mol. Cryst. Liq. Cryst.*, **296**, 53 (1997).
52. H. Fujiwara, Y. Misaki, T. Yamabe, T. Mori, H. Mori and S. Tanaka, *J. Mater. Chem.*, **10**, 1565 (2000).
53. J.-M. Fabre, S. Chakroune, A. Javidan, M. Calas, A. Souizi and L. Ouahab, *Synth. Met.*, **78**, 89 (1996).
54. A. J. Moore, L. M. Goldenberg, M. R. Bryce, M. C. Petty, J. Moloney, J. A. K. Howard, M. J. Joyce and S. N. Port, *J. Org. Chem.*, **65**, 8269 (2000).
55. H. Li, D. Zhang, Y. Yao, W. Xu, D. Zhu and Z. Wang, *J. Mater. Chem.*, **10**, 2063 (2000).
56. T. Ozturk, N. Saygili, S. Ozkara, M. Pilkington, C. R. Rice, D. A. Tranter, F. Turksoy and J. D. Wallis, *J. Chem. Soc., Perkin Trans.1*, 407 (2001).
57. Y. Ishikawa, T. Miyamoto, A. Yoshida, Y. Kawada, J. Nakazaki, A. Izuoka and T. Sugawara, *Tetrahedron Lett.*, **40**, 8819 (1999).
58. E. Aqad, A. Ellern, L. Shapiro and V. Khodorkovsky, *Tetrahedron Lett.*, **41**, 2983 (2000).
59. O. Neilands, S. Belyakov, V. Tilika and A. Edzina, *J. Chem. Soc., Chem. Commun.*, 325 (1995).
60. J. O. Jeppesen, K. Takimiya, N. Thorup and J. Becher, *Synthesis*, 803 (1999).
61. B. Domercq, T. Devic, M. Fourmigué, P. Auban-Senzier and H. Canadell, *J. Mater. Chem.*, **11**, 1570 (2001).
62. O. J. Dautel and M. Fourmigué, *J. Chem. Soc., Perkin Trans. 1*, 3399 (2001).
63. H. Fujiwara and H. Kobayashi, *Chem. Commun.*, 2417 (1999).
64. K. Takimiya, A. Oharuda, A. Morikami, Y. Aso and T. Otsubo, *Eur. J. Org. Chem.*, 3013 (2000); M. Kodani, K. Takimiya, Y. Aso, T. Otsubo, T. Nakayashiki and Y. Misaki, *Synthesis*, 1614 (2001).
65. K. Takimiya, N. Thorup and J. Becher, *Chem. Eur. J.*, **6**, 1947 (2000); K. Takimiya, A. Oharuda, Y. Aso, F. Ogura and T. Otsubo, *Chem. Mater.*, **12**, 2196 (2000); K. Takimiya, Y. Kataoka, Y. Aso, T. Otsubo, H. Fukuoka and S. Yamanaka, *Angew. Chem. Int. Ed.*, **40**, 1122 (2001).
66. K. Lerstrup, I. Johannssen and M. Jorgensen, *Synth. Met.*, **27**, B9 (1988).
67. J. Yamada, Y. Amono, S. Takasaki, R. Nakanishi, K. Matsumoto, S. Satoki and H. Anzai, *J. Am. Chem. Soc.*, **117**, 1149 (1995); J. Yamada, S. Satoki, S. Mishima, N. Akashi, K. Takahashi, N. Masuda, Y. Nishimoto, S. Takasaki and H. Anzai, *J. Org. Chem.*, **61**, 3987 (1996).
68. L. Giral, J.-M. Fabre and A. Gouasmia, *Tetrahedron Lett.*, **27**, 4315 (1986).
69. J.-M. Fabre, S. Chakroune, A. Javidan, L. Zanik, L. Ouahab, S. Golhen and P. Delhaes, *Synth. Met.*, **70**, 1127 (1995).
70. I. Sudmale, G. V. Tormos, V. Y. Khodorkovsky, A. S. Edzina, O. J. Neilands and M. P. Cava, *J. Org. Chem.*, **58**, 1355 (1993).
71. T. K. Hansen, M. R. Bryce, J. A. K. Howard and D. S. Yufit, *J. Org. Chem.*, **59**, 5324 (1993).
72. L. Binet, J.-M. Fabre and J. Becher, *Synthesis*, 26 (1997).

73. A. K. Gouasmia, J.-M. Fabre, L. Boudiba, L. Kaboub and C. Carcel, *Synth. Met.*, **120**, 809 (2001).
74. N. Svenstrup, K. M. Rasmussen, T. K. Hansen and J. Becher, *Synthesis*, 809 (1994).
75. K. Lerstrup, I. Johannsen and M. Jorgensen, *Synth. Met.*, **27**, B9 (1988).
76. M. Fourmigué, F. C. Krebs and J. Larsen, *Synthesis*, 509 (1993).
77. H. Mora, J.-M. Fabre, L. Giral and C. Montginoul, *Bull. Soc. Chim. Belg.*, **101**, 137 (1992); A. J. Moore, M. R. Bryce, A. S. Batsanov, J. C. Cole and J. A. K. Howard, *Synthesis*, 675 (1995).
78. J.-M. Fabre, J. Amouroux, B. Garreau and J. P. Legros, *Bull. Soc. Chim. Belg.*, **103**, 97 (1994).
79. M. R. Bryce, A. J. Moore, D. Lorcy, A. S. Dhindsa and A. Robert, *J. Chem. Soc., Chem. Commun.*, 470 (1990).
80. J. Hellberg, M. Moge, H. Schmitt and J. U. Von Schütz, *J. Mater. Chem.*, **5**, 1549 (1995); J. Hellberg and M. Moge, *Synthesis*, 198 (1996).
81. H.-J. Cristau, F. Darviche, E. Torreilles and J.-M. Fabre, *Tetrahedron Lett.*, **39**, 2103 (1998).
82. H.-J. Cristau, F. Darviche, M.-T. Babonneau, J.-M. Fabre and E. Torreilles, *Tetrahedron*, **55**, 13029 (1999).
83. H.-J. Cristau, F. Darviche, M.-T. Babonneau, J.-M. Fabre and E. Torreilles, *Tetrahedron Lett.*, **40**, 7219 (1999).
84. M. Jorgensen, K. A. Lerstrup and K. Bechgaard, *J. Org. Chem.*, **56**, 5684 (1991).
85. D. Lorcy, D. Guérin, K. Boubekeur, R. Carlier, P. Hascoat, A. Tallec and A. Robert, *J. Chem. Soc., Perkin Trans.1*, 2719 (2000).
86. H. Fujiwara, Y. Misaki, T. Yamabe, T. Mori, H. Mori and S. Tanaka, *J. Mater. Chem.*, **10**, 1565 (2000).
87. J. Yamada, S. Tanaka, J. Segawa, M. Hamasaki, K. Hagiya, H. Anzai, H. Nishikawa, I. Ikemoto and K. Kikuchi, *J. Org. Chem.*, **63**, 3952, (1998).
88. J. Yamada, S. Kawagishi, T. Shinomaru, H. Akutsu, S. Nakatsuji, H. Nishikawa, I. Ikemoto and K. Kikuchi, *Synth. Met.*, **120**, 787 (2001); J. Yamada, M. Watanabe, T. Toita, H. Akutsu, S. Nakatsuji, H. Nishikawa, I. Ikemoto and K. Kikuchi, *Chem. Commun.*, 1118 (2002).
89. J. Yamada, H. Nishikawa and K. Kikuchi, *J. Mater. Chem*, **9**, 617 (1999).
90. J. Yamada, M. Watanabe, H. Anzai, H. Nishikawa, I. Ikemoto and K. Kikuchi, *Angew. Chem. Int. Ed.*, **38**, 810 (1999).
91. J. Yamada, M. Watanabe, H. Akutsu, S. Nakatsuji, H. Nishikawa, I. Ikemoto and K. Kikuchi, *J. Am. Chem. Soc.*, **123**, 4174 (2001).
92. D. C. Green, *J. Chem. Soc., Chem. Commun.*, 161 (1977); D. C. Green, *J. Org. Chem.*, **44**, 1476 (1979).
93. H. Mora, J.-M. Fabre, L. Giral and C. Montginoul, *Bull. Soc. Chim. Belg.*, **101**, 741 (1992).
94. M. Gonzalez, N. Martín, J. L. Segura, J. Garín and J. Orduna, *Tetrahedron Lett.*, **39**, 3269 (1998).
95. A. S. Batsanov, M. R. Bryce, G. Cooke, A. S. Dhindsa, J. N. Heaton, J. A. K. Howard, A. J. Moore and M. C. Petty, *Chem. Mater.*, **6**, 1419 (1994).
96. K. Euzé, M. Fourmigué and P. Batail, *J. Mater. Chem.*, **9**, 2373 (1999).
97. J.-M. Fabre, J. Garín and S. Uriel, *Tetrahedron Lett.*, **32**, 6407 (1991); J.-M. Fabre, J. Garín and S. Uriel, *Tetrahedron*, **48**, 3983 (1992).
98. C. Wang, A. Ellern, V. Khodorkovsky, J. Bernstein and J. Y. Becker, *J. Chem. Soc., Chem. Commun.*, 983 (1994).
99. D. E. John, A. J. Moore, M. R. Bryce, A. S. Batsanov and J. A. K. Howard, *Synthesis*, 826 (1998).
100. J. Garín, J. Orduna, M. Saviron, M. R. Bryce, A. J. Moore and V. Morisson, *Tetrahedron*, **52**, 11063 (1996).
101. C. Guillot, P. Hudhomme, P. Blanchard, A. Gorgues, M. Jubault and G. Duguay, *Tetrahedron Lett.*, **36**, 1645 (1995).
102. G. Cooke, V. M. Rotello and A. Radhi, *Tetrahedron Lett.*, **40**, 8611 (1999); C. Guillot, P. Hudhomme, P. Blanchard, A. Gorgues, M. Jubault and G. Duguay, *Tetrahedron Lett.*, **36**, 1645 (1995).
103. T. Otsubo, Y. Kochi, A. Bitoh and F. Ogura, *Chem. Lett.*, 2047 (1994).
104. P. Hudhomme, S. G. Liu, D. Kreher, M. Cariou and A. Gorgues, *Tetrahedron Lett.*, **40**, 2927 (1999).
105. J. L. Segura, E. M. Priego, N. Martín, C. Luo and D. M. Guldi, *Org. Lett.*, **2**, 4021 (2000).
106. S. G. Liu, D. Kreher, P. Hudhomme, E. Levillain, M. Cariou, J. Delaunay, A. Gorgues, J. Vidal-Gancedo, J. Veciana and C. Rovira, *Tetrahedron Lett.*, **42**, 3717 (2001).
107. E. Allard, J. Delaunay, F. Cheng, J. Cousseau, J. Orduna and J. Garín, *Org. Lett.*, **3**, 3503 (2001).
108. J. Nakazaki, M. M. Matsushita, A. Izuoka and T. Sugawara, *Tetrahedron Lett.*, **40**, 5027 (1999).
109. M. Iyoda, K. Hara, C. R. V. Rao, Y. Kuwatani, K. Takimiya, A. Morikami, Y. Aso and T. Otsubo, *Tetrahedron Lett.*, **40**, 5729 (1999).
110. A. J. Moore, A. S. Batsanov, M. R. Bryce, J. A. K. Howard, V. Khodorkovsky, L. Shapiro and A Shames, *Eur. J. Org. Chem.*, 73 (2001).
111. M. R. Bryce, G. Cooke, W. Devonport, F. M. A. Duclairoir and V. M. Rotello, *Tetrahedron Lett.*, **42**, 4223 (2001).

112. A. Bitoh, Y. Kohchi, T. Otsubo, F. Ogura and K. Ikeda, *Synth. Met.*, **70**, 1123 (1995).
113. T. Otsubo, Y. Kochi, A. Bitoh and F. Ogura, *Chem. Lett.*, 2047 (1994).
114. D. E. John, A. J. Moore, M. R. Bryce, A. S. Batsanov, M. A. Leech and J. A. K. Howard, *J. Mater. Chem.*, **10**, 1273 (2000).
115. M. Iyoda, Y. Kuwatani, N. Ueno and M. Oda, *J. Chem. Soc., Chem. Commun.*, 158 (1992).
116. M. Iyoda, M. Fukuda, S. Sasaki and M. Yoshida, *Synth. Met.*, **70**, 1171 (1995).
117. C. Carcel, J.-M. Fabre, B. G. de Bonneval and C. Coulon, *New J. Chem.*, **24**, 919 (2001).
118. F. L. Derf, M. Sallé, N. Mercier, J. Becher, P. Richomme, A. Gorgues, J. Orduna and J. Garín, *Eur. J. Org. Chem.*, 1861 (1998).
119. L. Binet and J.-M. Fabre, *Synthesis*, 1179 (1997).
120. P. Bellon, E. Brulé, N. Bellec, K. Chamontin and D. Lorcy, *J. Chem. Soc., Perkin Trans. 1*, 4409 (2000).
121. C. Carcel and J.-M. Fabre, *Synth. Met.*, **120**, 747 (2001).
122. K. B. Simonsen, N. Thorup and J. Becher, *Synthesis*, 1399 (1997).
123. D. E. John, A. S. Batsanov, M. R. Bryce and J. A. K. Howard, *Synthesis*, 824 (2000).
124. H. Spanggaard, J. Prehn, M. B. Nielsen, P. Levillain, M. Allain and J. Becher, *J. Am. Chem. Soc.*, **122**, 9486 (2000).
125. J. Lau, P. Blanchard, A. Riou, M. Jubault, M. P. Cava and J. Becher, *J. Org. Chem.*, **62**, 4936 (1997).

2

Tetrachalcogenafulvalenes with Four Additional Heteroatoms

2.1 Introduction

Most organic conductors and superconductors are based on tetrachalcogenafulvalenes (TCFs) and metal 1,2-dichalcogenolenes.[1] TCFs serve as π-electron donors, and tetrathiafulvalene (TTF, Chart 2.1) is the first member of this series and one of the simplest symmetrical TCFs. TTF was synthesized for the first time in 1964,[2,3] although prior to this synthesis, DB-TTF (Chart 2.1) had already been reported.[4] Following the synthesis of TTF, selenium and tellurium analogues of TTF, *i.e.*, tetraselenafulvalene (TSF, Chart 2.1)[5] and tetratellurafulvalene (TTeF, Chart 2.1),[6] which are the two other simplest symmetrical TCFs, were synthesized. The discovery of electrical conductibility in charge-transfer (CT) complexes of some polycyclic aromatic compounds acting as π-donors with bromine or iodine, followed by the prediction that the CT complexes of sulfur-containing aromatic systems could exhibit good electrical conductivity[7] have led to the synthesis and study of a variety of TCFs. In fact, the CT complexes of TTF with bromine, chlorine, *etc.* were found to be good conductors.[8] It was also found that TTF with organic π-acceptors such as TCNQ (Chart 2.1)[9] or with metal 1,2-dichalcogenolenes such as Ni(dmit)$_2$ (dmit = 1,3-dithiole-2-thione-4,5-dithiolate, Chart 2.1)[10] forms CT complexes exhibiting good room-tempera-

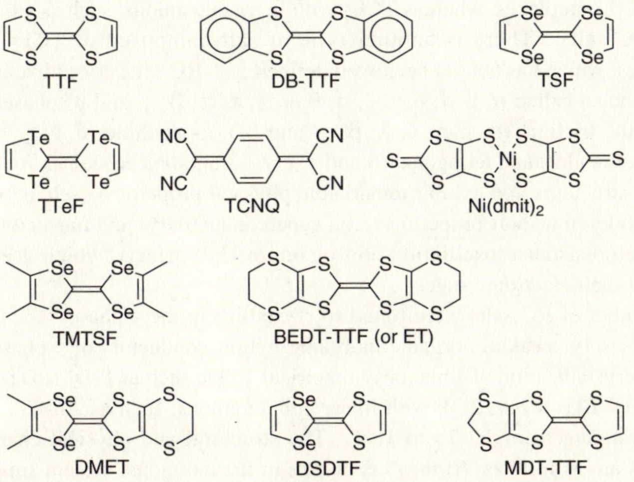

Chart 2.1

ture conductivity. Some of these CT complexes were metallic and underwent metal-to-insulator (MI) transition at low temperatures. In addition, the symmetrical π-donors TMTSF (Chart 2.1) and BEDT-TTF (or ET, Chart 2.1)[3b,11] have been found to give a number of conducting CT complexes and radical-cation (RC) salts (also called cation-radical salts), several of which underwent metal-to-superconductor transition at very low temperatures.[12,13] The first organic superconductor (TMTSF)$_2$PF$_6$, with a critical superconducting transition temperature (T_c) of 1.4 K under a pressure of 6.5 kbar, and the first ET-based organic superconductor (ET)$_2$ReO$_4$, with T_c = 2 K under a pressure of 4 kbar, were discovered in 1980[12] and 1982,[13] respectively. The superconducting salts based on TMTSF (the so-called Bechgaard salts) are quasi one-dimensional (1D) conductors, while the ET-based superconductors are regarded as quasi two-dimensional (2D) conductors. Furthermore, unsymmetrical donors, such as DMET (Chart 2.1),[14] which is a derivative of diselenadithiafulvalene (DSDTF, Chart 2.1), and MDT-TTF (Chart 2.1),[15] were found to give superconducting salts.

During the last two decades, a large number of conducting and superconducting materials based on symmetrical and unsymmetrical TCFs as well as M(dmit)$_2$ (M = metal) have been prepared and studied.[1,16–83] These systems are usually divided into the two categories labeled CT complexes and radical-ion (RI) salts.[1,76–78] In the case of CT complexes, a donor molecule (D) and an acceptor molecule (A) associate to form a stable complex through partial CT from D to A: $D^0 + A^0 \rightarrow D^{+\rho}A^{-\rho}$, where ρ denotes the amount of CT in a DA complex, and its value depends mainly on the ionization potential of D and the electron affinity of A, which can be estimated by the redox potentials of D and A in solutions.[76–78] In RI salts, a donor or an acceptor is first transformed into the corresponding RI *via* a chemical oxidation or reduction process, and then the resulting RI is spatially organized when molecular association with a counterion is feasible. A compound with $0 < \rho < 1$, leading to a mixed valence system, is a conducting material and obtained by the reaction of a donor having a moderate oxidation potential with an acceptor having a moderate reduction potential or by the reaction of a strong donor (or acceptor) and a weak acceptor (or donor). The combination of a strong donor and a strong acceptor leads to an ionic or a single valence complex ($\rho = 1$), whereas the combination of a weak donor and a weak acceptor gives rise to a neutral or a molecular complex in which no CT occurs. TCFs with organic π-acceptors, such as TCNQ, form CT complexes, whereas TCFs with inorganic anions, such as Cl$^-$, ClO$_4^-$, PF$_6^-$ and AuBr$_2^-$, give RC salts. There is another type of salt composed of TCFs and metal 1,2-dichalcogenolenes, which has not yet been well defined.[1, 19] RC salts crystallize in many different phases, such as the so-called α, β, γ, δ, ε, ζ, η, θ, κ, λ, τ, α', β', γ' and β'' phases, which belong, for example, to the triclinic (α, β, γ, ζ, λ, β', γ' and β''), monoclinic (δ, ε, ζ, η, κ and α'), orthorhombic (γ, δ, θ and κ) and tetragonal (θ and τ) crystallographic systems. RC salts have interesting electronic structures and exhibit remarkable physical properties, such as optical properties (*e.g.*, vibronic mode), transport properties (*e.g.*, superconductivity) and magnetotransport properties (*e.g.*, magnetoresistance oscillation and quantum Hall effect), which arise from the low dimensionality of their electronic states.

A small number of RC salts were found to crystallize in the τ-phase.[21–36] Organic conductors or, more precisely speaking, organic-inorganic hybrid conductors of τ-phase have been obtained by electrocrystallization of some unsymmetrical TTFs, such as P-DMEDT-TTF (Chart 2.2) and EDO-DMEDT-TTF (Chart 2.2), with linear counteranions, such as AuBr$_2^-$, AuI$_2^-$ and I$_3^-$, in stoichiometries ranging from 2:1.75 to 1:~1. The structural and physical characteristics of τ-phase conductors are as follows: (i) they crystallize in the tetragonal system simpler than the orthorhombic, monoclinic and triclinic systems; (ii) their structures predict star-like Fermi surfaces; (iii) with decreasing temperature, they exhibit metallic behavior and a change to semiconducting

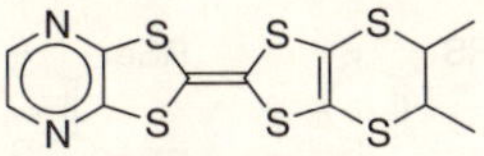

P-DMEDT-TTF 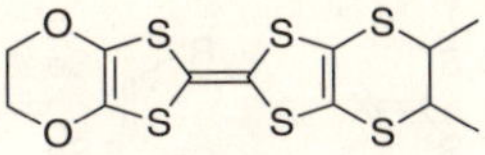   EDO-DMEDT-TTF

Chart 2.2

or insulating behavior at low temperatures; (iv) they exhibit broad ESR bands even at low temperatures; (v) they show negative Hall coefficients, indicating negative carriers (*i.e.*, electrons), negative magnetoresistances and positive magnetoreflectances at low fields; (vi) at higher field, they exhibit giant magnetoresistance oscillations and quantum Hall Plateaus.[21–36]

In this chapter, in addition to the synthesis and properties of TCFs with heteroatom-containing substituents, the formation and properties of their charged species, including their RCs, are reviewed. Special emphasis is placed on the unsymmetrical TCFs incorporating four additional heteroatoms and their charged species, which are components of τ-phase conductors. For the naming of derivatives of TCF, empirical or nonsystematic abbreviations rather than *Chemical Abstracts* or IUPAC nomenclature are used. Definitions of the abbreviations used herein have frequently appeared in the original papers and review articles.[1–83]

2.2 Preparation of Building Blocks

A variety of reactants and building blocks have been used for constructing TCFs. Elemental S, Se and Te, as well as many simple oxygen-, sulfur-, selenium- and nitrogen-containing compounds, such as CS_2, CSe_2, H_2Se, $HSCH_2CH_2SH$, $Me_2NC(=S)SNa$, 1,4-dioxane, $(Me_2N)_2C=S$ and 2,3-dichloropyrazine, have been used as reactants. Most of these reactants are commercially available. In contrast, most building blocks used for the synthesis of TCFs, *i.e.*, 2-oxo-1,3-dithioles (or 1,3-dithiol-2-ones), 2-oxo-1,3-diselenoles (or 1,3-diselenol-2-ones) and related compounds, were prepared by several methods.[1,2,5,14,15,28,30,37–75,82] Here, the preparation of 1,3-dichalcogenole-2-chalcogenones with more than one additional heteroatoms (O, S, Se and N) is mainly described. These compounds are listed in Chart 2.3, although most of them have been reported in the literature.[1,47,48,50,51,55]

Some building blocks, such as vinylenetrithiocarbonate (or 1,3-dithiole-2-thione, **1a′**, X = S) for TTF and 4,5-dimethyl-1,3-diselenole-2-selenone (**2b′**, X = Se) for TMTSF (Chart 2.1), are commercially available. Compounds **1** and **2** have been prepared by several cyclization procedures.[1–3,49,74] Vinylenetrithiocarbonate (**1a′**, X = S) was obtained in good yield by treating potassium trithiocarbonate [KS-C(=S)-SK] with 1,2-dichloroethyl ethyl ether [$ClCH_2CHCl(OEt)$], followed by elimination of ethanol promoted by *p*-toluenesulfonic acid. In addition, compound **1a′** and its selenium analogue **2a′** (X = Se) were prepared by thermolysis of 1,2,3-thiadiazole and 1,2,3-selenathiazole in the presence of CS_2 and CSe_2, respectively. Substituted derivatives of **1a′** and **2a′** were obtained from the corresponding β-keto-*N,N*-dialkylthiocarbonates, β-keto-*S*-alkylthiocarbonates and their Se-containing analogues by different cyclization procedures depending on the types of R′ group, heteroatom, *etc.* Compounds **1d′** and **2d′**, containing electron-withdrawing groups (R′ = CN, COOMe, CHO, *etc.*), were prepared by the reaction of the corresponding acetylenes with ethylenetrithiocarbonate and ethylenetriselenocarbonate, respectively.[1,41,43,46,50,51] Some compounds of types **1d′** and **2d′**, as well as compounds **11** and **12**, have functional groups which can be transformed into other functional groups. For example, (i) transformation of the CN group into the COOH group,[47,51] (ii) transformation of the ES group [E =

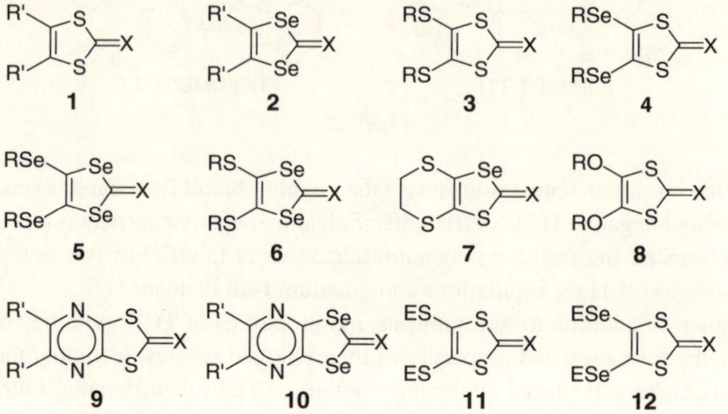

X = O, S, Se
a': R' = H, **b'**: R' = CH$_3$, **c'**: R'–R' = CH$_2$CH$_2$CH$_2$ *etc.*,
d': R' = CN, COOMe, CHO, *etc.*
a: R = CH$_3$, **b**: R–R = CH$_2$, **c**: R–R = CH$_2$CH$_2$, **d**: R–R = CH$_2$CH$_2$CH$_2$,
e:R–R = CH(CH$_3$)CH$_2$, **f**: R–R = CH(CH$_3$)CH(CH$_3$), **g**: R–R = CH=CH *etc.*
a: E = CH$_2$O(CH$_2$)$_2$SiMe$_3$, **b**: E = CH$_2$C$_6$H$_4$Ac-*p*,
c: E = CH$_2$CH$_2$CN, **e**: E = COPh *etc.*

Chart 2.3

CH$_2$O(CH$_2$)$_2$SiMe$_3$, CH$_2$C$_6$H$_4$Ac-*p*, CH$_2$CH$_2$CN, COPh, *etc.*] into the RS group *via* the corresponding thiolate (S$^-$),[48,53,62,70] *etc.*[74] have been reported. Also, compounds containing the vinylenedithio end group have been transformed into further extended compounds (see Schemes 2.2 and 2.13 for details).[46,51,60,61,74]

Compounds **3–5** have been prepared by treating the corresponding metal 1,2-dichalcogeno-lates such as (Bu$_4$N)$_2$Zn(dmit)$_2$ with alkyl halides such as CH$_3$I or with dibromoalkanes such as BrCH$_2$Br and BrCH$_2$CH$_2$Br.[39,40,42,45–48,50,51,59,69] Although (R$_4$N)$_2$Zn(dmit)$_2$ (R = Bu and Et) are commercially available, they can be obtained by chemical reduction of CS$_2$ with Na in DMF followed by addition of ZnCl$_2$ and R$_4$NBr (R = Bu and Et).[40,47,51] Their selenium analogues can also be obtained by a similar method using CSe$_2$ instead of CS$_2$.[47,51] An analogous compound

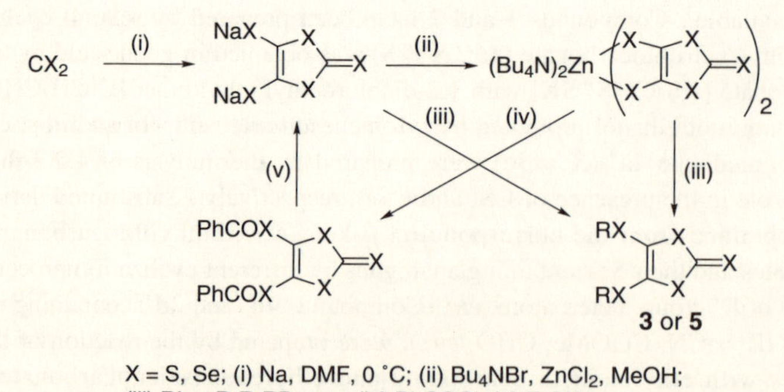

X = S, Se; (i) Na, DMF, 0 °C; (ii) Bu$_4$NBr, ZnCl$_2$, MeOH;
(iii) RI or BrR-RBr *etc.*; (iv) PhCOCl; (v) EtONa.

Scheme 2.1

(Bu$_4$N)$_2$Hg(dmit)$_2$ has been obtained by electrochemical reduction of CS$_2$ with Hg electrodes in DMF containing Bu$_4$NI.[10,37]   Scheme 2.1 outlines the procedures for the preparation of **3** or **5**. Procedures (i), (ii) and (iii) for the preparation of **3a–f**, procedures (i), (ii), (iv), (v) and (iii) for the preparation of **3g**, *etc.* have been utilized.[1,39,40,42,45–47,51,52,60)

Vinylenetrithiocarbonate (**1a′**, X = S) and its selenium analogue **2a′** (X = Se) have been used for the preparation of the corresponding zinc 1,2-dichalcogenolenes. Dilithiation of compounds with the vinylenedichalcogeno end group, in which each of two acidic hydrogen atoms bonded directly to the C=C bond is replaced by lithium, was achieved by using lithium diisopropylamide (LDA) as a metalation reagent at low temperature (–78 °C). To the resulting dilithiated product, elemental S or Se was added and the reaction mixture was allowed to warm up to 25 °C. Subsequent reaction with ZnCl$_2$ and Bu$_4$NBr led to a solid zinc 1,2-dichalcogenolate.[46,51,60,61) The procedures for preparations of **3–5** and further extended compounds are outlined in Schemes 2.2a–c. Starting from 2,3-dihydro-1,4-dithiin and its analogues, similar procedures were applied to the preparation of compounds of type **6**.[35,36,75)

An alternative method has been developed to prepare compounds of types **3** and **4**. The use of cyclic sulfates instead of the corresponding dibromoalkanes allowed the preparation of **3e,f**, as

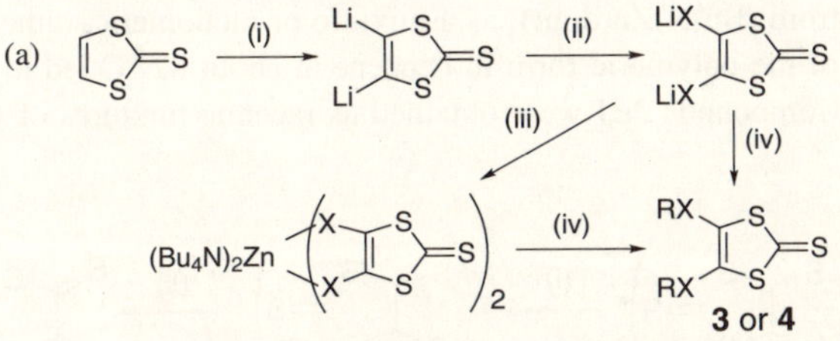

(i) LDA, THF, –78 °C; (ii) X (S, Se), THF, –78 °C → 25 °C;
(iii) Bu$_4$NBr, ZnCl$_2$, MeOH; (iv) RI or BrR-RBr.

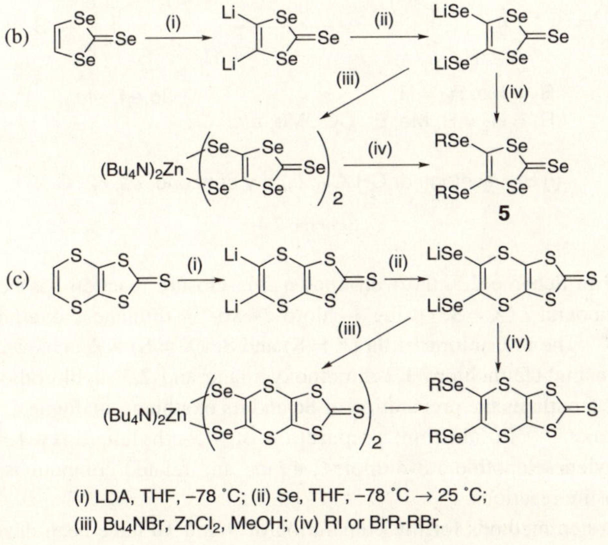

(i) LDA, THF, –78 °C; (ii) Se, THF, –78 °C → 25 °C;
(iii) Bu$_4$NBr, ZnCl$_2$, MeOH; (iv) RI or BrR-RBr.

Scheme 2.2

shown in Scheme 2.3.[35,36,63,66,67,71] Chiral compounds and their racemic mixtures as well as their selenium analogues were also prepared by this method. It has been reported that only the *trans* form of the cyclic sulfate reacts with the disodium dithiolate (X = S) shown in Scheme 2.3.[66]

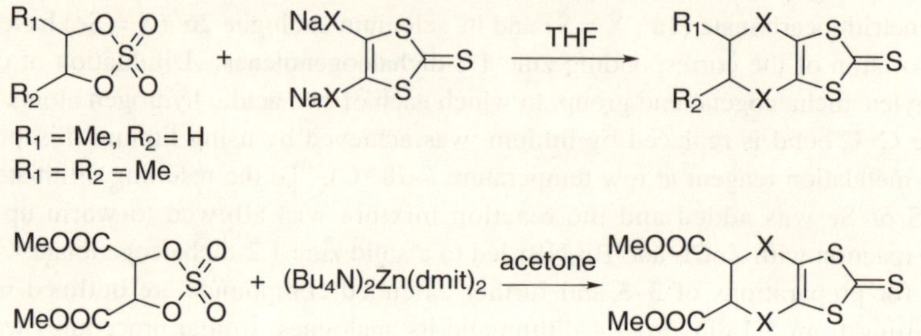

NaX
THF
$R_1 = Me, R_2 = H$
$R_1 = R_2 = Me$
MeOOC
acetone
$(Bu_4N)_2Zn(dmit)_2$

Scheme 2.3

Also, Diels-Alder cycloaddition reaction of the *trans*-alkenes with the 1,3-dithiole-2,4,5-trithione oligomer gave thiones **3c,e,f**,[28] *etc.*,[73,82b] as shown in Scheme 2.4. 1,3-Dithiole-2,4,5-trithione was obtained from $(Bu_4N)Zn(dmit)_2$ as a mixture of monomeric, dimeric and polymeric forms.[28] Thermolysis of the polymeric form in benzene at about 82 °C led to the generation of the monomeric form. Compounds **3e,f** were obtained as racemic mixtures of the corresponding enantiomers.

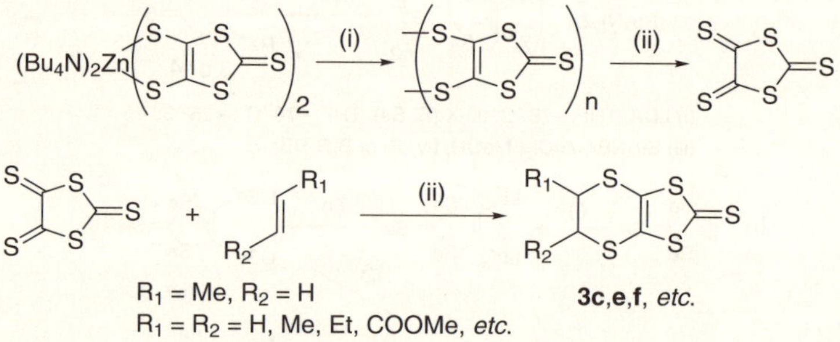

$(Bu_4N)_2Zn$
(i)
(ii)
$R_1 = Me, R_2 = H$
$R_1 = R_2 = H, Me, Et, COOMe, etc.$
3c,e,f, *etc.*

(i) Br_2, acetone or CH_2Cl_2; (ii) dry benzene, *ca.* 82 °C.

Scheme 2.4

As illustrated in Scheme 2.5, the well-known ring-closure reaction has been applied to the preparation of compound **7** (X = Se) using 3-chloro-2-oxo-1,4-dithiane and tetramethylthiourea as starting materials.[49] The preparation of **8a** (X = S) and **8c** (X = S) was achieved *via* similar ring-closure reactions using 1,2-dichloro-1,2-dimethoxyethane and 2,3-dichlorodioxane, respectively.[1,38,51] Scheme 2.6 outlines the preparation of **8c** and its oxathiane analogue 4,5-ethylenoxythio-1,3-dithiole-2-thione.[38a] In addition, preparation of 4,5-ethylenedioxy-1,3-selenothiole-2-thione,[38b] 4,5-ethyleneselenothio-1,3-dithiole-2-thione and related compounds[38e] proceeded *via* analogous ring-closure reactions.

Various short-step methods for the preparation of **9** and **10** have been developed,[54–58] some of which are outlined in Scheme 2.7. Compound **9a′** (X = S) was directly prepared by the treatment of 2,3-dichloropyrazine with potassium trithiocarbonate in DMF and converted into its

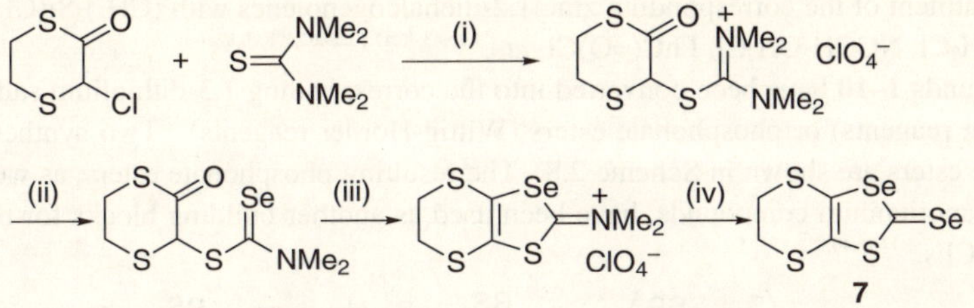

(i) HClO$_4$, EtOH; (ii) H$_2$Se, H$_2$O; (iii) H$_2$SO$_4$; then HClO$_4$;
(iv) H$_2$Se, MeOH-H$_2$O.

Scheme 2.5

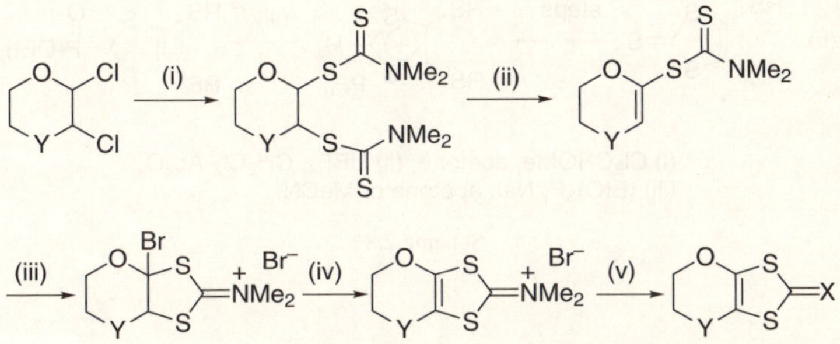

Y = O, S; (i) Bu$_4$N$^+$ $^-$SC(=S)NMe$_2$, MeCN; (ii) DMSO, 110 ˚C or
Et$_3$N, MeCN, 0 ˚C; (iii) Br$_2$, CH$_2$Cl$_2$, 0 ˚C; (iv)110 ˚C, 25 Torr;
(v) H$_2$Se, MeOH or NaSH, EtOH-AcOH, 25 ˚C.

Scheme 2.6

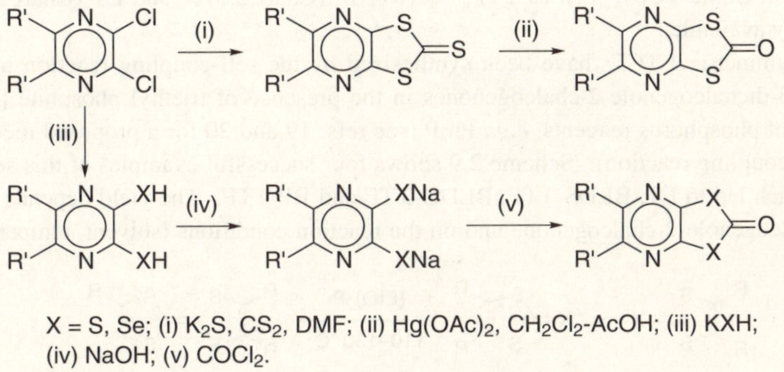

X = S, Se; (i) K$_2$S, CS$_2$, DMF; (ii) Hg(OAc)$_2$, CH$_2$Cl$_2$-AcOH; (iii) KXH;
(iv) NaOH; (v) COCl$_2$.

Scheme 2.7

oxo-analogue **9a'** (X = O), which was also obtainable by the three-step preparation *via* the reaction of 2,3-dichloropyrazine with KSH. The use of KSeH instead of KSH in this three-step preparation gave the selenium analogue of **9a'** (X = O), *i.e.*, **10a'** (X = O).

According to the procedures shown in Schemes 2.1 and 2.2, compounds **11** and **12** were prepared by treatment of the corresponding zinc 1,2-dichalcogenolenes with $(CH_3)_3Si(CH_2)_2OCH_2Cl$, p-$AcC_6H_4CH_2Cl$, $NCCH_2CH_2Cl$, $PhC(=O)Cl$, $etc.$[1,3b,43,46,47,53,59,62,69b,70,82c]

Compounds **1–10** have been converted into the corresponding 1,3-dithiolium salts, phosphoranes (Wittig reagents) or phosphonate esters (Wittig-Horner reagents). Two synthetic routes to phosphonate esters are shown in Scheme 2.8. The resulting phosphonate esters, as well as organotin and organotitanium compounds, have been used as another building blocks for the synthesis of various TCFs.[1,17,18,40,75]

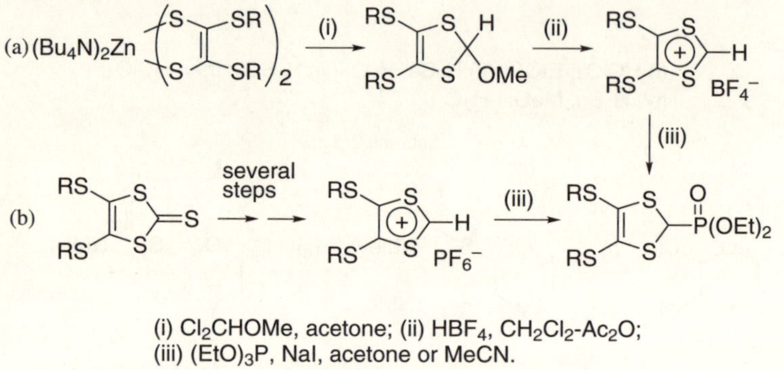

(i) Cl_2CHOMe, acetone; (ii) HBF_4, CH_2Cl_2-Ac_2O;
(iii) $(EtO)_3P$, NaI, acetone or MeCN.

Scheme 2.8

2.3   Synthesis of TCFs

One can design a wide variety of TCFs by combining two same or different 1,3-dichalcogenole-2-chalcogenones, $i.e.$, building blocks **1-12** shown in Chart 2.3 and their analogues. However, all possible combinations have not actually been attempted, and the number of TCFs giving τ-phase conductors is limited. In this section, some coupling reactions applied to the synthesis of TCFs are described. Some TCFs, such as TTF,[2,3] TMTSF (Chart 2.1)[11a] and ET (Chart 2.1),[11b,c] are commercially available.

Most symmetrical TCFs have been synthesized by the self-coupling reaction of the corresponding 1,3-dichalcogenole-2-chalcogenones in the presence of triethyl phosphite $[(EtO)_3P]$ or other trivalent phosphorus reagents, $e.g.$, Ph_3P (see refs. 19 and 20 for a proposed mechanism for this type of coupling reaction). Scheme 2.9 shows four successful examples of this self-coupling reaction, which led to ET, BEDS-TTF, BEDO-TTF and BP-TTF. The yield depends on the type of 1,3-dichalcogenole-2-chalcogenone and on the reaction conditions (solvent, temperature, $etc.$).

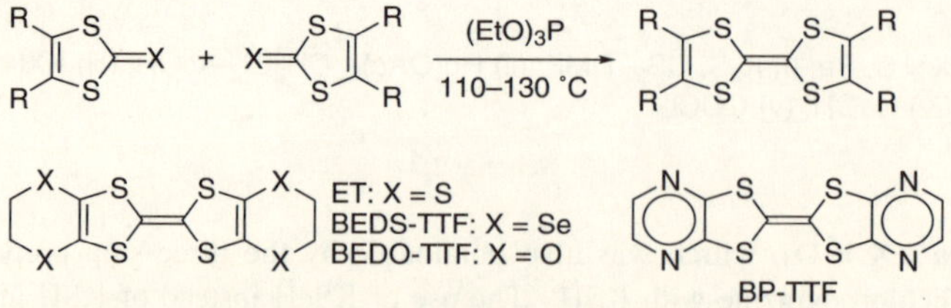

ET: X = S
BEDS-TTF: X = Se
BEDO-TTF: X = O

BP-TTF

Scheme 2.9

This coupling reaction is usually carried out under an inert atmosphere (N_2 or Ar).[1,39,40,47–57,59,60] In most cases, the best yield of the self-coupling product is obtained by using 2-oxo-1,3-dichalcogenoles (X = O in Chart 2.3) instead of the corresponding 2-thio- or 2-seleno-1,3-dichalcogenoles (X = S or Se in Chart 2.3). Conversion of 2-thio- or 2-seleno-1,3-dichalcogenoles into the corresponding 2-oxo-3-dichalcogenoles can be carried out by using, for example, mercury acetate [$Hg(OAc)_2$] in CH_2Cl_2/AcOH (v/v = 1/1).[42,45,46,49,56,57,71,73,74] Some trial experiments may be required to optimize the yield of the coupling product. Other building blocks, e.g., 1,3-dithiolium salts (see Scheme 2.8) and organotin compounds,[1,18,38,40,74] have been used for different types of coupling reactions with the aid of activating reagents, e.g., Et_3N[23,40] and Me_3Al.[18]

A good number of unsymmetrical TCFs have been synthesized by the cross-coupling reaction between the two different 1,3-dichalcogenole-2-chalcogenones in the presence of $(EtO)_3P$.[1,30,32,33,39,47–57,60] Useful applications of this cross-coupling reaction are found in the synthesis of P-DMEDT-TTF (Chart 2.2) and EDO-DMEDT-TTF (Chart 2.2) as well as the TTFs and DSDTFs shown in Scheme 2.10. The reaction conditions for obtaining these unsymmetrical donors in the best yields have been reported in the literature.[30,32,33,36,51,52,60] Also, the selenium analogues of EDT-TTF and EDO-MDT-TTF, i.e., EDT-DSDTF (Chart 2.4) and EDO-MDS-TTF (Chart 2.4), were synthesized by this kind of cross-coupling reaction.[43,44,51,52,58]

In the cross-coupling reaction, the product is usually accompanied by two self-coupling byproducts, and separation of the cross-coupling product from the mixture with silica gel column chromatography is essential. In the synthetic examples shown in Scheme 2.10, CH_2Cl_2 was used as the eluent solvent for the separation of the pyrazino-substituted TTF and DSDTF derivatives,

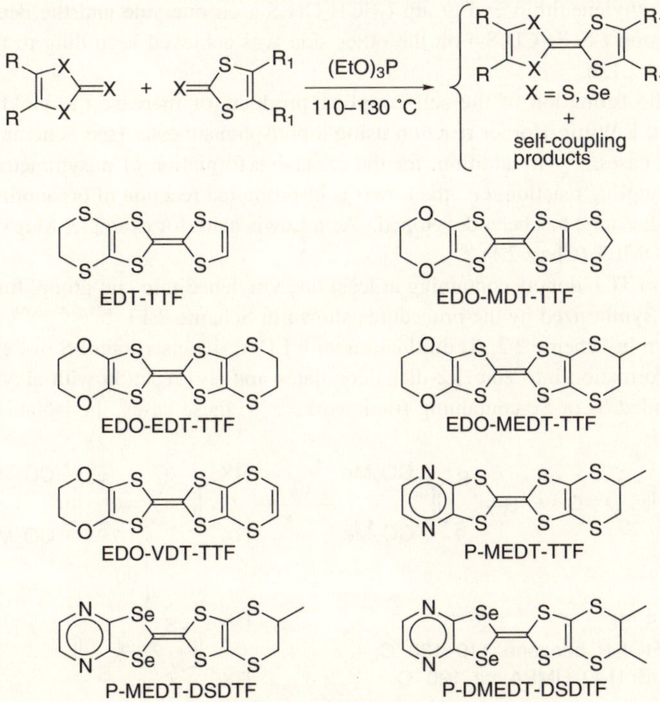

EDT-TTF

EDO-MDT-TTF

EDO-EDT-TTF

EDO-MEDT-TTF

EDO-VDT-TTF

P-MEDT-TTF

P-MEDT-DSDTF

P-DMEDT-DSDTF

Scheme 2.10

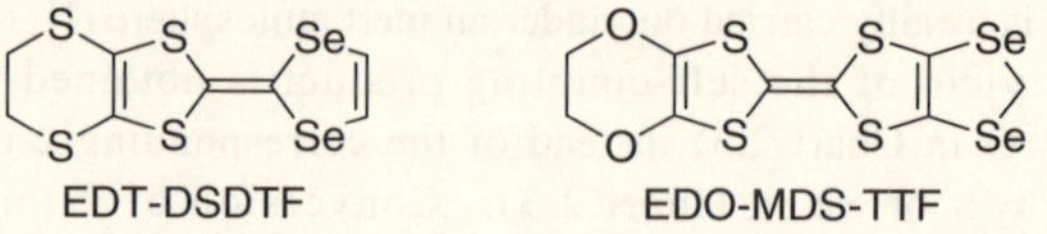

EDT-DSDTF   EDO-MDS-TTF

Chart 2.4

whereas separation of the ethylenedithio- and ethylenedioxy-substituted TTFs required CS_2 as the eluent solvent. It has been reported that the use of silica gel (230–400 mesh) with a mixed solvent hexane-CS_2 (v/v = 1/2) as the eluent solvent results in good separation of an unsymmetrical derivative of ET.[73] In each case, the value of the retardation factor (R_f) for the cross-coupling product (AB) is approximately given by the following equation:

$$R_f (AB) = \frac{1}{2} [R_f (AA) + R_f (BB)].\tag{2.1}$$

where AA and BB are the self-coupling products.[48] When the polarities between the cross-coupling product and the two self-coupling products differ considerably or moderately, the separation of the cross-coupling product by column chromatography is easily carried out. If this separation poses serious difficulties due to a small difference between the polarities of the cross- and self-coupling products, modification of the cross-coupling reaction is required. For example, as shown in Schemes 2.11 and 2.12a, building blocks with one or two polar functional groups, such as an ester, ether or nitril group, are first used for the cross-coupling reaction. Then, after separation of the desired cross-coupling products, removal of the polar functional group(s) attached is performed by appropriate methods.[1,20,41–43,45–51,53,58,59,61,62,70] Synthesis of unsymmetrical TCFs which have the ethylenedithio end group (-SCH$_2$CH$_2$S-) on one side and the deuterated ethylenedithio end group (-SCD$_2$CD$_2$S-) on the other side was achieved according to the procedures shown in Scheme 2.12b.

To avoid the formation of the self-coupling products or increase the yield of the cross-coupling product, a Wittig-Horner reaction using a phosphonate ester (see Scheme 2.8) has been utilized in some cases.[1,18] In addition, for the exclusive formation of unsymmetrical TCFs, the non-phosphite coupling reaction, *i.e.*, the Lewis acid-promoted reaction of organotin dithiolates or diselenolates with esters, has been developed. As a Lewis acid, for example, Me$_3$Al was used for the synthesis of DMET (Chart 2.1).[18]

Starting from TCF donors containing at least one vinylenedithio end group, further extended TCFs have been synthesized by the procedures shown in Scheme 2.13.[3b,35,43,48,60,80] Similar to the procedures shown in Scheme 2.2, (i) dilithiation with LDA, (ii) insertion of S or Se into the C–Li bond, (iii) transformation into zinc 1,2-dichalconolates and (iv) reaction with alkyl halides gave TCFs with extended S- or Se-containing framework.[60] In these cases, the isolation of zinc 1,2-

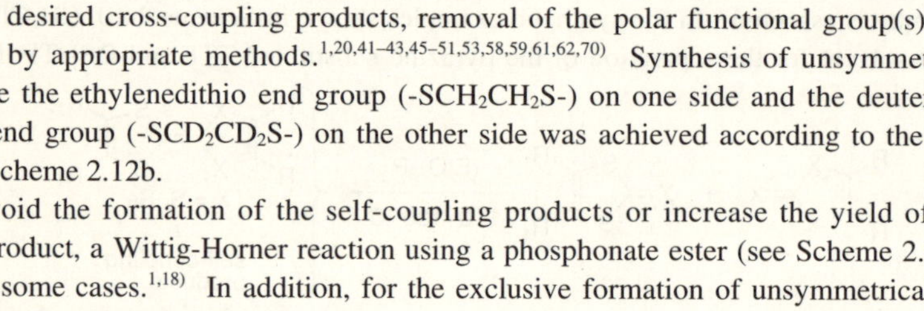

X = S, Se
(i) (EtO)$_2$P, benzene, 110–130 °C;
(ii) LiBr·H$_2$O, HMPA, *ca.* 120 °C.

Scheme 2.11

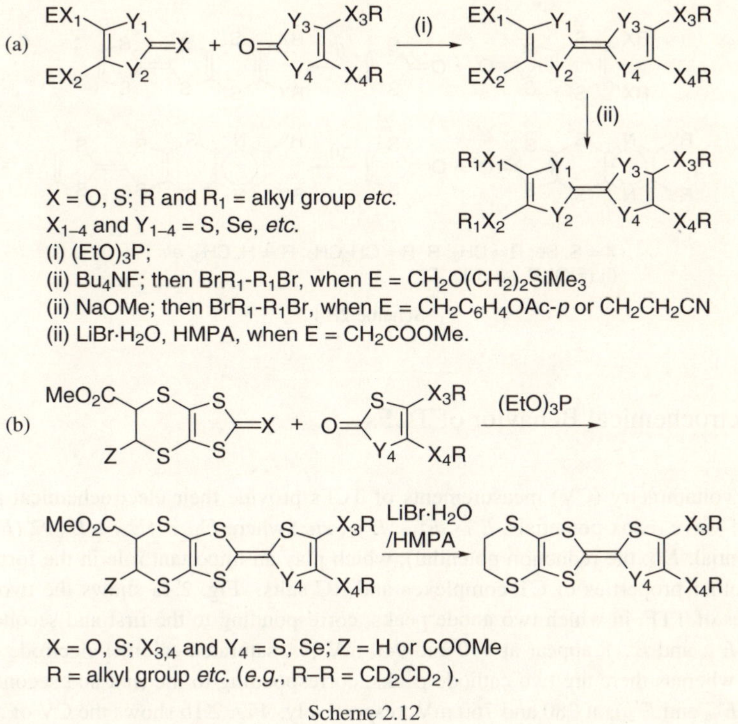
(a)
(i)
(ii)
X = O, S; R and R1 = alkyl group etc.
X1–4 and Y1–4 = S, Se, etc.
(i) (EtO)3P;
(ii) Bu4NF; then BrR1-R1Br, when E = CH2O(CH2)2SiMe3
(ii) NaOMe; then BrR1-R1Br, when E = CH2C6H4OAc-p or CH2CH2CN
(ii) LiBr·H2O, HMPA, when E = CH2COOMe.
(b)
(EtO)3P
LiBr·H2O
/HMPA
X = O, S; X3,4 and Y4 = S, Se; Z = H or COOMe
R = alkyl group etc. (e.g., R–R = CD2CD2).

Scheme 2.12

dichalcogenolates was necessary to obtain products of high purity, because direct reaction of dilithium dichalcogenolates, generated *via* insertion of a chalcogen atom into each C–Li bond, with alkyl halides resulted in deterioration in the purity of the products. Synthesis of the TTFs with extended π-conjugation from smaller TTFs with convertible functional groups has also been reported.[1] Scheme 2.14 outlines the synthesis of several TTFs containing four additional heteroatoms in only the outer hemisphere of the TTF core.[1,18,47,48,56,61] Moreover, the use of building blocks **3**–**8** (Chart 2.3) instead of 2-oxo-1,3-dithiole shown in Scheme 2.14 led to multi-heteroatom-containing TCFs.[18,47,48]

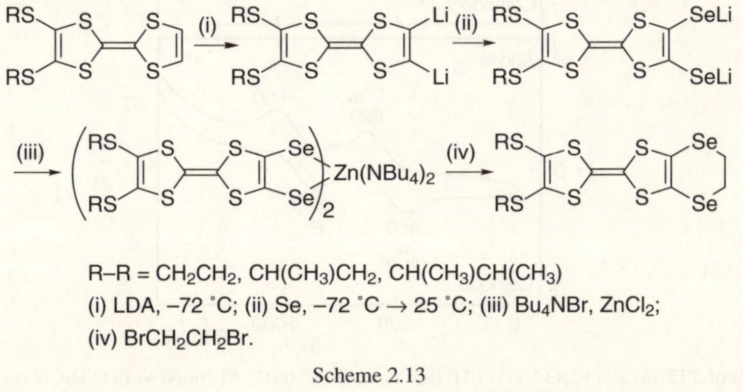
(i)
(ii)
(iii)
(iv)
Zn(NBu4)2
R–R = CH2CH2, CH(CH3)CH2, CH(CH3)CH(CH3)
(i) LDA, –72 °C; (ii) Se, –72 °C → 25 °C; (iii) Bu4NBr, ZnCl2;
(iv) BrCH2CH2Br.

Scheme 2.13

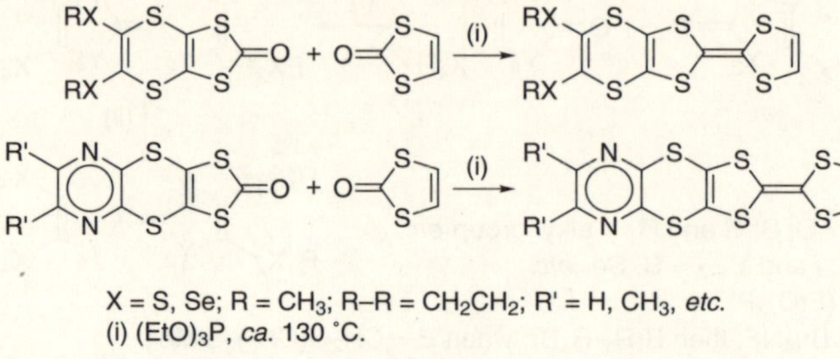

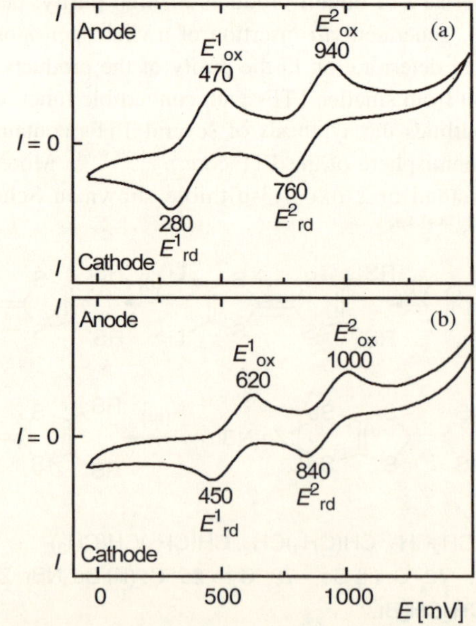

X = S, Se; R = CH$_3$; R–R = CH$_2$CH$_2$; R' = H, CH$_3$, *etc.*
(i) (EtO)$_3$P, *ca.* 130 °C.

Scheme 2.14

2.4   Electrochemical Behavior of TCFs

The cyclic voltammetry (CV) measurements of TCFs provide their electrochemical parameters, *i.e.*, the half-wave redox potentials, $E^1_{1/2}$, $E^2_{1/2}$, $E^3_{1/2}$, *etc.*, where $E^i_{1/2} = (E^i_{ox} + E^i_{rd})/2$ (E^i_{ox}: the oxidation potential, E^i_{rd}: the reduction potential), which play an important role in the formation, stability and other properties of CT complexes and RC salts. Fig. 2.1a shows the two reversible redox waves of TTF, in which two anode peaks, corresponding to the first and second oxidation potentials (E^1_{ox} and E^2_{ox}), appear at 470 and 940 mV [*vs.* saturated calomel electrode (SCE)], respectively, whereas there are two cathode peaks, corresponding to the first and second reduction potentials (E^1_{rd} and E^2_{rd}), at 280 and 760 mV, respectively. Fig. 2.1b shows the CV of a derivative of TTF, EDO-VDT-TTF (Scheme 2.10), measured under the same conditions.

Fig. 2.1   CVs of TTF (a) and EDO-VDT-TTF (b): V *vs.* SCE; 0.025 M Bu$_4$NPF$_6$ in C$_6$H$_5$CN containing 7 × 10^{-4} M TTF or EDO-VDT-TTF; Pt electrode; at room temperature; scan rate 100 mV s^{-1}.

Although the values of the redox potentials vary under different measurement conditions (temperature, solvent, electrolyte, *etc.*), the value of the first half-wave redox potential ($E^1_{1/2}$) indicates the donor (D) ability to release one electron according to the following equation:

$$D^0 \xrightarrow{-e} D^{\bullet+}.$$

The value of the second half-wave redox potential ($E^2_{1/2}$) represents the generation of a dication, as shown with the following equation:

$$D^{\bullet+} \xrightarrow{-e} D^{2+}.$$

The difference between the first and second half-wave redox potentials ($\Delta E = E^2_{1/2} - E^1_{1/2}$) corresponds to the on-site Coulombic repulsion involved in the formation of a dication.[1,2,76–79] In general, the value of $E^i_{1/2}$ of an unsymmetrical donor (AB) can be calculated from those of the two corresponding symmetrical donors (AA and BB) according to the following equation:

Table 2.1   Half-wave Redox, Oxidation and Reduction Potentials (mV) of TCFs[a]

Compound	Conditions[b]	$E^1_{1/2}$	$E^2_{1/2}$	ΔE	E^1_{ox}	E^2_{ox}	E^1_{rd}	E^2_{rd}
TM-TTF[c]	A	290	650	360				
TTF[d]	A	367	748	381				
	B	375	845	470	470	940	280	750
	C	319	701	382				
MDS-TTF[c]	B	405	790	385	490	870	320	710
BEDO-TTF[e]	A	435	699	264				
	B	415	810	395	490	900	340	720
BMDS-TTF[c]	B	(435)	(735)	(300)				
TMTSF[d]	A	440	720	280				
MDT-TTF[d]	B	455	832	377	535	915	375	750
	B	(451)	(829)	(378)	(532)	(910)	(370)	(748)
EDT-TTF[f]	A	(467)	(788)	(321)				
	B	460	885	425	545	970	375	800
	C	390	690	300				
	C	(402)	(720)	(318)				
TSF[d]	A	480	760	280				
EDO-MDT-TTF[f]	B	(466)	(811)	(395)				
EDO-EDT-TTF[f]	A	(501)	(764)	(263)				
	B	(467)	(858)	(391)				
EDO-DMEDT-TTF[g]	A	502	762	260				
	B	(468)	(853)	(388)				
BEDT-TTF[d]	A	567	829	262				
	B	520	907	387	600	980	440	835
	C	486	739	253				
BDMEDT-TTF[c]	B	(521)	(902)	(381)				
EDT-DMEDT-TTF[c]	B	522	897	375	595	970	450	825
BMDT-TTF[c]	B	527	812	285	595	880	460	745
EDO-VDT-TTF[f]	B	535	920	385	620	1000	450	840
P-MDS-TTF[c]	B	(677)	(1049)	(371)				
P-EDT-TTF[c]	B	720	1135	415	790	1220	650	1050
	C	680	1020	340				
	C	(688)	(955)	(267)				
P-DMEDT-TTF[g]	B	(721)	(1130)	(409)				
P-MDT-DSDTF[c]	C	690	960	270				
P-EDT-DSDTF[c]	C	740	1030	290				
BP-TTF[e]	B	(920)	(1363)	(443)				
	C	890	1170	280				
BDMP-TTF[c]	C	990	1280	290				

[a]The calculated values from equation (2.2) are given in parentheses. [b]A: 0.1 M Bu$_4$NPF$_6$ in CH$_3$CN, V *vs.* Ag/AgCl; B: 0.025 M Bu$_4$PF$_6$ in C$_6$H$_5$CN, V *vs.* SCE; C: 0.1 M Et$_4$NClO$_4$ in CH$_3$CN, V *vs.* SCE.[1,54,58,59,67,75,77] [c]See Chart 2.5 for the structural formula. [d]See Chart 2.1 for the structural formula. [e]See Scheme 2.9 for the structural formula. [f]See Scheme 2.10 for the structural formula. [g]See Chart 2.2 for the structural formula.

$$E^i_{1/2} (AB) = \frac{1}{2} [E^i_{1/2} (AA) + E^i_{1/2} (BB)]. \tag{2.2}$$

This equation is also applicable to the oxidation and reduction potentials (E^i_{ox} and E^i_{rd}). Table 2.1 summarizes the experimental $E^1_{1/2}$, $E^2_{1/2}$, ΔE, E^1_{ox}, E^2_{ox}, E^1_{rd} and E^2_{rd} values for selected TCFs. In several cases, the calculated values from equation (2.2) under various measurement conditions are given.[1,53,57,58,66,71,76] As can be seen from Table 2.1, the $E^1_{1/2}$ values increase on going from TM-TTF to BDMP-TTF. All the unsymmetrical TCFs incorporating four additional heteroatoms, listed in Table 2.1, show higher $E^1_{1/2}$ and $E^2_{1/2}$ values than those of TSF. In the CVs of TCFs with one or two appended heterocycles, more than two redox waves are observed.[1,18,76,80] These waves indicate the appearance of higher oxidized species, and probably arise from the additional heterocycle(s).

2.5   Role of Additional Functional Groups in the Synthesis and Properties of TCFs

There are two ways to obtain extended TCFs: one is the peripheral extension of the TCF core by substitution of the hydrogen atoms of the TCF molecule with other functional groups, while the other is the introduction of a π-linking group as a spacer group between the central C=C bond of TCF. The latter compounds are described elsewhere,[1,74,79] so the substituted TTFs belonging to the former category are described here. For example, substitution of the four hydrogen atoms on both sides of TTF with the CH$_3$, CH=CH-CH=CH, SCH$_2$CH$_2$S and N=CH-CH=N groups leads to the symmetrical donors TM-TTF (Chart 2.5), DB-TTF (Chart 2.1), ET (Chart 2.1) and BP-TTF (Scheme 2.9), respectively, whereas substitution of the two hydrogen atoms on either side of TTF with the same groups gives the unsymmetrical donors DM-TTF (Chart 2.6), B-TTF (Chart 2.6), EDT-TTF (Scheme 2.10) and P-TTF (Chart 2.6), respectively. Additional substituents attached to

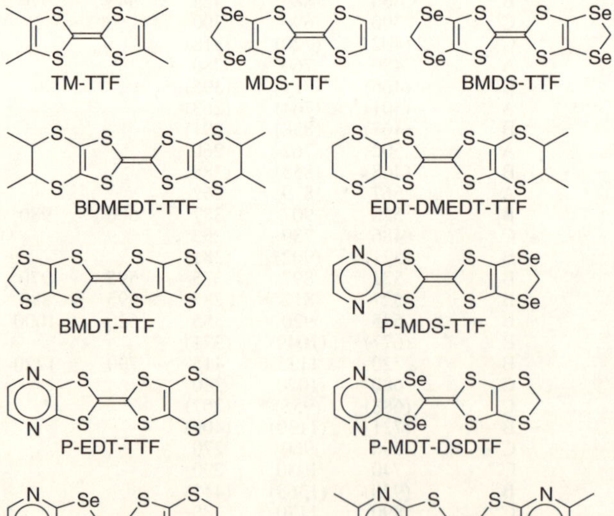

TM-TTF   MDS-TTF   BMDS-TTF

BDMEDT-TTF   EDT-DMEDT-TTF

BMDT-TTF   P-MDS-TTF

P-EDT-TTF   P-MDT-DSDTF

P-EDT-DSDTF   BDMP-TTF

Chart 2.5

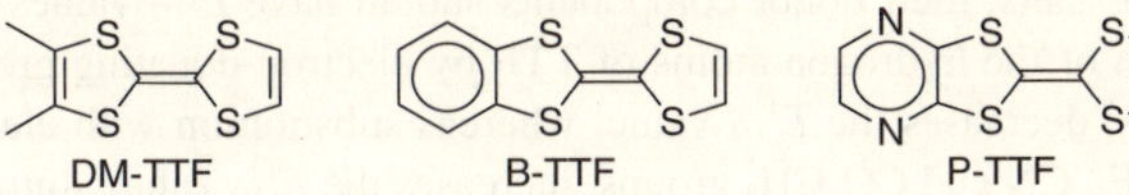

Chart 2.6

the TTF core generally affect the polarity, solubility, redox potentials, optical properties and other chemical or physical properties. Some examples are given in the following paragraphs.

As described in Section 2.3, a number of unsymmetrical TCFs have been obtained by the $(EtO)_3P$-promoted cross-coupling reaction of the two corresponding 1,3-dichalcogenol-2-ones followed by separation from the concomitant self-coupling byproducts by column chromatography when the R_f values of unsymmetrical TCFs are considerably or moderately different from those of the self-coupling products. For example, the R_f values of TTF, BMDT-TTF (Chart 2.5), BEDS-TTF (Scheme 2.9), BEDO-TTF (Scheme 2.9), BP-TTF (Scheme 2.9) and BMC-TTF (Chart 2.7) on a silica gel plate using CH_2Cl_2 as an eluent solvent are 0.75, 0.77, 0.55, 0.25, 0.08 and 0.05, respectively; that is, the polarities of the three latter donors are higher than those of the three former donors. This means that the additional substituents, $i.e.$, the OCH_2CH_2O, N=CH-CH=N and CO_2CH_3 groups, affect the R_f values. Consequently, unsymmetrical TTFs, such as MC-TTF (Chart 2.7), P-EDT-TTF (Chart 2.5) and EDO-EDS-TTF (Chart 2.7), were easily separated from the corresponding symmetrical TTFs by column chromatography because of a moderate difference between the R_f values of unsymmetrical and symmetrical TTFs. On the other hand, an easy separation of unsymmetrical TTFs, such as MDT-TTF (Chart 2.1) and EDT-EDS-TTF (Chart 2.7), was unsuccessful, even though the eluent solvent used for a silica gel column chromatography was changed. Therefore, modified cross-coupling reactions were required to obtain these unsymmetrical TTFs.

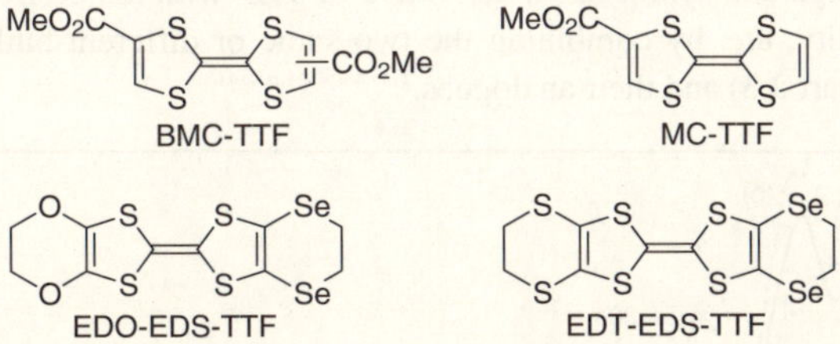

Chart 2.7

For the preparation of CT complexes and RC salts in solution, TCFs used as their donor components need to be soluble in organic solvents. For example, TTF and ET (Chart 2.1) are sufficiently soluble in CH_2Cl_2 and C_6H_5CN, while BEDS-TTF (Scheme 2.9) and BP-TTF (Scheme 2.9) are less soluble in these solvents. The solubility is increased by (i) replacement of the hydrogen atoms of TCF by the CH_3, $CH_3(CH_2)_n$, CN, CO_2CH_3 and $CH_2O(CH_2)_2SiMe_3$ groups, (ii) replacement of the XCH_2CH_2X (X = S and Se) groups by their oxygen analogue (X = O) and (iii) replacement of the N=CH-CH=N or SCH=CHS group by the XCH_2CH_2X (X = S or O) group.

It has been reported that TCFs with $E^1_{1/2}$ values close to or lower than that of TTF are oxidized by air in solution.[1,76)] On the other hand, symmetrical TCFs derived from the building blocks **9–12** (Chart 2.3) have high $E^1_{1/2}$ values and do not form a stable CT complex or RC salt. For example, the salts based on BP-TTF (Scheme 2.9) are unstable in moist air due to the low electron-donating ability of BP-TTF (see Table 2.1).[56)] Accordingly, for the preparation of stable

CT complexes and RC salts, their donor components should have $E^1_{1/2}$ values lower than that of BP-TTF. Substitution of the hydrogen atoms of TTF by electron-donating groups, such as alkyl and cycloalkyl groups, decreases the $E^1_{1/2}$ value, whereas substitution with electron-withdrawing groups, such as the CF_3, CN and CO_2CH_3 groups, increases the $E^1_{1/2}$ value. Introduction of sulfur- and oxygen-containing substituents instead of the hydrogen atoms increases the $E^1_{1/2}$ value, but decreases the ΔE ($E^2_{1/2} - E^1_{1/2}$) value, suggesting a decrease in the on-site Coulombic repulsion.[1,76–78] In addition, attachment of sulfur-, selenium-, tellurium- and nitrogen-containing substituents leads to the enhancement of intermolecular interaction in CT complexes and RC salts, but increases the $E^1_{1/2}$ value (see Table 2.1).[76,79–82] It should be noted that the extension of TCFs by π-linking spacer groups changes both the $E^1_{1/2}$ and ΔE values,[1,74,79] whereas substitution of the sulfur atoms in the TTF core with selenium or tellurium increases the $E^1_{1/2}$ value (see Table 2.1 and ref. 38). The TCFs listed in Table 2.1, except for BP-TTF (Scheme 2.9) and BDMP-TTF (Chart 2.5), were found to be good π-donors for the preparation of CT complexes and RC salts.

Additionally, the color of the TCF donors depends on the type of chalcogen atom in the TCF core and on the type of heteroatom-containing substituent appended to TCF.[20–22,24,38,41–62,74] For example, BP-TTF (Scheme 2.9), P-DMEDT-TTF (Chart 2.2), EDO-DMEDT-TTF (Chart 2.2) and BEDO-TTF (Scheme 2.9) are colored yellow ($\lambda_{max} = 402$ nm), brownish-red ($\lambda_{max} = 425$ nm),[57] red ($\lambda_{max} = ca.$ 480 nm)[50–52] and red ($\lambda_{max} = ca.$ 500 nm),[50,52] respectively. That is, the optical absorption (OA) spectra of TCFs in the UV-vis spectral region show a variety of absorption bands, and their respective positions, intensities and shapes differ. The absorption bands of unsymmetrical TTFs in solution appear between those of the two corresponding symmetrical ones. For example, Fig. 2.2 shows the OA spectrum of the unsymmetrical P-DMEDT-TTF (Chart 2.2) donor together with those of the symmetrical BDMEDT-TTF (Chart 2.5) and BP-TTF (Scheme 2.9) donors.

Taking into account the above-mentioned and other empirical rules as well as equations (2.1) and (2.2), one can design and synthesize a derivative of TCF with the desired polarity, solubility, electron-donating ability, *etc.* by combining the two same or different building blocks selected from among **1–10** (Chart 2.3) and their analogues.[1,47,48,50,51,55,74]

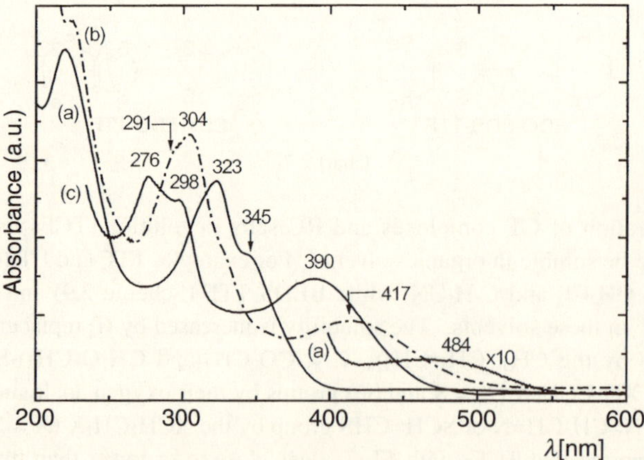

Fig. 2.2 OA spectra of BDMEDT-TTF (a), P-DMEDT-TTF (b) and BP-TTF (c) in CH_3CN at room temperature.

2.6 Formation and Properties of RCs and Other Charged Species

TCFs are easily oxidized by chemical and electrochemical methods to give RCs and other charged species in solution or in the solid state (*e.g.*, crystals and films).[1,19,22,48,76–78] The oxidation conditions affect what kinds of charged species and crystal structure (*i.e.*, α, β, γ, κ, λ, τ, *etc.*) are formed. For example, by chemical oxidation of TTF with chlorine gas in CCl_4 or CH_3CN, the RC of TTF ($TTF^{\cdot+}$) was obtained as a deep-purple crystal, and further oxidation with chlorine gas gave the yellow crystalline dication TTF^{2+}.[2,83] In general, RC ($D^{\cdot+}$) is paramagnetic, whereas neutral (D^0) and dicationic species (D^{2+}) are diamagnetic.[1,2,21,24,32,62,76,83] Charged species can also be obtained by oxidation with an organic acceptor, *e.g.*, TCNQ, or by electrochemical oxidation in the presence of an inorganic anion. Single crystals of CT complexes and RC salts are obtained mainly by the following two methods.

(i) Chemical methods: In the direct reaction method, donor and acceptor components are dissolved in a minimum amount of solvent. Then, to achieve crystallization, the resulting (saturated) solution is either cooled to a lower temperature to lower the solubility of the two components or warmed up to a higher temperature to evaporate the solvent. If this cooling or warming process is performed slowly, large single crystals of good quality could be obtained. When either or both components are not sufficiently dissolved in a solvent, they are mixed with a small portion of solvent in a stopped flask, and the mixture is allowed to stand until the reaction is completed.[48] If the reaction is promoted by heating or ultrasonic irradiation, the yield in this reaction can be improved in a shorter reaction time. In the diffusion method, a cell with two separated tubes connected by a common tube is employed. Donor and acceptor components are placed in two separate tubes, respectively, and the cell is filled with a solvent. When one component is volatile (*e.g.*, iodine vapor), only one tube containing the other component is filled with a solvent.

(ii) Electrochemical (galvanostatic and potentiostatic) methods: An H-shaped cell with a porosity frit and two platinum wire electrodes is usually used. A solution containing both a donor and a supporting electrolyte is added to the anode compartment of the cell, whereas, to the cathode compartment, only a solution of the same electrolyte is added.[1,19,33,77] In the galvanostatic method, crystals begin growing up on applying a constant current with low density (0.2–$1.0\ \mu A\ cm^{-2}$). The growth period varies from several days to some months, depending on the nature and purity of the materials, the concentration, the kind of solvent, the current density and the temperature. In most cases, stoichiometries of the resulting salts [D:A or D:anion] are 2:1 ($\rho = 0.5$, see Section 2.1) and 3:2.[1,14,15,19–21,43,44,57–59,62] Other stoichiometries, 1:~1,[20,22,39,44,51] 2:~1.75,[23–36] 3:1,[54,58] *etc.*,[1,10,19,43,48,51,58,59,73,77] are sometimes obtained. A salt with a 1:1 ($\rho = 1$) stoichiometry is obtained by applying a constant voltage (the potentiostatic method), while a salt with a 1:2 stoichiometry ($\rho = 2$) is formed under galvanostatic conditions in a solution containing an oxidizing agent. That is, the divergent formation of RC ($D^{\cdot+}$) and dication (D^{2+}) in the solid state can be attained by changing the electrochemical conditions.[77,83]

RCs are also formed by intermolecular CT from a neutral species to a dicationic species as follows:

$$D^{2+} + D^0 \rightarrow 2D^{\cdot+} \text{ (in solution and in the solid state)}$$

$$D_1^{2+} + 2D_2^0 \rightarrow D_1^0 2D_2^{\cdot+} \text{ (in the solid state).}$$

Treatment of RC and dicationic species with a reducing agent, *e.g.*, sodium hydrogen sulfite (NaHSO$_3$), generates the corresponding neutral species. It is considered that τ-phase conductors with a 2:~1.75 stoichiometry contain several charged species.[32] For example, it has been suggested by crystallographic and spectroscopic data that the τ-phase conductor based on EDO-DMEDT-TTF (Chart 2.2), τ-(EDO-DMEDT-TTF)$_2$AuBr$_2$(AuBr$_2$)$_y$, should consist of a mixture of charged species, expressed by the following formula: [2D^{2+} + 3D$^{\cdot+}$ + 3D^0][4X$^-$ + 4yX$^-$] or [1D^{2+} + 5D$^{\cdot+}$ + 2D^0][4X$^-$ + 4yX$^-$] *etc.*, where D = EDO-DMEDT-TTF, X = AuBr$_2$ and y = *ca.* 0.75. Similar formulas have been proposed for τ-phase conductors with other stoichiometries. Table 2.2 summarizes the peak positions in the OA spectra of charged species based on TTF, EDT-TTF (Scheme 2.10), ET (Chart 2.1), BDMEDT-TTF (Chart 2.5), EDO-DMEDT-TTF (Chart 2.2), BEDO-TTF (Scheme 2.9), EDO-VDT-TTF (Scheme 2.10) and P-DMEDT-TTF (Chart 2.2).[83] It was found that the OA spectral peak positions of charged species derived from unsymmetrical TTFs appear between those derived from the two corresponding symmetrical ones, as is observed for the neutral species. Fig. 2.3 shows, for example, the OA spectra of neutral and charged species of EDO-

Table 2.2   Peak Positions (nm) in the OA Spectra of Neutral and Charged Species[a]

Species	Experimental Value[83c] (Intensity[b])	Calculated Value[83a,b] (Oscillator Strength $\times$ 10^4)
TTF^0	305–316 (s), 361 (sh), 445 (w, b)	312 (561), 382(180), 452 (2)
TTF$^{\cdot+}$	336 (m), 400 (sh), 435 (s)	365 (3086), 381 (42), 451 (1)
	500 (sh), 579 (m, b)	523 (677) or > 362 (317)
TTF^{2+}	350 (?)	362 (3217)
EDT-TTF^0	309 (s)–330, 368 (sh), 440 (w, b)	
EDT-TTF$^{\cdot+}$	403 (s)–441–472, 543 (w)	
	788 (s, b)	
EDT-TTF^{2+}	600 (s, b)	
ET^0	321 (s)–345, 465–496 (w)	323 (364), 329 (1), 342 (881)
		358 (342), 451 (0)
ET$^{\cdot+}$	420 (w, sh), 456 (s)–482	416 (2401), 430 (96), 579 (1)
	580 (w), 952 (s, b)	1000 (1816) or 956 (1850)
ET^{2+}	710 (s, b)	
BDMEDT-TTF^0	323 (s)–345, 470 (w)	
BDMEDT-TTF$^{\cdot+}$	420 (sh), 457 (s)–481, 575 (w)	
	960 (s, b)	
BDMEDT-TTF^{2+}	707 (s, b)	
EDO-DMEDT-TTF^0	312–324 (s), 458–488 (w)	
EDO-DMEDT-TTF$^{\cdot+}$	425 (sh), 457 (s)–485, 596 (w)	
	937 (s, b)	
EDO-DMEDT-TTF^{2+}	649 (s, b)	
BEDO-TTF^0	316 (s)–337, 479–523 (w)	302 (3319), 324 (170), 331 (15)
		378 (1), 517 (0)
BEDO-TTF$^{\cdot+}$	435 (w, sh), 463 (s)–493	415 (2299), 436 (0), 596 (0)
	604 (w, b), 934 (m, b)	873 (1513)
BEDO-TTF^{2+}	568 (s, b)	
EDO-VDT-TTF^0	308 (sh)–323 (s)–350 (sh)	
	480 (w)	
EDO-VDT-TTF$^{\cdot+}$	455 (s)–481 (sh), 584 (w)	
	858 (s, b)	
EDO-VDT-TTF^{2+}	600 (s, b)	
P-DMEDT-TTF^0	291 (sh), 304 (s), 330 (sh)	
	417 (m)	
P-DMEDT-TTF$^{\cdot+}$	428 (s), 536 (sh), 890 (s, b)	
P-DMEDT-TTF^{2+}	712 (?)	

[a]In CH$_3$CN, at room temperature. [b]s = strong, sh = shoulder, m = medium, w = weak, b = broad, ? = poorly defined.

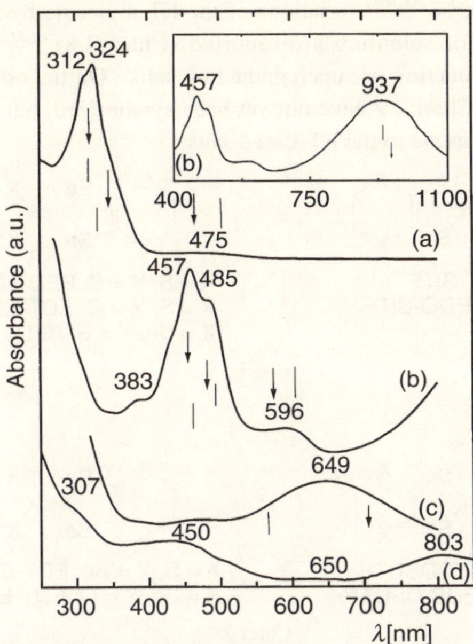

Fig. 2.3   OA spectra of (EDO-DMEDT-TTF)^0 (a), (EDO-DMEDT-TTF)$^{·+}$ (b) and (EDO-DMEDT-TTF)$^{2+}$ (c) in CH$_3$CN at room temperature and OA spectrum of a thin film based on τ-(EDO-DMEDT-TTF)$_2$AuBr$_2$(AuBr$_2$)$_{0.75}$ (d) at room temperature. Vertical bars indicate the peak positions in the OA spectra of the neutral, RC and dication of BEDO-TTF, whereas arrows indicate those of BDMEDT-TTF.

DMEDT-TTF (Chart 2.2) in solution and in the solid state.[24,32,83] The RC and dication, (EDO-DMEDT-TTF)$^{·+}$ and (EDO-DMEDT-TTF)$^{2+}$, were generated by oxidation of (EDO-DMEDT-TTF)^0 with bromine (or chlorine) in CH$_3$CN. The resulting charged species can be stabilized in the presence of ClO$_4^-$ or BF$_4^-$ anion in solution. The OA spectrum of a thin film of τ-(EDO-DMEDT-TTF)$_2$AuBr$_2$(AuBr$_2$)$_{0.75}$ on a quartz plate indicated the presence of a mixture of neutral and charged species. It has been reported that, in a concentrated solution containing RCs of TTF or tetrakis(methylthio)-TTF at low temperature,[83] as well as in quasi-1D conductors, 2D κ-phase conductors[1,11,12,15,16,19,22] and others,[75,76] the RCs form π-dimers (D$_2^{2+}$). The OA spectra in the spectral region from UV-vis to near IR give information on the intermolecular electronic behavior, and the spectrum of an oxidized donor molecule does not depend on the type of counteranion (Cl$^-$, Br$^-$, ClO$_4^-$, BF$_4^-$, *etc.*). There are, however, several cases where the electronic spectra ranging from far IR to near IR for RC salts in the solid state vary by a phase transition (*e.g.*, an MI transition) acompanying the formation of a different kind of intermolecular interaction;[1,25,29,77a] the same can be said for other physical properties of RC salts and CT complexes.[1,16,17,19]

2.7   Design of New Materials

ET (Chart 2.1) is considered to be the most important derivative of TCF with four additional heteroatoms, because this donor produces the largest number of superconducting salts. The ET-based salts are divided into several phases (α, β, γ, κ, λ, *etc.*), but none of them crystallizes in the τ-phase. In addition to BEDO-TTF (Scheme 2.9),[53] BEDS-TTF (Scheme 2.9)[3b,38e,60,80a] and

EDT-EDS-TTF (Chart 2.7),[48,73,70c,80b] variations of the ET molecule by replacement of the sulfur atom(s) with oxygen and/or selenium are reported (Chart 2.8).[49,62,75] Similar to ET, these analogous donors gave conducting or superconducting salts. On the other hand, the chalcogeno-analogues of ET shown in Chart 2.9 have not yet been synthesized, but they are expected to provide conducting salts comparable to the ET-based salts.

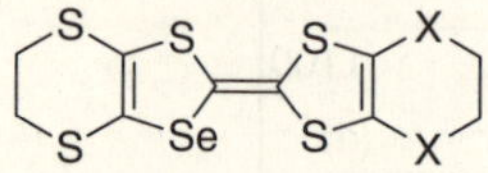

X = S: BEDT-StTF
X = O: EDT-EDO-StTF

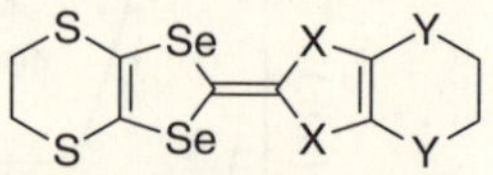

X = S, Y = S: BEDT-DSDTF
X = S, Y = O: EDT-EDO-DSDTF
X = Se, Y = S: BEDT-TSF

Chart 2.8

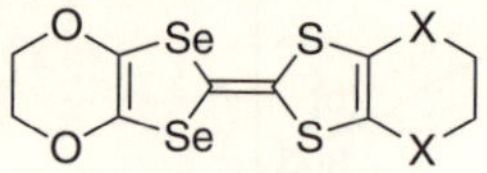

X = S: EDO-EDT-DSDTF
X = Se: EDO-EDS-DSDTF

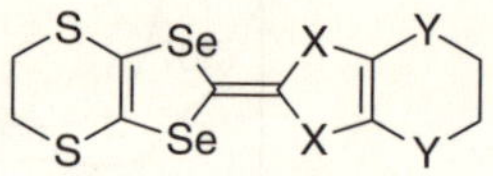

X = S, Y = Se: EDT-EDS-DSDTF
X = Se, Y = O: EDT-EDO-TSF

Chart 2.9

Besides EDO-MDT-TTF,[20,21,53] EDO-EDT-TTF[20,21,53] and EDO-VDT-TTF,[20,22,23,53] which are shown in Scheme 2.10, the racemic and chiral forms of EDO-MEDT-TTF (Scheme 2.10),[20,36] EDO-DMEDT-TTF (Chart 2.2),[23–32,36] EDT-DMEDT-TTF (Chart 2.5),[35,68b] P-MEDT-TTF (Scheme 2.10),[33,36] P-DMEDT-TTF (Chart 2.2)[23,25,26–28,31,34] and P-DMEDT-DSDTF (Scheme 2.10)[34] were found to form at least one τ-phase salt with a linear anion. All these donors are red and reddish-brown in color. Among them, EDO-MDT-TTF (Scheme 2.10) shows the lowest $E^i_{1/2}$ values, while the $E^i_{1/2}$ values of P-DMEDT-TTF (Chart 2.2) are highest (see Table 2.1). However, MDT-DMEDT-TTF (Chart 2.10) and VDT-DMEDT-TTF (Chart 2.10), which exhibit $E^i_{1/2}$ values between those of EDO-MDT-TTF (Scheme 2.10) and P-DMEDT-TTF (Chart 2.2), did not give the τ-phase salt.[60] It should be noted that, although EDO-VDT-TTF (Scheme 2.10) has $E^i_{1/2}$ values close to those of ET (see Table 2.1), this donor, unlike ET, did not produce a superconducting salt. As shown in Chart 2.11, analogues of the TCFs providing τ-phase salts have been synthesized,[35,36,51,52,59,60] but it is not clear whether these donors give the τ-phase conductor. Other analogues, several of which are, for example, shown in Chart 2.12, could be designed and synthesized. Generally speaking, it is possible that new TCF donors are obtained by the coupling reaction of building blocks of types **1–10** (Chart 2.3), but it is not sure that their RC salts will exhibit more interesting physical properties than those obtained from the well-known donors, ET (Chart 2.1), TMTSF (Chart 2.1), MDT-TTF (Chart 2.1), EDT-TTF (Scheme 2.10), P-DMEDT-TTF (Chart 2.2), etc.

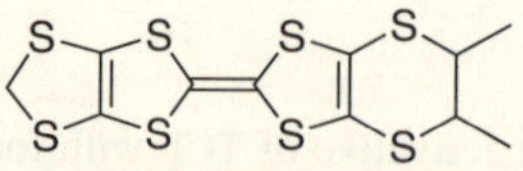

MDT-DMEDT-TTF

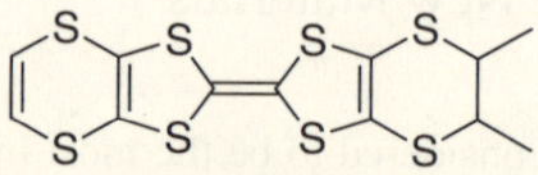

VDT-DMEDT-TTF

Chart 2.10

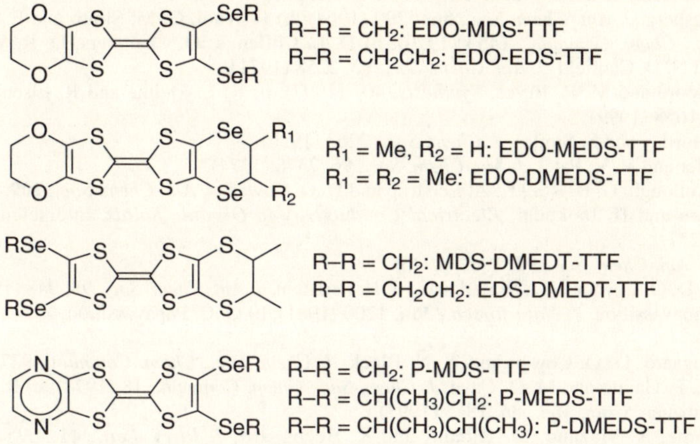

R–R = CH$_2$: EDO-MDS-TTF
R–R = CH$_2$CH$_2$: EDO-EDS-TTF

R$_1$ = Me, R$_2$ = H: EDO-MEDS-TTF
R$_1$ = R$_2$ = Me: EDO-DMEDS-TTF

R–R = CH$_2$: MDS-DMEDT-TTF
R–R = CH$_2$CH$_2$: EDS-DMEDT-TTF

R–R = CH$_2$: P-MDS-TTF
R–R = CH(CH$_3$)CH$_2$: P-MEDS-TTF
R–R = CH(CH$_3$)CH(CH$_3$): P-DMEDS-TTF

Chart 2.11

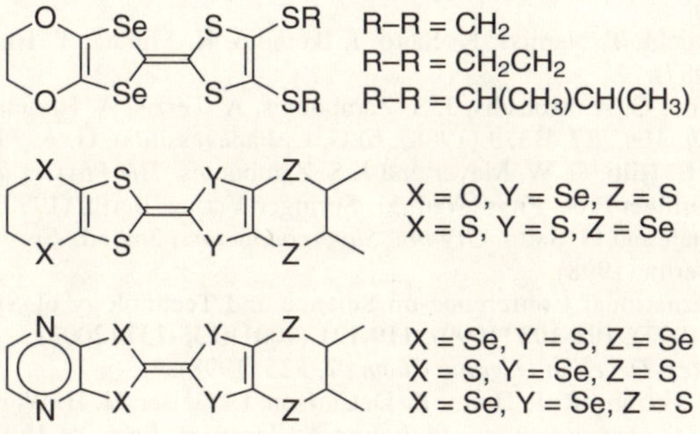

R–R = CH$_2$
R–R = CH$_2$CH$_2$
R–R = CH(CH$_3$)CH(CH$_3$)

X = O, Y = Se, Z = S
X = S, Y = S, Z = Se

X = Se, Y = S, Z = Se
X = S, Y = Se, Z = S
X = Se, Y = Se, Z = S

Chart 2.12

2.8   Summary

A number of TCFs with four additional heteroatoms have been synthesized by coupling reactions using mainly 1,3-dithiol-2-ones and their selenium analogues. Unsymmetrical TCFs (AB) exhibit R_f and $E^i_{1/2}$ values as well as OA spectral peaks midway between those of the two corresponding symmetrical ones (AA and BB). Successive oxidation of TCFs generates RCs, dications and other charged species in both solution and the solid state, and the OA spectral peaks of charged species of AB also appear between those of AA and BB. The RC salts based on TCFs crystallize in several phases (α, β, γ, *etc.*), but the τ-phase conductors are formed by the use of a limited number of TCFs. It is possible to undertake further studies of new TCFs as well as their CT complexes and RC salts by applying various building blocks, coupling reactions and empirical rules, which are described herein and elsewhere.

References

1. G. C. Papavassiliou, A. Terzis and P. Delhaes, *Handbook of Organic Conductive Molecules and*

Polymers, J. Wiley & Sons, Chichester (1997), Vol. 1, p. 151.

2. a) E. Klingsherg, *J. Am. Chem. Soc.*, **86**, 5290 (1964); b) F. Wudl, G. M. Smith and E. J. Hufnagel, *J. Chem. Soc., Chem. Commun.*, 1453 (1970); b) D. L. Coffen, J. Q. Chambers, D. R. Williams, P. E. Garrett and N. D. Canfield, *J. Am. Chem. Soc.*, **93**, 2258 (1971).

3. a) A. J. Moore and M. R. Bryce, *Synthesis*, 407 (1997); b) R. L. Meline and R. Elsenbaumer, *Synth. Met.*, **102**, 1658 (1999).

4. W. R. H. Hurtley and S. Smiles, *J. Chem. Soc.*, 2263 (1926).

5. E. M. Engler and V. V. Patel, *J. Am. Chem. Soc.*, **96**, 7376 (1974).

6. R. D. McCullough, G. B. Koc, K. A. Lerstrup and D. O. Cowan, *J. Am. Chem. Soc.*, **109**, 4115 (1987).

7. H. Akamatu and H. Inokuchi, *Electrical Conductivity in Organic Solids*, Interscience, New York (1960), p. 277.

8. F. Wudl, *J. Am. Chem. Soc.*, **97**, 1962 (1975).

9. J. Ferraris, D. O. Cowan, V. J. Walarka and J. H. Perlstein, *J. Am. Chem. Soc.*, **95**, 948 (1973).

10. a) G. C. Papavassiliou, *Z. Naturforsch.*, **36b**, 1200 (1981); b) G. C. Papavassiliou, *Z. Naturforsch.*, **37b**, 825 (1982).

11. a) K. Bechgaard, D. O. Cowan and A. N. Block, *J. Chem. Soc., Chem. Commun.*, 937 (1974); b) M. Mizuno, A. F. Gavito and M. D. Cava, *J. Chem. Soc., Chem. Commun.*, 18 (1978); c) R. L. Meline and R. L. Elsenbauer, *Synth. Met.*, **86**, 1845 (1997).

12. a) D. Jerome, A. Mazaud, M. Ribault and K. Bechgaard, *J. Phys. Lett.*, **41**, L95 (1980); b) K. Bechgaard, C. S. Jacobsen, K. Mortensen, M. J. Pederson and N. Thoup, *Solid State Commun.*, **33**, 1119 (1980).

13. S. S. Parkin, E. Engler, R. R. Schumaker, R. Lagier, V. Y. Lee, J. C. Scott and R. L. Greene, *Phys. Rev. Lett.*, **50**, 270 (1983).

14. K. Kikuchi, M. Kikuchi, T. Namiki, K. Saito, I. Ikemoto, K. Murata, T. Ishiguro and K. Kobayashi, *Chem. Lett.*, 931 (1987).

15. a) G. C. Papavassiliou, G. A. Mousdis, J. S. Zambounis, A. Terzis, A. Hountas, B. Hilti, C. W. Mayer and J. Pfeiffer, *Synth. Met.*, **27**, B379 (1988); b) G. C. Papavassiliou, G. A. Mousdis, V. Kakoussis, A. Terzis, A. Hountas, B. Hilti, C. W. Mayer and J. S. Zambounis, *The Physics and Chemistry of Organic Superconductors*, Springer Proc. Phys., Vol. 51, Springer-Verlag, Berlin (1990), p. 247.

16. T. Ishiguro, K. Yamaji and G. Saito, *Organic Superconductors*, 2nd ed., Springer Ser. Solid-State Sci., Vol. 88, Springer, Berlin (1998).

17. Proceedings of International Conference on Science and Technology of Synthetic Metals (ICSM), *Synth. Met.*, **84-86** (1997); **101-103** (1999); **119-121** (2001);**135-137** (2003).

18. J. Yamada, *Recent Res. Devel. in Organic Chem.*, **2**, 525 (1998).

19. J. M. Williams, J. R. Ferraro, R. J. Thorn, K. D. Carlson, U. Geiser, H. H. Wang, A. M. Kini and M.–H. Whangbo, *Organic Superconductors (Including Fullerenes)*, Prentice Hall, Englewood Cliffs, NJ (1992).

20. R. D. McCullough, M. A. Petruska and J. A. Belot, *Tetrahedron*, **55**, 9979 (1999).

21. G. C. Papavassiliou, D. J. Lagouvardos, V. Kakoussis, G. Mousdis, A. Terzis, A. Hountas, B. Hilti, C. Mayer, J. Zambounis, J. Pfeiffer and P. Delhaes, *Organic Superconductors*, Plenum Press, NY (1990), p. 367.

22. G. C. Papavassiliou, D. J. Lagouvardos, V. C. Kakoussis, A. Terzis, A. Hountas, B. Hilti, C. Mayer, J. S. Zambounis, J. Pfeiffer, M.-H. Whangbo, J. Ren and D. B. Kang, *Mat. Res. Soc. Symp. Proc.*, **247**, 535 (1992).

23. G. C. Papavassiliou, D. J. Lagouvardos, A. Terzis, C. P. Raptopoulou, B. Hilti, W. Hofherr, J. S. Zambounis, G. Rihs, J. Pfeiffer, P. Delhaes, K. Murata, N. A. Fortune and N. Shirakawa, *Synth. Met.*, **70**, 787 (1995).

24. G. C. Papavassiliou, D. J. Lagouvardos, J. S. Zambounis, A. Terzis, C. P. Raptopoulou, K. Murata, N. Shirakawa, L. Ducasse and P. Delhaes, *Mol. Cryst. Liq. Cryst.*, **285**, 83 (1996).

25. G. C. Papavassiliou, D. J. Lagouvardos, I. Koutselas, K. Murata, A. Graja, I. Olejniczak, J. S. Zambounis, L. Ducasse and J. P. Ulmet, *Synth. Met.*, **86**, 2043 (1997).

26. K. Murata, H. Yoshino, Y. Tsubaki and G. C. Papavassiliou, *Synth. Met.*, **94**, 69 (1998).

27. G. C. Papavassiliou, K. Murata, J. P. Ulmet, A. Terzis, G. A. Mousdis, H. Yoshino, A. Oda, D. Vignolles and C. P. Raptopoulou, *Synth. Met.*, **103**, 1921 (1999).

28. G. C. Papavassiliou, G. A. Mousdis and A. Papadima, *Z. Naturforsch.*, **55b**, 231 (2000).

29. I. Olejniczak, J. L. Musfeldt, G. C. Papavassiliou and G. A. Mousdis, *Phys. Rev.*, *B* **62**, 15634 (2001).

30. T. Konoike, K. Namba, T. Shinada, K. Sakaguchi, G. C. Papavassiliou, K. Murata and Y. Ohfune, *Synlett*, 1476 (2001).

31. K. Storr, L. Balicas, J. S. Brooks, D. Graf and G. C. Papavassiliou, *Phys. Rev.*, *B* **64**, 45107 (2001).

32. G. C. Papavassiliou, G. A. Mousdis, A. Terzis, C. Raptopoulou, K. Murata, T. Konoike and Y. Yoshino,

Synth. Met., **120**, 743 (2001) and unpublished work.

33. G. C. Papavassiliou, A. Terzis and C. P. Raptopoulou, *Z. Naturforsch.*, **56b**, 963 (2001).

34. T. Konoike, K. Iwashita, H. Yoshino, K. Murata, T. Sasaki and G. C. Papavassiliou, *Phys. Rev.*, *B* **66**, 245308 (2002); *J. Phys. IV France,* to be published.

35. G. C. Papavassiliou, G. A. Mousdis and G. C. Anyfandis, *Z. Naturforsch.*, **b**, to be published.

36. G. C. Papavassiliou, G. A. Mousdis, A. Terzis, C. Raptopoulou, K. Murata, T. Konoike, H. Yoshino, A. Graju and A. Lapinski, *Synth. Met.*, in press; *J. Phys. IV France*, to be published.

37. a) G. C. Papavassiliou, *Mol. Cryst. Liq. Cryst.*, **86**, 159 (1982); b) G. C. Papavassiliou, *J. Physique C3*, **44**, 1257 (1983).

38. a) J. Hellberg and M. Moge, *Synthesis*, 198 (1996); b) T. Imakubo and K. Kobayashi, *J. Mater. Chem.*, **8**, 1945 (1998); c) H. Nishikawa, Y. Misaki, T. Yamabe and M. Shiro, *Synth. Met.*, **102**, 1693 (1999); d) M. Iyoda, Y. Kuwatani, E. Ogura, K. Hara, H. Suzuki, T. Takano, K. Takeda, J. Takano, K. Ugawa, M. Yoshida, H. Matsuyama, H. Nishikawa, I. Ikemoto, T. Kato, N. Yoneyama, J. Nishijo, A. Miyazaki and T. Enoki, *Heterocycles*, **54**, 833 (2001); e) T. Takimiya, T. Jigami, M. Kawashima, M. Kodani, Y. Aso and T. Otsubo, *J. Org. Chem.*, **67**, 4218 (2002).

39. a) G. C. Papavassiliou, *Chem. Scr.*, **25**, 167 (1985); b) G. C. Papavassiliou, J. S. Zambounis and S. Y. Yiannopoulos, *Physica B*, **143**, 307 (1986).

40. G. C. Papavassiliou, J. S. Zambounis and S. Y. Yiannopoulos, *Chem. Scr.*, **37**, 261 (1987).

41. G. C. Papavassiliou, J. S. Zambounis, G. A. Mousdis, V. Gionis and S. Y. Yiannopoulos, *Mol. Cryst. Liq. Cryst.*, **156**, 269 (1988).

42. G. C. Papavassiliou, G. A. Mousdis, S. Y. Yiannopoulos and J. S. Zambounis, *Chem. Scr.*, **28**, 365 (1988).

43. G. C. Papavassiliou, G. A. Mousdis, S. Y. Yiannopoulos, V. C. Kakoussis and J. S. Zambounis, *Synth. Met.*, **27**, B373 (1988).

44. A. Terzis, A. Hountas, A. E. Underhill, A. Clark, B. Kaye, B. Hilti, C. Mayer, J. Pfeiffer, S. Y. Yiannopoulos, G. Mousdis and G. C. Papavassiliou, *Synth. Met.*, **27**, B97 (1988).

45. G. C. Papavassiliou, V. C. Kakoussis, E. A. Mousdis, J. S. Yiannopoulos and C. W. Mayer, *Chem. Scr.*, **29**, 71 (1989).

46. G. C. Papavassiliou, V. C. Kakoussis, J. S. Zambounis and G. A. Mousdis, *Chem. Scr.*, **29**, 123 (1989).

47. G. C. Papavassiliou, *Lower-Dimensional Systems and Molecular Electronics*, Plenum Press, NY (1990), **B248**, p. 143.

48. G. C. Papavassiliou, *Mat. Res. Soc. Symp. Proc.*, **247**, 523 (1992).

49. D. J. Lagouvardos and G. C. Papavassiliou, *Z. Naturforsch.*, **47b**, 898 (1992).

50. G. C. Papavassiliou, *Pure Appl. Chem.*, **62**, 483 (1990).

51. G. C. Papavassiliou, V. C. Kakoussis, D. J. Lagouvardos and G. A. Mousdis, *Mol. Cryst. Liq. Cryst.*, **181**, 171 (1990).

52. G. C. Papavassiliou, D. J. Lagouvardos, V. C. Kakoussis and G. A. Mousdis, *Z. Naturforsch.*, **45b**, 1216 (1990).

53. G. C. Papavassiliou, *Synth. Met.*, **41-43**, 2535 (1991).

54. G. C. Papavassiliou, S. Y. Yiannopoulos and J. S. Zambounis, *J. Chem. Soc., Chem. Commun.*, 820 (1986).

55. G. C. Papavassiliou, S. Y. Yiannopoulos and J. S. Zambounis, *Physica B*, **143**, 310 (1986).

56. G. C. Papavassiliou, S. Y. Yiannopoulos and J. S. Zambounis, *Chem. Scr.*, **27**, 265 (1987).

57. G. C. Papavassiliou, S. Y. Yiannopoulos, J. S. Zambounis, K. Kobayashi and K. Umemoto, *Chem. Lett.*, 1279 (1987).

58. G. C. Papavassiliou, V. Gionis, S. Y. Yiannopoulos, J. S. Zambounis, G. A. Mousdis, K. Kobayashi and K. Umemoto, *Mol. Cryst. Liq. Cryst.*, **156**, 277 (1988).

59. G. A. Mousdis, V. C. Kakoussis and G. C. Papavassiliou, *Low-Dimensional Systems and Molecular Electronics*, Plenum Press, NY (1990), **B248**, p. 181.

60. G. C. Papavassiliou, V. C. Kakoussis and D. J. Lagouvardos, *Z. Naturforsch.*, **46b**, 1269 (1991) and unpublished work.

61. G. C. Papavassiliou, D. J. Lagouvardos and V. C. Kakoussis, *Z. Naturforsch.*, **46b**, 1730 (1991).

62. G. C. Papavassiliou, D. J. Lagouvardos, A. Terzis, A. Hountas, B. Hilti, J. S. Zambounis, C. W. Mayer, J. Pfeiffer, W. Hofherr, P. Delhaes and J. Amiell, *Synth. Met.*, **55-57**, 2174 (1993).

63. J. D. Wallis, A. Karrer and J. D. Dunitz, *Helv. Chim. Acta*, **69**, 69 (1986).

64. A. Karrer, J. D. Wallis, J. D. Dunitz, B. Hilti, C. W. Mayer, M. Bukle and J. Pfeiffer, *Helv. Chim. Acta*, **70**, 942 (1987).

65. J. D. Wallis and J. D. Dunitz, *Acta Crystallogr.*, C**44**, 1037 (1988).

66. A. Izuoka, S. Matsumiya and T. Sugawara, *The Physics and Chemistry of Organic Superconductors*, Springer Proc. Phys., Vol. 51, Springer-Verlag, Berlin (1990), p. 379.

67. B. Chen, F. Deilacher, M. Hoch, H. J. Keller, P. Wu, P. Armbruster, R. Geiger, S. Kahlich and D. Schweitzer, *Synth. Met.*, **41-43**, 2101 (1991).
68. a) J. S. Zambounis, C. W. Mayer, K. Hauenstein, B. Hilti, W. Hofherr, J. Pfeiffer, M. Burkle and G. Rihs., *Adv. Mater.*, **4**, 33 (1992); b) J. S. Zambounis, C. W. Mayer, K. Hauenstein, B. Hilti, W. Hofherr, J. Pfeiffer, M. Burkle and G. Rihs, *Mat. Res. Soc. Symp. Proc.*, **247**, 509 (1992).
69. a) K. Swaminathan, P. J. Carrol, J. D. Singh and H. B. Singh, *Acta Crystallogr.*, C**50**, 1958 (1964); b) B. Chen, F. Deilacher, M. Hoch, H. J. Keller, P. Wu, S. Garner, S. Kahlich and D. Schweitzer, *Low-Dimensional Systems and Molecular Electronics*, Plenum Press, NY (1990), **B248**, p. 175.
70. a) J. S. Zambounis and C. W. Mayer, *Tetrahedron Lett.*, **32**, 2237 (1991); b) C. Gemmell, J. Kilburn, H. Ueek and A. E. Underhill, *Tetrahedron Lett.*, **33**, 3923 (1992); c) L. Binet, J.-M. Fabre and J. Becher, *Synthesis*, 26 (1997).
71. S. Matsumiya, A. Izuoka, T. Sugawara, T. Taruishi and Y. Kawada, *Bull. Chem. Soc. Jpn.*, **66**, 513 (1993).
72. S. Matsumiya, A. Izuoka, T. Sugawara, T. Taruishi, Y. Kawada and M. Tokumoto, *Bull. Chem. Soc. Jpn.*, **66**, 1949 (1993).
73. A. M. Kini, J. P. Parakka, U. Geiser, H.–H. Wang, F. Rivas, E. DiNino, S. Thomas, J. D. Dudek and J. M. Williams, *J. Mater. Chem.*, **9**, 883 (1999).
74. G. Schukat and E. Fanghänel, *Sulfur Rep.*, **18**, 1 (1996).
75. a) R. Kato, H. Kobayashi and A. Kobayashi, *Synth. Met.*, **41-43**, 2093 (1991); b) T. Naito, A. Miyamoto, H. Kobayashi, R. Kato and A. Kobayashi, *Chem. Lett.*, 1945 (1991); c) S. Aonuma, Y. Okano, H. Sawa, R. Kato and H. Kobayashi, *J. Chem. Soc., Chem. Commun.*, 1193 (1992); d) T. Imakubo, Y. Okano, H. Sawa and R. Kato, *J. Chem. Soc., Chem. Commun.*, 2493 (1995); e) T. Courcef, I. Malfant, K. Polhodnia and P. Cassoux, *New J. Chem.*, 585 (1998).
76. a) T. Akutagawa and G. Saito, *Bull. Chem. Soc. Jpn.*, **68**, 1753 (1995); b) H. Spangaard, J. Prehn, M. B. Nielsen, E. Levillain, M. Allain and J. Becher, *J. Am. Chem. Soc.*, **122**, 9486 (2000); c) E. Ribera, J. Veciana, E. Molins, I. Mata, K. Wurst and G. Rovira, *Eur. J. Org. Chem.*, **65**, 2867 (2000).
77. a) L.-K. Chou, M. A. Quijada, M. B. Clevenger, C. F. deOliveira, K. A. Abboud, D. B. Tanner and D. R. Talham, *Chem. Mater.*, **7**, 530 (1995); b) P. Batail, K. Boubeker, M. Formigue and J.-C. P. Gabriel, *Chem. Mater.*, **10**, 3005 (1998).
78. P. Delhaes, *Low-Dimensional Systems and Molecular Electronics*, Plenum Press, NY (1991), **B248**, p. 43.
79. G. C. Papavassiliou, Y. Misaki, K. Takahashi, J. Yamada, G. A. Mousdis, T. Sharahata and T. Ise, *Z. Naturforsch.*, **56b**, 297 (2001).
80. a) A. M. Kini, B. D. Gates, M. A. Beno and J. M. Williams, *J. Chem. Soc., Chem. Commun.*, 169 (1989); b) A. Otsuka, G. Saito, K. Ohfuchi and M. Konno, *Phosphorus, Sulfur, and Silicon*, **67**, 333 (1992); c) K. Tsutsui, K. Takimiya, Y. Aso and T. Otsubo, *Tetrahedron Lett.*, **38**, 7569 (1997); d) M. Moge, J. Hellberg, K. W. Tomroos, H. Schmitt and J.-U. von Schutz, *Synth. Met.*, **86**, 1877 (1997).
81. E. Ojima, H. Fujiwara, H. Kobayashi and M. Tokumoto, *Synth. Met.*, **120**, 887 (2001).
82. a) H. Tatemitsu, E. Nishikawa, Y. Sakata and S. Misuni, *J. Chem. Soc., Chem. Commun.*, 106 (1985); b) R. S. Medne, Y. Y. Katsens, I. L. Kraupsha and O. Y. Neiland, *Chem. Heterocycl. Compd.*, **27**, 1053 (1991); c) A. J. Moore and M. R. Bryce, *J. Chem. Soc., Chem. Commun.*, 1639 (1991); d) M. Salle, A. Gorgues, M. Jubault, K. Boubekeur and P. Batail, *Tetrahedron*, **48**, 3081 (1992); e) S. Ikegawa, K. Miyawaki, T. Nogami and Y. Shirota, *Bull. Chem. Soc. Jpn.*, **66**, 2770 (1993); f) P. Hudhomme, S. L. Moustarder, C. Durand, N. Gallego-Planas, N. Mercier, P. Blanchard, E. Levillain, M. A. Allain, A. Gorgues and A. Riou, *Chem. Eur. J.*, 7, 5070 (2001).
83. a) R. Andreu, J. Gavin and J. Orduna, *Tetrahedron*, **57**, 7883 (2001); b) V. Khodorkovsky, L. Shapiro, P. Krief, A. Shames, G. Mabon, A. Gorgues and M. Giffard, *Chem. Commun.*, 2736 (2001); c) G. C. Papavassiliou, unpublished work.

3

Halogenated TTFs

3.1 Introduction

Halogenated derivatives of conventional aromatic compounds, *e.g.*, bromobenzene, 2-bromothiophene, *etc.*, are important precursors to the more complex systems containing aromatic rings. The tetrathiafulvalene (TTF) molecule can be considered to be a pseudo-aromatic compound and halogenated TTFs were synthesized as raw materials in the early stages of TTF chemistry. Some oligomers and polymers containing a TTF skeleton have been synthesized using several types of iodinated TTFs as the key intermediates.

Direct use of halogenated TTFs to prepare conducting charge transfer salts is being studied continuously. Tetrachloro- and tetrabromo-TTFs were synthesized by the analogy of tetramethyltetrathiafulvalene (TMTTF) and tetramethyltetraselenafulvalene (TMTSF), because the chlorine or bromine atom could be approximated as an equivalent of the methyl group in terms of the size and shape of the substituent. Although the electron-withdrawing nature of the chlorine and bromine atoms prevented the formation of the conducting salts of tetrachloro- and tetrabromo-TTFs, some dichloro- and dibromo-TTF derivatives provided conductive charge transfer salts.

The most successful example of the utilization of halogenated TTFs is a series of iodinated derivatives. They were used to introduce the I···N type strong intermolecular interaction, *i.e.*, the *iodine bond*, for the crystal structure control of organic conductors, and a variety of charge transfer salts with unique supramolecular structure and high conductivity have been prepared. Recently, a supramolecular superconductor based on an iodine-bonded diselenadithiafulvalene (DSDTF) has also been reported.

This section summarizes the synthesis and properties of the halogenated TTFs, and the unique features of their charge transfer salts are reviewed.

3.2 Synthesis and Properties of Halogenated TTFs

Direct halogenation of the TTF skeleton using the conventional method cannot be attained because the oxidation reaction occurs easily due to the low oxidation potential as an electron donor.[1] One route for the introduction of halogen atom(s) into the TTF skeleton is halogenation *via* anionic intermediates generated by the metallation reaction in the last stage of their syntheses (Scheme 3.1, *Metallation Method*). This method was developed for the halogenation of the parent TTF and widely used for the synthesis of other halogenated TTFs. However, it is not always easy to prepare unsymmetrical TTF derivatives on a large scale in good yield. Therefore, a cross-coupling reaction using the halogenated 1,3-dithiole unit was investigated as a different route

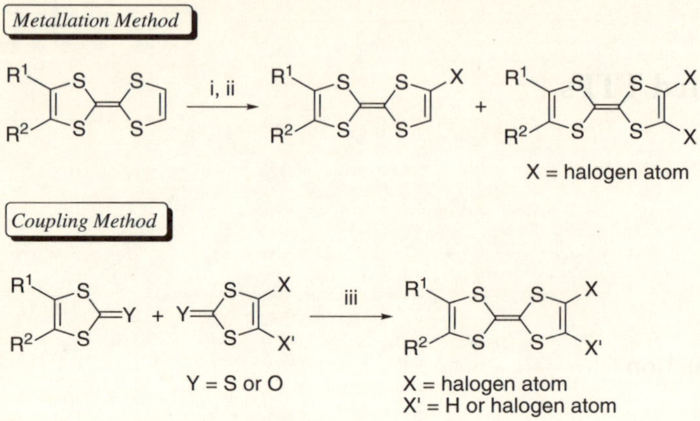

Metallation Method

X = halogen atom

Coupling Method

Y = S or O

X = halogen atom
X' = H or halogen atom

Scheme 3.1   Synthesis of halogenated TTFs. *Reagents and conditions*: i. metallating reagent (LDA *etc.*); ii. halogenating reagent (perfluorohexyl iodide, *etc.*); iii. P(OEt)$_3$.

other than the metallation method (Scheme 3.1, *Coupling Method*).

3.2.1 Chlorination and Bromination

The first synthesis of the chloro- and bromo-derivatives of the parent TTF were reported by Nakayama *et al.* in 1986.[2] Treatment of tetrathiafulvalenium lithium with Br$_2$ or *N*-chlorosuccinimide (NCS) afforded monohalogeno-TTF and a low yield of the mixture of two isomers of dihalogeno-TTFs (Scheme 3.2, 11% and 10% for chloride, 34% and 5% for bromide). According to the spectroscopic data of dihalogeno-TTFs, the positions of the halogenation were the 2,6- or 2,7-positions of TTF and no 2,3-substituted TTF was detected. Interestingly, treatment of 2-bromotetrathiafulvalene with lithium diisopropylamide (LDA) resulted in a *halogen dance* which afforded TTF, bromo-TTF, three isomers of dibromo-TTFs and tribromo-TTF. This type of *halogen dance* is similar to those observed for the halogenated thiophene derivatives.[3]

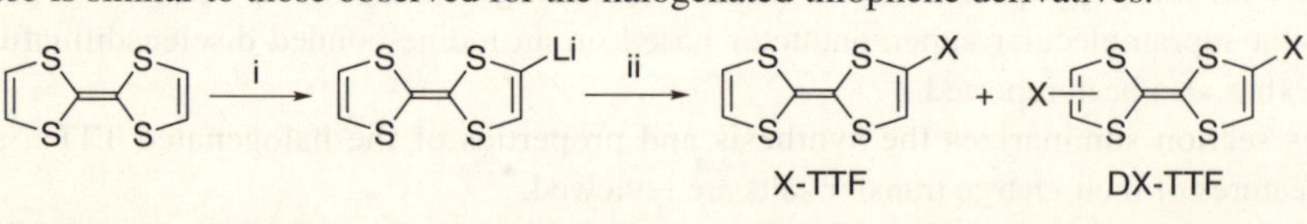

X-TTF   DX-TTF

Scheme 3.2   Synthesis of chloro- and bromo-TTFs (X = Cl or Br). *Reagents and conditions*: i. LDA; ii. Br$_2$ or NCS, –78 °C.

Tetrachloro-TTF (TCl-TTF) and tetrabromo-TTF (TBr-TTF) were prepared to compare the crystal and physical properties with those of TMTTF and TMTSF from the viewpoint of the isometric size and sphere shape of the halogen atom (the van der Waals radii of Cl and Br are 1.80 and 1.95 Å,[4] respectively): approximately the same size as the methyl group (*ca.* 2 Å radius pseudo-sphere shape). Syntheses of TCl-TTF and TBr-TTF were accomplished by quenching the tetralithiated intermediate with hexachloroethane and 1,2-dibromotetrachloroethane, respectively (Scheme 3.3[5]).

To date, several research groups have reported the synthesis of chloro- and bromo-derivatives of the parent TTF by the metallation method.[5–7] The product yield and distribution strongly

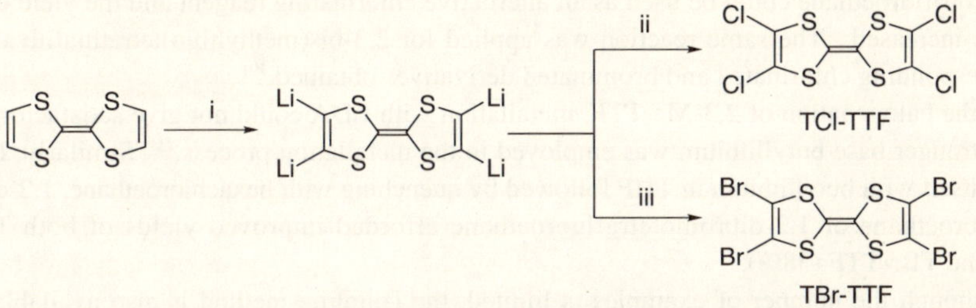

Scheme 3.3   Synthesis of TCl-TTF and TBr-TTF. *Reagents and conditions*: i. PhLi, −78 °C; ii. hexachloroethane; iii. 1,2-dibromotetrachloroethane.

Table 3.1   Chlorination and Bromination of TTF

	Metallating Reagent	Halogenating Reagent	Yields			Reference
			X-TTF	DX-TTF	TX-TTF	
Chlorination	LDA	NCS[a]	11	10[b]		[2]
	PhLi (5 eq.)	HCE[c] (5 eq.)			15	[5]
	LDA (1.1 eq.)	TsCl[d] (1.1 eq.)	48	4[e]		[7]
	LDA (4.4 eq.)	TsCl[d] (4.4 eq.)			30	[7]
	BuLi (4.5 eq.)	HCE[c] (2.5 eq.)			44	[11]
Bromination	LDA	Br$_2$	34	5[b]		[2]
	PhLi (5 eq.)	BCE[f] (5 eq.)			14	[5]
	LDA (1 eq.)	BCE[f] (1 eq.)	13	41[e]		[6]
	LDA (1.1 eq.)	TsBr[g] (1.1 eq.)	38	5[e]		[7]
	BuLi (4.5 eq.)	BCE[f] (2.5 eq.)			38	[11]

[a]NCS = *N*-Chlorosuccinimide. [b]Mixture of 2,6- and 2,7-isomers. [c]HCE = Hexachloroethane. [d]TsCl = *p*-Toluenesulfonyl chloride. [e]2,3-Isomer. [f]BCE = 1,2-Dibromotetrachloroethane. [g]TsBr = *p*-Toluenesulfonyl bromide.

depend on the combination of the metallating and halogenating reagents. Table 3.1 shows the selected results of the chlorination and bromination of the parent TTF molecule. LDA or butyllithium is generally preferable as a metallating reagent and poly-halogenated alkanes are often employed as halogenating reagents.

Unsymmetrical derivatives containing two halogen atoms on the 4,5-positions of unsymmetrical TTFs have also been synthesized by the metallation method. (Ethylenedithio)tetrathiafulvalene (EDT-TTF) and (ethylenedioxy)tetrathiafulvalene (EDO-TTF) derivatives were prepared by the reaction with LDA and hexachloroethane or 1,2-dibromotetrachloroethane (Scheme 3.4[8,9]).

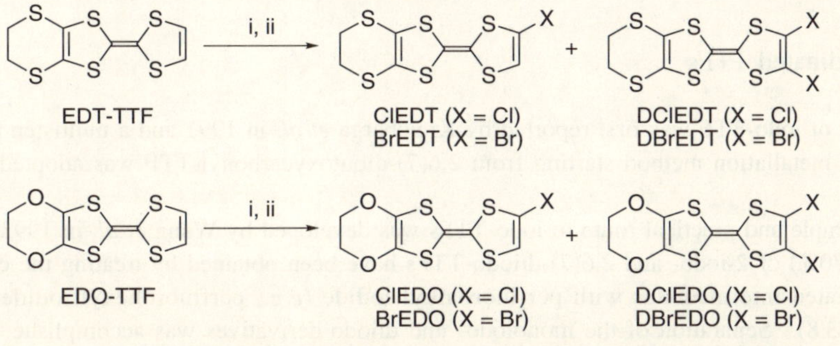

Scheme 3.4   Synthesis of chloro- and bromo-derivatives of EDT-TTF and EDO-TTF. *Reagents and conditions*: i. LDA; ii. hexachloroethane or 1,2-dibromotetrachloroethane.

Trichlorotrifluroethane could be used as an alternative chlorinating reagent and the yield of mono-chloride increased. The same reaction was applied for 2,3-bis(methylthio)tetrathiafulvalene and the corresponding chlorinated and brominated derivatives obtained.[10]

In the halogenation of 2,3-Me$_2$TTF, metallation with LDA could not give satisfactory results so the stronger base butyllithium was employed in the metallation process.[11] Similarly, lithiation of TTF itself with butyllithium in THF followed by quenching with hexachloroethane, 1,2-dibromo-tetrachloroethane or 1,2-dibromotetrafluoroethane afforded improved yields of both TCl-TTF (44%) and TBr-TTF (38%).

Although the number of examples is limited, the coupling method is also available for the synthesis of chloro- and bromo-TTFs. 4-Chloro-5-(methoxycarbonyl)-1,3-dithiole-2-thione was employed for the synthesis of 2,6(7)-dichloro-TTF (Scheme 3.5[12]) using the coupling method involving the decarbomethoxylation process. 4,5-Dibromo-1,3-dithiole-2-thione and 4-bromo-1,3-dithiole-2-thione were prepared from 1,3-dithiole-2-thione (Scheme 3.6[13]). Lithiation of 1,3-dithiole-2-thione followed by reaction with *p*-toluenesulfonyl bromide gave a mixture of 4,5-dibromo-1,3-dithiole-2-thione and 4-bromo-1,3-dithiole-2-thione, which could be separated by silica gel chromatography. 4,5-Dibromo-4',5'-bis(methylthio)-TTF was synthesized *via* P(OEt)$_3$-mediated cross-coupling of 4,5-dibromo-1,3-dithiole-2-thione and 4,5-bis(cyanoethylthio)-1,3-dithiol-2-one.

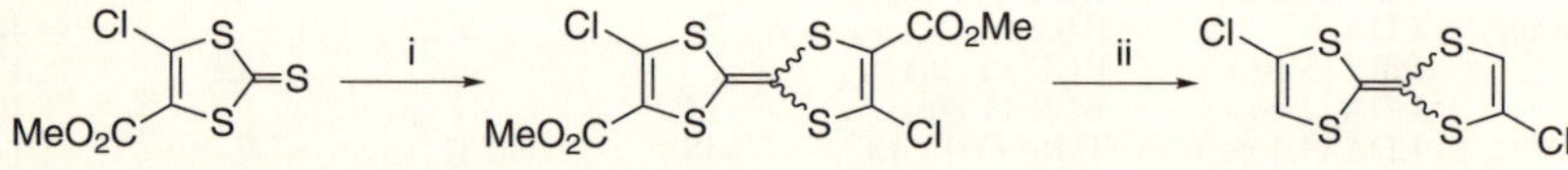

Scheme 3.5   Synthesis of 2,6(7)-dichloro-TTF *via* coupling method. *Reagents and conditions*: i. P(OMe)$_3$; ii. LiBr, HMPA.

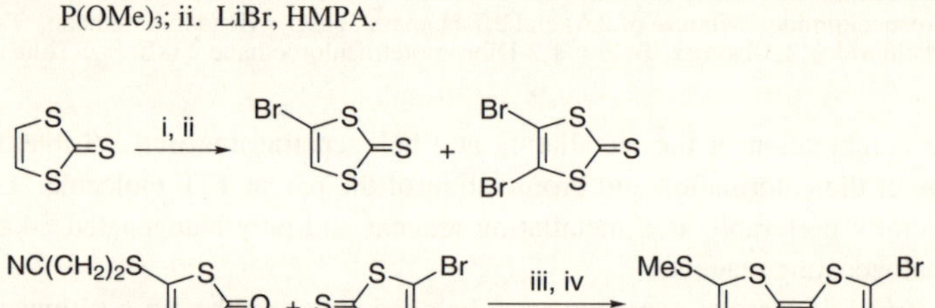

Scheme 3.6   *Reagents and conditions*: i. LDA, ether, −78 °C; ii. *p*-toluenesulfonyl bromide, −78 °C → −20 °C; iii. P(OEt)$_3$, toluene, reflux; iv. CsOH·H$_2$O, THF-MeOH, 20 °C, then MeI.

3.2.2 Iodinated TTFs

Synthesis of iodo-TTF was first reported by Kreicberga *et al.* in 1991 and a multistep procedure using the metallation method starting from 2,6(7)-dibutoxycarbonyl-TTF was adopted (Scheme 3.7[14]).

A simple and practical route to iodo-TTFs was developed by Wang *et al.* in 1993.[15] Good yields (~70%) of 2-iodo- and 2,6(7)-diiodo-TTFs have been obtained by treating the corresponding lithiated intermediates with perfluoroalkyl iodide (*e.g.*, perfluorohexyl iodide = PFHI) (Scheme 3.8). Separation of the monoiodo- and diiodo-derivatives was accomplished by silica gel chromatography using CS$_2$-light petroleum as an eluent. The X-ray crystal structure of diiodo-TTF has revealed that the iodination occurs primarily at the 2 and 6 (or 7) positions and

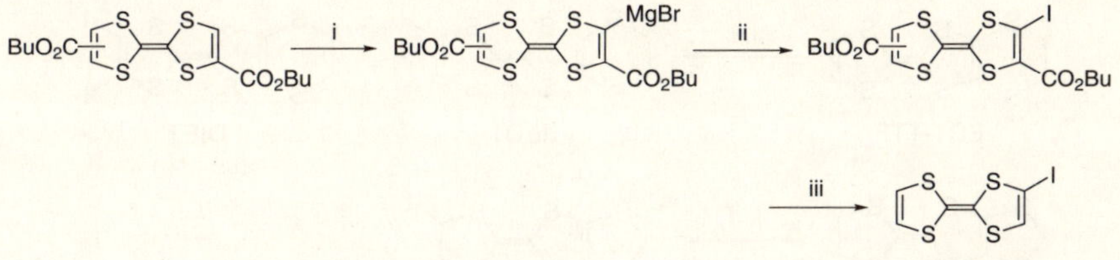

Scheme 3.7   Synthesis of iodo-TTF. *Reagents and conditions*: i. BuMgBr, –25 °C; ii. I$_2$; iii. NaOH, H$_2$O, Δ.

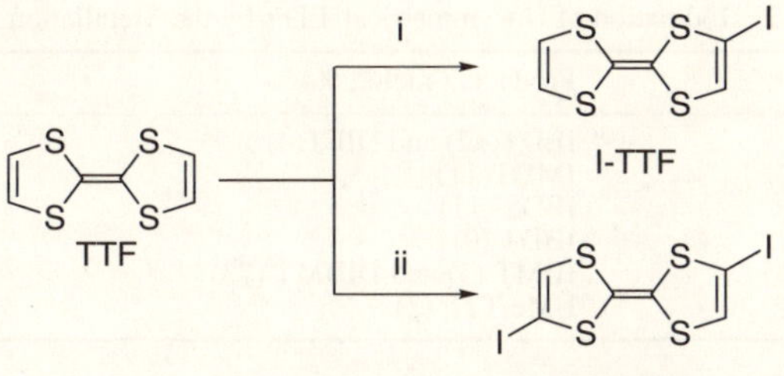

Scheme 3.8   Direct iodination of TTF. *Reagents and conditions*: i. LDA, –60 °C; ii. PFHI.

short I···I contacts were observed in the crystal of 2,6-diiodo-TTF (Fig. 3.1). Interestingly, the 2,3-diiodo-TTF was not detected using the PFHI as an iodinating reagent and it is remarkably different from the stereochemistry of chlorination and bromination of TTF. The difference in the iodinated position may be induced by the difference in the activating effect of halogen atoms and steric hindrance of the iodine atom. Reductive deiodination of iodo-TTF could be accomplished by using NaBH$_4$ as the reducing agent and this may be useful for the preparation of unsymmetrical TTFs containing a nonsubstituted 1,3-dithiole moiety.

Iodo-derivatives of the unsymmetrical donors EDT-TTF and (methylenedithio)tetrathiafulvalene (MDT-TTF) could be prepared using a similar metallation method (Scheme 3.9[16]). Using the smallest iodinating reagents, *i.e.*, iodine monochloride (ICl), the 4,5-diiodo-derivatives could be obtained, and it seems that the size of the iodinating reagent is an important factor in the stereochemistry of diiodo-derivatives. The same procedure is applicable to other unsymmetrical

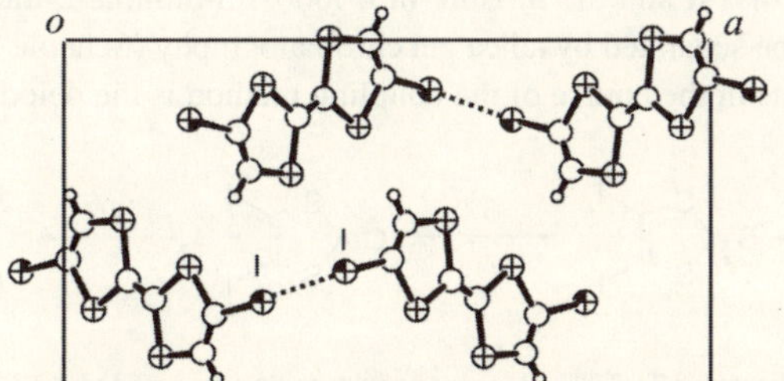

Fig. 3.1   Crystal structure of 2,6-diiodo-TTF. The dotted lines indicate short I···I contacts ($d_{\text{I···I}} = 3.69$ Å).

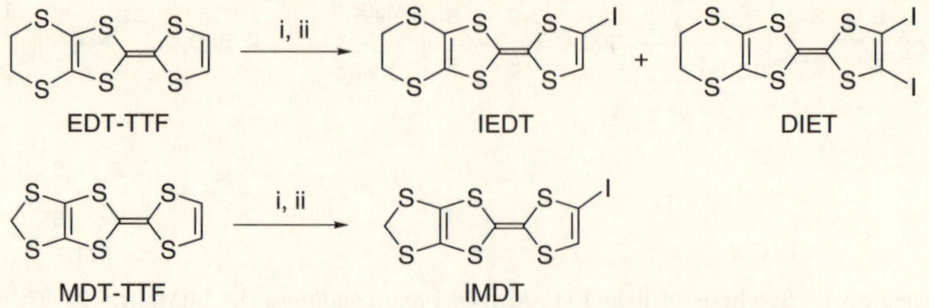

Scheme 3.9   Synthesis of iodinated derivatives of EDT-TTF and MDT-TTF.
Reagents and conditions: i. LDA, −78 °C; ii. ICl.

Table 3.2   Iodination of Unsymmetrical TTFs by the Metallation Method

Starting Material	Products (Yield^a, %)	Reference
EDT-TTF	IEDT(42) and DIET(4)	[16]
MDT-TTF	IMDT(11)	[16]
EDSe-TTF	IEDS (11)	[18]
BDT-TTP	IBDT (9)	[18]
BMT-TTF	IBMT (5) and DIBMT (73)	[10]
2,3-Me$_2$TTF	I$_2$Me$_2$TTF (2)	[11]

^{a}Isolated yield

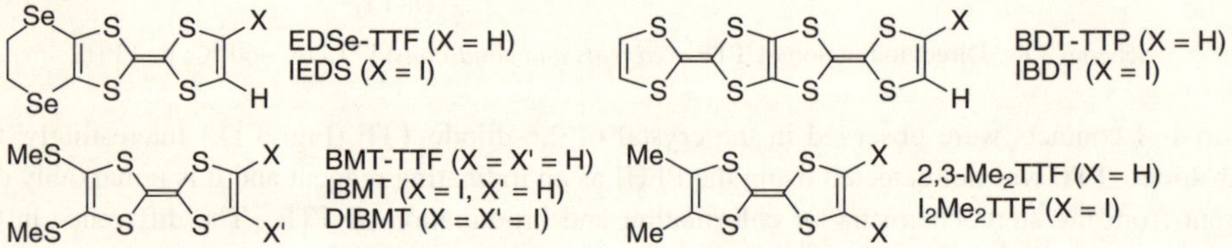

EDSe-TTF (X = H)
IEDS (X = I)

BDT-TTP (X = H)
IBDT (X = I)

BMT-TTF (X = X' = H)
IBMT (X = I, X' = H)
DIBMT (X = X' = I)

2,3-Me$_2$TTF (X = H)
I$_2$Me$_2$TTF (X = I)

TTFs, and the results of the iodination by the metallation method are summarized in Table 3.2.

It is not always easy to prepare the parent unsymmetrical TTFs, and other types of unsymmetrical iodo-TTFs were mainly prepared by the coupling method. Synthesis of 4,5-diiodo-1,3-dithiole-2-thione, which is the key intermediate for the coupling method, was first reported by Gompper *et al.* in 1994 and it was used for the synthesis of tetraiodo-TTF (TI-TTF) (Scheme 3.10[17]). LDA and *p*-toluenesulfonyl iodide (TsI) were used as the lithiating and iodinating reagents respectively. For the purpose of the iodination of 1,3-dithiole-2-thione, the more convenient reagents, PFHI or ICl is available instead of TsI and the mixture of 4-iodo-1,3-dithiole-2-thione and 4,5-diiodo-1,3-dithiole-2-thione could be separated by silica gel chromatography (Scheme 3.11[18]).

One of the problems in the course of the coupling method is the deiodination during the reac-

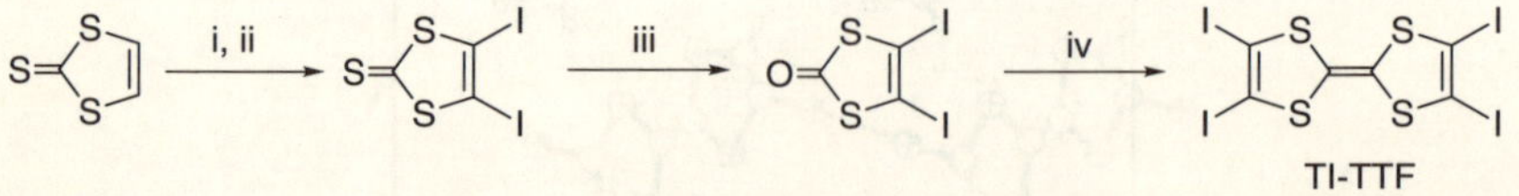

Scheme 3.10   Synthesis of TI-TTF. *Reagents and conditions*: i. LDA (2.1 eq.), THF, −78 °C, 3 h; ii. *p*-toluenesulfonyl iodide, −78 °C → −20 °C; iii. Hg(OAc)$_2$, AcOH-CHCl$_3$, room temperature; iv. P(OEt)$_3$, toluene, 110 °C, 16 h.

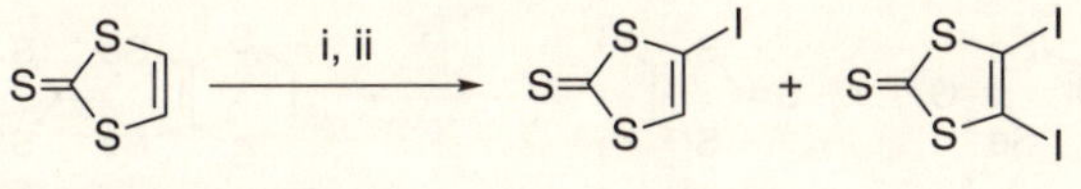

Scheme 3.11 Synthesis of 4-iodo- and 4,5-diiodo-1,3-dithiole-2-thiones.
Reagents and conditions: i. LDA, THF, –78 °C; ii. PFHI or ICl.

tion. Some research groups have reported *in situ* deiodination during the coupling reaction using 4,5-diiodo-1,3-dithiole-2-thione.[13,18] Particularly in the synthesis of TI-TTF, the randomness of the deiodination produces six isomers of the iodinated TTFs, and it is difficult to separate these isomers by conventional column chromatography. The mechanism of the deiodination reaction is not clear, although one possibility is the formation of an ylide-like intermediate between the iodinated TTFs and excess phosphite. This problem could be avoided using a smaller amount of P(OEt)$_3$ and shorter reaction time instead of a large excess of reagents.[19]

Using the above coupling method, iodinated TTFs containing an extended π-system, tetrathiapentalene (TTP) or TTF vinylogue, have also been synthesized in good yields (Scheme 3.12[18,20]).

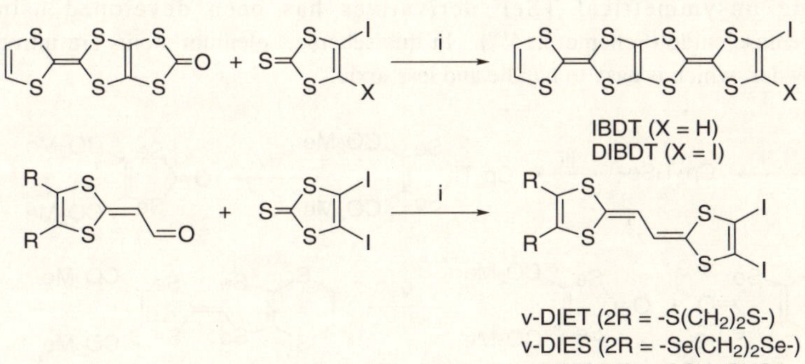

IBDT (X = H)
DIBDT (X = I)

v-DIET (2R = -S(CH$_2$)$_2$S-)
v-DIES (2R = -Se(CH$_2$)$_2$Se-)
v-DIVT (2R = -SCH=CHS-)

Scheme 3.12 Synthesis of iodinated TTFs containing an extended π-system. *Reagents and conditions*: i. P(OEt)$_3$, toluene, reflux (IBDT: 12%, DIBDT: 40%, v-DIET: 45%, v-DIES: 22%, v-DIVT: 63%).

3.2.3 DSDTFs and TSeFs

Introduction of the larger chalcogen atoms (selenium or tellurium) is one of the most effective and reliable methods of enhancing the conductivity of organic conductors based on TTFs. One of the merits of the sulfur-selenium substitution technique is the similarity of the molecular shape between TTF and its DSDTF or tetraselenafulvalene (TSeF) analogue, and it is promising for obtaining an isostructural salt with higher conductivity rather than the all-sulfur analogue. Several types of halogenated DSDTFs and TSeFs have been designed in this way.

Only two compounds are known as halogenated DSDTF derivatives. One is diiodo(ethylenedithio)diselenadithiafulvalene (DIETS) and the other is diiodo(pyrazino)diselenadithiafulvalene (DIPS). These were synthesized using the coupling method and the yields are modest to high as a cross-coupling reaction between the 1,3-dithiole and 1,3-diselenole units (Scheme 3.13[18,19,21]).

Syntheses of the halogenated TSeFs have been accomplished by two different routes. Iodo(ethylenedithio)tetraselenafulvalene (IETSe) and diiodo(ethylenedithio)tetraselenafulvalene

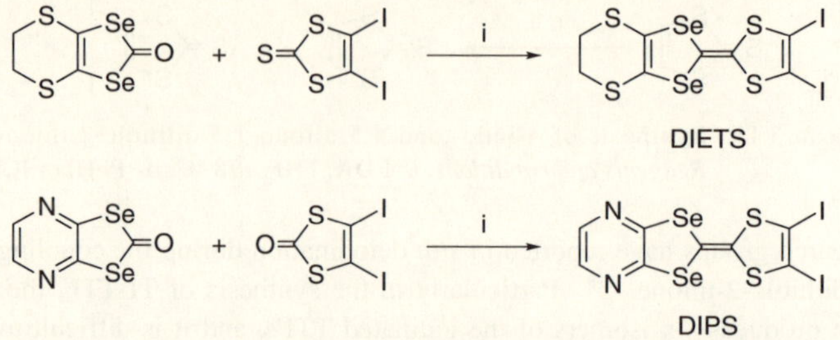

Scheme 3.13   Synthesis of DIETS and DIPS. *Reagents and conditions*: i.
P(OEt)$_3$, toluene, reflux (DIETS: 46%, DIPS: 49%).

(DIETSe) were prepared by the iodination of the parent (ethylenedithio)tetraselenafulvalene (EDT-TSeF). Usually, practical problems exist in the syntheses of unsymmetrical TSeF derivatives. The highly toxic CSe$_2$ is needed for the synthesis of an unsubstituted 1,3-diselenole unit. However, it is preferable not to use this reagent in the laboratory. A new CSe$_2$-free method of synthesizing unsymmetrical TSeF derivatives has been developed using known titanocene pentaselenide (Scheme 3.14[18]). In this scheme, selenium atoms are introduced from selenium powder, which is easy to handle and less toxic.

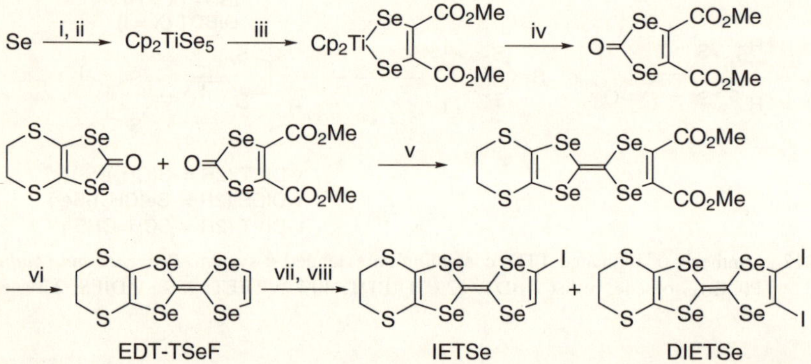

Scheme 3.14   Synthesis of IETSe and DIETSe. *Reagents and conditions*: i. LiEt$_3$BH,
THF; ii. Cp$_2$TiCl$_2$, reflux (63%); iii. MeO$_2$CC≡CCO$_2$Me, xylene, reflux
(84%); iv. (Cl$_3$CO)$_2$CO, THF, reflux (77%); v. P(OEt)$_3$, toluene, reflux
(19%); vi. LiBr·H$_2$O, HMPA, 110 °C (62%); vii. LDA, THF, −78 °C; viii.
ICl, −78 °C → room temperature (IETSe: 1%, DIETSe: 11%).

Tetra- and dihalogenated derivatives of the parent TSeF were prepared by the coupling reaction of the corresponding 1,3-diselenole-2-selones, which were synthesized by the one-pot reaction of trimethylsilylacetylene and CSe$_2$ (Scheme 3.15[22]). This coupling reaction affords the halogenated TSeF in a practical yield. However, the highly toxic and formidable reagent CSe$_2$ must be used in the first step of the synthesis of the 1,3-diselenole ring. The same synthetic route is also available for the synthesis of dimethyldihalo-TSeFs. In contrast to the synthesis of tetrahalogenated TTFs, treatment of the tetralithiated TSeF with hexachloroethane gave only 2,3-dichloro-TSeF and neither 2,6(7)-dichloro-TSeF nor tetrachloro-TSeF was detected.

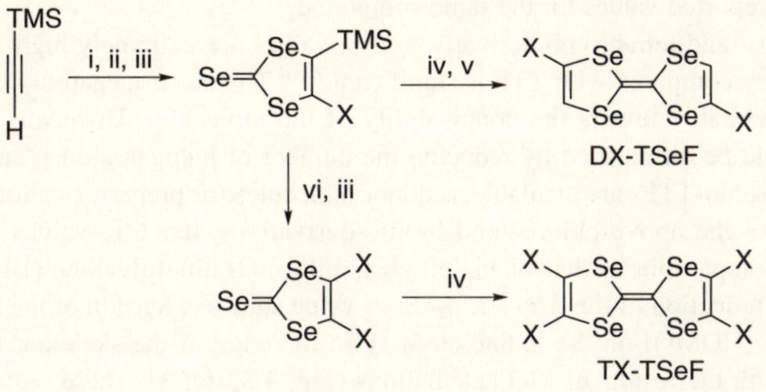

Scheme 3.15   Synthesis of halogenated TSeFs (X = Cl, Br or I). *Reagents and conditions*: i. BuLi; ii. Se, CSe$_2$; iii. C$_2$Cl$_6$, C$_2$F$_4$Br$_2$ or PFHI; iv. P (OEt)$_3$; v. KF; vi. Bu$_4$NF.

3.3   Electrochemical Properties of Halogenated TTFs

The donor ability of the halogenated TTFs could be examined by their redox potentials determined by cyclic voltammetry measurements. In general, the first redox potential $E^1_{1/2}$ approximately indicates the donor ability of the molecule and the ΔE (= $E^2_{1/2}$–$E^1_{1/2}$) value is related to the degree of on-site Coulombic repulsion. Table 3.3 summarizes the reported values of the redox potentials of halogenated TTFs calibrated by the $E^1_{1/2}$ value of the parent TTF. Due to the low solubility of the halogenated TTFs under the condition of cyclic voltammetry measurements and the differences in the criterion for the reversibility of the redox reaction, there exist some differ-

Table 3.3   Redox Potentials of Halogenated TTFs[a]

Donor	$E^1_{1/2}$ / V	$E^2_{1/2}$ / V	ΔE (= $E^2_{1/2}$–$E^1_{1/2}$)	Reference
TTF	0.00	0.42	0.42	
BEDT-TTF	0.16	0.48	0.32	
Cl-TTF	0.22	0.44	0.22	[7]
Br-TTF	0.21	0.42	0.21	[7]
I-TTF	0.21	0.46	0.25	[7]
	0.07	0.49	0.42	[11]
	0.07	0.47	0.40	[15]
	0.10	0.48	0.38	[23]
TCl-TTF	0.49	0.78	0.29	[7]
	0.53	0.73	0.20	[11]
TBr-TTF	0.45	0.79	0.34	[7]
	0.47	0.70	0.23	[11]
ClEDT	0.21	0.52	0.31	[8]
DClEDT	0.32	0.58	0.26	[8]
BrEDT	0.20	0.52	0.32	[8]
DBrEDT	0.30	0.57	0.27	[8]
IEDT	0.18	0.52	0.34	[16]
DIET	0.24	0.57	0.33	[16]
IMDT	0.17	0.47	0.30	[16]
IEDS	0.16	0.51	0.35	[18]
DIEDS	0.22	0.55	0.33	[18]
IETS	0.26	0.55	0.29	[18]
DIETS	0.32	0.59	0.27	[18]
DIEDO	0.16	0.51	0.35	[18]

[a]Potentials were calibrated by the $E^1_{1/2}$ value of the parent TTF.

ences among the reported values for the same compound.

In tetrachloro- and tetrabromo-derivatives, $E^1_{1/2}$ values are extremely high and the lowering of the donor ability compared with TTF is significant.[7,11] The electronegativity of the chlorine or bromine atom inevitably lowers the donor ability of the molecule. However, this lowering of donor ability could be diminished by reducing the number of halogen atoms, and some kinds of dichloro- and dibromo-TTFs are available as donor molecules for preparing cation radical salts.[10]

In contrast to the above chloro- and bromo-derivatives, the $E^1_{1/2}$ values of the iodinated derivatives are comparable to that of bis(ethylenedithio)tetrathiafulvalene (BEDT-TTF).[15,16,18] Furthermore, the reduction of the ΔE (= $E^2_{1/2}$–$E^1_{1/2}$) value shows extension of the highest occupied molecular orbital (HOMO) on the iodine atom(s) on the edge of the skeleton, and this result is also consistent with the results of MO calculations (Fig. 3.2, right). These reductions of the ΔE values are desirable for reducing the effect of on-site Coulombic repulsion and are also promising for the preparation of metallic charge transfer salts.

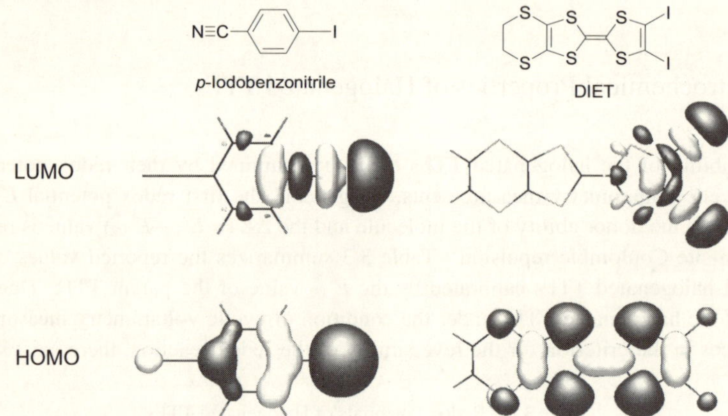

Fig. 3.2   Frontier orbitals of *p*-iodobenzonitrile (left) and DIETS (right) calculated by the MNDO-PM3 method.

3.4   Crystal Structure and Physical Properties of Charge Transfer Salts

It is well known that the molecular arrangement in the crystal is as important as the molecular design itself for exploring interesting electronic properties of organic conductors. Nevertheless, there have been few successful attempts to develop an active method for controlling molecular arrangement in organic conductors.[24]

In supramolecular chemistry, the hydrogen bond is the most popular interaction. However, the flexibility of the hydrogen bond hinders reliable control of the whole crystal structure of organic conductors. On the other hand, intermolecular interactions involving halogen atoms have recently gained increasing attention and are known as *halogen bonds*.[25] One of the most clear-cut examples of the halogen bond is the one-dimensional chain structure of *p*-iodobenzonitrile in the crystalline state (Fig. 3.3 [26]). Strong and directional I···N contacts (3.18 Å) much shorter than the van der Waals radii (3.53 Å) exist in the crystal of *p*-iodobenzonitrile. According to the MO calculations, the strong interaction could be recognized as an interaction between the lone pair on the nitrogen atom and pσ*-LUMO on the iodine atom (Fig. 3.2, left). In the crystal of organic con-

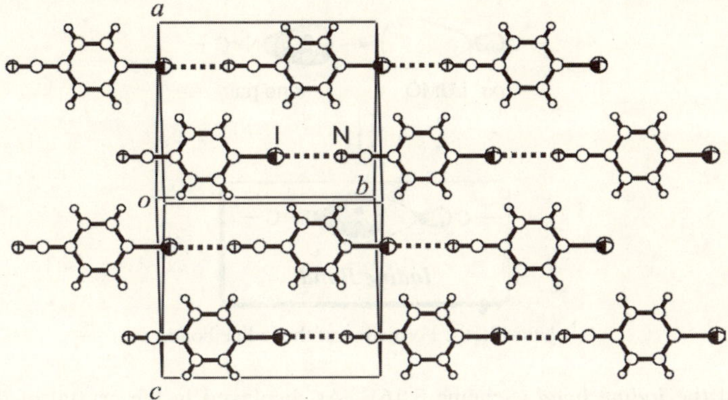

Fig. 3.3   One-dimensional chain structure of *p*-iodobenzonitrile in the crystal.
The intermolecular I···N distance is 3.18 Å.

ductors, metal-cyanate anions are often employed as the counterparts of TTFs, and halogenated TTFs contain halogen atoms directly connected to the pπ-conjugated system. Therefore, the halogenated TTFs should be good candidates for introducing the X···N type *halogen bond* to organic conductors.

3.4.1 Charge Transfer Salts of Chloro- and Bromo-TTFs

In the research on TCl-TTF and TBr-TTF, preparations of the conducting salts were attempted mainly to construct the isomeric crystal structures of TMTTF or TMTSF salts from the viewpoint of the isostructural framework of the donor molecules.[5] However, these attempts were unsuccessful because of the low donor ability due to the electronegativity of the halogen atoms.

Lowering of the electron donor ability could be diminished using fewer halogen atoms, and preparations of conducting salts based on the dichloro- and dibromo-derivatives have been reported. Most of the salts were, however, semiconductive or insulating at room temperature and few salts based on DClEDO and DBrEDO (Scheme 3.4) are metallic around room temperature.[10] Several types of halogen bonds based on the chlorine or bromine atoms were observed in these salts. However, their strengths are rather weak, and the range of the direction is too diverse to predict the molecular arrangement in the crystal. For this reason, weak halogen bonds based on the chlorine or bromine atoms are inadequate for effective control of the crystal structure of organic conductors.

3.4.2 Iodine Bond: A Special Case of the Halogen Bond

There are distinct differences between the iodine atom and the other halogen atoms. For example, the electronegativity of the halogen atom except the iodine atom is larger than that of the carbon atom. In contrast, the value of the electronegativity of the iodine atom (2.2) is smaller than that of the carbon atom (2.5)[27] and the introduction of the iodine atom does not suffer intrinsic donor ability. This has also been confirmed by cyclic voltammetry measurements, and the pπ orbital of the iodine atom contributes to the HOMO and extends it. Furthermore, the localization of the σ-type LUMO on the carbon(sp^2)-halogen bond is evident in the case of the iodine atom according to the MO calculations (Fig. 3.2). Therefore, the iodine-based halogen bond should more specifi-

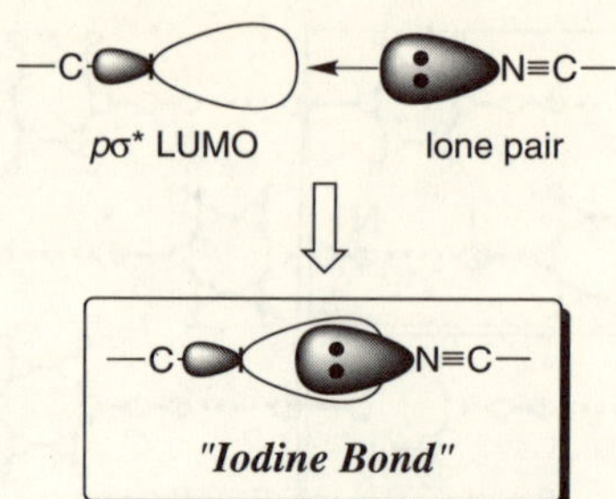

Scheme 3.16   Formation of the iodine bond.

cally be called the *iodine bond* (Scheme 3.16). As displayed in the crystal of the following iodo(ethylenedithio)tetrathiafulvalene (IEDT, Scheme 3.9) salts the I⋯N distance between the iodinated TTF and nitrogen atom on the counteranion is remarkably shorter than those of some neutral models of the halogen bond.

3.4.3 Cation Radical Salts of IEDT

(IEDT)$_2$[Ag(CN)$_2$] is the first example of a highly conducting salt that includes the iodine bond architecture in the crystal[16] (See Scheme 3.9 for the structural formula of IEDT). Temperature dependence of the resistivity is semiconductive from room temperature; however, the activation energy is small. Fig. 3.4 shows the donor⋯anion⋯donor type arrangement tailored by the strong I⋯N iodine bond. The distance of the I⋯N contact between the donor molecule and Ag(CN)$_2$ anion is 2.88 Å, which is almost 20% shorter than the sum of the van der Waals radii (3.53 Å). This value is much shorter than that of *p*-iodobenzonitrile (3.18 Å), indicating the existence of the strong iodine bond.

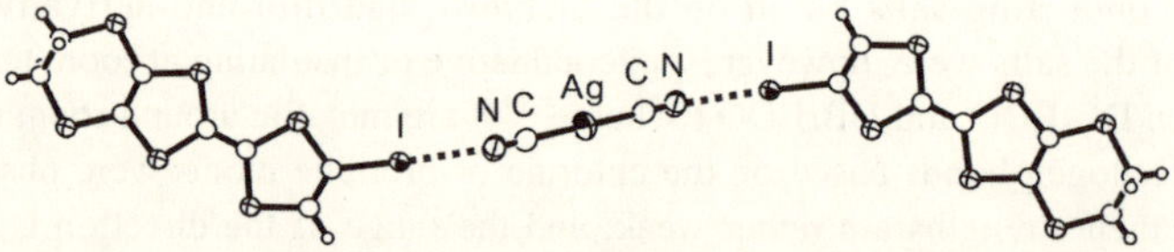

Fig. 3.4   Crystal structure of (IEDT)$_2$[Ag(CN)$_2$]. Dotted lines indicate the strong iodine bonds ($d_{\text{I}\cdots\text{N}} = 2.88$ Å).

The relationship between the number of the cyano groups on the counteranion and the iodine atoms on the donor molecule is also important to predict the motif of the iodine bond. Fig. 3.5 shows the crystal structure of (IEDT)$_4$[Pd(CN)$_4$] and the 4:1 donor⋯anion unit is constructed by the strong I⋯N iodine bonds.[28] Formation of the 4:1 unit could be recognized by the interaction between the donor molecule (one iodine atom) and the counteranion (four cyano groups). The differences in the directions of the iodine bonds will be discussed later in the paragraph on the diiodo(ethylenedithio)diselenadithiafulvalene (DIETS) salts. Temperature dependence of the resistivity of the Pd(CN)$_4$ salt is semi-metallic down to 130 K and room temperature conductivity is 2.5 S cm^{-1}.

When the counteranion is changed to the bromide anion, similar strong interactions were also observed (Fig. 3.6[16]). In this case, the counterpart of the iodine atom is the bromide anion and the

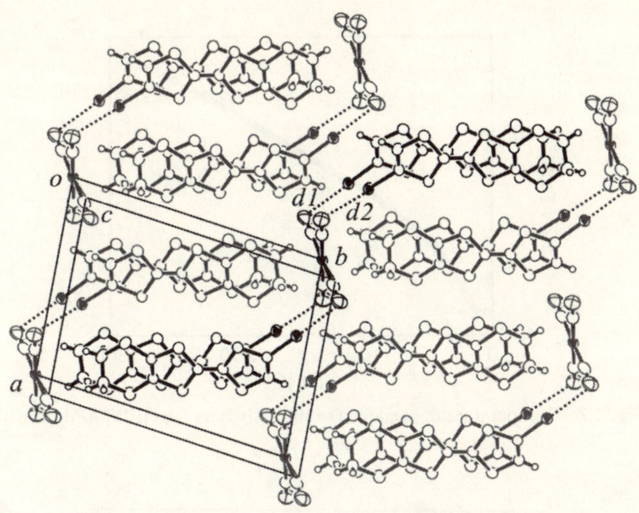

Fig. 3.5   4:1 Donor⋯anion unit in the crystal of (IEDT)$_4$[Pd(CN)$_4$]. The short I⋯N distances are: $d1$ = 3.075(7) Å, $d2$ = 3.144(7)Å.

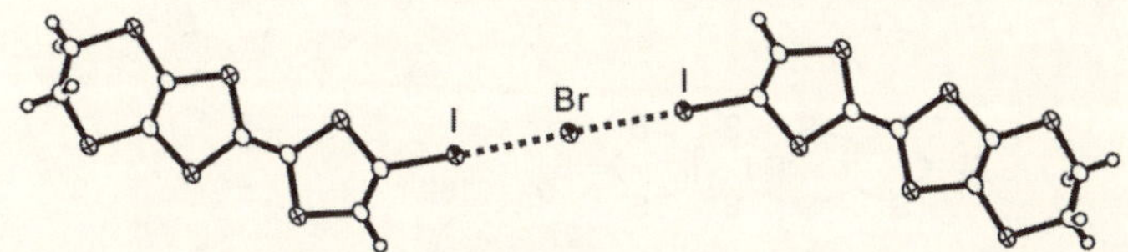

Fig. 3.6   Crystal structure of (IEDT)$_2$Br ($d_{I⋯Br}$ = 3.21 Å). Dotted lines indicate the I⋯Br type iodine bonds.

distance between the iodine atom and the bromine atom is 3.21 Å. It is also 20% shorter than the sum of the van der Waals radii (3.83 Å), showing that the iodine bond architecture is also possible using the halide anions.

3.4.4 Donor-acceptor Complexes of Iodinated TTFs

As well as the TTF-TCNQ (TCNQ = 7,7,8,8-tetracyanoquinodimethane) complex, donor-acceptor type charge transfer complexes attract much attention from the viewpoint of multicomponent organic conductors.

The TCNQ salts with 4-iodo-bis(methylthio)tetrathiafulvalene and 4-iodo-5-methyl-bis(methylthio)tetrathiafulvalene were prepared and their electrical conductivities measured.[13] X-ray structure analyses of these salts revealed stair-like stacks of alternating donor and acceptor molecules, *i.e.*, mixed stacking form, and their electrical conductivities were almost insulative (10^{-7}–10^{-8} S cm^{-1}). The TCNQF$_4$ salt of TI-TTF is rather conductive (0.2 S cm^{-1}), but the crystal structure is unknown.

The first metallic salt involving an iodinated-TTF was developed using the metal dithiolene complex Pd(dmit)$_2$ (dmit = 2-thioxo-1,3-dithiole-4,5-dithiolato) as the counteranion.[29] As shown in Fig. 3.7, the room temperature conductivity of (IEDT)[Pd(dmit)$_2$] reaches 400 S cm^{-1} and the

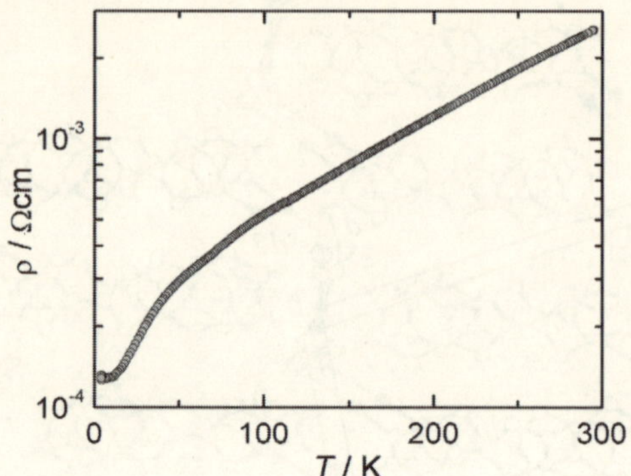

Fig. 3.7   Temperature dependence of the resistivity for (IEDT)[Pd(dmit)$_2$].

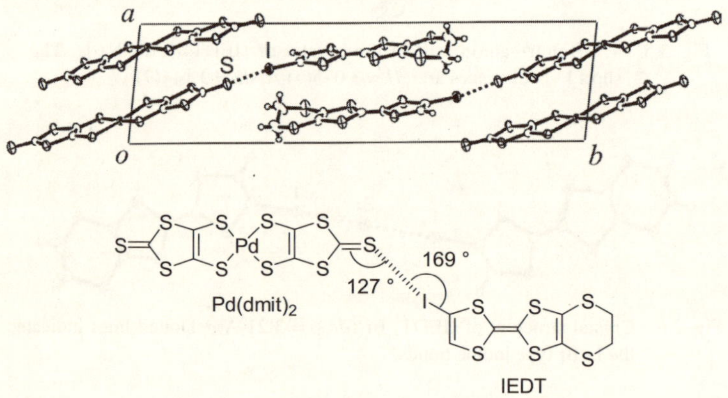

Fig. 3.8   Crystal structure of (IEDT)[Pd(dmit)$_2$]. The intermolecular short I$\cdots$S
distance is 3.308(4) Å.

temperature dependence of the resistivity shows metallic behavior down to 1.3 K. Fig. 3.8 shows
the crystal structure of (IEDT)[Pd(dmit)$_2$] and the shortest iodine$\cdots$sulfur distance between the
IEDT and Pd(dmit)$_2$ molecules is 3.308(4) Å. It is 13% shorter than the sum of the van der Waals
radii (3.78 Å). The strong I$\cdots$S interaction also affects the C=S double bond distances in the dmit
ligand; the one which interacts with the IEDT molecule is longer than the other (1.61 and 1.66 Å,
respectively). Bond angles of C=S$\cdots$I (127 °) and S$\cdots$I-C (169 °) are consistent with the directions
of the lone pair of sulfur atoms and the pσ*-LUMO on the C-I single bond; therefore, the I$\cdots$S
contact is one of the varieties of the iodine bond.

It has been reported that the parent EDT-TTF forms the Pd(dmit)$_2$ salt, which is isostructural
to the IEDT salt. However, the EDT-TTF complex shows metal-semiconductor transition at low
temperature.[30] Tight-binding band calculations gave some information about the differences in
the stability of the metallic states between the EDT-TTF and IEDT salts. Although the IEDT salt
has two types of closed Fermi surfaces associated with the frontier orbitals of IEDT and
Pd(dmit)$_2$, it should be noted that there exist small but significant values between the IEDT and
Pd(dmit)$_2$ layers (about one third of the inter-column interactions within each layer). This

suggests that one of the origins of the stable metallic state of the IEDT salt may be an increase in the dimensionality of the electronic structure through the I···S iodine bond.

3.4.5 Donor···anion Network in DIETS Salts

To confirm the effect of the sulfur-selenium substitution on the enhancement of the conductivity and stabilize the metallic state at low temperature, cation radical salts of DIETS (Scheme 3.13) were prepared and their physical properties examined.[31]

Temperature dependent resistivities of the $M(CN)_4$ [M = Ni, Pd, Pt] salts are shown in Fig. 3.9. They are metallic down to 80–100 K and then undergo moderate metal-semiconductor transitions. Comparing the conducting behavior of cation radical salts based on the all-sulfur analogue of DIETS, diiodo(ethylenedithio)tetrathiafulvalene (DIET [16] Scheme 3.9), the room temperature conductivities are much higher and the sulfur-selenium substitution in the parent TTF skeleton has effectively improved the electrical conductivity. Tight-binding band calculation indicates that there exist two pairs of quasi two-dimensional opened Fermi surfaces, which is consistent with the metallic behavior of the $M(CN)_4$ salts (Fig. 3.10).

X-ray diffraction study for the $Pd(CN)_4$ salt revealed that the donor molecules are packed in a so-called β-type arrangement. As shown in Fig. 3.10, the strong iodine bonds among the two iodine atoms on the donor molecule and four cyano groups on the counteranion construct an infinite chain of anions and quadruple donors. There are two kinds of iodine bonds [$d_{I···N}$ = 2.95(1) Å and 3.11(1) Å, respectively] between DIETS and $Pd(CN)_4$ and both interactions are 12-16% shorter than the sum of the van der Waals radii. Furthermore, the I···N-C angles [143(1) ° and 106.7(8) °] suggest that these two iodine bonds are of different types from that of the *p*-iodobenzonitrile. In the $Pd(CN)_4$ anion, the hybridization state of the atomic orbital of the nitrogen atom might be changed from sp to sp^2 as illustrated in Fig. 3.11. In sp hybridization, the number of the lone pair on the nitrogen atom is one and the direction is parallel to the CN triple bond as well as *p*-iodobenzonitrile. On the other hand, in the sp^2 state, there are two lone pairs and each direction is bent about 120 ° towards the fundamental CN triple bond. Concerning the I···N-C angle shown above, the hybridization state of the nitrogen atom is nearly sp^2 and the iodine bonds are made of

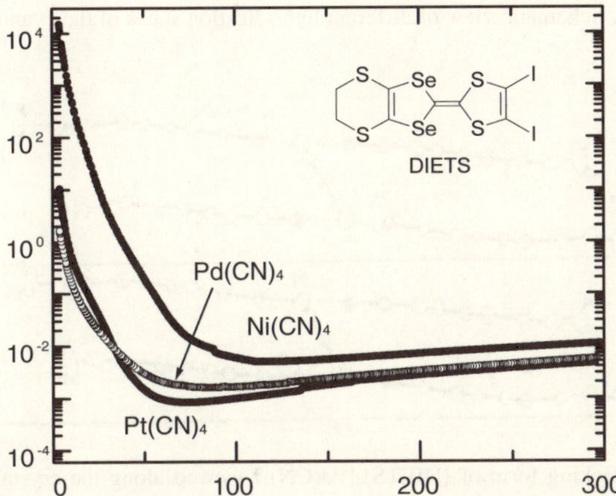

Fig. 3.9   Temperature dependence of the resistivity for $(DIETS)_4[M(CN)_4]$ (M = Ni, Pd, Pt).

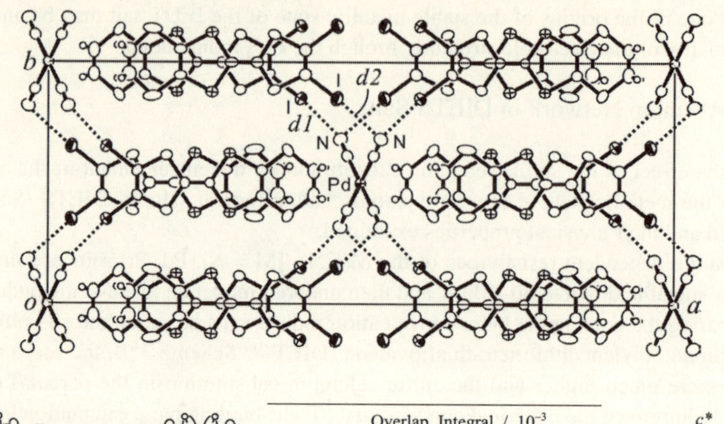

	Overlap Integral / 10^{-3}
p	-14.1
q	-9.19
r	-1.92
s	-1.94
t	-1.96
u	-1.94

Fig. 3.10   Crystal and electronic structure of $(DIETS)_4[Pd(CN)_4]$. The short I$\cdots$N distances are: $d1 = 2.95(1)$ Å, $d2 = 3.11(1)$ Å.

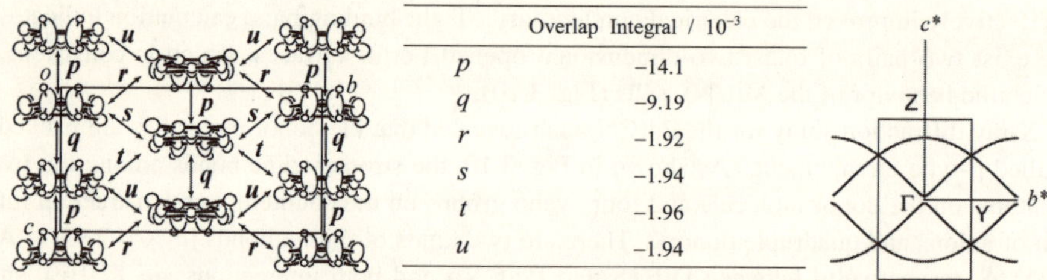

Fig. 3.11   Schematic view of different hybridization states of the cyano group.

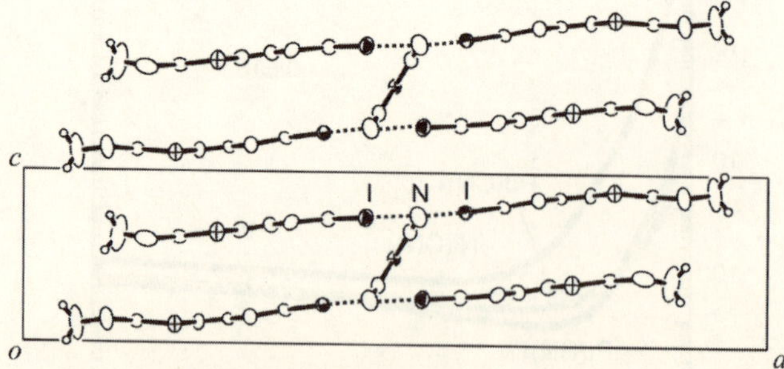

Fig. 3.12   Stacking form of $(DIETS)_4[Pd(CN)_4]$ viewed along the crystallographic b-axis. Donor molecules are arranged in the same direction by the strong iodine bond (dotted lines).

the lone pair on the sp^2-like nitrogen atom and the localized LUMO on the donor molecule. The strong iodine bond also affects the stacking direction of the donor molecules. Usually, unsymmetrical π-donors are stacked alternatively in the column. In this case, however, the donors are arranged in the same direction within the column and we can easily understand that this feature derives from the strong I$\cdots$N iodine bonds (Fig. 3.12).

3.4.6   FeCl$_4$ and FeBr$_4$ Salts of DIETS and DIETSe

In order to explore the possibility of the unique iodine bond architecture, the introduction of a tetrahedral FeX$_4$ (X = Cl or Br) anion has been examined.[28,32] In supramolecular chemistry, tetrahedral components are effectively used to construct three-dimensional intermolecular contacts. On the other hand, the molecular conductor that contains both itinerant π-electrons and localized magnetic moments is an attractive material, and if the central atom of the tetrahedral anion has a localized magnetic moment, donor$\cdots$anion contacts would bring about not only specific molecular arrangements but also an interaction pathway between the itinerant π-electrons and the localized spins. Temperature dependence of the resistivity for (DIETS)$_2$FeCl$_4$ is metallic down to 4.2 K. (DIETS)$_2$FeBr$_4$ and (DIETSe)$_2$FeCl$_4$ are isomorphous and show metal-semiconductor transitions around 15 K and 11 K, respectively (Fig. 3.13). In the crystals of these 2:1 salts, the three-dimensional donor$\cdots$anion network is tailored by the characteristic iodine bond as shown in Fig. 3.14. The low yield of the metallic 2:1 salt and co-crystallization of the 1:1 insulating salt prevented the magnetic measurements of these FeX$_4$ salts, unfortunately, and ferromagnetic interaction through the iodine bond was first observed in the crystal of charge transfer salts using the metal dithiolene complexes as follows.

3.4.7   Metallic and Ferromagnetic Salts (DIEDO)$_2$[M(mnt)$_2$] (M = Ni, Pt)

Charge transfer salts (DIEDO)$_2$[M(mnt)$_2$] [DIEDO = diiodo(ethylenedioxy)tetrathiafulvalene; mnt = maleonitriledithiolate; M = Ni, Pt] were prepared by galvanostatic oxidation.[33] These two salts

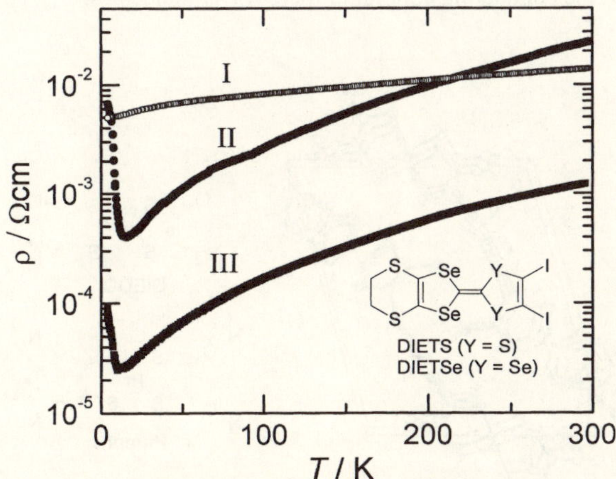

Fig. 3.13   Temperature dependence of the resistivity for (DIETS)$_2$FeCl$_4$ (I), (DIETS)$_2$FeBr$_4$ (II) and (DIETSe)$_2$FeCl$_4$ (III).

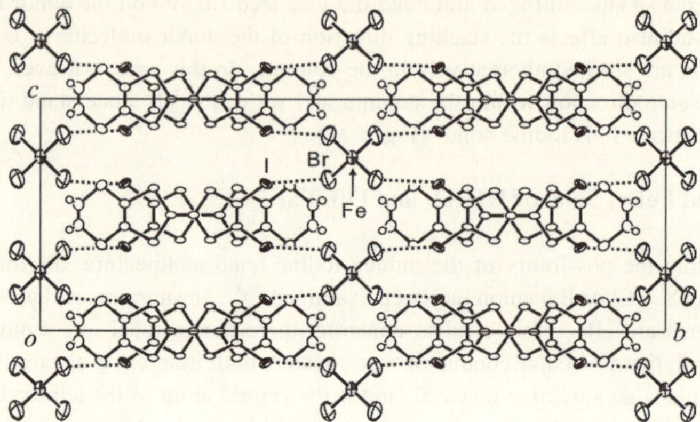

Fig. 3.14   Crystal structure of (DIETS)$_2$FeBr$_4$, which is isomorphous with (DIETSe)$_2$FeCl$_4$.
The short I···Br distance is 3.629(1) Å.

are isostructural and consist of one-dimensional chains of DIEDO and M(mnt)$_2$ aligned parallel to each other (Fig. 3.15). The strength of the I···N iodine bond between the DIEDO and M(mnt)$_2$ molecules (3.03-3.05 Å) is almost the same as that observed in the crystal of (iodo-TTF)[Ni(mnt)$_2$] (3.04 Å).[34]

According to the temperature dependence of the magnetic susceptibility, localized spins of the M(mnt)$_2$ molecules behave as a one-dimensional ferromagnet. The susceptibility could be fitted with the one-dimensional Heisenberg model for the Ni(mnt)$_2$ salt (J = 18 K) and the Ising model for the Pt(mnt)$_2$ salt (J = 20 K). The origin of the ferromagnetic interactions is the orthogonality of the molecular orbitals of M(mnt)$_2$ and this may be induced by the unique molecular arrangement tailored by the characteristic iodine bond. Temperature dependence of the resistivity for both salts is metallic down to 110 K and shows metal-insulator transition around 90 K due to the low dimensionality of the conducting column of the DIEDO molecules.

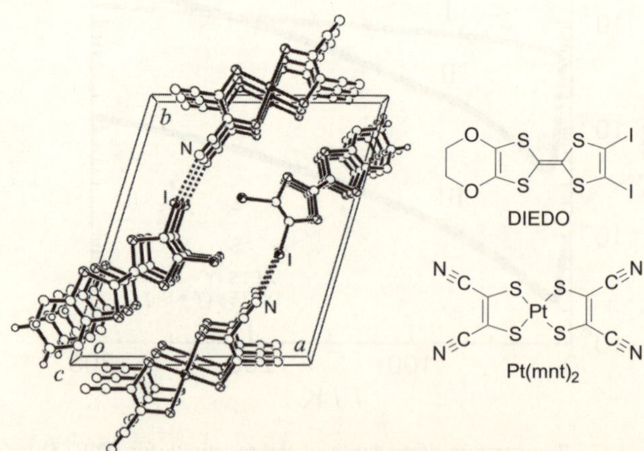

Fig. 3.15   Molecular arrangement of (DIEDO)$_2$[Pt(mnt)$_2$]. Dotted lines
indicate the I···N iodine bonds ($d_{\text{I···N}}$ = 3.05 Å).

Although the metallic state is unstable at low temperature and the ferromagnetic interaction through the I···N iodine bond is not completely coexistent with the metallic state, these $M(mnt)_2$ salts displayed the potential ability of the iodine bonding architecture to construct a novel electronic system including both itinerant π-electrons and localized spins in organic conductors.[35]

3.4.8 Hexagonal System of DIPS Salts

Like the donor···anion interaction, donor···donor interaction is also controllable using the iodine bond. DIPS (Scheme 3.13) was designed to introduce both iodine and nitrogen atoms into the TTF skeleton.[21] Hexagonal rod-shape crystals of $(DIPS)_3(anion)(solvent)_x$ (anion = PF_6^-, BF_4^-; solvent = chlorobenzene, dichloromethane, trichloroethane) were obtained by galvanostatic oxidation. All of the salts crystallize into a hexagonal $P6_3/mcm$ space group and are isostructural according to single crystal X-ray analyses. This is the first example of a hexagonal system in TTF-based organic conductors. The hexagonal lattice is constructed by the alternative repetition of the triangular units of the DIPS molecules tailored by the strong I···N iodine bond, which is almost 20% shorter than the sum of the van der Waals radii [$d_{I···N}$ = 2.879(6) Å, Fig. 3.16]. The C-I···N angle is almost linear [178.3(2)°] being in good accordance with the direction of the pσ*-LUMO along the carbon-iodine single bond and the lone pair on the pyrazine ring. The topology of the donor network should be recognized as the three-dimensional combination of: (i) the strong and directional iodine bond parallel to the molecular plane and (ii) the columnar stack favored by the planar π-donors.

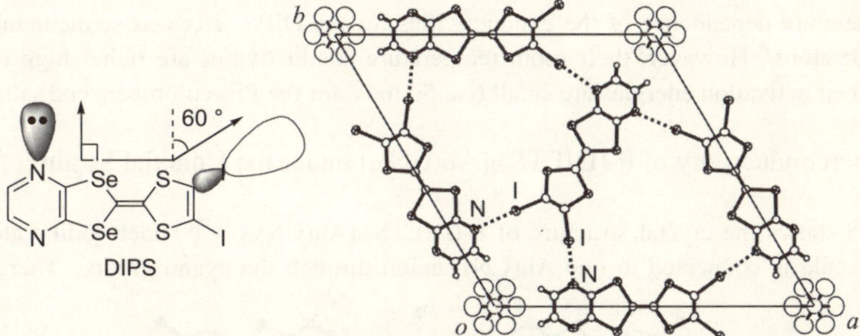

Fig. 3.16   Molecular structure of DIPS (left) and crystal structure of $(DIPS)_3(PF_6)(solvent)_x$ viewed along the crystallographic c-axis (right). Crystalline solvent molecules are omitted for clarity. The shortest I···N intermolecular distance (dotted line) is 2.879(6) Å.

Figure 3.17 shows the extended donor network viewed along the stacking direction of the donor molecules. There are two types of unidimensional channels parallel to the donor columns. One is surrounded by the edges of the six donor molecules (channel site A) and the other is surrounded by the sides of the three donor molecules (channel site B). Cross sections of the channel sites A and B are triangular (side lengths are $ca.$ 8 Å), and the counteranion is included in channel site A and the crystalline solvent is included in channel site B. Comparing the structures of the salts including the same counteranion but a different solvent molecule, the difference in the crystalline solvent does not change the entire crystal system and the hexagonal arrangement is identical. The guest-independent channel structure is similar to those of the inorganic "zeolite" systems, [36] and the DIPS salts must be good candidates for a new class of molecular-based

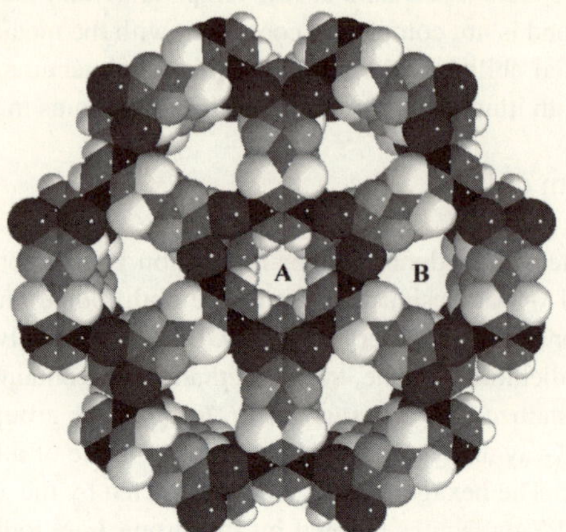

Fig. 3.17   Extended hexagonal donor-column arrangement
of the (DIPS)$_3$(anion)(solvent)$_x$ crystal system.

microporous material, namely "conductive organic zeolites." Preliminary experiments on the
thermogravimetry of the PF$_6$-dichloromethane salt showed that the crystalline solvent could be
evaporated around 110 °C and the weight loss of the sample stopped at around 6%. This value is
in accordance with the calculated weight percent of the dichloromethane molecule.

Temperature dependence of the conductivities for the DIPS salts was semiconductive from
room temperature. However, their room temperature conductivities are rather high ($\sigma_{rt} \sim 10$ S
cm^{-1}) and their activation energies are small (*ca.* 50 meV for the PF$_6$-chlorobenzene salt).

3.4.9   Superconductivity of θ-(DIETS)$_2$[Au(CN)$_4$] under the Uniaxial Strain

Figure 3.18 shows the crystal structure of θ-(DIETS)$_2$[Au(CN)$_4$].[19,37] Each iodine atom on the
donor molecule is connected to one Au(CN)$_4$ anion through the cyano group. There exists an

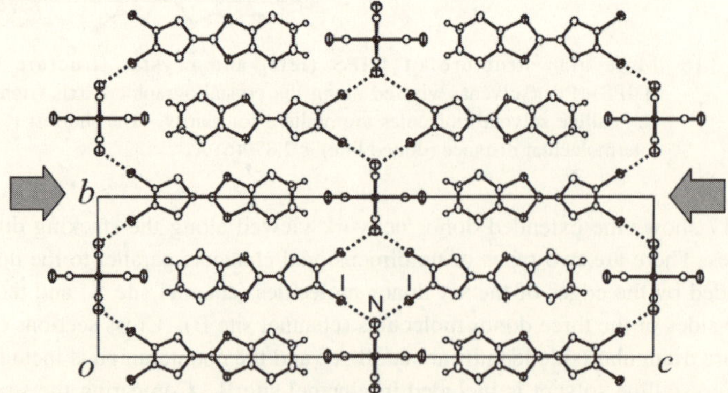

Fig. 3.18   Crystal structure of θ-(DIETS)$_2$[Au(CN)$_4$] viewed along the
crystallographic *a*-axis. The short I···N distance is 3.018(7) Å
(dotted lines) and thick arrows indicate the direction of the
uniaxial strain which induces superconductivity.

infinite donor⋯anion chain along the side-by-side direction as well as the crystal of (DIETS)$_4$-[Pd(CN)$_4$]. The I⋯N intermolecular distance is 3.018(7) Å, which is 15% shorter than the sum of van der Waals radii. Fig. 3.19 (left) shows the end-on view of the conducting layer. The adjacent columns consist of donor molecules inclined towards opposite directions and this "herring bone" type packing within the donor layer belongs to the so-called θ-type. The strong iodine bond connects every other column, and the planar Au(CN)$_4$ anions also construct a one-dimensional column. The direction of the anion column is parallel to the donor column because the donors and anions are linked by the strong iodine bond.

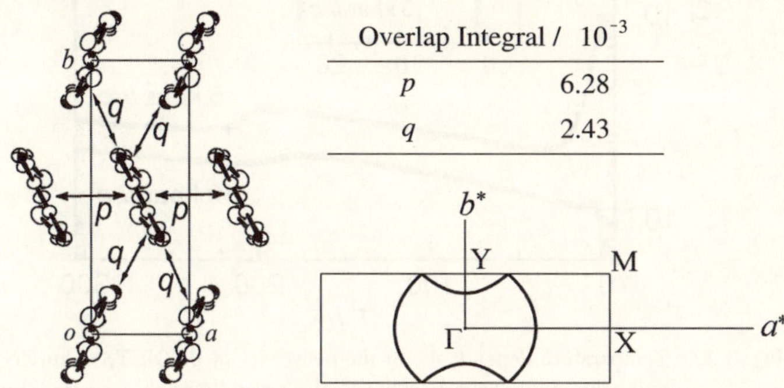

Fig. 3.19   θ-Type molecular arrangement within the donor layer (left) and the calculated Fermi surface of θ-(DIETS)$_2$[Au(CN)$_4$] (right).

According to the band calculations, the calculated Fermi surface has a cross section of a two-dimensional distorted circle (Fig. 3.19, right). At ambient pressure, however, the resistivity of the Au(CN)$_4$ salt shows a sharp semimetal-insulator transition around 226 K and is similar to the θ-type salts of BEDT-TTF.[38] On the other hand, the metal-insulator transition is gradually suppressed by the hydrostatic pressure, in contrast to the known θ-type salts. This inverse effect of hydrostatic pressure was thought to be induced by the strong iodine bond and this was confirmed by uniaxial pressure experiments as follows.

One of the features of the iodine bond is its anisotropic character. However, the pressure which is usually applied to organic conductors is hydrostatic, that is, isotropic. Therefore, anisotropic pressure is desirable to investigate the nature of the iodine bond. From this point of view, uniaxial pressure was applied to θ-(DIETS)$_2$[Au(CN)$_4$] along the three crystal axes. Fig. 3.20 shows the temperature dependence of the resistivity under the uniaxial strain along the crystallographic c-axis, which corresponds to the interlayer direction between the donors and counteranions. Under moderate pressure up to 5 kbar, the semimetal-insulator transition around 226 K was easily suppressed. At 10 kbar, an abrupt drop in the resistivity was observed around 8 K, which is characteristic of the onset of superconducting transition. The existence of superconductivity was confirmed by the recovery of resistance in the magnetic field and the onset temperature of the superconducting transition was determined to be 8.6 K.[19] To date, all of the transition temperatures (T_c's) of the known superconductors using unsymmetrical TTFs are under 5 K and the T_c of 8.6 K is currently the record transition temperature. The metal-insulator transition around 226 K was also gradually suppressed under uniaxial strain within the conducting layer, but the pressure effect within the donor-layer is similar to that of hydrostatic pressure and no superconductivity was observed. The role of the iodine bond is not clear at the present time. However, there is no doubt that the strong iodine-nitrogen interaction affects the manner of distortion under pressure and plays an important role in the superconducting state under uniaxial strain.

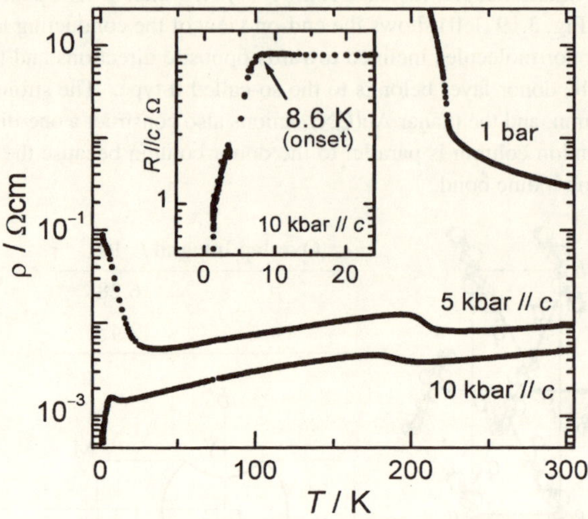

Fig. 3.20   Temperature dependence of the resistivity of θ-(DIETS)₂[Au(CN)₄] under uniaxial strain parallel to the crystallographic *c*-axis. Inset shows the superconducting transition at 8.6 K (onset) under 10 kbar.

3.5   Summary and Outlook

Although research on halogenated TTFs started almost 15 years after the discovery of the high conductivity of the TTF-TCNQ complex, synthetic methods for the halogenation have been developed during the last decade and halogenated derivatives of most known TTFs can be synthesized using one or more synthetic routes. The halogen atom has a unique nature unlike alkyl or alkylthio substituents which are popular in TTF chemistry. The electronic feature of halogen atoms, *i.e.,* the electron-withdrawing nature, sometimes prevents the formation of conducting charge transfer salts. However, the introduction of one or two halogen atoms is not intrinsic to the donor ability. Furthermore, recent advances in supramolecular chemistry have drawn attention to the halogen bond as a powerful tool for designing unique crystal structures of organic materials, and the halogenated TTFs are unique building blocks for supramolecular crystals with attractive electronic properties.

Several examples of supramolecular organic conductors based on the halogenated TTFs have been discussed in this chapter. Among these, the first hexagonal system of an organic conductor (DIPS)₃(anion)(solvent)ₓ and a supramolecular superconductor θ-(DIETS)₂[Au(CN)₄] are especially promising as a new wave of TTF-based organic conductors. The highly symmetrical system of the DIPS salts is interesting not only as an organic conductor but also as a magnetic material such as a spin frustration system, a zeolite-like organic material with electrical conductivity, including, *conductive organic zeolite* and others. The large anisotropy of the pressure effect of θ-(DIETS)₂[Au(CN)₄] also suggests some kind of relationship between the iodine bond and the electronic state and will open a new architecture for the design of organic superconductors, possibly with higher transition temperatures.

References

1. F. Wudl, G. M. Smith and E. J. Hufnagel, *Chem. Commun.*, 1453 (1970); B. A. Scott, S. J. La Placa, J. B. Torrence, B. D. Silverman and B. Welber, *J. Am. Chem. Soc.*, **99**, 6631 (1977).
2. J. Nakayama, N. Toyoda and M. Hoshino, *Heterocycles*, **24**, 1145 (1986).
3. J. F. Bunnett, *Acc. Chem. Res.*, **5**, 139 (1972) and references cited therein.
4. A. Bondi, *J. Phys. Chem.*, **68,** 441 (1964).
5. M. Jørgensen and K. Bechgaard, *Synthesis*, 207 (1989).
6. J. Y. Becker, J. Bernstein, S. Bittner, L. Shahal and S. S. Shaik, *J. Chem. Soc., Chem. Commun.*, 92 (1991).
7. M. R. Bryce and G. Cooke, *Synthesis*, 263 (1991).
8. U. Kux, H. Suzuki, S. Sasaki and M. Iyoda, *Chem. Lett.*, 183 (1995); B. Domercq, T. Devic, M. Fourmigue, P. Auban-Senzier and E. Canadell, *J. Mater. Chem.*, **11**, 1570 (2001).
9. M. Iyoda, E. Ogura, T. Takano, K. Hara, Y. Kuwatani, T. Kato, N. Yoneyama, J. Nishijo, A. Miyazaki and T. Enoki, *Chem. Lett.*, 680 (2000).
10. M. Iyoda, Y. Kuwatani, K. Hara, E. Ogura, H. Suzuki, H. Ito and T. Mori, *Chem. Lett.*, 599 (1997).
11. C. Wang, J. Y. Becker, J. Bernstein, A. Ellern and V. Khodorkovsky, *J. Mater. Chem.*, **5**, 1559 (1995).
12. R. Andreu, M. J. Blesa, J. Garín, A. López, L. Orduna and M. Savirón, *Synth. Met.*, **86**, 1897 (1997).
13. A. S. Batsanov, M. R. Bryce, A. Chesney, J. A. K. Howard, D. E. John, A. J. Moore, C. L. Wood, H. Gershtenman, J. Y. Becker, V. Y. Khodorkovsky, A. Ellern, J. Bernstein, I. F. Perepichka, V. Rotello, M. Gray and A. O. Cuello, *J. Mater. Chem.*, **11**, 2181 (2001).
14. J. Kreicberga, A. Edzina, R. Kampare and O. Neilands, *Zh. Org. Khim.*, **25**, 1456 (1989).
15. C. Wang, A. Ellern, V. Khodorkovsky, J. Bernstein and J. Y. Becker, *J. Chem. Soc., Chem. Commun.*, 983 (1994).
16. T. Imakubo, H. Sawa and R. Kato, *Synth. Met.*, **73**, 117 (1995).
17. R. Gompper, J. Hock, K. Polborn, E. Dormann and H. Winter, *Adv. Mater.*, **7**, 41 (1995).
18. T. Imakubo, H. Sawa and R. Kato, *Synth. Met.*, **86**, 1883 (1997).
19. T. Imakubo, N. Tajima, M. Tamura, R. Kato, Y. Nishio and K. Kajita, *J. Mater. Chem.*, **12**, 159 (2002).
20. T. Imakubo, T. Iijima, K. Kobayashi and R. Kato, *Synth. Met.*, **120**, 899 (2001).
21. T. Imakubo, T. Maruyama, H. Sawa and K. Kobayashi, *Chem. Commun.*, 2021 (1998).
22. K. Takimiya, Y. Kataoka, A. Morikami, Y. Aso and T. Otsubo, *Synth. Met.*, **120**, 875 (2001).
23. V. Khodorkovskii, A. Edzifna, O. Neilands, *J. Mol. Electron.*, **5**, 33 (1989).
24. For examples of the applications of the hydrogen bond to organic conductors, see: P. Blanchard, K. Boubekeur, M. Sallé, G. Duguay, M. Jubault, A. Gorgues, J. D. Martin, E. Canadell, P. Auban-Senzier, D. Jerome and P. Batail, *Adv. Mater.*, **4**, 579 (1992); A. S. Batsanov, M. R. Bryce, G. Cooke, A. S. Dhindsa, J. N. Heaton, J. A. K. Howard, A. J. Moore and M. C. Petty, *Chem. Mater.*, **6**, 1419 (1994); A. Dolbecq, M. Formigué, P. Batail and C. Coulon, *Chem. Mater.*, **6**, 1413 (1994); O. Neilands, S. Belyakov, V. Tilika and A. Edzina, *J. Chem. Soc., Chem. Commun.*, 325 (1995).
25. For a review of the halogen bond, see: P. Metrangolo and G. Resnati, *Chem. Eur. J.*, **7**, 2511 (2001) and references cited therein.
26. E. O. Schlemper and D. Britton, *Acta. Crystallogr.*, **18**, 419 (1965); G. R. Desiraju and R. L. Harlow, *J. Am. Chem. Soc.*, **111**, 6757 (1989).
27. A. L. Allred and E. G. Rochow, *J. Inorg. Nucl. Chem.*, **5**, 264 (1958).
28. T. Imakubo, H. Sawa and R. Kato, *Synth. Met.*, **86**, 1847 (1997).
29. T. Imakubo, H. Sawa and R. Kato, *J. Chem. Soc., Chem. Commun.*, 1097 (1995).
30. A. Kobayashi, A. Sato, K. Kawano, T. Naito, H. Kobayashi and T. Watanabe, *J. Mater. Chem.*, **5**, 1671 (1995).
31. T. Imakubo, H. Sawa and R. Kato, *J. Chem. Soc., Chem. Commun.*, 1667 (1995).
32. T. Imakubo, H. Sawa and R. Kato, *Mol. Cryst. Liq. Cryst.*, **285**, 27 (1996).
33. J. Nishijo, E. Ogura, J. Yamaura, A. Miyazaki, T. Enoki, T. Takano, Y. Kuwatani and M. Iyoda, *Solid State Commun.*, **116**, 661 (2000).
34. A. S. Batsanov, A. J. Moore, N. Robertson, A. Green, M. R. Bryce, J. A. K. Howard and A. E. Underhill, *J. Mater. Chem.*, **7**, 387 (1997).
35. For examples of metallic organic conductors containing paramagnetic anions, see: M. Kurmoo, A. W. Graham, P. Day, S. J. Coles, M. B. Hursthouse, J. L. Caulfield, J. Singleton, F. L. Pratt, W. Hayes, L. Ducasse and P. Guionneau, *J. Am. Chem. Soc.*, **117**, 12209 (1995); H. Kobayashi, H. Tomita, T. Naito, A. Kobayashi, F. Sakai, T. Watanabe and P. Cassoux, *J. Am. Chem. Soc.*, **118**, 368 (1996); E. Coronado, J. R. Galán-Mascarós, C. J. Gómez-Garcia and V. Laukhin, *Nature*, **408**, 447 (2000).
36. W. M. Meier and D. H. Olson, *Atlas of Zeolite Structure Types*, Butterworth-Heinemann, London

(1992).
37. T. Imakubo, A. Miyake, H. Sawa and R. Kato, *Synth. Met.*, **120**, 927 (2001).
38. For the θ-type salts, see: H. Mori, S. Tanaka and T. Mori, *Phys. Rev., B* **57**, 12023 (1998) and references cited therein.

4

Oxygen Analogues of TTFs

4.1   Introduction

The representative compound in this category is BEDO-TTF (BO). Since the first synthesis of this peculiar donor molecule, the TTF derivatives have been synthesized by incorporating oxygen atoms to the outside the TTF skeleton and within the TTF moiety itself. Very few numbers of the latter class of compounds have been reported so far. This chapter describes the TTF derivatives and the charge-transfer (CT) complexes in both categories.

It is noted here that the van der Waals (vdW) radii proposed by Bondi[1] are used in discussing the intermolecular atomic contacts. To introduce this chapter, the relationship between the transition energy of the CT band ($h\nu_{CT}$) and the redox properties of the component molecules in a complex is described. The concept originally arose from the exploration of the neutral-ionic (NI) transition system by Torrance and his coworkers.[2] The lowest CT band transition energy of a solid CT complex is expressed using the donor ionization potential (I_p), the acceptor electron affinity (E_A), the charge of an electron (e), the distance between donor and acceptor molecules (a) and the Madelung constant (α) as follows.

For the complex having neutral ground state: $h\nu_{CT}^{N} = (I_p - E_A) - <e^2/a>$ (4.1)

For the complex having ionic ground state: $h\nu_{CT}^{I} = (2\alpha-1)<e^2/a> - (I_p - E_A)$ (4.2)

where $< >$ indicates the average in the solid. Substituting the I_p and E_A by the redox potentials of the donor (D) and acceptor (A), $I_p - E_A$ is expressed as

$$I_p - E_A = \Delta E_{redox} + \Delta G \tag{4.3}$$

where ΔE_{redox} and ΔG represent the difference between the first oxidation potential (E_{ox}^{1}) of D and the first reduction potential (E_{red}^{1}) of A measured under the same conditions and the solvation energy, respectively. Based on ΔG of 3.9 eV and the observed $h\nu_{CT}^{N}$ and $h\nu_{CT}^{I}$, Torrance and his coworkers obtained the values of $<e^2/a>$ and $\alpha<e^2/a>$. As a result, the CT transition energy is given in the unit of wave number as follows.

$$h\nu_{CT}^{N} = 8066 \times \Delta E_{redox} + 4033 \ (cm^{-1}) \tag{4.4}$$

$$h\nu_{CT}^{I} = -8066 \times \Delta E_{redox} + 7259.4 \ (cm^{-1}) \tag{4.5}$$

The "theoretical" linear relationships afford a V-shaped line in the plot of $h\nu_{CT}$ vs. ΔE_{redox} as shown in Fig. 4.1. In principle, this type of plot allows the assessment of the ionicity of the complex based on the readily available experimental data, and Torrance and his coworkers found

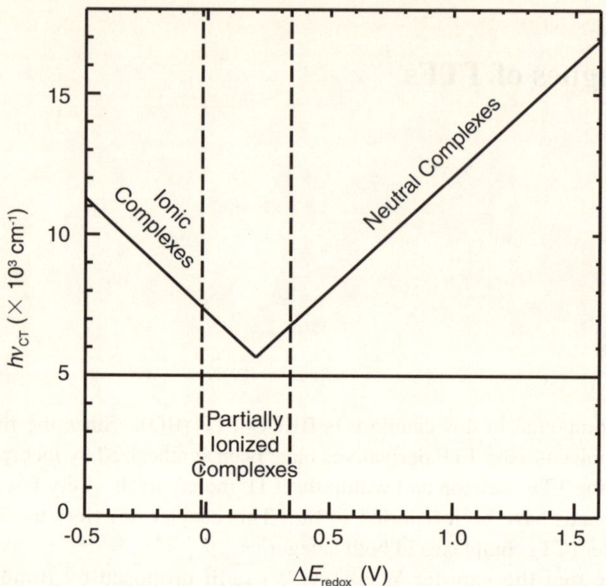

Fig. 4.1   Torrance's V-shape plot (hv_{CT} *vs.* ΔE_{redox}). CT complexes having alternating stack structures with ionic and neutral ground states are expected to be located on the left and right sides of the V-shaped line. The conductive complexes having segregated stack structures with the partially charge-transferred ground states, however, are located in the bottom part ($hv_{CT} \leq 5 \times 10^3$ cm⁻¹).

the first NI transition system applying this concept.[2] However, the application should be done with caution. This concept only regards the CT transition between D and A molecules. In the ionic region, the CT transition between the same chemical species, *e.g.*, $D^{\cdot+} + D^{\cdot+} \rightarrow D^{2+} + D$, is often observed instead of the one between D and A. It should also be noted that CT complexes prepared from donors, such as aromatic amines, TTF derivatives and aromatic hydrocarbons, and acceptors, such as TCNQ, quinone derivatives, aromatic carboxylic acid anhydrides and TCNE, are used to derive Equations (4.4) and (4.5).

Later, the application of this diagram to classify the CT complexes in a family was carried out.[3] Since metallic CT complexes consist of partially charged component molecules in the segregated column or two-dimensional(2D) layered structure in general, they exhibit low energy CT absorption due to the transition of $D^{\cdot+} + D \rightarrow D + D^{\cdot+}$ and/or $A^{\cdot-} + A \rightarrow A + A^{\cdot-}$ in the region below *ca.* 5×10^3 cm⁻¹ (shown by the horizontal line in Fig. 4.1). Caution is required here because the appearance of such a low energy transition is necessary but not sufficient for the formation of a metallic CT complex, since even an isolated dimer of donor molecules having a unit charge can give the same type of CT transition in the same energy range.

For comparison with the typical low dimensional organic metal system, two vertical lines are often added at $\Delta E_{redox} = -0.02$ and $+0.34$ V between which the complexes of the TTF-TCNQ family afford partial CT ground states (indicated by dashed lines in Fig. 4.1).[4] Strictly speaking, ΔG depends on the series of CT complexes concerned and a smaller value by 0.14–0.17 eV (*i.e.*, shifting the V-shape to the right side by 0.14–0.17 V) is suggested for the TTF-TCNQ series, although such an adjustment is not generally carried out in the application of this type of plot.[5]

Even with the restrictions noted above, the plot shown in Fig. 4.1 for the CT complexes based on a donor or an acceptor serves to help understand the nature of the component molecule described in the following sections. In this chapter, this type of plot is referred to as "Torrance's V-shape plot."

The chemical formulas which are not explained in the succeeding sections are summarized in Chart 4.1. The composition of a complex is expressed by the ratio of donor:acceptor (or anion):solvent (if included), and ΔE is used to express the difference between the first and second redox potentials ($E_{1/2}^2 - E_{1/2}^1$).

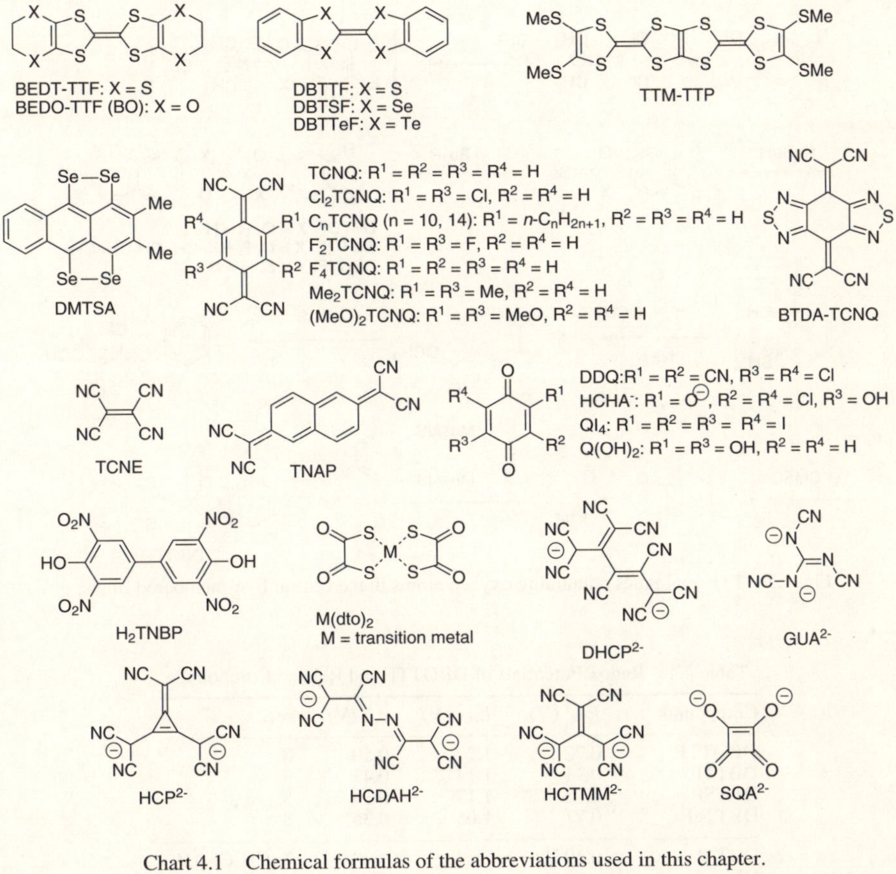

BEDT-TTF: X = S
BEDO-TTF (BO): X = O

DBTTF: X = S
DBTSF: X = Se
DBTTeF: X = Te

TTM-TTP

DMTSA

TCNQ: $R^1 = R^2 = R^3 = R^4 = H$
Cl$_2$TCNQ: $R^1 = R^3 = Cl$, $R^2 = R^4 = H$
C$_n$TCNQ (n = 10, 14): $R^1 = n\text{-}C_nH_{2n+1}$, $R^2 = R^3 = R^4 = H$
F$_2$TCNQ: $R^1 = R^3 = F$, $R^2 = R^4 = H$
F$_4$TCNQ: $R^1 = R^2 = R^3 = R^4 = H$
Me$_2$TCNQ: $R^1 = R^3 = Me$, $R^2 = R^4 = H$
(MeO)$_2$TCNQ: $R^1 = R^3 = MeO$, $R^2 = R^4 = H$

BTDA-TCNQ

TCNE

TNAP

DDQ: $R^1 = R^2 = CN$, $R^3 = R^4 = Cl$
HCHA$^-$: $R^1 = O^-$, $R^2 = R^4 = Cl$, $R^3 = OH$
QI$_4$: $R^1 = R^2 = R^3 = R^4 = I$
Q(OH)$_2$: $R^1 = R^3 = OH$, $R^2 = R^4 = H$

H$_2$TNBP

M(dto)$_2$
M = transition metal

DHCP^{2-}

GUA^{2-}

HCP^{2-}

HCDAH^{2-}

HCTMM^{2-}

SQA^{2-}

Chart 4.1   Chemical formulas of the abbreviations used in this chapter.

4.2   Tetraoxafulvalene (TOF) and Related Compounds

The oxygen-substituted derivatives at the central five-membered rings are the most poorly reported group among the TTF families. Although DBTOF was reported in 1988,[6] clear data on this compound were not published until 2000.[7] The latter paper contradicted the synthesis described in the former. DBTOF is labile to acidic conditions, although the paper of 1988 stated that this donor molecule was prepared by a coupling reaction under strongly acidic conditions. Historically, the first compound in this category is the DBOTTF synthesized in 1996.[8] The synthetic routes and the known compounds in this class are summarized in Fig. 4.2.[7-9] In the case of DBOTF, the *trans* and *cis* isomers regarding the position of oxygen (or sulfur) atoms are separated by HPLC.

The redox potentials are reported as summarized in Table 4.1. Comparing the electron-donating abilities among them and to other DBTTF derivatives, the order is determined as DBTTeF ≈ DBTTF > DBOTTF ≥ DBTSF ≈ *cis*-DBOTF > *trans*-DBOTF > DBTOF > DNTOF. So far only one CT complex is reported in this category of donor molecules. For the 1:1 complex

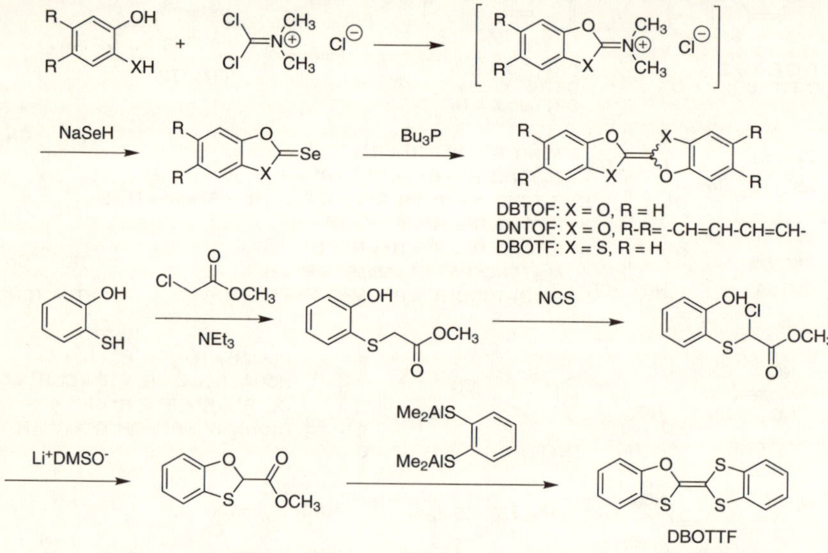

DBTOF: X = O, R = H
DNTOF: X = O, R–R= -CH=CH-CH=CH-
DBOTF: X = S, R = H

DBOTTF

Fig. 4.2   TTF analogues containing oxygen atoms in the central five-membered rings.

Table 4.1   Redox Potentials of DBOTTF and Related Compounds

Compound	$E_{1/2}^1$ (V)	$E_{1/2}^2$ (V)	ΔE (V)	ref.
DBOTTF	0.72	1.22	0.50	8
DBTTF	0.71	1.14	0.43	8
DBTSF	0.78	1.17	0.39	8
DBTTeF	0.71	1.05	0.36	8
DBTTF	0.49 [pa]			7 (CH$_2$Cl$_2$), 9
DBTOF	0.61 [pa]			7 (CH$_2$Cl$_2$), 9
trans-DBOTF	0.56 [pa]			9
cis-DBOTF	0.47 [pa]			9
DBTOF	0.58 [pa]			7 (THF)
DNTOF	0.63 [pa]			7 (THF)

pa: Anodic peak potential. Measurement conditions: in ref. 8, Bu$_4$NBF$_4$/ CH$_2$Cl$_2$, V *vs.* Ag/Ag$^+$; in ref. 7, 0.2 M Bu$_4$NBF$_4$/ CH$_2$Cl$_2$ or THF, V *vs.* SCE; in ref. 9, in CH$_2$Cl$_2$, V *vs.* Ag/Ag$^+$.

of *trans*-DBOTF with TCNQ, the degree of CT is estimated as 0.8 based on the IR spectra (CN stretching), although other data are not reported.[9]

The crystal structures of DBTOF, DNTOF and both isomers of DBOTF have been established and the details of the former two reported. Both DBTOF and DNTOF show almost planar molecular structures while the analogues with heavier chalcogen atoms, DBTTF and DBTSF, exhibit somehow chair-shaped molecular configurations.[10] Oxygen derivatives slip along the molecular longitudinal axis to form stacking columns. The donor molecules in a column and the neighboring ones are crystallographically parallel to each other. While the packing patterns of DBTTF and DBTSF are characterized by a columnar structure with a small molecular slip and the intercolumnar chalcogen···chalcogen contacts, respectively, intermolecular CH···O contacts are observed in the crystal structures of DBTOF and DNTOF. The intermolecular atomic distances between H and O [2.70(4) Å for DBTOF and 2.6–2.7 Å for DNTOF] are comparable to the sum of the vdW radii (2.72 Å).

4.3 BEDO-TTF and (MeO)₄TTF

The report of the first synthesis of BO was published in 1989 by Wudl and his coworkers.[11] The characteristic feature of this donor molecule is a strong tendency to afford metallic complexes. Up to now, about 100 kinds of CT complexes with organic and inorganic counter components have been reported, and the metallic behaviors were confirmed for at least 64 species. However, only two complexes showed superconductivity. The origin of these peculiar properties is understood as the self-assembling nature of this donor molecule described below. In this section, the structural properties of the conductive complexes of BO are mainly described along with those of (MeO)₄TTF, which is the sole tetrakis(alkoxy)TTF derivative other than BO appearing in the regular journals. It is noteworthy that improved reaction conditions applicable to the synthesis of BO have recently been reported.[12]

4.3.1 Redox Properties and CT Complexes with Organic Acceptor Molecules

As donor molecules, a comparison of the results of the photo electron spectra and the redox properties for BO, BEDT-TTF and TTF showed somewhat complicated features due to the difficulty of the vaporization of the former two; BO is the strongest donor in vapor phase.[13] The situation was reviewed and the redox data, which are directly related to the complex formation, reexamined later.[14] As summarized in Table 4.2, BO showed intermediate strength between TTF and BEDT-TTF and a small ΔE similar to that of BEDT-TTF.

The systematic investigation of the complexes mainly with organic acceptor molecules along with the organic closed-shell anions proved the self-assembling property of BO. In 1996, the preparation of 30 kinds of CT complexes with 29 organic acceptor molecules was reported.[14] Only two of the complexes afforded single crystals, while other complexes were obtained as

Table 4.2 Redox Potentials of TTF, BO and BEDT-TTF
in 0.1 M Bu₄NBF₄/ CH₃CN (V *vs.* SCE)[14]

Compound	$E_{1/2}^{1}$ (V)	$E_{1/2}^{2}$ (V)	ΔE (V)
TTF	0.37	0.62	0.25
BO	0.43	0.69	0.26
BEDT-TTF	0.53	0.78	0.25

powder samples. As summarized in Torrance's V-shape plot (Fig. 4.3), most of the complexes exhibited $h\nu_{CT}$'s below 5×10^3 cm^{-1}; however, their single crystals were not obtained. The complexes indicated by open circles in Fig. 4.3 showed metallic temperature dependence of the conductivity (σ) even when measured on compressed pellets [σ_{RT} (σ at room temperature) = 14–170 S cm^{-1}]. The distribution of the metallic complexes indicates that the acceptor molecule of which $\Delta E_{redox} \leq 0.56$ V generally affords the metallic complex (Fig. 4.3). The ΔE_{redox} values of the metallic complexes ranged exactly from –0.29 to +0.56 V. The regime is more than twice that of the quasi-one-dimensional (1D) TTF-TCNQ system (0.38 V).

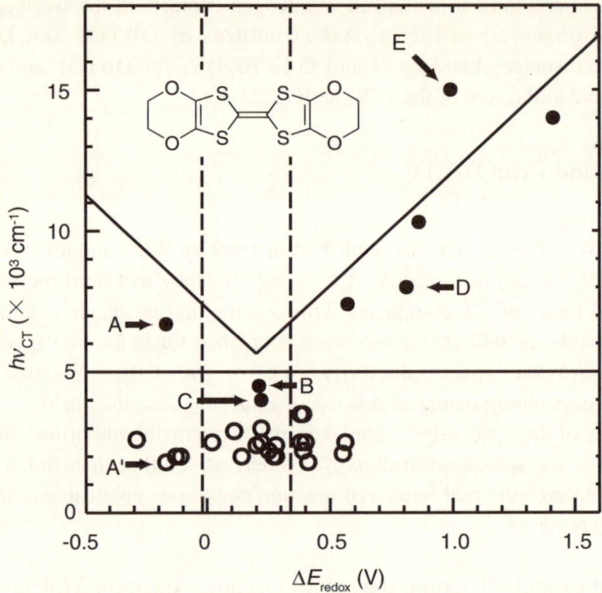

Fig. 4.3   Torrance's V-shape plot for BO complexes. Metallic and insulating (semiconducting) complexes were indicated by open and closed circles, respectively. The labeled complexes are; A: (BO)(F$_4$TCNQ), A': (BO)$_9$(F$_4$TCNQ)$_5$(THF)$_4$, B: (BO)(TCNQ), C: (BO)$_9$(C$_{14}$TCNQ)$_4$(H$_2$O)$_2$, D: (BO)[Q(OH)$_2$]$_2$ and E: (BO)(H$_2$TNBP).

It is also noteworthy that BO rarely produces polymorphous complexes. Among the complexes in Fig. 4.3, only those labeled A (1:1 F$_4$TCNQ complex) and A' [9:5:4(THF) F$_4$TCNQ complex] are exceptional cases. In the former, BO was ionized completely to +1 and hence the complex was an insulator ($\sigma_{RT} = 7.9 \times 10^{-8}$ S cm^{-1}). In the region of $\Delta E_{redox} \leq 0.56$ V, two other complexes showed semiconductive behavior. The complex labeled B is 1:1 TCNQ complex ($\sigma_{RT} = 8.1 \times 10^{-2}$ S cm^{-1}), in which the charge on BO was expected to be inhomogeneously distributed based on the IR spectra. However, a later study to determine the charge on this donor could not confirm the charge disproportion, and the origin of the nonmetallic behavior is not yet clear.[15] The other is (BO)$_9$(C$_{14}$TCNQ)$_4$(H$_2$O)$_2$ labeled C ($\sigma_{RT} = 9.9 \times 10^{-1}$ S cm^{-1}), in which the bulky alkyl group was assigned to the origin which inhibited the metallic property. Additionally, it should be mentioned that the stoichiometries of the complexes in this region are frequently apart from simple ones, *e.g.*, 2:1, 1:1 and 3:2, and include the solvent of crystallization as exemplified by the F_4TCNQ and C_{14}TCNQ complexes mentioned above.

The conducting property of the neutral complex in Fig. 4.3 was negligible. The complexes with the acceptor molecules of $\Delta E_{redox} \geq 0.57$ V were insulators ($\sigma_{RT} = 10^{-7}$–10^{-10} S cm^{-1}). Among them, the crystal structures of two complexes showed novel features. The crystal of $(BO)[Q(OH)_2]_2$ labeled D consisted of DAA-type alternating stacks. The crystallographic (120) plane contained the side-by-side type infinite BO array and a chain made of $Q(OH)_2$ connected with the OH$\cdots$O hydrogen bonds, in which the molecular planes of both components were almost parallel to each other. The intermolecular S$\cdots$S distance between the donor molecules of 3.20 Å was extraordinarily short compared to those observed in the complexes of TTF derivatives (> 3.3 Å; the sum of the vdW radii = 3.60 Å). The origin of this exceptional intermolecular close contact was assigned to the existence of the C-H$\cdots$O contacts between the donor and acceptor molecules and the difference in their molecular sizes. The intermolecular distance between donor molecules was understood to be adjusted by the periodicity of the acceptors. $(BO)(H_2TNBP)$ is indicated by the label E in Fig. 4.3. The transition energy of the CT band [$h\nu_{CT} = 15 \times 10^3$ cm^{-1} (shoulder)] was clearly different from the expected value and shifted to a higher energy side. In the crystal, no specific intermolecular interactions were observed except the O$\cdots$HO contact with just the distance of the sum of the vdW radii (O$\cdots$O = 3.04 Å) between the donor and acceptor. This complex is regarded to be a kind of clathrate compound. The reason why the CT band appeared in the high-energy region was assigned to the lack of effective CT interactions.

As a donor-acceptor type complex, a peculiar BO complex was prepared recently. $(BO)(Cl_2TCNQ)$ is an ionic complex in which the degree of CT was estimated to be almost complete.[16] The crystal structure consists of uniform columns of alternating donor and acceptor molecules. The intermolecular overlap integrals between a pair of donor and acceptor within and among the columns were comparable ($s = 0.74$ and 0.43×10^{-3}, respectively). This complex showed temperature dependence of the magnetic susceptibility obeying the Curie-Weiss law ($C \approx 0.744$ emu K mol^{-1}, $\theta \approx 21$ K). The Curie constant (C) corresponded to a spin concentration of 1/2 spins per component molecule, which is consistent with the feature of completely ionized and weakly interacting donor and acceptor molecules in this complex. At around 120 K, the spin concentration showed an abrupt decrease with hysteresis. However, no evidence could be found for the dimerization of the donor-acceptor pair, the formation of the superlattice or for the repartition of the charge among the component molecules. This may be a new type of structural transition given the hysteretic behavior, but the mechanism is not yet known.

4.3.2 Self-assembling Nature of BO — I₃-type Packing Pattern

It should be noted that the systematic investigation described above proved not only the strong tendency of BO to afford the metallic complexes but also that the metallic states are generally realized even in the severely disordered system of the powder form. The origin of these peculiar properties was definitely revealed by the analysis of radical cation salts of BO with organic closed-shell anions. The idea was proposed in the early 1990s. The dianion of HCTMM afforded a milestone complex of $(BO)_5(HCTMM)(PhCN)_2$ in which BO is +0.4 charged.[14,17] In this complex, BO formed a two-dimensional donor layer, the packing motif of which corresponds to the β"-phase of BEDT-TTF complexes as shown in Fig. 4.4, while the counter components were completely disordered to inhibit the determination of the atomic positions in the crystal structure analysis. This means that no specific interactions between donor and counter components affect the packing pattern of the donor molecules; hence the packing pattern observed in this complex is regarded as the result of the self-assembling of partially oxidized BO in the 2D space sandwiched

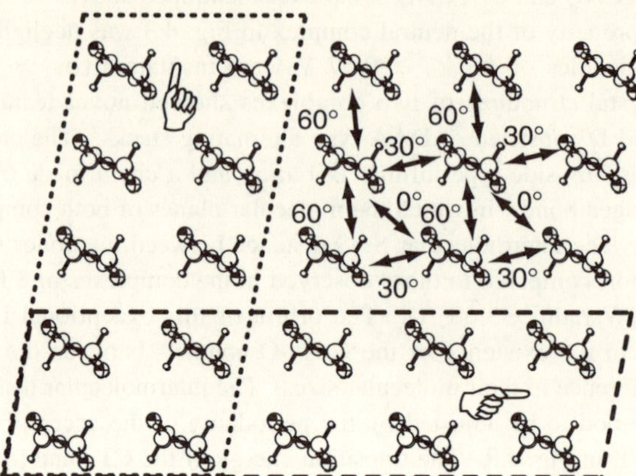

Fig. 4.4   Donor longitudinal axis projection of the donor layer in
(BO)₅(HCTMM)(PhCN)₂ (I₃-type packing). The forefingers at the
top and bottom indicate the stacking (60°) and side-by-side (30°)
directions of BO molecules, respectively. The areas surrounded by
dotted and dashed lines correspond to the structural units of HCP-
and M(CN)₄-type packings, respectively.

by negatively charged sheets. In fact, exactly the same situation was found in
$(BO^{0.4+})_{10}[C(CN)_3^-]_4(H_2O)_3$ in which BO formed the 2D donor layer isostructural to that in the
HCTMM complex.[14] This packing pattern of BO is denoted as I₃-type after $(BO)_{2.4}I_3$, the crystal
structure of which had been reported as the first example of BO complexes, showed this packing
pattern.[18] It should be noted that the counteranion of I_3^- in this complex formed a
crystallographic lattice, the periodicity of which was incommensurate with that of BO.[18b]
Reviewing the crystal structures published before, in and after those in reference 14, the I₃-type is
the most common packing motif of BO complexes, regardless of whether the counter components
showed disorder.

   In the I₃-type packing, a donor is surrounded by six neighboring BO molecules. A quasi-
columnar stack is formed along the 60° direction with respect to the molecular plane. The side-
by-side direction corresponds to the 0° direction and the neighboring molecules are located on the
same plane. From the viewpoint of overlap integrals, the strongest intermolecular interactions
take place along the 0° and 30° directions ($s = 10$–15×10^{-3}), while the interactions are weak
along the 60° direction ($s = 3$–6×10^{-3}). The former two intermolecular interactions afford the
2D electronic structures to the I₃-type BO complexes. In fact, most of the BO complexes of I₃-
type maintain the metallic state down to the lowest or close to the lowest temperature measured
(Table 4.3). Structurally, the relative positions of the neighboring donor molecules are almost the
same in the complexes of this category. The geometrical parameters D, δ and ε defined in Fig.
4.5 are 3.4–3.6 Å, 59–62° and 87–90°, respectively.[14] The most striking feature in this packing
pattern is the existence of the weak hydrogen bond network along the 60° direction, which is the
axis with the smallest intermolecular overlap integrals. Although the network consists of aliphatic
hydrogen and ether oxygen with approximated distances of 2.5–2.7 Å (the sum of the vdW radii =
2.72 Å)[17b] indicating only a small stabilization energy,[19] this intermolecular interaction can not be
ignored due to the number of contributing sites. Also, the pattern of the side-by-side
intermolecular atomic contacts is noteworthy. The sulfur atoms embedded in the inner

Table 4.3 Selected Data of BO Complexes

Counter Component(s)	Composition[a]	Charge on BO	Notes[b]	ref.	ref. for other data
I$_3$-type					
I$_3$	2.4 : 1	0.417	σ_{RT} = 100–280 S cm^{-1}(S), $R_{rt}/R_{1.2K*}$ = 250 (S), $T_{\sigma max}$ = 1.2 K*	19	18a, 65
Cu$_2$(NCS)$_3$	3 : 1	0.333	black crystals (β_m-phase), Superconductor (T_c = 1.06 K (rf penetration onset))	29	24
Ag(CN)$_2$	2 : 1	0.50	needle-shaped, Crystal structure was solved on a twinned crystal	66	
AuBr$_2$			thin plate, σ_{RT} = 68 S cm^{-1}(S), T_{MI} = 260 K, E_a = 0.07 eV (220-80 K)	67	
AuI$_2$	2 : 1	0.50	thin plates, σ_{RT} = 10 S cm^{-1}(S), $T_{\sigma max}$ = 15 K*	68	
Hg$_{1.9}$Br$_{6.8}$	5 : 1	0.60	parallel-epipedic + needles, σ_{RT} = 20–200 S cm^{-1}(S), $T_{\sigma max}$ = 20 K	69	
HgBr$_4$	9 : 2 : 5(DCE)	0.444	mirror-like surfaces, σ_{RT} = 5–50 S cm^{-1}(S), $T_{\sigma max}$ = 20 K, the evaporation of solvent was observed.	69	
CsHg(SCN)$_4$	5 : 2	0.40	σ_{RT} = 2–10 S cm^{-1} (S), $T_{\sigma max}$ ≈ 85 K, ρ_{300K}/ρ_{85K} = 2.5	70	
CF$_3$SO$_3$	2 : 1 : 0.5 (THF)	0.50	irregular shape plates, σ_{RT} ≥ 100 Scm^{-1}, semiconductor (abrupt decrease at 250 K)	28	
ClO$_4$	2 : 1	0.50	σ_{RT} = 100 S cm^{-1} (S), $T_{\sigma max}$ ≈ 200 K, T_{MI} ≈ 170 K (cooling), 210 (warming)	67, 68	
ReO$_4$	2 : 1 : 1(H$_2$O)	0.50	black plate-like crystals, σ_{RT} ≈ 140 S cm^{-1} (S), Superconductor below *ca.* 2.5 K. Other phase transitions were observed in a higher temperature region.	30	
Br[MnBr$_2$ (H$_2$O)$_4$] (H$_2$O)	2 : 1	0.50	black plates, σ_{RT} = 200–300 S cm^{-1} (S//c), 40–60 S cm^{-1} (S//a-c), $T_{\sigma max}$ = 1.3 K*	71	
[Fe(CN)$_5$NO]	4 : 1	0.50	σ_{RT} = 40–100 S cm^{-1} (S), 50 S cm^{-1} (S//a), $T_{\sigma max}$ = 1.3 K*	72, 73	
Cl	2 : 1 : x(H$_2$O)		$\rho_{293K}/\rho_{1.3K}$ ≈ 15, Degree of CT was discussed based on SdH	25	
Br	2 : 1 : 3(H$_2$O)	0.50	black elongated plates, σ_{RT} = 1×10^2 S cm^{-1} (S//c)	61	
HCTMM	5 : 1 : 2 (PhCN)	0.40	black plates, σ_{RT} = 2.8×10 S cm^{-1} (S//a+2c), $T_{\sigma max}$ = 5 K, σ_{5K}/σ_{RT} = 40, σ_{RT} = 7.0 × 10 S cm^{-1} (S//2a-c), $T_{\sigma max}$ = 5 K, σ_{5K}/σ_{RT} = 30	14	17a, b
C(CN)$_3$	10 : 4 : 3(H$_2$O)	0.50	black needles, σ_{RT} = 1.1×10^2 S cm^{-1} (S//c), $T_{\sigma max}$ = 1.3 K*, σ_{5K}/σ_{RT} = 19, $h\nu_{CT}$ = 2.5 × 10^3 cm^{-1} (KBr)	14	
SQA	4 : 1 : 6(H$_2$O)	0.50	black needles, σ_{RT} = 1.7 × 10^2 S cm^{-1} (S//c), $T_{\sigma max}$ = 1.4 K*, σ_{5K}/σ_{RT} = 46, $h\nu_{CT}$ = 2.1 × 10^3 cm^{-1} (KBr)	14	
HCHA	2 : 1	0.50	black needle-like crystals, σ_{RT} = 100–200 S cm^{-1} (S), $T_{\sigma max}$ = 1.3 K*	71	
GUA	4 : 1 : 1(H$_2$O)	0.50	shiny black elongated plates, σ_{RT} = 11.5 S cm^{-1} (S), σ_{8K} = 38.5 S cm^{-1} (S)	52	
HCP-type					
HCP	5 : 1 : 0.2 (PhCN)	0.40	black needles, conducting plane = bc, σ_{RT} = 1.8 × 10 S cm^{-1}, E_a = 0.10 eV (S//c), σ_{RT} = 4.1 × 10^{-1} S cm^{-1}, E_a = 0.05 eV (S//b), $h\nu_{CT}$ = 3.0 × 10^3 cm^{-1}	14	
HCDAH	6 : 1	0.333	black prisms, conducting plane = ac, σ_{RT} = 0.10–16.7 S cm^{-1} (S)	21	
DHCP-type					
DHCP	5 : 1 : 2(THF)	0.40	black plates, σ_{RT} = 100 S cm^{-1} (S)	23	
M(CN)$_4$-type					
Pt(CN)$_4$	4 : 1 : 1(H$_2$O)	0.50	thin plates with a parallelogram, $\sigma_{4K}/\sigma_{293K}$ ≈ 30 (S), $T_{\sigma max}$ = 1.5 K*	22	

(continued)

Table 4.3

Counter Component(s)	Composition[a]	Charge on BO	Notes[b]	ref.	ref. for other data
κ-type					
CF_3SO_3	2 : 1	0.50	regularly shaped plates, $\sigma_{RT} \geq 100$ S cm^{-1} (semiconductor).	28	
Cl-type					
Cl	2 : 1 : 3(H$_2$O)	0.50	black rhombic-like plates, $\sigma_{RT} = 50$–100 S cm^{-1} (S), $T_{\sigma max} \approx 15$ K, $R(80$ kbar$)/R(0$ kbar$) \approx 8$, SdH	48	24, 45, 49
Other types					
I$_3$	1 : 1	1.0	black plates, $\sigma_{RT} = 1 \times 10^{-6}$ S cm^{-1}, $E_a = 0.43$ eV (P)	50	
	1 : 2	2.0	black plates, $\sigma_{RT} < 10^{-9}$ S cm^{-1} (S)	50	
Ni(dto)$_2$	2 : 1	1.00	black plates, $\sigma_{RT} = 1.0 \times 10^{-5}$ S cm^{-1}, $E_a = 0.52$ eV (S)	74	
Pd(dto)$_2$	2 : 1	1.00	black blocks, $\sigma_{RT} = 1.0 \times 10^{-7}$ S cm^{-1}, $E_a = 0.50$ eV (S)	74	

[a] BO:(Anion or Acceptor):solvent (if included). [b] σ_{xK}, ρ_{xK}: conductivity and resistivity at x K, respectively (RT: room temperature, S: Measured on single crystals, P: Measured on compressed powder samples). $T_{\sigma max}$: The temperature at maximum conductivity (* indicates the lowest temperature measured). R_{xK}: The resistance at x K. T_c: Phase transition temperature to a superconductor. T_{MI}: Metal-insulator transition temperature. E_a: Activation energy of conductivity. SdH: Shubnikov-de Haas oscillation. $h\nu_{CT}$: Transition energy of CT-band in KBr.

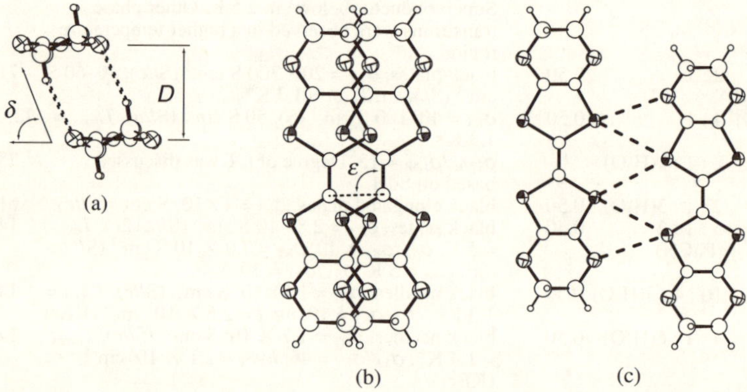

Fig. 4.5   Neighboring donor molecules in the I$_3$-type packing. (a) A pair in the 60° direction projected along the molecular longitudinal direction. (b) A pair in the 60° direction projected onto the molecular plane. (c) A donor pair in the 0° direction projected onto the molecular plane. D: intermolecular separation along the stacking axis, δ, ε: tilted angles of the stacking direction with the molecular plane and with the central C=C bond. In (a) and (c), intermolecular atomic contacts shorter than the sum of the vdW radii are shown by dotted and dashed lines, respectively.

five-member rings, at which the atomic orbital coefficients in the highest occupied molecular orbital (HOMO) are concentrated, form contacts shorter than the sum of the vdW radii. Compared to BEDT-TTF, whose outer six-member rings are larger than the inner five-member rings, the similar size of the outer six-member and inner five-member rings allows this type of intermolecular contact.[18a]   The packing pattern of the I$_3$-type is understood as a result of the appropriate molecular shape of BO to allow the formation of the side-by-side intermolecular atomic contacts between the inner sulfur atoms and the construction of the weak hydrogen bond network along the stacking (60°) direction. Moreover, it should be noted that the ethylene groups show the eclipsed conformation even though the calculated activation energy for the conformation change from the eclipsed form to the staggered one was smaller than that of BEDT-TTF; hence

BO is more flexible than BEDT-TTF.[20]

4.3.3 Other Packing Patterns of Conductive BO Complexes

Besides the I_3-type, five other kinds of donor packing patterns have been reported for the conducting BO complexes. Among them, three motifs are regarded as modified I_3-type and the remaining two are exceptional cases, with only one example each having been reported.

In a paper on a systematic investigation of the BO complexes, HCP-type packing was reported.[14] In (BO)$_5$(HCP)(PhCN)$_{0.2}$, the donor molecules formed a columnar stack along the b-axis (Fig. 4.6a). The tilt of the molecular planes of BO changed with every two columns, although they were parallel to each other in alignment. This type of packing pattern was observed in the α''-BEDT-TTF complexes. To avoid confusion in the naming of the packing pattern occurring in the BEDT-TTF complexes, the donor arrangement was denoted as the HCP-type for the BO complexes. The donor-packing pattern in the neighboring columns with parallel donor molecules was isostructural to that in the I_3-type complexes. The neighboring couples of the "I_3-type" double columns are related by a two-fold screw axis parallel to the columnar axis and the lateral intermolecular interactions were weak ($s \approx 8 \times 10^{-3}$) compared to that within the parallel donor pair. As a result, the calculated electronic band structure showed a rather 1D feature as described below. The isostructural donor layer was found in (BO)$_6$(HCDAH).[21] In both cases of HCP-type complexes, the superlattice formed by the counter component prohibited the metallic

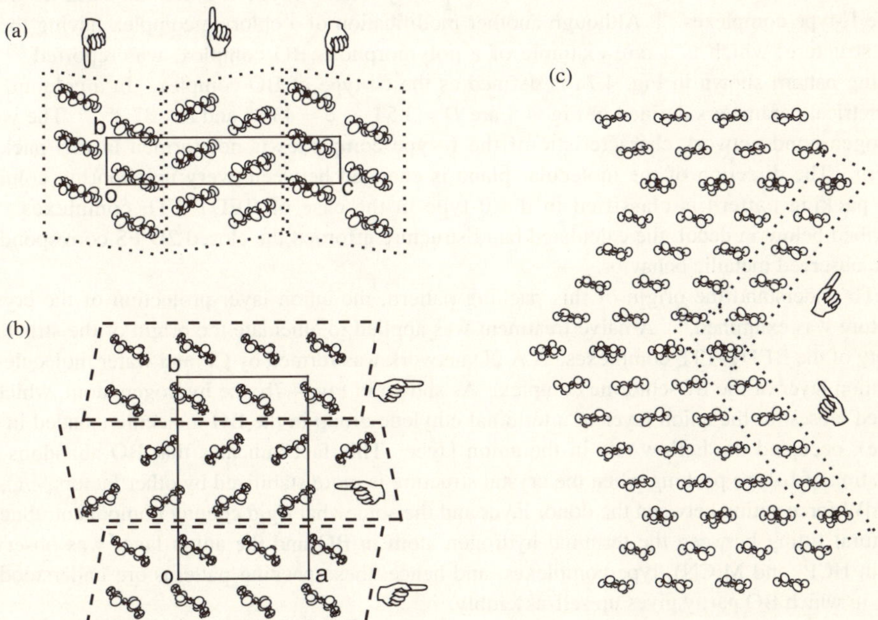

Fig. 4.6   Donor layer structures of the I_3-type analogy of the BO complexes. (a) The a-axis projection of (BO)$_5$(HCP)(PhCN)$_{0.2}$. (b) The c-axis projection of the crystal structure of (BO)$_4$[Pt(CN)$_4$](H$_2$O). (c) The donor layer of (BO)$_5$(DHCP)(THF)$_2$. The areas surrounded by dotted and dashed lines and the forefingers correspond to the unit isostructural to the I_3-type packing and their relative orientations, respectively.

behavior of the complexes, although the calculated band structure based on the averaged crystal structure afforded the Fermi surface (FS).

On the other hand, the donor layer in $(BO)_4[Pt(CN)_4](H_2O)$ is understood to consist of two rows along the 30° direction in the I_3-type packing; this is denoted as $M(CN)_4$–type packing pattern in this chapter (Fig. 4.6b).[22] In this case, the structural units of the double ribbons are related by the two-fold screw axis parallel to the donor longitudinal axis, and hence the molecular plane changes the tilting direction between the pairs of rows. The intermolecular interactions across the border between the I_3-type units ($s \approx 9$–13×10^{-3}) are smaller than that within a unit. The calculated band structure, however, contained the closed 2D FS. In fact, the metallic state of the $Pt(CN)_4$ complex was retained down to the lowest temperature measured (1.5 K).

Another example of modified I_3-type complex was recently reported, although details were not given.[23] In $(BO)_5(DHCP)(THF)_2$, the donor molecules formed a columnar structure in which the neighboring molecules were located in the 60° direction with respect to the molecular plane. This arrangement is isostructural to that in the I_3-type packing. However, the column was bent every five BO molecules to form a zigzag column (Fig. 4.6c).

The origin of the formation of these modified I_3-type packing patterns is understood to be the cooperative combination of the self-assembling nature of BO and the stabilization of the crystal structure by the structural fitting between the donor and anion layers, the latter of which is described below. In these crystal structures, BO partly gives up but retains in some measure the most profitable packing pattern.

In the chlorine complex, BO formed a completely different type of columnar stack from that in the I_3-type complexes.[24] Although another modification of a chlorine complex having the I_3-type structure, which is a rare example of a polymorphous BO complex, was reported,[25] the packing pattern shown in Fig. 4.7a is defined as the Cl-type of BO complex. In a column, the geometrical parameters defined in Fig. 4.5 are $D = 3.54$ Å, $\delta = 43.8°$ and $\varepsilon = 87.3°$.[14] The weak hydrogen bond network characteristic of the I_3-type complexes is not formed in this packing pattern. The direction of the molecular plane is changed between every neighboring column. This packing pattern is classified in the θ-type in the case of BEDT-TTF complexes. As described below in detail, the calculated band structure afforded the closed 2D FS corresponding to the observed metallic behavior.

To understand the origin of this packing pattern, the anion layer projection of the crystal structure was examined.[26] A naive treatment was applied to elucidate the origin of the structural variety of the BEDT-TTF complexes.[27] A 2D network was formed by Cl^- and water molecules in the anion layer of the BO-chlorine complex. As shown in Fig. 4.7b, the hydrogen atom, which is located closest to the anion layer of a terminal ethylene group in the BO molecule (shaded in the figure), occupied the hollow site in the anion layer. This fact indicates that BO abandons the formation of I_3-type packing when the crystal structure is more stabilized by other factors, e.g., by the vdW force acting between the donor layer and the somewhat rigid counter component. Such a structural fitting between the terminal hydrogen atom in BO and the anion layer was observed also in HCP- and $M(CN)_4$-type complexes, and hence, these packing patterns are understood as those in which BO partly gives up self-assembly.

The κ-type packing, in which the face-to-face donor dimer is orthogonally arranged in a layer, is a structure well known as being one of the superconducting phases of BEDT-TTF complexes. Of the BO complexes, only one, $(BO)_2(CF_3SO_3)$, has been reported to show this type of donor packing.[28] Unlike the BEDT-TTF complexes, this κ-type BO complex exhibited semiconducting behavior, while the calculated band structure showed a feature similar to those of the superconducting κ-type BEDT-TTF complexes. The origin of the conducting behavior is not

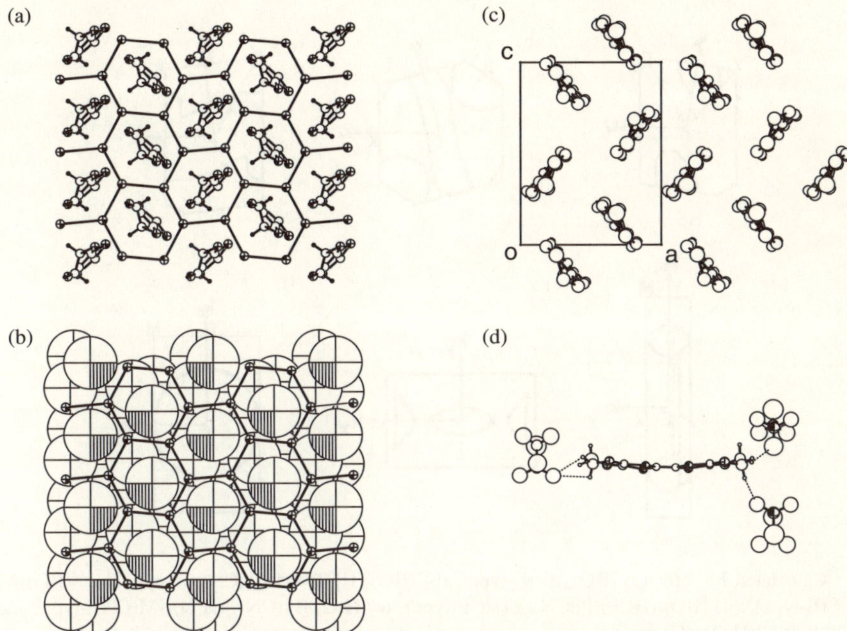

Fig. 4.7   Crystal structure of (BO)$_2$Cl(H$_2$O)$_3$ (left) and κ-(BO)$_2$(CF$_3$SO$_3$) (right). The atomic coordinates of the former complex were taken from ref. 24, although a different composition was stated. (a) The *ab*-plane projection of the crystal structure of chlorine complex. (b) Hydrogen atoms are drawn as spheres of the vdW radius (1.2 Å) in the same crystal structure to (a). The hydrogen atom among an ethylene group closest to the anion layer is shaded. (c) The *b*-axis projection of the crystal structure. (d) A donor molecule and the anion molecules, which give the shorter intermolecular atomic contacts than the sum of the vdW radii (indicated by dashed lines).

yet clear, despite a thorough discussion about the dipolar nature of the anion and anion layer. It should be noted that, in the crystal structure, the donor molecule was slightly distorted to show the boat form and short atomic contacts of 2.50–2.63 Å were observed between the terminal hydrogen atoms of the donor and the oxygen atoms of the anion (Fig. 4.7d).

4.3.4 Superconductors and Fermiology of BO Complexes

As noted above, BO forms an electronic 2D conducting layer. The Fermiology was started by the tight binding approximated band calculation based on the extended Hückel molecular orbital calculation. The shape of the FS of an I$_3$-type complex is understood to be a cylindrical tube or tubes of a folded cylinder according to the shape of the unit cell, as shown in Fig. 4.8. A typical 2D FS was calculated for (BO)$_{2.4}$I$_3$. In the case of (BO)$_5$(HCTMM)(PhCN)$_2$, the elliptical FS's in the extended Brillouin zone expression were folded to give closed and open FS's. Occasionally, the folding and the interactions between the basis bands afford 1D FS as calculated for (BO)$_2$AuBr$_2$. Although the M(CN)$_4$- and Cl-type packing patterns afford the 2D electronic structures, the calculated FS for HCP-type was 1D. As mentioned above, the crystal structures of both examples in this category contained a superlattice due to the counter components. In the case of (BO)$_5$(HCP)(PhCN)$_{0.2}$, the periodicity of the superlattice was as much as five times that of

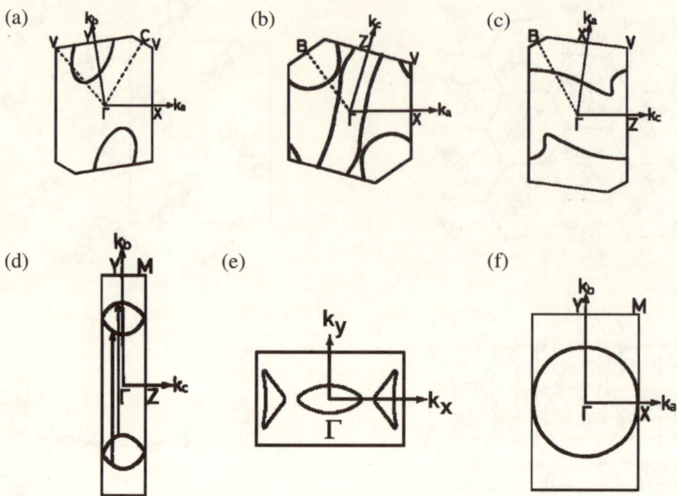

Fig. 4.8   Calculated FS's for (a) $(BO)_{2.4}I_3$ (I_3-type), (b) $(BO)_5(HCTMM)(PhCN)_2$ (I_3-type), (c) $(BO)_2AuBr_2$ (I_3-type), (d) $(BO)_5(HCP)(PhCN)_{0.2}$ (HCP-type), (e) $(BO)_4[Pt(CN)_4](H_2O)$ [$M(CN)_4$-type] and (f) $(BO)_2Cl(H_2O)_3$ (Cl-type).
(The figures (a)-(d) and (f) are taken from ref. 14. The FS drawn in (e) was obtained by the calculation applying the same AO parameters to those in ref. 14 and the atomic coordinates corresponding to those in ref. 22)

the donor lattice along the stacking direction (b-axis). As indicated in Fig. 4.8d, the translation of two FS's by $(3/5)b^*$ effectively superimposes them to another pair (nesting). Since nominally the band is 4/5 filled due to the charge of 2/5+ on each BO molecule, the gap formation at the wave numbers of $b^*/5$ and the multiples was assigned as being the origin of the semiconducting nature.

As for experimental studies on Fermiology, the tendency to concentrate on the superconductors is also the case for BO complexes as in the case of BEDT-TTF and other analogous complexes. The first superconductor based on BO, $(BO)_3Cu_2(NCS)_3$ was reported in 1990 by an American group.[29] The superconductivity was confirmed by a radio frequency (rf) penetration depth measurement with an onset temperature of 1.06 K. Although the superconducting transition was confirmed by the complex susceptibility (χ_{AC}) measurement conducted by a German group [T_c = 1.1 K (onset), 0.6 K (midpoint)],[24] the fact that an experimental investigation of the Fermiology of this complex has not yet been conducted may be due to the difficulty in obtaining single crystals of adequate quality. The poor quality of the crystals obtained was noted in the first report.[29]

As for the Fermiology of BO complexes, the most thoroughly studied, and at same time the most complicated, sample is $(BO)_2ReO_4(H_2O)$. The superconductivity of this complex was first described in a report by a German group along with other phase transitions observed in the transport property measurements.[30] On cooling the crystal, a distinct jump and subsequent faint jump to less and more conductive states were observed at 213 and 80–90 K, respectively. Further cooling caused a transition to a semiconducting state at around 35 K and an abrupt decrease in the resistance at 2.5 K (onset). All these transitions except for the last one were also detected in the thermopower measurement. The transition at the lowest temperature was understood to be a superconducting one, although a finite resistance was observed even at 0.5 K. In fact, AC susceptibility measurement supported this assignment with the onset temperature of the phase transition (T_c) of 0.9 K. Soon after the publication of this paper, the conducting and superconducting properties were confirmed by a Russian group.[31] This group observed the hysteretic behavior of the highest temperature metal-to-metal (MM) transition which takes place at 205–220 K, the higher T_c of 3.5 K (onset) observed in the conductivity measurement and the anisotropic nature of the conduction (anisotropy in the conducting plane = 2–4; however, the resistance is 10^3–10^4 times greater along the interplanar direction). Also, the mechanism of the 35 K transition was deduced to be the formation of a spin density wave (SDW) based on the result of the microwave conductivity measurement. Later the origin of the 205–220 K MM transition was assigned to the reorientation of the ReO_4 anion along with the possibility of a reversible loss and regain of water molecules, which had a negligible effect on the donor layer structure.[32]

Some groups independently performed band structure calculations based on the different unit cells, although the same crystal structure was investigated. Fig. 4.9 shows two of the results.[33] Both calculations afforded small pockets of closed FS's. The areas of the electron and hole pockets were estimated to be 1.5% and 3% of the first Brillouin zone (FBZ) based on "cell-1"[34] and to be ~1.7% and ~3.4% based on "cell-2," [35] respectively, and hence the complex was considered to be a semi-metal. The experimental values were obtained by the analysis of Shubnikov-de Haas (SdH) oscillation. At 0.5 K applying the magnetic field up to 24 T, the oscillation was observed as the superposition of two fundamental ones with frequencies of $F_1 =$ (37 ± 3) and $F_2 = (76 \pm 2)$ T, which correspond to the sizes of the closed pockets of 0.7% and 1.5% of FBZ.[35] The cyclotron carrier mass values corresponding to F_1 and F_2 were determined to be (1.15 ± 0.1) m_0 and (0.90 ± 0.05) m_0, where m_0 is the mass of electron.[34] Similar results were obtained by another group along with information on pressure dependence.[36] The pressure dependence of the T_c, dT_c/dP, was determined to be 0.2 K/kbar. This value is one order of magnitude lower than those of BEDT-TTF superconductors and understood to be the result of the competition between the SDW and superconducting states. To understand the discrepancy between the calculated and observed area sizes of the pockets, investigations under higher magnetic field have been carried out. Based on the data taken at 1.9 K up to 52 T, the third component with a frequency of $F_4 = (150 \pm 10)$ T was claimed to be needed to fit the observed SdH oscillation.[37] Further, an oscillation series with a larger frequency of $F_H = (460 \pm 10)$ T was observed.[38] The area of the closed FS derived from this frequency (9% of FBZ) corresponded to that obtained from the angular dependent magnetoresistance oscillation (AMRO) measurement,

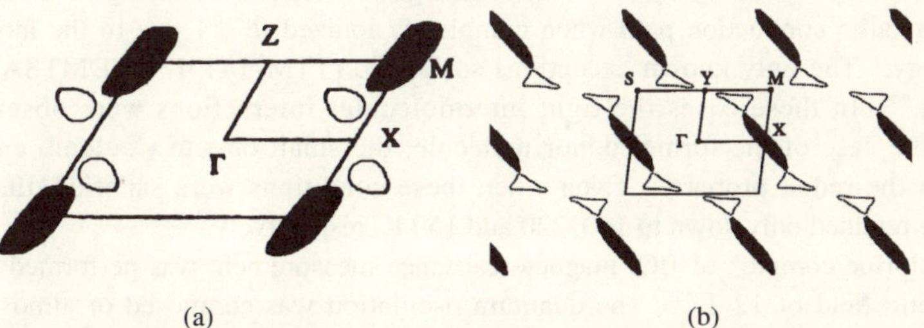

(a)   (b)

Fig. 4.9   Calculated FS's for $(BO)_2ReO_4(H_2O)$ (a) based on "cell-1" and (b) on "cell-2." [33] In both cases, the electron and hole pockets are indicated by open and closed areas, respectively.

although the calculated band structure does not contain the corresponding closed FS.[39]   To account for these complicated results, the occurrence of a magnetic field-induced phase transition was suggested.   However, a systematic investigation is needed to understand the nature of this complex.[40]

In the case of $(BO)_5[CsHg(SCN)_4]_2$, clear results were obtained for the SdH oscillation.[41] Based on the triclinic crystal structure,[42] the calculated band structure afforded the closed and open FS's (Fig. 4.10).   The area of the closed orbital denoted α was 19% of FBZ.   The SdH and de Haas-van Alphen (dHvA) oscillations were observed at 1.5 K up to 14 T.   The spectrum consisted of four frequencies of oscillations: $F_1 \approx 650$, $F_2 \approx 2600$, $F_3 \approx 3200$, $F_4 \approx 3850$ T. F_1 and F_4 corresponded to cross sections, whose areas were 16% and 100% of FBZ, and assigned to the closed FS (α-orbit) and the one produced by the magnetic breakdown (β-orbit), respectively. Two additional frequencies were assigned to the forbidden orbits of $F_2 \approx F_4 - 2F_1$ and $F_3 \approx F_4 - F_1$.   The effective mass values were estimated corresponding to F_1 to F_4 to be $m_1 = (1.6 \pm 0.1) \, m_0$, $m_2 = (0 \pm 0.3) \, m_0$, $m_3 = (1.5 \pm 0.1) \, m_0$ and $m_4 = (3.0 \pm 0.3) \, m_0$, respectively.

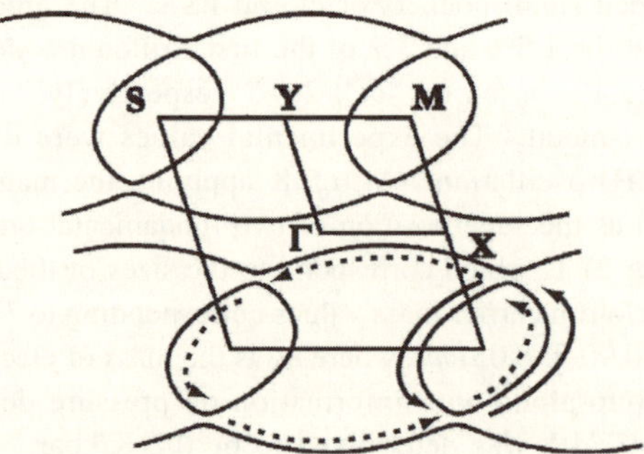

Fig. 4.10   Calculated FS for $(BO)_5[CsHg(SCN)_4]_2$.   Solid and dotted arrows indicate the cyclotron paths denoted as α and β, respectively.

For the chlorine complex of BO, the first report claimed the composition to be $(BO)Cl(H_2O)$ despite the metallic behavior which persisted down to 30 mK.[24]   This conclusion was based mainly on crystal structure analysis.   Since the Cl^- and water molecules formed a 2D network (Fig. 4.7) randomly occupying each site, the result of the population analysis (refinement of the site occupancy factors) was the main factor to determine the composition, although it is difficult to obtain the exact stoichiometry by this method.   In general, however, an organic molecule can not give the metallic conduction path when completely ionized to ± 1 due to the large on-site Coulomb energy.   The only known exceptions so far are $(TTM-TTP)I_3$,[43] $(DMTSA)NO_3$ and $(DMTSA)BF_4$.[44]   In these cases the tight intermolecular interactions were observed, and especially in the case of the former donor molecule, the small on-site Coulomb energy was expected from the redox property.   Even when these conditions were satisfied, the metallic behaviors were retained only down to 160, 220 and 160 K, respectively.

On the chlorine complex of BO, magnetoresistance measurement was performed at 0.48 K up to a magnetic field of 12 T.[45]   The quantum oscillation was composed of almost a single fundamental frequency of $F = 5000$ T.   This period corresponded to 53% of the area of FBZ.   To give this size of cross section, the charge on BO should be +0.5 based on the calculated band structure, and hence the composition of this complex was estimated to be $(BO)_2Cl(H_2O)_3$.   The shape of the FS was experimentally determined to confirm the degree of CT on BO.[46]   Further, the same authors examined the pressure effect on the conductivity and detected a loss of metallic behavior above 4 kbar and a loss of semiconducting behavior above 10 kbar.[47] Although the size of the cross section was confirmed also by another research group,[48] there still remained the ambiguity of the composition. The composition of $(BO^{0.5+})_2 \, Cl^-_{1.28} \, (H_3O^+)_{0.28} \, (H_2O)_{2.44}$ was estimated based on population analysis in low temperature X-ray structural analysis and the IR spectra of 3320–4000 cm^{-1}.[49] Although this composition satisfies the structural analysis within the experimental error, the conclusion must be viewed with caution since the IR peak assigned to H_3O^+ was a weak one at 3250 cm^{-1}, in which region the O-H stretching absorption of neutral water molecule can also appear when the intermolecular hydrogen bond is formed.

4.3.5 Fundamental Data of BO in the Complexes

Although only the redox properties of BO are discussed in Section 4.3.1, fundamental data have been reported over a dozen years from the first synthesis of this donor molecule. Among them, the preparation of the I_3 complexes, in which BO was oxidized to 1+ and 2+, afforded useful information.[50] In Section 4.1, the electronic spectra of the donor-acceptor type complexes were mentioned qualitatively. The electronic spectra of BO in various oxidation states in solution and in solid (KBr) were compared in order to assign and predict the spectra of the complexes as shown in Fig. 4.11. Summarizing the results, the transition energy ($\times \, 10^3$ cm^{-1}) and the assignments are as follow: A (1.9–3.0): $BO^0 + BO^+ \rightarrow BO^+ + BO^0$, B (7–9): $BO^+ + BO^+ \rightarrow BO^{2+} + BO^0$, C (12–14); intramolecular transition of BO^+ (plausibly, next HOMO to HOMO), D ($\approx$ 16), E ($\approx$ 20), F ($\approx$ 22); intamolecular transition of BO^+, G (29.9 in solution), H (31–31.8); intramolecular transition of BO^0, while the band I (24–27) and J (31–31.8) are assigned to the intramolecular transiton of I_3^-. [14,50]

The estimation of the charge on BO in a complex was a difficult problem. The comparison of the bond lengths is a familiar method in the field of the TTF derivatives. However, as summarized in Table 4.4, the variation is not so large.[50] Furthermore, in general, BO complexes afford poor quality crystals that give bond lengths with poor accuracy. Also, the vibrational spectra of IR absorption did not show adequate dependence on the charge on BO.[14] Although the application to the BEDT-TTF complexes is not adequate, the Raman spectra afforded a fairly accurate relationship between the spectral peak energies and the charge.[15] The peaks denoted v_3 and v_2, which are assigned to the totally symmetric stretching vibrations of the central and ring C=C bonds of BO, respectively, showed the shifts according to the charge on this donor molecule. The charge ρ is expressed as

$$\rho = (1524.9 - v_3 \, (cm^{-1}))/109.0 \pm 0.05, \qquad \rho = (1660.8 - v_2 \, (cm^{-1}))/74.1 \pm 0.1.$$

Although the peak corresponding to v_3 is more easily detected in general and gives a more accurate ρ than v_2, both peaks should be checked so as not to assign wrong peaks due to the superposition of those from the counter components in a complex. Also, it is noteworthy that this method is applicable not only to single crystal materials but also to other forms of the samples, *e.g.*, powder samples and even complexes embedded in a polymer matrix. Moreover, a recent investigation based on this methodology proved that BO tends to ionize the counter component as much as possible to form the partially oxidized state of BO itself.[51]

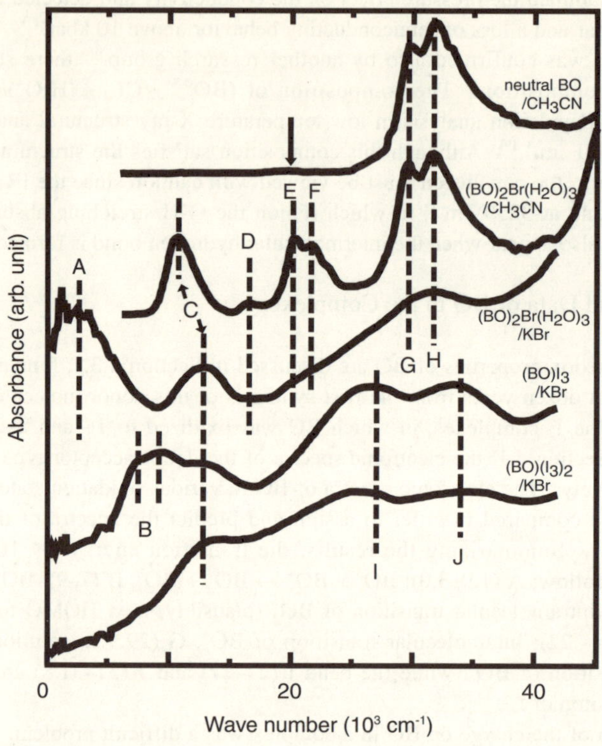

Fig. 4.11   The optical absorption spectra of BO in various oxidation states in CH_3CN solutions and in KBr pellets.
[taken from S. Horiuchi, *Doctoral Thesis*, Kyoto University (1997)]

Table 4.4   Mean Dimensions of BO in Various Oxidation States

Compound	BO	$(BO)_{2.4}I_3$	$(BO)I_3$	$(BO)(I_3)_2$
Oxidation state	0	+0.42	+1	+2
a (Å)	1.357	1.347	1.398	1.427
b (Å)	1.762	1.741	1.723	1.702
c (Å)	1.754	1.733	1.727	1.702
d (Å)	1.333	1.312	1.350	1.420
e (Å)	1.368	1.365	1.352	1.309

As for the magnetism of BO complexes, the principal values of the ESR g-tensor were reported, although sufficient experimental data are not available to discuss the absolute value of the magnetic susceptibility arising from the radical cation.[52] Similar to the case of BEDT-TTF, the principal values are governed by the molecular orientation and found to be $g_{xx} = 2.004(1)$, $g_{yy} = 2.000(1)$ and $g_{zz} = 2.014(1)$, where the principal axes are parallel to the transverse and longitudinal molecular axes and perpendicular to the molecular plane, respectively.

4.3.6 Thin Films Composed of BO Complexes

Since the metallic states of BO complexes persist even in severely disordered systems, it is natural to examine the formation of the conducting films with these complexes. This will be beneficial to both basic and applied science. The first example was reported by Schweitzer and his coworkers, *i.e.*, vapor phase deposition films prepared in a high vacuum of 10^{-8} mbar.[53] Ground crystals of (BO)$_{2.4}$I$_3$ were sublimed at a source temperature of 130–150 °C onto the substrate, the temperature of which was maintained at 200–300 K. The film deposited on a quartz glass consisted of at least three directions of the orientations of the complex with $\sigma_{RT} < 10^{-1}$ S cm^{-1} and an activation energy (E_a) of 50 meV, while that prepared on the (100) surface of a silicon wafer exhibited only one orientation of the crystals in X-ray diffraction spectra. The best quality of the film was obtained when NaCl was used as the substrate. This film was most conductive among those examined ($\sigma_{RT} < 1$ S cm^{-1}, $E_a = 20$ meV). The orientation of the complex was deduced from the X-ray diffraction data, which indicated the stacking axis (60° direction) of the I$_3$-type complex to be parallel to the NaCl surface. This film was mechanically stable and could be removed from the substrate without damage.

The first metallic film based on BO complexes was obtained from the water-air interface by Nakamura and his coworkers.[54] The Langmuir-Blodgett (LB) film was prepared by the deposition of the 1:1 mixture of (BO)$_{10}$(C$_{10}$TCNQ)$_4$(H$_2$O) and icosanoic acid (n-C$_{19}$H$_{39}$COOH) on a quartz-poly(ethyleneterephthalate) (PET) substrate. These films are the first examples of LB film which show metallic behavior without secondary treatment after deposition. The conductivity measurement exhibited metallic behavior down to *ca.* 250 K ($\sigma_{RT} = 10$ S cm^{-1}). Although the round maximum in the plot of conductivity *vs.* temperature could be regarded as the behavior of a narrow gap semiconductor, thermoelectric power measurement proved its metallic nature at least down to *ca.* 50 K. The structure of the film was investigated by IR and ESR techniques, the latter of which also supported the metallic behavior of this film down to *ca.* 50 K.[55] The orientations of the longitudinal axes of the donor and the TCNQ moiety in the acceptor molecules were deduced to be almost perpendicular to the surface of the substrate. Later, the conducting behavior was analyzed by physicists utilizing the concept of percolation conduction in which the coexistence of the metallic, semiconducting and insulating domains in a film was assumed.[56] Not only the TCNQ derivative having a long alkyl chain but the derivative with small substituents was also found to afford metallic LB films. Treatment of a 1:1 mixture of (BO)$_2$[(MeO)$_2$TCNQ] and icosanoic acid similar to that of the above films afforded the deposited films. In this case, metallic behavior was observed down to *ca.* 150 K by the conductivity measurement ($\sigma_{RT} = 11.3$ S cm^{-1}).[57]

A striking LB film was prepared by Ohnuki and Izumi in 1997. They deposited the LB film of 2:1 and 1:1 mixtures of BO and behenic acid (C$_{21}$H$_{43}$COOH) on glass or CaF$_2$.[58] These films showed metallic behavior down to 14 K, the lowest temperature observed ($\sigma_{RT} = 40$ S cm^{-1}). The structure was determined by X-ray diffraction measurements to be a Y-type film in which the donor and behenic acid form double layered structures individually and the double layers are stacked alternately. The mechanism of the ionization of BO in the film formation process is deduced to be the deprotonation of the carboxylic acid based on the IR spectra of the resultant film. Recently, they observed negative magnetoresistance in a similar LB film prepared from BO and stearic acid (C$_{17}$H$_{35}$COOH) which showed metallic behavior down to 120 K.[59] The negative magnetoresistance was interpreted in the weak localization of a 2D electronic system to fit the observed data from 1.7 to 16.7 K under a magnetic field of up to 7 T.

The doping of CT complexes into polymer films is one of the methods for preparing conducting films. For BO complexes, the reticulate doped polymer (RDP) films were prepared to afford a unique material which shows metallic behavior with transparent appearance. After the publication of short papers, Ulanski and his coworkers reported a full description of the film.[60] The preparation of the conducting films was initiated by the casting of the hot o-dichlorobenzene solution of polycarbonate (PC) containing $ca.$ 1 wt% of BO at 120°C. The resultant film was subjected to the vapor of an organic solvent containing iodine or bromine. In this procedure, conducting BO complexes were formed near the surface of the PC films. By choosing the solvent, the concentration of the halogen and the swelling period (time for subjecting the film), conducting films with a surface resistivity (ρ) of 1×10^3 $\Omega/\square$ were prepared. The bulk ρ concerning the thickness of the conducting layer was $ca.$ 10^{-3} to 10^{-2} Ω cm. The best films showed metallic behavior down to at least 15 K and about 100 K when iodine and bromine were used as halogen molecules, respectively. The most striking feature is not the conducting behavior itself but the near transparency of the bromine-doped films. These films were almost transparent to normal incident light to the film surface. This peculiarity was understood as the result of the adequate orientation of the crystallites of the BO complexes near the film surface.[61] As illustrated in Fig. 4.12, the direction of the crystal growth, $i.e.$, the orientation of the conducting plane, was expected to be parallel to the film surface. In each BO complex crystallite the longitudinal molecular axis is arranged perpendicular to the conducting plane (film surface). In fact, when the incident light was irradiated from a direction slanted to the film surface, the optical absorption in the visible region (intramolecular next HOMO to HOMO transition of BO radical cation) partly regained its intensity. X-ray diffraction experiments confirmed this interpretation.

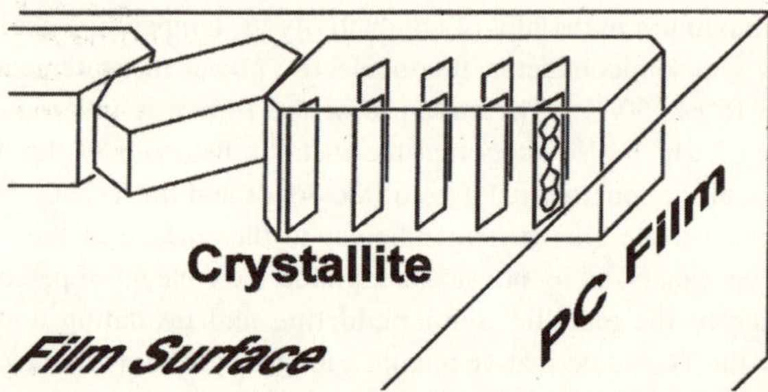

Fig. 4.12   The estimated structure of the RDP films of BO complexes in a PC film.

4.3.7 Some Specific Complexes of BO

In addition to the complexes mentioned above, two complexes of BO deserve to be briefly mentioned. Due to the strong tendency to give metallic complexes, investigations incorporating the functional counter components were carried out.

The BO complexes with rare-earth complex anions were prepared to examine the formation of π-f composite metal.[62] The BO complexes with $[Ln(SCN)_6]^{2-}$ (Ln = Ho, Er, Yb and Y) showed metallic behavior, although the specific interactions concerning the physical properties between the donor molecule and the counter ion were not detected. The introduction of a magnetic counter component was also examined. The complex with the bimetallic oxalate, $(BO)_3[FeCr(ox)_3]$ $(H_2O)_{3.5}$ was prepared as powder samples.[63] In this case, the anion layer showed a ferromagnetic transition at around 11 K. Although this complex is the second example to show metallic behavior in the ferromagnetic phase, no distinct effect of the magnetic transition was observed on the conducting property. Only an anomaly of the ESR parameters was detected below the ferromagnetic transition temperature. These observations indicate the difficulty in modulating metallic behavior by counter components in BO complexes.

4.3.8 (MeO)$_4$TTF

(MeO)$_4$TTF is the unique TTF derivative having four alkoxy substituents directly bonded to the TTF skeleton. Although the redox properties seem to be superior to those of BO, only one experimental paper has been published.[64] It was claimed that $E_{1/2}^1$ is comparable to that of BO, while the ΔE was 0.06 V smaller. As for the complex formation, only [(MeO)$_4$TTF](TCNQ) was reported to show conductive behavior [σ_{RT} = 1.7 S cm^{-1}, E_a = 21 meV (> 160 K) and 19 meV (< 160 K) measured on a compressed pellet]. Due to the difficulty in preparing the starting materials to synthesize other tetrakis(alkoxy)TTF derivatives, systematic investigations on this category of donor molecules have not yet been carried out.

4.4 DBTTFs Having Alkoxy Groups

To improve the solubility of DBTTF and its strength as a donor molecule, Inayoshi and her coworkers substituted the outermost hydrogen atoms with alkoxy groups. The redox properties are summarized in Table 4.5 and indicate that all the alkoxy-substituted derivatives are stronger donors than DBTTF, even though they are weaker than BO.[5,75]

For the neutral single component crystals, X-ray structure analysis has been performed to show a variety of packing patterns. The schematic representation of the packing pattern is summarized in Fig. 4.13. In (MeO)$_4$DBTTF, a molecule is surrounded by the orthogonally tilted neighbors. Substituting two vicinal methoxy groups in this compound by a methylenedioxy group, the packing pattern is modified to consist of a head-to-tail type face-to-face dimer, which resembles the κ-type packing in BEDT-TTF complexes. Other DBTTF derivatives crystallize to form herring-bone type columnar stacks. Even though there are several variations, all of them show slippage along the longitudinal molecular axis between the neighboring molecules in the transverse direction. It should be noted that intermolecular C-H···O contacts shorter than the sum of the vdW radii were observed in all cases even though the respective packing patterns are different.[5,75,76]

Table 4.5 Redox Potentials of DBTTF and Related Compounds in 0.1 M Bu$_4$NClO$_4$/ CH$_2$Cl$_2$ (V *vs.* SCE)[75]

Compound	R^1 or R^1-R^1	R^2 or R^2-R^2	$E_{1/2}^1$ (V)	$E_{1/2}^2$ (V)	ΔE (V)
(MeO)$_4$DBTTF	CH$_3$O-	CH$_3$O-	0.41	0.79	0.38
(MeO)$_2$(MDO)DBTTF	CH$_3$O-	-OCH$_2$O-	0.43	0.81	0.38
(MeO)$_2$(EDO)DBTTF	CH$_3$O-	-O(CH$_2$)$_2$O-	0.44	0.85	0.41
(MDO)$_2$DBTTF	-OCH$_2$O-	-OCH$_2$O-	0.48	0.83	0.35
(EDO)(MDO)DBTTF	-O(CH$_2$)$_2$O-	-OCH$_2$O-	0.49	0.85	0.36
(EDO)$_2$DBTTF	-O(CH$_2$)$_2$O-	-O(CH$_2$)$_2$O-	0.50	0.90	0.40
DBTTF	H	H	0.64	1.06	0.42

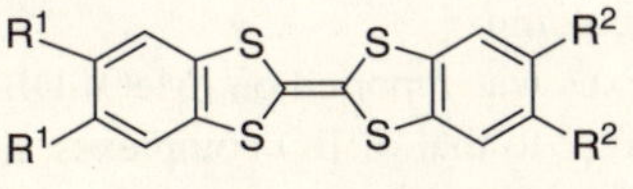

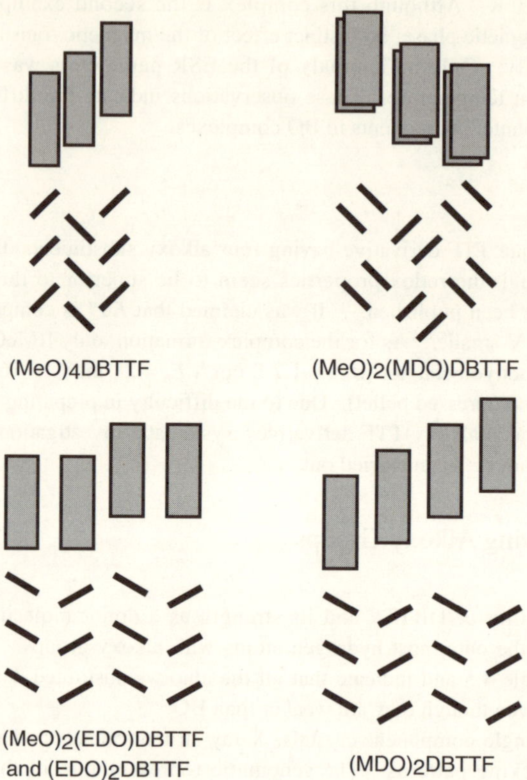

Fig. 4.13   Schematic representation of the packing pattern of the neutral single-component crystals of DBTTF derivatives. In each category, the lower and upper panels show the projection along the donor longitudinal axis and the top view of the former, respectively.

Although more than 50 CT complexes based on the donor molecules in this category including some conducting complexes with σ_{RT} of the order of 10^0 S cm^{-1} have been reported, no metallic behavior has been observed and very few crystal structures have been solved. For (MeO)$_4$DBTTF, two kinds of TCNQ complexes were prepared in the early days.[77] The 1:1 complex synthesized by mixing the solution of each component showed the highly conducting nature of $\rho_{RT} = 1.5$ Ω cm ($E_a = 66$ meV: measured on the compressed pellet), though the sample was prepared only as powder and no structural data are available. Judging from the Raman spectra, the degree of CT in this complex is estimated to be ~0.6, which is consistent with the strength of (MeO)$_4$DBTTF as a donor molecule. On the other hand, needle-shaped single crystals of [(MeO)$_4$DBTTF]$_2$(TCNQ) were prepared by an electrocrystallization method. The structure consists of uniform face-to-face stacks of donor molecules, the columnar direction of which is parallel to the acceptor molecular plane. The relatively high resistivity of $\rho_{RT} = 67$ Ω cm ($E_a = 113$ meV) and the semiconducting nature are interpreted as a result of the existence of molecules with different charges in a donor column.

Recently, a series of complexes was reported on (MeO)$_4$DBTTF.[78] Although some of them showed unusual composition similar to that of BO complexes, no metallic complexes have been obtained yet; D:A = 5:6 for TNAP, 3:4:2(DCE) for DDQ, 4:3:1(DCE) for TCNE, 1:1:0.25(DCE) for Br, 4:1:1(THF) for AuBr$_2$, 6:5:3(DCE) for SbF$_6$, 3:2:1(DCE) for Cu(NCS)$_2$, 1:1 for I$_3$, Ag(CN)$_2$, ReO$_4$ and PF$_6$ complexes where DCE indicates 1,2-dichloroethane. Of these, the crystal structure of [(MeO)$_4$DBTTF]$_3$Cu(NCS)$_2$·DCE has been solved. In the crystal structure, two kinds of crystallographically independent donor molecules formed a face-to-face stacked column and the Cu(NCS)$_2$ anion was located in the transverse direction of the donor molecule. The charge on copper was confirmed to be +1 by ESR measurement for this complex. It should be noted here that the copper(I) ion and thiocyanate anion formed a discrete anion of [Cu(NCS)$_2$]$^-$, while this combination is well known to afford a polymeric anion, an infinite chain in κ-(BEDT-TTF)$_2$Cu(NCS)$_2$ and a 2D network in β_m-(BO)$_3$Cu$_2$(NCS)$_3$. In the donor column, the longitudinal axis of (MeO)$_4$DBTTF was rotated by 15° between the neighboring columns. Short C-H···O contacts (2.46–2.60 Å) were found between the donor columns, while short intermolecular S···S contacts (3.40–3.49 Å) along with intermolecular C-H···O contacts (2.70 Å) barely shorter than the sum of the vdW radii were observed with in a column. Even though the interplanar distances were almost uniform (3.43 and 3.45 Å) within a column, the overlap integrals showed significant alternation (31:8). Due to the latter feature, the calculated band structure clearly showed a gap between the occupied and unoccupied levels. Moreover, a comparison of the intramolecular bond lengths suggested the occurrence of a charge disproportion between the two kinds of donor molecules. Both of these features coincide with the observed semiconducting behavior (ρ_{RT} = 7.8 × 10^3 Ω cm, E_a = 0.20 eV).

Compared to the other derivatives, (EDO)$_2$DBTTF has been more thoroughly investigated because of its similarity to BO: combined with 37 organic acceptor molecules, 40 CT complexes were prepared.[5] Similar to BO, the Torrance V-shape plot for this donor molecule shows that (EDO)$_2$DBTTF affords highly conductive complexes in the region of $\Delta E_{redox} \leq 0.53$ V, which is wider than that of the quasi-1D TTF-TCNQ family, though none of them showed metallic behavior (Fig. 4.14). It should be noted that three kinds of acceptor molecules [TCNQ, Me$_2$TCNQ and (MeO)$_2$TCNQ] afforded polymorphism near the boundary between the neutral and partial CT regions. These results indicate that (EDO)$_2$DBTTF shows a weaker self-assembling property than BO. In fact, the isostructural complexes of [(EDO)$_2$DBTTF](Me$_2$TCNQ) and [(EDO)$_2$DBTTF][(MeO)$_2$TCNQ], the only complexes whose crystal structures have been published among the complexes based on this donor, showed the alternating columnar structure, while BO formed metallic complexes with the same or even relatively weaker acceptor molecules in the scale of ΔE_{redox}.

The origin of the poor ability of (EDO)$_2$DBTTF to afford metallic complexes is suspected to be the lack of side-by-side S···S contacts along the donor transverse directions. In fact, a comparison of the molecular shapes in the neutral donor crystals revealed that the hydrogen atoms on the benzene rings can prohibit this type of intermolecular interactions (Table 4.6).

The optical properties were examined for the (EDO)$_2$DBTTF complexes. Comparison between the spectra measured on the KBr pellet and in the methanol solution of [(EDO)$_2$DBTTF]Br$_{2.1}$(H$_2$O)$_{0.8}$ along with the data collected in the above study revealed the approximate transition energy of the low lying bands of the complexes.

A-band (D$^{\cdot+}$ + D $\rightarrow$ D + D$^{\cdot+}$): 2.0–3.8 × 10^3 cm^{-1}

B-band (D$^{\cdot+}$ + D$^{\cdot+}$ $\rightarrow$ D^{2+} + D): 10.5 × 10^3 cm^{-1}

C-band (intramolecular transition in D$^{\cdot+}$): 13.4 × 10^3 cm^{-1}.

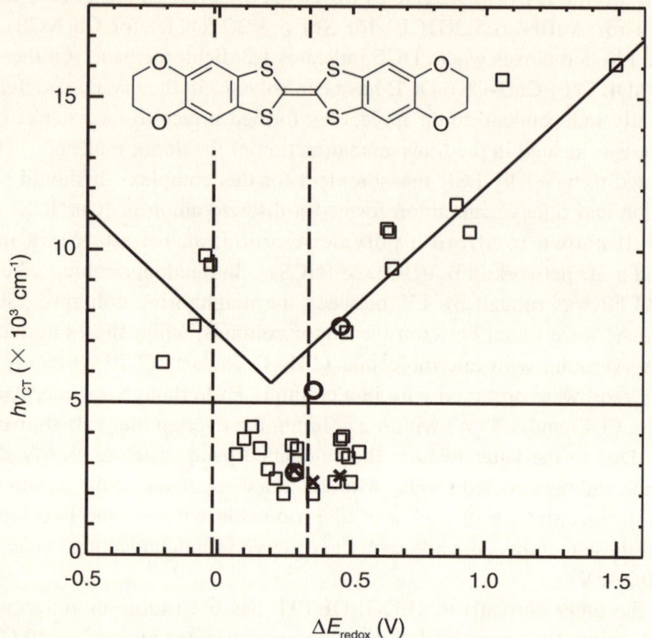

Fig. 4.14   Torrance's V-shape plot for (EDO)$_2$DBTTF complexes. A polymorphism with the same acceptor
molecules is indicated by the paired circle and cross at the same $\Delta E_{\mathrm{redox}}$ value.

Table 4.6   Comparison of the Molecular Dimension of
(EDO)$_2$DBTTF with Those of BO and BEDT-TTF

Compound	a (Å)	b (Å)	c (Å)	d (Å)
BO	10.1	2.90	2.99	–
BEDT-TTF	11.8	3.54	2.96	–
(EDO)$_2$DBTTF	16.0	2.84	2.97	4.48

The large transition energy of the B-band indicates that the effective on-site Coulomb energy of
(EDO)$_2$DBTTF is larger than those of BO (7.8–9.8 $\times$ 10^3 cm^{-1}) and BEDT-TTF (5.5 $\times$ 10^3 cm^{-1}).
Also, from this viewpoint, this donor molecule is inadequate to form metallic complexes, even
though its strength as an electron donor is improved as designed.

4.5   EDO-TTF and Its Derivatives

The unsymmetric derivatives of BO have been investigated from just after the first synthesis of this donor molecule. The fundamental skeleton is regarded to be EDO-TTF (Fig. 4.15; $R^1 = R^2 =$ H). By 1990, the syntheses of most of the derivatives had been reported.[79,80] As in the cases of other unsymmetric donor molecules, the $E_{1/2}^1$ appeared about midway between the parent symmetrical donor molecules. The ΔE, which is the measure of the on-site Coulomb energy of a molecule, also remained of a magnitude similar to those of the parent molecules.[79]

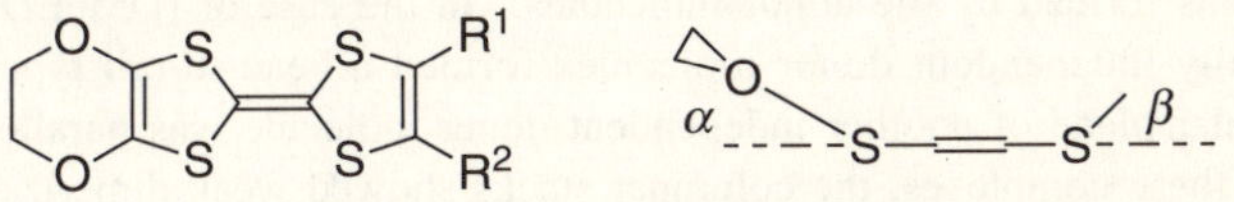

Fig. 4.15   EDO-TTF derivatives (left) and the respective dihedral angles
(α and β) around the sulfur-to-sulfur axes in EDO-TTF (right).

Although a number of EDO-TTF derivatives have been reported, a systematic investigation and/or deep examination of specified complexes have been carried out on only a few cases. For example, the derivatives whose chemical structures are indicated in Fig. 4.15 as $R^1 = R^2 =$ SCH_2CH_2OH,[81] SCH_2CH_2CN[81]; R^1-R^2 = -S(CH$_2$)$_3$S-,[80b] -Se(CH$_2$)$_3$Se-,[80b] are reported only to give conductive TCNQ complexes, respectively. Also, for some of the members of this class only the synthesis or appearance as intermediates of synthetic procedures have been reported. The derivatives of $R^1 = R^2 = CO_2Me$, CO_2Et, SCH_3; $R^1 = H$, $R^2 = Me$, CO_2H, CO_2Me, $CONH_2$, $CONHMe$, $CONMe_2$, $COCl$; R^1-R^2 = -SCH$_2$OCH$_2$S- are known compounds without reports regarding their complexes.

Judging from the redox potentials, the parent EDO-TTF is a stronger donor but shows a larger on-site Coulomb energy compared to BO due to the smaller size of the π-moiety (Table 4.7). The neutral EDO-TTF crystallized in the monoclinic orange parallelepipeds.[82] In the crystal, the bent donor molecule ($\alpha = 8°$ and $\beta = 17°$ in Fig. 4.15) formed a head-to-tail type face-to-face dimer. The neighboring dimers were arranged orthogonally to each other, similar to BEDT-TTF in the neutral crystal. Although three papers have been published concerning the complex formation, no metallic complexes have been reported and even the compositions have not been examined so far.[80a,83,84] Only recently have some of the complexes with TCNQ derivatives been examined systematically to show that this donor exhibits no self-assembling property and affords the conductive complexes ($\sigma_{RT} > 10^{-1}$ S cm^{-1}) in the same ΔE_{redox} region as that of the quasi-1D TTF-TCNQ family.[85]

The substitution of the vinyl proton by the electron-donating methyl group is expected to reinforce the donor strength. Although MeEDO-TTF ($R^1 = CH_3$, $R^2 = H$ in Fig. 4.15) was unexpectedly reported to be weaker than EDO-TTF,[86] DMeEDO-TTF ($R^1 = R^2 = CH_3$ in Fig.

Table 4.7   Redox Potentials of EDO-TTF Derivatives and
Related Compounds

Compound	$E_{1/2}^1$ (V)	$E_{1/2}^2$ (V)	ΔE (V)	ref.[a]
DMeEDO-TTF	0.33	0.63	0.30	83
TTF	0.34	0.70	0.36	83
MeEDO-TTF	0.40	0.77[b]		86
EDO-TTF	0.36	0.69	0.33	83
	0.38	0.70	0.32	86
BO	0.39	0.65	0.26	83

[a] Measurement conditions: in ref. 83, 0.1 M Bu$_4$NClO$_4$/ CH$_3$CN, V vs. SCE; in ref. 86, 0.1 M Bu$_4$NClO$_4$/ CH$_3$CN, V vs. Ag/AgCl.
[b] Peak oxidation potential (irreversible peak).

4.15) is as strong as TTF itself, but the on-site Coulomb energy is lower (Table 4.7).[83] For the latter donor molecule, highly conductive 2:1 radical cation salts were prepared with PF_6^-, AsF_6^-, $CF_3SO_3^-$, NO_3^- and ReO_4^- (σ_{RT} = 5–140 S cm^{-1}) mainly as needles, while BF_4^-, Br^- and ClO_4^- afforded poor conductors (σ_{RT} = 10^{-3}–10^{-6} S cm^{-1}).[87] Among them, the crystal structures of ReO_4^- and NO_3^- complexes were solved. In (DMeEDO-TTF)$_2$ReO$_4$, the head-to-tail type columnar stack was formed by the donor molecules. In the case of (DMeEDO-TTF)$_2$NO$_3$, the crystallographically independent donor molecules formed a head-to-tail type columnar stack, while the molecular plane of another independent doner molecule was parallel to this stacking axis. In both of these complexes, the columnar stacks showed weak dimerization of the donor molecules (interplanar distances were 3.50 and 3.57 Å in the former and 3.52 and 3.48 Å in the latter). Although the crystal structure is not yet known, (DMeEDO-TTF)$_2$PF$_6$ was reported to show an abrupt phase transition at 130 K.

Some of the EDO-TTF derivatives having an additional aromatic ring have been synthesized, but the complexes have not been investigated well so far. For the TCNQ complexes, only qualitative descriptions are available. The complex with benzene-fused EDO-TTF (R^1-R^2 = -CH=CH-CH=CH- in Fig. 4.15) is an insulator,[88] while those with pyridine (-N=CH-CH=CH-)-, pyrazine (-N=CH-CH=N-)-, dimethylpyrazine (-N=CMe-CMe=N-)- and thiophene (=CH-S-CH=)- fused compounds are conducting solids.[80b] As for the radical cation salts, only the preparation of the conducting semiconductors has been reported without a description of their composition; the benzene-fused EDO-TTF afforded I_3 (σ_{RT} = 0.13 S cm^{-1}) and AuI_2 (σ_{RT} = 3.4 × 10^{-2} S cm^{-1}) complexes and the dimethylpyrazine-fused EDO-TTF gave I_3 complex (σ_{RT} = 21.1 S cm^{-1}).[89]

The extension of EDO-TTF using heterocycles has been examined. For the derivatives having the five- and seven-membered alkylenedithio or alkylenediseleno rings [R^1-R^2 = -X(CH$_2$)$_n$X- , n = 1, 3, X = S or Se in Fig. 4.15], only the TCNQ complexes have been reported to be conductive,[80b] with the exception of the reports on the radical cation salts of EOMT-TTF (R^1-R^2 = -SCH$_2$S- in Fig. 4.15).[89] Along with the τ-pase I_3 complexes, EOMT-TTF afforded semiconducting complexes with I_3^- (σ_{RT} = 5 S cm^{-1}), AuI^- (σ_{RT} = 0.17 S cm^{-1}), $AuBr_2^-$ (σ_{RT} = 0.36 S cm^{-1}) and Cu(NCS)$_2^-$ (σ_{RT} = 0.002 S cm^{-1}). Among these, the I_3 complex was claimed to have β'-type packing, although the details have not been reported.

As for the ethylenedichalcogeno-fused EDO-TTF derivatives, EOET and EOES are known (Fig. 4.16). Both of them afforded the conductive TCNQ complexes.[80b] EOET has been well investigated from the viewpoint of a hybrid molecule of BEDT-TTF and BO and the synthesis was reported very early.[79,80b,90] The crystal structure of the neutral molecule was reported to be

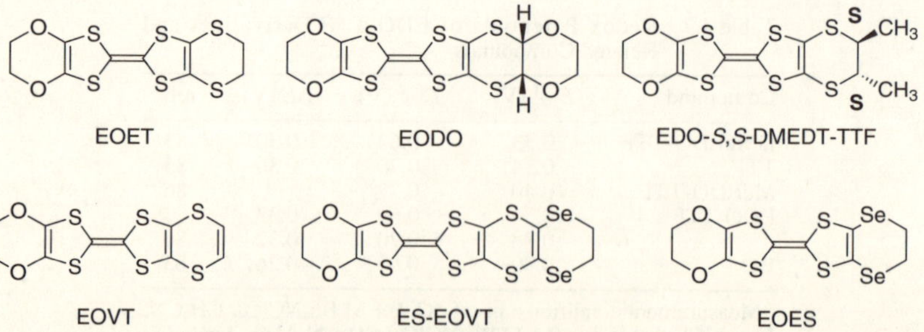

Fig. 4.16   EOET and its analogues.

isostructural to that of BEDT-TTF.[90] Also, this donor molecule is known to be the source of the τ-phase complexes. The 2:1 $Cu(NCS)_2$ [σ_{RT} = 29 S cm^{-1}, T_{MI} (metal-insulator transition temperature) = *ca.* 25 K][91] and 4:1 $Pt(CN)_4$ (semiconductor with σ_{RT} = 2–3 S cm^{-1})[92] complexes were reported. Although other radical cation salts have not been widely examined yet, the crystal structures of the 4:1 Hg_3Br_8 (σ_{RT} = 1–5 S cm^{-1})[93] and 2:1 AuI_2 (semiconductor with σ_{RT} = 4 × 10^{-3} S cm^{-1})[89] complexes were reported. The AuI_2 complex was denoted as β'-type; however, the figure in the paper seems to show the α'-type packing of BEDT-TTF complexes. Recently, two modifications of $AuBr_2$ complexes were obtained.[94] One is β"-$(EOET)_2AuBr_2$, which exhibited metallic behavior down to 1.6 K (σ_{RT} = 60 S cm^{-1}). Another 2:1 modification was a semiconductor [σ_{RT} = 5 × 10^{-3} S cm^{-1}, E_a = 0.22 eV (260–310 K)] the crystal structure of which was isostructural to that of α'-$(BEDT-TTF)_2AuBr_2$.

The EOET complexes with organic acceptor molecules were systematically investigated.[95] Although most of the complexes were not obtained as single crystals, the compositions, optical spectra in KBr pellets and transport properties (mostly measured on compressed pellets) were examined along with the redox properties of the acceptor molecules. As a result, 36 kinds of CT complexes were prepared with 30 acceptor molecules which are categorized into five groups as shown in Fig. 4.17. It is interesting to compare this figure with Fig. 4.3 for BO complexes. In the case of EOET, also, the conductive complexes were prepared with the acceptor molecules having a wide variety of strengths: –0.25 ≤ ΔE_{redox} ≤ 0.60 V (group A). Despite this similarity to BO, only the 2:1:1(H_2O) complex with F_2TCNQ exhibited metallic behavior down to a low temperature (T_{MI} = 15 K). Other complexes showed semiconductive behavior below 100–250 K or even at RT (8 complexes each). The complexes of group B were located in the ΔE_{redox} region

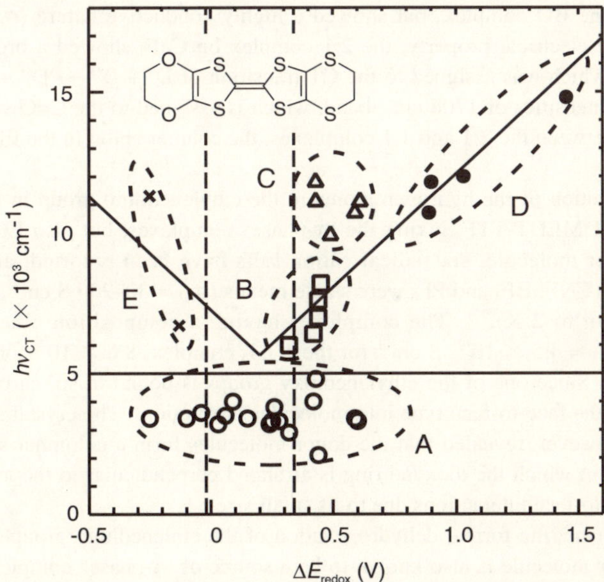

Fig. 4.17 Torrance's V-shape plot for EOET complexes. The ellipses of the dashed lines group the CT complexes of (A) partially ionic, (B) insulator with alternating stacks, (C) neutral clathrate with alternating stacks, (D) completely neutral and (E) completely ionic complexes.

which is included in that of group A. Based on the fact that they gave a larger $h\nu_{CT}$ than group A and showed an insulating nature ($\sigma_{RT} < 10^{-8}$ S cm^{-1}), they were assigned to be partially ionic complexes having alternating stacks, although the ionicities of EOET were estimated to be small (≤ 0.3 for three complexes). In fact, the crystal structure analyses of 2:1 (MeO)$_2$TCNQ and 1:1 and 1:2 BTDA-TCNQ complexes revealed that they consiste of expected columnar structures. Along with the completely neutral alternating stacked complexes (group D) and fully ionic ones (group E), another group C was formed. This group showed a significant deviation of $h\nu_{CT}$ from the estimated value and IR spectra as the superposition of the neutral component molecules. The crystal structure of a complex belonging to this group (2:1 QI$_4$ complex) assigned this group as a clathrate complex in which no specific intermolecular interactions such as heteroatomic contacts, overlapping of π-orbitals or Coulomb attractive interactions contribute to the stabilization of the crystal structure. The number of complexes belonging to this group is as many as six, while only one example has been reported among the BO complexes (1:1 H$_2$TNBP complex). Hence, the tendency to give a clathrate complex is regarded as one of the characteristics of EOET. Also, it is noteworthy that EOET afforded polymorphism with five kinds of acceptor molecules. These features, especially the instability of the metallic state of group A and the existence of groups B and C, indicate that EOET shows a weaker self-assembling nature than BO.

The results reported in this paper demonstrate that the introduction of larger chalcogen atoms into BO reduces the self-assembling property and hence the ability to give stable metallic complexes similar to the discussion in Section 4.4.

Additionally, it is noteworthy that EOET has been employed as the donor molecule of conducting LB films.[96] The 2:1 and 1:1 mixtures of EOET and behenic acid afforded the Y-type films deposited on CaF$_2$ or glass substrate. The 2:1 complex on a glass was less conductive than that consisting of the BO complex, but showed a highly conductive nature ($\sigma_{RT} = 1.0$ S cm^{-1}). Consistent with the electrical property, the 2:1 complex on CaF$_2$ showed a broad absorption at around 3800 cm^{-1} which was assigned to the CT transition of $D^0 + D^{\cdot+} \rightarrow D^{\cdot+} + D^0$. Comparing the IR absorption intensities of 1708 cm^{-1} band, which is assigned to the C=O stretching band of the COOH group between the 2:1 and 1:1 complexes, the counter anion in the film was presumed to be C$_{21}$H$_{43}$COO$^-$.

Further substitution of the hydrogen atoms in the ethylenedithio group in EOET has led to not only EDO-S,S-DMEDT-TTF, giving the "τ-phase" complexes, but also EODO (Fig. 4.16). For the latter donor molecule, six radical cation salts have been reported, among which 2:1 complexes with Au(CN)$_2$, BF$_4$ and PF$_6$ were stable metals ($\sigma_{RT} = 30$–200 S cm^{-1}, no MI transition was observed down to 2 K).[97] The complexes having a composition other than 2:1 were semiconductors ($\sigma_{RT} = 4.5 \times 10^{-7}$ S cm^{-1} for the 1:1 I$_3$ complex, 8.6×10^{-3} for 3:2 AuI$_2$, 8.4×10^{-3} for 1:1 ClO$_4$). Since one of the ethylenedioxy groups is bound to sp^3 carbon atoms, it was expected to hinder the face-to-face type intermolecular interaction. The crystal structure analysis of (EODO)$_2$BF$_4$, however, revealed that the donor molecules form a columnar structure with the head-to-tail dimers in which the dioxane ring is arranged perpendicular to the π-moiety but does not inhibit the face-to-face interactions due to its small size.

On the other hand, the formal dehydrogenation of the ethlenedithio group in EOET derives EOVT. This donor molecule is also known to be a source of "τ-phase" complexes. Along with these peculiar complexes, others have been reported, $e.g.$, the TCNQ complex has been reported as a conducting solid.[80b] As the I$_3$ complexes, 2:1 β'- and 1:1 γ'-phases have been reported as semiconductors ($\sigma_{RT} = 10.4$ and 0.2 S cm^{-1}, respectively) without structural information. The complexes with AuBr$_2$ ($\sigma_{RT} = 11$–22 S cm^{-1}) and Cu(SCN)$_2$ ($\sigma_{RT} = 1.5$ S cm^{-1}) were also semiconductors, but the compositions are not clear yet.[89,91] This family of vinylene compounds has an ethylenedichalcogeno-extended member. ES-EOVT was prepared in relatively good yield by a coupling reaction (19%), although the complex formation has not yet been reported.[98]

EOES has not been as well examined as EOET. This selenium derivative is a strong donor molecule comparable to BO (Table 4.8),[99] but only three complexes have been reported. The TCNQ and I_3 complexes were reported to be conducting and insulating, respectively.[80b,99] $(EOES)_2AuBr_2$ is reported to show metallic behavior down to 1.5 K ($\rho_{RT} = 0.05\ \Omega$ cm).[99] The crystal structure was reported to be isostructural to that of β''-$(BEDT-TTF)_2AuBr_2$, in which EOES are arranged alternately along the side-by-side direction to gain effective intercolumn interactions.

Table 4.8   Comparison of Redox Potentials of TOST Analogues in 0.1 M Bu_4NBF_4/ PhCN (V *vs.* $Ag/AgNO_3$)[99]

Compound	Y1	X1	X2	Y2	$E_{1/2}^{1}$ (V)	$E_{1/2}^{2}$ (V)	ΔE (V)
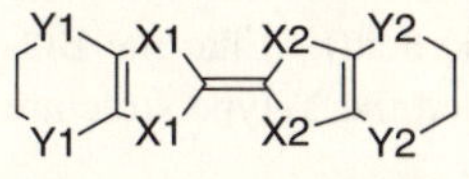 BO	O	S	S	O	0.02	0.34	0.32
BEDT-TTF	S	S	S	S	0.13	0.45	0.32
BS	Se	S	S	Se	0.09	0.43	0.34
BETS	S	Se	Se	S	0.30	0.55	0.25
TOST	O	S	Se	S	0.13	0.44	0.31
EOES	O	S	S	Se	0.04	0.38	0.34

4.6   Sulfur-selenium Analogues of EDO-TTF and Its Derivatives

As the "parent" compounds in this class, EDO-DTSTF and EDO-DTDSF (Fig. 4.18) have been reported. Although the TCNQ complexes of these donor molecules were reported to be conductive,[80b] the preparation of the radical cation salts were examined only on the latter donor (with ClO_4^-, ReO_4^-, PF_6^-, NO_3^-, $CF_3SO_3^-$) to give the powder samples. Among them, at least, the ReO_4 and CF_3SO_3 complexes were insulating ($\sigma_{RT} = 5 \times 10^{-8}$ and 7×10^{-4} S cm^{-1},

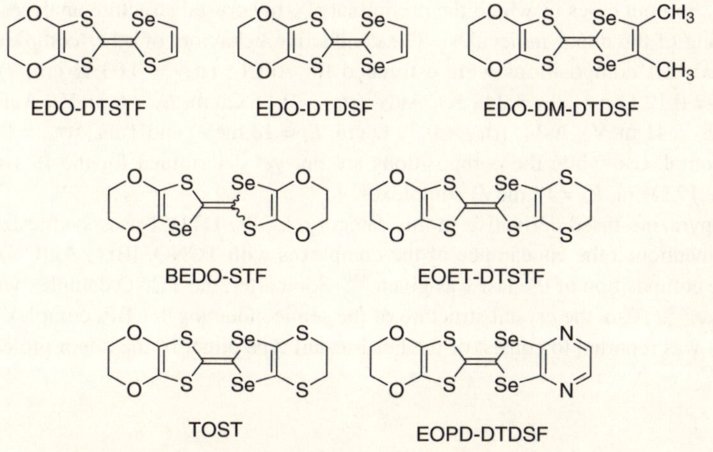

EDO-DTSTF   EDO-DTDSF   EDO-DM-DTDSF

BEDO-STF   EOET-DTSTF

TOST   EOPD-DTDSF

Fig. 4.18   Sulfur-selenium analogues of EDO-TTF and its derivatives.

respectively).[84] Introduction of the methyl groups amended the transport property. EDO-DM-DTDSF afforded not only the TCNQ complex (σ_{RT} = 6 S cm^{-1}) but also the conducting radical cation salts with ClO$_4^-$, ReO$_4^-$, PF$_6^-$, CF$_3$SO$_3^-$ and AuBr$_2^-$ (σ_{RT} = 44, 200, 45, 200 and 8 S cm^{-1}, respectively).[100] Although the structural aspects are not reported yet, it is claimed that the ReO$_4$ complex showed polymorphism; another phase of σ_{RT} = 10^{-1} S cm^{-1} was obtained.

As a BO-like molecule, BEDO-STF was synthesized as a mixture of *cis* and *trans* isomers regarding the relative position of two selenium (or sulfur) atoms.[101] The cyclic voltammetry (CV) measurement showed that $E_{1/2}^1$ of this donor was midway between BEDT-TTF and BO, while the ΔE was smaller by 0.03 V than those of these reference donors. Based on this property, the formation of metallic complexes was anticipated and in fact the 2:1 AuBr$_2$ and GaCl$_4$ radical cation salts were metallic down to 4.2 K. The crystal structure of the latter complex was solved to show the κ-type donor packing in which the selenium and sulfur atoms randomly occupied the chalcogen sites in the fulvalene skeleton.

As for BEDO-STF, the formation of the LB films was also examined on the mixture with behenic acid.[102] Contrary to the case of BO, only the 1:1 mixture could be transferred onto the solid substances among the 0.5:1, 1:1 and 1.5:1 mixtures (donor:behenic acid). Like the BO-behenic acid LB films, the X-ray diffraction and IR measurements revealed the Y-type structure and ionization of carboxylic acid in the deposited films.

EOET-DTSTF and TOST are the EOET analogues of the selenium-containing donor molecules. Although only the synthesis was reported for the former,[103] the latter afforded not only the conducting TCNQ complex[80b] but also the metallic radical cation salts.[99] The redox properties of TOST are understood by the general tendency of unsymmetric TTF derivatives. As summarized in Table 4.8, TOST showed intermediate properties between BO and BETS, although the proportionality for ΔE is not sufficient. This tendency, which is also observed for EOES described in Section 4.5, may be due to the small deviation in ΔE values among the compounds compared in this table.

The crystal structure of neutral TOST consists of head-to-tail type dimers, in which *ca.* 19% of the molecules are packed oppositely along the longitudinal axis. The arrangement of the dimers was isostructural to that of BEDT-TTF. Among the reported complexes, 2:1 FeCl$_4^-$ and GaCl$_4^-$ radical cation salts showed metallic behavior down to 4.2 K (ρ_{RT} = 1.9 and 0.31 Ω cm, respectively), in both cases of which the preliminary X-ray crystal structure analyses revealed the κ-type packing of the donor molecules. The conducting behaviors of other complexes have also been reported; 2:1 compositions were estimated for AuCl$_2^-$ (ρ_{RT} = 0.03 Ω cm, T_{MI} = 194 K), AuBr$_2^-$ (σ_{RT} = 0.12 Ω cm, T_{MI} = 242 K), AuI$_2^-$ (σ_{RT} = 0.11 Ω cm, E_a = 6 meV), Au(CN)$_2^-$ (σ_{RT} = 0.19 Ω cm, E_a = 41 meV), AsF$_6^-$ (σ_{RT} = 0.39 Ω cm, E_a = 13 meV) and TaF$_6^-$ (σ_{RT} = 0.38 Ω cm, E_a = 8 meV) complexes, while the compositions are not yet determined for the I$_3^-$ (insulator) and PF$_6^-$ (ρ_{RT} = 0.19 Ω cm, E_a = 18 meV) complexes.

As the pyrazine-fused derivative in this category, EOPD-DTDSF was synthesized. The first report only mentioned the appearance of the complexes with TCNQ, IBr$_2^-$, Ag(CN)$_2^-$, AuI$_2^-$ and BF$_4^-$, but the composition of the last was given.[80a] Soon after, the TCNQ complex was claimed to be conductive.[80b] Also, the crystal structure of the semiconducting 1:1 BF$_4$ complex (σ_{RT} = 3.3 $\times$ 10^{-4} S cm^{-1}) was reported to consist of the head-to-tail type dimer of the donor molecules and the counter ion.[89]

4.7 Ethylenoxythio-substituted TTFs

The substitution of one of the oxygen atoms in an ethylenedioxy group in the BO molecule has been initiated by the synthesis of BEOT-TTF by Hellberg and his coworkers.[104] As shown in Fig. 4.19, this donor molecule is a structural isomer of EOET and consists of E- and Z- geometrical isomers. Although the coexistence of the geometrical isomers was proved by ^{13}C-NMR measurement, they have not been isolated from each other. As for the CT complexes of this donor, the preparation of 2:1 radical cation salts with ClO_4^-, BF_4^- and PF_6^- and the AsF_6^- complex with unknown composition were reported. The physical properties were examined ($\sigma_{RT} = ca.$ 5 S cm^{-1}) only on (BEOT-TTF)$_2$ClO$_4$, and a Lorentzian ESR signal having a line width of $ca.$ 25 G was observed at RT with the intensity corresponding to 0.4 spins per molecular unit.

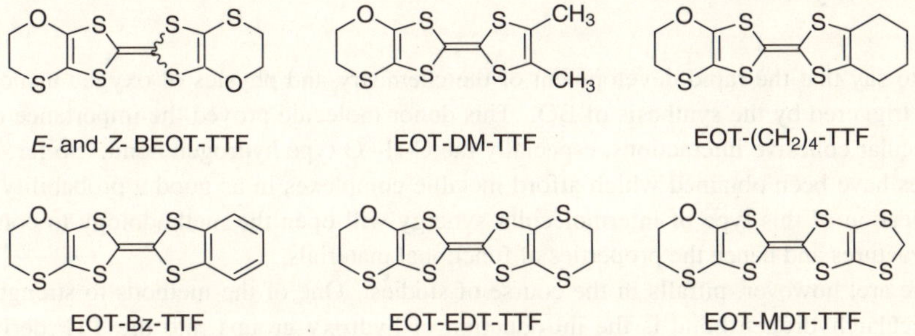

E- and Z- BEOT-TTF   EOT-DM-TTF   EOT-(CH$_2$)$_4$-TTF

EOT-Bz-TTF   EOT-EDT-TTF   EOT-MDT-TTF

Fig. 4.19   Ethylenoxythio-TTF derivatives.

Subsequently, the ethylenoxythio-substituted TTF derivatives were prepared to give a series of donor molecules of EOT-DM-TTF,[105–107] EOT-(CH$_2$)$_4$-TTF,[105] EOT-Bz-TTF,[105,107] EOT-EDT-TTF[106–108] and EOT-MDT-TTF[106] (Fig. 4.19). With the exception of EOT-Bz-TTF, the electrochemical oxidation of these donor molecules was easier than that of BEDT-TTF (Table 4.9).

Besides BEOT-TTF, the preparation of the CT complexes was examined on two derivatives. EOT-DM-TTF has afforded radical cation salts with 7 counter ions so far, all of which are reported to have 2:1 compositions; NO$_3$ ($\sigma_{RT} = 100$ S cm^{-1}, semimetallic down to $ca.$ 180 K), ClO$_4$ ($\sigma_{RT} = 20$ S cm^{-1}, $T_{MI} = 160$ K), ReO$_4$ ($\sigma_{RT} = 0.1$ S cm^{-1}, semiconductor), PF$_6$ [δ-phase ($\sigma_{RT} = 0.05$ S cm^{-1}) and ϕ-phase ($\sigma_{RT} = 300$ S cm^{-1}, $T_{MI} = 120$ K)], AsF$_6$ ($\sigma_{RT} = 390$ S cm^{-1}, $T_{MI} = 150$ K), SbF$_6$ ($\sigma_{RT} = 480$ S cm^{-1}, $T_{MI} = 170$ K) and CF$_3$SO$_3$ ($\sigma_{RT} = 1$ S cm^{-1}) salts.[105,106] In the case of the PF$_6$ complexes, it is reported that only the ESR and conducting behaviors were different between the δ- and ϕ-phases despite the visual similarity. Among these highly conducting complexes, the

Table 4.9   Comparison of Redox Potentials of TOST Analogues in 0.15 M Bu$_4$NPF$_6$/ CH$_2$Cl$_2$ (V $vs.$ SCE)

Compound	$E_{1/2}^1$ (V)	$E_{1/2}^2$ (V)	ΔE (V)	ref.
EOT-DM-TTF	0.29	0.79	0.50	106
EOT-(CH$_2$)$_4$-TTF	0.31	0.81	0.50	105
BEOT-TTF	0.39	0.81	0.42	106
EOT-EDT-TTF	0.43	0.85	0.42	106
EOT-MDT-TTF	0.45	0.86	0.41	106
EOT-Bz-TTF	0.48	0.95	0.47	105
BEDT-TTF	0.46	0.89	0.43	106

crystal structure of the NO_3 complex was reported briefly; short intermolecular atomic contacts between the oxygen in NO_3 and sulfur in EOT group were noted.

EOT-EDT-TTF is the most BEDT-TTF-like hybrid molecule among the BO analogues. In fact the crystal structure of the neutral molecule is isostructural to that of BEDT-TTF.[108] Interestingly, no disorder of the oxygen atom was reported, while disorder of the ethylene groups was detected. In contrast to EOT-DM-TTF, the composition of the EOT-EDT-TTF complex depended on the counter ion.[106,108] The most popular composition was 2:1 observed in ClO_4 (σ_{RT} = 3.2 S cm^{-1}), I_3 (σ_{RT} = 1 S cm^{-1}), AsF_6 (σ_{RT} = 4 S cm^{-1}, semiconductor) and SbF_6 (σ_{RT} = 30 S cm^{-1}, semiconductor) complexes. Other compositions were observed in ReO_4 (3:2, σ_{RT} = 10 S cm^{-1}, T_{MI} = 180 K) and $Ag_2(CN)_3$ (1:1, σ_{RT} = 0.1 S cm^{-1}) complexes along with the $Cu[N(CN)_2]Cl$ complex (σ_{RT} = 0.05 S cm^{-1}), the composition of which has not yet been determined.

4.8   Summary and Outlook

It is fair to say that the rapid development of the chemistry and physics of oxygen analogues of TTF was triggered by the synthesis of BO. This donor molecule proved the importance of weak intermolecular cohesive interactions, especially the C-H$\cdots$O type hydrogen bond. So far, no TTF derivatives have been obtained which afford metallic complexes in as good a probability as BO. The application of this type of intermolecular synergy will open the methodology to control the crystal structures and hence the properties of functional materials.

There are, however, pitfalls in the course of studies. One of the methods to strengthen the intermolecular hydrogen bond is the introduction of hydroxy groups into the TTF derivatives. This strategy was examined by synthesizing hydroxymethyl-substituted TTF derivatives.[109] The CT complexes were not conducting in general and analysis proved that the strong hydrogen bond among the component molecules resulted in non-uniform segregated columnar structures and charge separation among the donor molecules. As one method to introduce oxygen atoms in TTF derivatives, the fusion of the furan ring is a possibility. In fact, [3,4]furan-fused TTF derivatives were synthesized and it was proven that the introduction of this heterocyle reduced their ability as electron donors.[110]

Oxygen is an interesting atom to control the properties of a molecule as exemplified by biological molecules. Although caution is required, further application of the oxygen atom will bring new features to TTF chemistry and physics.

References

1. A. Bondi, *J. Phys. Chem.*, **68**, 441 (1964).
2. J. B. Torrance, J. E. Vazquez, J. J. Mayerle and V. Y. Lee, *Phys. Rev., B* **46**, 253 (1981).
3. For example, K. Nakasuji, *Pure Appl. Chem.*, **62**, 477 (1990).
4. G. Saito and J. P. Ferraris, *Bull. Chem. Soc. Jpn.*, **53**, 2141 (1980).
5. T. Senga, K. Kamoshida, L. A. Kushch, G. Saito, T. Inayoshi and I. Ono, *Mol. Cryst. Liq. Cryst.*, **296**, 97 (1997).
6. O. G. Safiev, D. V. Nazarov, V. V. Zorin and D. L. Rukhmankulov, *Khim. Geterotsikl. Soedin.*, 852 (1988) [*Chem. Abs.*, 110:212656].
7. K. Tanaka, K. Yoshida, T. Ishida, A. Kobayashi and T. Nogami, *Adv. Mater.*, **12**, 661 (2000).
8. S. T. D'Arcangelis and D. O. Cowan, *Tetrahedron Lett.*, **37**, 2931 (1996).
9. T. Nogami, K. Tanaka, T. Ishida and A. Kobayashi, *Synth. Met.*, **120**, 755 (2001).
10. K. Lerstrup, M. Lee, F. M. Wiygul, T. J. Kistenmacher and D. O. Cowan, *J. Chem. Soc., Chem. Commun.*, 294 (1983).
11. T. Suzuki, H. Yamochi, G. Srdanov, K. Hinkelmann and F. Wudl, *J. Am. Chem. Soc.*, **111**, 3108 (1989).
12. M. Iyoda, Y. Kuwatani, E. Ogura, K. Hara, H. Suzuki, T. Takano, K. Takeda, J. Takano, K. Ugawa, M.

Yoshida, H. Matsuyama, H. Nishikawa, I. Ikemoto, T. Kato, N. Yoneyama, J. Nishijo, A. Miyazaki and T. Enoki, *Heterocycles*, **54**, 833 (2001).

13. a) K. Hinkelmann, K. K. Liou and F. Wudl, *J. Chem. Soc., Chem. Commun.*, 1744 (1989); b) D. L. Lichitenberger, R. L. Johnston, K. Hinkelmann, T. Suzuki and F. Wudl, *J. Am. Chem. Soc.*, **112**, 3302 (1990).

14. S. Horiuchi, H. Yamochi, G. Saito, K. Sakaguchi and M. Kusunoki, *J. Am. Chem. Soc.*, **118**, 8604 (1996).

15. O. Drozdova, H. Yamochi, K. Yakushi, M. Uruichi, S. Horiuchi and G. Saito, *J. Am. Chem. Soc.*, **122**, 4436 (2000).

16. T. Hasegawa, T. Mochida, R. Kondo, S. Kagoshima, Y. Iwasa, T. Akutagawa, T. Nakamura and G. Saito, *Phys. Rev., B* **62**, 10059 (2000).

17. a) H. Yamochi, T. Nakamura, G. Saito, T. Sugano and F. Wudl, *Synth. Met.*, **42**, 1741 (1991); b) H. Yamochi, S. Horiuchi, G. Saito, M. Kusunoki, K. Sakaguchi, T. Kikuchi and S. Sato, *Synth. Met.*, **56**, 2096 (1993).

18. a) F. Wudl, H. Yamochi, T. Suzuki, H. Isotalo, C. Fite, H. Kasmai, K. Liou, G. Srdanov, P. Coppens, K. Maly and A. Frost-Jensen, *J. Am. Chem. Soc.*, **112**, 2461 (1990); b) I. Cisarova, K. Maly, X. Bu, A. Frost-Jensen, P. Sommer-Larsen and P. Coppens, *Chem. Mater.*, **3**, 647 (1991).

19. M.-H. Whangbo, D. Jung, J. Ren, M. Evain, J. J. Novoa, F. Mota, S. Alvarez, J. M. Williams, M. A. Beno, A. M. Kini, H. H. Wang and J. R. Ferraro, *The Physics and Chemistry of Organic Superconductors*, Springer Proc. Phys., Vol. 51, Springer-Verlag, Berlin (1990), p. 262.

20. H. Yamochi and G. Saito, *Synth. Met.*, **120**, 863 (2001).

21. H. Yamochi, T. Tsutsumi, T. Kawasaki and G. Saito, *Mat. Res. Symp. Proc.*, **488**, 641 (1998).

22. E. I. Zhilyaeva, R. N. Lyubovskaya, S. A. Torunova, S. V. Konovalikhin, O. A. Dyachenko and R. B. Lyubovskii, *Synth. Met.*, **80**, 91 (1996).

23. S. Sekizaki, A. Konsha, H. Yamochi and G. Saito, *Mol. Cryst. Liq. Cryst.*, **376**, 207 (2002).

24. D. Schweitzer, S. Kahlich, I. Heinen, S. E. Lan, B. Nuber, H. J. Keller, K. Winzer and H. W. Helberg, *Synth. Met.*, **56**, 2827 (1993).

25. R. P. Shibaeva, S. S. Khasanov, B. Z. Narymbetov, L. V. Zorina, A. V. Bazhenov, N. D. Kushch, E. B. Yagubskii and E. Canadell, *Synth. Met.*, **102**, 1650 (1999).

26. H. Yamochi, K. Tsutsumi, T. Kawasaki and G. Saito, *Synth. Met.*, **103**, 2004 (1999).

27. H. Yamochi, T. Komatsu, N. Matsukawa, G. Saito, T. Mori, M. Kusunoki and K. Sakaguchi, *J. Am. Chem. Soc.*, **115**, 11319 (1993).

28. M. Fettouhi, L. Ouahab, D. Serhani, J.-M. Fabre, L. Ducasse, J. Amiell, R. Canet and P. Delhaes, *J. Mater. Chem.*, **3**, 1101 (1993).

29. M. A. Beno, H. H. Wang, A. M. Kini, K. D. Carlson, U. Geiser, W. K. Kwok, J. E. Thompson, J. M. Williams, J. Ren and M.-H. Whangbo, *Inorg. Chem.*, **29**, 1599 (1990).

30. S. Kahlich, D. Schweitzer, I. Heinen, S. E. Lan, B. Nuber, H. J. Keller, K. Winzer and H. W. Helberg, *Solid State Commun.*, **80**, 191 (1991).

31. a) L. I. Buravov, A. G. Khomenko, N. D. Kuschch, V. N. Laukhin, A. I. Schegolev, E. B. Yagubskii, L. P. Rozenberg and R. P. Shibaeva, *J. Phys. I*, **2**, 529 (1992); b) V. N. Laukhin, L. I. Buravov, A. G. Khomenko, N. D. Kushch, A. I. Schegolev, E. B. Yagubskii, L. P. Rozenberg and R. P. Shibaeva, *Synth. Met.*, **56**, 2877 (1993).

32. a) S. S. Khasanov, B. Z. Narymbetov, L. V. Zorina, L. P. Rozenberg, R. P. Shibaeva, N. D. Kushch, E. B. Yagubskii, R. Rousseau and E. Canadell, *Eur. Phys. J., B* **1**, 419 (1998); b) S. S. Khasanov, L. V. Zorina, R. P. Shibaeva, N. D. Kushch, E. B. Yagubskii, R. Rousseau, E. Canadell, Y. Barrans, J. Gaultier and D. Chasseau, *Synth. Met.*, **103**, 1853 (1999).

33. The first report of $(BO)_2ReO_4(H_2O)$ from Germany gave the unit cell parameters of the space group $P2_1/n$, $a = 12.16(1)$, $b = 34.05$, $c = 8.091(4)$ Å, $\beta = 123.44(4)°$, $V = 2795.5$ Å^3 and $Z = 4$.[30] This cell choice is referred as "cell-1" in this chapter. A subsequent paper from Russia applied the following parameters of the space group $P2_1/n$, $a = 8.072(3)$, $b = 10.230(4)$, $c = 34.012(9)$ Å, $\gamma = 98.00(3)°$, $V = 2781(3)$ Å^3 and $Z = 4$, which is referred to as "cell-2" in this chapter.[31] The relationship between these cell choices are described in ref. 35. The donor layer is parallel to the *ac*- and *ab*-plane in "cell-1" and "cell-2," respectively. The axes of *a*, *b* and *c* in "cell-1" correspond to *a*, *c* and (*a+b*) in "cell-2."

34. D. Schweitzer, E. Balthes, S. Kahlich, I. Heinen, H. J. Keller, W. Strunz, W. Biberacher, A. G. M. Jansen and E. Steep, *Acta Phys. Pol., A* **87**, 749 (1995).

35. S. Kahlich, D. Schweitzer, C. Rovira, J. A. Paradis, M.-H. Whangbo, I. Heinen, H. J. Keller, B. Nuber, P. Bele, H. Brunner and R. P. Shibaeva, *Z. Phys., B* **94**, 39 (1994).

36. M. V. Kartsovnik and V. N. Laukhin, *J. Phys. I*, **6**, 1753 (1996).

37. A. Audouard, V. N. Laukhin, C. Proust, L. Brossard, N. D. Kushch and S. Askenazy, *Physica B*, **246**, 117 (1998).

38. C. Proust, A. Audouard, M. Honold, M.-S. Nam, V. N. Laukhin, L. Brossard, E. H. Haanappel, J. Singleton and N. D. Kushch, *Synth. Met.*, **103**, 2040 (1999).

39. C. Proust, M. Honold, V. N. Laukhin, M.-S. Nam, A. Audouard, L. Brossard, E. Canadell, J. Singleton, J. P. Legros and N. D. Kushch, *Synth. Met.*, **103**, 1938 (1999).

40. C. Proust, A. Audouard, V. Laukhin, L. Brossard, M. Honold, M.-S. Nam, E. Haanappel, J. Singleton and N. Kushch, *Eur. Phys. J., B* **21**, 31 (2001).

41. R. B. Lyubovskii, S. I. Pesotskii, M. Gener, R. Rousseau, E. Canadell, J. A. A. J. Perenboom, V. I. Nizhankovskii, E. I. Zhilyaeva, O. A. Bogdanova and R. N. Lyubovskaya, *J. Mater. Chem.*, **12**, 483 (2002).

42. E. I. Zhilyaeva, O. A. Bogdanova, R. N. Lyubovskaya, R. B. Lyubovskii, K. A. Lyssenko and M. Y. Antipin, *Synth. Met.*, **99**, 169 (1999).

43. T. Mori, H. Inokuchi, Y. Misaki, T. Inabe, H. Mori and S. Tanaka, *Bull. Chem. Soc. Jpn.*, **67**, 661 (1994).

44. K. Takimiya, A. Ohnishi, Y. Aso, T. Otsubo, F. Ogura, K. Kawabata, K. Tanaka and M. Mizutani, *Bull. Chem. Soc. Jpn.*, **67**, 766 (1994).

45. T. Mori, K. Oshima, H. Okuno, K. Kato, H. Mori and S. Tanaka, *Phys. Rev., B* **51**, 11110 (1995).

46. T. Mori, S. Ono, H. Mori and S. Tanaka, *J. Phys. I*, **6**, 1849 (1996).

47. S. Hanazato, T. Mori, H. Mori and S. Tanaka, *Synth. Met.*, **102**, 1770 (1999).

48. E. I. Zhilyaeva, S. A. Torunova, R. N. Lyubovskaya, S. V. Konovalikhin, O. A. Dyachenko, R. B. Lyubovskii and S. I. Pesotskii, *Synth. Met.*, **83**, 7 (1996).

49. R. P. Shibaeva, S. S. Khasanov, B. Z. Narymbetov, L. V. Zorina, L. P. Rozenberg, A. V. Bazhenov, N. D. Kushch, E. B. Yagubskii, C. Rovira and E. Canadell, *J. Mater. Chem.*, **8**, 1151 (1998).

50. S. Horiuchi, H. Yamochi, G. Saito and K. Matsumoto, *Mol. Cryst. Liq. Cryst.*, **284**, 357 (1996).

51. O. Drozdova, H. Yamochi, K. Yakushi, M.Uruichi and G. Saito, *Mol. Cryst. Liq. Cryst.*, **376**, 135 (2002).

52. B. H. Ward, G. E. Granroth, J. B. Walden, K. A. Abboud, M. W. Meisel, P. G. Rasmussen and D. R. Talham, *J. Mater. Chem.*, **8**, 1373 (1998).

53. J. Moldenhauer, U. Niebling, T. Ludwig, B. Thoma, D. Schweitzer, W. Strunz, H. J. Keller, P. Bele and H. Brunner, *Mol. Cryst. Liq. Cryst.*, **284**, 161 (1996).

54. a) T. Nakamura, G. Yunome, R. Azumi, M. Tanaka, M. Yumura, M. Matsumoto, S. Horiuchi, H. Yamochi and G. Saito, *Synth. Met.*, **57**, 3853 (1993); b) T. Nakamura, G. Yunome, R. Azumi, M. Tanaka, H. Tachibana, M. Matsumoto, S. Horiuchi, H. Yamochi and G. Saito, *J. Phys. Chem.*, **98**, 1882 (1994).

55. K. Ikegami, S. Kuroda, T. Nakamura, R. Azumi, G. Yunome, M. Matsumoto, S. Horiuchi, H. Yamochi and G. Saito, *Synth. Met.*, **71**, 1909 (1995).

56. K. Ogasawara, T. Ishiguro, S. Horiuchi, H. Yamochi, G. Saito and Y. Nogami, *J. Phys. Chem. Solids*, **58**, 39 (1997).

57. K. Ogasawara, T. Ishiguro, S. Horiuchi, H. Yamochi and G. Saito, *Jpn. J. Appl. Phys.*, **35**, L571 (1996).

58. H. Ohnuki, T. Noda, M. Izumi, T. Imakubo and R. Kato, *Phys. Rev., B* **55**, R10225 (1997).

59. Y. Ishizaki, M. Izumi, H. Ohnuki, K. Kalita-Lipinska, T. Imakubo and K. Kobayashi, *Phys. Rev., B* **63**, 134201 (2001).

60. J. K. Jeszka, A. Tracz, A. Sroczynska, M. Kryszewski, H. Yamochi, S. Horiuchi, G. Saito and J. Ulanski, *Synth. Met.*, **106**, 75 (1999).

61. S. Horiuchi, H. Yamochi, G. Saito, J. K. Jeszka, A. Tracz, A. Sroczynska and J. Ulanski, *Mol. Cryst. Liq. Cryst.*, **296**, 365 (1997).

62. M. Tamura, K. Yamanaka, Y. Mori, Y. Nishio, K. Kajita, H. Mori, S. Tanaka, J. Yamaura, T. Imakubo, R. Kato, Y. Misaki and K. Tanaka, *Synth. Met.*, **120**, 1041 (2001).

63. H. Yamochi, T. Kawasaki, Y. Nagata, M. Maesato and G. Saito, *Mol. Cryst. Liq. Cryst.*, **376**, 113 (2002).

64. Y. Misaki, H. Nishikawa, K. Nomura, T. Yamabe, H. Yamochi, G. Saito, T. Sato and M. Shiro, *J. Chem. Soc., Chem. Commun.*, 1410 (1992).

65. V. Petricek, K. Maly, P. Coppens, X. Bu, I. Cisarova and A. Frost-Jensen, *Acta Cryst.*, A**47**, 210 (1991).

66. R. Li, V. Petricek, I. Cisarova and P. Coppens, *Acta Crystallogr.*, B**51**, 798 (1995).

67. M. A. Beno, H. H. Wang, K. D. Carlson, A. M. Kini, G. M. Frankenbach, J. R. Ferraro, N. Larson, G. D. McCabe, J. Thompson, C. Purnama, M. Vashon, J. M. Williams, D. Jung and M.-H. Whangbo, *Mol. Cryst. Liq. Cryst.*, **181**, 145 (1990).

68. M. A. Beno, A. M. Kini, U. Geiser, H. H. Wang, K. D. Carlson and J. M. Williams, *The Physics and Chemistry of Organic Superconductors,* Springer Proc. Phys., Vol. 51, Springer-Verlag, Berlin (1999), p. 369.

69. E. I. Zhilyaeva, R. N. Lyubovskaya, S. V. Konovalikhin, O. N. Dyachenko and R. B. Lyubovskii, *Synth.*

Met., **94**, 35 (1998).

70. E. I. Zhilyaeva, O. A. Bogdanova, R. N. Lyubovskaya, R. B. Lyubovskii, K. A. Lyssenko and M. Y. Antipin, *Synth. Met.*, **99**, 169 (1999).

71. S. Horiuchi, H. Yamochi, G. Saito and K. Matsumoto, *Synth. Met.*, **86**, 1809 (1997).

72. L. Kushch, L. Buravov, V. Tkacheva, E. Yagubskii, L. Zorina, S. Khasanov and R. Shibaeva, *Synth. Met.*, **102**, 1646 (1999).

73. L. V. Zorina, S. S. Khasanov, R. P. Shibaeva, M. Gener, R. Rousseau, E. Canadell, L. A. Kushch, E. B. Yagubskii, O. O. Drozdova and K. Yakushi, *J. Mater. Chem.*, **10**, 2017 (2000).

74. G. Saito, H. Izukashi, M. Shibata, K. Yoshida, L. A. Kushch, T. Kondo, H. Yamochi, O. O. Drozdova, K. Matsumoto, M. Kusunoki, K. Sakaguchi, N. Kojima and E. B. Yagubskii, *J. Mater. Chem.*, **10**, 893 (2000).

75. T. Inayoshi, K. Sato, I. Miyazaki, S. Yokoyama and I. Ono, *Mol. Cryst. Liq. Cryst.*, **309**, 251 (1998).

76. T. Inayoshi and I. Ono, *Synth. Met.*, **110**, 153 (2000).

77. T. Inayoshi, I. Ono, S. Matsumoto and O. Mitsunobu, *Mol. Cryst. Liq. Cryst.*, **285**, 89 (1996).

78. T. Inayoshi, H. Komatsu, I. Ono and S. Matsumoto, *Synth. Met.*, **128**, 333 (2002).

79. T. Mori, H. Inokuchi, A. M. Kini and J. M. Williams, *Chem. Lett.*, 1279 (1990).

80. a) G. C. Papavassiliou, V. C. Kakoussis, D. J. Lagouvardos and G. A. Mousidis, *Mol. Cryst. Liq. Cryst.*, **181**, 171 (1990); b) G. C. Papavassiliou, *Synth. Met.*, **42**, 2535 (1991).

81. L. Binet, C. Montginoul, J.-M. Fabre, L. Ouahab, S. Golhen and J. Becher, *Synth. Met.*, **86**, 1825 (1997).

82. C. Meziere, M. Fourmigue and J.-M. Fabre, *C. R. Acad. Sci., Ser. IIc: Chim.*, **3**, 387 (2000).

83. J.-M. Fabre, D. Serhani, K. Saoud, S. Chakroune and M. Hoch, *Synth. Met.*, **60**, 295 (1993).

84. A. Javidan, M. Calas, A. K. Gousmia, J.-M. Fabre, L. Ouahab and S. Golhen, *Synth. Met.*, **86**, 1811 (1997).

85. A. Ota, H. Yamochi and G. Saito, *Mol. Cryst. Liq. Cryst.*, **177**, 135 (2002).

86. A. Miyazaki and T. Enoki, *Synth. Met.*, **120**, 939 (2001).

87. J.-M. Fabre, S. Chakroune, A. Javidan, L. Zanik, L. Ouahab, S. Golhen and P. Delhaes, *Synth. Met.*, **70**, 1127 (1995).

88. G. C. Papavassiliou, D. J. Lagouvardos, V. C. Kakoussis and G. A. Mousdis, *Z. Naturforsch.*, **45b**, 1216 (1990).

89. A. Terzis, A. Hountas, B. Hilti, G. Mayer, J. S. Zambounis, D. Lagouvardos, V. Kakoussis, G. Mousdis and G. C. Papavassiliou, *Synth. Met.*, **42**, 1715 (1991).

90. A. M. Kini, T. Mori, U. Geiser, S. M. Budz and J. M. Williams, *J. Chem. Soc., Chem. Commun.*, 647 (1990).

91. G. C. Papavassiliou, D. Lagouvardos, V. Kakoussis, G. Mousdis, A. Terzis, A. Hountas, B. Hilti, C. Mayer, J. Zambounis, J. Pfeiffer and P. Delhaes, *Organic Superconductivity*, Plenum, New York (1990), p. 367.

92. L. Ducasse, G. Mousdis, M. Fettouhi, L. Ouahab, J. Amiell and P. Delhaes, *Synth. Met.*, **56**, 1995 (1993).

93. S. V. Konovalikhin, G. V. Shilov, E. I. Zhilyaeva, S. A. Torunova and R. N. Lyubovskaya, *Russ. J. Coord. Chem.*, **26**, 89 (2000).

94. T. Aoki, G. Saito, H. Yamochi and M. Maesato, *Mol. Cryst. Liq. Cryst.*, **376**, 201 (2002).

95. a) H. Sasaki, T. Kondo, K. Kamoshida, Y. Sacho and G. Saito, *Synth. Met.*, **102**, 1626 (1999); b) G. Saito, H. Sasaki, T. Aoki, Y. Yoshida, A. Otsuka, H. Yamochi, O. O. Drozdova, K. Yakushi, H. Kitagawa and T. Mitani, *J. Mater. Chem.*, **12**, 1640 (2002).

96. H. Ohnuki, T. Noda, M. Izumi, T. Imakubo and R. Kato, *Supramol. Sci.*, **4**, 413 (1997).

97. a) J. Yamada, H. Akutsu, S. Nakatsuji, H. Nishikawa, I. Ikemoto and K. Kikuchi, *Mol. Cryst. Liq. Cryst.*, **356**, 253 (2001); b) K. Kikuchi, S. Ikeda, H. Nishikawa, T. Kodama, I. Ikemoto and J. Yamada, *Synth. Met.*, **120**, 901 (2001).

98. G. C. Papavassiliou, D. J. Lagouvardos and V. C. Kakoussis, *Z. Naturforsch.*, **46b**, 1730 (1991).

99. a) T. Imakubo, Y. Okano, H. Sawa and R. Kato, Reizo, *J. Chem. Soc., Chem. Commun.*, 2493 (1995); b) Y. Okano, K. Yamamoto, T. Imakubo, H. Sawa and R. Kato, *Synth. Met.*, **86**, 1829 (1997).

100. J.-M. Fabre, A. Javidan, L. Binet, J. Ramos and P. Delhaes, *Synth. Met.*, **86**, 1889 (1997).

101. T. Imakubo and K. Kobayashi, *J. Mater. Chem.*, **8**, 1945 (1998).

102. H. Ohnuki, M. Nagata, Y. Ishizaki, T. Imakubo, K. Kobayashi, R. Kato and M. Izumi, *Synth. Met.*, **102**, 1699 (1999).

103. D. J. Lagouvardos and G. C. Papavassiliou, *Z. Naturforsch., B: Chem. Sci.*, **47**, 898 (1992).

104. a) J. Hellberg, M. Moge, D. Bauer and J.-U. von Schuetz, *J. Chem. Soc., Chem. Commun.*, 817 (1994); b) J. Hellberg, M. Moge, D. Bauer and J.-U. von Schuetz, *Synth. Met.*, **70**, 1135 (1995).

105. J. Hellberg, M. Moge, H. Schmitt and J.-U. von Schuetz, *J. Mater. Chem.*, **5**, 1549 (1995).

106. M. Moge, J. Hellberg, K. W. Tornroos, H. Schmitt and J.-U. von Schutz, *Synth. Met.*, **86**, 1877 (1997).
107. a) F. Darviche, M. T. Babonneau, H. J. Cristau, E. Torreilles and J.-M. Fabre, *Synth. Met.*, **102**, 1662 (1999); b) H. J. Cristau, F. Darviche, M.-T. Babonneau, J.-M. Fabre and E. Torreilles, *Tetrahedron*, **55**, 13029 (1999).
108. M. Moge, J. Hellberg, K. W. Toernroos and J.-U. von Schuetz, *Adv. Mater.*, **8**, 807 (1996).
109. T. Inoue, H. Yamochi, G. Saito and K. Matsumoto, *Synth. Met.*, **70**, 1139 (1995).
110. Y. Siquot, P. Frere, T. Nozdryn, J. Cousseau, M. Salle, M. Jubault, J. Orduna, J. Garin and A. Gorgues, *Tetrahedron Lett.*, **38**, 1919 (1997).

5

Selenium Analogues of TTFs

5.1 Introduction

The importance of the selenium analogues of tetrathiafulvalenes (TTFs) in the field of organic conductors and superconductors as exemplified by TSF and TMTSF is widely recognized (See Table 5.1 for list of abbreviations). The science of organic superconductors was initiated by the

Table 5.1   Abbreviated Names for Selenium Analogues of TTFs Used in This Chapter

abbreviation	compound
TSF	tetraselenafulvalene
TMTSF	tetramethyltetraselenafulvalene
BEDSe-TTF	bis(ethylenediseleno)tetrathiafulvalene
DMET	dimethyl(ethylenedithio)diselenadithiafulvalene
BEDSe-TSF	bis(ethylenediseleno)tetraselenafulvalene
BETS	bis(ethylenedithio)tetraselenafulvalene
TMET-STF	trimethylene(ethylenedithio)diselenadithiafulvalene
MDT-TSF	methylenedithiotetraselenafulvalene
BMDSe-TTF	bis(methylenediseleno)tetrathiafulvalene
BPDSe-TTF	bis(propylenediseleno)tetrathiafulvalene
MDSe-TTF	methylenediselenotetrathiafulvalene
EDSe-TTF	ethylenediselenotetrathiafulvalene
PDSe-TTF	propylenediselenotetrathiafulvalene
STF	diselenadithiafulvalene
EDS-EDT-TTF	ethylenediseleno(ethylenedithio) tetrathiafulvalene
CHET-STF	cyclohexenyl(ethylenedithio)diselenadithiafulvalene
BPDT-TSF	bis(propylenedithio)tetraselenafulvalene
DMET-TSF	dimethyl(ethylenedithio)tetraselenafulvalene
EDT-TSF	ethylenedithiotetraselenafulvalene
PEDT-TSF	pyrazino(ethylenedithio)tetraselenafulvalene
EDST	ethylenedithiodiselenadithiafulvalene
BEDT-STF	bis(ethylenedithio)diselenadithiafulvalene
DED	thiadiazole(ethylenedisulfanyl)diselenadithiafulvalene
EDTS	ethylenedithiodiselenadithiafulvalene
TMVT	trimethylene(vinylenedithio)diselenadithiafulvalene
DMVT	dimethyl(vinylenedithio)diselenadithiafulvalene
EDVT	ethylenedithio(vinylenedithio)diselenadithiafulvalene
DSDTeF	diselenaditellurafulvalene
BES-TTF	bis(ethyleneseleno)tetrathiafulvalene
BES-TSF	bis(ethyleneseleno)tetraselenafulvalene
BPT-TSF	bis(propylenethio)tetraselenafulvalene
PDT-TSF	propylenedithiotetraselenafulvalene
BMDT-TSF	bis(methylenedithio)tetraselenafulvalene
MDSe-TSF	methylenediselenotetraselenafulvalene
EDSe-TSF	ethylenediselenotetraselenafulvalene
BMDSe-TSF	bis(methylenediseleno)tetraselenafulvalene

discovery of the first organic superconductor, (TMTSF)$_2$PF$_6$, in 1979.[1] The substitution of sulfur atoms in the TTF framework by selenium atoms is an effective way to enhance intermolecular interaction and dimensionality, stabilizing the metallic state of the resulting radical cation salts.[2] Thus the development of new seleno-analogues of TTFs has been a major focus of interest in synthetic TTF chemistry. However, replacement of the desired sulfur atoms in TTF derivatives by selenium atoms was not so easy owing to lack of suitable building blocks and synthetic methods. The major selenium reagents frequently used in the synthesis of the seleno-analogues of TTFs are listed in Table 5.2, together with the sulfur counterparts. As seen in the table, selenium reagents are more expensive and less available than their sulfur counterparts, and furthermore, their high toxicity and bad smell are always problematic in synthetic experiments.

Table 5.2   Representative Selenium Reagents and Their Sulfur Counterparts

Reagent	Price^a (x 10^2 JPY^b)	Reagent	Price^a (x 10^2 JPY^b)
Se	112 /250 g	S	33 /500 g
CSe$_2$	–^{c}	CS$_2$	42 /1000 mL
SeO$_2$	175 /500 g	SO$_2$	–^{c}
H$_2$Se	–^{c}	H$_2$S	486 /227 g
MeSe-SeMe	46 /1 g	MeS-SMe	154 /1000 mL
KSeCN	139 /50 g	KSCN	35 /100 g

^{a}Prices on Aldrich website.
^{b}JPY = Japanese Yen (1 USD = *ca.* 130 JPY in April 2002)
^{c}Not commercially available.

Until around 1990, few selenium analogues of TTFs, such as the parent TSF,[3,4] TMTSF,[5] BEDSe-TTF,[6] DMET[7] and BEDSe-TSF[8](Fig. 5.1), were reported. The synthetic methods used were very diverse, *e.g.*, Engler's and Cava's syntheses of TSF *via* 1,3-diselenole-2-selone and 2-methylene-1,3-diselenole, respectively, (Scheme 5.1a), the synthesis of TMTSF *via* a seleno-carbamate intermediate (Scheme 5.1b) and reduction of carbon diselenide to prepare 1,3-diselenole-2-selone-4,5-diselenolate (dsis), which was converted into BEDSe-TSF (Scheme 5.1c). These methods are very useful for the synthesis of specific compounds, but mostly inapplicable to the synthesis of different types of compounds or even slightly modified derivatives.

In the last decade, effective reactions and protocols to construct and connect the building blocks of the seleno-analogues of TTF have been newly developed. Specifically, these reactions and protocols are roughly classified into the following five categories:

1) protection/deprotection protocol of tetrachalcogenafulvalene-chalcogenolate anions,

TSF   TMTSF   BEDSe-TTF

DMET   BEDSe-TSF

Fig. 5.1   Seleno-analogues of TTFs developed before the 1990's.

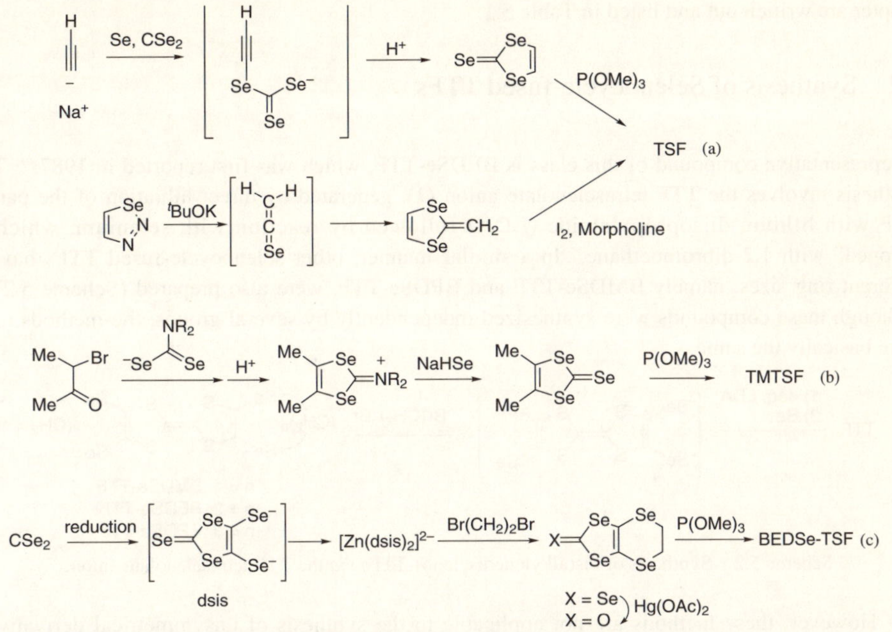

Scheme 5.1 Initial approaches used in the synthesis of (a) TSF, (b) TMTSF and (c) BEDSe-TSF.

2) titanocene method for the synthesis of 1,3-diselenole-2-chalcogenone derivatives,
3) Me$_3$Al-promoted cross-coupling reaction,
4) one-pot synthesis of 1,3-diselenole-2-selone derivatives,
5) outer thio- or seleno-cycle formation *via* intramolecular transalkylation reaction on chalcogen atom(s).

Using these procedures and their combinations, a range of new seleno-analogues of TTFs have been efficiently synthesized, involving important electron donors such as BETS,[9] TMET-STF[10] and MDT-TSF[11](Fig. 5.2), which produce superconducting radical cation salts. This chapter reviews the novel synthetic procedures that have stimulated and promoted research on the selenium analogues of TTFs.

Since the aim of this chapter is not to cover the seleno-analogues of TTFs completely, it is advisable for readers, if necessary, to refer to other reviews, for example, a comprehensive review of the synthetic TTF chemistry written by Schukat and Fanghänel[12] and Cowan's chapter in Patai's book describing selenium-containing electron donors developed earlier.[13] Two types of selenium analogues of TTFs, *i.e.*, halogen-substituted analogues and those containing oxygen, are described in Chapters 3 and 4, respectively.

As noted above, the abbreviations of the selenium analogues of TTFs described in this

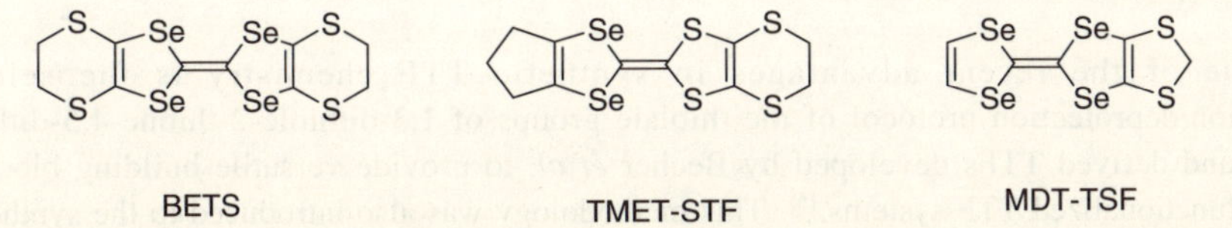

BETS   TMET-STF   MDT-TSF

Fig. 5.2 Advanced selenium-containing TTF type donors forming superconducting radical cation salts.

chapter are written out and listed in Table 5.1.

5.2 Synthesis of Selenocycle-fused TTFs

A representative compound of this class is BEDSe-TTF, which was first reported in 1987.[6] The synthesis involves the TTF tetraselenolate anion (**1**), generated by direct lithiation of the parent TTF with lithium diisopropylamide (LDA) followed by reaction with selenium, which is "capped" with 1,2-dibromoethane. In a similar manner, other selenocycle-fused TTFs having different ring sizes, namely BMDSe-TTF and BPDSe-TTF, were also prepared (Scheme 5.2).[14] Although these compounds were synthesized independently by several groups, the methods used were basically the same.

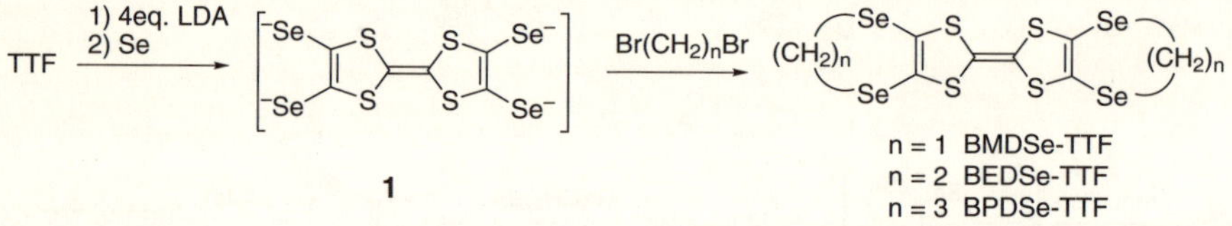

Scheme 5.2   Synthesis of bis(alkylenediseleno)-TTFs *via* the TTF tetraselenolate anion.

However, these methods are not applicable to the synthesis of unsymmetrical derivatives. Instead of the direct functionalization of TTF, the introduction of selenium atoms into 1,3-dithiole-2-thione (**2**) was developed (Scheme 5.3).[15] 1,3-Dithiole-2-thione dilithiated with LDA was reacted with selenium, and the resulting 1,3-dithiole-2-thione-4,5-diselenolate (dsit) was trapped as a zinc complex. The outer selenocycle was conveniently constructed by capping with a dibromoalkane, and the resulting 1,3-dithiole-2-chalcogenones (**3** and **4**) were utilized in phosphite-mediated cross-coupling reactions with 4,5-dimethoxycarbonyl-1,3-dithiole-2-thione (**5**). In this way, a series of monoselenocycle-fused TTFs (MDSe-TTF, EDSe-TTF, and PDSe-TTF) were prepared by Papavassiliou *et al.*[15c]

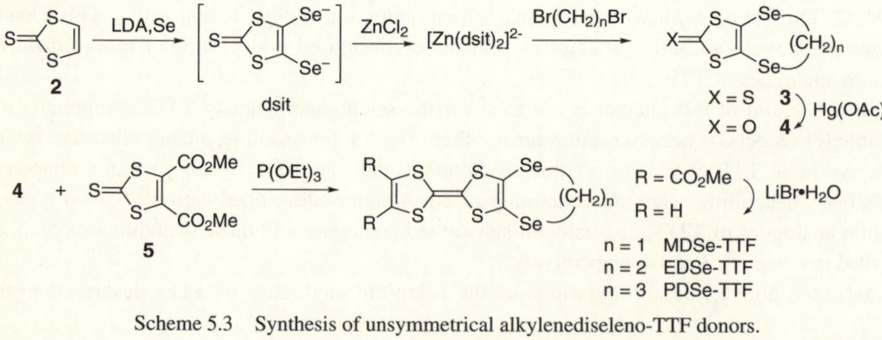

Scheme 5.3   Synthesis of unsymmetrical alkylenediseleno-TTF donors.

One of the recent advantages in synthetic TTF chemistry is the selective protection/deprotection protocol of the thiolate groups of 1,3-dithiole-2-thione-4,5-dithiolate (dmit) and derived TTFs developed by Becher *et al.* to provide versatile building blocks for highly functionalized TTF systems.[16] This methodology was also introduced to the synthesis of selenium analogues of TTFs; the TTF and STF derivatives **7** and **10** bearing thiolate moieties protected with cyanoethyl groups were developed. As shown in Scheme 5.4, **7** was synthesized through a phosphite-mediated cross-coupling between the thione **3** (n = 2) and the ketone **6**, while, on the other hand, the synthesis of **10** was effected by a pseudo-Wittig condensation between the phosphonim salt **8** and the selenonium salt **9**.[17] These compounds are useful intermediates for the synthesis of EDS-EDT-TTF and DMET.

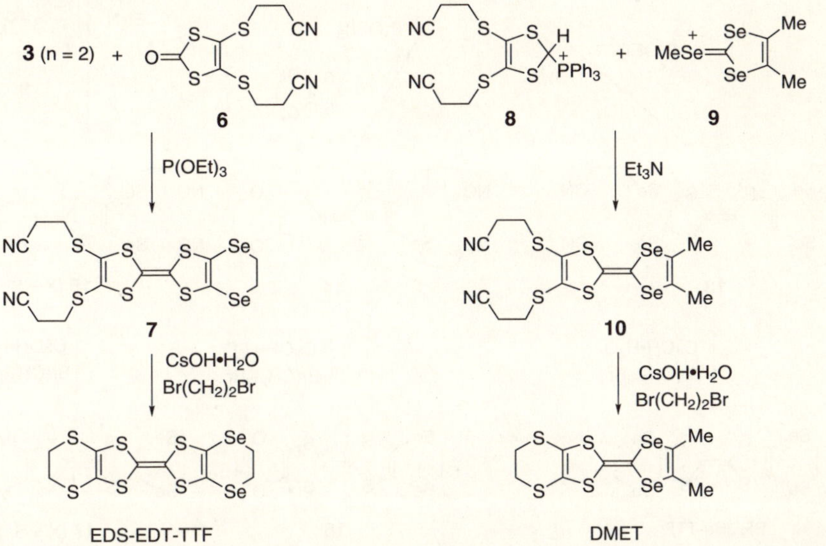

Scheme 5.4   Use of protected TTF- and STF-thiolates in the synthesis of EDS-EDT-TTF and DMET.

The cyanoethyl-protected dsit, 4,5-bis(cyanoethylseleno)-1,3-dithiole-2-thione (**11**), and its oxo derivative **12** were also synthesized as shown in Scheme 5.5.[18] The latter compound was readily converted into the corresponding tetrakis(cyanoethylseleno)-TTF (**13**). Treatment of **13** with sodium ethoxide or cesium hydride monohydrate followed by a reaction with 1,2-dibromoethane afforded BEDSe-TTF in a moderate yield. The cyanoethyl-protected dsit substructure was also introduced into unsymmetrical TTF derivatives **14** and **15** (X = S) and a derivative of STF **15** (X = Se) *via* a phosphite-promoting cross-coupling or a pseudo-Wittig condensation as shown in Scheme 5.5. Effective construction of the outer ethylenediseleno moiety to provide unsymmetrical TTFs **16**, **17** (X = S) and a derivative of STF **17** (X = Se) was achieved in good yields by the well-established deprotection/realkylation protocol as in the case of TTF thiolates.

5.3   Titanocene Method: Synthesis of TSFs and STFs

Transition metal complexes containing polychalcogenide chelates have been recognized since the late 1960s,[19] and the synthesis of 4,5-bis(methoxycarbonyl)-1,3-diselenole-2-thione (**19e**, X = S) using dicyclopentadienyltitanium (II) pentaselenide (Cp$_2$TiSe$_5$) was reported in 1982 (Scheme 5.6).[20] Kato *et al.* first employed this type of titanocene complex intermediates **18** in the synthesis of selenium-containing electron donors such as BETS,[21] TMET-STF[22] and CHET-STF.[23]

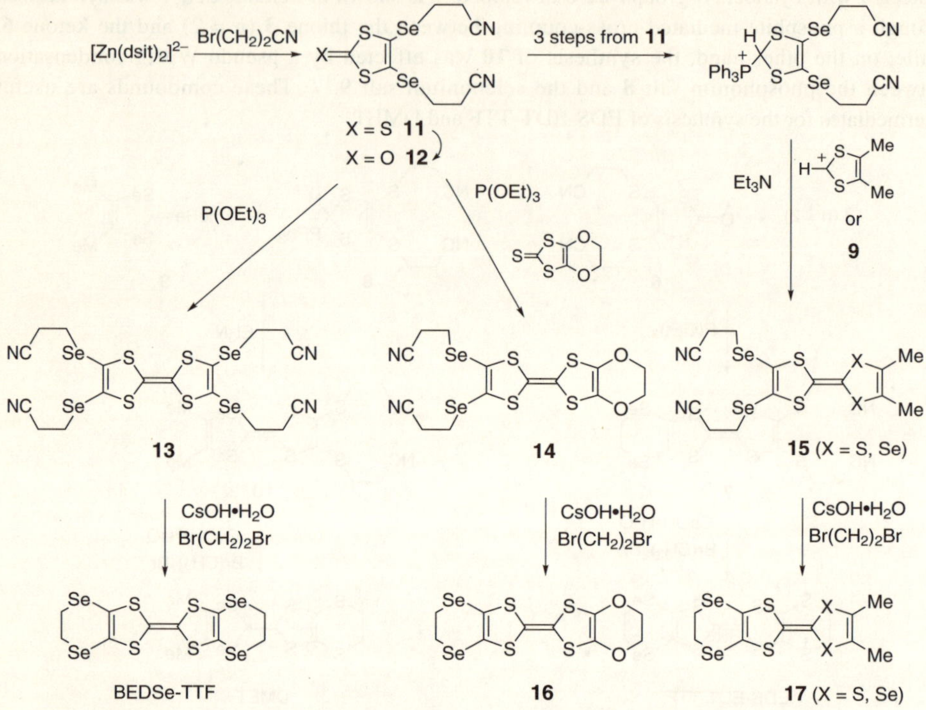

Scheme 5.5   Syntheses of seleno-analogues of TTFs *via* protected TTF- and STF- selenolates.

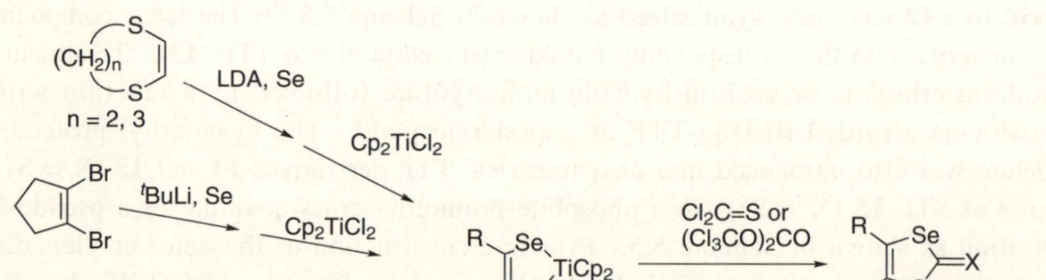

Scheme 5.6   Syntheses of 1,3-diselenole-2-chalcogenones **19** from titanocene complexes **18**.

One great merit of the titanocene method is that it avoids the use of highly toxic and less accessible H_2Se or CSe_2 as a selenium source. There are two ways to introduce selenium atoms into titanocene complexes **18** (Scheme 5.6). The first is a reaction of lithiated intermediates with selenium powder producing diselenolate anions, which are trapped with dicyclopentadienyltitanium (II) dichloride, leading to **18a–c**. The second is the use of Cp_2TiSe_5: the cyclohexyl-substituted 1,2,3-selenadiazole and dimethyl acetylenedicarboxylate[24] are reacted with Cp_2TiSe_5, giving **18d** and **18e**, respectively.

The titanocene complexes **18** were successfully converted to the corresponding 1,3-diselenole-2-chalcogenones **19** in good yields by the action of triphosgene or thiophosgene. 1,3-Diselenole-2-chalcogenone derivatives thus obtained were widely used in the syntheses of not only symmetrical BETS and BPDT-TSF,[25] but also unsymmetrical TSF and STF derivatives, as shown in Fig. 5.3[22–24,26]. Based on these selenium-containing donors, vast numbers of organic metals have been synthesized. It should be noted that three donors developed by this method, BETS, DMET-TSF and TMET-STF, are important electron donors capable of producing organic superconductors.

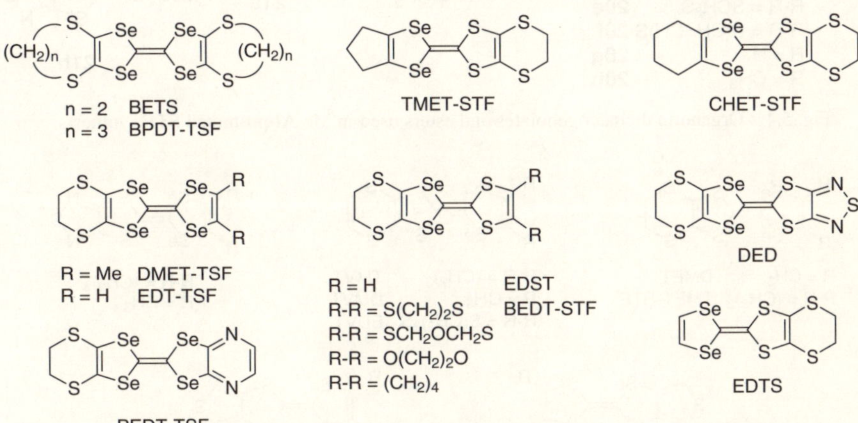

Fig. 5.3   Seleno-analogues of TTFs synthesized *via* the titanocene method.

5.4   Synthesis of STFs *via* Me₃Al-promoted Reaction

Most conventional synthesis of unsymmetrical tetrachalcogenafulvalenes **22** including STFs rely on phosphite-promoted cross-coupling reactions as the key step.[12] The yields of the desired cross-coupling products are, however, usually not high for statistical reasons, and furthermore, the purification is so difficult as to require tedious chromatographic separation from symmetrical coupling products. Thus, selective cross-coupling reactions for unsymmetrical tetrachalcogenafulvalenes have been developed.[27] Among such reactions, a Me₃Al-promoted reaction between the organotin dichalcogenolate **20** and the ester **21** developed by Yamada *et al.* is the most suitable for the synthesis of various STF derivatives (Scheme 5.7).[28]

As shown in Fig. 5.4, many organotin dichalcogenolates **20a–h** and esters **21a–h** having different substituents were readily derived from the corresponding 1,3-dichalcogenol-2-ones, and their versatile combinations made it possible to synthesize various kinds of STF derivatives

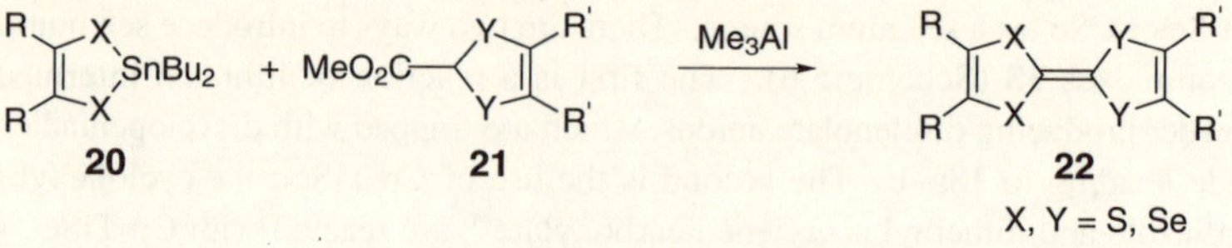

Scheme 5.7   Synthesis of STF derivatives *via* Me$_3$Al-promoted reaction.

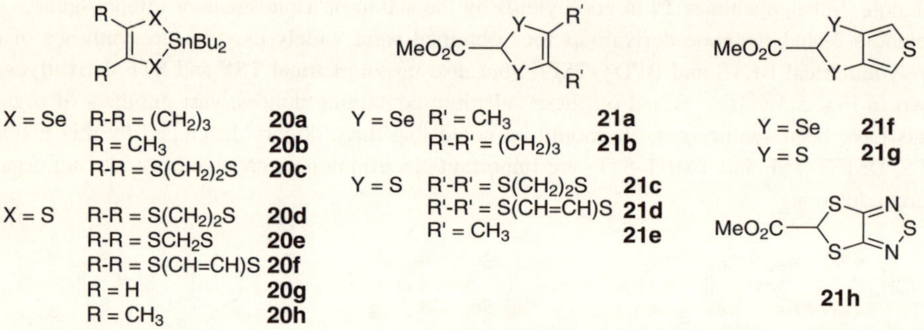

Fig. 5.4   Organotin dichalcogenolates and esters used in Me$_3$Al-promoted STF synthesis.

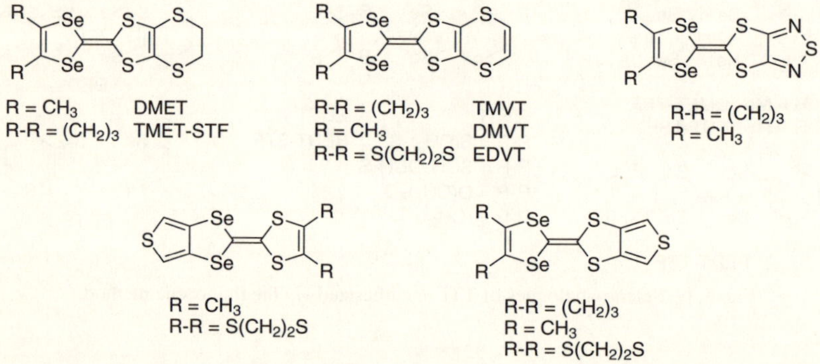

Fig. 5.5   STF derivatives synthesized by Me$_3$Al-promoted coupling reactions.

illustrated in Fig. 5.5 in moderate yields.[29]

Several important STF derivatives such as DMET can be synthesized by this method. Compared to the classical method *via* a phospite-promoted cross-coupling reaction,[7] this method has apparent advantages in terms of high selectivity and easy purification.

5.5   One-pot Synthesis of 1,3-Diselenole-2-selones

In the first synthesis of TSF reported by Engler *et al.*, the key intermediate, 1,3-diselenole-2-selone, was synthesized by a cyclization reaction of sodium acetylide with selenium and carbon diselenide (CSe$_2$) in liquid ammonia (Scheme 5.1a).[3] This one-pot reaction is straightforward, but had not been taken up until recently because of its low yield (15–25%) and requirement of highly toxic and less available CSe_2. We focused on its hidden potential and reexamined it with the following experimental modifications. The reaction medium was changed from liquid ammonia to THF and the substrate from sodium acetylide to lithium trimethysilylacetylide *in situ* generated from trimethysilylacetylene (**23**, R^1 = SiMe₃) and *n*-butyllithium. These modifications were very effective not only in making the experimental operation easy but also in enhancing the isolated yield of 1,3-diselenole-2-selone (**25**, R^1 = H) to 94%. In this reaction, the trimethylsilyl group is detached during work-up under the aqueous alkaline condition. Furthermore, monosubstituted acetylene derivatives bearing phenyl-, hexyl-, methylthio-, tetrahydropyranyl (THP)-protected hydroxymethy- and acetal-protected formyl-substituents were used as substrates of this reaction to form the corresponding monosubstituted 1,3-diselenole-2-selones (**25**) in good to excellent yields (Scheme 5.8).[30]

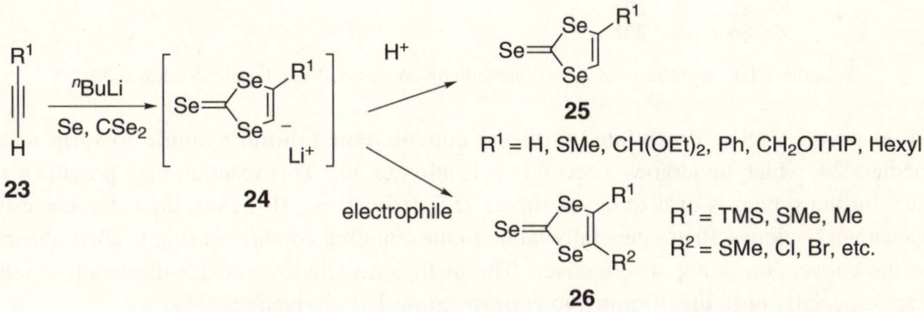

Scheme 5.8   One-pot formation of 1,3-diselenole-2-selones **25** and **26**.

This procedure is also applicable to the synthesis of bis-functionalized 1,3-diselenole-2-selone derivatives (**26**). In the one-pot reaction, a cyclic vinyl anion **24** is formed as the reaction product before quenching, which can thus be functionalized with appropriate electrophiles such as alkylthiocyanates and halogenating reagents, giving various bis-substituted 1,3-diselenole-2-selones.[31]

This methodology has been successfully applied to the synthesis of a selenium-tellurium hybrid system. Replacement of selenium by tellurium in the initial step of the one-pot reaction of lithium trimethylsilylacetylide gave 1,3-selenatellurole-2-selone (**27**) in 83% yield. This hybrid heterocycle was readily converted into DSDTeF (Scheme 5.9).[32]

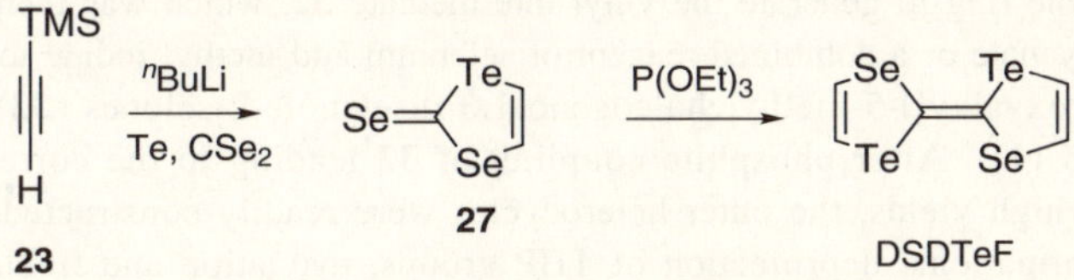

Scheme 5.9   Synthesis of DSDTeF.

During further study by our group on the scope of the one-pot reaction, we discovered an interesting tandem cyclization. When lithium trimethylsilylacetylide was treated successively with selenium, CSe_2 and α,ω-bis(chalcogenocyanato)alkanes, the major products were 4,5-alkylenedichalcogeno-1,3-diselenole-2-selones (**30a–f**). As shown in Scheme 5.10, all the methylene-, ethylene- and propylene-dichalcogeno derivatives were isolated in high yields.[33] By this reaction, 4,5-bis(methylthio)-1,3-diselenole-2-selone (**26**) was also synthesized using excess methylthiocyanate as an alkyl-sulfurating reagent. In the reaction mechanism, it is assumed that

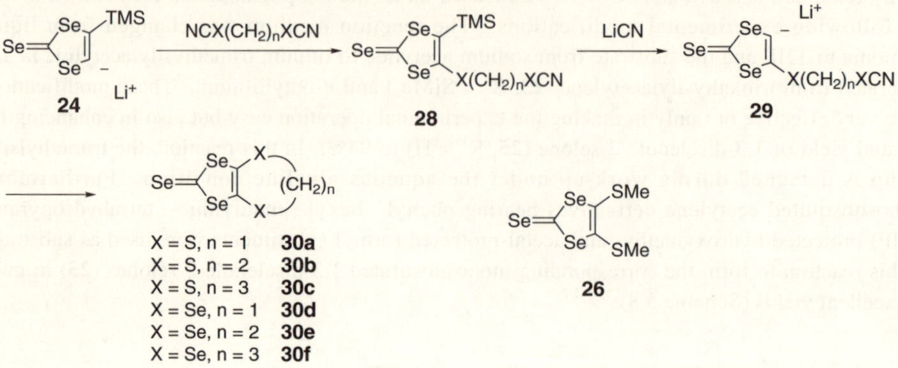

Scheme 5.10   Synthesis of 4,5-alkylenedichalcogeno-1,3-diselenole-2-selones **30**.

X = S, n = 1 **30a**
X = S, n = 2 **30b**
X = S, n = 3 **30c**
X = Se, n = 1 **30d**
X = Se, n = 2 **30e**
X = Se, n = 3 **30f**

the initial product **28** is desilylated with the concomitant lithium cyanide to form another intermediate **29**, which undergoes a second cyclization to **30**. This reaction may provide a rapid approach for the synthesis of heterocycle-fused TSF derivatives. However, the heterocycle-fused 1,3-diselenole-2-selones **30** are generally labile to the coupling condition using trialkyl phosphite, so that the conversion is not so effective. The methylenedichalcogeno-1,3-diselenole-2-selones **30a,d** are especially unstable, forming no corresponding TSF derivatives.

5.6   Outer Thio- and Seleno-cycle Formation *via* Transalkylation Reaction on Chalcogen Atom(s)

As seen in the last section, two different substituents are conveniently introduced in 1,3-diselenole-2-selones by the one-pot synthesis. We tried to apply this method to the synthesis of heterocycle-fused TSFs by introducing appropriate functional groups, which can be transformed to the heterocyclic rings. As a result, the following reaction sequence turned out to be very effective.

THP-protected 3-butyn-1-ol and 4-pentyn-1-ol (**31**) were employed in the one-pot synthesis of the 1,3-diselenole ring to generate the vinyl intermediate **32**, which was then functionalized with methyl thiocyanate or a combined reagent of selenium and methyl iodide to give the THP-protected 4-(hydroxyalkyl)-5-methylchalcogeno-1,3-diselenole-2-selones (**33**) in reasonable yields (Scheme 5.11). After phosphite coupling of **33** leading to the corresponding TSF derivatives **34** in high yields, the outer heterocycles were readily constructed by a series of functional transformations: deprotection of THP groups, tosylation and finally ring-closing reaction *via* transalkylation on the chalcogen atoms, giving the ethylenechalcogeno- and propyrenechalcogeno-substituted TSF derivatives **37** and **38**.[34]

By this method, various TTF and STF derivatives **39** and **40**, illustrated in Fig. 5.6, were obtained from the same starting materials using different chalcogen reagents in the one-pot reaction.[35]

The great versatility of this synthetic method is demonstrated by developing a series of selenium variants **42** of bis(ethylenedithio)tetrathiafulvalene (BEDT-TTF) from the sulfur-containing acetylene **41**, as shown in Scheme 5.12. Thus, it follows that this procedure makes it possible to introduce selenium atoms selectively at desired positions of the BEDT-TTF

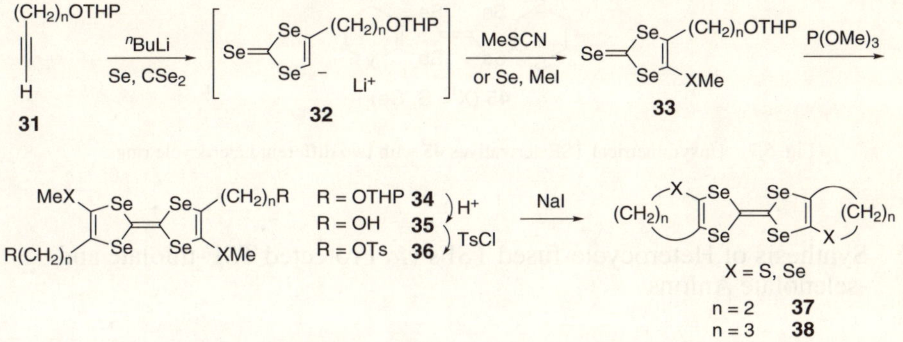

Scheme 5.11   Synthesis of heterocycle-fused TSF derivatives **37** and **38**.

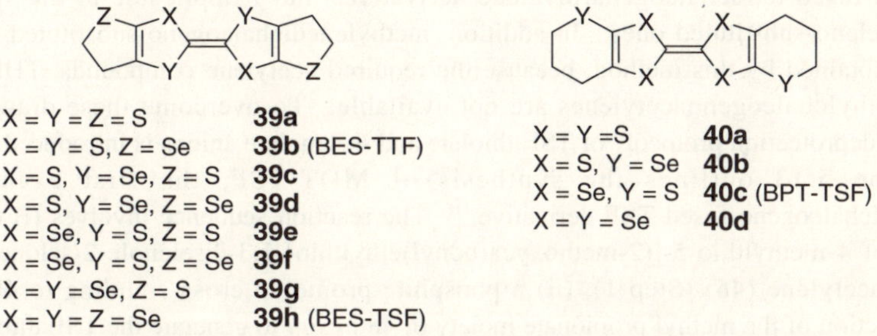

X = Y = Z = S   **39a**
X = Y = S, Z = Se   **39b** (BES-TTF)
X = S, Y = Se, Z = S   **39c**
X = S, Y = Se, Z = Se   **39d**
X = Se, Y = S, Z = S   **39e**
X = Se, Y = S, Z = Se   **39f**
X = Y = Se, Z = S   **39g**
X = Y = Z = Se   **39h** (BES-TSF)

X = Y = S   **40a**
X = S, Y = Se   **40b**
X = Se, Y = S   **40c** (BPT-TSF)
X = Y = Se   **40d**

Fig. 5.6   Ethylenechalcogeno- and propylenechalcogeno-substituted TTF type donors **39** and **40** synthesized by transalkylation reactions.

X = Y = Z = S
X = Y = S, Z = Se
X = S, Y = Se, Z = S
X = S, Y = Se, Z = Se
X = Se, Y = S, Z = Se
X = Y = Se, Z = S
X = Y = Z = Se

X, Y = S or Se

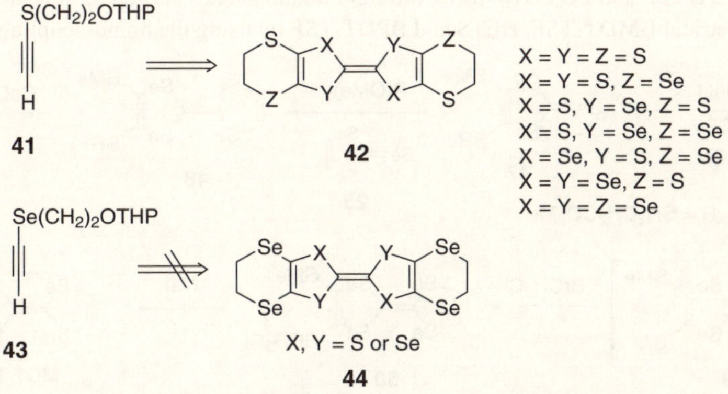

Scheme 5.12   Synthesis of selenium variants **42** of BEDT-TTF.

framework.[36)] However, the synthetic attempts for BEDSe-type derivatives **44** starting from the selenium-containing acetylene **43** were unsuccessful owing to the difficult construction of the ethylenediseleno moiety *via* transalkylation reaction (Scheme 5.12).

Finally, unsymmetrical complicated derivatives **45** were also synthesized.[37)] In this case, for the sake of efficient purification of the cross-coupling products, one of the protecting groups in the 1,3-diselenole-2-selones is replaced by a less polar trialkylsilyl group than the THP group.

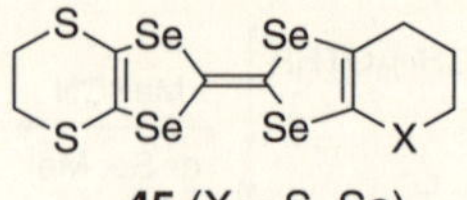

45 (X = S, Se)

Fig. 5.7   Unsymmetrical TSF derivatives **45** with two different heterocycle rings.

5.7   Synthesis of Heterocycle-fused TSFs *via* Protected TSF-thiolate and -selenolate Anions

As discussed in the above section, the one-pot method is very powerful in synthesizing versatile heterocycle-fused tetrachalcogenafulvalene derivatives, but inapplicable to the synthesis of ethylenediseleno-substituted ones. In addition, methylenedichalcogeno-substituted derivatives cannot be obtained by this method, because the required acetylene compounds, THP-protected hydroxymethylchalcogenoacetylenes are not available. To overcome these drawbacks, the "protection/deprotection protocol" of TSF-thiolate and -selenolate anions is introduced.[38]

Scheme 5.13 outlines the synthesis of MDT-TSF, the first example of a methylenedichalcogeno-fused TSF derivative.[11] The reaction sequence involves (i) the one-pot formation of 4-methylthio-5-[(2-methoxycarbonyl)ethylthio]-1,3-diselenole-2-selone (**47**) from methylthioacetylene (**46**) (Step 1), (ii) a phosphite-promoted cross-coupling to **48** (Step 2), (iii) deprotection of the methyl propionate moiety of **48** *in situ* to generate the TSF thiolate **49** and subsequent realkylation with bromochloromethane to **50** (Step 3) and (iv) a ring-closing reaction *via* transalkylation reaction on the outer sulfur atom (Step 4).

The further usefulness of the combination of these reactions is exemplified by the successful synthesis of EDT-TSF and PDT-TSF using different dihaloalkanes at Step 3. It is also possible to prepare symmetrical BMDT-TSF, BETS and BPDT-TSF by using the homo-coupling at Step 2.[39]

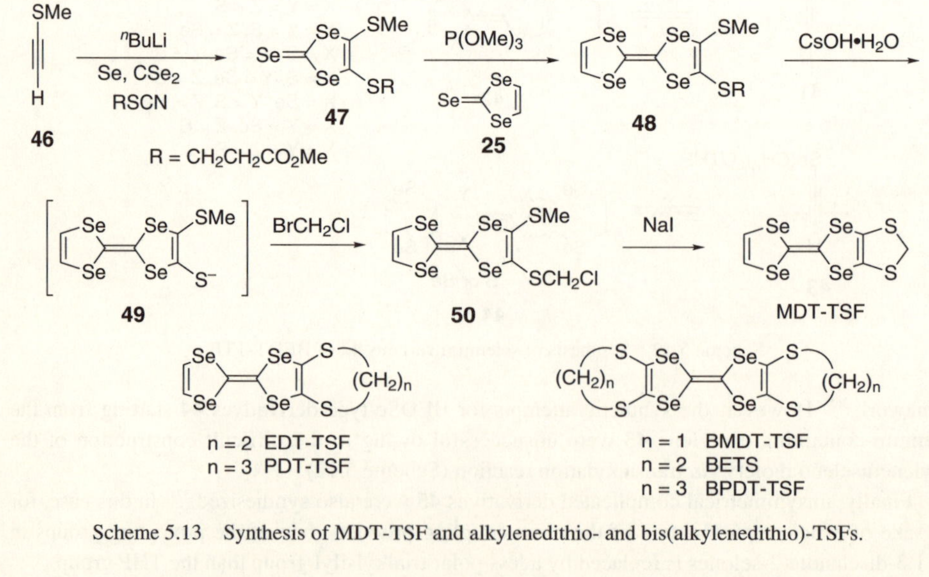

Scheme 5.13   Synthesis of MDT-TSF and alkylenedithio- and bis(alkylenedithio)-TSFs.

For the synthesis of alkylenediseleno-substituted TSFs, we developed a dsis version of "protection/deprotection protocol" as illustrated in Scheme 5.14.[40a] The key intermediate, dsis anion, readily prepared by dilithiation of 1,3-diselenole-2-selone (**25**) with LDA followed by addition of selenium,[40] was once trapped as 4,5-bis[(2-methoxycarbonyl)ethylseleno]-1,3-diselenole-2-selone (**51**), and its trimethylphophite-promoted coupling reaction afforded unsymmetrical- or symmetrical-protected TSF selenolates (**52** or **53**) in reasonable yields. The final outer selenocycle formation was readily achieved in the same manner as for the case of TTF thiolates,[16] giving MDSe-TSF, EDSe-TSF, BMDSe-TSF and BEDSe-TSF. Although BEDSe-TSF was reported earlier by Engler *et al.* (Scheme 5.1c),[8] the total yield of the present method (*ca.* 40% from CSe_2) exceeds the previous one (*ca.* 10%).

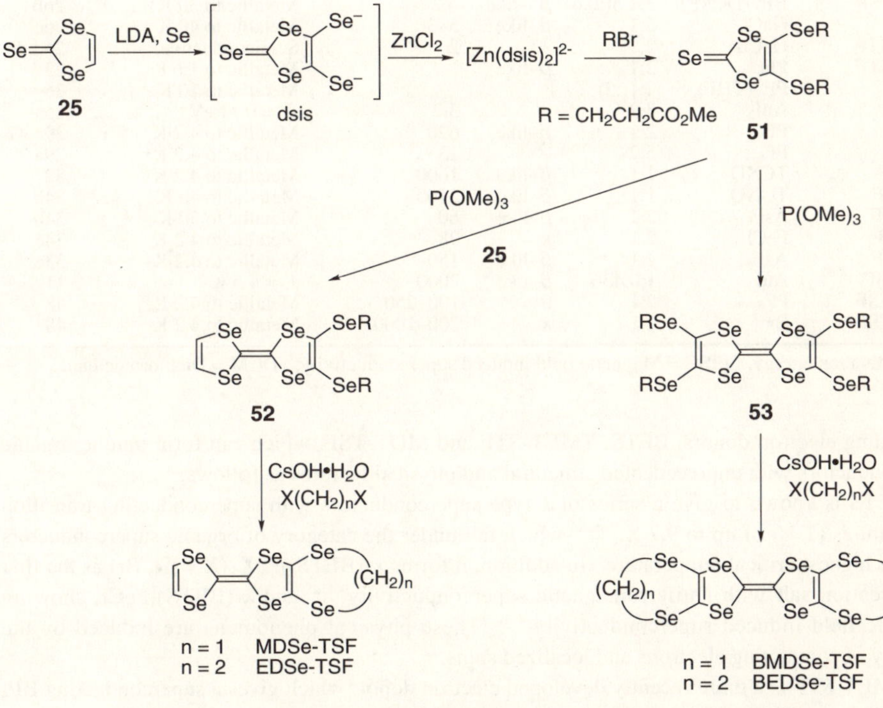

Scheme 5.14   Synthesis of alkylenediseleno- and bis(alkylenediseleno)-TSFs.

5.8   Conducting Salts of New Selenium Analogues of TTFs

Many of the selenium analogues of TTFs presented in this chapter have been utilized in the formation of highly conducting charge-transfer(CT) salts, and representative examples are listed in Table 5.3. Generally speaking, there is a tendency for the STF and TSF derivatives, irrespective of the kind of the chalcogen atoms in the outer heterocycles, to produce metallic salts that remain stable down to very low temperatures, whereas TTF derivatives with the outer selenocycles can give either semiconductive salts or metallic salts with metal-insulator transitions. It should also be emphasized that among the seleno-analogues of TTFs, there are three

Table 5.3   Representative Radical Cation Salts Based on Seleno-analogues of TTF

Donor	Anion	D:A	Structure	σ_{rt} / S cm^{-1}	Remarks	Ref.
BEDSe-TTF	I$_3$	2:1	β	1	Metallic to 260 K	41
BEDSe-TTF	I$_2$Br	2:1	β'	3×10^{-3}	$E_a^a = 0.09$ eV	41
BEDSe-TTF	CuN(CN)$_2$Br	2:1	κ	0.1	$T_c = 7.5$ K (0.15 GPa)	42
MDSe-TTF	Au(CN)$_2$	2:1	θ	12–41	Metallic to 15 K	43
EDS-EDT-TTF	Ag$_2$(CN)$_3$	2:1	κ	3.0×10^{-4}	$E_a = 0.12$ eV	44
BETS	GaCl$_4$	2:1	λ		$T_c = 8$ K	45a
BETS	GaBr$_{1.5}$Cl$_{2.5}$	2:1	λ		$T_c = 9.7$ K (3 kbar)	45b
BETS	FeBr$_4$	2:1	κ	60	$T_c = 1.1$ K	45c,d
BETS	FeCl$_4$	2:1	λ	20	FISC^b	45c,e
DMET-TSF	AuI$_2$	2:1	β–like		$T_c = 0.58$ K	26a
DMET-TSF	I$_3$	2:1	β–like		$T_c = 0.4$ K	26a
TMET-STF	BF$_4$	2:1		50	$T_c = 4.1$ K	10
PEDT-TSF	BF$_4$ (DCM^c)	3:1.5(:0.6)	β''–like	6.2	Metallic to 30 K	26b
EDST	GaCl$_4$	2:1	β–like	5–30	Metallic to 40 K	26c
BEDT-STF	GaCl$_4$	2:1	λ	2	Semiconductive	26d,46
CHET-STF	PF$_6$	2:1	β–like		Metallic to 1.6 K	23
DED	PF$_6$ (THF)	2:1(:2)			Metallic to 20 K	26g
DMVT	AuI$_2$	3:2	–	3.2	$E_a = 0.11$ eV	26a
TMVT	PF$_6$	2:1	β–like	620	Metallic to 4.2 K	29a,47
EDVT	PF$_6$	5:2	–	13	Metallic to 4.2 K	29a
DSDTeF	TCNQ	1:1	β–like	1600	Metallic to 4.2 K	32
BES-TSF	TCNQ	1:1	β–like	2700	Metallic to 40 K	34b
BES-TTF	AsF$_6$	2:1	β–like	60	Metallic to 50 K	34b
BPT-TSF	FeCl$_4$	2:1	κ	98	Metallic to 4.2 K	34c
BPT-TSF	AsF$_6$	2:1	β–like	150	Metallic to 4.2 K	33c
MDT-TSF	AuI$_2$	1:0.436	β–like	2000	$T_c = 4.5$ K	11
MDSe-TSF	PF$_6$	2:1	θ	100–250	Metallic to 4.2 K	48
MDSe-TSF	Br	2:1	κ	200–1000	Metallic to 4.2 K	48

$^a E_a$ = Activation energy. ^{b}FISC = Magnetic field induced superconductivity. ^{c}DCM = Dichloromethane.

outstanding electron donors, BETS, TMET-STF and MDT-TSF, which can form unique organic superconductors with unprecedented structural and physical features as follows.

BETS is known to give a series of λ-type superconductors with superconducting transition temperatures (T_c's) of up to 9.7 K,$^{9,45a,b)}$ which fall under the category of organic superconductors with the highest critical temperature. In addition, it forms κ-(BETS)$_2$FeX$_4$ (X = Cl, Br) as the first radical cation salt with antiferromagnetic superconductivity$^{9,45d)}$ and λ-(BETS)$_2$FeCl$_4$ showing magnetic field-induced superconductivity.$^{45e)}$ These physical phenomena are induced by the interplay of conducting electrons and localized spins.

TMET-STF is a more recently developed electron donor, which gives a superconducting BF$_4$ salt with $T_c = 4.1$ K.$^{10)}$ Its crystal structure is unique in that there are two different donor arrangements; one is two dimensional and the other is quasi-one dimensional. Thus, (TMET-STF)$_2$BF$_4$ is a rare organic superconductor with conductive multi channels.

Finally, MDT-TSF is the newest selenium-including electron donor forming superconductors.$^{11)}$ The superconducting (MDT-TSF)(AuI$_2$)$_{0.436}$ salt with $T_c = 4.5$ K shows a special crystal structure, which consists of conducting donor layers based on uniform donor stacks and anion layers incommensurate with the donor lattice, leading to the nonstoichiometric composition of the donor to anion (Fig. 5.8).

These unique structural and physical features observed for selenium-containing organic superconductors are rarely found in the TTF-based radical cation salts. It is thus safe to say that the introduction of selenium atoms is an important chemical modification of TTF-based electron donors for the development not only of metallic states but also of unique structural and physical properties of conducting radical cation salts.

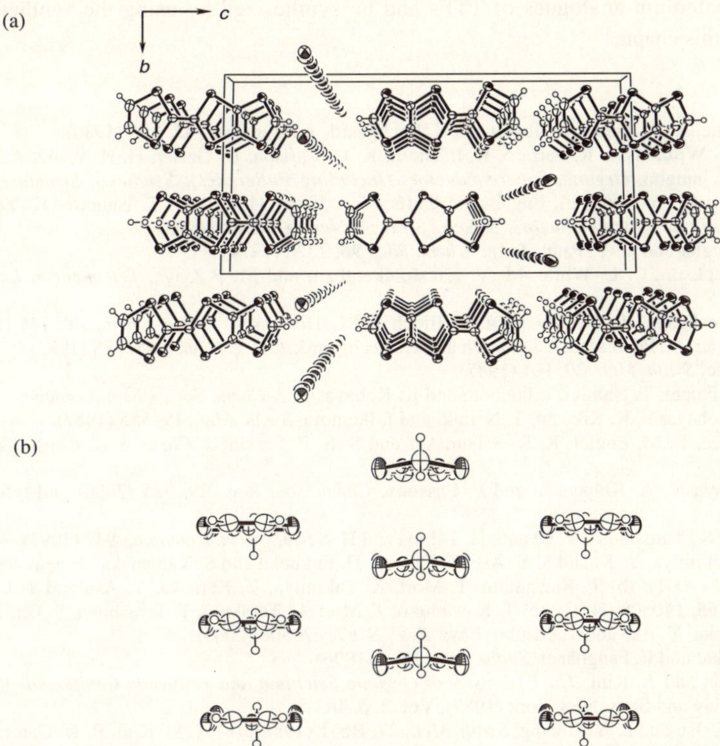

Fig. 5.8   Crystal structure of (MDT-TSF)(AuI$_2$)$_{0.436}$: (a) view along the a-axis, (b) side view of the conducting donor sheet.

5.9   Summary and Outlook

The above is an overview of recent advances in the synthetic chemistry of selenium analogues of TTFs. These advances are roughly divided into three classes: straightforward approaches to selenium-containing 1,3-dichalcogenole building blocks, innovation of the ensuing coupling reactions and, if necessary, the ready formation of the outer heterocyclic rings. By combining these reactions, almost all kinds of seleno-analogues of TTFs are now accessible. It should be emphasized that the combined methods offer ready access not only to new compounds but also to known compounds more easily than previous methods. It is natural that each synthetic method has its own scope and limitation, so the appropriate synthetic method must be selected to obtain the desired selenium donor. For example, although the synthetic method for MDT-TSF illustrated in Scheme 5.13 is very useful for the synthesis of heterocycle-fused TSF derivatives, it is not applicable to the synthesis of cycloalkene-annulated compounds such as TMET-STF and CHET-STF, which can be effectively synthesized by the titanocene method.

Not all the compounds described in this chapter have been examined throughout as electron donors for the conducting and superconducting CT salts. The authors believe that the key compounds of the next generation in the field of organic (super)conductors will be discovered among the selenium analogues of TTFs and be synthesized by using the synthetic methods described in this chapter.

References

1. D. Jerome, A. Mazaud, M. Ribault and K. Bechgaard, *J. Phys. Lett.*, **41**, L95 (1980).
2. (a) J. M. Williams, J. R. Ferraro, R. J. Thorn, K. D. Carlson, U. Geiser, H. H. Wang, A. M. Kini and M.-H. Whangbo, *Organic Superconductors (Including Fullerenes): Synthesis, Structure, Properties, and Theory*, Prentice Hall, Englewood Cliffs, New Jersey (1992); (b) T. Ishiguro, K. Yamaji and G. Saito, *Organic Superconductors,* 2nd ed., Springer-Verlag, Berlin (1998).
3. E. M. Engler and V. V. Patel, *J. Am. Chem. Soc.*, **96**, 7376 (1974).
4. Y. A. Jackson, C. L. White, M. V. Lakshmikantham and M. P. Cava, *Tetrahedron Lett.*, **28**, 5635 (1987).
5. (a) K. Bechgaard, D. O. Cowan, A. N. Bloch and L. Henriksen, *J. Org. Chem.*, **40**, 746 (1975); (b) A. Moradpour, V. Peyrussan, I. Johansen and K. Bechgaard, *J. Org. Chem.*, **48**, 388 (1983).
6. V. Y. Lee, *Synth. Met.*, **20**, 161 (1987).
7. (a) K. Kikuchi, T. Namiki, I. Ikemoto and K. Kobayashi, *J. Chem. Soc., Chem. Commun.*, 1472 (1986); (b) K. Kobayashi, K. Kikuchi, T. Namiki and I. Ikemoto, *Synth. Met.*, **19**, 555 (1987).
8. V. Y. Lee, E. M. Engler, R. R. Schumaker and S. S. P. Parkin, *J. Chem. Soc., Chem. Commun.*, 235 (1983).
9. H. Kobayashi, A. Kobayashi and P. Cassoux, *Chem. Soc. Rev.*, **29**, 325 (2000) and references cited therein.
10. R. Kato, K. Yamamoto, Y. Okano, H. Tajima and H. Sawa, *Chem. Commun.*, 947 (1997).
11. (a) K. Takimiya, Y. Kataoka, Y. Aso, T. Otsubo, H. Fukuoka and S. Yamanaka, *Angew. Chem. Int. Ed.*, **40**, 1122 (2001); (b) T. Kawamoto, T. Mori, K. Takimiya, Y. Kataoka, Y. Aso and T. Otsubo, *Phys. Rev., B* **65**, 140508 (2002); (c) T. Kawamoto, T. Mori, C. Terakura, T. Terashima, S. Uji, K. Takimiya, Y. Kataoka, Y. Aso and T. Otsubo, *Phys. Rev., B* **67**, 020508 (2003).
12. G. Schukat and E. Fanghänel, *Sulfur Rep.*, **18**, 1 (1996).
13. D. Cowan and A. Kini, *The Chemistry of Organic Selenium and Tellurium Compounds* (S. Patai ed.), John Wiley and Sons, New York (1987), Vol. 2, p. 463.
14. (a) S.-Y. Hsu and L. Y. Chiang, *Synth. Met.*, **27**, B651 (1988); (b) A. M. Kini, B. D. Gates, M. A. Beno and J. M. Williams, *J. Chem. Soc., Chem. Commun.*, 169 (1989).
15. (a) R.-M. Olk, A. Röhr, B. Olk and E. Hoyer, *Z. Chem.*, **28**, 304 (1988); (b) P. J. Nigrey, *J. Org. Chem.*, **53**, 201 (1988); (c) G. C. Papavassiliou, V. C. Kakoussis, J. S. Zambounis and G. A. Mousdis, *Chem. Scr.*, **29**, 123 (1989); (d) J. Garín, J. Orduna, M. Savirón, M. R. Bryce, A. J. Moore and V. Morisson, *Tetrahedron*, **52**, 11063 (1996).
16. (a) N. Svenstrup, K. M. Rasmussen, T. K. Hansen and J. Becher, *Synthesis*, 809 (1994); (b) K. B. Simonsen, N. Svenstrup, J. Lau, O. Simonsen, P. Mørk, G. J. Kristensen and J. Becher, *Synthesis*, 407 (1996).
17. L. Binet, J.-M. Fabre, C. Montginoul, K. B. Simonsen and J. Becher, *J. Chem. Soc., Parkin Trans. 1*, 783 (1996).
18. L. Binet, J.-M. Fabre and J. Becher, *Synthesis*, 26 (1997).
19. H. Köpf, B. Block and M. Schmidt, *Chem. Ber.*, **101**, 272 (1968).
20. C. M. Bolinger and T. B. Rauchfuss, *Inorg. Chem.*, **21**, 3947 (1982).
21. (a) R. Kato, H. Kobayashi and A. Kobayashi, *Synth. Met.*, **42**, 2093 (1991); (b) T. Courcet, L. Malfant, K. Pokhodnia and P. Cassoux, *N. J. Chem.*, **22**, 585 (1998).
22. Y. Okano, H. Sawa, S. Aonuma and R. Kato, *Synth. Met.*, **70**, 1161 (1995).
23. Y. Okano, M. Iso, Y. Kashimura, J. Yamaura and R. Kato, *Synth. Met.*, **102**, 1703 (1999).
24. T. Imakubo, H. Sawa and R. Kato, *Synth. Met.*, **86**, 1883 (1999).
25. R. Kato, S. Aonuma, Y. Okano, H. Sawa, H. Kobayashi and A. Kobayashi, *Synth. Met.*, **55–57**, 2084 (1993).
26. (a) R. Kato, S. Aonuma, Y. Okano, H. Sawa, M. Tamura, M. Kinoshita, K. Oshima, A. Kobayashi, K. Bun and H. Kobayashi, *Synth. Met.*, **61**, 199 (1993); (b) E. Ojima, H. Fujiwara and H. Kobayashi, *Adv. Mater.*, **11**, 459 (1999); (c) A. Sato, E. Ojima, H. Kobayashi and A. Kobayashi, *J. Mater. Chem.*, **9**, 2365 (1999); (d) T. Naito, H. Kobayashi and A. Kobayashi, *Bull. Chem. Soc. Jpn.*, **70**, 107 (1997); (e) A. Kobayashi, R. Kato, T. Naito and H. Kobayashi, *Synth. Met.*, **55–57**, 2078 (1993); (f) N. Sakurai, H. Mori, S. Tanaka and H. Moriyama, *Chem. Lett.*, 1191 (1999); (g) T. Naito, H. Kobayashi, A. Kobayashi and A. E. Underhill, *Chem. Commun.*, 521 (1996).
27. M. Fourmigué, F. C. Krebs and J. Larsen, *Syntheis*, 509 (1993).
28. J. Yamada, Y. Amano, S. Takasaki, R. Nakanishi, K. Matsumoto, S. Satoki and H. Anzai, *J. Am. Chem.*

Soc., **117**, 1149 (1995).
29. (a) J. Yamada, S. Satoki, H. Anzai, K. Hagiya, M. Tamura, Y. Nishio, K. Kajita, E. Watanabe, M. Konno, T. Sato, H. Nishikawa and K. Kikuchi, *Chem. Commun.,* 1955 (1996); (b) J. Yamada, S. Satoki, S. Mishima, N. Akashi, K. Takahashi, N. Masuda, Y. Nishimoto, S. Takasaki and H. Anzai, *J. Org. Chem.,* **61**, 3987 (1996); (c) J. Yamada, *Recent Res. Devel. in Organic Chem.*, **2**, 525 (1998); (d) J. Yamada, H. Nishikawa and K. Kikuchi, *J. Mater. Chem.*, **9**, 617 (1999).
30. K. Takimiya, A. Morikami and T. Otsubo, *Synlett*, 319 (1997).
31. K. Takimiya, Y. Kataoka, A. Morikami, Y. Aso and T. Otsubo, *Synth. Met.*, **120**, 875 (2001).
32. (a) K. Takimiya, A. Morikami, Y. Aso and T. Otsubo, *Chem. Commun.,* 1925 (1997); (b) A. Morikami, K. Takimiya, Y. Aso and T. Otsubo, *J. Mater. Chem.*, **11**, 2431 (2001).
33. A. Morikami, K. Takimiya, Y. Aso and T. Otsubo, *Org. Lett.*, **1**, 23 (1999).
34. (a) T. Jigami, K. Takimiya, Y. Aso and T. Otsubo, *Chem. Lett.*, 1091 (1997); (b) T. Jigami, K. Takimiya, T. Otsubo and Y. Aso, *J. Org. Chem.*, **63**, 8865 (1998); (c) T. Jigami, M. Kodani, S. Murakami, K. Takimiya, Y. Aso and T. Otsubo, *J. Mater. Chem.*, **11**, 1026 (2001).
35. M. Kodani, S. Murakami, T. Jigami, K. Takimiya, Y. Aso and T. Otsubo, *Heterocycles*, **54**, 225 (2001).
36. K. Takimiya, T. Jigami, M. Kawashima, M. Kodani, Y. Aso and T. Otsubo, *J. Org. Chem.*, **67**, 4218 (2002).
37. K. Takimiya, S. Murakami, M. Kodani, Y. Aso and T. Otsubo, *Mol. Cryst. Liq. Cryst.*, **380**, 189 (2002).
38. K. Takimiya, A. Oharuda, A. Morikami, Y. Aso and T. Otsubo *Eur. J. Org. Chem.*, 3013 (2000).
39. K. Takimiya, Y. Kataoka, N. Niihara, Y. Aso and T. Otsubo, *J. Org. Chem.*, **68**, 5217 (2003).
40. (a) M. Kodani, K. Takimiya, Y. Aso, T. Otsubo, T. Nakayashiki and Y. Misaki, *Synthesis*, 1614 (2001); (b) H. Poleschner, R. Radeglia, J. Fuchs, *J. Organomet. Chem.*, **427**, 213 (1992).
41. H. H. Wang, L. K. Montgomery, U. Geiser, L. C. Porter, K. D. Carlson, J. R. Ferraro, J. M. Williams, C. S. Cariss, R. L. Rubinstein, J. R. Whitworth, M. Evain, J. J. Novoa and M.-H. Whangbo, *Chem. Mater.*, **1**, 140 (1989).
42. J. Sakata, H. Sato, A. Miyazaki, T. Enoki, Y. Okano and R. Kato, *Solid State Commun.*, **108**, 377 (1998).
43. H. Mori, I. Hirabayashi, S. Tanaka, T. Mori, Y. Maruyama and H. Inokuchi, *Solid State Commun.*, **88**, 411 (1993).
44. S. Golhen, L. Ouahab, A. Lebeuze, M. Bouayed, P. Delhaes, Y. Kashimura, R. Kato, L. Binet and J.-M. Fabre, *J. Mater. Chem.*, **9**, 387 (1999).
45. (a) H. Kobayashi, T. Udagawa, H. Tomita, K. Bun, T. Naito and A. Kobayashi, *Chem. Lett.*, 1559 (1993); (b) H. Tanaka, A. Kobayashi, A. Sato, H. Akutsu and H. Kobayashi, *J. Am. Chem. Soc.*, **121**, 760 (1999); (c) H. Kobayashi, H. Tomita, T. Naito, A. Kobayashi, F. Sakai, T. Watanabe and P. Cassoux, *J. Am. Chem. Soc.*, **118**, 368 (1996); (d) H. Fujiwara, E. Fujiwara, Y. Nakazawa, B. Z. Narymbetov, K. Kato, H. Kobayashi, A. Kobayashi, M. Tokumoto and P. Cassoux, *J. Am. Chem. Soc.*, **123**, 306 (2001); (e) S. Uji, H. Shinagawa, T. Terashima, T. Yakabe, Y. Terai, M. Tokumoto, A. Kobayashi, H. Tanaka and H. Kobayashi, *Nature*, **410**, 908 (2001).
46. H. Mori, T. Okano, M. Kamiya, M. Haemori, H. Suzuki, S. Tanaka, M. Tamura, Y. Nishio, K. Kajita, M. Kodani, K. Takimiya, T. Otsubo and H. Moriyama, *Synth. Met.*, **120**, 979 (2001).
47. H. Nishikawa, T. Sato, T. Kodama, I. Ikemoto, K. Kikuchi, H. Anzai and J. Yamada, *J. Mater. Chem.*, **9**, 693 (1999).
48. M. Kodani, A. Takamori, K. Takimiya, Y. Aso and T. Otsubo, *J. Solid State Chem.*, **168**, 582 (2002).

6

TTFs with Organic Stable Radicals

6.1 Introduction

Tetrathiafulvalene (TTF) derivatives are representative donor components of organic conductors which have been well documented as correlated electron systems. In such systems, Coulombic interaction between electrons makes a large contribution to physical properties. Electronic spin also plays an important role, as seen in the spin-density-wave state or in the antiferromagnetic Mott insulator state. When a localized spin is put into a sea of conduction electrons, several types of interactions are expected. In the case of rare earth metal systems, such a situation may be rationalized in terms of the Kondo effect, RKKY (Ruderman, Kittel, Kasuya and Yoshida) interaction or a heavy electron system. Correlation between itinerant and localized electrons will produce a variety of physical phenomena. Concerning organic conductors, various conduction behaviors have been found by the use of inorganic counterions with magnetic moments. Furthermore, a genuine organic system may be realized if a stable organic radical group is introduced to the constituent donor molecule or acceptor, as shown in Fig. 6.1.

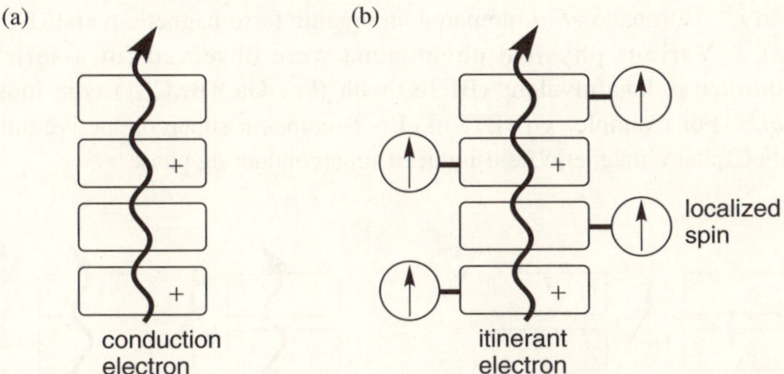

Fig. 6.1   Schematic drawing of (a) an organic conductor composed of TTF
derivatives and (b) an organic magnetic conductor composed of
TTF derivatives carrying stable organic radicals.

The development of organic magnetic conductors is also desirable from the viewpoint of molecular magnetism.[1] More than twenty genuine organic ferromagnets have been discovered, though their transition temperatures (T_c's) are very low.[2] In order to achieve a high T_C organic ferromagnet, the utilization of itinerant electrons may be necessary.[3] As the first step of this approach, an itinerant electron and a localized spin should be coupled ferromagnetically. For this purpose, the efficient design of a donor-radical molecule is important. When the itinerant electron migrates intermolecularly, it will align all the localized spins which are ferromagnetically coupled with the itinerant electron as depicted in Fig. 6.1b. Such a mechanism resembles the double-exchange system seen in metal oxide compounds such as $La_{1-x}Sr_xMnO_3$.[4] Although $LaMnO_3$ is a Mott insulator and antiferromagnet, it becomes conducting and the local spins on Mn sites are aligned ferromagnetically through conduction electrons when it is doped with Sr. Since Mn in $La_{1-x}Sr_xMnO_3$ is of a high-spin state, localized spins (electrons in the t_{2g} orbital) are ferromagnetically coupled with electrons in the e_g orbital. When this system is partially doped, down spin electrons in the e_g orbital can migrate along the metal array, aligning the localized spins in the same direction at each metal site. If this type of spin alignment system is achieved in organic magnetic materials, the transition temperature may exceed room temperature.

TTF derivatives are also well known as a source of stable cation radicals. A TTF derivative with stable organic radicals is expected to be a redox-active switchable spin system. Organic materials can be designed on the molecular level, and will be used as unimolecular devices. Since magnetoresistant devices have been applied in recording facilities, studies on correlated electron systems for device applications have been further developed. For device application of TTF-based switchable spin systems or conductive ferromagnetic systems, studies on TTFs carrying stable organic radicals will be fruitful.

6.2   Organic Conductors Incorporating Inorganic Spins

Day *et al.* prepared the $FeCl_4$ salt of bis(ethylenedithio)tetrathiafulvalene (ET) and showed its conducting and magnetic behavior.[5] Since then, several organic conductors incorporating inorganic spins have been reported (Fig. 6.2).[6-9] Among such magnetic conductors, correlation between conductivity and magnetism was observed in some cases. Enoki *et al.* reported the physical properties of $(ET)_3CuBr_4$, which showed a simultaneous change in conductivity and magnetic property.[6] Coronado *et al.* prepared an organic ferromagnetic metal $(ET)_3[MnCr(ox)_3]$ (ox = oxalate).[7] Various physical phenomena were observed for a series of salts of bis(ethylenedithio)tetraselenafulvalene (BETS) with $(Fe_{1-x}Ga_x)(Br_yCl_{4-y})$ type ions prepared by Kobayashi *et al.*[8] For example, κ-$(BETS)_2FeBr_4$ becomes a superconductive antiferromagnet, and λ-$(BETS)_2FeCl_4$ has a magnetic field-induced superconducting phase.

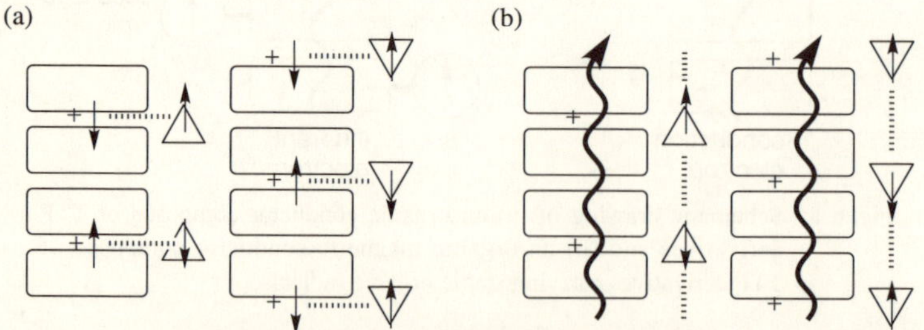

Fig. 6.2   Schematic drawing of a donor-magnetic anion system:(a) when the interaction between conduction electron and localized spin is dominant, (b) when the interaction between localized spins is dominant.

6.3   Magnetic Conductors Composed of Dimeric Donors

6.3.1 Spin Alignment Based on Degenerated Molecular Orbitals

McConnell proposed a model for intermolecular spin alignment derived from spin-polarization.[10] The model was experimentally supported using cyclophanedicarbene derivatives.[11] Most organic ferromagnets are based on this mechanism. McConnell also proposed the possibility of spin alignment in a charge-transfer (CT) complex between a donor with degenerated SOMOs (singly occupied molecular orbitals) and an acceptor.[12] The spin alignment in an alternate stack of donors and acceptors may be achieved by the contribution of the reverse CT state, in which two electrons in the degenerate SOMOs are in parallel (Fig. 6.3a).

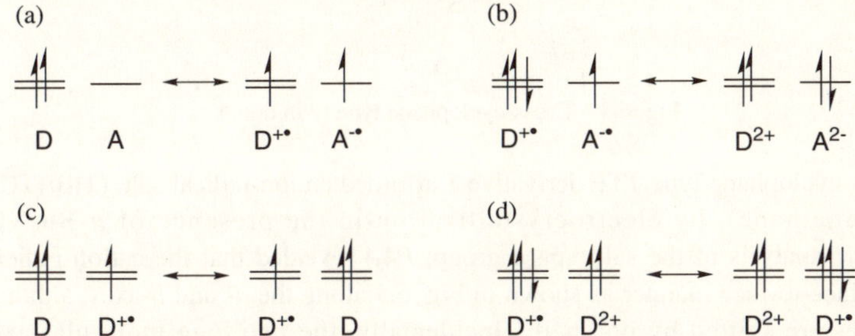

Fig. 6.3   Models of spin alignment in CT complexes based on (a) a triplet donor, (b) a triplet dication diradical of a donor, (c) partially oxidized triplet donors and (d) partially generated triplet dication diradical of a donor.

The latter model was revised by Breslow as shown in Fig. 6.3b.[13] If a donor molecule has two degenerated HOMOs (highest occupied molecular orbitals), its dication diradical species should be in the triplet state. Such donor molecules were designed based on high symmetry, and, in fact, triplet species were observed. However, CT complexes of the donors did not exhibit any ferromagnetic interactions, proving that the symmetry of the donors is lowered due to the Jahn-Teller distortion.[14] High-spin polycation polyradical species can also be achieved by the use of π-topology.[15] For example, a 1,3,5-triaminobenzene derivative was proposed as a building block for organic ferromagnetic metals.

If a columnar stacking of donors with degenerated SOMOs is partially oxidized, two unpaired electrons residing in SOMOs are ferromagnetically coupled, and the intermolecular spin alignment may be achieved along the migration of the electrons, as proposed by Wudl (Fig. 6.3c).[16]

6.3.2 Ion-radical Salt of the Cross-cyclophane Type Twin Donor

A similar situation should occur when a ferromagnetically connected dication diradical of a dimeric donor is prepared (Fig. 6.3d). Although such a dimeric donor may be achieved by the use of π-topology,[17] detailed results have not been reported. Another approach has been proposed. The designed molecule is a cross-cyclophane type twin donor (**1**) in which two equivalent donor units are connected with four trimethylenedithio chains (Fig. 6.4).[18] In this case, two types of interactions are expected: through-space and through-bond interactions. While through-space interaction always leads to an antiferromagnetic coupling, through-bond interaction depends on the number of carbon atoms: an odd number causes ferromagnetic interaction, while an even number is antiferromagnetic. In the twin donor **1**, the through-bond interaction is considered to be ferromagnetic and the antiferromagnetic through-space interaction is expected to be negligibly small, since the two donor units are fixed by long alkyl chains in an orthogonal orientation.

1

Fig. 6.4   Cross-cyclophane type twin donor.

The cross-cyclophane type TTF derivative **1** afforded an ion-radical salt, (**1**)Br(TCE)$_2$ (TCE = 1,1,2-trichloroethane), by electrocrystallization in the presence of n-Bu$_4$NBr. X-ray crystallographic analysis of the salt (space group: $P4_1$) revealed that the cation radicals of **1** are arranged in a face-to-face manner as shown in Fig. 6.5 along the a- and b-axes, although the long molecular axes are shifted by one half. Incidentally, the two long molecular axes in **1** are equivalent according to the symmetry of the crystal structure. Inside the cavity created by the surrounding donors, a bromide ion and two solvent molecules of TCE are incorporated per one donor molecule.

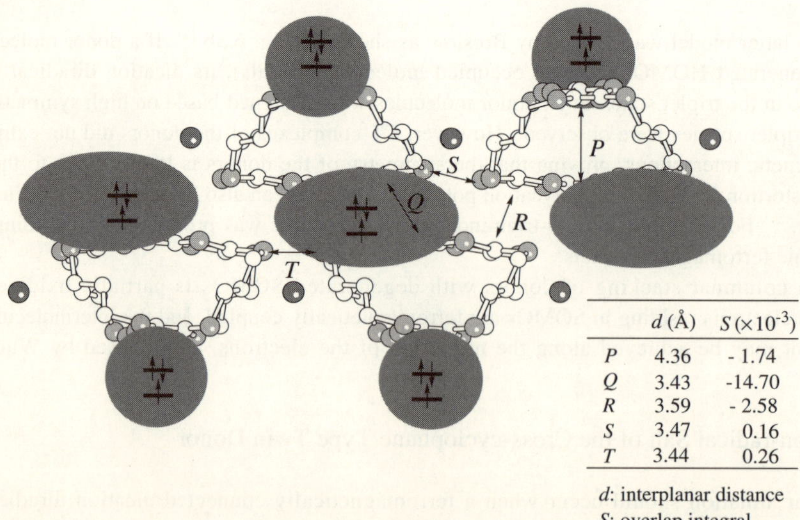

	d (Å)	S ($\times 10^{-3}$)
P	4.36	- 1.74
Q	3.43	-14.70
R	3.59	- 2.58
S	3.47	0.16
T	3.44	0.26

d: interplanar distance
S: overlap integral

Fig. 6.5   Molecular alignment of donors and calculated overlap integrals in (**1**)Br(TCE)$_2$.

The conductivities of the salt measured by a four-probe method were 1.0×10^{-4} S cm^{-1} along the *a*- or *b*-axis and 5.5×10^{-5} S cm^{-1} along the *c*-axis. The temperature dependence of the conductivity indicated that the salt exhibits semiconducting behavior, and the activation energy was estimated to be 0.3 eV.

The magnetic property of the single crystal of the salt was measured by a SQUID magnetometer with increasing temperature from 2 K to 300 K (Fig. 6.6). The χT value is 0.375 emu K mol^{-1} at room temperature, indicating that each donor molecule carries one unpaired electron. The temperature dependence of the χT value was analyzed based on the model of a one-dimensional ferromagnetic Heisenberg chain with antiferromagnetic couplings with adjacent chains ($z = 4$). The ferromagnetic interaction was evaluated to be $J = +1.6$ K, and the antiferromagnetic interaction to be $J = -0.1$ K. Judging from the calculated overlap integrals shown in Fig. 6.5, the unpaired electrons are localized in a region where the HOMO and the SOMO of the TTF donor planes interact intermolecularly (along Q). Two unpaired electrons are, therefore, considered to reside across the cyclophane donor **1**, although the donor exists as a cation radical. Thus the ferromagnetic interaction detected here may originate from the ferromagnetic coupling mediated *via* one donor molecule.

Although the degree of the ferromagnetic coupling in (**1**)Br(TCE)$_2$ is weak, the coexistence of conductivity and magnetism of the ion-radical salt observed here is of significance in designing an organic ferromagnetic metal.

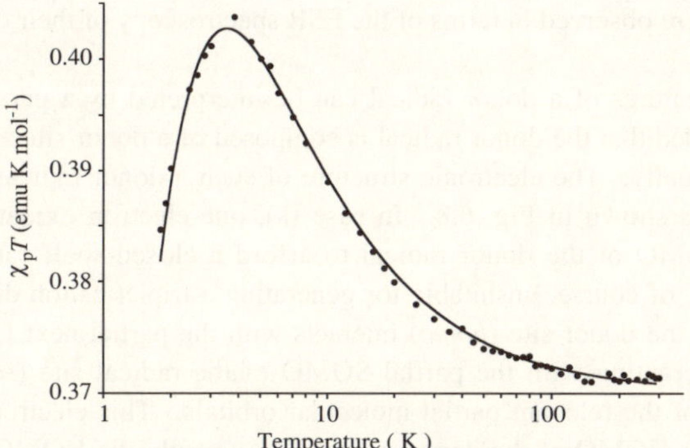

Fig. 6.6   Temperature dependence of the magnetic susceptibility of (**1**)Br(TCE)$_2$.

6.4   Switchable Spin System Composed of Spin-polarized Donors

6.4.1 Spin-polarized Donors

Trimethylenemethane (TMM) is a well-known ground state triplet diradical, and the ferromagnetic coupling between two unpaired electrons in TMM is significantly large.[19] TMM can be regarded as the combination of an allyl radical and a methyl radical. If the allyl radical part and the methyl radical part are replaced with a nitronyl nitroxide (NN) and the cation radical of a donor, respectively, the resultant cation diradical of donor-NN can be regarded as a hetero-

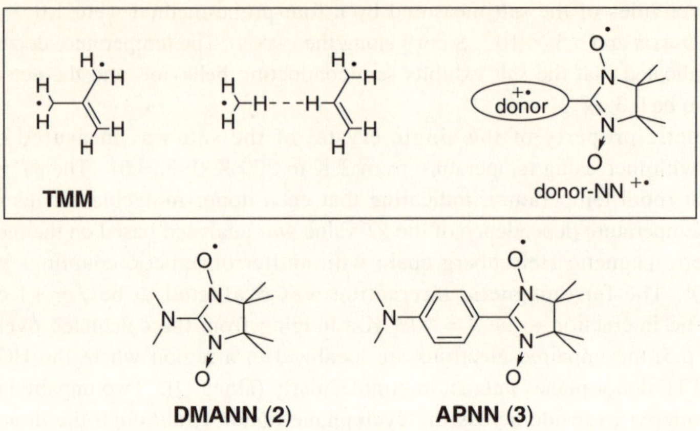

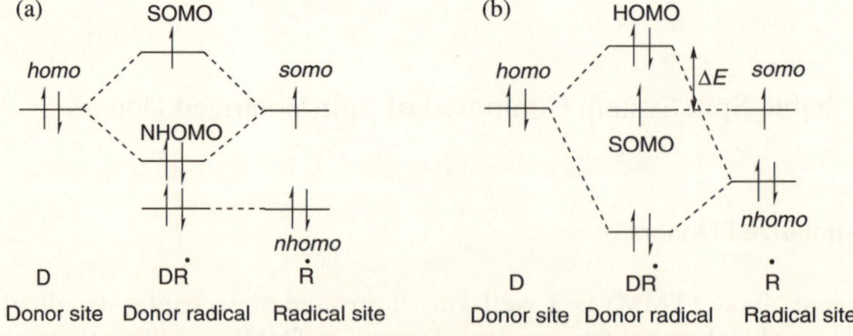

Fig. 6.7   Trimethylenemethane (**TMM**) and its hetero-analogues. Prototypical
spin-polarized donors, dimethylamino nitronyl nitroxide (**2**) and
dimethylaminophenyl nitronyl nitroxide (**3**).

analogue of TMM (Fig. 6.7). As prototype compounds, dimethylamino nitronyl nitroxide
(DMANN, **2**) and dimethylaminophenyl nitronyl nitroxide (APNN, **3**) were prepared, and ground
state triplet species were observed in terms of the ESR spectroscopy of their one-electron oxidized
state.[20]

The electronic features of a donor radical can be interpreted by a perturbational molecular
orbital method, provided that the donor radical is composed of a donor site and a radical site, and
that they interact mutually. The electronic structure of such a donor radical may be represented
by either (a) or (b) as shown in Fig. 6.8. In case (a), one-electron oxidation will remove the
electron from the SOMO of the donor radical to afford a closed-shell cationic species. This
electronic structure is, of course, unsuitable for generating a triplet cation diradical. In case (b),
the partial HOMO of the donor site (*homo*) interacts with the partial next HOMO of the radical
site (*nhomo*), not interacting with the partial SOMO of the radical site (*somo*) because of the
symmetry mismatch of the relevant partial molecular orbitals. This electronic interaction raises
the energy level of the HOMO of the donor radical. As a result, the HOMO is placed above the
SOMO. Such an exotic electronic structure can be maintained if the on-site Coulombic repulsion

Fig. 6.8   Comparison of the electronic configurations in the donor-radical interaction:
(a) a conjugated π-radical, (b) a spin-polarized donor.

of the SOMO is larger than the orbital energy difference (ΔE) between the HOMO and the SOMO. If the SOMO', which is derived from the HOMO upon one-electron oxidation, and the SOMO are of space-sharing types with each other, two unpaired electrons residing in the SOMO and the SOMO' should interact ferromagnetically to afford a triplet cation diradical. Since this situation is depicted as a spin-polarized electronic structure in the UHF (unrestricted Hartree-Fock) description, donor radicals in case (b), including **2** and **3**, are designated as "spin-polarized donors."

It should be noted that SOMO and SOMO' are not in the same energy level. Most high-spin molecules are based on degenerated molecular orbitals, but cation diradicals of the spin-polarized donors are different. As a result, the spin-polarized donors are free from the effect of the Jahn-Teller distortion observed in Breslow's CT complex crystal.

6.4.2 Design of TTF-based Spin-polarized Donors

In order to obtain a conductive and magnetic material, the TTF skeleton is desirable for the donor part of the spin-polarized donor. As a prototypical donor radical, a TTF derivative carrying an NN group (**4**) was prepared.[21] Although a triplet ESR spectrum was observed in the iodine-doped sample of **4**, the triplet signal was found to be a thermally populated one ($J \sim -100$ K). The reason for the antiferromagnetic interaction may be explained as follows. Since the nitroxide group in NN is located close to the sulfur atom of the TTF skeleton, the radical site may be twisted from the donor plane. The twisting causes a significant decrease in the coefficients of the HOMO at the radical site, breaking the condition of the space-sharing types of the SOMOs.

In order to remove such steric repulsion, a *p*-phenylene group was inserted between the donor site and the radical site (**5**). Furthermore, sulfur-extended derivatives (**6-8**) were prepared in order to enhance the kinetic stability and to increase the intermolecular interaction (Fig. 6.9).[22] Among these donor radicals, *p*-phenylene type derivatives (**5-7**) were prepared as shown in Fig. 6.10. The key step is the palladium-catalyzed coupling of the stannylated TTF derivatives and bromo- or iodobenzaldehyde.

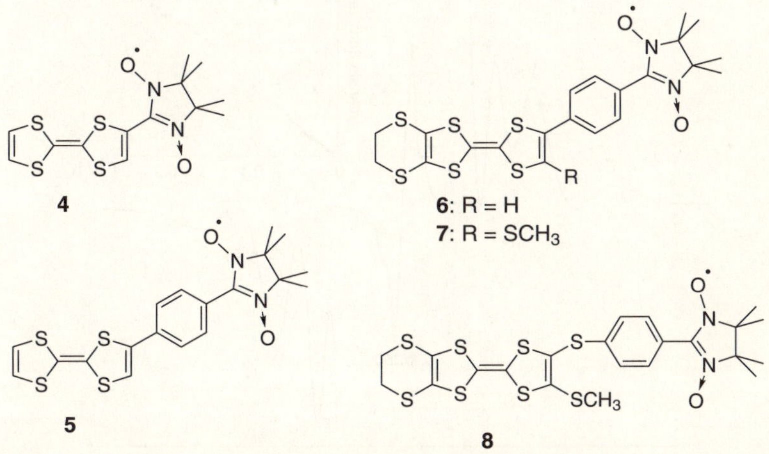

Fig. 6.9   Designed TTF-based spin-polarized donors.

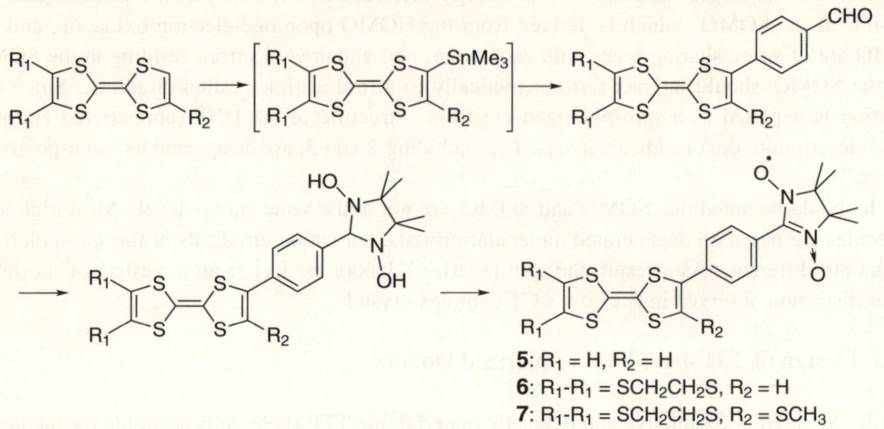

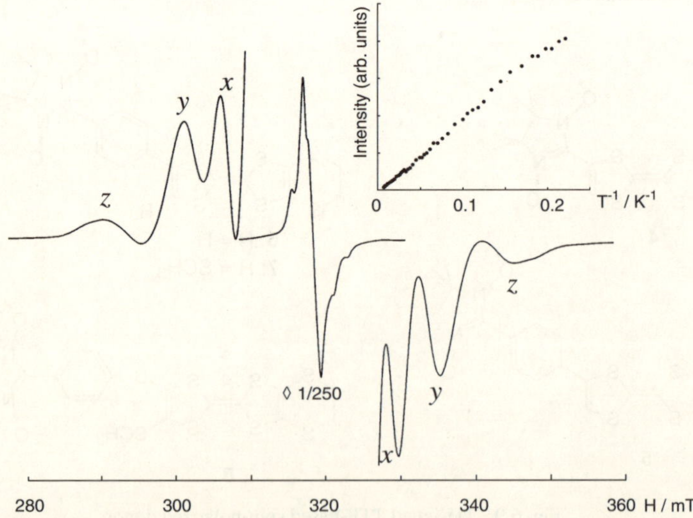

Fig. 6.10   Synthetic scheme of TTF-based spin-polarized donors.

The cyclic voltammograms of these donor radicals were recorded to show the first redox potentials at **5**: +0.44 V, **6**: +0.54 V, **7**: +0.62 V, **8**: +0.68 V (*vs.* Ag/AgCl in 0.1 M Bu_4NClO_4/CH_2Cl_2), suggesting a significant electronic interaction between the donor and radical moieties. The lower oxidation potentials of these donor radicals, compared with the parent donors, suggest that the HOMO of the donor radical is raised due to the interaction between the *homo* of the donor part and the *nhomo* of the radical part. ESR spectra of the iodine-doped **5-8** showed triplet signals derived from the cation diradical species (Fig. 6.11). The temperature dependence of the triplet signals revealed that the triplet is the ground state of these cation diradicals. This suggests that the insertion of a *p*-phenylene group decreases the twisting between the donor and radical units considerably.

Fig. 6.11   ESR spectrum of the oxidized species of **7** at 5 K. Inset shows the temperature dependence of the triplet signal intensity.

6.4.3 Spin-polarized Polyradical Donors

As an extension of the spin-polarized donor, spin-polarized polyradical donors are designed as depicted in Fig. 6.12.[23] Since these radical units are coupled loosely in the neutral state, the unpaired electrons behave paramagnetically. However, once the core unit is singly oxidized, the generated unpaired electron delocalizes over the entire molecule, aligning these fluctuating local spins to the same direction, giving rise to a high-spin species. The point here is that one can generate high-spin species by removing only one electron.

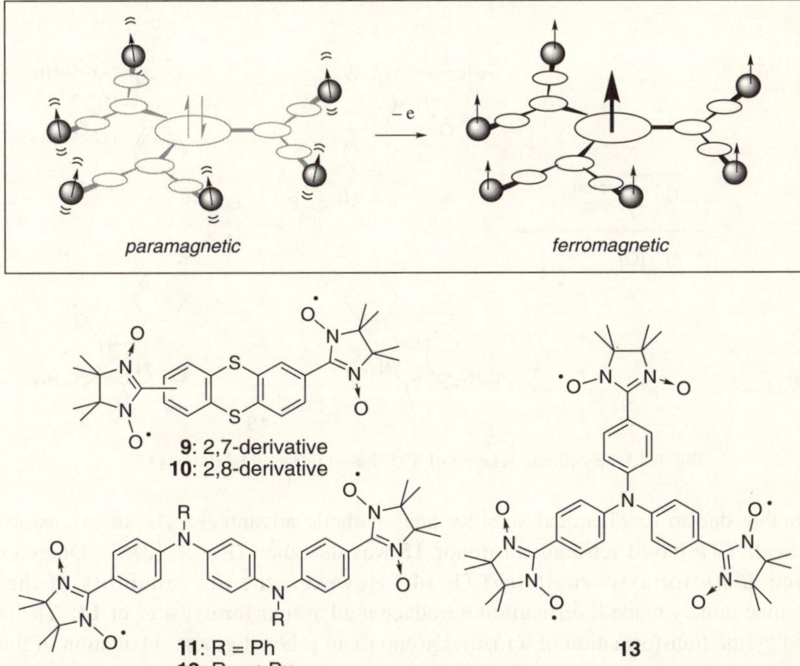

Fig. 6.12   Schematic drawing of a switchable spin system composed of spin-polarized polyradical donors. Examples are depicted below.

Thianthrene or *N,N'*-diphenyl-*para*-phenylenediamine is used as an oxidizable core for the diradical, and NN groups are introduced into the aromatic rings (**9-12**). The magnetic measurements of the diradicals **9** and **10** showed that the magnetic interactions between two NN groups were negligibly small. ESR spectra of the oxidized donor radicals in matrices afforded fine-structure signals due to the quartet species. Their Curie plots suggest that the quartet is the ground state for the cation triradicals. The triradical with a triphenylamine skeleton (**13**) also provided a corresponding result. It should be noted that the high-spin state of the oxidized species is achieved free from the π-topology rule.

6.4.4 Preparation of TTF-based Tetraradical Donors

Among TTF-based spin-polarized donors **5-8** (Fig. 6.9), the *p*-phenylenethio derivative **8** is the

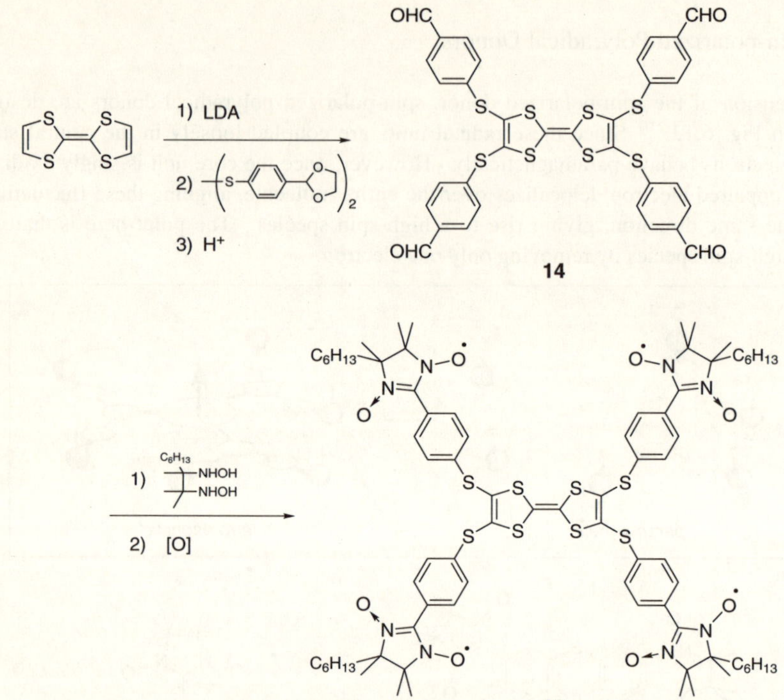

Fig. 6.13   Synthetic scheme of TTF-based tetraradical donor (**15**).

most favorable due to its chemical stability and synthetic advantage. As an extension of this
donor radical, TTF-based tetraradical donor **15** was designed (Fig. 6.13).[24) One can easily
prepare tetrakis(*p*-formylphenylthio)TTF **14**. However, the low solubility of the cyclic
hydroxylamine moiety made it difficult to introduce it all at four formyl sites of **14**. The problem
was solved by the transformation of a methyl group to an *n*-hexyl group. Oxidation of the cyclic
hydroxylamine derivative obtained afforded the tetraradical **15**.

6.5   Assembled TTF Derivatives Carrying Stable Radicals

6.5.1 Nitroxide-based Donor Radicals

TTF derivatives carrying a TEMPO (tetramethylpiperidine-*N*-oxyl) radical have been prepared by
some groups.[25) Since a TEMPO radical can survive under synthetic reaction conditions, amino-
or oxo-TEMPO was used directly for the preparation of these donor radicals (Fig. 6.14).
Although it may be difficult to introduce a large ferromagnetic interaction between the donor and
radical sites, this approach is advantageous because the TTF site gets only a small perturbation.
Several CT complexes or ion-radical salts of these donor radicals have been reported. Among
these donor radicals, the tetrathiapentalene derivative **18** was expected to afford a good
conducting material based on its π-extended structure. Fujiwara *et al.* reported the physical

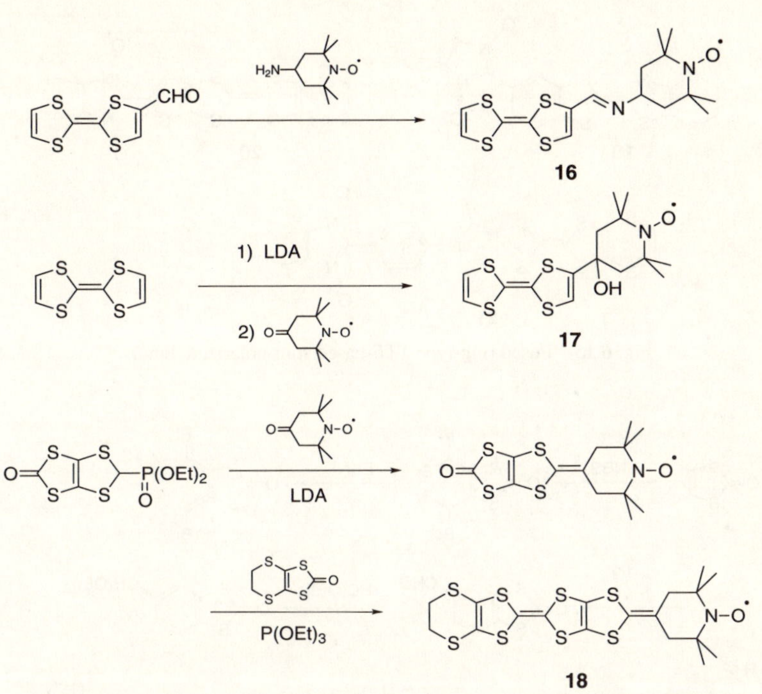

Fig. 6.14   Synthetic schemes of TTF derivatives carrying a TEMPO radical.

properties of the Au(CN)$_2$ salt of the donor radical **18**. The cyclic voltammogram of **18** showed four oxidation waves at +0.56, +0.88, +1.17 and +1.77 V *vs.* Ag/AgCl. Compared with the corresponding piperidine analogue, which showed three redox waves at +0.55, +0.87 and +1.74 V, the third oxidation potential (+1.17 V) of **18** can be attributed to the oxidation of the TEMPO radical. Based on X-ray crystallographic analysis of the Au(CN)$_2$ salt, the donor radical **18** in the salt was revealed to have 1.5 valence, although the nitroxide part is intact. The salt is a semiconductor with an activation energy of 0.20 eV. Temperature dependence of the magnetic susceptibility of the salt is roughly fitted to the Bleaney-Bowers expression and the S-T gap is estimated to be $2J \sim 200$ cm^{-1}. This indicates the existence of strong antiferromagnetic interaction.

6.5.2 Ion-radical Salt of Spin-polarized Donors

Although several TTF-based spin-polarized donors have been prepared, no crystalline ion-radical salts have been obtained. In order to improve the crystallinity of donor radicals, fused ring-type donor radicals **19-21** were prepared (Fig. 6.15).[26] The synthetic scheme of the donor radical **19**, which incorporates a fused benzene ring, is shown in Fig. 6.16. The key step is the formation of the benzene ring through the Diels-Alder reaction.

In fact, these fused ring-type donor radicals exhibited good crystallinity, and their molecular structures were all determined by X-ray crystallography. For example, the molecular structure of the donor radical **19** is shown in Fig. 6.17.[27] The intramolecular interaction of the donor radical depends on the twist angle between the donor and radical sites: the small dihedral angle of **19**,

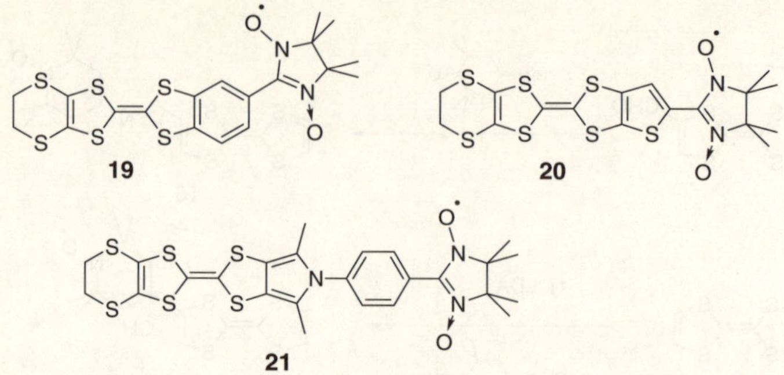

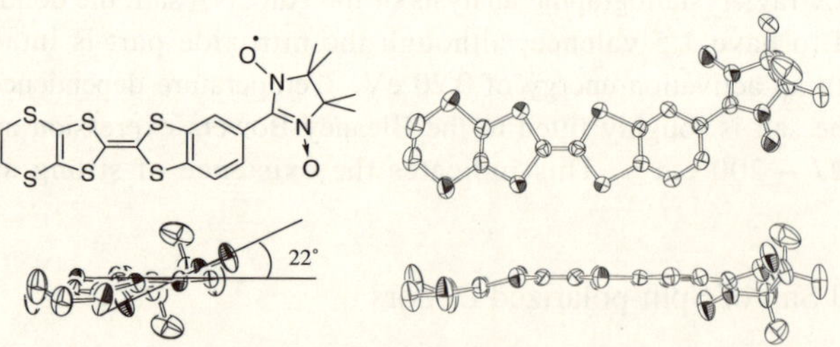

19   **20**

21

Fig. 6.15   Fused ring-type TTF-based spin-polarized donors.

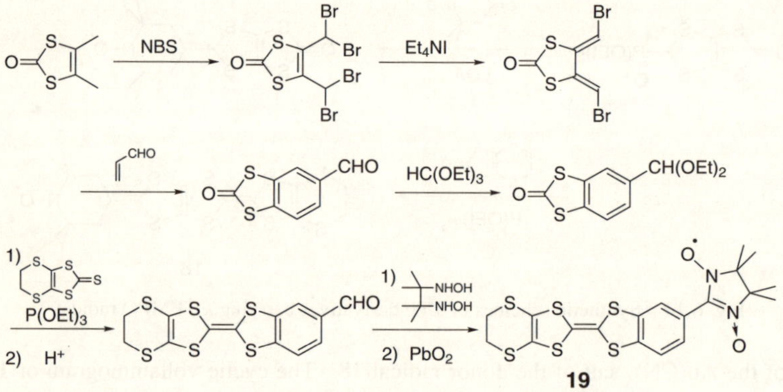

Fig. 6.16   Synthetic scheme of the fused ring-type donor radical **19**.

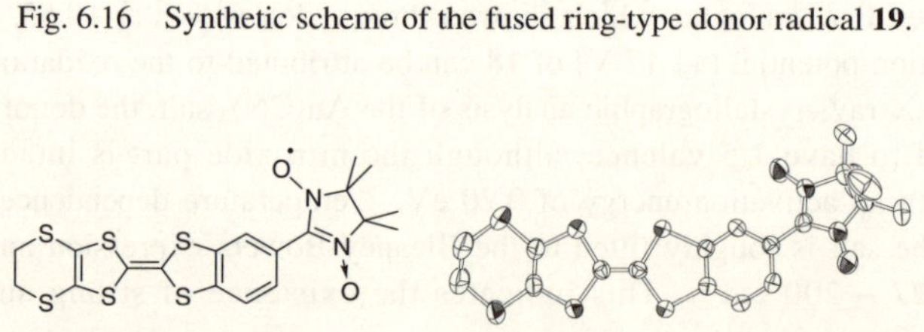

Fig. 6.17   Molecular structure of **19**.

only 22°, suggests an effective π-interaction. The coplanarity of the donor radical is expected to be improved especially in the singly oxidized state.

The trigonal unit cell of the crystal of **19** contains eighteen crystallographically equivalent molecules, as shown in Fig. 6.18. When the unit cell is viewed along the *c*-axis, each donor molecule forms a face-to-face dimer with inversion symmetry. These dimers are each correlated by a threefold rotatory-inversion axis, forming a propeller-like six-membered ring structure. The

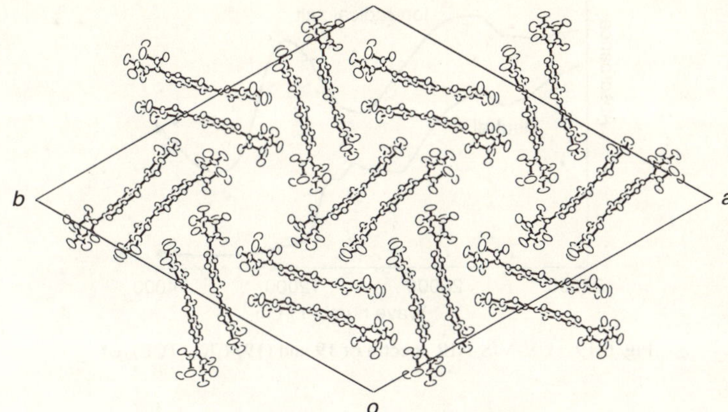

Fig. 6.18   Crystal structure of **19**.

donor molecules are arranged side-by-side along the *c*-axis by translation. Triangular prismatic structures are constructed along the threefold screw axes at the (1/3, 0), (2/3, 0), (0, 1/3), (0, 2/3), (1/3, 1/3) and (2/3, 2/3) positions in the *ab* plane.

Temperature dependence of the magnetic susceptibility of the crystals of **19** is well reproduced by the Curie-Weiss law with a positive Weiss temperature of 0.48 K. This suggests ferromagnetic interaction. Judging from the crystal structure, the ferromagnetic interaction is considered to be derived from the CH⋯ON type contacts between the neighboring NN moieties of the translated molecules along the *c*-axis [C-O: 3.60(1) Å] and the rotatory-inverted ones within the *ab* plane [C-O: 3.38(1), 3.52(1) Å].

Galvanostatic electrocrystallization of the donor radical **19** was performed in a 1 mM solution of **19** in 1,1,1-trichloroethane (TCE)-tetrahydrofuran (9:1 v/v) using 5 mM of *n*-Bu$_4$NClO$_4$ as the supporting electrolyte with a constant current of 0.7 μA for a period of one week. Very thin, black plate-like crystals were obtained, and the composition of the salt was estimated by elemental analysis to be (**19**)$_2$ClO$_4$(TCE)$_{0.5}$. It should be noted that the salt is stable even at room temperature for at least one month despite the rather limited kinetic stability of **19**$^{+\bullet}$ in solution.

The UV-VIS-NIR spectrum of the salt showed absorption maxima at 456 and 876 nm, together with a broad absorption band due to the interband transition extending from 1300 nm to the IR region (Fig. 6.19). The IR spectrum of the salt showed a sharp peak at 1350 cm^{-1}, which corresponds to the $\nu_{\text{N-O}}$ absorption of the neutral **19** observed at 1348 cm^{-1}.

The conductivity of the single crystal of the salt at room temperature was 1×10^{-2} S cm^{-1}, and the salt turned out to be a semiconductor. The Arrhenius plot of the resistivity of the salt was linear down to 150 K, and the activation energy was evaluated to be 0.16 eV.

Temperature dependence of the paramagnetic susceptibility was almost reproduced by the Curie-Weiss law with a Curie constant (*C*) of 0.73 emu K mol^{-1} and a Weiss temperature of -5.2 K (Fig. 6.20). The Curie constant is close to the theoretical value (*C* = 0.75 emu K mol^{-1}) for two independent *S* = 1/2 spins per chemical formula, suggesting that each donor radical carries one paramagnetic spin as depicted in Fig. 6.20b.

The ESR spectrum of the polycrystalline sample of the salt showed a narrow peak at *g* = 2.0068 with a peak-to-peak linewidth of 0.8 mT at room temperature. The ESR signals of the powdered neutral **19** and phenyl NN were observed at *g* = 2.0070 and 2.0068, respectively,

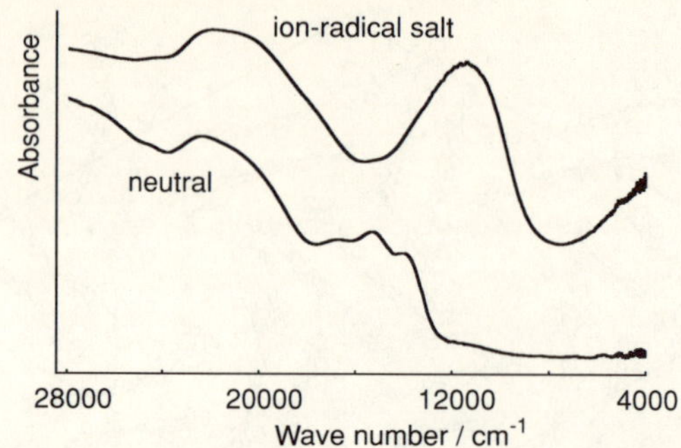

Fig. 6.19   UV-VIS-NIR spectra of **19** and **(19)**$_2$ClO$_4$(TCE)$_{0.5}$.

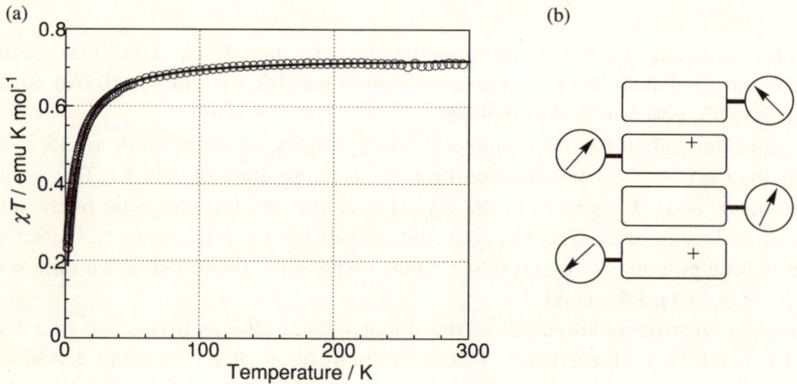

Fig. 6.20   (a) Temperature dependence of the magnetic susceptibility of **(19)**$_2$ClO$_4$(TCE)$_{0.5}$.
(b) Schematic drawing of the spin configuration in the ion-radical salt.

whereas that of ET in its ion-radical salt has been reported to show a relatively large anisotropy [g = 2.0125, 2.0022 for (ET)$_2$ClO$_4$(TCE)$_{0.5}$].[28] Accordingly, it is assumed that the ESR signal of the salt comes from the NN group. Temperature dependence of the ESR signal intensity also obeyed the Curie-Weiss law and no thermal excitation was recognized.

According to the magnetic data, each donor radical has only one paramagnetic spin in spite of the 2:1 ratio of donor to counterion. This suggests that the generated π-delocalizing spin on the TTF skeleton does not contribute to the magnetic susceptibility of the salt. Although the π-spins on the donor sites are conductive, they are considered to be coupled strongly in an antiferromagnetic manner, and the density of the charge carrier is very low in this salt, as seen from the conductivity data. Therefore, the spin polarization of the π-spin caused by the local spin on the radical site cannot be transmitted to the local spin on the adjacent donor radical.

6.6   Summary and Outlook

Since the strategy for the spin alignment of localized spins on π-radical units through spin-polarized conduction electrons has been established, one may design a ferromagnetic molecular wire which passes a spin-polarized current along a π-conjugating polymer carrying π-radical units as pendant groups. A hole-doped polypyrrole is known to exhibit high conductivity on the basis of the polaronic mechanism. If a pyrrole-based spin-polarized donor is used as the building block of a conducting polymer, the electron transmitted along the chain is spin-polarized to β, aligning all the localized spins on the pendant groups to α. Such a spin-polarized magnetic wire is indispensable to the quantum spin device.

A unimolecular quantum spin device is one of the most challenging targets for organic spin systems. If a spin-polarized donor is used as the building block of a conducting polymer, polaronic conduction electrons on the π-conjugated chain may couple ferromagnetically with the localized spin on the radical site on the basis of a double-exchange mechanism. As a spin-aligning unit for the above purpose, N-(p-NN-phenyl)pyrrole (NN = nitronyl nitroxide) was selected because it was found to afford a ground state triplet species upon one-electron oxidation.

In order to incorporate the spin-aligning unit into a molecular wire, N-(p-NN-phenyl)pyrrole carrying quaterthiophenes at both α positions of the pyrrole ring has been prepared.[29] Thiophene oligomers carrying stable radical groups at every several monomer units will also be synthesized. These oligomers may be regarded as ferromagnetic wires to pass a spin-polarized current, if both ends of the oligomer are connected to electrodes. Such a ferromagnetic wire is expected to be a crucial part of quantum spin devices.

References

1. For a recent overview of molecular magnetism see: *Molecular Magnetism - New Magnetic Materials*, Kodansha & Gordon and Breach, Tokyo (2000); *Magnetic Properties of Organic Materials*, Marcel Dekker, New York (1999).
2. M. Kinoshita, P. Turek, M. Tamura, K. Nozawa, D. Shiomi, Y. Nakazawa, M. Ishikawa, M. Takahashi, K. Awaga, T. Inabe and Y. Maruyama, *Chem. Lett.*, 1225 (1991); R. Chiarelli, M. A. Novak, A. Rassat and J. L. Tholence, *Nature*, **363**, 147 (1993); T. Nogami, K. Tomioka, T. Ishida, H. Yoshikawa, M. Yasui, F. Iwasaki, H. Iwamura, N. Takeda and M. Ishikawa, *Chem. Lett.*, 29 (1994); T. Sugawara, M. M. Matsushita, A. Izuoka, N. Wada, N. Takeda and M. Ishikawa, *J. Chem. Soc. Chem. Commun.*, 1723 (1994); J. Cirujeda, M. Mas, E. Molins, F. L. Panthou, J. Laugier, J. G. Park, C. Paulsen, P. Rey, C. Rovira and J. Veciana, *J. Chem. Soc. Chem. Commun.*, 709 (1995); A. Caneschi, F. Ferraro, D. Gatteschi, A. Lirzin, M. A. Novak, E. Rentschler and R. Sessoli, *Adv. Mater.*, **7**, 476 (1995); K. Mukai, K. Konishi, K. Nedachi and K. Takeda, *J. Magn. Magn. Mater.*, **140–144**, 1449 (1995); K. Togashi, R. Imachi, K. Tomioka, H. Tsuboi, T. Ishida, T. Nogami, N. Takeda and M. Ishikawa, *Bull. Chem. Soc. Jpn.*, **69**, 2821 (1996); M. M. Matsushita, A. Izuoka, T. Sugawara, T. Kobayashi, N. Wada, N. Takeda and M. Ishikawa, *J. Am. Chem. Soc.*, **119**, 4369 (1997).
3. T. Sugawara, *J. Synth. Org. Chem. Jpn*, **47**, 306 (1989); K. Yamaguchi, H. Namimoto, T. Fueno, T. Nogami and Y. Shirota, *Chem. Phys. Lett.*, **166**, 408 (1990); A. Izuoka, R. Kumai, T. Tachikawa and T. Sugawara, *Mol. Cryst. Liq. Cryst.*, **218**, 213 (1992).
4. M. Imada, A. Fujimori and Y. Tokura, *Rev. Mod. Phys.*, **70**, 1039 (1998).
5. T. Mallah, C. Hollis, S. Bott, M. Kurmoo, P. Day, M. Allan and R. H. Friend, *J. Chem. Soc. Dalton Trans.*, 859 (1990).
6. J. Yamaura, K. Suzuki, Y. Kaizu, T. Enoki, K. Murata and G. Saito, *J. Phys. Soc. Jpn.*, **65**, 2645 (1996); T. Enoki, J. Yamaura and A. Miyazaki, *Bull. Chem. Soc. Jpn.*, **70**, 2005 (1997).
7. E. Coronado, J. R. Galán-Mascarós, C. J. Gómez-García and V. Laukhin, *Nature*, **408**, 447 (2000).
8. H. Kobayashi, H. Tomita, T. Naito, A. Kobayashi, F. Sakai, T. Watanabe and P. Cassoux, *J. Am. Chem. Soc.*, **118**, 368 (1996); H. Kobayashi, A. Sato, E. Arai, H. Akutsu, A. Kobayashi and P. Cassoux, *J. Am.*

Chem. Soc., **119**, 12392 (1997); H. Akutsu, E. Arai, H. Kobayashi, H. Tanaka, A. Kobayashi and P. Cassoux, *J. Am. Chem. Soc.*, **119**, 12681 (1997); H. Akutsu, K. Kato, E. Ojima, H. Kobayashi, H. Tanaka, A. Kobayashi and P. Cassoux, *Phys. Rev.*, *B* **58**, 9294 (1998); L. Brossard, R. Clerac, C. Coulon, M. Tokumoto, T. Ziman, D. K. Petrov, V. N. Laukhin, M. J. Naughton, A. Audouard, F. Goze, A. Kobayashi, H. Kobayashi and P. Cassoux, *Eur. Phys. J. B*, **1**, 439 (1998); E. Ojima, H. Fujiwara, K. Kato and H. Kobayashi, *J. Am. Chem. Soc.*, **121**, 5581 (1999); H. Kobayashi, A. Kobayashi and P. Cassoux, *Chem. Soc. Rev.*, **29**, 325 (2000); H. Fujiwara, E. Fujiwara, Y. Nakazawa, B. Z. Narymbetov, K. Kato, H. Kobayashi, A. Kobayashi, M. Tokumoto and P. Cassoux, *J. Am. Chem. Soc.*, **123**, 306 (2001); S. Uji, H. Shinagawa, T. Terashima, T. Yakabe, Y. Terai, M. Tokumoto, A. Kobayashi, H. Tanaka and H. Kobayashi, *Nature*, **410**, 908 (2001); H. Fujiwara, H. Kobayashi, E. Fujiwara and A. Kobayashi, *J. Am. Chem. Soc.*, **124**, 6816 (2002): B. Zhang, H. Tanaka, H. Fujiwara, H. Kobayashi, E. Fujiwara and A. Kobayashi, *J. Am. Chem. Soc.*, **124**, 9982 (2002).

9. M. Kurmoo, A. W. Graham, P. Day, S. J. Coles, M. B. Hursthouse, J. L. Caulfield, J. Singleton, F. L. Pratt, W. Hayes, L. Ducasse and P. Guionneau, *J. Am. Chem. Soc.*, **117**, 12209 (1995); P. Day and M. Kurmoo, *J. Mater. Chem.*, **7**, 1291 (1997); R. Kumai, A. Asamitsu and Y. Tokura, *Chem. Lett.*, 753 (1996); H. Mori, N. Sakurai, S. Tanaka and H. Moriyama, *Bull. Chem. Soc. Jpn.*, **72**, 683 (1999); M. Iwamatsu, T. Kominami, K. Ueda, T. Sugimoto, T. Tada, K. Nishimura, T. Adachi, H. Fujita, F. Guo, S. Yokogawa, H. Yoshino, K. Murata and M. Shiro, *J. Mater. Chem.*, **11**, 385 (2001); T. Kominami, T. Matsumoto, K. Ueda, T. Sugimoto, K. Murata, M. Shiro and H. Fujita, *J. Mater. Chem.*, **11**, 2089 (2001); S. Rashid, S. S. Turner, P. Day, J. A. K. Howard, P. Guionneau, E. J. L. McInnes, F. E. Mabbs, R. J. H. Clark, S. Firth and T. Biggs, *J. Mater. Chem.*, **11**, 2095 (2001); T. Naito, T. Inabe, K. Takeda, K. Awaga, T. Akutagawa, T. Hasegawa, T. Nakamura, T. Kakiuchi, H. Sawa, T. Yamamoto and H. Tajima, *J. Mater. Chem.*, **11**, 2221 (2001); J. Yamada, T. Toita, H. Akutsu, S. Nakatsuji, H. Nishikawa, I. Ikemoto and K. Kikuchi, *Chem. Commun.*, 2538 (2001).

10. H. M. McConnell, *J. Chem. Phys.*, **39**, 1910 (1963).

11. A. Izuoka, S. Murata, T. Sugawara and H. Iwamura, *J. Am. Chem. Soc.*, **107**, 1786 (1985); A. Izuoka, S. Murata, T. Sugawara and H. Iwamura, *J. Am. Chem. Soc.*, **109**, 2631 (1987).

12. H. M. McConnell, *Proc. R. A. Welch Found. Conf. Chem. Res.*, **11**, 144 (1968).

13. R. Breslow, *Pure Appl. Chem.*, **54**, 927 (1982).

14. J. S. Miller, D. A. Dixon, J. C. Calabrese, C. Vazquez, P. J. Krusic, M. D. Ward, E. Wasserman and R. L. Harlow, *J. Am. Chem. Soc.*, **112**, 381 (1990).

15. H. Fukutome, A. Takahashi and M. Ozaki, *Chem. Phys. Lett.*, **133**, 34 (1987); M. M. Murray, P. Kaszynski, D. A. Kaisaki, W. Chang and D. A. Dougherty, *J. Am. Chem. Soc.*, **116**, 8152 (1994); K. Yoshizawa, A. Chano, A. Ito, K. Tanaka, T. Yamabe, H. Fujita and J. Yamauchi, *Chem. Lett.*, 369 (1992); K. Yoshizawa, M. Hatanaka, H. Ago, K. Tanaka and T. Yamabe, *Bull. Chem. Soc. Jpn.*, **69**, 1417 (1996); K. R. Stickley and S. C. Blackstock, *J. Am. Chem. Soc.*, **116**, 11576 (1994); K. R. Stickley, T. D. Selby and S. C. Blackstock, *J. Org. Chem.*, **62**, 448 (1997); M. M. Wienk and R. A. J. Janssen, *Chem. Commun.*, 267 (1996); M. M. Wienk and R. A. J. Janssen, *J. Am. Chem. Soc.*, **118**, 10626 (1996); M. M. Wienk and R. A. J. Janssen, *J. Am. Chem. Soc.*, **119**, 4492 (1997); R. J. Bushby and K. M. Ng, *Chem. Commun.*, 659 (1996); R. J. Bushby, D. R. McGill, K. M. Ng and N. Taylor, *J. Chem. Soc. Perkin Trans. 2*, 1405 (1997); R. J. Bushby and D. Gooding, *J. Chem. Soc. Perkin Trans. 2*, 1069 (1998); K. Sato, M. Yano, M. Furuichi, D. Shiomi, T. Takui, K. Abe, K. Itoh, A. Higuchi, K. Katsuma and Y. Shirota, *J. Am. Chem. Soc.*, **119**, 6607 (1997).

16. E. Dormann, M. J. Nowak, K. A. Williams, R. O. Angus Jr. and F. Wudl, *J. Am. Chem. Soc.*, **109**, 2594 (1987).

17. M. Iyoda, M. Fukuda, M. Yoshida and S. Sasaki, *Chem. Lett.*, 2369 (1994).

18. J. Tanabe, T. Kudo, M. Okamoto, Y. Kawada, G. Ono, A. Izuoka and T. Sugawara, *Chem. Lett.*, 579 (1995); J. Tanabe, G. Ono, A. Izuoka, T. Sugawara, T. Kudo, T. Saito, M. Okamoto and Y. Kawada, *Mol. Cryst. Liq. Cryst.*, **296**, 61 (1997); A. Izuoka, J. Tanabe, T. Sugawara, Y. Kawada, R. Kumai, A. Asamitsu and Y. Tokura, *Mol. Cryst. Liq. Cryst.*, **306**, 265 (1997); T. Sugawara and A. Izuoka, *Mol. Cryst. Liq. Cryst.*, **305**, 41 (1997); T. Sugawara, *Mol. Cryst. Liq. Cryst.*, **334**, 257 (1999); A. Izuoka, J. Tanabe, T. Sugawara, T. Kudo, T. Saito and Y. Kawada, *Chem. Lett.*, 910 (2002).

19. P. Dowd, *Acc. Chem Res.*, **5**, 242 (1972); W. T. Borden and E. R. Davidson, *J. Am. Chem. Soc.*, **99**, 4587 (1977); J. A. Berson, *Acc. Chem. Res.*, **11**, 446 (1978); P. G. Wenthold, J. Hu, R. R. Squires and W. C. Lineberger, *J. Am. Chem. Soc.*, **118**, 475 (1996).

20. H. Sakurai, R. Kumai, A. Izuoka and T. Sugawara, *Chem. Lett.*, 879 (1996); R. Kumai, H. Sakurai, A. Izuoka and T. Sugawara, *Mol. Cryst. Liq. Cryst.*, **279**, 133 (1996); H. Sakurai, A. Izuoka and T. Sugawara, *J. Am. Chem. Soc.*, **122**, 9723 (2000).

21. R. Kumai, M. M. Matushita, A. Izuoka and T. Sugawara, *J. Am. Chem. Soc.*, **116**, 4523 (1994).

22. J. Nakazaki, M. M. Matsushita, A. Izuoka and T. Sugawara, *Tetrahedron Lett.*, **40**, 5027 (1999).

23. T. Sugawara, *Mol. Cryst. Liq. Cryst.*, **334**, 257 (1999); A. Izuoka, M. Hiraishi, T. Abe, T. Sugawara, K. Sato and T. Takui, *J. Am. Chem. Soc.*, **122**, 3234 (2000).
24. G. Harada, T. Jin, A. Izuoka, M. M. Matsushita and T. Sugawara, *Tetrahedron Lett.*, **44**, 4415 (2003).
25. T. Sugano, T. Fukasawa and M. Kinoshita, *Synth. Met.*, **41–43**, 3281 (1991); T. Sugimoto, S. Yamaga, M. Nakai, M. Tsuji, H. Nakatsuji and N. Hosoito, *Chem. Lett.*, 1817 (1993); T. Ishida, K. Tomioka, T. Nogami, H. Iwamura, K. Yamaguchi, W. Mori and Y. Shirota, *Mol. Cryst. Liq. Cryst.*, **232**, 99 (1993); S. Nakatsuji, N. Akashi, Y. Komori, K. Suzuki, T. Enoki and H. Anzai, *Mol. Cryst. Liq. Cryst.*, **279**, 73 (1996); S. Nakatsuji, A. Takai, K. Nishikawa, Y. Morimoto, N. Yasuoka, K. Suzuki, T. Enoki and H. Anzai, *Chem. Commun.*, 275 (1997); S. Nakatsuji and H. Anzai, *J. Mater. Chem.*, **7**, 2161 (1997); S. Nakatsuji, A. Takai, K. Nishikawa, Y. Morimoto, N. Yasuoka, K. Suzuki, T. Enoki and H. Anzai, *J. Mater. Chem.*, **9**, 1747 (1999); H. Fujiwara and H. Kobayashi, *Chem. Commun.*, 2417 (1999).
26. Y. Ishikawa, T. Miyamoto, A. Yoshida, Y. Kawada, J. Nakazaki, A. Izuoka and T. Sugawara, *Tetrahedron Lett.*, **40**, 8819 (1999).
27. J. Nakazaki, Y. Ishikawa, A. Izuoka, T. Sugawara and Y. Kawada, *Chem. Phys. Lett.*, **319**, 385 (2000).
28. T. Enoki, K. Imaeda, M. Kobayashi, H. Inokuchi and G. Saito, *Phys. Rev.*, *B* **33**, 1553 (1986).
29. J. Nakazaki, M. M. Matsushita, A. Izuoka and T. Sugawara, *Mol. Cryst. Liq. Cryst.*, **306**, 81 (1997); J. Nakazaki, I. Chung, M. M. Matsushita, T. Sugawara, R. Watanabe, A. Izuoka and Y. Kawada, *J. Mater. Chem.*, **13**, 1011 (2003); J. Nakazaki, I. Chung, R. Watanabe, T. Ishitsuka, Y. Kawada, M. M. Matsushita and T. Sugawara, *Int. Electr. J. Mol. Des.*, **2**, 112 (2003).

7

TTFs as Ligands of Metal Complexes

7.1   Introduction

Bis(dithiolene) metal complexes with sulfur-rich ligands have been synthesized and their electrical conducting properties investigated. The Ni, Pd and Pt complexes with malononitriledithiolato (mnt) or isotrithionedithiolato (dmit) groups (Scheme 7.1) were used as acceptors to give their highly conducting alkali metal and ammonium salts as well as their charge-transfer (CT) salts with TTF. Almost all the salts were semiconducting, but some salts exhibited metallic and even superconductivity.[1,2] In contrast, the metal complexes with 5,6-dihydro-1,4-dithiin-2,3-dithiolato (dddt) groups worked as donors, and in one of their ClO_4 and BF_4 salts metallic conductivity was observed.[3]

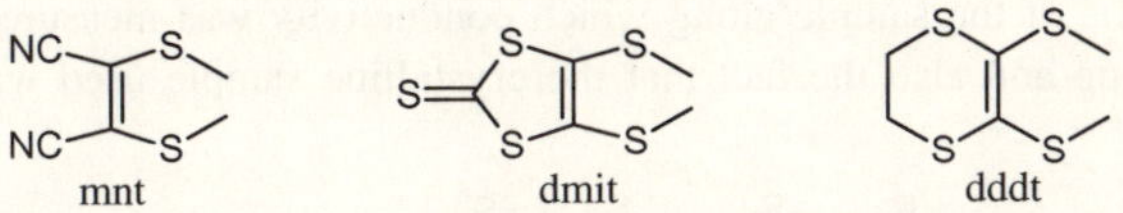

mnt   dmit   dddt

Scheme 7.1

Very recently, more attention has been directed toward neutral metal complexes with sulfur-rich ligands, since all organic metals and superconductors obtained so far are based on more than two components and there has till quite recently been no known organic single-component metal or superconductor. The finding that neutral Au complexes with benzenedithiolato,[4,5] 2,3-dihydro-4,5-thiophenedithiolato[6] and 2,3-[6] and 3,4-thiophenedithiolato groups[6] (Scheme 7.2) exhibited comparatively high electrical conductivities (σ_{RT}'s) at room temperature raised expectations that the first organic single-component metal and even superconductor might emerge from one of the neutral metal complexes with sulfur-rich ligands. In particular, a tetrathiafulvalenyldithiolato (TTFdithiolato) group was a most promising candidate, since it was a highly extended π-electron system involving many sulfur atoms to bring about strong intermolecular interaction. Indeed, high electrical conductivities were observed for several neutral metal complexes with TTFdithiolato groups, and eventually the first organic single-component metal was discovered from the Ni complex. This chapter mainly reviews the synthesis of several neutral metal (Ni, Pd, Pt, Cu, Au, Hg, V, Mn) complexes with TTF-dithiolato and -diselenolato groups and their electrical conducting/magnetic properties.

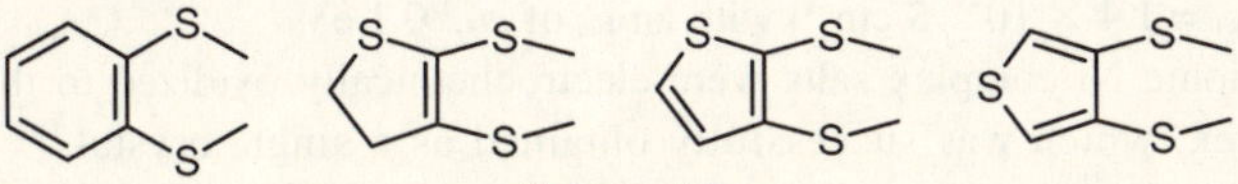

Scheme 7.2

7.2   Ni, Pd and Pt Complexes with TTFdithiolato Ligands

DitrifluoromethylTTFdithiolato (Scheme 7.3) dianion was reacted with $Ni(OAc)_2$ in CH_3OH, followed by treatment with O_2 to give the first neutral bis(ditrifluoromethylTTFdithiolato) Ni complex as a green-black platelet.[7]   The σ_{RT} was not reported.   However, afterwards new neutral Ni complexes with dimethylthio-, diethylthio- and di(n-butylthio)-TTFdithiolato groups[8] were prepared by chemical (I_2 and O_2) or electrochemical oxidation of the corresponding dianionic complexes, and the compressed pellets of the neutral complexes exhibited comparatively high σ_{RT}'s of 10^{-4}–10^{-1} S cm^{-1}.   The crystal structure of the neutral bis(diethylthioTTFdithiolato) Ni complex was successfully analyzed (see Fig. 7.1), in which (1) the whole molecular skeleton except for the four alkylthio groups is essentially planar, (2) the molecules stack in a slightly slipped manner with an interplanar distance of $ca.$ 3.4 Å and (3) there is the shortest S···S contact with a distance of 3.51 Å between the molecules in the neighboring columns.   Electrical conductivity (σ) was measured for the single crystal of the neutral bis(diethylthioTTFdithiolato) Ni complex ($\sigma_{RT} = 2.7 \times 10^{-2}$ S cm^{-1}) and found to drop steadily with decreasing temperature. This shows a semiconducting behavior with an activation energy (E_a) of 0.10 eV.   Nevertheless, in the measurement of the compressed pellet a metallic temperature dependence of σ was observed between 275 and 300 K, although below 275 K it changed to a semiconducting behavior with an E_a of 0.09 eV.   This difference in electrical conducting behavior between the single crystal and the compressed pellet of the same neutral Ni complex may reflect the anisotropy of the material since the axis of the sample along which conductivity was measured is not necessarily the highest conducting and also the fact that the crystalline sample used was not a true single crystal.

R = CH_3, CF_3, SMe, SEt, S(n-Bu)

R, R = -$(CH_2)_3$-, -$O(CH_2)_2O$-, -$S(CH_2)_2S$-, -$S(CH_2)_3S$-

Scheme 7.3

A neutral Ni complex with two propylenedithioTTFdithiolato groups was also prepared.[9] The precursors, mono- and di-anionic complex salts, were first isolated by a procedure similar to that described above.   The NMe_4^+ and PPh_4^+ salts of the monoanionic Ni complex were isolated as a single crystal.   The crystal structures are shown in Figs. 7.2 and 7.3.   For both salts the geometry around the Ni atoms is almost square planar.   The anionic Ni complex molecules are stacked in a slightly slipped fashion to form a one-dimensional column.   The adjacent columns are connected by transverse short S···S contacts.   The NMe_4^+ salt of the monoanionic Ni complex was a semiconductor ($\sigma_{RT} = 1.4 \times 10^{-3}$ S cm^{-1}) with an E_a of $ca.$ 0.1 eV.

The monoanionic Ni complex salts were electrochemically oxidized to the corresponding neutral Ni complex, which was successfully obtained as a single crystal.[10]   The neutral Ni complex has a molecular structure such that the geometry around the Ni atom is square planar and

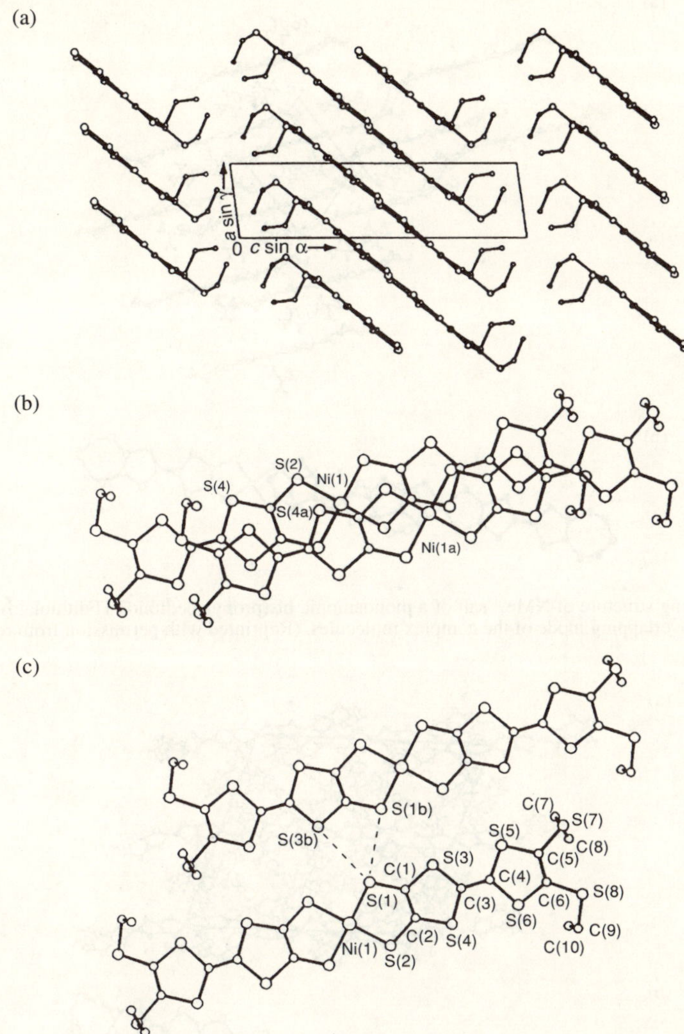

Fig. 7.1   (a) Packing structure viewed down to the *b* direction, and two adjacent molecules separated by a unit translation in the (b) *a* and (c) *b* directions for a neutral bis(diethylthioTTFdithiolato) Ni complex. (Reprinted with permission from ref. 8)

the eight sulfur atoms of the propylenedithiolatoTTFdithiolato group are essentially coplanar except for the bent terminal propylene group (see Fig. 7.4). The molecules form one-dimensional columns having short intermolecular transverse S···S contacts. The detailed overlap modes between the molecules along (a) the stacking direction, (b) the transverse direction and (c) the long axis of the molecule are shown in Fig. 7.5. The crystal exhibited the highest σ_{RT} of 7 S cm^{-1} among the neutral metal complex-based organic conductors. The electrical resistivity ($\rho = 1/\sigma$) exhibited a monotonic increase down to 60 K with the E_a of 0.03 eV (see Fig. 7.6). Below 60 K the ρ still increased with lowering temperature, but the E_a became larger. From comparison

(a)

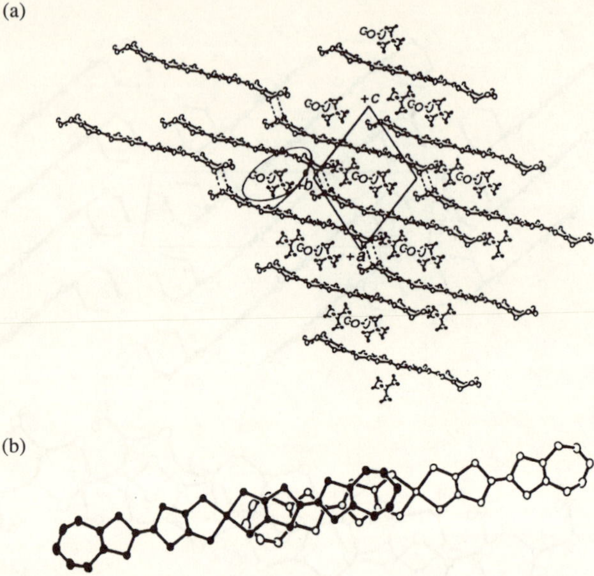

(b)

Fig. 7.2   (a) Packing structure of NMe$_4^+$ salt of a monoanionic bis(propylenedithioTTFdithiolato) Ni complex and (b) overlapping mode of the complex molecules. (Reprinted with permission from ref. 9)

(a)

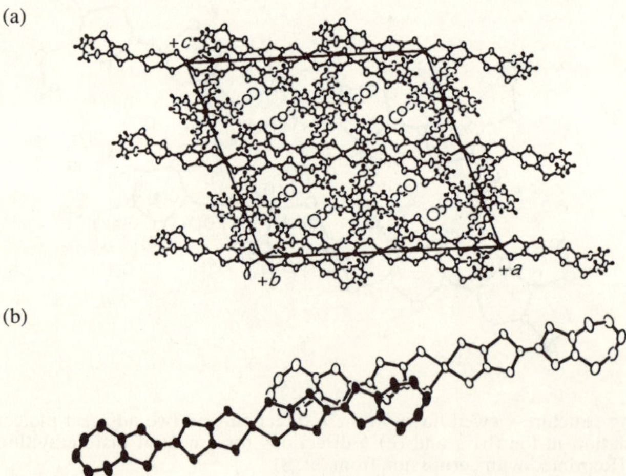

(b)

Fig. 7.3   (a) Packing structure of PPh$_4^+$ salt of a monoanionic bis(propylenedithioTTFdithiolato) Ni complex and (b) overlapping mode of the complex molecules. (Reprinted with permission from ref. 9)

between the crystal structures at room temperature and 30 K, it was shown that the main cause of the phase transition in the electrical conducting state at around 60 K is due to the development of a superlattice accompanying the crystal structure change. A gradual suppression of the phase transition was observed with increasing pressure, but no conversion of the semiconductivity to metallic behavior occurred.

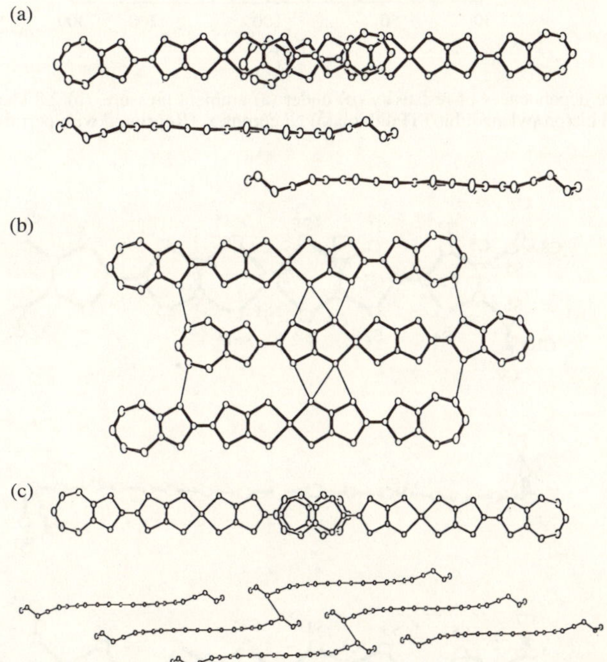

Fig. 7.4   Molecular structure of a neutral bis(propylenedithioTTFdithiolato) Ni complex: (a) top and (b) side views. (Reprinted with permission from ref. 10)

Fig. 7.5   (a) Stacking structure of neutral bis(propylenedithioTTFdithiolato) Ni complex molecules, (b) molecular arrangement along the transverse direction and (c) the overlapping mode along the long axis of the molecule. (Reprinted with permission from ref. 10)

The monoanionic and neutral Ni complexes with TTFdiselenolato groups were also prepared.[11] Bis(2'-cyanoethylseleno)diethylthioTTF was reacted with NMe₄OH in the presence of N(n-C₆H₁₃)₄Br in MeOH, followed by the addition of nickel acetylacetonate to give an N(n-C₆H₁₃)₄⁺ salt of the monoanionic bis(diethylthioTTFdiselenolato) Ni complex in 89% yield. The corresponding monoanionic bis(diethylthioTTFdithiolato) Ni complex was obtained in 37% yield by using bis(2'-cyanoethylthio)diethylthioTTF as the starting material in a similar procedure. The single crystals obtained by recrystallization from CS_2 were examined by X-ray structure analysis. The molecular and crystal structures of the two monoanionic Ni complexes are shown

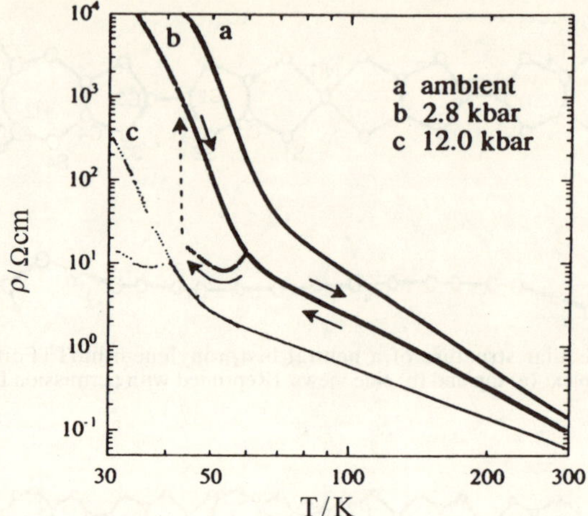

Fig. 7.6   Temperature dependences of resistivity (ρ) under (a) ambient pressure, (b) 2.8 kbar and (c) 12.0 kbar for a neutral bis(propylenedithioTTFdithiolato) Ni complex. (Reprinted with permission from ref. 10)

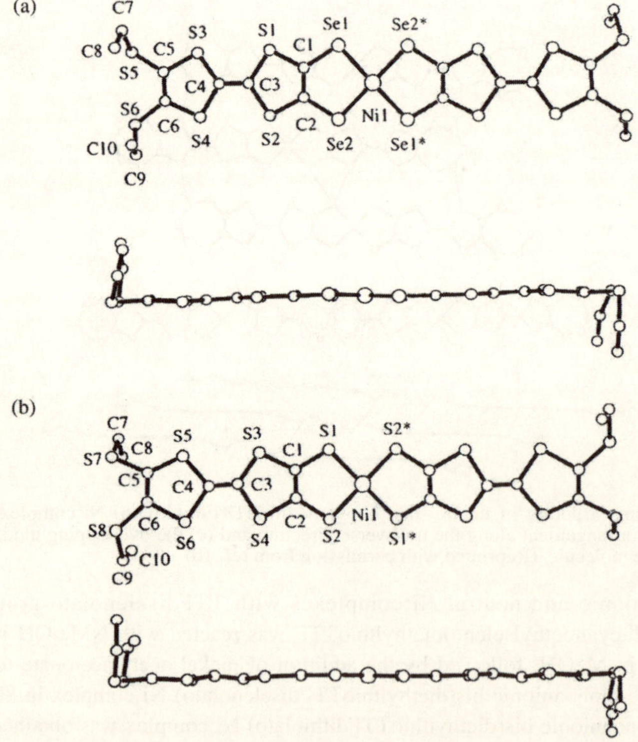

Fig. 7.7   Molecular structures of monoanionic (a) bis(diethylthioTTFdiselenolato) and (b) bis(diethylthioTTFdithiolato) Ni complexes. (Reprinted with permission from ref. 11)

in Figs. 7.7 and 7.8. In both cases the geometry around the Ni atom is essentially a square plane. Each of the TTF moieties involving two S and two Se atoms or four S atoms has high planarity, so that the entire molecular skeleton, except for the four Et groups, is almost coplanar. Each of the monoanionic Ni complex molecules is surrounded by four $N(n\text{-}C_6H_{13})_4^+$ ions, and there is no close contact between the molecules. Two different arrangements of the molecules are present, and they are almost perpendicular to each other.

Each of the monoanionic bis(diethylthioTTFdiselenolato) and bis(diethylthioTTFdithiolato) Ni complex molecules possesses one $S = 1/2$ spin. The ESR spectra of the microcrystals measured at 297 K showed two signals: $g_{//} = 2.105$ and $g_{\perp} = 2.060$ for the former salt, and $g_{//} = 2.069$ and $g_{\perp} = 2.024$ for the latter salt. These $g_{//}$ and $g_{\perp}$ values are very close to those of monoanionic Ni complexes with small π-conjugated bis(ethylenediselenolato) and bis(ethylenedithiolato) ligands, so that the spins residing in the monoanionic bis(diethylthioTTFdiselenolato) and bis(diethylthioTTFdithiolato) Ni complexes appear to be preferentially localized in the vicinity of the Ni atom. In addition, SQUID magnetic measurements showed the interaction between the spins of the monoanionic Ni molecules to be very weakly antiferromagnetic, as expected from the crystal structures with almost no contact between the neighboring molecules. In correspondence to the ESR and SQUID measurement

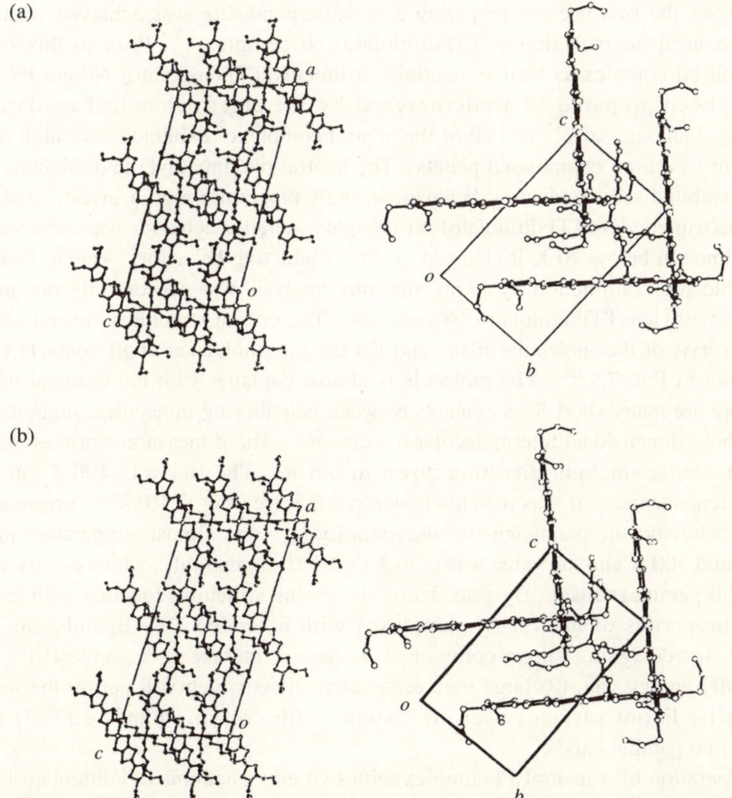

Fig. 7.8   Packing structures of $N(n\text{-}C_6H_{13})_4^+$ salts of monoanionic (a) bis(diethylthioTTFdiselenolato) and (b) bis(diethylthioTTFdithiolato) Ni complexes. (Reprinted with permission from ref. 11)

results, the σ_{RT}'s of the single crystals were very low ($< 10^{-10}$ S cm^{-1}).

The conversion of the monoanionic bis(diethylthioTTF-dithiolato or -diselenolato) Ni complex salt to the corresponding neutral compound was again carried out by a diffusion method in which I_2 vapor was gradually added to a CS_2 solution of the monoanionic complex salt at room temperature to obtain the single crystal. The black microcrystal was isolated in 67% yield, but unfortunately the crystal could not undergo X-ray structure analysis because it was too small. From the magnetic and ESR results, almost all the spins are distributed to the TTF ligands and can contribute to the electrical conduction, while only a very small amount of spin remains within the Ni-Se or Ni-S vicinity. As expected from the above spin distribution, bis(diethylthioTTF-diselenolato) and bis(diethylthioTTFdithiolato) Ni complexes exhibited considerably high σ_{RT}'s of 2.8 and 5.0 $\times$ 10^{-2} S cm^{-1} on the compressed pellets, respectively. It should be noted that the σ_{RT} of the Se compound is higher by *ca.* 60 times than that of the S compound, although such a comparison between the σ_{RT}'s on the compressed pellets involves some problems. The temperature dependence of ρ in the Se compound is semiconducting, but the E_a is very small (14 meV). The fact that the Se atoms bound to Ni exert a stronger effect on the electrical conduction than do the corresponding S atoms may be related to the much greater orbital overlap between the TTFdiselenolato groups than between the TTFdithiolato groups within and between columns.

After numerous attempts to prepare single component-based organic metallic conductors and superconductors, the first success preparing a metallic conductor was achieved with the single crystal of a neutral bis(trimethyleneTTFdithiolato) Ni complex.[12a] Prior to this success, the present neutral Ni complex as well as neutral bis(dimethylTTFdithiolato) Ni and Pd complexes had already been prepared as a microcrystal by the electrochemical oxidation of the corresponding dianionic salts,[13] and all of the neutral complexes exhibited very high σ_{RT}'s of *ca.* 100–400 S cm^{-1} even on compressed pellets. The neutral bis(dimethylTTFdithiolato) Ni and Pd complexes exhibited semiconducting behavior in the temperature region investigated, while for the neural bis(trimethyleneTTFdithiolato) Ni complex metallic behavior was observed between 70–300 K, although below 70 K it changed to semiconducting behavior. Among them, a single crystal suitable for examination by X-ray structure analysis was successfully obtained for the neutral bis(trimethyleneTTFdithiolato) Ni complex. The crystal structures viewed along (a) the perpendicular axis of the molecular plane and (b) the intermolecular short contacts along the *a* axis are shown in Fig. 7.9.[12b] The molecule is almost coplanar with the terminal trimethylene groups. There are many short S···S contacts between neighboring molecules, suggesting that the system has three-dimensional intermolecular interactions. The ρ measurement was made closely along the molecule-stacking direction down to 0.6 K. The σ_{RT} was 400 S cm^{-1}, and the temperature dependence of σ was metallic down to 0.6 K (see Fig. 7.10).[12a] Corresponding with the metallic behavior, the paramagnetic susceptibility (χ_p) was almost temperature-independent between 0.5 and 300 K and the value was (2.6–3.1) $\times$ 10^{-4} emu mol^{-1}, which is very close to the value for Pauli paramagnetism. By considering the crystal structures together with the electrical conducting properties of neutral Ni complexes with extended TTF ligands, the following requirements for designing single-component molecular metals were suggested: (1) a small HOMO-LUMO energy gap, (2) large transverse interactions which will permit the formation of (semi-)metallic Fermi surfaces even in systems with crossing bands and (3) molecular arrangement into parallel bands.

The preparation of a neutral Pt complex with two ethylenedithioTTFdithiolato ligands was also attempted by the reaction of ethylenedithioTTFdithiolate with *cis*-PtCl$_2$(SMe)$_2$, but only a mixture of the neutral and monoanionic bis(ethylenedithioTTFdithiolato) Pt complexes with a σ_{RT} of 0.2 S cm^{-1} was isolated.[14]

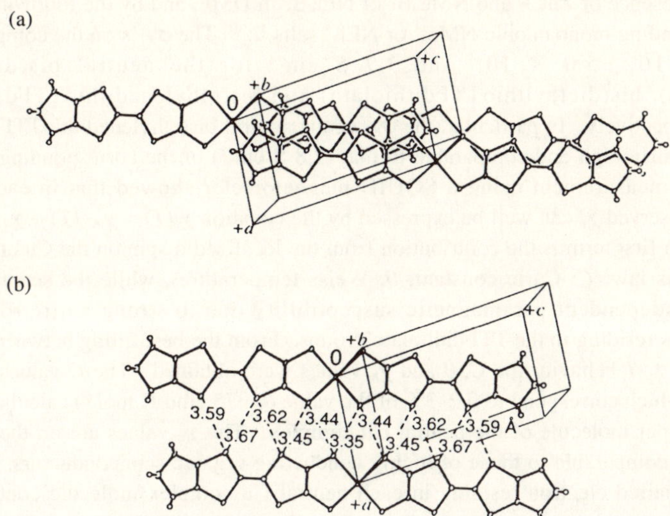

Fig. 7.9   Molecular packing viewed along (a) the perpendicular axis of the molecular plane and (b) intermolecular short contacts along the a axis for a neutral bis(trimethyleneTTFdithiolato) Ni complex. (Reprinted with permission from ref. 12b)

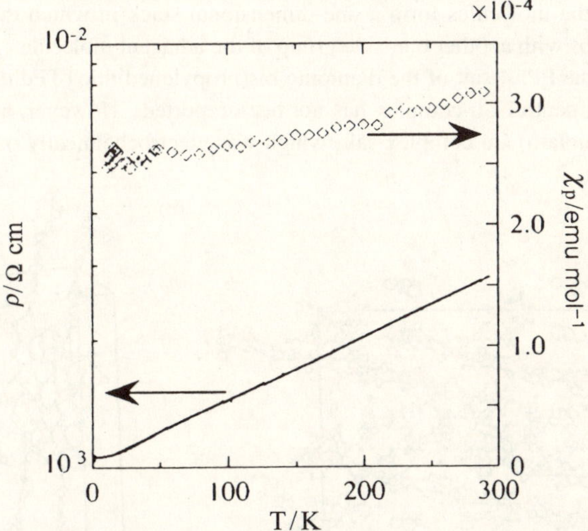

Fig. 7.10   Temperature dependences of resistivity (ρ) and paramagnetic susceptibility (χ_p) for a neutral bis(trimethyleneTTFdithiolato) Ni complex. (Reprinted with permission from ref. 12a)

7.3   Cu Complexes with TTFdithiolato Ligands

Neutral Cu complexes with two dimethylthio-, diethylthio- and ethylenedithio-TTFdithiolato groups were obtained as powdered samples by the reaction of the TTFdithiolato dianions with

$CuCl_2$ in the presence of $ZnCl_2$ and NMe_4Br or NEt_4Br in DMF, and by the following I_2 oxidation of the corresponding monoanionic NMe_4^+ or NEt_4^+ salts.[15,16] The σ_{RT}'s on the compressed pellets were 7.8×10^{-2}, 5.0×10^{-4} and 3.7 S cm^{-1} for the neutral bis(dimethylthio-TTFdithiolato), bis(diethylthioTTFdithiolato) and bis(ethylenedithioTTFdithiolato) Cu complexes, respectively. In particular, the σ_{RT} of the neutral bis(ethylenedithioTTFdithiolato) Cu complex was fairly high and comparable to that (2.8 S cm^{-1}) of the corresponding Au complex. The magnetic measurement using a SQUID magnetometer showed that in each neutral Cu complex the observed χ_p can well be expressed by the equation $\chi_p(T) = \chi_{pCu}(T) + \chi_\pi = C/(T - \theta) + \chi_\pi$, in which the first term is the contribution from the localized d spin on the Cu atom and obeys the Curie-Weiss law (C: Curie constant; θ: Weiss temperature), while the second term is the temperature-independent paramagnetic susceptibility due to strong antiferromagnetically interacting spins residing in the TTFdithiolato groups. From the best fitting between the observed and theoretical χ_p-T behavior the C, θ and χ_π values were obtained. The C values are 10^{-3}–10^{-2} emu K mol^{-1}, which correspond to 0.2–3% of the value (0.375 emu K mol^{-1}) calculated as one $S = 1/2$ spin entity per molecule of the neutral Cu complex. The χ_π values are on the order of 10^{-4} emu mol^{-1} and comparable to those of highly conducting organic semiconductors. Accordingly, almost one unpaired electron residing in each neutral Cu complex molecule contributes to the electrical conduction and only a very small amount is located on the Cu atom.

When using the propylenedithioTTFdithiolato group, the PPh_4^+ salt of the dianionic Cu complex obtained by a procedure similar to that described above was isolated as a single crystal.[9] As shown in Fig. 7.11, the dianionic Cu complex molecule has a distorted tetrahedral geometry around the Cu atom, and the dihedral angle between the planes of the TTFdithiolato groups is 54.2°. As a result, the molecules form a one-dimensional stack in which the one-side group of one molecule overlaps with another one-side group of the adjacent molecule.

Conversion of the PPh_4^+ salt of the dianionic bis(propylenedithioTTFdithiolato) Cu complex to the corresponding neutral Cu complex has not been reported. However, another monoanionic bis(dimethylTTFdithiolato) Cu complex salt, which was electrochemically oxidized with a small

(a)   (b)

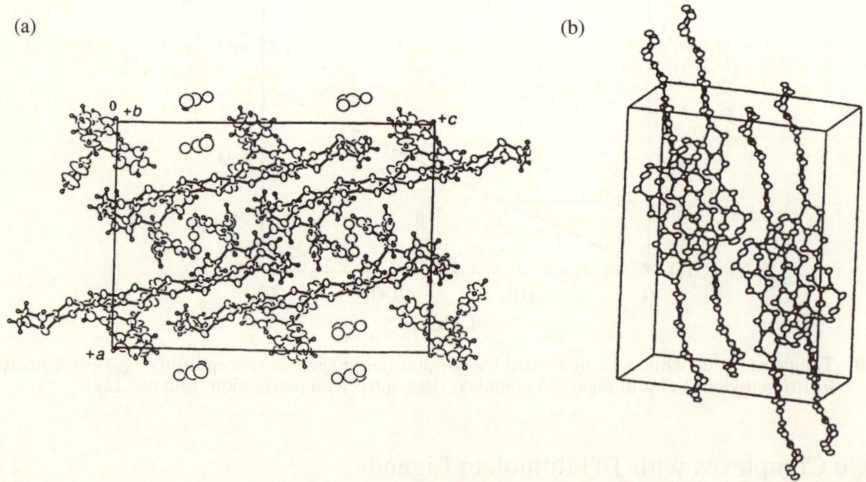

Fig. 7.11   (a) Packing structure of PPh_4^+ salt of a monoanionic bis(propylenedithioTTFdithiolato) Cu complex and (b) overlapping mode of the complex molecules. (Reprinted with permission from ref. 9)

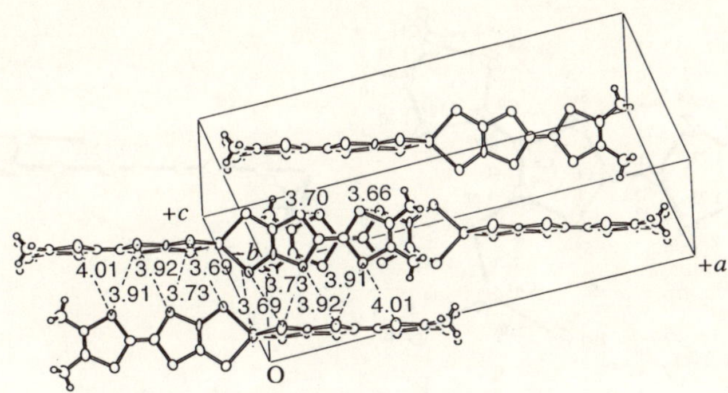

Fig. 7.12   Stacking structure of a neutral bis(dimethylTTFdithiolato) Cu complex.
(Reprinted with permission from ref. 17)

and constant current to give the corresponding neutral Cu complex as a single crystal, has been prepared for the first time.[17] Its crystal structure is shown in Fig. 7.12. Each of the dimethylTTFdithiolato groups is almost coplanar, but arranged almost perpendicular to each other. One half of the dimethylTTFdithiolato groups overlaps with those in neighboring dimethylTTFdithiolato groups to form two π stacks perpendicular to each other. The σ_{RT} on the single crystal was 1.2–3.5 S cm^{-1} and the temperature dependence of σ was semiconducting with an E_a of *ca.* 60 meV in the temperature range of 200 to 300 K. This neutral Cu complex also possessed a considerable amount of spin. From magnetic measurements the C was 0.327 emu K mol^{-1}, which corresponds to *ca.* 90% of the value (0.375 emu K mol^{-1}) calculated as one Cu(II) (S = 1/2) spin entity. The θ was –4.2 K, so that the interaction between the Cu d spins is very weak and antiferromagnetic, as expected from the fairly long distance (5.96 Å) between the nearest Cu atoms.

7.4   CuBr₂ Complexes with TTF Derivatives

Another neutral molecular π/d system was also prepared using TTF derivatives, dimethylthio- and ethylenedithio-tetrathiafulvalenothioquinone-1,3-dithiolemethides (**1a,b**, Scheme 7.4), which formed a 1:1 complex with CuBr₂ through coordination with the C=S group.[18] The molecular structures of **1a** • CuBr₂ and **1b** • CuBr₂ are shown in Fig. 7.13. The Cu atom of the CuBr₂ moiety is bound to the thiocarbonyl S atom of **1a** or **1b**. For **1a** • CuBr₂ the molecular skeleton is almost planar except for two methyl groups and the CuBr₂ moiety, while the entire molecular skeleton of **1b** • CuBr₂ has high planarity. The geometry around the Cu atom is a distorted tetrahedron in

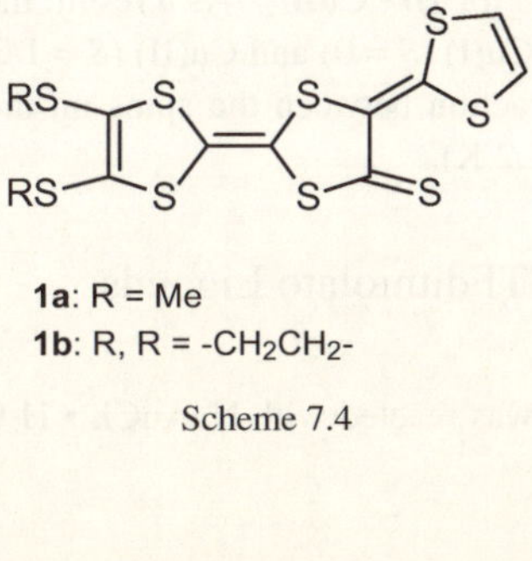

1a: R = Me
1b: R, R = -CH₂CH₂-

Scheme 7.4

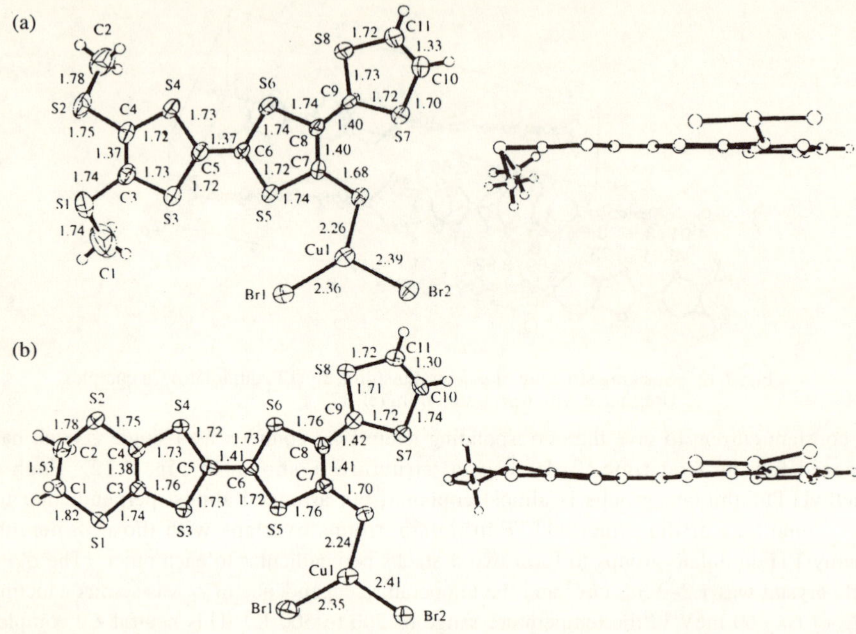

Fig. 7.13   Molecular structures of (a) **1a** • CuBr$_2$ and (b) **1b** • CuBr$_2$. (Reprinted with permission from ref. 18)

both CuBr$_2$ complexes. The C=S distances are 1.68 and 1.70 Å for **1a** • CuBr$_2$ and **1b** • CuBr$_2$, respectively, which are slightly longer than those of **1a,b**. The crystal structures of both CuBr$_2$ complexes are shown in Fig. 7.14. For **1a** • CuBr$_2$ two neighboring molecules form a tight dimer through close contact between the Cu atom of one molecule and one methylthio S atom of the other molecule. The neighboring dimers have a fairly short interplanar distance of 2.72 Å, but scarcely any effective overlap. On the other hand, the column of **1b** • CuBr$_2$ has a quasi-uniform stacking structure composed of a unit of two crystallographically independent molecules [A(A")] and A']. The overlap between A and A' is good while A' and A" have inferior contact.

The σ_{RT}'s on the single crystals of **1a** • CuBr$_2$ and **1b** • CuBr$_2$ were 2.1 × 10^{-5} and 4.0 S cm^{-1}, respectively, and the temperature dependence of σ for **1b** • CuBr$_2$ showed semiconducting behavior with a small E_a of 0.18 eV. The σ_{RT}'s obtained can be readily expected from their stacking structures. However, it must also be taken into consideration that the desirable degree of intramolecular electron transfer from **1a** or **1b** to the CuBr$_2$ moiety, in producing the appropriate amount of hole carriers into the **1a**- or **1b**-stacked columns, is responsible for the σ_{RT}'s of both CuBr$_2$ complexes. By magnetization measurement, the degree of electron transfer was estimated to be 88% for **1a** • CuBr$_2$ and 58% for **1b** • CuBr$_2$. As a result, the Cu atom of each CuBr$_2$ moiety becomes an intermediate state of Cu(I) ($S = 0$) and Cu(II) ($S = 1/2$) of 88:12 for **1a** • CuBr$_2$ and of 58:42 for **1b** • CuBr$_2$. The interaction between the spins on the Cu atoms was very weak and antiferromagnetic ($\theta = $ -1.9 and -1.2 K).

7.5   Au Complexes with TTFdithiolato Ligands

The ethylenedithioTTFdithiolate was reacted with NaAuCl$_4$ • H$_2$O in the presence of N(n-Bu)$_4$Br

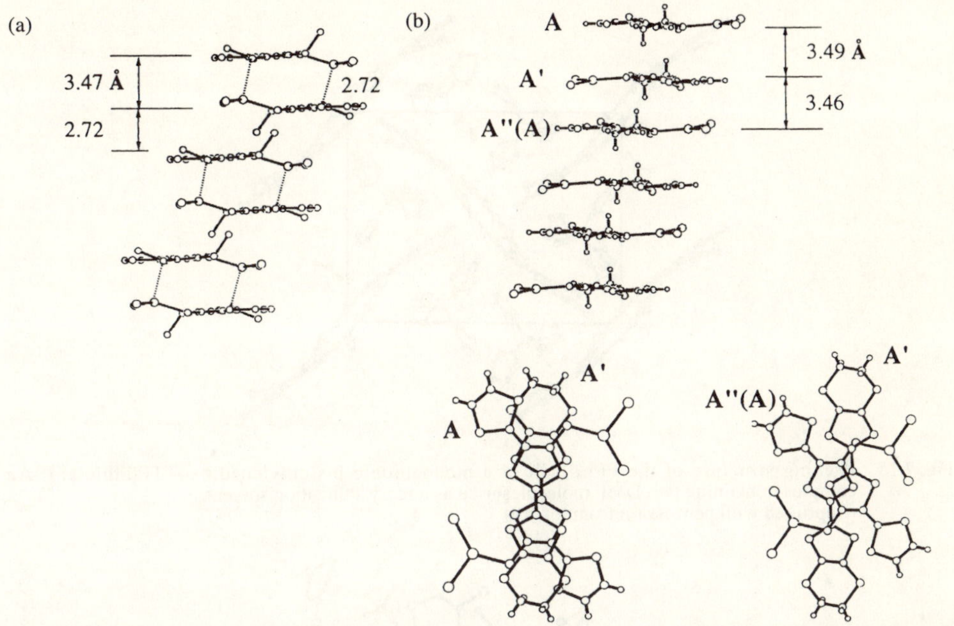

(a)

3.47 Å

2.72

(b)

A

A'

A''(A)

3.49 Å

3.46

2.72

A'

A

A''(A)

A'

Fig. 7.14   Stacking structures of (a) **1a** • CuBr$_2$ and (b) **1b** • CuBr$_2$. (Reprinted with permission from ref. 18)

in MeOH/CH$_2$Cl$_2$ to give the corresponding monoanionic N(n-Bu)$_4^+$ salt, which was converted to the neutral complex by treatment with I$_2$ in PhCN.[14]   The σ_{RT} on the compressed pellet of the neutral complex was 2.8 S cm^{-1}.

The crystal structure of the monoanionic Au complex was determined in the case of the PPh$_4^+$ salt of a monoanionic bis(ethylenedioxoTTFdithiolato) Au complex, which was obtained as brown plates by keeping the corresponding N(n-C$_6$H$_{13}$)$_4^+$ salt in DMF containing an excess of PPh$_4$Br at room temperature.[19]   As shown in Fig. 7.15, the monoanionic molecule as a whole takes a chair conformation, but the geometry around the Au atom is almost planar.   The monoanionic molecules are orthogonally arranged with each other in the bc plane, while they form a sheet-like network along the a axis.   There are many side-by-side S$\cdots$S contacts between the adjacent molecules along the a axis.   Other monoanionic salts of unsubstituted and dimethylthio-substituted bis(TTFdithiolato) Au complexes were also prepared, but their single crystals were not obtained.   The reaction of these monoanionic salts with TCNQ gave CT complexes with different compositions of Au complex:TCNQ = 2:1, 4:1, 6:1, which exhibited high σ_{RT}'s of 5–30 S cm^{-1} with very low E_a's of 0.024–0.051 eV.

7.6   Hg Complexes with TTFdithiolato Ligands

By the reaction of diethylthio- and di(n-butyl)thio-TTFdithiolate with Hg(OAc)$_2$ in MeOH, the corresponding dianionic bis(TTFdithiolato) Hg complexes were obtained.[20]   In the case of the NMe$_4^+$ salt of a dianionic bis(diethylthioTTFdithiolato) Hg complex, its single crystal was isolated for X-ray structure analysis.   Fig. 7.16 shows the crystal structure of the dianion.   The diethylthioTTFdithiolato units are approximately tetrahedrally coordinated around the central Hg

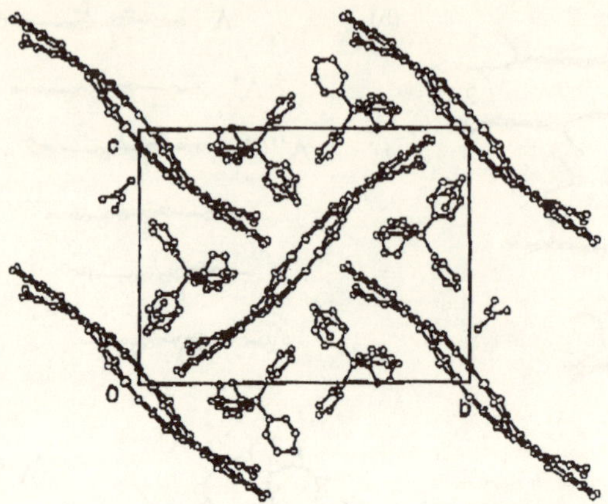

Fig. 7.15   Packing structure of the PPh$_4^+$ salt of a monoanionic bis(ethylenedioxoTTFdithiolato) Au complex containing two DMF molecules used as a recrystallization solvent. (Reprinted with permission from ref. 19)

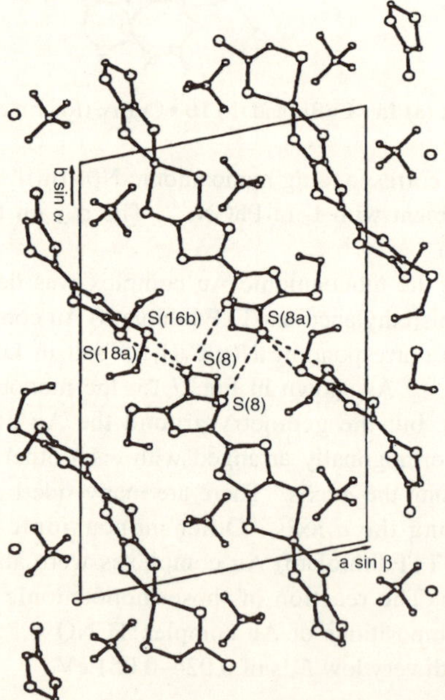

Fig. 7.16   Packing structure of the NMe$_4^+$ salt of a dianionic bis(diethylthioTTFdithiolato) Hg complex. (Reprinted with permission from ref. 20)

atom. The addition of a stoichiometric amount of I$_2$ to the solution of the dianionic diethylthioTTFdithiolato Hg salt in CH$_3$CN at room temperature resulted in the formation of the corresponding neutral Hg complex as a brown precipitate. Compressed pellets of the powdered dianionic and neutral complexes exhibited very low σ_{RT}'s of $< 10^{-7}$ and 10^{-5} S cm^{-1}, respectively.

7.7   V Complexes with TTFdithiolato Ligands

Neutral V complexes with three dialkylthioTTFdithiolato groups [dialkyl = dimethyl, di(*n*-butyl) and ethylene] were very recently prepared by the I_2 oxidation of NEt_4^+ salts of the corresponding monoanionic V complexes obtained by the reaction of $2NEt_4^+$ salts of bis(dialkylthioTTF dithiolato) Zn complexes with VCl_3 in DMF.[21]   The NEt_4^+ salt of the monoanionic tris(diethylthioTTFdithiolato) V complex was isolated as black crystals by recrystallization from CS_2, and X-ray structure analysis was successfully performed. As shown in Fig. 7.17, the geometry around the V atom is close to trigonal prismatic, and each of the three diethylthioTTFdithiolato groups have almost a planar molecular skeleton except for two ethyl groups. Fig. 7.18 shows the crystal structure projected down to the *ac* plane. The monoanionic V complex molecules are arranged to form quadratic layers in which cavities of three different sizes are present. The NEt_4^+ ions are incorporated in the smallest cavity and CS_2 molecules in the two

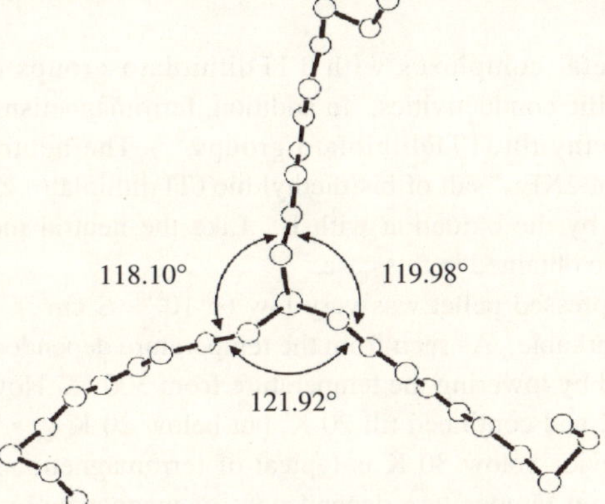

Fig. 7.17   Molecular structure of a monoanionic tris(diethylthioTTFdithiolato) V complex. (Reprinted with permission from ref. 21)

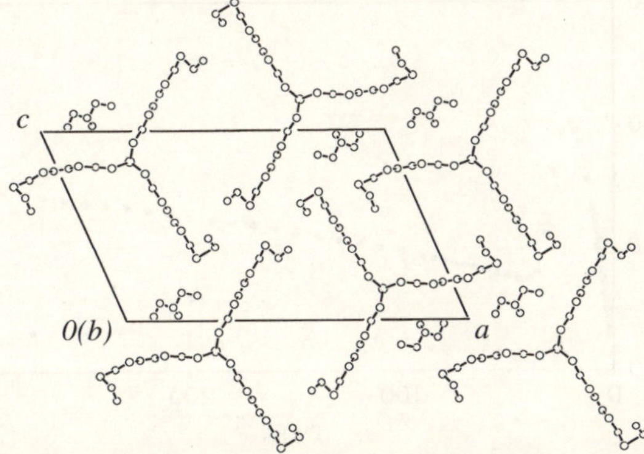

Fig. 7.18   Packing structure projected down to the *ac* plane for NEt_4^+ salt of a monoanionic tris(diethylthioTTFdithiolato) V complex. CS_2 molecules are omitted for clarity. (Reprinted with permission from ref. 21)

other cavities. As a result there is no $\pi \cdots \pi$ contact between the diethylthioTTFdithiolato groups. On the other hand, the X-ray structure analysis of the neutral V complexes has not yet been successful despite many attempts to obtain single crystals. However, it is assumed that the molecular structures of the neutral V complexes are similar to that of the monoanionic tris(diethylthioTTFdithiolato) V complex, and that their stacking structures also form a quadratic layer with much tighter contact between the dialkylthioTTFdithiolato groups.

From ESR and SQUID measurements the monoanionic V complexes are diamagnetic, so that the V atom is in the V(V) ($S = 0$) state. On the other hand, the neutral V complexes involve unpaired electrons being mainly distributed at the dialkylthioTTFdithiolato groups in contrast to the neutral tris(dddt) V complex, V(dddt)$_3$, with an unpaired electron center at the V atom. Corresponding to the absence or presence and the preferential distribution of the unpaired electrons, the σ_{RT}'s were $< 10^{-10}$ S cm^{-1} for the monoanionic V complexes, 10^{-8}–10^{-4} S cm^{-1} for the neutral V complexes and $< 10^{-10}$ S cm^{-1} for the neutral V(dddt)$_3$.

7.8   Mn Complexes with TTFdithiolato Ligands

A number of neutral metal complexes with TTFdithiolato groups obtained so far exhibit semiconducting and metallic conductivities. In addition, ferromagnetism emerged from a neutral MnCl$_2$ complex with diethylthioTTFdithiolato groups.[22] The neutral MnCl$_2$ complex was prepared by the reaction of 2NEt$_4^+$ salt of bis(diethylthioTTFdithiolato) Zn complex with MnCl$_2 \cdot$ 4H$_2$O in DMF, followed by the oxidation with I$_2$. Like the neutral metal complexes thus far, single crystals could not be obtained in this case.

The σ_{RT} on the compressed pellet was very low ($< 10^{-10}$ S cm^{-1}). However, the magnetic properties were most remarkable. As seen from the temperature dependence of $\chi_p \cdot T$ in Fig. 7.19, $\chi_p \cdot T$ gradually decreased by lowering the temperature from 300 K. However, abrupt increase in $\chi_p \cdot T$ occurred near 30 K and continued till 20 K, but below 20 K $\chi_p \cdot T$ again decreased. This kind of $\chi_p \cdot T$ vs. T behavior below 30 K is typical of ferromagnetic spin ordering, which was confirmed by three different temperature dependences of magnetization, i.e., field-cooled, zero-field-cooled and remnant magnetizations (FCM, ZFCM and RM) in Fig. 7.20. The onset temperature of ferromagnetic phase transition was determined to be 22 K from the ZFCM and

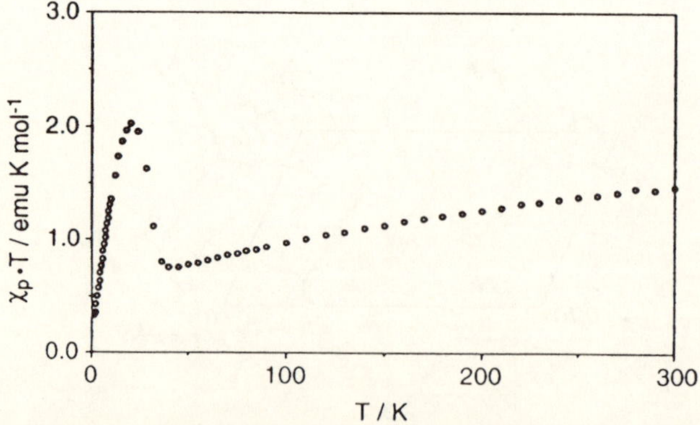

Fig. 7.19   Temperature dependence of $\chi_p \cdot T$ for a neutral bis(diethylthioTTFdithiolato) MnCl$_2$ complex. (Reprinted with permission from ref. 22)

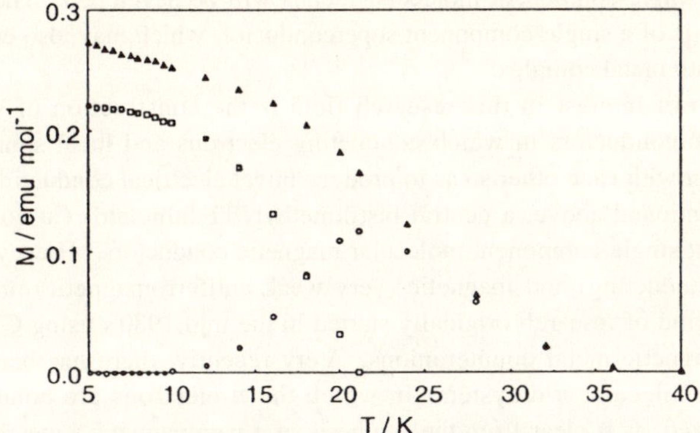

Fig. 7.20   Temperature dependences of magnetization for a neutral bis(diethylthioTTFdithiolato) MnCl$_2$ complex: ($\triangle$) FCM, ($\bigcirc$) ZFCM and ($\square$) RM. (Reprinted with permission from ref. 22)

RM experiments. Nevertheless, further details on the magnetism and spin interaction in this neutral MnCl$_2$ complex must wait until magnetic measurement on single crystals is achieved.

7.9   Neutral Metal Complex Oligomers with TTFdithiolato Groups

TTFtetrathiolate produced *in situ* by the treatment of bis(carbonyldithio)TTF with NaOMe/MeOH or MeLi/THF was reacted with several transition metal (Ni, Pd, Cu, Pt, Fe) salts.[7] The solvent-insoluble and amorphous powders with a metal:ligand ratio of *ca.* 1:1 were isolated; these are assumed to have an oligomer structure with a repeating TTF-metal bis(dithiolene) composition (Scheme 7.5). The neutral Ni complex oligomer was found to have a surprising high σ_{RT} of 30 S cm^{-1}. However, the other metal oligomers displayed lower σ_{RT}'s: Pd, *ca.* 10^{-3} S cm^{-1}; Cu, *ca.* 0.1 S cm^{-1}; Pt, 10^{-2} S cm^{-1}; Fe, *ca.* 10^{-5} S cm^{-1}.

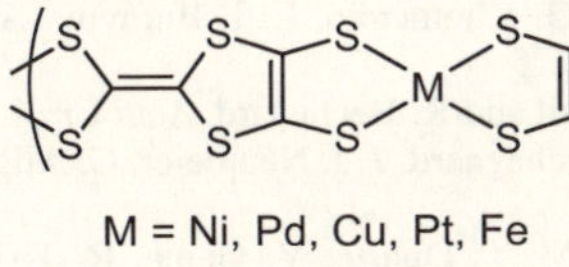

M = Ni, Pd, Cu, Pt, Fe

Scheme 7.5

7.10 Summary and Outlook

The common guiding principle for the design of molecular metals and superconductors has so far been based on achieving partial filling of the conduction band through partial CT from donor to acceptor molecules in the CT complexes or through partial oxidation (reduction) of donor (acceptor) molecules. However, the discovery of a single-component molecular metal using a neutral bis(trimethyleneTTFdithiolato) Ni complex[12] has not only countered this principle, but also extended an approach to the development of molecular metals and superconductors. In the near future further single-component molecular metals will be developed. The next challenging target is the synthesis of a single-component superconductor, which may also emerge from a new neutral TTFdithiolato metal complex.

Another focus of interest in this research field is the construction of single-component molecular magnetic conductors in which conducting electrons and local spins are expected to significantly interact with each other so as to produce novel electrical conducting and/or magnetic properties. As mentioned above, a neutral bis(dimethylTTFdithiolato) Cu complex[17] and **1b** • $CuBr_2$[18] are the first single-component molecular magnetic conductors. However, their electrical conducting (semiconducting) and magnetic (very weak antiferromagnetic) properties were not remarkable. This kind of research originally started in the mid 1980's using CT salts of π donor molecules with magnetic-metal counteranions. Very recently, there has been a great deal of progress made in molecular π/d systems in which the π electrons are conducting and the d electrons are localized, as is clear from the synthesis of a paramagnetic superconductor based on β''-(BEDT-TTF)$_4$ • [(H$_2$O)M(C$_2$O$_4$)]$_3$ • PhCN (BEDT-TTF = bis(ethylenedithio)tetrathiafulvalene; M = Fe, Cr),[23,24] an antiferromagnetic superconductor based on κ-(BETS)$_2$ • FeBr$_4$ (BETS = bis(ethylenedithio)tetraselenafulvalene),[25,26] a ferromagnetic metal based on (BEDT-TTF)$_2$ • [MnCr(C$_2$O$_4$)]$_3$[27] and a magnetic field-induced ferromagnetic superconductor based on λ-(BETS)$_2$ • FeCl$_4$.[28] However, the simultaneous appearance of antiferro- or ferro-magnetism and metallic or super-conductivity is not always the result of the interaction between the conducting π electrons and the local d spins involved in the CT salts other than λ-(BETS)$_2$ • FeCl$_4$[29] and κ-(BETS)$_2$ • FeBr$_4$,[25,26] in which significant interaction begins to occur at temperatures lower than 10 K. Because of these circumstances, an active study will be conducted using new neutral magnetic-metal complexes with sulfur-rich ligands and CT salts with magnetic-metal counteranions to develop molecular magnetic conductors with both semi-, metallic- or super-conductivity and ferromagnetism due to strong interaction between conducting π electrons and local d spins.

References

1. P. Cassoux and L. Valade, *Inorganic Materials,* 2nd ed., Wiley, Chichester (1996) p. 1.
2. P. Cassoux and J. S. Miller, *Chemistry of Advanced Materials*, Wiley-VCH, New York (1998) p. 19.
3. E. B. Yagubskii, A. I. Kotov, A. G. Khomenko, L. I. Buravov, A. I. Schegorev and R. P. Shibaeva, *Synth. Met.*, **46**, 255 (1992).
4. G. Rindorf, N. Thorup, T. Bjørnholm and K. Bechgaard, *Acta Cryst.*, C**46**, 1437 (1990).
5. N. C. Schiødt, T. Bjørnholm, K. Bechagaard, J. J. Neumeier, C. Allgeier, C. S. Jacobsen and N. Thorup, *Phys. Rev.*, B **53**, 1773 (1996).
6. D. Belo, H. Alves, E. B. Lopes, M. T. Duarte, V. Gama, R. T. Henriques, M. Almeida, A. Pérez-Benitez, C. Rovira and J. Veciana, *Chem. Eur. J.*, **7**, 511 (2001).
7. N. M. Rivera and E. M. Engler, *J. Chem. Soc., Chem. Commun.*, 184 (1979).
8. N. L. Narvor, N. Robertson, T. Weyland, J. D. Killburn, A. E. Underhill, M. Webster, N. Svenstrup and J. Becher, *Chem. Commun.*, 1363 (1996).
9. M. Kumasaki, H. Tanaka and A. Kobayashi, *J. Mater. Chem.*, **8**, 301 (1998).
10. A. Kobayashi, H. Tanaka, M. Kumasaki, H. Torii, B. Narymbetov and T. Adachi, *J. Am. Chem. Soc.*, **121**, 10763 (1999).
11. K. Ueda, Y. Kamata, M. Iwamatsu, T. Sugimoto and H. Fujita, *J. Mater. Chem.*, **9**, 2979 (1999).
12. (a) H. Tanaka, Y. Okano, H. Kobayashi, W. Suzuki and A. Kobayashi, *Science*, **291**, 285 (2001).
     (b) A. Kobayashi, H. Tanaka and H. Kobayashi, *J. Mater. Chem.*, **11**, 2078 (2001).
13. H. Tanaka, H. Kobayashi and A. Kobayashi, *Synth. Met.*, **120**, 1037 (2001).
14. M. Nakano, A. Kuroda and G. Matsubayashi, *Inorg. Chim. Acta*, **254**, 189 (1997).
15. K. Ueda, M. Goto, T. Sugimoto, S. Endo, N. Toyota, K. Yamamoto and H. Fujita, *Synth. Met.*, **85**, 1679 (1997).
16. K. Ueda, M. Goto, M. Iwamatsu, T. Sugimoto, S. Endo, N. Toyota, K. Yamamoto and H. Fujita, *J. Mater. Chem.*, **8**, 2195 (1998).

17. H. Tanaka, H. Kobayashi and A. Kobayashi, *J. Am. Chem. Soc.*, **124**, 10002 (2002).
18. M. Iwamatsu, T. Kominami, K. Ueda, T. Sugimoto, T. Adachi, H. Fujita, H. Yoshino, Y. Mizuno, K. Murata and M. Shiro, *Inorg. Chem.*, **39**, 3810 (2000).
19. Y. Misaki, Y. Tani, M. Taniguchi, T. Maitani, K. Tanaka and K. Bechgaard, *Mol. Cryst. Liq. Cryst.*, **343**, 59 (2000).
20. N. L. Narvor, N. Robertson, E. Wallace, J. P. Killburn, A. E. Underhill, P. N. Bartlett and M. Webster, *J. Chem. Soc., Dalton Trans.*, 823 (1996).
21. T. Yoneda, Y. Kamata, K. Ueda, T. Sugimoto, M. Shiro, H. Yoshino and K. Murata, *Synth. Met.*, **135-136**, 573 (2003).
22. K. Ueda, Y. Kamata, T. Yoneda, T. Matsumoto and T. Sugimoto, *Mol. Cryst. Liq. Cryst.*, **376**, 417 (2002).
23. M. Kurmoo, A. W. Graham, P. Day, S. J. Coles, M. B. Hursthouse, J. L. Caulfield, J. Singleton, F. L. Pratt, W. Hayes, L. Ducasse and P. Guionneau, *J. Am. Chem. Soc.*, **117**, 12209 (1995).
24. L. Martin, S. S. Turner, P. Day, F. E. Mabbs and E. J. L. McInnes, *Chem. Commun.*, 1367 (1997).
25. E. Ojima, H. Fujiwara, K. Kato, H. Kobayashi, H. Tanaka, A. Kobayashi, M. Tokumoto and P. Cassoux, *J. Am. Chem. Soc.*, **121**, 5581 (1999).
26. H. Fujiwara, E. Fujiwara, Y. Nakazawa, B. Z. Narymbetov, K. Kato, H. Kobayashi, A. Kobayashi, M. Tokumoto and P. Cassoux, *J. Am. Chem. Soc.*, **123**, 306 (2001).
27. E. Coronado, J. R. Galán-Mascarós, C. J. Gómez-Garcia and V. Laukhin, *Nature*, **408**, 447 (2000).
28. S. Uji, H. Shinagawa, T. Terashima, T. Yakabe, Y. Terai, M. Tokumoto, A. Kobayashi, H. Tanaka and H. Kobayashi, *Nature*, **410**, 908 (2001).
29. H. Kobayashi, H. Tomita, T. Naito, A. Kobayashi, F. Sakai, T. Watanabe and P. Cassoux, *J. Am. Chem. Soc.*, **118**, 368 (1996).

II Dimeric TTFs

8

Bi- and Bis-TTFs

8.1 Introduction

There is currently considerable interest in the dimeric tetrathiafulvalene (TTF) molecules and higher oligomers.[1-3] Dimeric TTFs **1-4** linked by a σ-bond, a conjugated π-system, a single chalcogen atom or an alkyl chain may have intramolecular through-bond or through-space interaction between two TTFs. As shown in Scheme 8.1, dimeric TTFs display multistage redox behavior, and hence can be expected to reveal unique structures and physical properties in their charge-transfer (CT) complexes and ion radical salts.[4,5] For example, bi-TTF **1** is a four-electron redox system, and the cation radical, dication and further oxidized species $\mathbf{1}^{\cdot+}$, $\mathbf{1}^{2+}$, $\mathbf{1}^{\cdot3+}$ and $\mathbf{1}^{4+}$ can exist as a localized or delocalized form. In addition, dimeric and oligomeric TTFs may be employed either for the enhancement of the dimensionality by the extended π-conjugation or for controlling the stoichiometry and band filling in the corresponding conductive complexes.[6,7]

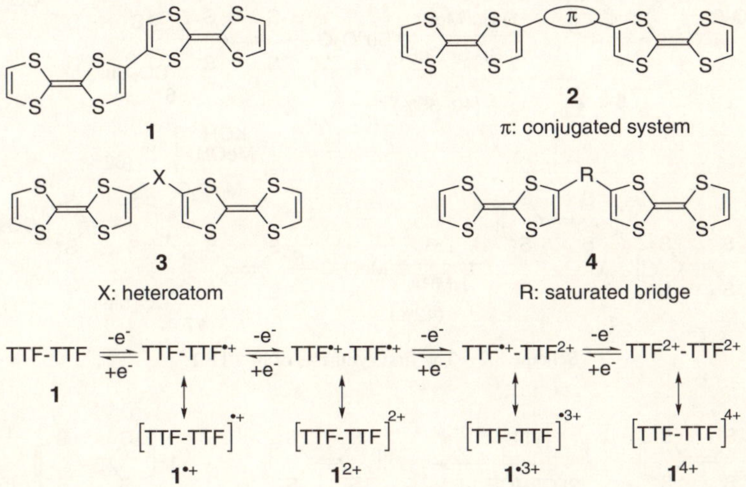

1

2

π: conjugated system

3

X: heteroatom

4

R: saturated bridge

$$\text{TTF-TTF} \underset{+e^-}{\overset{-e^-}{\rightleftharpoons}} \text{TTF-TTF}^{\cdot+} \underset{+e^-}{\overset{-e^-}{\rightleftharpoons}} \text{TTF}^{\cdot+}\text{-TTF}^{\cdot+} \underset{+e^-}{\overset{-e^-}{\rightleftharpoons}} \text{TTF}^{\cdot+}\text{-TTF}^{2+} \underset{+e^-}{\overset{-e^-}{\rightleftharpoons}} \text{TTF}^{2+}\text{-TTF}^{2+}$$

1

$$\left[\text{TTF-TTF}\right]^{\cdot+} \quad \left[\text{TTF-TTF}\right]^{2+} \quad \left[\text{TTF-TTF}\right]^{\cdot3+} \quad \left[\text{TTF-TTF}\right]^{4+}$$

$\mathbf{1}^{\cdot+}$ $\mathbf{1}^{2+}$ $\mathbf{1}^{\cdot3+}$ $\mathbf{1}^{4+}$

Scheme 8.1

Recently, dimeric TTFs have come to be regarded as building blocks in supramolecular chemistry, and large "belt-shape" molecules, cage molecules and "tweezers-like" molecules have been investigated in order to realize redox sensors, conducting organic magnets and novel crystal structures.[8-10] Although a variety of dimeric TTFs have been prepared and investigated, this chapter focuses mainly on the chemistry of conjugated dimeric TTFs **1-3** with or without a spacer group.

8.2   Bi-TTF and Its Derivatives

8.2.1 Synthesis

Among dimeric TTFs, bi-TTF **1** is a feasible candidate for constructing organic metals with high dimensionality based on a π-expanded donor. It provides the opportunity to control the stoichiometry, band filling and molecular assembly in the desired conductive complexes.[6] Although bi-TTF **1** was first reported in 1982 in an abstract for a conference, no physical and spectral data were presented.[11] The second synthesis of **1** with spectral characterization was reported by Neilands *et al.* in 1989 using the copper-catalyzed coupling of the Grignard reagent and decarboxylation (Scheme 8.2).[12] Thus, the reaction of **5** with butylmagnesium bromide followed by treatment with bromine or $CuBr_2$ produced **6** in 40–56% yields. Transesterification of **6** with KOH in methanol afforded **7**, which was converted into **1** using decarboxylation with LiBr in HMPA.

For the synthesis of **1**, Ullmann coupling of iodo-TTF **8** was reported by Becker *et al.* (Scheme 8.3).[13] Besides **1** (22%), they reported the formation of a trimer **10** in 1–3% yields. The production of **10** may be due to a small amount of diiodo-TTF **9** contaminating **8** employed for the Ullmann reaction.

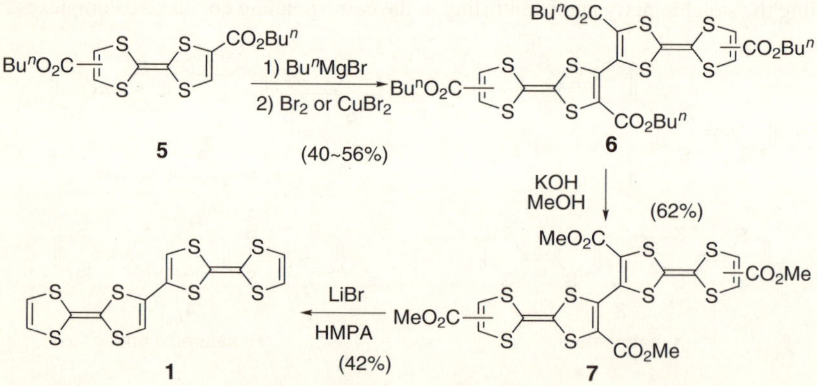

Scheme 8.2   The first synthesis of bi-TTF **1**.

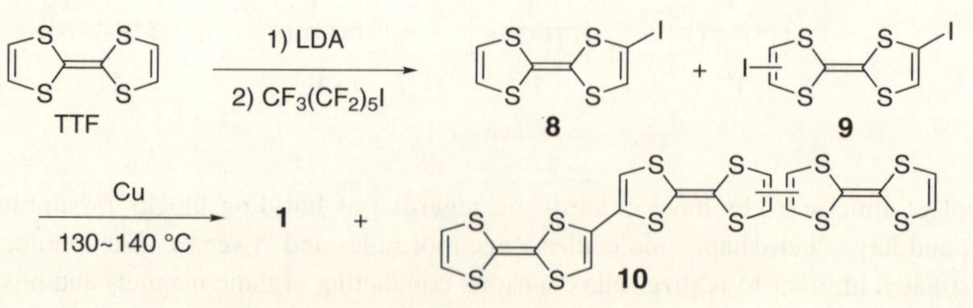

Scheme 8.3   Synthesis of **1** together with **10**.

In the course of our studies on the synthesis of functionalized π-electron systems using transition metal complexes,[14] we carried out an efficient synthesis of **1** by the copper- or palladium-catalyzed homo-coupling of trimethylstannyl-TTF **11** (Scheme 8.4).[15,16] TTF was first converted into **11** in 86% yield. The transition metal-catalyzed homo-coupling of **11** is summarized in Table 8.1. We found Pd(OAc)$_2$ to be an efficient catalyst for the homo-coupling of **11**. Thus, the reaction of **11** with a stoichiometric amount of Pd(OAc)$_2$ in benzene at room temperature produced **1** in 67% yield, together with the recovered TTF (20%). Although the reaction of **11** with PdCl$_2$(CH$_3$CN)$_2$ in HMPA or PdCl$_2$(PPh$_3$)$_2$ in benzene afforded **1** in 62% or 58% yield, no TTF was recovered in the former reaction or the reaction needed a much longer time in the latter case. The homo-coupling reaction of tributylstannylbenzene with a catalytic amount of Pd(OAc)$_2$ in the presence of t-BuOOH (excess) was reported to produce the corresponding biphenyl in good yields.[17] However, a similar reaction of **11** under similar conditions produced **1** in 8% yield, together with 72% yield of TTF (protodestannylation product). In addition, the Cu(OAc)$_2$-mediated homo-coupling of **11** produced **1** in 70% yield.

Scheme 8.4   Coupling of trimethylstannyl-TTF **11**.

Table 8.1   Transition Metal-catalyzed Homo-coupling of **11**

Compound	Metal catalyst	Solvent	Time (h)	Product	Yield(%)
11	PdCl$_2$(CH$_3$CN)$_2$	HMPA	2	**1**	62
11	PdCl$_2$(PPh$_3$)$_2$	Benzene	20	**1**	58
11	Pd(OAc)$_2$	Benzene	2	**1**	67
11	Pd(OAc)$_2$-t-BuOOH	Benzene	7	**1**	8[a]
11	Cu(NO$_3$)$_2$·3·H$_2$O	THF	1	**1**	70

[a]10 Mol% of Pd(OAc)$_2$ and 2 equiv of t-BuOOH were used.

For the synthesis of alkylthio derivatives of **1**, Tatemitsu *et al.* reported the tetrakis(alkylthio)-bi-TTFs **14a-c** using the cross-coupling method (Scheme 8.5).[18,19] Thus, the reactions of **12a-c** with **13** in the presence of P(OEt)$_3$ at 140 °C yielded the 2:1 coupling products **14a-c**. When the reaction was carried out in refluxing benzene, the 1:1 coupling products **15b,c** were obtained in 37 and 31% yields, respectively. The homo-coupling reactions of **15b,c** with P(OEt)$_3$ in refluxing toluene afforded the corresponding trimers **16b,c**.

Tetramethyl-bi-TTF **17** was first prepared using multi-step synthesis.[20] Recently, Bryce *et al.* reported a new methodology for the synthesis of bi-TTF derivatives.[21,22] As shown in Scheme 8.6, the reactions of **18a-d** with copper(I) thiophene-2-carboxylate (CuTC) in 1-methylpyrrolidin-2-one (NMP) at 20 °C proceeded smoothly to produce the corresponding coupling products **19a-d** in 75, 65, 80 and 72% yields, respectively. Decarboxylation of **19d** with LiBr in DMF at 140 °C afforded **14a** in 59% yield. In the case of **18e**, however, the Ullmann coupling of **18e** with Cu powder produced **19e** in 62% yield.[22]

To synthesize bi-TTF derivatives **14a**, **26** and **27**, we applied the palladium-catalyzed homo-coupling similar to the synthesis of **1**.[16,23] As shown in Scheme 8.7, BMT-TTF **20**, EDT-TTF **21** and EDO-TTF **22** were converted into the corresponding tin(IV) derivatives **23-25** in 55, 48 and 29% yields, respectively. Although the reactions of **23-25** with Cu(NO$_3$)$_2$ • 3H$_2$O in THF were

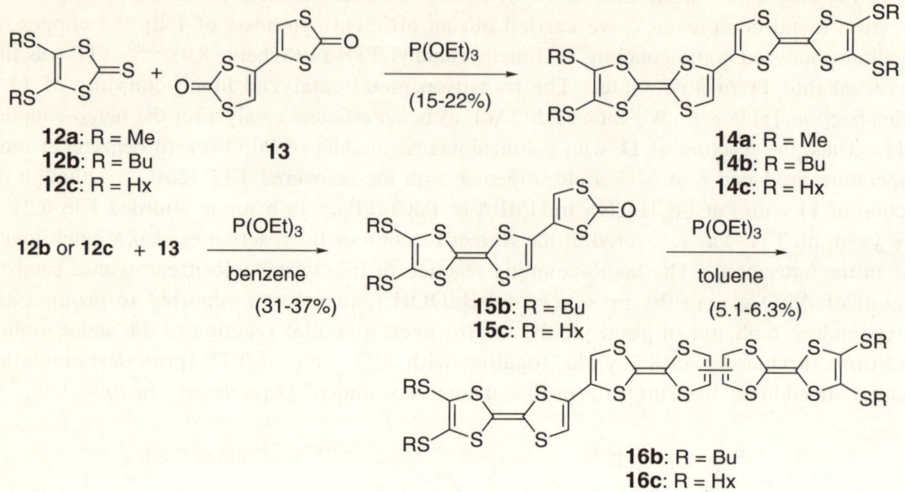

12a: R = Me
12b: R = Bu
12c: R = Hx

13

P(OEt)$_3$

(15-22%)

14a: R = Me
14b: R = Bu
14c: R = Hx

12b or 12c + **13**

P(OEt)$_3$

benzene

(31-37%)

15b: R = Bu
15c: R = Hx

P(OEt)$_3$

toluene

(5.1-6.3%)

16b: R = Bu
16c: R = Hx

Scheme 8.5   Synthesis of **14** and **16**.

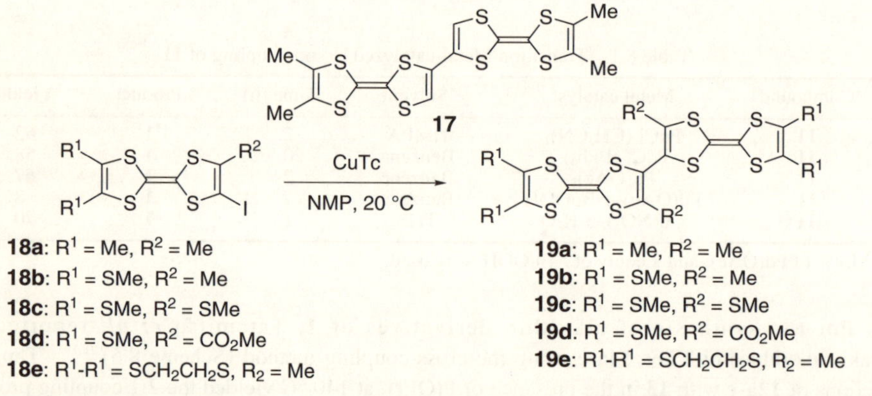

17

18a: R^1 = Me, R^2 = Me
18b: R^1 = SMe, R^2 = Me
18c: R^1 = SMe, R^2 = SMe
18d: R^1 = SMe, R^2 = CO$_2$Me
18e: R^1-R^1 = SCH$_2$CH$_2$S, R$_2$ = Me

CuTc

NMP, 20 °C

19a: R^1 = Me, R^2 = Me
19b: R^1 = SMe, R^2 = Me
19c: R^1 = SMe, R^2 = SMe
19d: R^1 = SMe, R^2 = CO$_2$Me
19e: R^1-R^1 = SCH$_2$CH$_2$S, R$_2$ = Me

Scheme 8.6   Coupling of iodo-TTFs with copper(I) thiophene-2-carboxylate (CuTc).

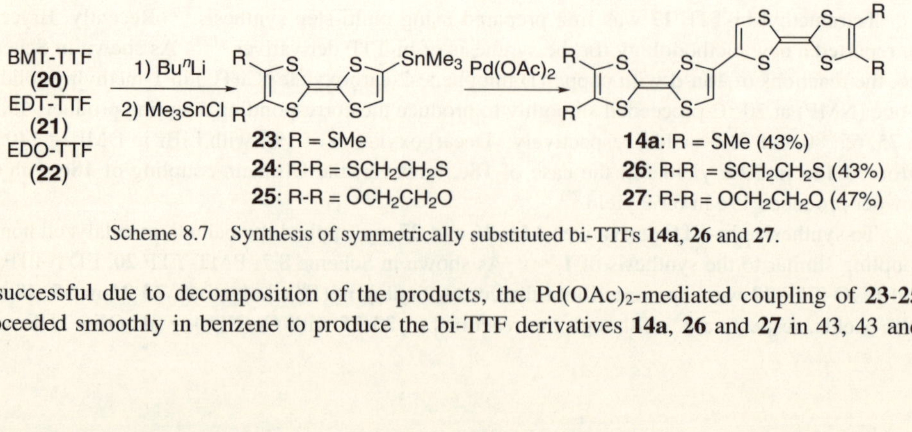

BMT-TTF
(20)

EDT-TTF
(21)

EDO-TTF
(22)

1) Bu^{n}Li

2) Me$_3$SnCl

23: R = SMe
24: R-R = SCH$_2$CH$_2$S
25: R-R = OCH$_2$CH$_2$O

Pd(OAc)$_2$

14a: R = SMe (43%)
26: R-R = SCH$_2$CH$_2$S (43%)
27: R-R = OCH$_2$CH$_2$O (47%)

Scheme 8.7   Synthesis of symmetrically substituted bi-TTFs **14a**, **26** and **27**.

unsuccessful due to decomposition of the products, the Pd(OAc)$_2$-mediated coupling of **23-25** proceeded smoothly in benzene to produce the bi-TTF derivatives **14a**, **26** and **27** in 43, 43 and

47% yields, respectively. All reactions form protodestannylation products, *i.e.*, the starting materials **20-22**, which can be used again for the synthesis of organotin(IV) compounds **23-25**.

The transition metal-catalyzed coupling of organozinc species derived from TTF can also be employed for the synthesis of symmetrically and unsymmetrically substituted bi-TTF derivatives (Scheme 8.8).[16,24] Thus, a reddish-brown solution of TTFZnCl **28** was prepared from the monolithiated TTF with 1.2 equiv of anhydrous $ZnCl_2$ in THF, and **28** can be expected to be a versatile reagent. Although TTFLi is unstable at 0 °C to disproportionate to TTF and $TTFLi_2$,[25] the zinc species **28** is stable at 0 °C in solution and shows no decomposition or disproportionation. As shown in Scheme 8.8, 0.5 equiv of $PdCl_2(PPh_3)_2$ was added to a solution of **28** in THF at -78 °C, and the resulting solution was stirred at -10 °C for 1 h and then at room temperature for 1 h to produce bi-TTF **1** in 57% yield based on the consumed TTF (13% of the recovered TTF). Similarly, the homo-coupling of zinc species **29** and **30** with $PdCl_2(PPh_3)_2$ afforded **14a** and **26** in 70 and 80% yields, respectively. For the cross-coupling, the reaction of **28** (1.7–2 equiv) with the iodo-TTF derivatives **31-33**[26-29] (1 equiv) in the presence of $Pd(PPh_3)_4$ (10 mol%) in THF led to the unsymmetrical bi-TTFs **34-36** in 88, 70 and 20% yields, respectively. The low yield of **36** is presumably due to the difficulty in isolating the product from the reaction mixture.

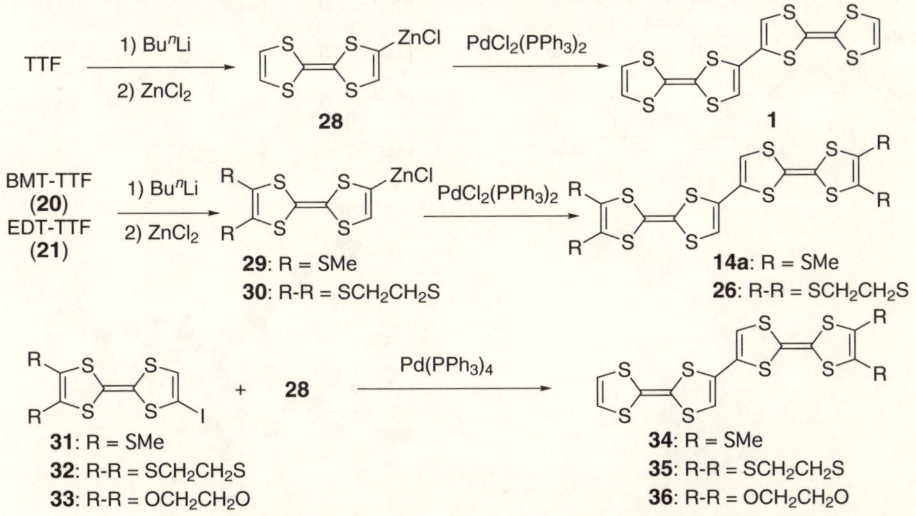

Scheme 8.8   Synthesis of bi-TTF **1** and its derivatives using **28**.

Bi-tetraselenafulvalene (bi-TSF) **38** has been synthesized to investigate its structure and donor ability.[30] As shown in Scheme 8.9, the reaction of TSF with LDA, followed by treatment with chlorotrimethylstannane afforded **37**. The reaction of **37** with $Pd(OAc)_2$ or $Cu(NO_3)_2 \cdot 3H_2O$ produced **38** in 33% or 39% yield, together with the recovered TSF (47% or 56%).

Our synthetic procedure for dimeric and oligomeric TTFs was applied to the synthesis of ter-

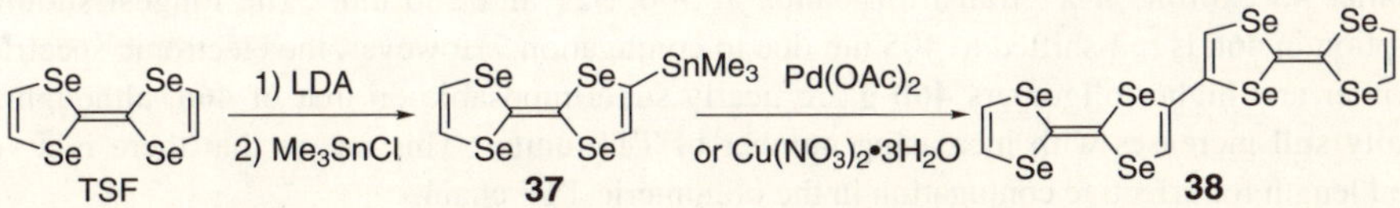

Scheme 8.9   Synthesis of bi-TSF **38**.

and quarter-TTFs **41** and **42** (Scheme 8.10).[24] Thus, the reaction of **39** with butyllithium followed by $ZnCl_2$ formed the organozinc intermediate **40**. The cross-coupling of **40** with **18c** in the presence of $Pd(PPh_3)_4$ produced **41**, whereas the homo-coupling of **40** using a stoichiometric amount of $PdCl_2(PPh_3)_2$ resulted in the formation of **42**. The PM3 calculations of **41** and **42** suggest nonplanar structures with a zigzag conformation (Scheme 8.10). The two TTF units at the terminal positions in **41** cannot interact in a face-to-face manner; however, the two terminal TTFs in **42** may interact with each other by a spiral arrangement.

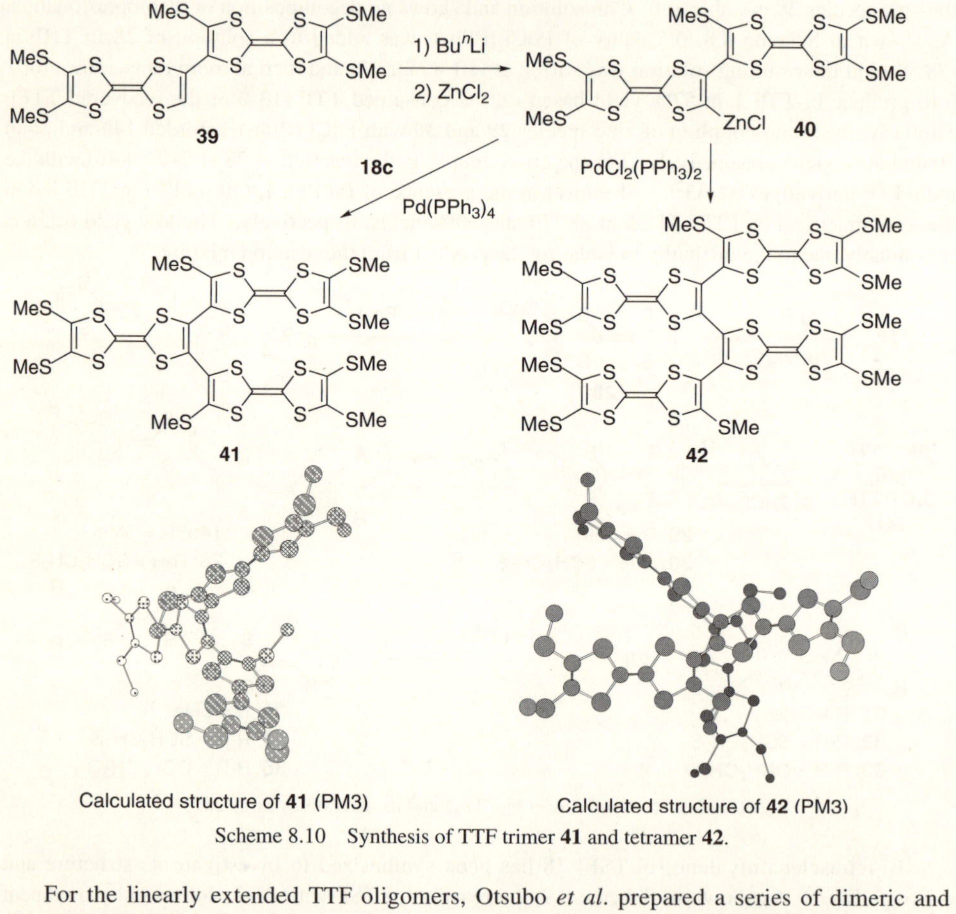

Calculated structure of **41** (PM3)

Calculated structure of **42** (PM3)

Scheme 8.10   Synthesis of TTF trimer **41** and tetramer **42**.

For the linearly extended TTF oligomers, Otsubo *et al.* prepared a series of dimeric and oligomeric TTFs **46a-g** (Scheme 8.11).[31] Although the oligomers **46b-g** comprise various isomeric mixtures due to the *E/Z* configurations, the structures of these oligomers were well characterized by MS and NMR spectroscopy. The electronic absorption spectrum of the monomer **43** exhibits π-π* transition bands at 306, 321 and 380 nm. The longest shoulder absorption in **46a** is red-shifted to 405 nm due to conjugation. However, the electronic spectra of the trimer and higher oligomers **46b-g** are nearly superimposable on that of **46a**, although the intensity still increases with increasing number of TTF units. This means that there is a very limited length for effective conjugation in the oligomeric TTF chain.

Interestingly, treatment of **47a** with iodine in DMF resulted in the formation of bis(thioxotetrathiafulvalenylidene) **48** in 63% yield (Scheme 8.12).[32] Compound **48** shows good

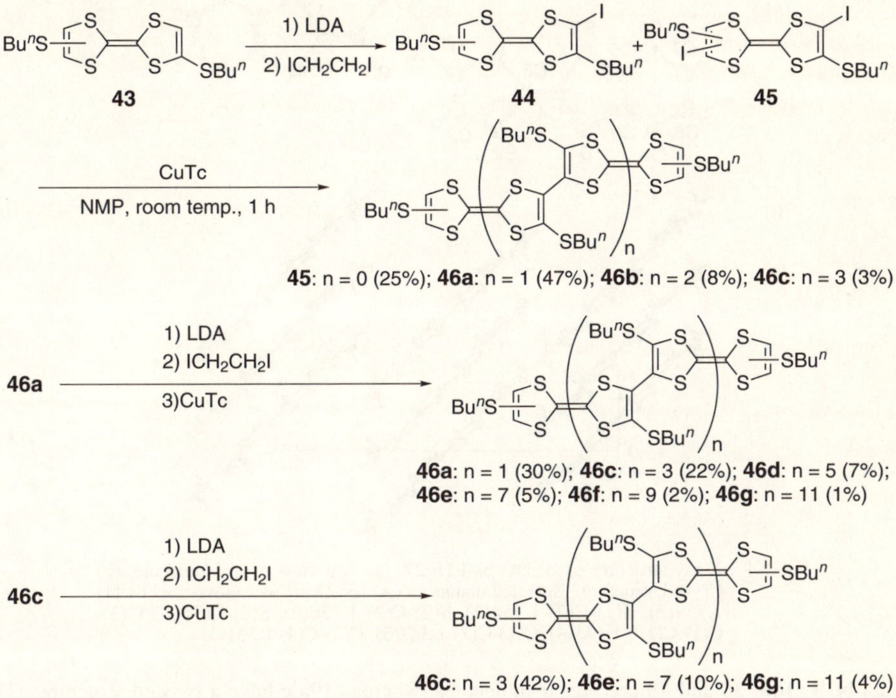
1) LDA
2) ICH2CH2I
43
44
45
CuTc
NMP, room temp., 1 h
45: n = 0 (25%); 46a: n = 1 (47%); 46b: n = 2 (8%); 46c: n = 3 (3%)
46a
1) LDA
2) ICH2CH2I
3)CuTc
46a: n = 1 (30%); 46c: n = 3 (22%); 46d: n = 5 (7%);
46e: n = 7 (5%); 46f: n = 9 (2%); 46g: n = 11 (1%)
46c
1) LDA
2) ICH2CH2I
3)CuTc
46c: n = 3 (42%); 46e: n = 7 (10%); 46g: n = 11 (4%)

Scheme 8.11 Synthesis of TTF oligomers up to the decamer.

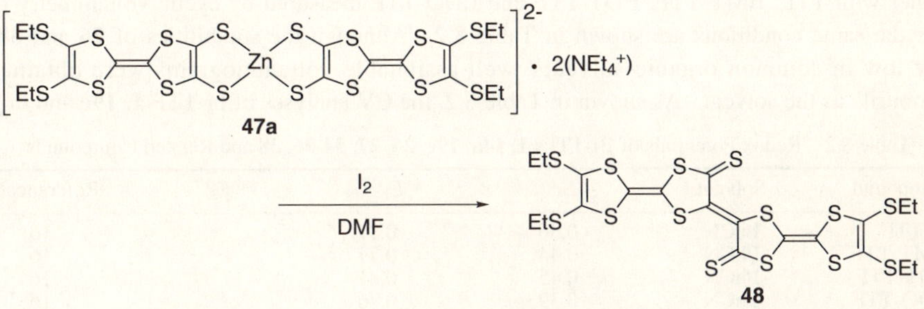
47a
I2
DMF
48

Scheme 8.12 Formation of bis(thioxotetrathiafulvalenylidene) **48**.

donor ability comparable to that of BEDT-TTF. In addition, **48** has the longest absorption maximum at 1118 nm ($\varepsilon = 6.03 \times 10^3$), reflecting its planar structure.

8.2.2 Structures and Properties

The structures of **1**, **14a**, **19a**,**c**, **27**, **38** and **48** have been determined by X-ray analyses.[13,16,21,22,32] As shown in Fig. 8.1, **27** has a crystallographic S_2 symmetry and the two-fold axis passing through the C(1)-C(1*) bond between two TTF units. The central $S_2C_2S_2$ moiety of each TTF unit is exactly planar, and both TTF units have an almost planar structure with a zigzag conformation.

(a)

(b)

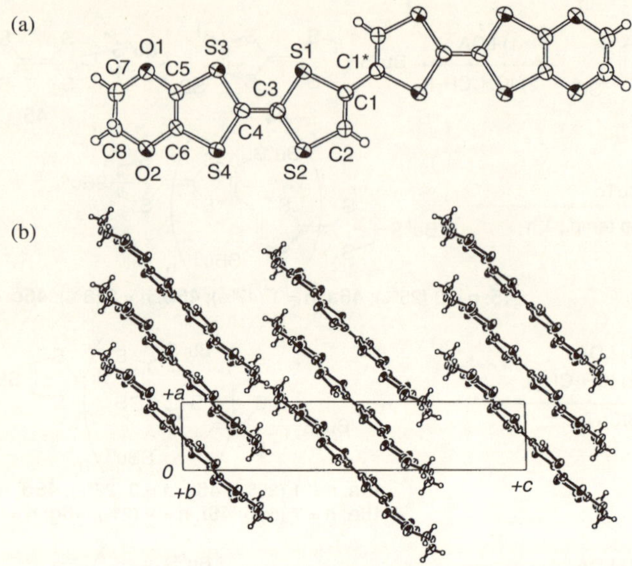

Fig. 8.1   X-ray structure of BEDO-bi-TTF **27**. (a) Top view of the molecule. (b) Crystal structure. Selected distances (Å) for **27** are as follows: S(1)-C(1) 1.759(3), S(1)-C(3) 1.758(2), S(2)-C(2) 1.730(3), S(2)-C(3) 1.751(3), C(1)-C(1)* 1.441(4), C(1)-C(2) 1.337(3), C(3)-C(4) 1.331(3).

Similarly, **1**, **14a**, **38** and **48** are planar molecules, whereas **19a,c** have a twisted structure. The torsion angles around the central bond in **19a,c** are 54° and 89°, respectively.

The oxidation potentials of **1**, **14a**, **19c**, **26**, **27**, **34-36**, **38** and 4-phenyl-TTF (4-Ph-TTF), together with TTF, BMT-TTF, EDT-TTF and EDO-TTF measured by cyclic voltammetry (CV) under the same conditions are shown in Table 8.2. Although the solubilities of **26** and **38** are fairly low in common organic solvents, well-assignable voltammograms were obtained in benzonitrile as the solvent. As shown in Table 8.2, the CV analyses of bi-TTF **1**, **19c** and bi-TSF

Table 8.2   Redox Potentials of Bi-TTFs **1**, **14a**, **19c**, **26**, **27**, **34-36**, **38** and Related Compounds[a]

Compound	Solvent	$E_1^{1/2}$	$E_2^{1/2}$	$E_3^{1/2}$	Reference(s)
TTF	PhCN	0.36	0.77		16
BMT-TTF	PhCN	0.44	0.77		16
EDT-TTF	PhCN	0.45	0.81		16
EDO-TTF	PhCN	0.39	0.76		16
1[b]	CH₃CN	0.45	0.65	0.97	12
1	PhCN	0.43	0.84		15,16
14a	CH₂Cl₂	0.50	0.59	0.85	19
14a	PhCN	0.52	0.63	0.87	16
19c[c]	CH₃CN	0.53	0.60	0.77	22
19c	PhCN	0.58	0.85		24
26	PhCN	0.52	0.62	0.87	16
27	PhCN	0.46	0.55	0.87	16
34	PhCN	0.47	0.62	0.86	16
35	PhCN	0.46	0.68	0.89	16
36	PhCN	0.42	0.54	0.86	16
38	PhCN	0.45	0.71[d]		26
4-Ph-TTF	PhCN	0.38	0.80		12,16

[a]Conditions: Bu₄NClO₄; 100 mV s⁻¹; rt; potentials referred to SCE unless otherwise stated.
[b]Reference: Ag/AgCl. [c]Reference: decamethylferrocene. [d]The anodic potential.

38 indicate two redox waves in benzonitrile as the solvent,[15,13] whereas the oxidation of **1** and **19c** in acetonitrile was reported as a three-step process.[12,13] By contrast, **14a**, **26**, **27** and **34-36** in benzonitrile show three redox waves, in which the third redox process seems to be two-electron oxidation. 4-Ph-TTF shows a slightly higher first oxidation potential ($E^1_{1/2}$ = 0.38 V) than TTF ($E^1_{1/2}$ = 0.36 V),[15] whereas bi-TTF **1** has a much higher first oxidation potential ($E^1_{1/2}$ = 0.43 V) than TTF, presumably due to the larger electron-withdrawing effect of the TTF moiety as compared to the phenyl group. Although the bond formation between two TTF molecules at the 4-position causes the first oxidation potentials to shift positively by 0.03–0.12 V, the oxidation potentials ($E^1_{1/2}$) of **1**, **14a**, **26**, **27** and **34-36** are much lower than that ($E^1_{1/2}$ = 0.55 V) of BEDT-TTF, which is a well-known donor. In the case of bi-TSF **38**, this compound shows lower oxidation potentials than TSF. Therefore, bi-TTFs and bi-TSF can be expected to show good donor ability.

The redox behavior of bi-TTFs and bi-TSF suggests that the through-bond interaction between the two TTF or TSF units is weak in the neutral, cation radical and dicationic species derived from bi-TTFs and bi-TSF,[33] although the face-to-face through-space interaction between the two TTF units of the bis(EDT-TTF)-substituted naphthalene is known to be strong enough to form a mixed valence state or charge-delocalized π-dimer in the cation radical or dicationic species.[34] There are three possible resonance structures in the cation radical **1**[·+] (Fig. 8.2). The first electronic structure shown in Fig. 8.2a has a partial delocalization of the cation radical charge (Class I). In the case of nonplanar bi-TTFs, the weak conjugation through the central pivot bond reduced the interaction of the two TTF units to form the Class I state. The second structure shown in Fig. 8.2b is an insufficiently delocalized state (Class II). The major cation radical charge is localized on one TTF unit, whereas the other can participate in delocalization to accept a slight charge density. The third one is a strongly or fully delocalized state as shown in Fig. 8.2c (Class III). In the case of the cation radicals derived from bi-TTFs and bi-TSF, no cation radicals in the Class III state have been reported until now.

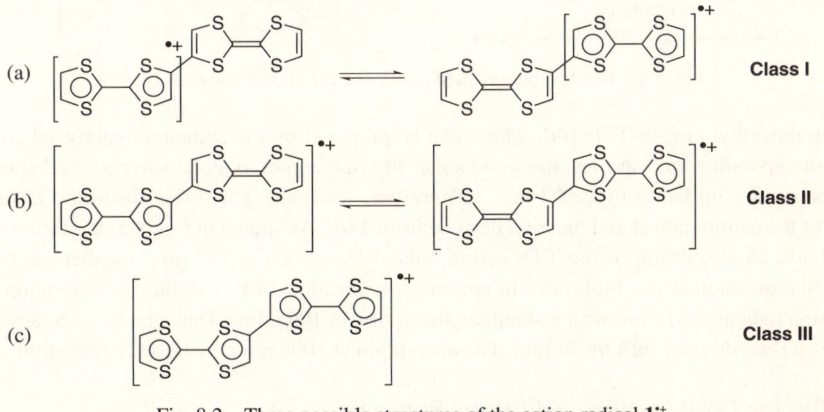

Fig. 8.2   Three possible structures of the cation radical **1**[·+].

The extended HMO calculations of **1** show that the electron density in the highest occupied molecular orbital (HOMO) is mainly located in the left part of **1** (Fig. 8.3a). On the other hand, the density in the next HOMO (NHOMO) is located in the right part of **1** (Fig. 8.3b). Thus, the extended HMO calculations of **1** indicate the Class II state for **1**[·+]. Similarly, the weak interaction in the cation radicals of **14a**, **19c**, **26**, **27**, **34-36** and **38** results in formation of the Class I or Class II state,[5] although the cation radicals derived from the face-to-face fixed bis-TTFs

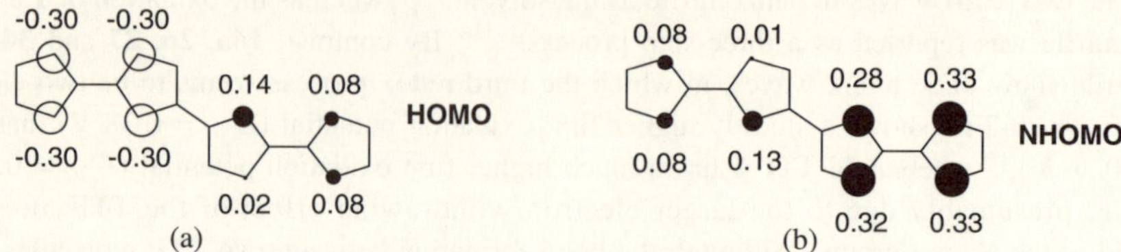

Fig. 8.3   Electron density in HOMO (a) and NHOMO (b) ($\Delta E = 0.08$ eV) of **1** calculated by extended HMO.

form a strongly delocalized (Class III) species.[34]   The cation radical **19e**[.+] seems to possess a Class I state, presumably due to its nonplanar structure with a torsion angle of 77° around the central σ-bond.[22]   In the electrochemical oxidation of bi-TTFs and bi-TSF, π-dimer formation or aggregation of cation radicals can be excluded due to their low solubilities in acetonitrile and benzonitrile.

The intramolecular interaction of the two cation radical parts in the dication **1**[2+] has been investigated very little because **1**[2+] is nearly insoluble in common organic solvents and forms no single crystals for X-ray analysis. Torrance *et al.* presumed a Davydov red shift for a head-to-tail oriented TTF[.+] dimer (Fig. 8.4).[35]   Thus, the two TTF[.+] aligned in a head-to-tail manner make the dipole interaction attractive; hence a lower energy is sufficient to excite this transition. Since the interaction between the two cation radical parts in **1**[2+] seems to be weak, **1**[2+] can be regarded as a model system for the head-to-tail oriented TTF[.+] dimer.

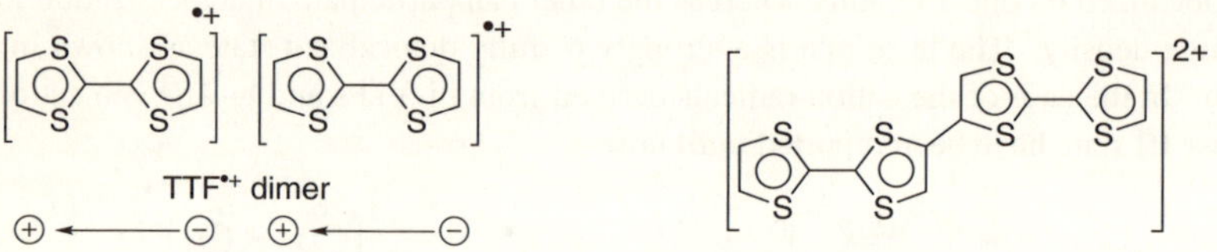

Fig. 8.4   Head-to-tail oriented dimer and the bi-TTF dication **1**[2+].

Tetrakis(ethylthio)-bi-TTF **14d**, which can be prepared by the palladium-catalyzed coupling of the corresponding zinc species, has good solubility in common organic solvents, and shows the redox behavior similar to that of **14a**.[33]   Therefore, we investigated the electronic absorption spectra of the cation radical and dication derived from **14d**. As shown in Fig. 8.5, the cation radical **14d**[.+] shows an absorption of the TTF cation radical ($S_o \rightarrow S_1$) at 722 nm, together with a very broad CT absorption at *ca.* 1400 nm. In contrast, the dication **14d**[2+] exhibits an absorption of the TTF cation radical at 816 nm with a shoulder absorption at 1098 nm. Thus, the $S_o \rightarrow S_1$ absorption indicates a Davydov red shift by 94 nm. The absorption at 1098 nm may be a CT transition.

8.2.3 Electric Conductivities and Crystal Structures

Planar bi-TTFs formed CT complexes and cation radical salts.[16,23]   As shown in Table 8.3, the CT complexes derived from **1** and **27** with TCNQ showed fairly high room-temperature conductivities (4.8 and 3.6 S cm^{-1}, respectively), and **1**•TCNQ was metallic down to 160 K. However, the **26**•DDQ complex is a semiconductor (0.84 S cm^{-1}). Although the I$_3^-$ salt of **1** and the ClO$_4^-$ salt of **14a** were semiconductors, the I$_3^-$, AuI$_2^-$ and BrI$_2^-$ salts of **26** exhibited high conductivities (125, 778 and 80 S cm^{-1}) with metallic behavior down to 240, 285 and 240 K,

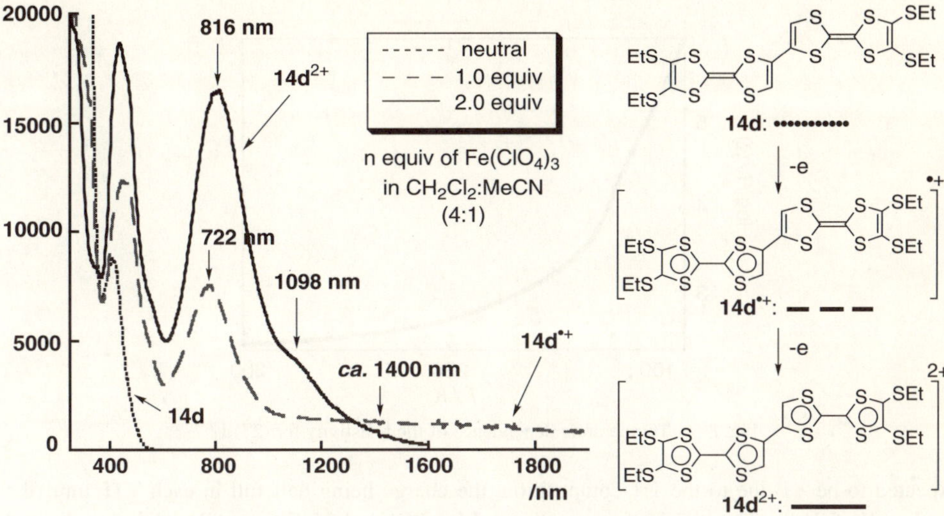

Fig. 8.5   Electronic absroption spectra of **14d**, **14d**$^{•+}$ and **14d**$^{2+}$.

Table 8.3   Electrical Conductivities of CT Complexes and Radical Cation Salts Derived from **1**, **14a**, **19a**, **19e**, **26**, **27** and **38**

Donor	Acceptor	Solvent	D:A	σ_{rt}/S cm^{-1}
1	TCNQ	PhCl	1:1	4.8 (metallic)
1	I$_3$	TCE^a	3:2	7.6
14a	ClO$_4$	THF	1:1	9.4 x 10^{-2}
19a	TCNQ	CH$_3$CN	1:1	1.2 x 10^{-2}
19e	ClO$_4$	TCE^a	1:1	1.6 x 10^{-3}
26	DDQ	benzene	2:1	8.4 x 10^{-1}
26	I$_3$	THF	2:1	125 (metallic)
26	AuI$_2$	THF	1:1	778 (metallic)
26	BrI$_2$	PhCl	1:1	80 (metallic)
27	TCNQ	PhCl	3:1	3.6 (E_a = 72 meV)
27	ClO$_4$	THF	1:1	6.9 (E_a = 35 meV)
27	I$_3$	PhCl	3:1	8.0 (metallic)
38	ClO$_4$	CS$_2$-PhCl	1:1	1.2 (semiconductive)
38	PF$_6$	CS$_2$-PhCl	3:2	4.2 (semiconductive)

^{a}1,1,2-Trichloroethane.

respectively, being transformed at lower temperatures to semiconductors with small activation energies. Furthermore, the ClO$_4^-$ salt of **27** indicated semiconductive resistive temperature dependence (6.9 S cm^{-1}) with a small activation energy (35 meV), whereas the I$_3^-$ salt of **27** was metallic (8.0 S cm^{-1}) at room temperature. Interestingly, fully substituted bi-TTFs (**19a,e**) form semiconducting CT complexes and cation radical salts, although these bi-TTFs possess nonplanar structures.[22] As shown in Fig. 8.6, (**27**)$_3$I$_3$ showed metallic behavior at 300 K and semi-metallic conductivity down to 135 K, exhibiting a typical temperature dependence of the resistivities for the cation radical salts of bi-TTFs.

In order to clarify the crystal packings of novel cation radical salts derived from **1**, **14a**, **26** and **27**, the crystal structures of **14a•**ClO$_4$ and **27•**ClO$_4$ were determined by the X-ray diffraction method.[16] As shown in Figs. 8.7 and 8.8, the cation radical salts of **14a** and **27** show unique crystal packings, reflecting a dimeric TTF structure. The net charge on a donor molecule is

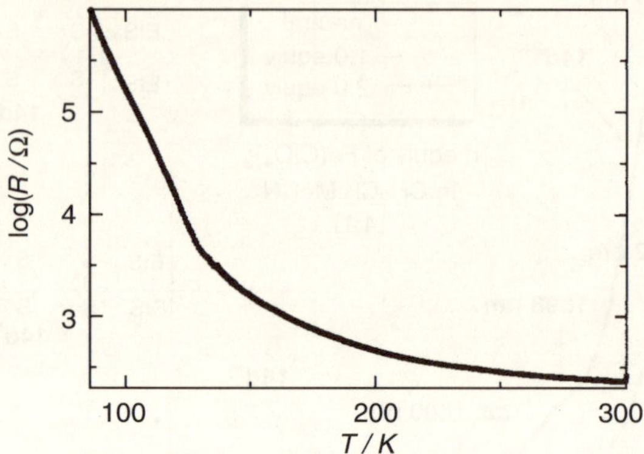

Fig. 8.6   Temperature dependence of the resistivity for $(27)_3I_3$.

expected to be +1 due to the 1:1 composition, the charge being half full in each TTF unit (the band is 3/4 filled). The bi-TTF frameworks in 14a•ClO$_4$ and 27•ClO$_4$ exhibit slightly bent and planar structures, the maximum atomic deviations from the least-square plane of the bi-TTF units excepting the substituents being 0.46 and 0.18 Å, respectively.

As shown in Fig. 8.7, the TMT-bi-TTF molecules are stacked face-to-face to form a dimeric structure (Fig. 8.7a), and the dimers are arranged in the so-called β′-type structure to form a conducting path along the c axis. There are four intra- and interstack interactions between the cation radicals (Fig. 8.7a). Although 14a•ClO$_4$ has a semiconducting property (9.4 x 10^{-2} S cm^{-1}), the overlap integral analysis was examined. Since the C(3)-C(4) [1.394(9) Å] and C(11)-C(12) [1.343(9) Å] bonds have single and double bond characters, respectively, the two TTF units in the donor molecule have different oxidation states, $i.e.$, neutral and cation radical states on the A and B parts, respectively (Fig. 8.7b). The two face-to-face distances observed between TMT-bi-TTFs along the c axis (Figs. 8.7a-c) are 3.50 and 3.55 Å, corresponding to the two overlaps (II and I), respectively. There are two types of side-by-side interactions between TMT-bi-TTFs, one of which is strong (IV in Fig. 8.7c), whereas the other is weak (III in Fig. 8.7b). Thus, each cation radical unit and each neutral unit in the donor molecules form a column structure, and the former shows strong side-by-side interaction (IV in Figs. 8.7a,c). In the calculations of the overlap integrals, the two TTF parts (A and B in Fig. 8.7b) of a single molecule are treated as if they were two independent molecules, although the corresponding molecular orbitals are originally HOMO and NHOMO, as shown in Fig. 8.3. Thus, each of the I to IV interactions in Fig. 8.7a has A-A, A-B, B-A and B-B combinations. When the interaction is connected by an inversion center, A-B and B-A are the same. In the case of 14a•ClO$_4$, the energy difference between HOMO and NHOMO is large (0.13 eV), indicating that the intramolecular charge separation is significant (nearly A^{0}B$^{•+}$) in accordance with the difference in the bond distances. Therefore, the B part in the TMT-bi-TTF molecule constructs a dimerized half-filled band, resulting in the relatively poor conductivity of the Mott insulating state.

In contrast to 14a•ClO$_4$, the cation radical salt 27•ClO$_4$ has a fairly high conductivity (6.9 S cm^{-1}) with a small activation energy (35 meV). The C(3)-C(4) and C(11)-C(12) bonds have the same distance [1.37(1) Å] as shown in Fig. 8.8a, indicating a mixed-valence state. The cation radicals form a conducting sheet in the ac plane as well as a segregated column. There are five intra- and interstack interactions between the cation radicals (Fig. 8.8e). Interestingly, the two

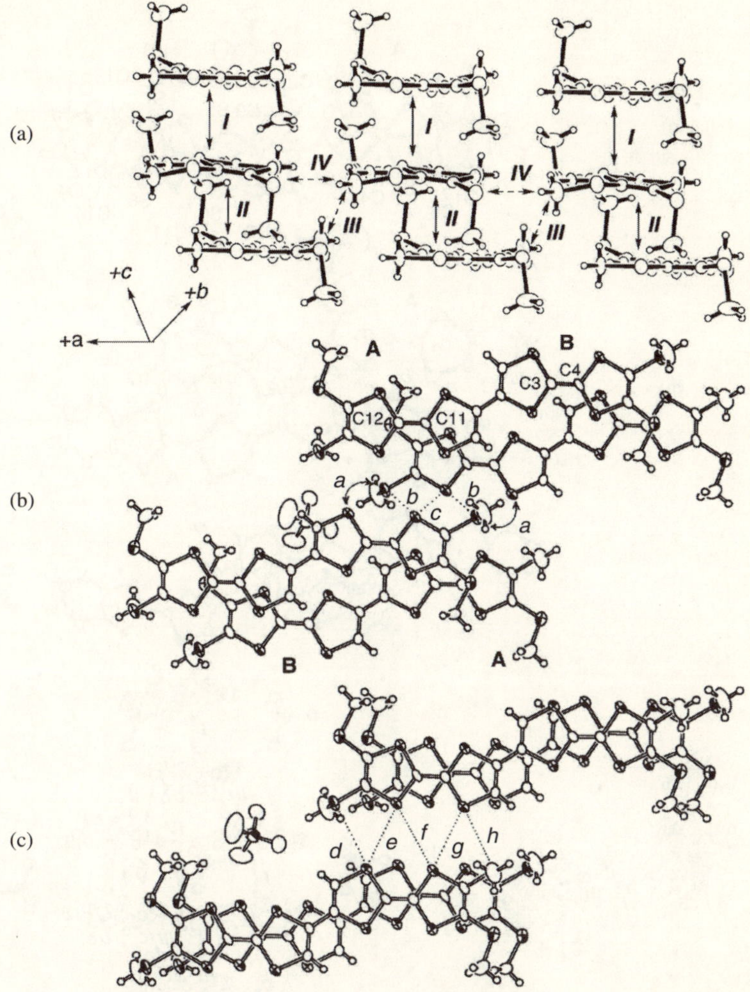

Fig. 8.7 The intra- and interstack overlaps of **14a**•ClO$_4$. The inter-columnar S···S distances are indicated by dotted lines. (a) The overlaps in **14a**•ClO$_4$; the overlap integrals (x 10^{-3}), *I*: A-A = -8.22; A-B = B-A = -18.05; B-B = 4.81, *II*: A-A = 9.18; A-B = B-A = -3.38; B-B = -2.22, *III*: A-A = 2.53; A-B = B-A = 0.26; B-B = 0.00, *IV*: A-A = -3.58; A-B = 9.81; B-A = 0.50; B-B = 0.91. (b) The overlaps *II* and *III*; inter-columnar S···S distances (*a* = 3.91, *b* = 3.79, *c* = 3.97 Å) with 3.55 Å of the face-to-face stacking. (c) The overlaps *I* and *III*; inter-columnar S···S distances (*d* = 3.64, *e* = 3.57, *f* = 3.79, *g* = *h* = 3.50 Å) with 3.50 Å of the face-to-face stacking.

overlaps (*I* and *II*) are made up of the whole molecules, *i.e.*, Figs. 8.8a,b, whereas the other two (*III* and *IV*) are constructed by the interaction between the TTF moieties of the adjacent cation radicals, *i.e.*, Figs. 8.8c,d. The side-by-side interstack interaction *V* made up of the slipped whole molecules is weak and may be neglected in the conducting interaction. The donor packing in **27**•ClO$_4$ is also regarded as a β′-like dimerized structure. Although this dimer structure resembles that of β′-(BEDT-TTF)$_2$ICl$_2$,[36] the interdimer interactions are strong enough to lead to a two-dimensional (2D) conducting sheet. In the molecular orbital calculations, NHOMO appears

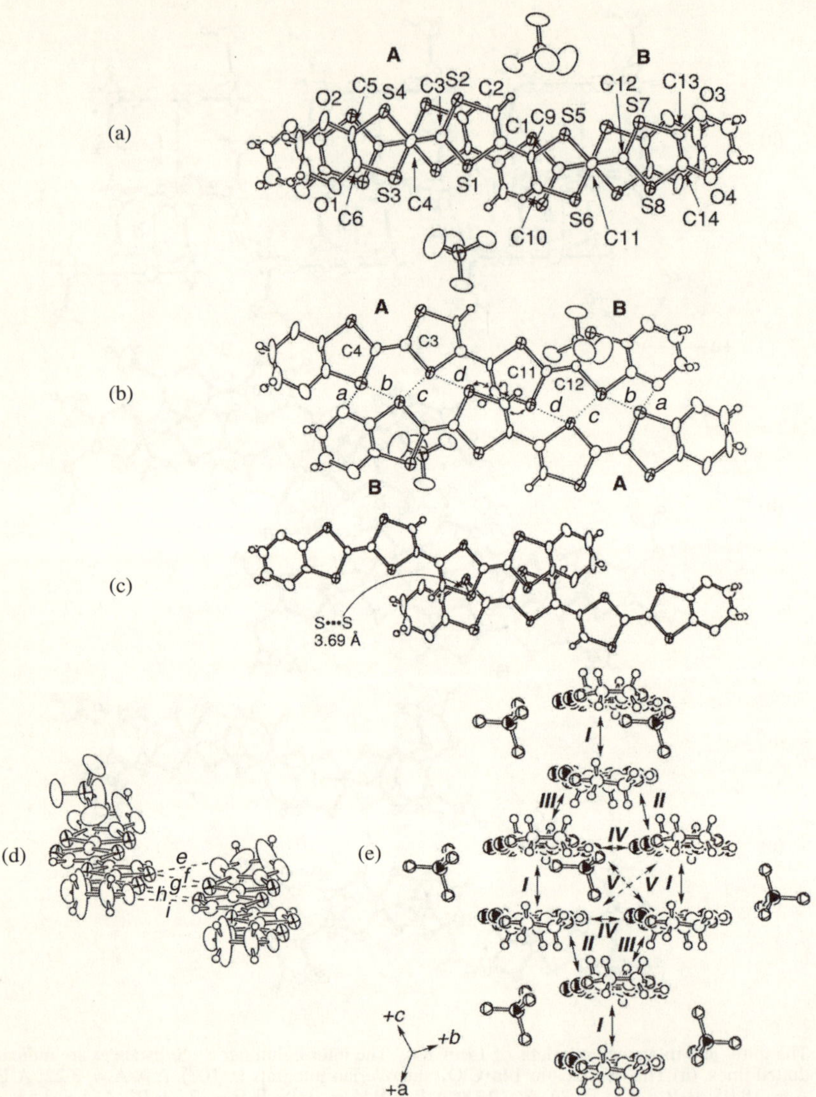

Fig. 8.8   The intra- and interstack overlaps of **27•ClO₄**. The inter-columnar S···S or S···O distances are indicated by dotted lines and the intra-column short S···S contacts are indicated by solid lines. (a) The overlap *I* and two TTF parts (A and B). (b) The overlap *II* and inter-columnar S···S and S···O distances (a = 3.48, b = 3.78, c = 3.70, d = 3.83 Å). (c) the overlap *III*. (d) The overlap *IV* and inter-columnar S···S and S···O distances (e = 3.36, f = 3.45, g = 3.59, h = 3.46, i = 3.41 Å). (e) The overlap integrals (x 10^{-3}), *I*: A-A = 9.31; A-B = B-A = -22.70; B-B = -14.11, *II*: A-A = -4.02; A-B = B-A = -1.97; B-B = -0.75, *III*: A-A = 2.53; A-B = B-A = -11.05; B-B = -5.62, *IV*: A-A = 1.66; A-B = 0.78; B-A = -10.88; B-B = -4.10, *V*: A-A = 0.05; A-B = B-A = -0.40; B-B = 2.48.

very close to the HOMO level (ΔE = 0.08 eV). NHOMO and HOMO are mainly located on the left (A) and right (B) halves of the molecule in Fig. 8.8a, respectively, and have the correct symmetry of TTF HOMO. The A and B units within the same molecule have the interaction of the order of $1/2\Delta E = 0.04$ eV, *i.e.*, 4 x 10^{-3} as the orbital overlap; however, this weak intramolecular interaction plays no important role in the conducting interaction, and the A and B parts can be regarded as two independent molecules. Thus, each "intermolecular interaction" based on the A and B parts has four kinds of interactions (A-A, A-B, B-A and B-B) like those in **14a•ClO₄**. From the crystal symmetry ($P\bar{1}$), one unit cell contains two donors. As shown in Fig. 8.8b, two donor molecules, which are connected by the interaction *I*, form a dimer unit. The face-to-face distance in Fig. 8.8a is 3.40 Å, and the overlap integral of the conduction orbitals seems to be fairly large (*I*: A-B = B-A = -22.7 x 10^{-3}). Because these two molecules are connected by an inversion center, the A part is located above the B part of the other molecule so that there is strung A-B interaction. There are considerable interdimer interactions *II* and *III*, spreading in the *ac* plane; this makes a 2D conducting sheet. Because one donor has two TTF parts, this conducting sheet has a double sheet structure stacking along the *b* axis. There are no short S···S contacts less than the sum of the van der Waals radii (3.70 Å) in the side-by-side overlap in Fig. 8.8b, but the six S···S contacts (3.70-3.83 Å) can contribute to make a conducting path (*II*: A-B = B-A = -11.1 x 10^{-3}). The face-to-face interaction (*III*) and the side-by-side interaction (*IV*) seem to be strong through the S···S short contacts (*III*: 3.69 Å in Fig. 8.8c; *IV*: 3.45, 3.46 and 3.59 Å in Fig. 8.8d). Although the interactions *III* and *IV* are mediated by only one TTF part, these interactions are moderately strung (*III*: A-A = -4.0 x 10^{-3}; *IV*: B-A = -10.9 x 10^{-3}). Consequently, **27•ClO₄** shows a fairly high conductivity with a small activation energy, reflecting a 2D band structure in contrast to the one-dimensional β′-phase.

8.3   Conjugated Bis-TTFs Linked by π-Systems

8.3.1 Synthesis

In order to design superior electronic donors with a reduced on-site Coulombic repulsion and an increased dimensionality, the conjugated dimeric TTFs linked by ethylene, acetylene and diacetylene spacers **49**, **50**, **51** and **52** were investigated. For the synthesis of **49**,[37] reductive coupling of 4-formyl-TTF **53** with a low-valent titanium reagent produced **49** in 56% yield. On the other hand, **50** was prepared by the Sonogashira coupling of **8** with **54** in 22% yield (Scheme 8.13).[37]

Recently, the synthesis of **51** and **52** has been carried out using the palladium-catalyzed couplings (Scheme 8.14).[33,38] For the Sonogashira coupling of iodo-TTFs, the reaction proceeded

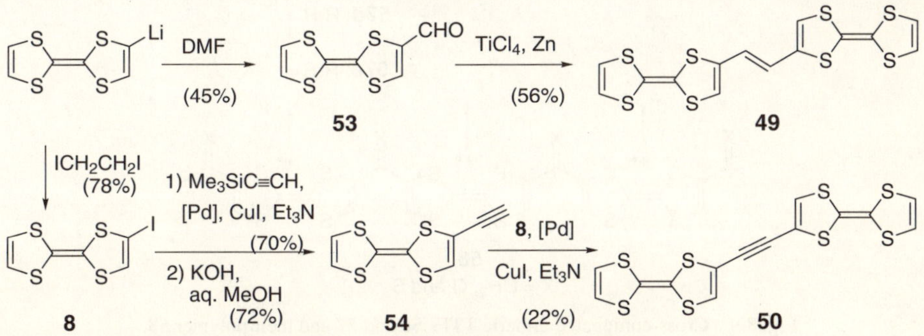

Scheme 8.13   Synthesis of **49** and **50**.

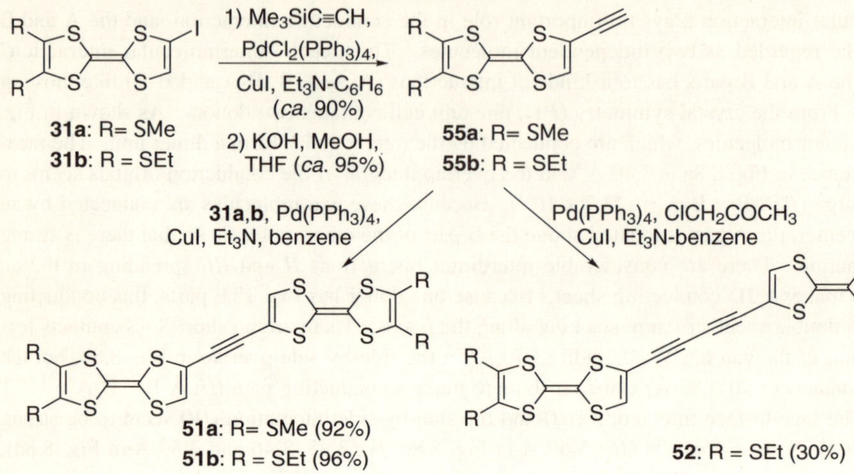

31a: R= SMe
31b: R = SEt

1) Me$_3$SiC≡CH,
   PdCl$_2$(PPh$_3$)$_4$,
   CuI, Et$_3$N-C$_6$H$_6$
   (*ca.* 90%)
2) KOH, MeOH,
   THF (*ca.* 95%)

55a: R= SMe
55b: R = SEt

31a,b, Pd(PPh$_3$)$_4$,
CuI, Et$_3$N, benzene

Pd(PPh$_3$)$_4$, ClCH$_2$COCH$_3$
CuI, Et$_3$N-benzene

51a: R= SMe (92%)
51b: R = SEt (96%)

52: R = SEt (30%)

Scheme 8.14   Synthesis of **51** and **52**.

smoothly under Krause conditions.[39] Thus, the cross-coupling of **55a** with **31a** in the presence of Pd(PPh$_3$)$_4$ and CuI in Et$_3$N-benzene at 65 °C afforded **51a** in 92% yield, whereas a similar reaction of **55b** with **31b** produced **51b** in 96% yield. In contrast, treatment of **55b** with Pd(PPh$_3$)$_4$, CuI and α-chloroacetone[40] in Et$_3$N-benzene at room temperature led to the homo-coupling product **52** in 30% yield.

The cross-conjugated dimeric TTFs **56**[41] and **57**[42] provide the possibility of a stable triplet ground state, and they can be regarded as simple model compounds for the high-spin system **58**

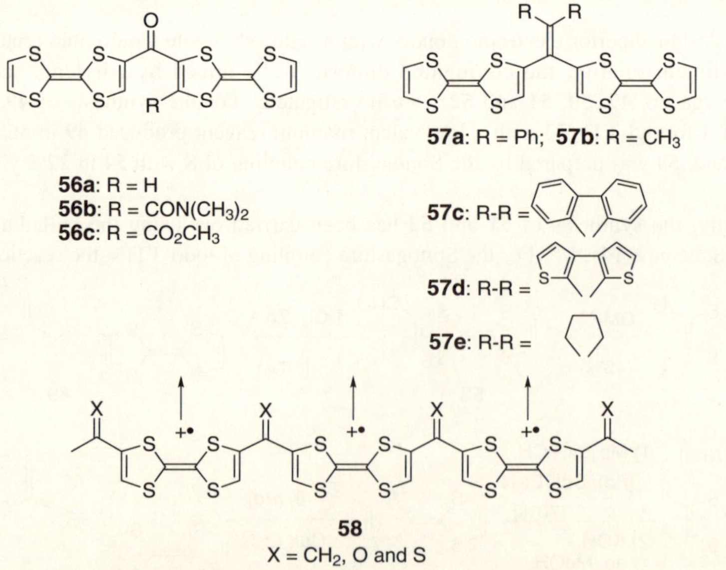

56a: R = H
56b: R = CON(CH$_3$)$_2$
56c: R = CO$_2$CH$_3$

57a: R = Ph; **57b**: R = CH$_3$

57c: R-R =

57d: R-R =

57e: R-R =

58

X = CH$_2$, O and S

Fig. 8.9   Cross-conjugated dimeric TTFs **56** and **57** and their polymer **58**.

(Fig. 8.9). Mizouchi *et al.* theoretically predicted that the radical cation of **58** should show a high-spin ground state.[43)]

For the synthesis of **56a-c**, monolithiated TTF was reacted with Me_2NCOCl (0.5 equiv) to afford **56a** (9.4%) and **56b** (6.5%) (Scheme 8.15).[41)] **56a** was also obtained by the reaction of **7** with 1,1'-carbonyldiimidazole in 5.6% yield. On the other hand, **56c** was prepared in 23% yield by using MeOCOCl in analogy to Green's method.[25)]

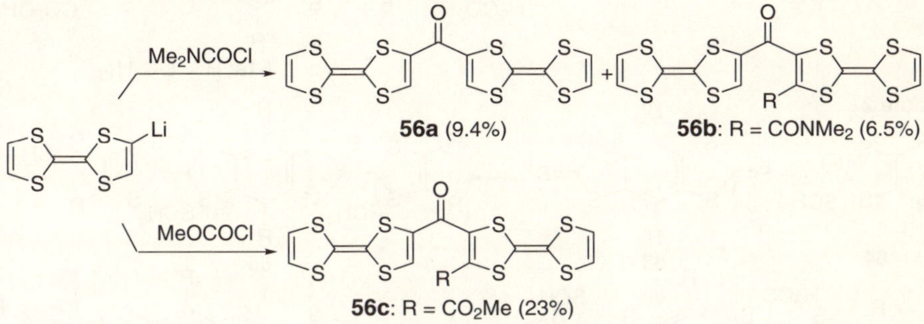

Me2NCOCl
MeOCOCl
Li
56a (9.4%)
56b: R = CONMe2 (6.5%)
56c: R = CO2Me (23%)

Scheme 8.15 Synthesis of **56a-c**.

As shown in Scheme 8.16, trimethylstannyl-TTF **11** was found to be an effective synthon to construct the 1,1-(tetrathiafulvalenyl)ethylene framework **57**.[42)] Thus, the Stille reaction of **11** (2 equiv) with 1,1-dibromo-2,2-diphenylethylene **59a** in the presence of $Pd(PPh_3)_4$ (0.1 equiv) proceeded smoothly in refluxing benzene to afford **57a** in 57% yield. Similarly, the palladium-catalyzed reactions of **11** with **59b-e** in refluxing benzene or toluene gave the corresponding products in 26, 52, 41 and 28% yields, respectively. Because of the low solubilities of **57a,c** in common organic solvents, **57a,c** were converted into the hexakis(methylthio) derivatives **60a,c** in a known procedure.

Dimeric TTFs linked by aromatic systems were studied by many groups independently. Müllen *et al.* investigated oligomeric systems of the benzene-fused bis-TTFs **61a-d** (Scheme

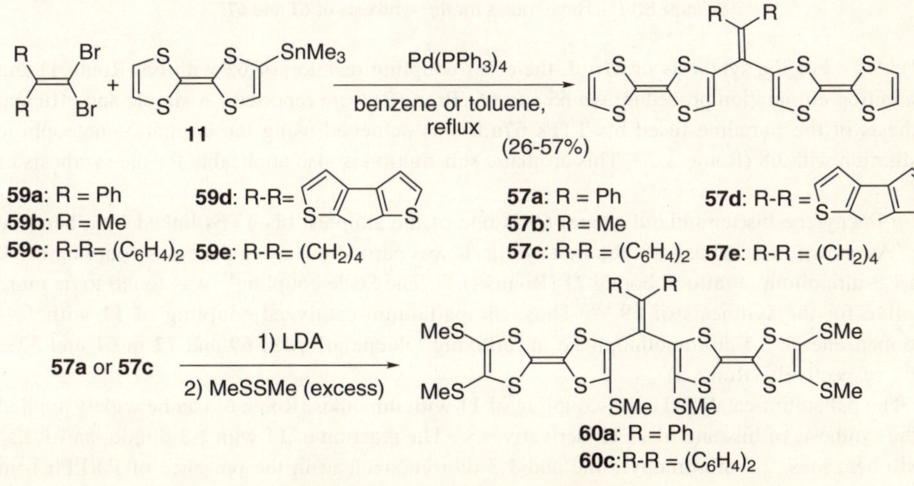

SnMe3
11
Pd(PPh3)4
benzene or toluene,
reflux
(26-57%)
59a: R = Ph
59b: R = Me
59c: R-R= (C6H4)2
59d: R-R=
59e: R-R= (CH2)4
57a: R = Ph
57b: R = Me
57c: R-R = (C6H4)2
57d: R-R =
57e: R-R = (CH2)4
57a or 57c
1) LDA
2) MeSSMe (excess)
60a: R = Ph
60c:R-R = (C6H4)2

Scheme 8.16 Synthesis of **57** and **60**.

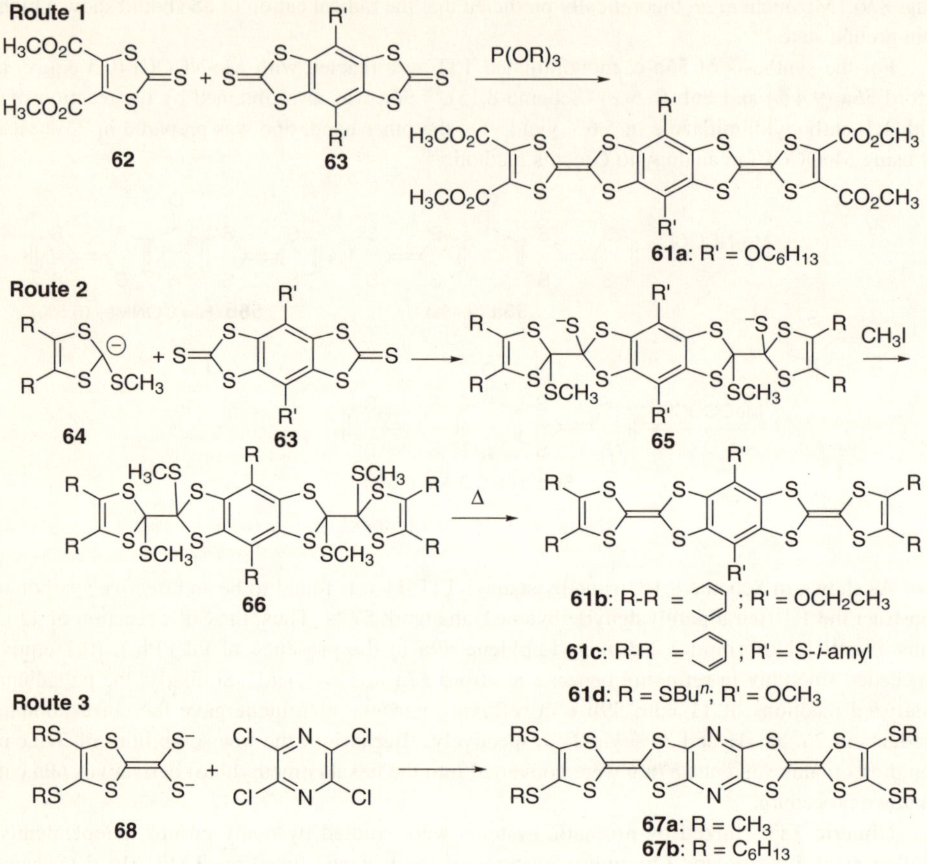

Scheme 8.17   Three routes for the synthesis of **61** and **67**.

8.17).[1,44-46] For the synthesis of **61a-d**, the cross-coupling reaction of **62** with **63** (Route 1) and the addition-elimination procedure *via* **65** and **66** (Route 2) were reported. A simple and efficient synthesis of the pyradine-fused bis-TTFs **67a,b** was achieved using the aromatic nucleophilic substitution with **68** (Route 3).[47] This aromatic substitution is also applicable for the synthesis of **61**.[4]

p-Phenylene-bis(tetrathiafulvalene) **69** is one of the simplest bis-TTFs linked by a benzene ring. As shown in Scheme 8.18, the first synthesis was carried out using the cross-coupling of **70** with 1,3-dithiolium tetrafluoroborate **71** (Route 4).[48] The Stille coupling[49] was found to be more effective for the synthesis of **69**.[15] Thus, the palladium-catalyzed coupling of **11** with 1,4-diiodobenzene or 2,5-dibromothiophene in refluxing toluene afforded **69** and **72** in 61 and 52% yields, respectively (Route 5).

The palladium-catalyzed cross-coupling of **11** with dihalides (Route 6) can be widely applied for the synthesis of bis- and tris-TTF derivatives.[50] The reaction of **11** with 1,3-diiodo- and 1,3,5-triiodo-benzenes, 2,6-dibromopyridine and 1,3-dibromoazulene in the presence of Pd(PPh₃)₄ in refluxing toluene afforded **73-76** in good yields (Scheme 8.19). However, similar reactions of **11** with 1,8-diiodonaphthalene and 2,10- and 2,7-dibromo-1,6-methano[10]annulenes gave **77a,78**

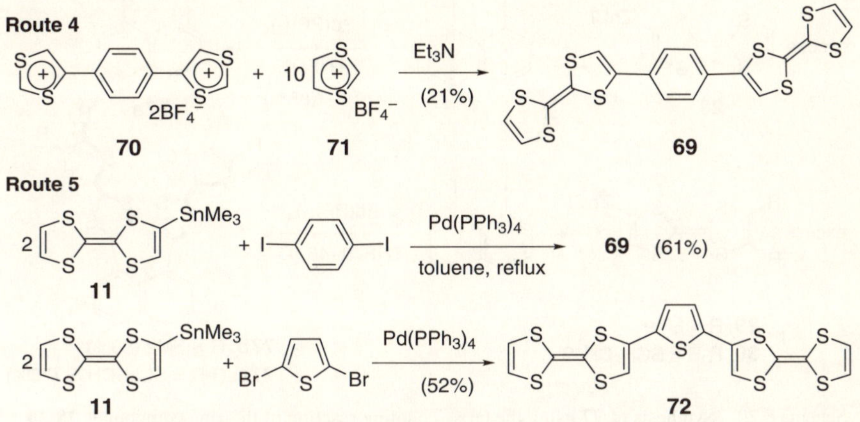

Route 4

70 + 10 **71** → (Et₃N, 21%) → **69**

Route 5

2 **11** + I–⬡–I → (Pd(PPh₃)₄, toluene, reflux) → **69** (61%)

2 **11** + Br–thiophene–Br → (Pd(PPh₃)₄, 52%) → **72**

Scheme 8.18   Synthesis of phenylene-bis-TTF and its analogue.

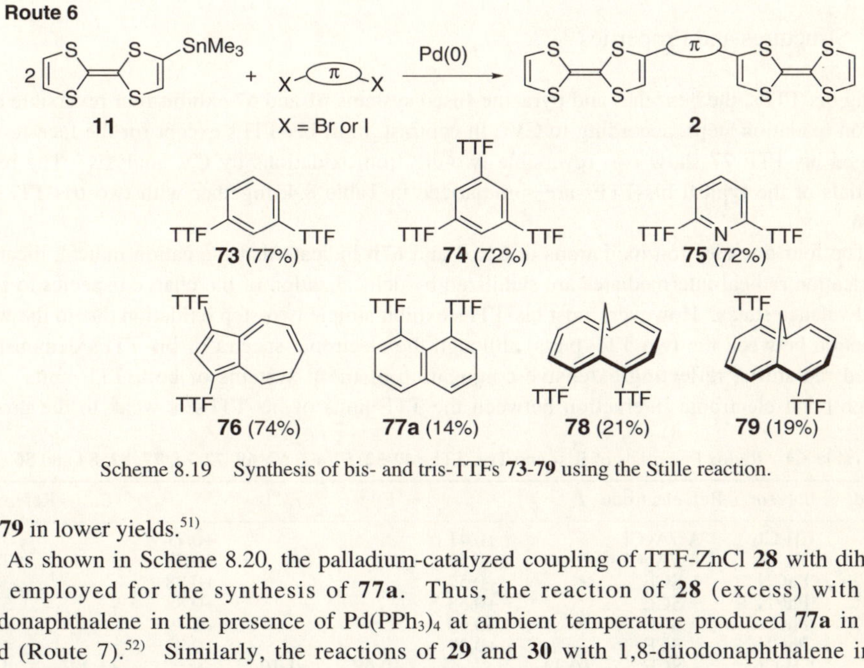

Route 6

2 **11** + X–π–X (X = Br or I) → (Pd(0)) → **2**

73 (77%) **74** (72%) **75** (72%)

76 (74%) **77a** (14%) **78** (21%) **79** (19%)

Scheme 8.19   Synthesis of bis- and tris-TTFs **73-79** using the Stille reaction.

and **79** in lower yields.[51]

As shown in Scheme 8.20, the palladium-catalyzed coupling of TTF-ZnCl **28** with dihalide was employed for the synthesis of **77a**. Thus, the reaction of **28** (excess) with 1,8-diiodonaphthalene in the presence of Pd(PPh₃)₄ at ambient temperature produced **77a** in 85% yield (Route 7).[52] Similarly, the reactions of **29** and **30** with 1,8-diiodonaphthalene in the presence of Pd(PPh₃)₄ afforded **77b,c** in 59 and 78% yields, respectively.[53]

Synthesis of dumbbell-type bis-TTFs **82** and **84** was carried out using a wittig reaction to construct the molecular framework (Route 8) as shown in Scheme 8.21. Thus, a series of bis-TTFs **82a-c** and **84a,b** were prepared by the reaction of **53** and **80a** with **81** and **83a,b** in the presence of lithium ethoxide.[54,55] Using a similar methodology, Martín et al. prepared **86a-c** to investigate conducting and magnetic properties of their cation radical species.[56a] The titanium-mediated self-condensation of **87** was applied to prepare **88**.[57]

Route 7

28

29: R = SMe
30: R-R = SCH$_2$CH$_2$S

77a

77b: R = SMe (59%)
77c: R-R = SCH$_2$CH$_2$S (78%)

Scheme 8.20   Synthesis of **77** using the cross-coupling reaction of the zinc compounds **28-30** with 1,8-diiodonaphthalene.

8.3.2 Structures and Properties

Among bis-TTFs, the benzene- and pyrazine-fused systems **61** and **67** exhibit four reversible one-electron oxidation steps, according to CV. In contrast, most bis-TTFs except for the face-to-face arranged bis-TTF **77** show two reversible two-electron oxidations by CV analysis. The redox potentials of the typical bis-TTFs are summarized in Table 8.4, together with two tris-TTFs **74** and **86**.

The four one-electron oxidations in **61c,d** and **67b** indicate that the cation radical, dication, and trication radical intermediates are stabilized by delocalization of the charged species to form mixed valence states. However, most bis-TTFs exhibit simple two-step oxidation due to the weak interaction between the two TTF parts, although the electronic spectra of bis-TTFs demonstrate marked red shifts, reflecting extensive conjugation of the π systems of both TTF units. The through-bond electronic interaction between the TTF units of bis-TTFs is weak in the ground

Table 8.4   Redox Potentials of Bis- and Tris-TTFs **49-52, 57, 61, 67, 69, 72-74, 77, 82, 84** and **86**

Compd.	Solvent	Ref. electrode	$E_1^{1/2}$	$E_2^{1/2}$	$E_3^{1/2}$	$E_4^{1/2}$	Reference
49	CH$_2$Cl$_2$	Ag/AgCl	+0.41			+0.71	37
50	CH$_2$Cl$_2$	Ag/AgCl	+0.56			+0.78	37
51b	PhCN	SCE	+0.57			+0.88	33
52b	PhCN	SCE	+0.58			+0.88	33
57a	PhCN	SCE	+0.40			+0.79	42
57b	PhCN	SCE	+0.39			+0.72	42
61c	CH$_2$Cl$_2$	SCE	+0.44	+0.68	+1.10	+1.29	45
61d	CH$_2$Cl$_2$	SCE	+0.39	+0.61	+1.07	+1.23	48
67b	CH$_2$Cl$_2$	SCE	+0.49	+0.71	+1.24	+1.50	48
69	PhCN	SCE	+0.35			+0.82	15
72	PhCN	SCE	+0.38			+0.81	15
73	PhCN	SCE	+0.40			+0.81	50
74	PhCN	SCE	+0.29			+0.77	50
77a	PhCN	SCE	+0.33	+0.45		+0.89	53
82a	PhCN	Ag/AgCl	+0.56			+0.93	54
84a	CH$_2$Cl$_2$	Ag/AgCl	+0.39			+0.67	55
86a	CH$_2$Cl$_2$	SCE	+0.48			+0.85	56

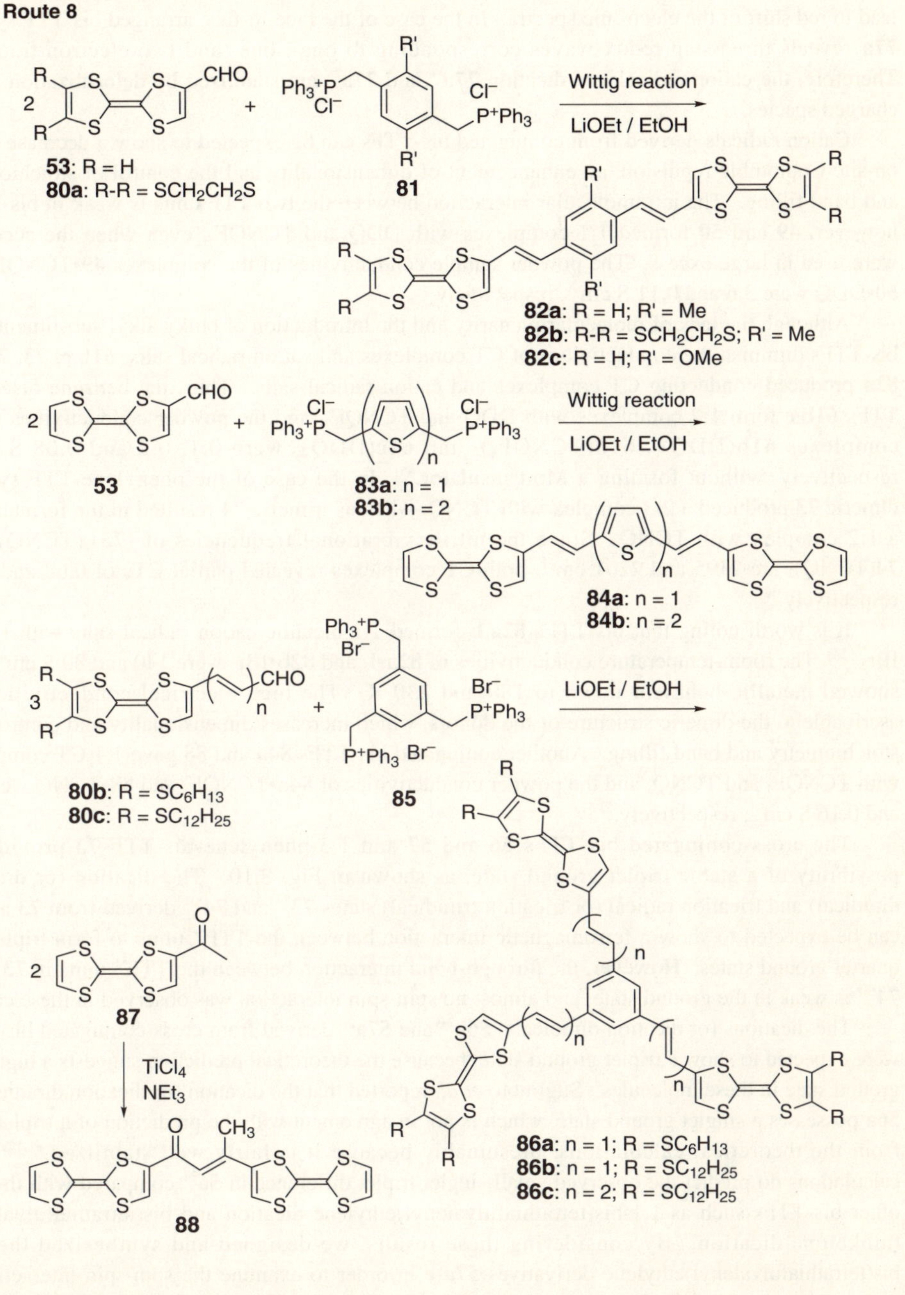

Route 8

82a: R = H; R' = Me
82b: R-R = SCH₂CH₂S; R' = Me
82c: R = H; R' = OMe

53: R = H
80a: R-R = SCH₂CH₂S

81

Wittig reaction
LiOEt / EtOH

2

53

83a: n = 1
83b: n = 2

84a: n = 1
84b: n = 2

Wittig reaction
LiOEt / EtOH

80b: R = SC₆H₁₃
80c: R = SC₁₂H₂₅

85

LiOEt / EtOH

86a: n = 1; R = SC₆H₁₃
86b: n = 1; R = SC₁₂H₂₅
86c: n = 2; R = SC₁₂H₂₅

87

TiCl₄
NEt₃

88

Scheme 8.21   Synthesis of bis-TTFs **82**, **84**, **86** and **88** using wittig reactions.

state, because the coefficients of atomic orbitals at the 4,5-positions are small in HOMO, as shown in Fig. 8.3a. However, the coefficients at the same 4,5-positions in LUMO are large to lead to red shift of the electronic spectra. In the case of the face-to-face arranged bis-TTF system, **77a** reveals three-step redox waves corresponding to one-, one- and two-electron transfers. Therefore, the cation radical and dication **77a**$^{\cdot+}$ and **77a**$^{2+}$ are stabilized by delocalization of the charged species.

Cation radicals derived from conjugated bis-TTFs can be expected to show a decrease of the on-site Coulombic repulsion, an enhancement of dimensionality and the control of stoichiometry and band filling. The intramolecular interaction between the two TTF units is weak in bis-TTFs; however, **49** and **50** formed 1:1 complexes with DDQ and TCNQF$_4$, even when the acceptors were used in large excess. The powder sample conductivities of the complexes **49**•TCNQF$_4$ and **50**•DDQ were 3.6 and 0.11 S cm^{-1}, respectively.[2]

Although the lack of molecular planarity and the introduction of bulky alkyl substituents into bis-TTFs diminish the conductivities of CT complexes and cation radical salts, **61b,c**, **73**, **74** and **82a** produced conducting CT complexes and cation radical salts. Thus, the benzene-fused bis-TTFs **61b,c** form 1:2 complexes with DDQ and TCNQF$_4$, and the powder conductivities of the complexes **61b**(DDQ)$_2$, **61b**(TCNQF$_4$)$_2$ and **61c**(DDQ)$_2$ were 0.1, 0.2 and 0.08 S cm^{-1}, respectively, without forming a Mott insulator.[45] In the case of the phenylene-TTF system, dimeric **73** produced a 2:3 complex with TCNQ, whereas trimeric **74** resulted in the formation of a 1:2 complex with TCNQ. Since the nitrile vibrational frequencies of (**73**)$_2$(TCNQ)$_3$ and **74**(TCNQ)$_2$ are 2195 and 2201 cm^{-1}, both CT complexes revealed partial CTs of 0.68 and 0.59, respectively.[58]

It is worth noting that bis-TTFs **82a,b** formed 1:1 metallic cation radical salts with I$_3^-$ and IBr$_2^-$.[54] The room-temperature conductivities of **82a**•I$_3$ and **82b**•IBr$_2$ were 140 and 80 S cm^{-1}, and showed metallic behavior down to 180 and 230 K. The high electrical conductivities are ascribable to the dimeric structure of the donors, which increases dimensionality and controls the stoichiometry and band filling. Another conjugated bis-TTFs **84a** and **88** gave 1:1 CT complexes with TCNQF$_4$ and TCNQ, and the powder conductivities of **84a**•TCNQF$_4$ and **88**•TCNQ were 0.1 and 0.16 S cm^{-1}, respectively.[55,57]

The cross-conjugated bis-TTFs **56** and **57** and 1,3-phenylene-bis-TTF **73** provide the possibility of a stable triplet ground state, as shown in Fig. 8.10. The dication (or dication diradical) and trication radical (or trication triradical) states **73**$^{2+}$ and **74**$^{\cdot3+}$ derived from **73** and **74** can be expected to show a ferromagnetic interaction between the TTF$^{\cdot+}$ units to form triplet and quartet ground states. However, the through-bond interaction between the TTF$^{\cdot+}$ units in **73**$^{2+}$ and **74**$^{\cdot3+}$ is weak in the ground state, and almost no spin-spin interaction was observed in these cases.

The dications (or dication diradicals) **56a**$^{2+}$ and **57a**$^{2+}$ derived from cross-conjugated bis-TTFs were expected to show a triplet ground state, because the theoretical prediction suggests a high-spin ground state in these molecules. Sugimoto *et al.* reported that the dication (or dication diradical) of **56a** possesses a singlet ground state, which is not in agreement with the prediction of a triplet state from the theoretical calculations, presumably because it is fairly well stabilized.[41,59] The calculations do predict the observed small singlet-triplet difference in **56**$^{2+}$ compared with those in other bis-TTFs such as 1,1-bis(tetrathiafulvalenyl)ethylene dication and bis(tetrathiafulvalenyl) thioketone dication. By considering these results, we designed and synthesized the 1,1-bis(tetrathiafulvalenyl)ethylene derivatives **57a-e** in order to examine the spin-spin interaction in their oxidation states.[42,60] The redox processes of **57a,b** in Table 8.4 display no separation in either of the two-electron transfers in their CVs, presumably due to the weak interaction between the two TTF moieties in these molecules. This weak interaction reflects the twisted structure, which was

56a^{2+}

Triplet ground state?

57a^{2+}

Triplet ground state?

73^{2+}

Triplet ground state?

74$^{\cdot 3+}$

Quartet ground state?

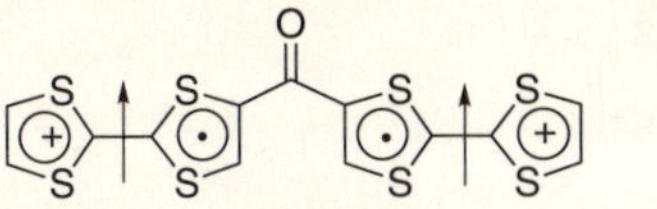

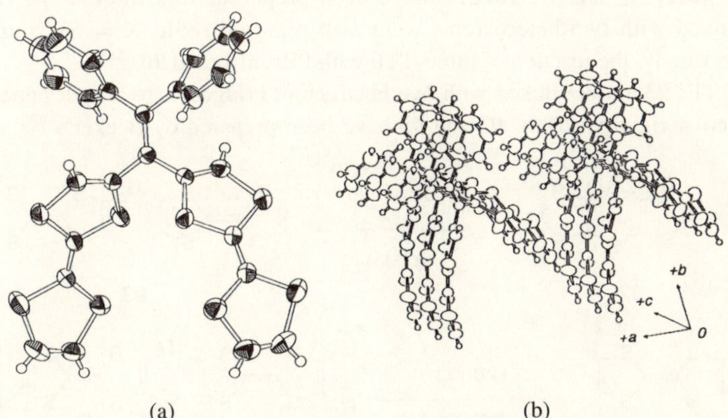

Fig. 8.10 Possible interaction between TTF cation radicals.

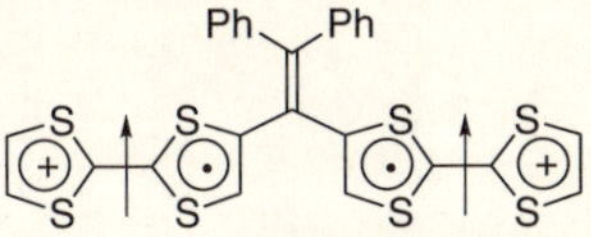

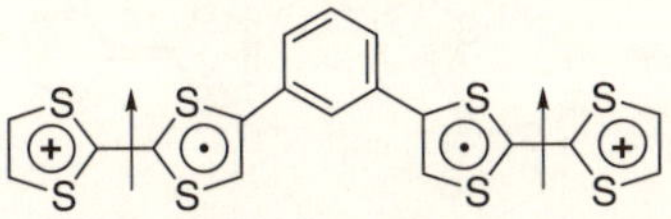

elucidated by X-ray analysis of **57a** (Fig. 8.11). Treatment of **57a** with iodine produced the corresponding cation radical salt, which shows interesting magnetic behavior. The ESR spectrum of the powdered sample of the cation radical salt of **57a** consists of a broad signal at the center and a fine structure (g = 2.0074, D = 108 G at 163 K) together with a signal of $\Delta m_s = 2$ corresponding to a triplet species. The distance between the two spins, as estimated from the D value by the point-dipole approximation (6.36 Å), seems to be comparable to the distance between the two TTF units in the radical salt. This result paves the way towards the synthesis of polymeric analogues such as **58** (Fig. 8.9), which exhibits a more preferential triplet ground state.

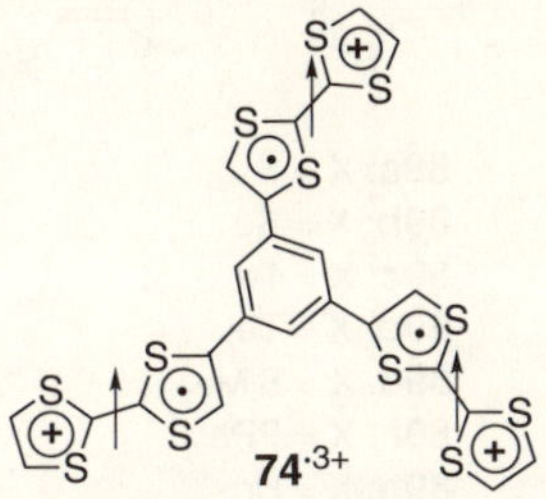

(a) (b)

Fig. 8.11 Molecular (a) and crystal (b) structures of **57a**.

8.4 Conjugated Bis-TTFs Linked by Heteroatoms

Main group elements such as sulfur, selenium, tellurium, silicon, phosphorus and mercury have

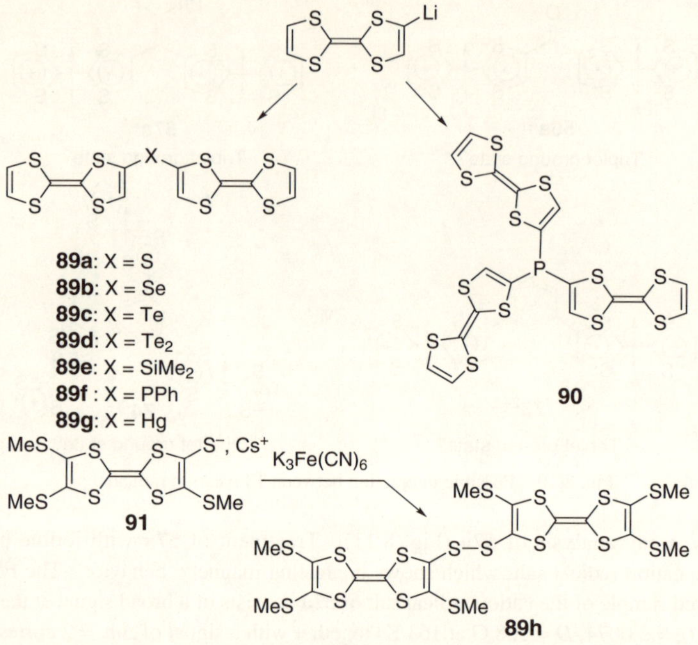
Li
89a: X = S
89b: X = Se
89c: X = Te
89d: X = Te2
89e: X = SiMe2
89f : X = PPh
89g: X = Hg
90
MeS
S⁻, Cs⁺
SMe
91
K3Fe(CN)6
89h

Scheme 8.22 Bi-TTFs **89** and **90** linked with various heteroatoms.

been employed to link the TTF units, and a number of dimeric and trimeric TTF derivatives have been prepared. The heteroatom-bridged bis-TTFs involving S (**89a**),[61] Se (**89b**),[61] Te (**89c**),[62,63] Si (**89e**),[64] P (**89f**)[64,65] and Hg (**89g**)[64] have been prepared from lithio-TTF. The bis-TTF derivatives linked with two heteroatoms were also reported (**89h**: X = $-S_2-$[66] and **89d**: X = $-Te_2-$[67]). Interestingly, the reaction of lithio-TTF with PBr_3 afforded **90**.[65,68]

The bis-TTFs **93** and **95** linked with two heteroatom bridges were also designed in order to enhance molecular rigidity. Thus, **93** and **95** have been prepared by Becker's[69,70] and Saito's[71]

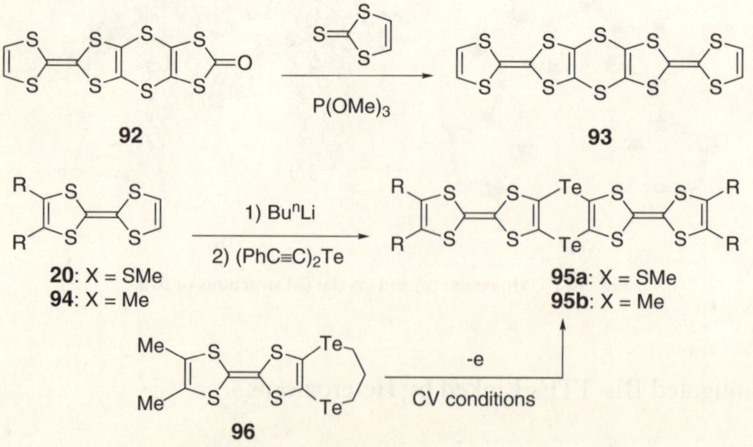
P(OMe)3
92
93
1) BunLi
2) (PhC≡C)2Te
20: X = SMe
94: X = Me
95a: X = SMe
95b: X = Me
96
-e
CV conditions

Scheme 8.23 Bi-TTFs **93** and **95** linked with two heteroatom bridges.

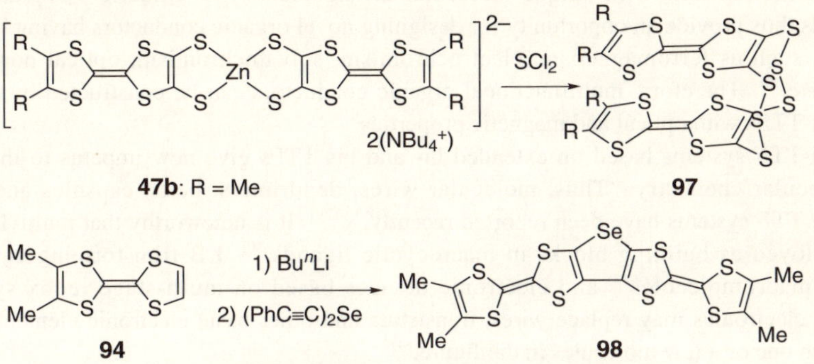

Scheme 8.24   Formation of bi-TTFs **97** and **98**.

groups and by Becker's[72,73] and Kobayashi's[74] groups, respectively. The use of heteroatoms as spacers has an additional advantage in that they may participate in intra- and interstack interactions in the donor system, and the resulting high dimensionality of the donor molecules would serve to suppress the Peierls distortion leading to the metal-insulating transition characteristic of one-dimensional conducting systems.

Two interesting reactions for the formation of bis-TTFs **97** and **98** were reported (Scheme 8.24). Thus, the reaction of **47b** with SCl_2 afforded a unique cyclophane-like **97**.[75] On the other hand, the reaction of **94** with butyllithium, followed by treatment with bis(phenylacetylenyl) selenide resulted in the formation of **98**.[76]

X-ray analyses of the sulfur-containing compounds **89a** and **93** and the tellurium-containing compounds **89c,d** and **95b** revealed remarkably close 2D chalcogen networks in the crystal structures. Although **89f,h** and **90** showed two two-electron redox waves in their CVs, the other compounds indicated three-stage redox waves, reflecting the sequential formation of the cation radical and dicationic species, followed by a further two-electron oxidation to the trication radical and tetracationic species.

Extended Hückel calculations for **89a,e,f,g** indicated that through-space and through-bond orbital overlaps between the two TTF moieties are both negligible, being unable to account for the observed redox trends.[64] It has thus been concluded that the dominant interaction is ascribed to Coulombic repulsion between the two TTF units, depending mainly on the nature of the bridging atom and the distances between the two redox centers. Despite extensive work on these main group element-containing dimeric TTF donors, surprisingly few conductive complexes have been reported: **89d•**TCNQ, 0.3 S cm^{-1}; **89d•**(TCNQ)$_2$, 8.4 S cm^{-1};[77] **95b•**TCNQ, 6 S cm^{-1}; **95b•**DMTCNQ, 6 S cm^{-1};[72] **95b•**[Au(CN)$_2$]$_{0.42}$, 13 S cm^{-1};[74] **98•**TCNQ, 32 S cm^{-1}.[76]

8.5   Summary and Outlook

Dimeric TTFs and related compounds summarized in this chapter are a relatively new class of donors playing an important role in the basic science of organic conductors. It has been found that some show intramolecular electronic interactions between the TTF units and that these depend on the nature of the spacer groups which can control the stoichiometry of the desired complexes. In addition, they have versatile molecular shapes such as linear, crooked, butterfly, star-shaped and helical structures, which can lead to the formation of a variety of CT complexes and cation radical salts with unique molecular arrangements.[2,10] Dimeric TTFs and related compounds thus provide an opportunity for designing novel organic conductors having interesting properties such as ferromagnetism, electrochromism, solvatochromism, optical nonlinearity, chirality, *etc.*[78] Therefore, multifunctional organic conductors can be constructed by endowing bi- and bis-TTFs with optical and magnetic properties.

Multi-TTF systems based on extended bi- and bis-TTFs give new impetus to the field of supramolecular chemistry. Thus, molecular wires, dendrimers, belts, capsules and shuttles containing TTF systems have been reported recently.[3,10,79] It is noteworthy that multi-TTFs have been employed as building blocks in macrocyclic ligands,[80] LB film-forming amphiles,[81] dendritic macromolecules[82] and electronic devices based on multi-stage redox systems.[83] Molecular electronics may replace wires, transistors and other basic electronic elements such as silicon with one or a few molecules in the future.[84]

References

1. M. Adam and K. Müllen, *Adv. Mater.*, **6**, 439 (1994).
2. T. Otsubo, Y. Aso and K. Takimiya, *Adv. Mater.*, **8**, 203 (1996).
3. (a) J. Roncali, *J. Mater. Chem.*, **7**, 2307 (1997); (b) J. Becher, L. Lau and P. Mørk, *Electronic Materials: The Oligomer Approach*, Wiley-VCH, Weinheim (1998), p. 198.
4. K. Lahlil, A. Moradpour, C. Bowlas, F. Menou, P. Cassoux, J. Bonvoisin, J.-P. Launay, G. Dive and D. Dehareng, *J. Am. Chem. Soc.*, **117**, 9995 (1995).
5. H. Spanggaard, J. Prehn, M. B. Nielsen, E. Levillain, M. Allain and J. Becher, *J. Am. Chem. Soc.*, **122**, 9486 (2000).
6. K. Bechgaard, K. Lerstrup, M. Jørgensen, I. Johannsen and J. Christensen, *The Physics and Chemistry of Organic Superconductors*, Springer Pro. Phys., Springer-Verlag, Berlin (1990), p. 349.
7. M. R. Bryce, *J. Mater. Chem.*, **5**, 1481 (1995).
8. M. R. Bryce, *J. Mater. Chem.*, **10**, 589 (2000).
9. M. B. Nielsen, C. Lomholt and J. Becher, *Chem. Soc. Rev.*, **29**, 153 (2000).
10. J. L. Segura and N. Martín, *Angew. Chem., Int. Ed.*, **40**, 1372 (2001).
11. V. Y. Lee, R. R. Schumaker, E. M. Engler and J. J. Mayerle, *Mol. Cryst. Liq. Cryst.*, **86**, 317 (1982).
12. Y. N. Kreitsberga, A. S. Édzhinya, R. B. Kampare and O. Y. Neiland, *Zh. Org. Khim.*, **25**, 1456 (1989); *J. Org. Chem. USSR*, 1312 (1989).
13. J. Y. Becker, J. Bernstein, A. Ellern, H. Gershtenman and V. Khodorkovsky, *J. Mater. Chem.*, **5**, 1557 (1995).
14. M. Iyoda, Y. Kuwatani, M. Oda, Y. Kai, N. Kanehisa and N. Kasai, *Angew. Chem., Int. Ed.*, **29**, 1062 (1990), and references cited therein.
15. M. Iyoda, Y. Kuwatani, N. Ueno and M. Oda, *J. Chem. Soc., Chem. Commun.*, 153 (1992).
16. M. Iyoda, K. Hara, E. Ogura, T. Takano, M. Hasegawa, M. Yoshida, Y. Kuwatani, H. Nishikawa, K. Kikuchi, I. Ikemoto and T. Mori, *J. Solid State Chem.*, **168**, 597 (2002).
17. S. Kanemoto, S. Matsubara, K. Oshima, K. Utimoto and H. Nozaki, *Chem. Lett.*, 5 (1987).
18. K. Sako, M. Kusakabe, T. Watanabe and H. Tatemitsu, *Synth. Met.*, **71**, 1949 (1995); K. Sako, M. Kusakabe, T. Watanabe, H. Takamura, T. Shinmyozu and H. Tatemitsu, *Mol. Cryst. Liq. Cryst.*, **296**, 31 (1997).
19. H. Tatemitsu, E. Nishikawa, Y. Sakata and S. Misumi, *Synth. Met.*, **19**, 565 (1987).
20. K. Lerstrup, M. Jørgensen, I. Johannsen and K. Bechgaard, *The Physics and Chemistry of Organic Superconductors*, Springer Pro. Phys., Springer-Verlag, Berlin (1990), p. 383.
21. D. E. John, A. J. Moore, M. R. Bryce, A. S. Batsanov and J. A. K. Howard, *Synthesis*, 826 (1998).
22. D. E. John, A. J. Moore, M. R. Bryce, A. S. Batsanov, M. L. Leech and J. A. K. Howard, *J. Mater. Chem.*, **10**, 1273 (2000).
23. M. Iyoda, E. Ogura, K. Hara, Y. Kuwatani, H. Nishikawa, T. Sato, K. Kikuchi, I. Ikemoto and T. Mori, *J. Mater. Chem.*, **9**, 335 (1999).
24. M. Iyoda, K. Hara, Y. Kuwatani and S. Nagase, *Org. Lett.*, **2**, 2217 (2000).
25. D. L. Green, *J. Org. Chem.*, **44**, 1476 (1979).
26. M. Iyoda, Y. Kuwatani, K. Hara, E. Ogura, H. Suzuki, H. Ito and T. Mori, *Chem. Lett.*, 599 (1997).
27. U. Kux, H. Suzuki, S. Sasaki and M. Iyoda, *Chem. Lett.*, 183 (1995).
28. Y. Kuwatani, E. Ogura, H. Nishikawa, I. Ikemoto and M. Iyoda, *Chem. Lett.*, 817 (1997).
29. M. Iyoda, Y. Kuwatani, E. Ogura, K. Hara, H. Suzuki, T. Takano, K. Takeda, J. Takano, K. Ugawa, M.

Yoshida, H. Matsuyama, H. Nishikawa, I. Ikemoto, T. Kato, N. Yoneyama, J. Nishijo, A. Miyazaki and T. Enoki, *Heterocycles*, **54**, 833 (2001).

30. M. Iyoda, K. Hara, C. R. V. Rao, Y. Kuwatani, K. Takimiya, A. Morikami, Y. Aso and T. Otsubo, *Tetrahedron Lett.*, **40**, 5729 (1999).

31. T. Kageyama, S. Ueno, K. Takimiya, Y. Aso and T. Otsubo, *Eur. J. Org. Chem.*, 2983 (2001).

32. K. Ueda, Y. Kamata, T. Kominami, M. Iwamatsu and T. Sugimoto, *Chem. Lett.*, 987 (1999).

33. M. Iyoda, M. Hasegawa, J. Takano, K. Hara and Y. Kuwatani, *Chem. Lett.*, 590 (2002).

34. M. Iyoda, M. Hasegawa, Y. Kuwatani, H. Nishikawa, K. Fukui, S. Nagase and G. Yamamoto, *Chem. Lett.*, 1146 (2001).

35. J. B. Torrance, B. A. Scott, B. Welber, F. B. Kaufman and P. E. Seiden, *Phys. Rev., B* **19**, 730 (1979).

36. T. Mori and H. Inokuchi, *Solid State Commun.*, **62**, 525 (1987).

37. T. Otsubo, Y. Kochi, A. Bitoh and F. Ogura, *Chem. Lett.*, 2047 (1994).

38. K. Hara, M. Hasegawa, Y. Kuwatani and M. Iyoda, unpublished results.

39. S. Thorand and N. Krause, *J. Org. Chem.*, **63**, 8551 (1998).

40. R. Rossi, A. Carpita and C. Bigelli, *Tetrahedron Lett.*, **26**, 523 (1985).

41. T. Sugimoto, S. Yamaga, M. Nakai, H. Nakatsuji, J. Yamauchi, H. Fujita, H. Fukutome, A. Ikawa, H. Mizouchi, Y. Kai and N. Kanehisa, *Adv. Mater.*, **5**, 741 (1993).

42. M. Iyoda, S. Sasaki, M. Miura, F. Fukuda and J. Yamauchi, *Tetrahedron Lett.*, **40**, 2807 (1999).

43. (a) H. Mizouchi, A. Ikawa and H. Fukutome, *Mol. Cryst. Liq. Cryst.*, **233**, 127 (1993); (b) H. Mizouchi, A. Ikawa and H. Fukutome, *J. Am. Chem. Soc.*, **117**, 3260 (1995).

44. (a) M. Adam, A. Bohnen, V. Enkelmann and K. Müllen, *Adv. Mater.*, **3**, 600 (1991); (b) U. Scherer, Y.-J. Shen, M. Adam, W. Bietsch, J. U. von Schütz and K. Müllen, *Adv. Mater.*, **5**, 109 (1993).

45. (a) R. Weger, N. Beye, E. Fanghänel, U. Scherer, R. Wirschem and K. Müllen, *Synth. Met.*, **53**, 353 (1993); (b) S. Frenzel and K. Müllen, *Synth. Met.*, **80**, 175 (1996).

46. (a) M. Adam, P. Wolf, H.-J. Räder and K. Müllen, *J. Chem. Soc., Chem. Commun.*, 1624 (1990); (b) M. Adam, U. Schere, Y.-J. Shen and K. Müllen, *Synth. Met.*, **55-57**, 2108 (1992).

47. K. Lahlil, A. Moradpour, C. Merienne and C. Bowlas, *J. Org. Chem.*, **59**, 8030 (1994).

48. M. L. Kaplan, R. C. Haddon and F. Wudl, *J. Chem. Soc., Chem. Commun.*, 388 (1977).

49. J. K. Stille and B. L. Groh, *J. Am. Chem. Soc.*, **109**, 813 (1987).

50. M. Iyoda, M. Fukuda, M. Yoshida and S. Sasaki, *Chem. Lett.*, 2369 (1994).

51. U. Kux and M. Iyoda, *Chem. Lett.*, 2327 (1994).

52. M. Hasegawa, Y. Kuwatani and M. Iyoda, unpublished results.

53. For the related reaction, see; M. Iyoda, T. Kondo, K. Nakao, K. Hara, Y. Kuwatani, M. Yoshida and H. Matsuyama, *Org. Lett.*, **2**, 2081 (2000).

54. (a) M. Mizutani, K. Tanaka, K. Ikeda and K. Kawabata, *Synth. Met.*, **46**, 201 (1992); (b) K. Ikeda, K. Kawabata, K. Tanaka and M. Mizutani, *Synth. Met.*, **55-57**, 2007 (1993); (c) K. Kawabata, T. Tanaka, M. Mizutani and N. Mori, *Synth. Met.*, **70**, 1141 (1995).

55. A. Bitoh, Y. Kohchi, T. Otsubo, F. Ogura and K. Ikeda, *Synth. Met.*, **70**, 1123 (1995).

56. (a) A. González, J. L. Segura and N. Martín, *Tetrahedron Lett.*, **41**, 3083 (2000); (b) R. Gómez, J. L. Segura and N. Martín, *Org. Lett.*, **2**, 1585 (2000).

57. J. Besançon, A. Radecki-Sudre, J. Szymoniak and J. Padiou, *J. Organomet. Chem.*, **429**, 335 (1992).

58. J. S. Chappell, A. N. Bloch, W. A. Bryden, M. Maxfield, T. O. Poehler and D. O. Cowan, *J. Am. Chem. Soc.*, **103**, 2442 (1981).

59. I. Ueda, M. Iwamatsu and T. Sugimoto, *J. Synth. Org. Chem. Jpn.*, **56**, 755 (1998).

60. J. Yamauchi, T. Aoyama, K. Kanemoto, S. Sasaki and M. Iyoda, *J. Phys. Org. Chem.*, **13**, 197 (2000).

61. M. R. Bryce, G. Cooke, A. S. Dhindsa, D. J. Ando and M. B. Hursthouse, *Tetrahedron Lett.*, **33**, 1783 (1992).

62. J. Y. Becker, J. Bernstein, S. Bittner, J. A. R. P. Sarma and L. Shahal, *Tetrahedron Lett.*, **29**, 6177 (1988).

63. J. Y. Becker, J. Bernstein, S. Bittner and S. S. Shaik, *Pure Appl. Chem.*, **62**, 467 (1990).

64. M. Fourmigué and Y.-S. Huang, *Organometallics*, **12**, 797 (1993).

65. M. Fourmigué and P. Batail, *Bull. Soc. Chim. Fr.*, **129**, 29 (1992).

66. P. Leriche, M. Giffard, A. Riou, J.-P. Majani, J. Cousseau, M. Jubault, A. Gorgues and J. Becher, *Tetrahedron Lett.*, **37**, 5115 (1996).

67. J. Y. Becker, J. Bernstein, M. Dayan and L. Shahal, *J. Chem. Soc., Chem. Commun.*, 1048 (1992).

68. M. Fourmigué and P. Batail, *J. Chem. Soc., Chem. Commun.*, 1370 (1991).

69. V. Y. Khodorkovsky, J. Y. Becker and J. Bernstein, *Synth. Met.*, **55-57**, 1931 (1993).

70. E. Aqad, J. Y. Becker, J. Bernstein, A. Ellern, V. Khodorkovsky and L. Shapiro, *J. Chem. Soc., Chem. Commun.*, 2775 (1994).

71. Y. Yamochi and G. Saito, *Synth. Met.*, **85**, 1467 (1997).

72. C. Wang, A. Ellern, V. Khodorkovsky, J. Y. Becker and J. Bernstein, *J. Chem. Soc., Chem. Commun.*, 2115 (1994).
73. C. Wang, A. Ellern, J. Y. Becker and J. Bernstein, *Tetrahedron Lett.*, **35**, 8489 (1994).
74. E. Ojima, H. Fujiwara, H. Kobayashi and A. Kobayashi, *Adv. Mater.*, **11**, 1527 (1999).
75. H. Fujiwara, E. Arai and H. Kobayashi, *J. Mater. Chem.*, **8**, 829 (1998).
76. C. Wang, A. Ellern, J. Y. Becker and J. Bernstein, *Adv. Mater.*, **7**, 644 (1995).
77. (a) J. Y. Becker, J. Bernstein, S. Bittner, Y. Giron, E. Harlev, L. Kaufmanorenstein, D. Peleg, L. Shahal and S. S. Shaik, *Synth. Met.*, **42**, 2523 (1991); (b) J. Y. Becker, J. Bernstein, M. Dayan, A. Ellern and L. Shahal, *Adv. Mater.*, **6**, 758 (1994).
78. (a) K. Uchida, G. Matsuda, Y. Aoi, K. Nakayama and M. Irie, *Chem. Lett.*, 1071 (1999); (b) R. Gómez, J. L. Segura and N. Martín, *Org. Lett.*, **2**, 1585 (2000); (c) M. Iyoda, K. Hara, M. Hasegawa and Y. Kuwatani, submitted for publication.
79. (a) D. Solooki, T. C. Parker, S. I. Khan and Y. Rubin, *Tetrahedron Lett.*, **39**, 1327 (1998); (b) T. Yamamoto, *Bull. Chem. Soc. Jpn.*, **72**, 621 (1999); (c) J. Becher, T. Brimert, J. O. Jeppesen, J. Z. Pedersen, R. Zubarev, T. Bjørnholm, N. Reitzel, T. R. Jensen, K. Kjaer and E. Levillain, *Angew. Chem., Int. Ed.*, **40**, 2497 (2001); (d) H. A. de Cremiers, G. Clavier, F. Ilhan, G. Cooke and V. M. Rotello, *J. Chem. Soc., Chem. Commun.*, 2232 (2001).
80. (a) T. Otsubo and F. Ogura, *Bull. Chem. Soc. Jpn.*, **58**, 1343 (1985); (b) M. Badri, J. P. Majoral, F. Gonce, A.-M. Caminade, M. Salle and A. Gorgues, *Tetrahedron Lett.*, **31**, 6343 (1990); (c) S. Yunoki, K. Takimiya, Y. Aso and T. Otsubo, *Tetrahedron Lett.*, **38**, 3017 (1997); (d) M. B. Nielsen, Z.-T. Li and J. Becher, *J. Mater. Chem.*, **7**, 1175 (1997).
81. (a) T. Nakamura, *Handbook of Organic Conductive Molecules and Polymers*, J.Wiley & Sons, Chichester (1997), Vol.1, p. 727; (b) S. Flink, F. C. J. M. van Veggel and D. N. Reinhoudt, *Adv. Mater.*, **12**, 1315 (2000).
82. (a) M. R. Bryce, W. Devonport and A. J. Moore, *Angew. Chem., Int. Ed.*, **33**, 1761 (1994); (b) Y. L. Bras, M. Sallé, P. Leríche, C. Mingotaud, P. Richomme and J. Møller, *J. Mater. Chem.*, **7**, 2393 (1997); (c) F. L. Derf, E. Levillain, G. Trippé, A. Gorgues, M. Sallé, R.-M. Sebastían, A.-M. Caminade and J.-P. Majoral, *Angew. Chem., Int. Ed.*, **40**, 224 (2001); (d) A. Beeby, M. R. Bryce, C. A. Christensen, G. Cooke, F. M. A. Duclairoir and V. M. Rotello, *Chem. Commun.*, 2950 (2002).
83. M. R. Bryce, *J. Mater. Chem.*, **10**, 589 (2000).
84. (a) M. A. Reed and J. M. Tour, *Sci. Am.*, **282**, 86 (2000); (b) R. F. Service, *Science*, **285**, 314 (1999); (c) J. K. Gimzewski and C. Joachim, *Science*, **283**, 1683 (1999); (d) D. Goldhaber-Gordon, M. S. Montemerlo, J. C. Love, G. J. Opiteck and J. C. Ellenbogen, *Prod. IEEE*, **85**, 521 (1997).

9

Cyclophane-type TTFs and TSFs

9.1　Introduction

It is widely recognized that the solid-state properties of molecular complexes depend highly on their crystal structures and electronic states. Highly conductive complexes generally meet the following two important requirements: an interactive stacking structure and a mixed valence state of the component donor or acceptor species.[1-4] The synthesis of such complexes is still considered to be formidable if not serendipitous. It is thus ideal and desirable, if possible, to control the solid state structures and properties of molecular complexes based on the design of the component molecules. One such trial in the molecular modification of tetrathiafulvalene (TTF) has been done with dimeric TTFs in which two TTF units are connected by one or more spacer linkages.[5-7] This approach is based on the idea that an electronic interaction between the two TTF units in a through-space or through-bond manner may afford the chance of controlling the crystal structures and electronic states of the derived complexes.[8-10] Although a variety of dimeric TTFs have been studied heretofore, most of them are of the simple single linkage type (**1**, Chart 9.1)[11-22] and double linkage type (**2**),[23-25] which do not necessarily ensure any close interactions between the two TTF units.

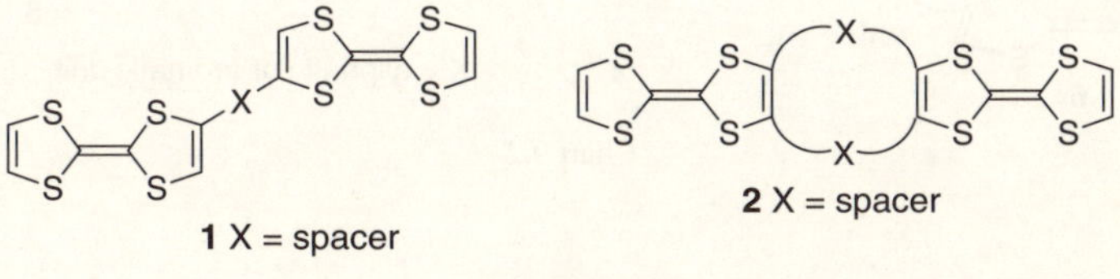

Chart 9.1

More intriguing are dimeric TTFs of the cyclophane type, in which the two TTF units are stacked by two or more spacer bridges and forced to strongly interact with each other by the close fixation of the two TTF units. Such double-layered tetrathiafulvalenophanes (TTF phanes) are very useful in understanding the stacking interactions of conducting TTF complexes, and can form unique molecular complexes which are different in crystal structures and solid state properties from those based on usual monomeric TTF donors. In addition, they are of particular interest as model compounds of κ-type superconductors. The crystal structures of κ-type superconductors with high critical temperatures are characterized by the close packing of paired TTF units, although those of highly conducting molecular complexes of TTF and related donor molecules generally comprise segregated stacks of the π-systems with a face-to-face arrangement along the stacks.[26,27]

There are various types of double-layered TTF phanes depending on the bridge number and bridge mode, *e.g.,* double-bridged TTF phane (**3**), quadruple-bridged TTF phane (**4**) of parallel stacking type and quadruple-bridged TTF phane (**5**) of crisscross stacking type(Chart 9.2). In addition, triple-layered TTF phanes (**6**) and double-layered tetraselenafulvalenophanes (TSF phanes, **7**) have been developed. This chapter reviews the synthesis and characteristics of these electron donors from the viewpoint of their unique molecular structures, electrochemical properties and conductive complexes. Most of these TTF phanes are composed of short alkylene bridges to ensure electronic interactions among the TTF units. On the other hand, Becher's group of Odense University has developed the macrocyclic TTF compounds of cyclophane types using polyether bridges.[28-31] These compounds are very interesting in terms of inclusion and supramolecular chemistry, which are described in detail in Chapter 15 of this volume. In addition, there are a number of simple TTF phanes (**8**) which include a single TTF unit in the macrocycles. These compounds are of considerable interest rather from the standpoints of bent TTFs caused by short bridges[32-35] and of interactive TTFs with other π-systems.[36-40] These topics are not discussed in this chapter.

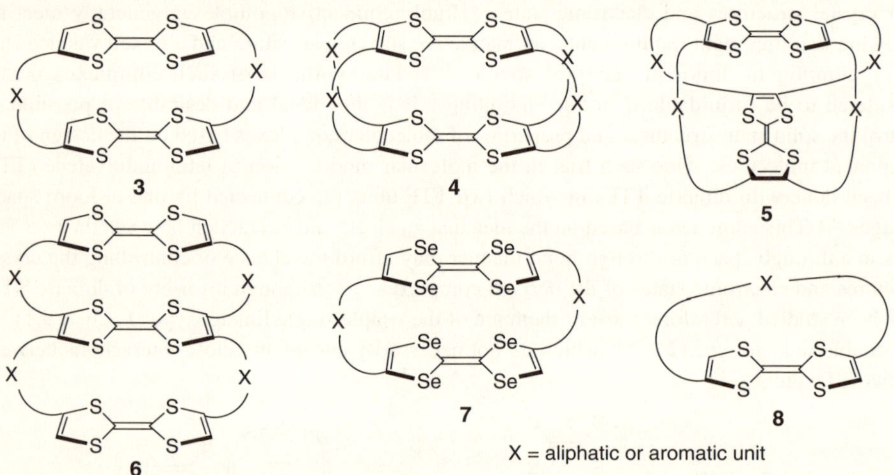

X = aliphatic or aromatic unit

Chart 9.2

9.2   Double-bridged TTF Phanes

For double-bridged TTF phanes, some structural isomers mainly resulting from the *cis* and *trans* substitutions are possible: 1) "paracyclophane-like" compounds with diagonal 4,4- and 5',5'-linkages, in which there is the possibility of a stereoisomer with parallel TTF axes and a stereoisomer with crossed TTF axes; 2) "metacyclophane-like" compounds with 4,5- and 4',5'-linkages; 3) "metaparacyclophane-like" compounds with 4,4- and 4',5'-linkages. As the first examples of double-bridged TTF phanes, Staab and coworkers reported [2.2]- and [3.3]-TTF phanes (**9**, Chart 9.3) in 1980.[41] The structure of **9a** was described as having an *anti*-metacyclophane-like conformation, where the TTF units deviate considerably from a planar arrangement. Since then, many double-bridged TTF phanes **10**,[36] **11**[42] and **12**[43] with longer bridges were also prepared, but little information on their structures and complexation was

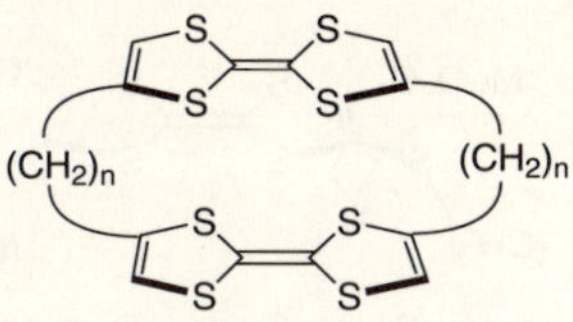

9a n = 2
9b n = 3

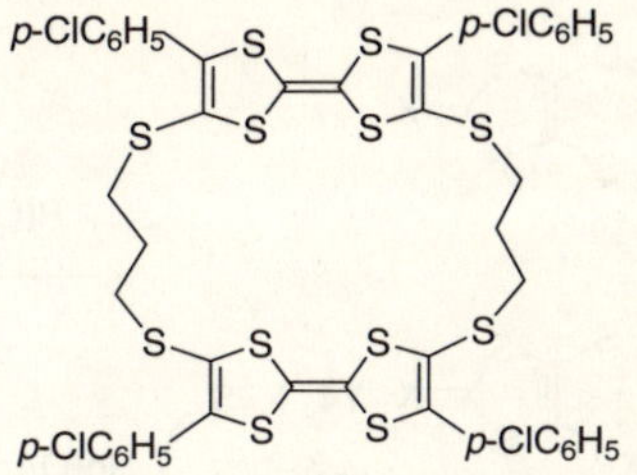

p-ClC$_6$H$_5$ *p*-ClC$_6$H$_5$ *p*-ClC$_6$H$_5$ *p*-ClC$_6$H$_5$

10

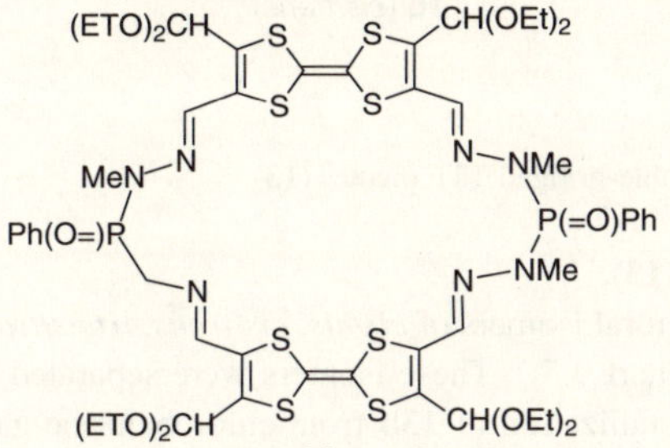

(ETO)$_2$CH   CH(OEt)$_2$
MeN—N   N—NMe
Ph(O=)P   P(=O)Ph
N—NMe
(ETO)$_2$CH   CH(OEt)$_2$

11

MeS   SMe   MeS   SMe

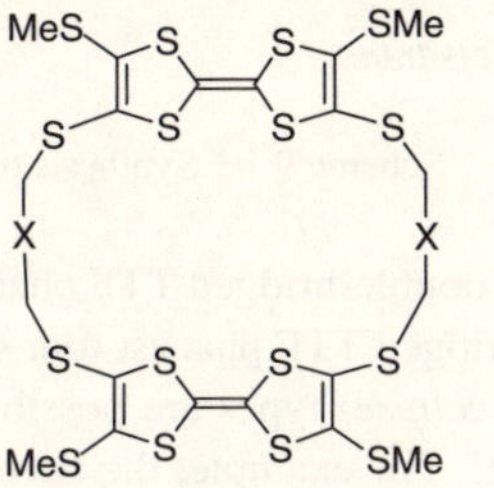

12

X = -C≡C-   **a**
-C≡C-C≡C-   **b** 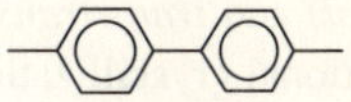
    c

Chart 9.3

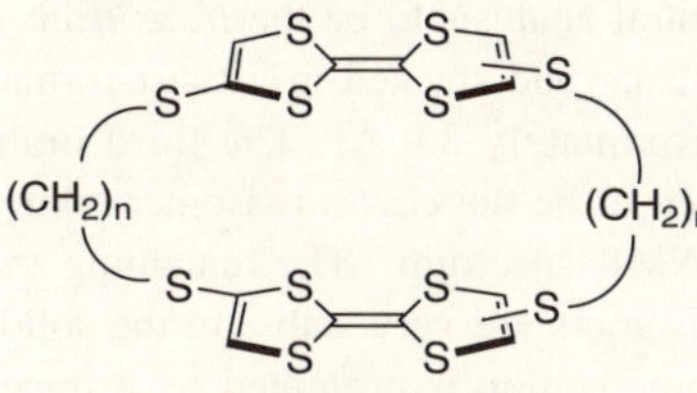

13a n = 1
13b n = 2
13c n = 3
13d n = 4

Chart 9.4

reported.

We studied a series of double-bridged TTF phanes (**13**, Chart 9.4) as an ideal model of interactive TTF dimers.[44-46)] The flexible structures depending on different bridge lengths must allow the involvement of many regio- and stereoisomers. These compounds provide important information not only on what kind of TTF phanes can form conductive complexes but also on what kind of regio- and stereoisomers can serve as the most active species in the complexes.

The synthesis of **13** was carried out as shown in Scheme 9.1. The starting materials, 4,4-(alkylenedithio)bis(5-methoxycarbonyl-1,3-dithiole-2-thione)s (**14**) were readily prepared according to the Cava's method[15)] and converted with mercury acetate into the ketones (**15**) in almost quantitative yield. The self-coupling reactions of the ketones promoted by triisopropyl phosphite gave the tetrakis(methoxycarbonyl)-TTF phanes (**16**), which existed as a mixture of *cis/trans* structural isomers as expected. Because of the mutual isomerization of these isomers, separation at this stage was skipped and the tetraester mixtures were decarboxylated with

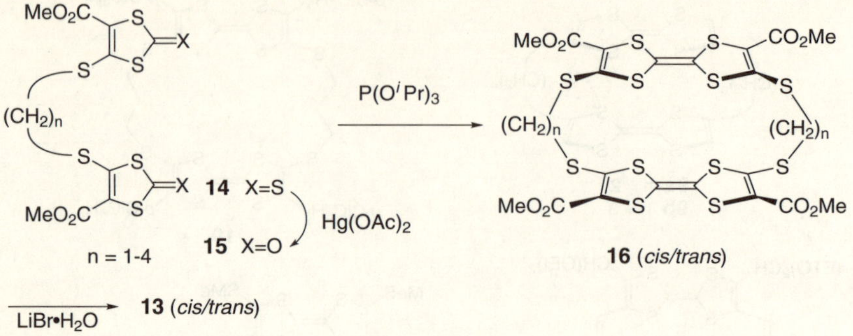

Scheme 9.1   Synthesis of double-bridged TTF phanes (**13**).

LiBr•H$_2$O to the desired double-bridged TTF phanes (**13**).

For these double-bridged TTF phanes, four structural isomers of *cis/cis*, *cis/trans*, *trans/trans (twist)* and *trans/trans (eclipse)* types are possible(Chart 9.5). These isomers were separated by fractional crystallization. For example, the first crystallization of **13b** from chlorobenzene gave yellow plates, which were characterized as the *cis/cis* isomer by X-ray crystallographic analysis (Fig. 9.1). The TTF moieties are almost planar and take a step-by-step *anti*-conformation reminiscent of [2.2]metacyclophane.[47,48] The second orange prisms from carbon disulfide were also determined by X-ray structural analysis to be the *trans/trans (twist)* isomer (Fig. 9.2). The two TTF moieties are almost planar and stacked in a twist formation; the average transannular distance between them is approximately 3.9 Å. The third orange fine needles from carbon disulfide-hexane were determined to be the *cis/trans* isomer taking into consideration two kinds of fulvalenic singlets in its ^{1}H NMR spectrum. The remaining *trans/trans (eclipse)* isomer was not detected. These structural isomers are very stable in the solid state, but isomerize with one another in solution.[32] The isomerization is promoted by a trace of acidic media contained in solution, and retarded by the addition of pyridine or triethylamine.

The three isomers of **13c** were also isolated, and only the *trans/trans (twist)* isomer gave a suitable orange crystal for X-ray analysis. As shown in Fig. 9.3, this structure is unlike that of the *trans/trans (twist)* isomer of **13b** (Fig. 9.2): the TTF moieties are not stacked and, despite

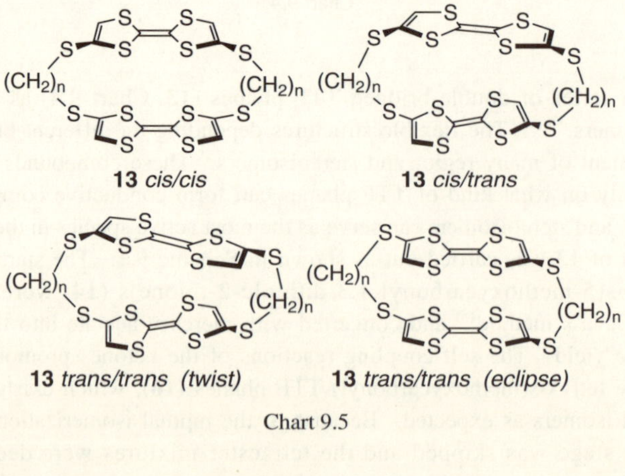

Chart 9.5

prolongation of the bridge chains, slightly nonplanar.

In the separation of **13d** with the longest bridge chains, the *cis/cis* isomer and the *trans/trans (twist)* isomer were isolated as pure crystals, and both structures have large cavities (about 7.0 Å distance across) inside the two TTF moieties, which can accommodate one carbon disulfide or chloroform molecule, as shown in Figs. 9.4 and 9.5.

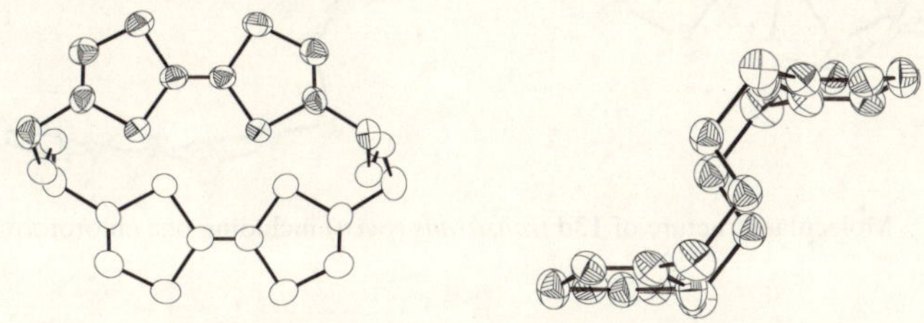

Fig. 9.1   Molecular structure of **13b** *cis/cis*.

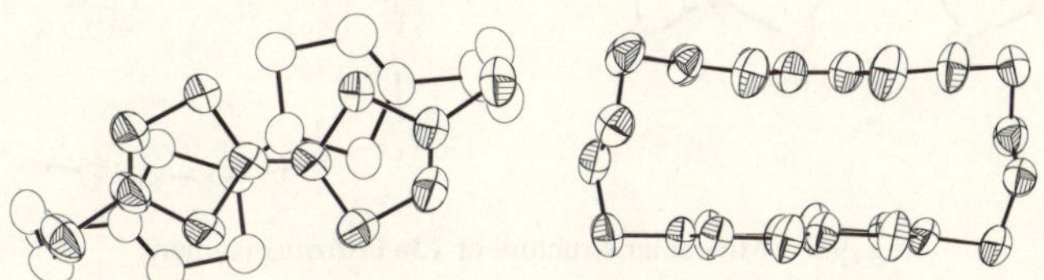

Fig. 9.2   Molecular structure of **13b** *trans/trans (twist)*.

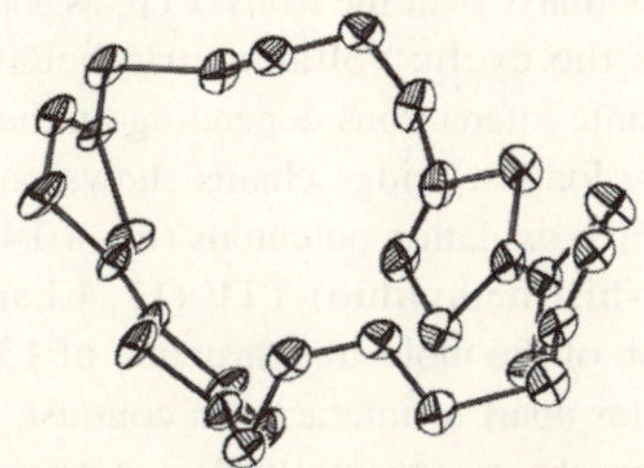

Fig. 9.3   Molecular structure of **13c** *trans/trans (twist)*.

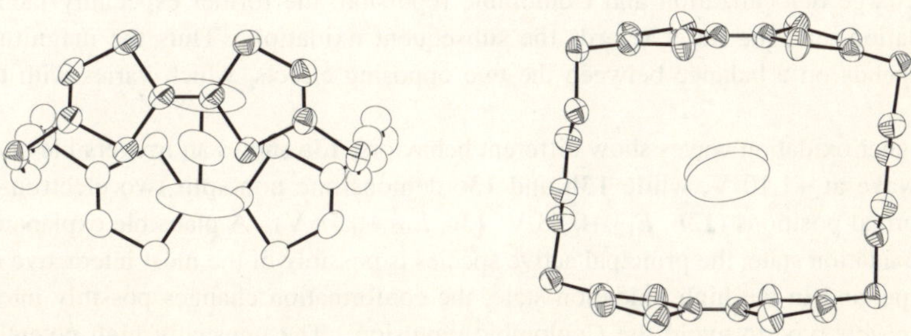

Fig. 9.4   Molecular structure of **13d** *cis/cis* including one carbon disulfide molecule.

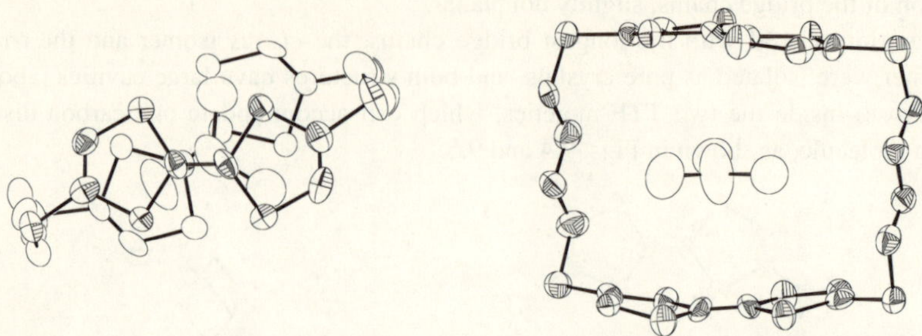

Fig. 9.5   Molecular structure of **13d** *trans/trans (twist)* including one chloroform molecule.

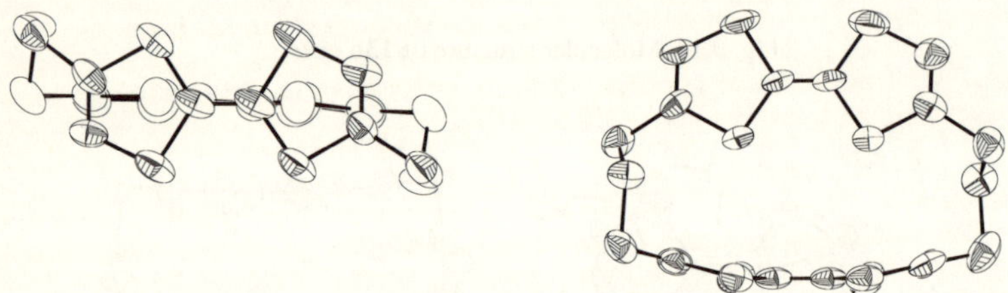

Fig. 9.6   Molecular structure of **13a** *cis/trans* isomer.

The *cis/cis* and *cis/trans* isomers were separated out by recrystallization of **13a** with the shortest bridge chains. X-ray crystallographic analysis of the *cis/trans* isomer revealed that the *cis*-TTF moiety stands perpendicularly over the *trans*-TTF, as shown in Fig. 9.6.

Figure 9.7 demonstrates the cyclic voltammetric behaviors of **13a-d**, which provide information on different electronic interactions depending on the length of the bridge chain. The voltammogram of **13d** with the longest bridge chains shows only two degenerated two-electron oxidation waves, whose half-wave oxidation potentials ($E_{1/2}$ +0.45 and +0.83 V *vs.* Ag/AgCl) are the same as those of 4,4'(5')-bis(methylthio)-TTF (**17**, Chart 9.6). This indicates that, in accordance with the observation of the molecular structure of **13d** in the crystal (Fig. 9.4 or 9.5), the two TTF moieties are too far apart to interact. In contrast, the voltammograms of the other TTF phanes with shorter bridge chains apparently demonstrate a splitting in the first oxidation process due to a through-space electronic interaction: **13a**, $E_{1/2}$ +0.35 and +0.48 V; **13b**, $E_{1/2}$ +0.41 and +0.56 V; **13c**, $E_{1/2}$ +0.40 and +0.50 V. The electronic interaction is explained in terms of two factors of charge delocalization and Coulombic repulsion: the former especially facilitates the initial oxidation, and the latter retards the subsequent oxidation. Thus the magnitude of the splitting depends on a balance between the two opposing effects, which varies with the bridge length.

The higher oxidation waves show different behaviors; **13a** shows an irreversible one-electron oxidation wave at +1.10 V, while **13b** and **13c** demonstrate non-split two electron-oxidation waves at normal positions (**13b**, $E_{1/2}$ +0.83 V; **13c**, $E_{1/2}$ +0.78 V). A plausible explanation is that in the low oxidation state, the principal active species is possibly of the most interactive *trans/tans (eclipse)* type, and in the high oxidation state, the conformation changes possibly into the non-interactive *cis/cis* type to avoid the Coulombic repulsion. The unusually high potential of **13a** indicates that its high oxidation species, even of the *cis-cis* type, suffers a strong Coulombic repulsion.

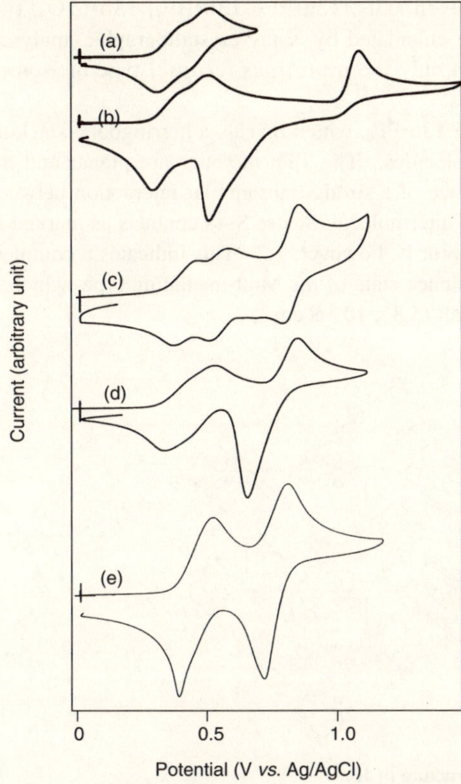

Fig. 9.7   Cyclic voltammograms of (a) **13a** (0-0.65 V scan), (b) **13a** (0-1.5 V scan), (c) **13b**, (d) **13c** and (e) **13d** in benzonitrile.

17 (*cis/trans*)

Chart 9.6

Table 9.1   Electrical Conductivities of Radical Cation Salts of Double-bridged TTF Phanes (**13**)

Salt	D:A	$\sigma/\text{S cm}^{-1}$
13a•PF$_6$	1:2	5.3×10^{-4}
13a•AsF$_6$	1:2	4.6×10^{-2}
13a•BF$_4$	1:1	8.9×10^{-2}
13a•ClO$_4$	2:3	1.1×10^{-2}
13a•I$_3$	3:4	6.8×10^{-2}
13b•BF$_4$	3:4	1.5
13b•ClO$_4$	3:4	2.0×10^{-1}
13b•I$_3$ (plate)	1:1.8	28
13b•I$_3$ (needle)	1:2	5.8×10^{-7}
13c•PF$_6$	1:1	2.1×10^{-2}
13c•I$_2$Br	1:1	3.0×10^{-3}
13d•I$_3$	1:2	2.0×10^{-4}

Electrocrystallization of **13** gave crystals of their radical cation salts, which are summarized in Table 9.1. No matter which donor isomer was used, the same radical cation salt was obtained.

This can be understood by speculating that the isomerization readily takes place during electrolysis and the most favorable cation species forms predominantly.

The crystal structures of the following seven salts, **13a•PF₆**, **13b•BF₄**, **13b•ClO₄**, two dimorphoric **13b•I₃**, **13c•PF₆** and **13c•I₂Br**, were elucidated by X-ray crystallographic analyses. All the salts except the two **13b•I₃** salts contain only the *trans/trans (eclipse)* type of isomer, corresponding to the most favorable cation species.

Figure 9.8 shows the crystal structure of the **13a•PF₆**, which reveals a herringbone stacking structure of the *trans/trans (eclipse)* dimeric molecules. The TTF moieties are planar and the stacking distance is 3.42 Å, indicating the existence of a strong transannular interaction between the two stacked TTF units. In addition, there are intermolecular close S⋯S contacts as marked in Fig. 9.8. The composition ratio of donor-to-acceptor is, however, 1:2. This indicates a complete charge transfer on each TTF moiety, that is, a valence state of the Mott-insulating type, which is thus responsible for the low conductivity of this salt (5.3×10^{-4} S cm^{-1}).

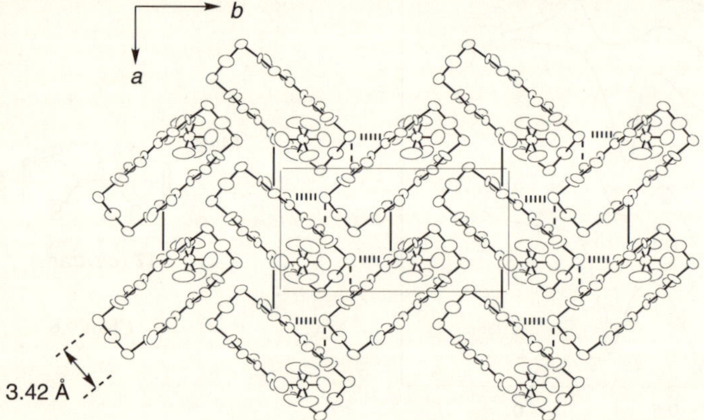

Fig. 9.8   Crystal structure of **13a•PF₆**.

The BF₄⁻ and ClO₄⁻ salts of **13b** are crystallographically isostructural with the same stoichiometry of 3:4 (D:A) and contain the donor columns comprising two kinds of monocationic and dicationic donor molecules in a ratio of 2:1. Although both are of the *trans/trans (eclipse)* type, the donor in the monocationic state has a completely overlapped conformation with a distance of 3.4 Å, while the donor in the dicationic state has a sliding conformation with a distance of 3.7 Å. It is suggested that the involvement of the different cationic species in the donor column hinders the formation of a conduction path along the column. The PF₆⁻ and I₂Br⁻ salts of **13c** have similar crystal structures, which comprise the stacking columns of the *trans/trans (eclipse)* isomers. The stacking distance of the two TTF units is long (*ca.* 3.90 Å), which explains the modest conductivities of these salts.

The electrocrystallization of **13b** provided two kinds of crystal salts with I₃⁻. The plate crystal showed a high room temperature conductivity of 28 S cm^{-1} and behaved metallic down to 210 K. On the other hand, the needle crystal was nearly insulating (5.8×10^{-7} S cm^{-1}). X-ray analyses of both crystals revealed different crystal packing from each other, although in both cases the donor species is of the *cis/cis* type, unlike that of the above BF₄ and ClO₄ salts. In the highly conductive plate crystal, there are two kinds of I₃⁻ anions: one forms a channel-like assembly with disordering, and the other is included in the cavity of the donor assembly. The TTF unit is stacked with a TTF unit of another molecule instead of by intramolecular contact (Fig. 9.9a). The stacked donor pair interacts with another pair *via* close intermolecular S⋯S contacts to form a diagonal interactive network (Fig. 9.9b). This network thus must correspond to a

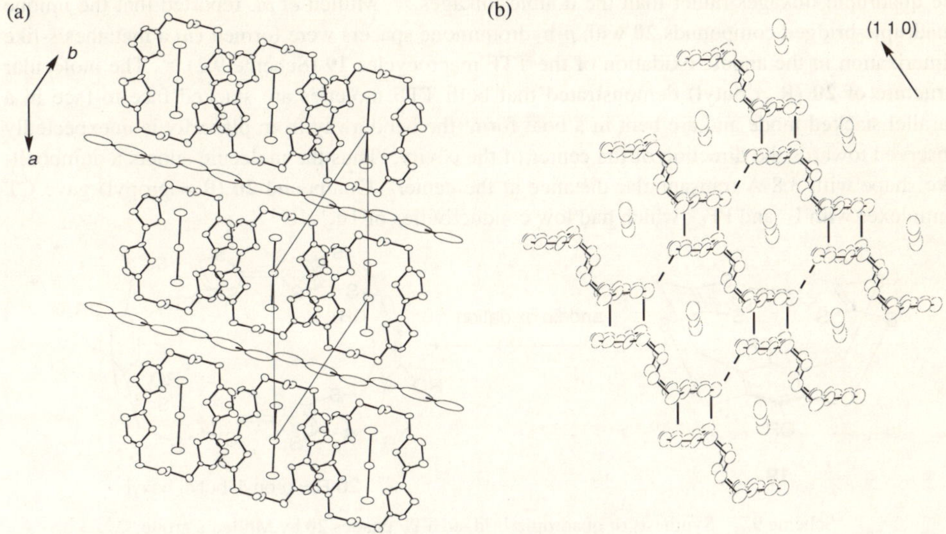

Fig. 9.9   Crystal structure of **13b•I₃** (plate).

conduction path in the salt. The needle crystal also comprises the donor species of the *cis/cis* type; however, the packing mode is quite different from that of the plate crystal. The I_3^- anions included in the donor assembly interfere with the intermolecular interaction of the donors, thus allowing no conduction path.

Bryce and coworkers reported the synthesis of different double-bridged TTF phanes **18**(Chart 9.7).[49] Both molecular structures of the neutral **18a** and its ClO_4 salt show no overlapping of the two TTF units.[50] Nevertheless, the ClO_4 salt has a relatively high conductivity of 10^{-2} S cm^{-1}. Although **18a** can complex iodine, the resulting I_8 salt is insulating.

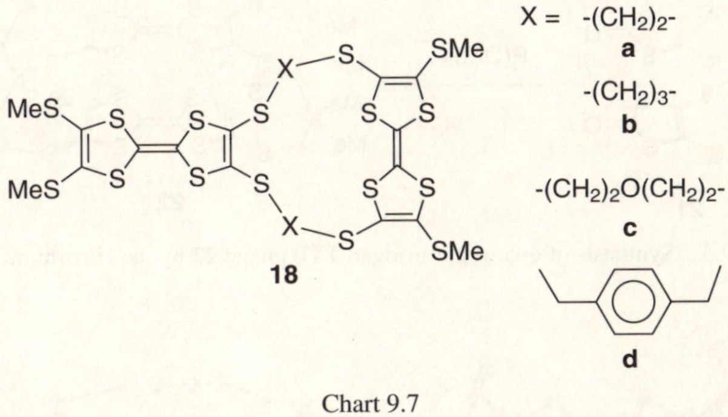

Chart 9.7

9.3   Quadruple-bridged TTF Phanes of Parallel-stacked Type

Since the study of the above double-bridged TTF phanes is complicated by the involvement of many regioisomers, it is expected that a fixed sandwich TTF phane could be constructed by using the quadruple linkages rather than the double linkages.[51] Müllen *et al.* reported that the unique quadruple-bridged compounds **20** with *p*-hydroquinone spacers were formed *via* a metathesis-like dimerization in the anodic oxidation of the TTF macrocycles **19** (Scheme 9.2).[52] The molecular structure of **20** (R = butyl) demonstrated that both TTF moieties are stacked face-to-face in a parallel stacked mode and are bent in a boat form; the bend away from planarity is unexpectedly observed towards the direction of the center of the cavity. Thus the molecule adopts a dumbbell-like shape with 3.8 Å transannular distance at the center. Compound **20** (R = propyl) gave CT complexes with I_3^- and PF_6^-, which had low conductivities of 10^{-3}–10^{-4} S cm^{-1}.

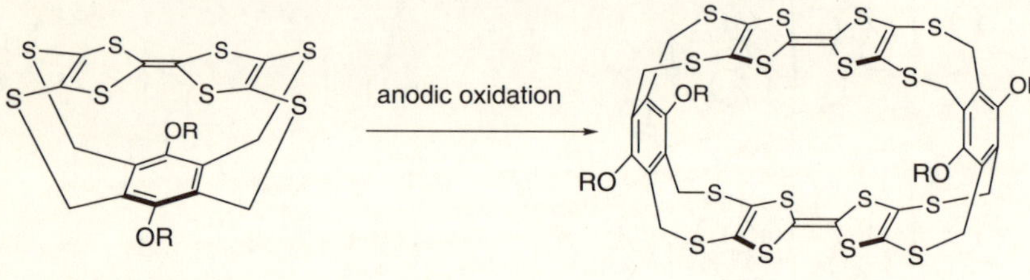

Scheme 9.2   Synthesis of quadruple-bridged TTF phanes **20** by Müllen's group.

We succeeded in the synthesis of the quadruple-bridged TTF phane **22** *via* a homo-coupling reaction of **21** (Scheme 9.3).[53] The molecular structure of **22** showed that the TTF moieties are also stacked face-to-face, but are bent slightly to the outside in contrast to that of **20** (Fig. 9.10). The bridged alkylenedithio chains are perpendicular to the TTF planes so as to make a small cavity in the molecular center. Thus, the two TTF units do not interact with each other. Compound **22** demonstrated two two-electron redox waves; the first and second half-wave oxidation potentials (+0.68 and +1.00 V *vs.* Ag/AgCl) are elevated as compared with the corresponding potentials of tetrakis(propylthio)-TTF (**23**, Chart 9.8) (+0.61 and +0.82 V). This result is explainable by Coulombic repulsion between both TTF units in the oxidation states.

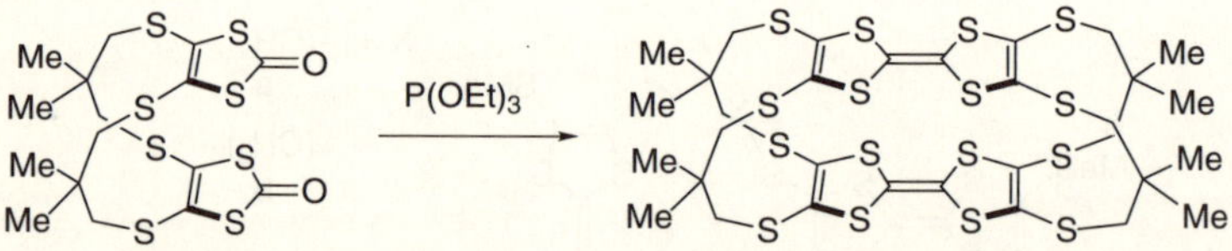

Scheme 9.3   Synthesis of quadruple-bridged TTF phane **22** by the Hiroshima group.

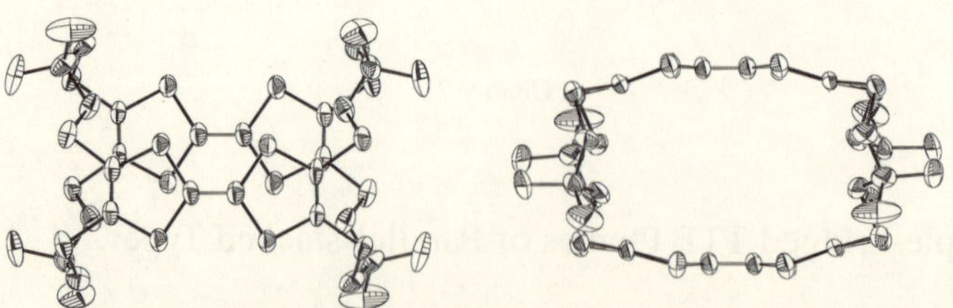

Fig. 9.10   Molecular structure of quadruple-bridged TTF phane **22.**

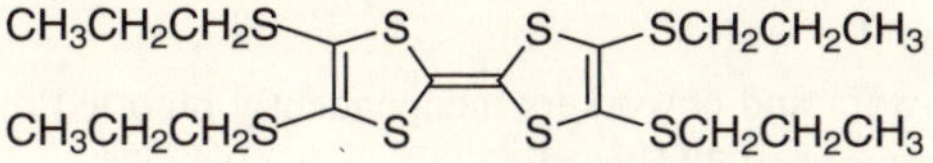

23

Chart 9.8

Compound **22** formed semiconductive complexes of 10^{-5} S cm^{-1} with TCNQF$_4$, DDQ or iodine.

Very recently Izuoka and Kawada of Ibaraki University succeeded in the synthesis of the quadruple-bridged TTF phane **25** with shorter bridge chains following a similar approach (Scheme 9.4).[54]   The molecular structure is severely deformed into a spherical shape, where the transannular distance around the central double bonds is over 6 Å.

As other examples of quadruple-bridged TTF phanes, Becher and coworkers reported compounds **26** (Chart 9.9) with well-defined large cavities.[55]   The crystal structure of **26c** reveals that it crystallizes with two chloroform molecules inside its cavity.

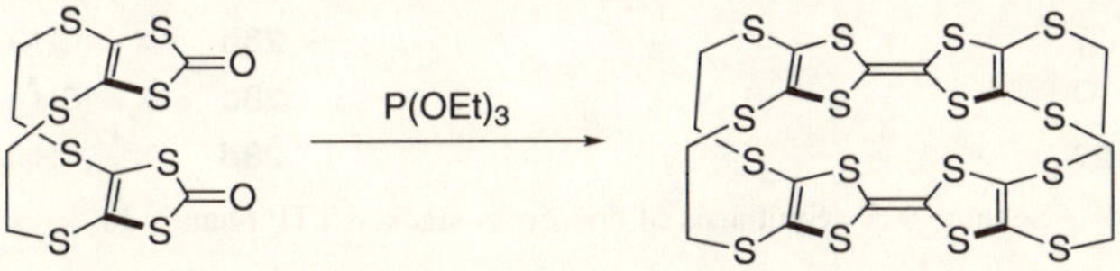

24   **25**

Scheme 9.4   Synthesis of quadruple-bridged TTF phane **25** by the Ibaraki group.

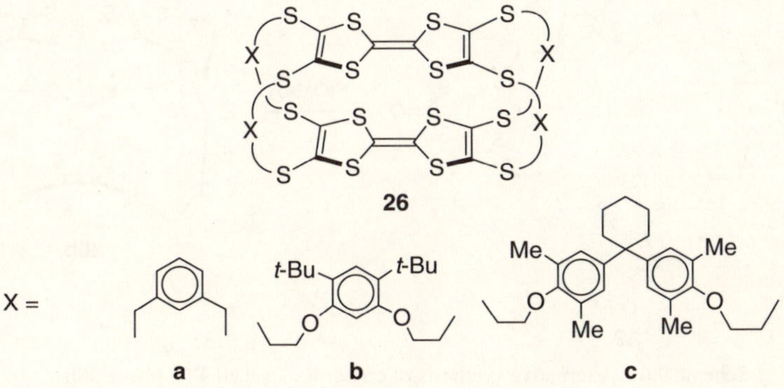

26

X =

a   b   c

Chart 9.9

9.4   Quadruple-bridged TTF Phanes of Crisscross-stacked Type

Although the crystal structures of highly conductive molecular complexes usually comprise stacked columns of TTFs in an approximately parallel arrangement, an unusual crisscross overlap was reported for trimeric TTFs of the semiconductive salts (TTF)$_3$W$_6$O$_{19}$ and (TTF)$_3$Mo$_6$O$_{19}$.[56] An important question is how the crisscross-stacked TTF phanes can contribute to electrical conduction when compared with the parallel-stacked TTF phanes.   In order to elucidate this point, we prepared a series of novel crisscross-stacked TTF phanes **28** *via* an intramolecular coupling reaction of **27** (Scheme 9.5).[57,58]

Kawada, Izuoka, Sugawara and coworkers independently reported the one-step synthesis of **28b** from the macrocyclic compound **29** according to Scheme 9.6.[59,60]

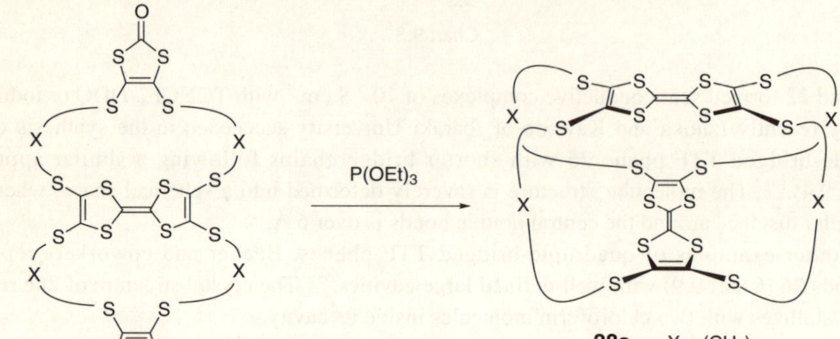

P(OEt)$_3$

28a	X = (CH$_2$)$_2$
28b	X = (CH$_2$)$_3$
28c	X = (CH$_2$)$_4$
28d	X = (CH$_2$)$_5$

27

Scheme 9.5   Synthesis of crisscross-stacked TTF phanes **28**.

P(OMe)$_3$

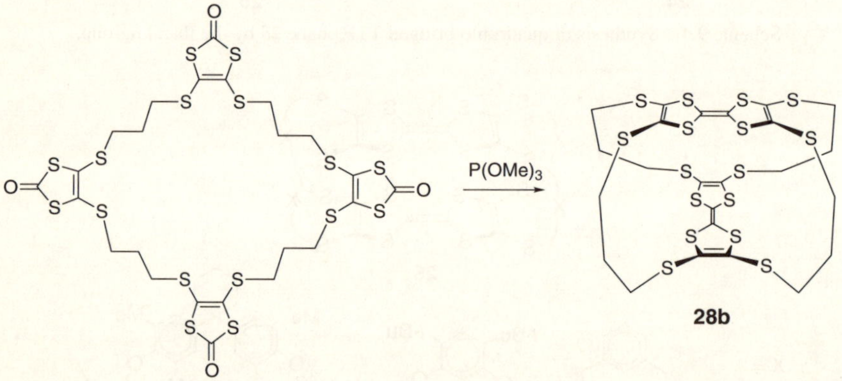

29

28b

Scheme 9.6   Alternative synthesis of crisscross-stacked TTF phane **28b** .

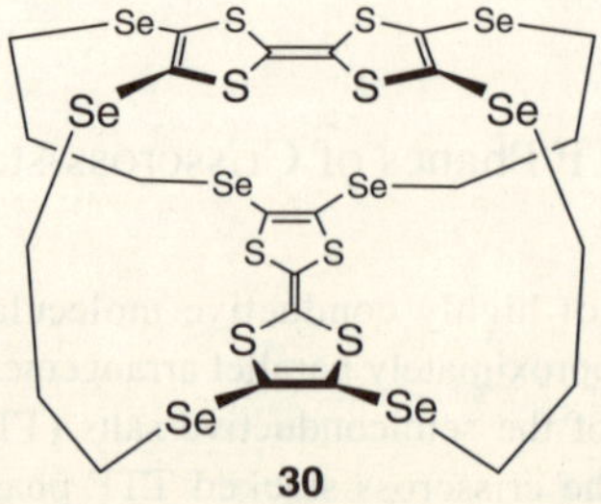

30

Chart 9.10

Furthermore, the Ibaraki group also succeeded in the synthesis of the analogue **30**(Chart 9.10) containing selenium atoms in the bridge chains.[61]

The unique molecular structures of **28** were characterized by X-ray crystal analyses, which confirmed that two TTF moieties overlap each other in a crisscross manner by the four bridged linkages.[57,58]   The molecular structure of the TTF phane **28a** with the shortest ethylenedithio bridges is depicted in Fig. 9.11.   The TTF units are severely bent into a boat shape, and the dihedral angles defined by the central tetrathiaethylene part and the outer tetrathiaethylene part in the TTF unit range from 27° to 45°.   As a result, the two TTF units are rather separated from each other, although they are linked by the short bridges.   In the trimethylenedithio-bridged compound **28b**, the bent angles are reduced to 16–29°.[60]   In the tetramethylenedithio bridged compound **28c**, the strain is expected to be less.   Nevertheless, the molecular structure of **28c** demonstrates that the bent angles, 24–40°, are somewhat larger than those of **28b**.   This unusual phenomenon is explainable in terms of the conformational restriction of the tetramethylene unit in the bridges. Nearly all the methylenes are forced to take torsional gauche conformations to make the bridge rigid and stretched, so that it stands perpendicular to the TTF moieties against release of the strain.   On the other hand, the molecular structure of **28d** with pentamethylenedithio bridges is fairly free from such strain: all the bridged methylenes take *anti* conformations, and the dihedral angles in the TTF units are only 6–17°.   In addition, **28d** has a cavity whose size is large enough to include a chloroform molecule.

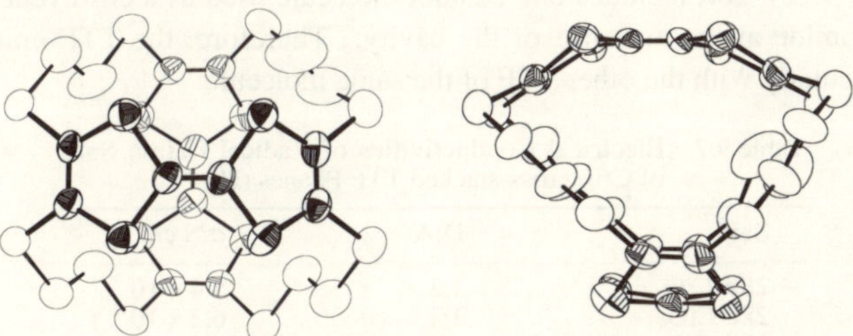

Fig. 9.11   Molecular structure of **28a**.

The cyclic voltammogram of the most strained molecule **28a** shows only one irreversible oxidation wave at an extraordinarily high potential (+0.94 V).   This is because the severe bending of the TTF moieties can not stabilize the oxidation state, as reminiscent of the bent TTF derivative **31**(Chart 9.11) with a high oxidation potential (+1.01 V).   On the other hand, the cyclic voltammogram of **28b** has four reversible redox waves, each of which corresponds to a one-electron process.   The splitting of all the waves clearly indicates that the two TTF moieties strongly interact with each other.   The first half-wave oxidation potential (+0.57 V) is slightly higher than that of tetrakis(methylthio)-TTF (**32**, Chart 9.11) (+0.49 V), but the second one (+0.70 V) is markedly higher.   This indicates that the splitting is ascribed to the intramolecular Coulombic repulsion of the resulting dicationic TTF species.   Similar Coulombic repulsion is also responsible for the splitting and high potentials of the third (+0.98 V) and fourth (+1.11 V) oxidation as compared with the second wave of **32** (+0.72 V).   In contrast, **28c** shows only two reversible redox waves (+0.60 V and +0.89 V), each of which corresponds to a two-electron process.   Both the first and second half-wave oxidation potentials are somewhat higher than the respective ones of **32**, suggesting the bent structure of the TTF moieties.   The equivalent redox behavior of the two TTF moieties indicates that there exists no Coulombic interaction between the two TTF moieties of **28c**.   Rather unusual phenomena were found for the redox process of **28d**:

31   **32**

Chart 9.11

the first redox process occurs with a two-electron process, but the half-wave oxidation potential (+0.46 V) is a little lower than that of **32**. Probably the monocationic species of the flexible compound **28d** can take a favorable conformation for transannular charge delocalization. On the other hand, the second oxidation process splits into two waves with high potentials (+0.74 V and +0.85 V), also demonstrating the emergence of Coulombic repulsion in the higher oxidation state.

Although **28a** afforded no crystalline complexes as expected from the cyclic voltammetry, the others formed crystalline radial cation salts. As summarized in Table 9.2, the conductivities of all the salts are of the order of 10^{-6}–10^{-7} S cm^{-1}, independent of the donor and the counterion species, suggesting that the crisscross stacking arrangement of TTF phanes is not suitable for electrical conduction. The crystal structure of the **28d**•Br complex shows the reason for its low conductivity (Fig. 9.12): **28d** includes one ethanol molecule used as a cosolvent inside the cavity and a counter bromide at the entrance of the cavity. Therefore, the TTF unit of **28d** can not interact intramolecularly with the other TTF of the same molecule.

Table 9.2   Electrical Conductivities of Radical Cation Salts
of Crisscross-stacked TTF Phanes (**28**)

Salt	D:A	σ/S cm^{-1}
28b•ClO$_4$	1:1	5.4×10^{-7}
28c•ClO$_4$	1:1	6.2×10^{-7}
28d•ClO$_4$	1:2	7.3×10^{-7}
28b•PF$_6$	3:2	1.1×10^{-6}
28c•PF$_6$	1:1	1.3×10^{-6}
28d•PF$_6$	2:3	3.1×10^{-7}
28d•Br	1:1	1.4×10^{-6}

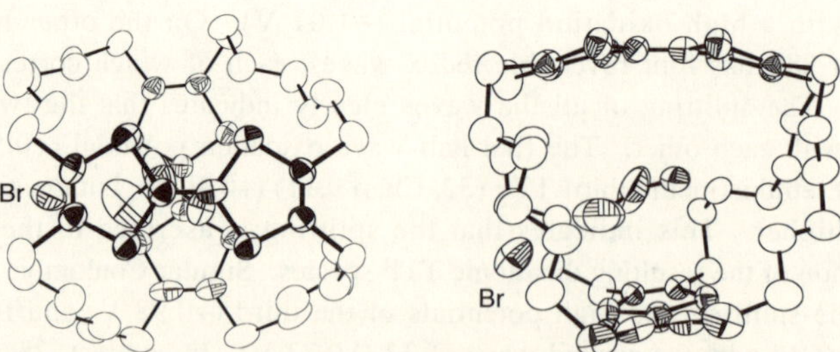

Fig. 9.12   Molecular structure of 28d•Br including one ethanol molecule.

9.5   Triple-layered TTF Phanes

Triple-layered TTF phanes would be probably more interesting than double-layered TTF phanes, if all the TTF units can contribute to the through-space electronic interaction. However, little is known so far about triple-layered TTF phanes, although macrocyclic cage molecules surrounded by three TTF bridges have been reported.[62] We thus studied two isomeric triple-layered TTF phanes **33** and **34**(Chart 9.12).[63]

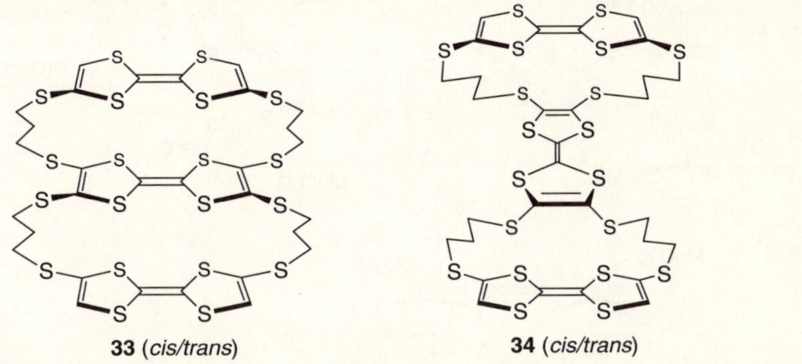

33 (*cis/trans*)   **34** (*cis/trans*)

Chart 9.12

The synthesis of **33** of the parallel-oriented type was carried out by a selective deprotection of the derivative of double-layered TTF phane **35** followed by coupling with **36** and then decarboxylation of the tetraester of triple-layered TTF phane **37**, as shown in Scheme 9.7. The

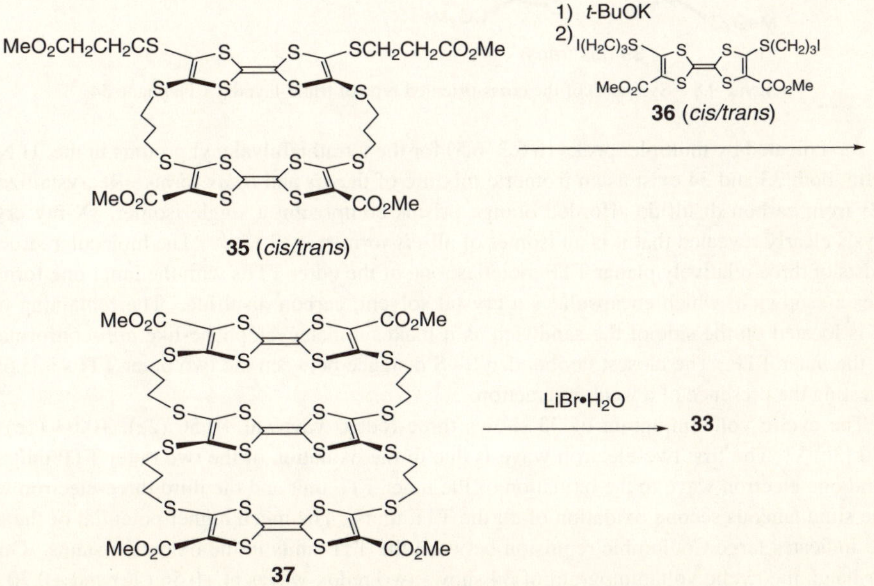

35 (*cis/trans*)

36 (*cis/trans*)

37

Scheme 9.7   Synthesis of the parallel-oriented type of triple-layered TTF phane **33**.

synthetic strategy using the protected TTF tetrathiolate was developed by Becher [64,65] and beautifully applied to the syntheses of numerous macrocyclic TTF compounds.[28-31]

For the synthesis of **34** of the cross-oriented type, we adopted a synthetic strategy *via* simultaneous construction of the two outer TTF units from the tetra substituted TTF **38**, as shown in Scheme 9.8. The intramolecular double-coupling reaction of **38** induced by triethyl phosphite has the possibility of providing not only the intermediate **39** but also the parallel-oriented intermediate **37**. However, only the isomer **39** was obtained and decarboxylated to the cross-oriented triple-layered TTF phane **34**.

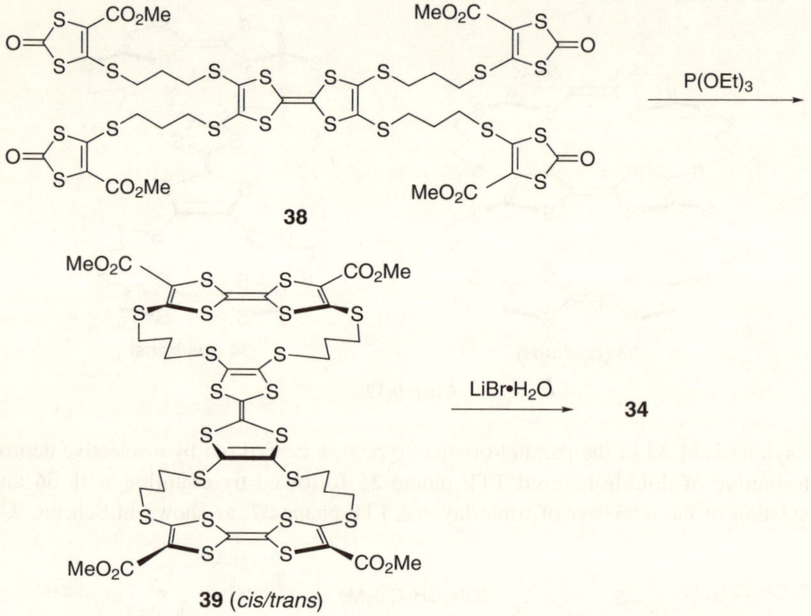

Scheme 9.8   Synthesis of the cross-oriented type of triple-layered TTF phane **34**.

As indicated by multiplet peaks (δ 6.3–6.5) for the tetrathiafulvalenyl protons in the ^{1}H NMR spectra, both **33** and **34** exist as an isomeric mixture of the *cis* and *trans* forms. Recrystallization of **33** from carbon disulfide afforded orange prisms comprising a single isomer. X-ray crystal analysis clearly revealed that it is an isomer of all-*cis* form (Fig. 9.13).[63] The molecular structure consists of three relatively planar TTF moieties; one of the outer TTFs with the inner one formally makes a sandwich, which encapsulates a crystal solvent, carbon disulfide. The remaining outer TTF is located on the side of the sandwich as it makes a metacyclophane-like *anti*-conformation with the inner TTF. The closest nonbonded S···S distance between the two outer TTFs is 3.69 Å, suggesting the presence of a weak interaction.

The cyclic voltammogram of **33** shows three redox waves at +0.50 (2e), +0.63 (1e) and +0.80 (3e) V. The first two-electron wave is due to the oxidation of the two outer TTF units, the second one-electron wave to the oxidation of the inner TTF unit and the third three-electron wave to the simultaneous second oxidation of all the TTF units. The much higher potential of the third wave indicates large Coulombic repulsion between the TTF units in the oxidation states. On the other hand, the cyclic voltammogram of **34** shows two redox waves at +0.56 (2e) and +0.70 (4e) V. The first two-electron wave is due to the oxidation of the outer TTF units, and unlike the case

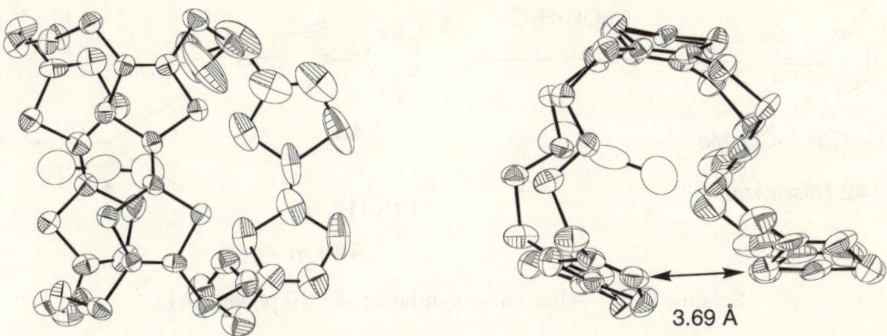

Fig. 9.13   Molecular structure of **33** including one carbon disulfide molecule.

of **33**, the second wave appears as coalescence with the subsequent oxidation. This means that the Coulombic repulsion between the TTF units of **34** is very small.

9.6   Double-bridged TSF Phanes

The high conductivities of radical cation salts of TTFs are induced by strong intermolecular nonbonded interactions of the sulfur atoms incorporated in the TTF moieties. Since the nonbonded interaction of selenium is generally more effective than that of sulfur,[66] we expected TSF phanes (**41**) to be better potential electron donors. The first synthesis of the TSF phanes was accomplished by trimethyl phosphite-induced coupling reaction of **40**, as shown in Scheme 9.9.[67] However, this coupling reaction is problematic because of poor yield (9–13%) and irreproducibility.

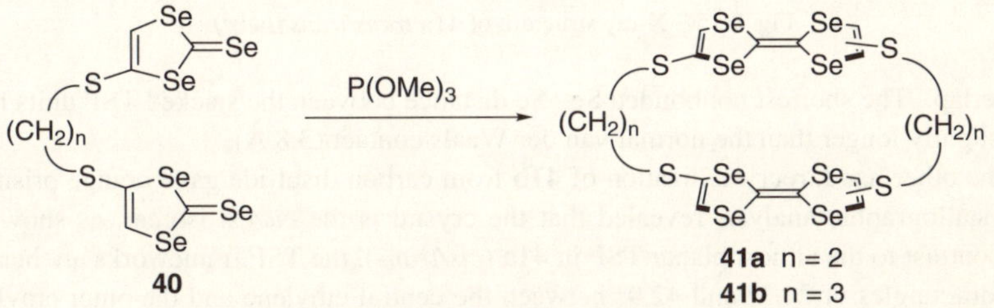

Scheme 9.9   The first synthesis of TSF phanes (**41**).

We devised an alternative synthetic method for the TSF phanes, based on the four-molecule coupling of two TSF dithiolates (**43**) and two α,ω-dibromoalkanes (**44**) at the crucial cyclization step (Scheme 9.10).[68] This improved approach based on the deprotection/realkyation protocol of **42** is apparently more advantageous than the previous synthetic route of the TSF phanes (Scheme 9.9) in terms of not only yield (20–25%) but also reproducibility.

As observed for the double-bridged TTF phanes, the double-bridged TSF phanes **41** also exist as a mixture of three geometrical isomers of different bonding types, *i.e.*, *cis/cis*, *cis/trans*, and *trans/trans (twist)*. Recrystallization of **41a** from carbon disulfide-benzene gave red columnar crystals of the *cis/trans* isomer, and from chlorobenzene, pink plate-like crystals of the *trans/trans (twist)* isomer. Their molecular structures were revealed by X-ray crystallographic analyses as shown in Figs. 9.14 and 9.15. The two TSF units are almost planar and stacked with

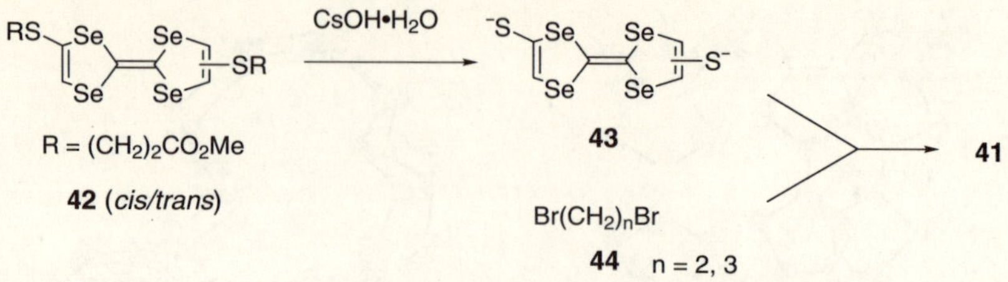

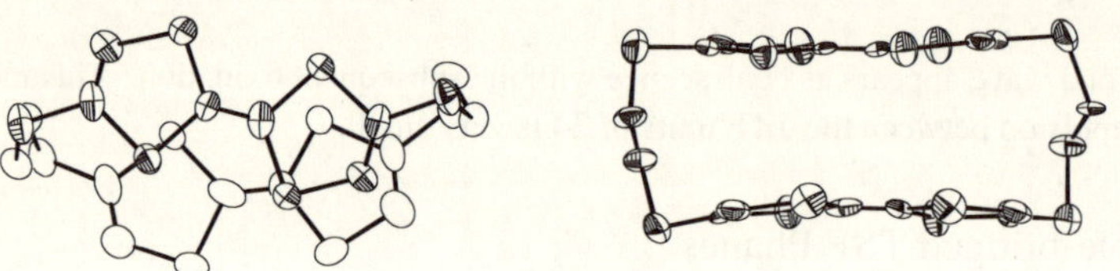

RS
Se
Se
Se
Se
SR
R = (CH2)2CO2Me
42 (cis/trans)
CsOH•H2O
⁻S
Se
Se
Se
Se
S⁻
43
Br(CH2)nBr
44 n = 2, 3
41

Scheme 9.10 Alternative synthesis of TSF phanes (**41**).

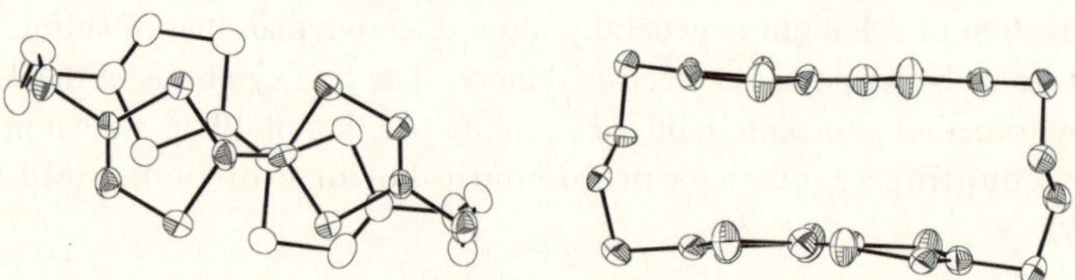

Fig. 9.14 X-ray structure of **41a** *cis/trans*.

Fig. 9.15 X-ray structure of **41a** *trans/trans (twist)*.

partial overlap. The shortest nonbonded Se···Se distance between the stacked TSF units is 3.92 Å, which is slightly longer than the normal van der Waals contact (3.8 Å).

On the other hand, recrystallization of **41b** from carbon disulfide gave orange prisms whose X-ray crystallographic analysis revealed that the crystal is the *cis/cis* isomer, as shown in Fig. 9.16. In contrast to the almost planar TSF in **41a** (*cis/trans*), the TSF frameworks are heavily bent with dihedral angles of 24.3° and 42.9° between the central ethylene and the outer ethylene part. The molecule takes a step-by-step *anti* conformation, where the two TSF units do not overlap with each other and there is no close nonbonded Se···Se contact.

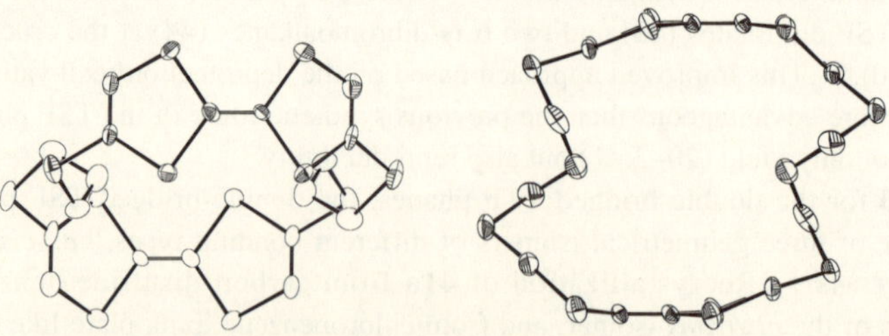

Fig. 9.16 Molecular structure of **41b** (*cis/cis*).

The cyclic voltammogram of the TSF phane (**41a**) showed two reversible one-electron redox waves ($E_{1/2}$ +0.47 and +0.67 V) and one quasi-reversible two-electron redox wave ($E_{1/2}$ +0.96 V), and the voltametric pattern is very similar to that of the TTF phane (**13b**) with the same ethylenedithio bridges. The split of the first and second redox processes is ascribed to a marked through-space interaction between the two TSF units, and the higher potential shift of the combined third and fourth oxidation wave also supports the presence of a large Coulombic repulsion in the tri- and tetra-cationic states. On the other hand, the cyclic voltammogram of **41b** (n = 3) with longer trimethylenedithio bridges showed only two broad two-electron redox waves at +0.51 and +0.93 V. Unlike the TTF phane (**13c**) with the same bridges, the two TSF units do not interact so much as to split the first and second redox processes.

Electrocrystallization of the TSF phanes gave some conductive radical cation salts, whose properties are summarized in Table 9.3. Their conductivities vary in the range of 10^{-3}–10^0 S cm^{-1} at room temperature with semiconductive temperature dependence.

Table 9.3   Electrical Conductivities of Radical Cation Salts
of Double-bridged TSF Phanes (**41**)

Salt	D:A	σ/S cm^{-1}
41a•ClO$_4$	3:2	3.5
41a•BF$_4$	3:4	4 x 10^{-2}
41a•Au(CN)$_2$	1:1:1[a]	2.8 x 10^{-3}
41b•AsF$_6$	2:3	1.3 x 10^{-3}
41b•Au(CN)$_2$	1:1	2 x 10^{-1}
41b•Au(CN)$_2$	1:1:1[b]	53

[a]Chlorobenzene is included from the solvent.
[b]2-Propanol is included.

The ClO$_4$ salt of **41a** (3.5 S cm^{-1}) is more conductive by one order of magnitude than the corresponding radical cation salt of the TTF phane **13b** (0.2 S cm^{-1}), suggesting the effect of the selenium substitution. However, the BF$_4$ salt of **41a** (0.04 S cm^{-1}) is not so conductive as that of **13b** (1.5 S cm^{-1}). Fig. 9.17 shows the crystal packing structure of the BF$_4$ salt of **41a**, which contains two donor molecules in a composition ratio of 2:1 in a stacking column. Both the donor molecules adopt the *trans/trans (eclipse)* form. The major component is monocationic and keeps an interactive through-space distance of 3.54 Å between the two TSF units. On the other hand, the minor component is dicationic, and due to the larger Coulombic repulsion, the through-space distance is extended to 4.13 Å, where the two TSF units are too far to interact. Presumably the conduction path along the column direction is interrupted at the dicationic position, resulting in the relatively low conductivity for the **41a** •BF$_4$ salt.

The Au(CN)$_2$ salts of **41b** gave two morphogenic crystals of plate and needle. Both showed a 1: 1 (D:A) composition ratio, but the needle crystal included 2-propanol from solvent. The plate salt is semiconductive (0.2 S cm^{-1}) with a relatively small activation energy, whereas the needle salt shows a very high conductivity of 53 S cm^{-1}. This value is much larger than any of the radical cation salts obtained from the double-bridged TTF phanes. Fig. 9.18 shows the crystal structure of the needle salt. The donor takes the *trans/trans* form with slight sliding along the long axis direction of the molecule to construct a stacking column with other donors. Judging from the 1:1 (D:A) ratio, the formal charge of the donor is monocationic, which means that each TSF is in a half-oxidized state suitable for charge migration. The shortest intramolecular and intermolecular Se⋯Se contacts along the columnar direction are 3.87 and 3.84 Å, respectively, approximately comparable to the van der Waals contact (3.8 Å) of two selenium atoms. In addition, there are close side-by-side Se⋯Se contacts (3.68 Å) between the neighboring donor columns, forming a sheet-like interaction array. This sheet-like interaction array probably contributes to the high conductivity of this salt.

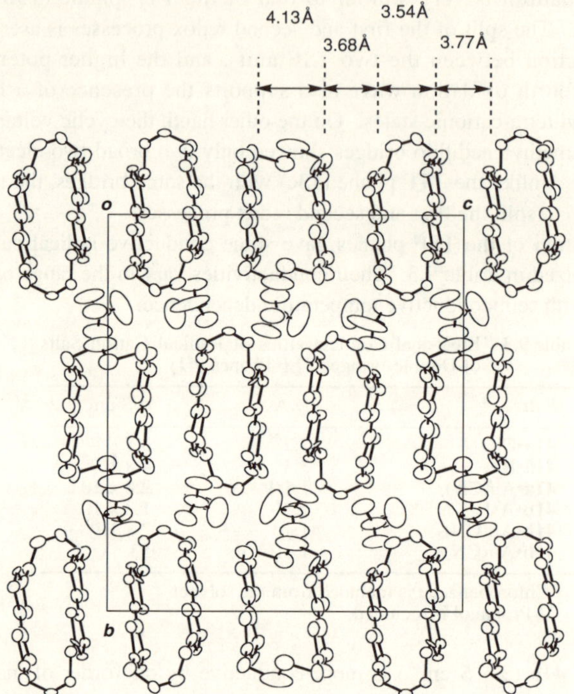

Fig. 9.17   Crystal structure of **41a•BF₄**.

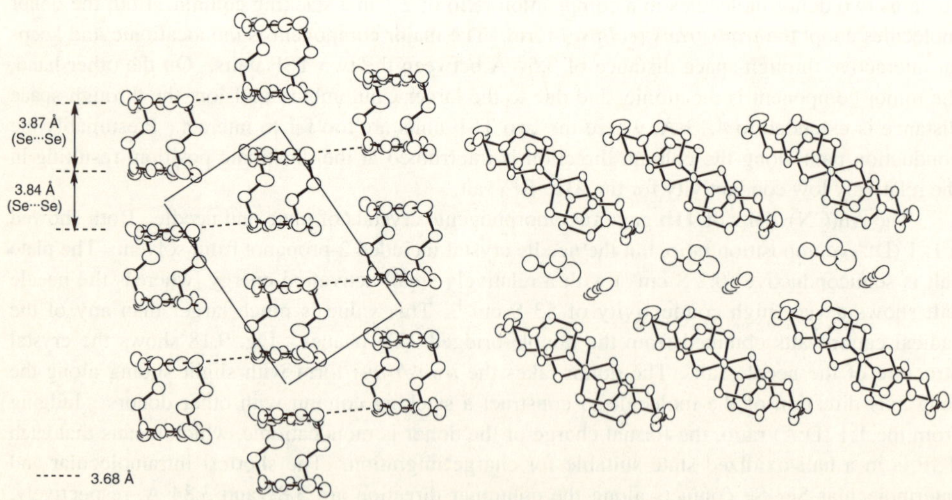

Fig. 9.18   Crystal structure of **41b•Au(CN)₂**.

9.7   Summary and Outlook

This chapter has presented many examples of TTF phanes, two or more TTF units of which interact in a through-space mode, depending largely on the number of the bridges, the length of the bridge chains and the stacking modes. These TTF phanes can actually form a variety of unique molecular complexes based on the versatile stacked structures of the donor molecules. It is evident that the electronic interactions of the TTF phanes serve to control the solid-state structures or electronic structures of their molecular complexes. However, the TTF phanes do not necessarily form highly conductive molecular complexes. It is also true that although they are expected to be model compounds for κ-type superconductors, dimeric TTF phanes have not provided any superconductors yet. One of the drawbacks in the TTF phanes is the very easy bending of the TTF skeleton, especially when the bridge chains are shortened. The bent TTF phanes thus reduce their capability as electron donors for forming conductive complexes. Instead, they tend to form large cavities inside, which can include a solvent molecule or hopefully a counterion in their molecular complexes. From the viewpoint of supramolecular assembly, TTF phanes may find practical use as molecular electronic materials with switching and sensory functions.[7]

References

1. J. B. Torrance, *Acc. Chem. Res.*, **12**, 79 (1979).
2. G. Saito and J. B. Ferraris, *Bull. Chem. Soc. Jpn.*, **53**, 2141 (1980).
3. D. O. Cowan, A. Kini, L.-Y. Chiang, K. Lerstrup, D. R. Talham, T. O. Poehler and A. N. Bloch, *Mol. Cryst. Liq. Cryst.*, **86**, 1741 (1982).
4. P. Delhaes, *Mol. Cryst. Liq. Cryst.*, **96**, 229 (1983).
5. M. Adam and K. Müllen, *Adv. Mater.*, **6**, 439 (1994).
6. T. Otsubo, Y. Aso and K. Takimiya, *Adv. Mater.*, **8**, 203 (1996).
7. J. L. Segura and N. Martín, *Angew. Chem. Int. Ed.*, **40**, 1372 (2001).
8. M. L. Kaplan, R. C. Haddon and F. Wudl, *J. Chem. Soc., Chem. Commun.*, 388 (1977).
9. K. Bechgaard, K. Lerstrup, M. Jørgensen, I. Johannsen, J. Cristensen and J. Larsen, *Mol. Cryst. Liq. Cryst.*, **181**, 161 (1990).
10. A. Izuoka, R. Kumai, T. Tachikawa and T. Sugawara, *Mol. Cryst. Liq. Cryst.*, **218**, 213 (1992).
11. M. Jørgensen, K. A. Lerstrup and K. Bechgaard, *J. Org. Chem.*, **56**, 5684 (1991).
12. N. Thorup, G. Rindorf, K. Lerstrup and K. Bechgaard, *Synth. Met.*, **42**, 2423 (1991).
13. M. R. Bryce, G. J. Marshallsay and A. J. Moore, *J. Org. Chem.*, **57**, 4859 (1992).
14. A. Izuoka, R. Kumai and T. Sugawara, *Chem. Lett.*, 285 (1992).
15. I. V. Sudmale, G. V. Tormos, V. Y. Khodorkovsky, A. S. Edzina, O. J. Neilands and M. P. Cava, *J. Org. Chem.*, **58**, 1355 (1993).
16. R. P. Parg, J. D. Kilburn, M. C. Petty, C. Pearson and T. G. Ryan, *Synthesis*, 613 (1994).
17. R. P. Parg, J. D. Kilburn and T. G. Ryan, *Synth. Met.*, **70**, 1163 (1995).
18. A. Dolbecq, K. Boubekeur, P. Batail, E. Canadell, P. Auban-Senzier, C. Coulon, K. Lerstrup and K. Bechgaard, *J. Mater. Chem.*, **5**, 1707 (1995).
19. M. Iyoda, H. Suzuki and U. Kux, *Tetrahedron Lett.*, **36**, 8259 (1995).
20. I. Sudmale, A. Puplovskis, A. Edzima, O. Neilands and V. Khodorkovsky, *Synthesis*, 750 (1997).
21. S. Booth, E. N. K. Wallace, K. Singhal, P. N. Bartlett and J. D. Kilburn, *J. Chem. Soc., Perkin Trans. 1*, 1467 (1998).
22. M. Fourmigué, C. Mézière, E. Canadell, D. Zitoun, K. Bechgaard and P. Auban-Senzier, *Adv. Mater.*, **11**, 766 (1999).
23. T. Tachikawa, A. Izuoka, R. Kumai, T. Sugawara and Y. Sugawara, *Solid State Commun.*, **82**, 19 (1992).
24. T. Tachikawa, A. Izuoka and T. Sugawara, *J. Chem. Soc., Chem. Commun.*, 1227 (1993).
25. A. Izuoka, T. Tachikawa, T. Sugawara, Y. Saito and H. Shinohara, *Chem. Lett.*, 1049 (1992).
26. J. M. Williams, J. R. Ferraro, R. J. Thorn, K. D. Carlson, U. Geiser, H. H. Wang, A. M. Kini and M.-H.

Whangbo, *Organic Superconductors*, Prentice-Hall, New Jersey (1992).
27. T. Ishiguro, K. Yamaji and G. Saito, *Organic Superconductors*, Springer-Verlag, Berlin (1998).
28. M. B. Nielsen and J. Becher, *Lieb. Ann.*, 2177 (1997).
29. T. Jørgensen, T. K. Hansen and J. Becher, *Chem. Soc. Rev.*, **23**, 41 (1994).
30. J. Becher, Z.-T. Li, P. Blanchard, N. Svenstrup, J. Lau, M. B. Nielsen and P. Leriche, *Pure Appl. Chem.*, **69**, 465 (1997).
31. M. B. Nielsen, C. Lomholt and J. Becher, *Chem. Soc. Rev.*, **29**, 153 (2000).
32. K. Boubekeur, C. Lenoir, P. Batail, R. Carlier, A. Tallec, M.-P. Le Paillard, D. Lorcy and A. Robert, *Angew. Chem., Int. Ed. Engl.*, **33**, 1379 (1994).
33. C. Wang, M. R. Bryce, A. S. Batsanov and J. A. K. Howard, *Chem. Eur. J.*, **3**, 1679 (1997).
34. T. Finn, M. R. Bryce, A. S. Batsanov and J. A. K. Howard, *Chem. Commun.*, 1835 (1999).
35. H. A. Staab, J. Ippen, C. Tao-pen, C. Krieger and B. Starker, *Angew. Chem. Int. Ed. Engl.*, **19**, 66 (1980).
36. F. Bertho-Thoraval, A. Robert, A. Souizi, K. Boubekeur and P. Batail, *J. Chem. Soc., Chem. Commun.*, 843 (1991).
37. C. A. Christensen, A. S. Batsanov, M. R. Bryce and J. A. K. Howard, *J. Org. Chem.*, **66**, 3313 (2001).
38. R. M. Moriarty, A. Tao, R. Gilardi, Z. Song and S. M. Tuladhar, *Chem. Commun.*, 157 (1998).
39. K. B. Simonsen, K. Zong, R. D. Rogers, M. P. Cava and J. Becher, *J. Org. Chem.*, **62**, 679 (1997).
40. K. B. Simonsen, N. Thorup, M. P. Cava and J. Becher, *Chem. Commun.*, 901 (1998).
41. J. Ippen, C. Tao-pen, B. Starker, D. Schweitzer and H. A. Staab, *Angew. Chem. Int. Ed. Engl.*, **19**, 67 (1980).
42. M. Bardri, J. P. Majoral, F. Gonce, A.-M. Caminade, M. Salle and A. Gorgues, *Tetrahedron Lett.*, **31**, 6343 (1990).
43. K. B. Simonsen, N. Thorup and J. Becher, *Synthesis*, 1399 (1997).
44. K. Takimiya, Y. Aso, F. Ogura and T. Otsubo, *Chem. Lett.*, 735 (1995).
45. K. Takimiya, Y. Aso and T. Otsubo, *Synth. Met.*, **86**, 1981 (1997).
46. K. Takimiya, A. Oharuda, Y. Aso, F. Ogura and T. Otsubo, *Chem. Mater.*, **12**, 2196 (2000).
47. C. J. Brown, *J. Chem. Soc.*, 3278 (1953).
48. Y. Kai, N. Yasuoka and N. Kasai, *Acta Crystllogr.*, B**33**, 754 (1977).
49. D. E. John, A. S. Batsanov, M. R. Bryce and J. A. K. Howard, *Synthesis*, 824 (2000).
50. A. S. Batsanov, D. E. John, M. R. Bryce and J. A. K. Howard, *Adv. Mater.*, **10**, 1360 (1998).
51. X. Yang, P. Wu and D. Zhu, *Hecheng Huaxue*, **1**, 141 (1993); *Chem. Abst.*, **121**, 108745c (1994).
52. M. Adam, V. Enkelmann, H.-J. Räder, J. Röhrich and K. Müllen, *Angew. Chem., Int. Ed. Engl.*, **31**, 309 (1992).
53. K. Matsuo, K. Takimiya, Y. Aso, T. Otsubo and F. Ogura, *Chem. Lett.*, 523 (1995).
54. T. Ishizuka, T. Ishikawa, M. Fujisaku, K. Itoh, T. Tejima, A. Izuoka and Y. Kawata, 31st Symposium on Structural Organic Chemistry, Yamaguchi, Japan, Oct. 27-28, 2001, Abstract p. 36.
55. K. B. Simonsen, N. Svenstrup, J. Lau, N. Thorup and J. Becher, *Angew. Chem. Int. Ed.*, **38**, 1417 (1999).
56. L. Ouahab, J.-F. Halet, O. Peña, J. Padiou, D. Grandjean, C. Garrigou-Lagrange and P. Delhaes, *J. Chem. Soc., Dalton Trans.*, 1217 (1992).
57. K. Takimiya, Y. Shibata, K. Imamura, A. Kashihara, Y. Aso, T. Otsubo and F. Ogura, *Tetrahedron Lett.*, **36**, 5045 (1995).
58. K. Takimiya, K. Imamura, Y. Shibata, Y. Aso, F. Ogura and T. Otsubo, *J. Org. Chem.*, **62**, 5567 (1997).
59. J. Tanabe, T. Kudo, M. Okamoto, Y. Kawada, G. Ono, A. Izuoka and T. Sugawara, *Chem. Lett.*, 579 (1995).
60. J. Tanabe, G. Ono, A. Izuoka, T. Sugawara, T. Kudo, T. Saito, M. Okamoto and Y. Kawada, *Mol. Cryst. Liq. Cryst.*, **296**, 61 (1997).
61. J. Tanabe, A. Izuoka, T. Sugawara, T. Kudo and Y. Kawada, 27th Symposium on Structural Organic Chemistry, Osaka, Japan, Oct. 8-9,1997, Abstract p. 27.
62. P. Blanchard, N. Svenstrup and J. Becher, *Chem. Commun.*, 615 (1996).
63. S. Yunoki, K. Takimiya, Y. Aso and T. Otsubo, *Tetrahedron Lett.*, **38**, 3017 (1997).
64. K. B. Simonsen, N. Svenstrup, J. Lau, O. Simonsen, P. Mørk, G. J. Kristensen and J. Becher, *Synthesis*, 407 (1996).
65. K. B. Simonsen and J. Becher, *Synlett*, 1211 (1997).
66. D. Cowan and A. Kini, *The Chemistry of Organic Selenium and Tellurium Compounds*, J. Wiley & Sons, Chichester (1987), Vol. 2, p. 463.
67. K. Takimiya, A. Oharuda, A. Morikami, Y. Aso and T. Otsubo, *Angew. Chem. Int. Ed.*, **37**, 619 (1998).
68. K. Takimiya, A. Oharuda, A. Morikami, Y. Aso and T. Otsubo, *Eur. J. Org. Chem.*, 3013 (2000).

10

Bis-fused TTFs –Tetrathiapentalene Donors

10.1 Introduction

In the search for new organic metals and organic superconductors, tetrathiafulvalene (TTF) and its derivatives have played a dominant role.[1] In fact, they have yielded a large number of organic conductors retaining metallic conductivity down to low temperatures, and most molecular superconductors have been prepared from TTF derivatives. Recently, considerable attention has been focused on dimeric (and oligomeric) TTF systems.[2] Such donors have the advantage that the on-site Coulombic repulsion in the dicationic states is considerably reduced by the delocalization of two positive charges over four (or more) redox-active 1,3-dithiol-2-ylidenes (DTs), while two positive charges must localize in two DT units in the case of TTF. Thus, stabilization of the metallic state can be expected in dimeric TTF systems. On the other hand, realization of multi-dimensional electronic structures is very important for the development of organic materials retaining metallic conductivity down to low temperatures and of superconductors, because conventional one-dimensional (1D) conductors represented as (TTF)(TCNQ) inherently exhibit metal-to-insulator transition. The long molecular skeleton of dimeric TTF systems raises high expectations for the production of two-dimensional (2D) conductors, although there are several obstacles which must be overcome.

The characteristic of TTF derivatives, which give organic superconductors, is a ladder-like array of chalcogen atoms. Thus, a 2D (or quasi 1D) electronic structure can be realized thanks to the so-called side-by-side interaction through chalcogen atoms as is observed in BEDT-TTF conductors (Fig. 10.1).[3] In this context, 2,5-bis(1,3-dithiol-2-ylidene)-1,3,4,6-tetrathiapentalene (BDT-TTP or simply TTP), in which two TTF units are directly fused with each other, is of particular interest for the preparation of stable metals down to low temperatures because it has a "ladder-like array" of sulfur atoms similar to BEDT-TTF (Fig. 10.1).

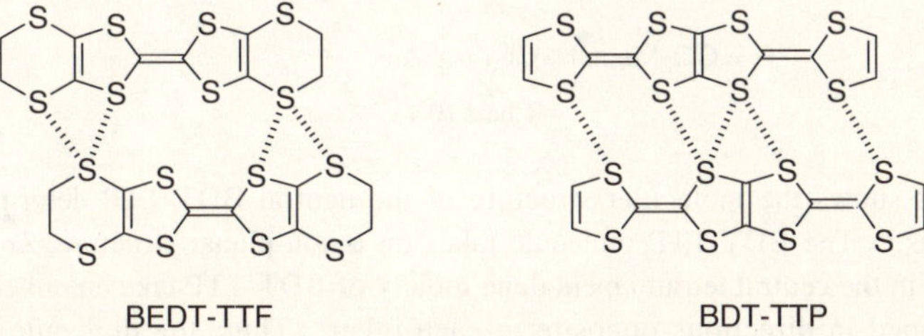

Fig. 10.1   Side-by-side interaction in BEDT-TTF and BDT-TTP conductors.

10.2   Synthesis and Properties of Bis-fused TTFs

Schumaker and coworkers first reported the synthesis of BDT-TTP derivatives (**1**) possessing strongly electron-withdrawing substituents such as cyano and trifluoromethyl groups in 1980 by the cross-coupling reaction between 1,3,4,6-tetrathiapentalene-2,5-dione (**2**) and 1,3-dithiole-2-thione derivatives (**3**) in P(OEt)$_3$ (Chart 10.1).[4] Their interest in fused TTF donors focused on the higher homologues of TTF, because a 2D metal based on BEDT-TTF was not yet reported at that time. Unfortunately, the details of their structures and properties, particularly crystal structure, redox behavior and formation of conducting salts, were not mentioned at all. Furthermore, Schumaker's method has not been applied to the synthesis of other BDT-TTP derivatives probably due to the difficulty in optimizing the reactions conditions of cross-coupling using **2**. In this connection, the establishment of a general synthetic method for BDT-TTP derivatives has been desired for a long time. A versatile route for the synthesis of BDT-TTP derivatives was developed early in 1990, as shown in Scheme 10.1.[5,6] The feature of this synthetic route is the construction of unsymmetrical TTF (**4**) followed by the formation of 1,3-dithiole ring by means of deprotection of TTF dithiolate followed by reaction with (Cl$_3$CO)$_2$CO. The resulting 1,3-dithiol-2-one fused with TTF (**5**) can be cross-coupled with **3** by using P(OMe)$_3$ in refluxing toluene. However, the cross-coupling reaction of the final step does not proceed at all when the unsubstituted or alkyl-substituted **3** is used. This is in contrast to the fact that the first cross-coupling reaction easily proceeds regardless of the substituents on compounds **3**. Fortunately, the unsubstituted derivative has been obtained by demethoxycarbonylation using LiBr·H$_2$O in HMPA. The 2-cyanoethyl group[7] can also be utilized as the protecting group for thiolate instead of the *p*-acetoxybenzyl group.[8] On the other hand, the cross-coupling reaction for the synthesis of selenium analogues ST-TTP and BDS-TTP derivatives (**6**) has been carried out by another combination of 1,3-dithiole-2-thiones fused with TTF (or STF) (**7**) and 1,3-diselenol-2-ones (**8**) (Scheme 10.2).[9] More than eighty kinds of BDT-TTP derivatives and selenium analogues have been synthesized by the above method so far.[10-28] On the other hand, Yamada and coworkers have developed a new synthetic strategy for **5** (Scheme 10.3).[29] In this procedure, one of the TTF frameworks is constructed *via* the Me$_3$Al-promoted reaction of organotin compound **9** prepared from **2** with esters **10**, although the P(OEt)$_3$-mediated cross coupling is still required because of the extremely low solubilities of **5** and **11** in common organic solvents.

R = CO$_2$Me, CN, CF$_3$

Chart 10.1

Figure 10.2 shows the molecular structure of the neutral BDT-TTP determined by X-ray structure analysis.[6] The BDT-TTP molecule takes on a non-planar structure, *i.e.*, the two five-membered rings in the central tetrathiapentalene moiety of BDT-TTP take an envelope form, with C2 and C8 flapped in directions opposite to each other. Thus, the molecule adopts a chair conformation. The dihedral angle between the two planes formed by S1-S2-C1-C2-S3-S4 (or S5-S6-C7-C8-S7-S8) and S3-S4-C5-C6-S5-S6 is 27.17° (or 27.33°).

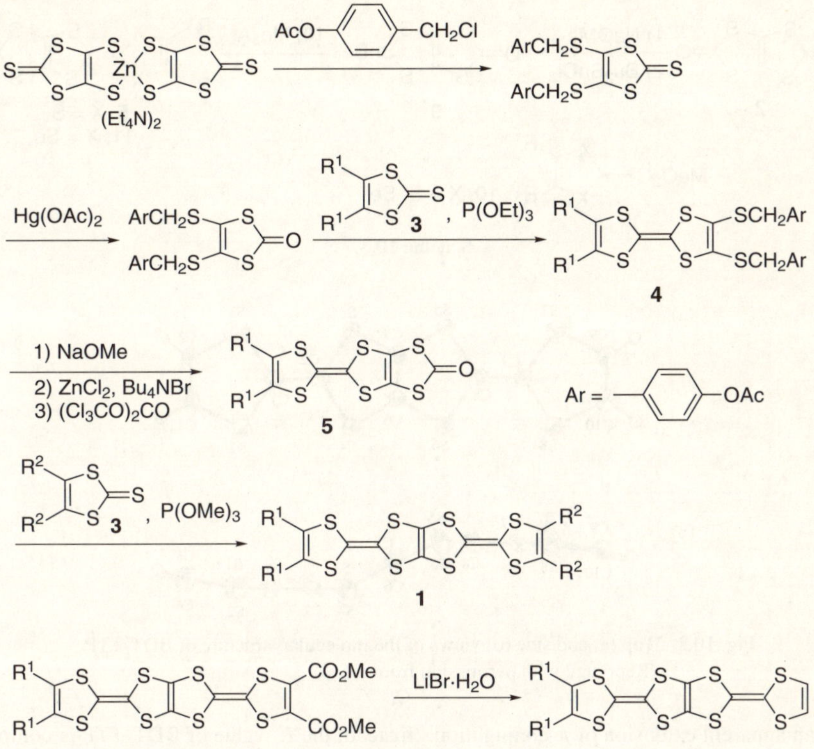

Scheme 10.1

The redox potentials of BDT-TTP derivatives measured by cyclic voltammetry (CV) are summarized in Table 10.1. The CVs of most BDT-TTP derivatives are composed of four pairs of redox waves in correspondence with the presence of four redox-active sites of DT units, while the selenium analogue ST-TTP (**18**, Chart 10.2) shows three pairs of waves, in which the most positive one corresponds to a two-electron transfer. The first redox potential (E_1) of BDT-TTP ($E_1 = 0.44$ V *vs.* SCE, in PhCN) is somewhat higher by *ca.* 0.1 V than that of TTF (0.35 V) in

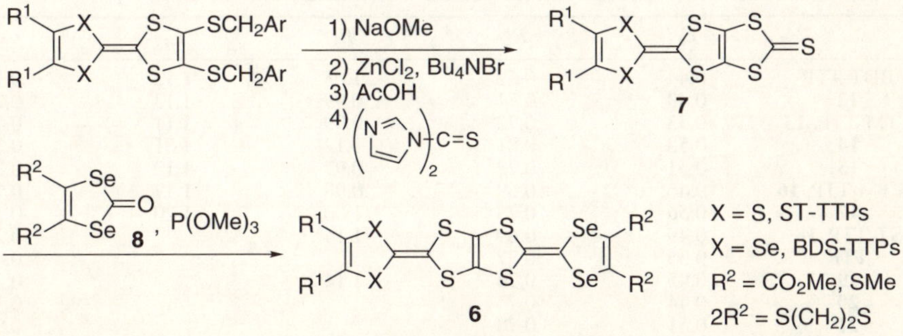

Scheme 10.2

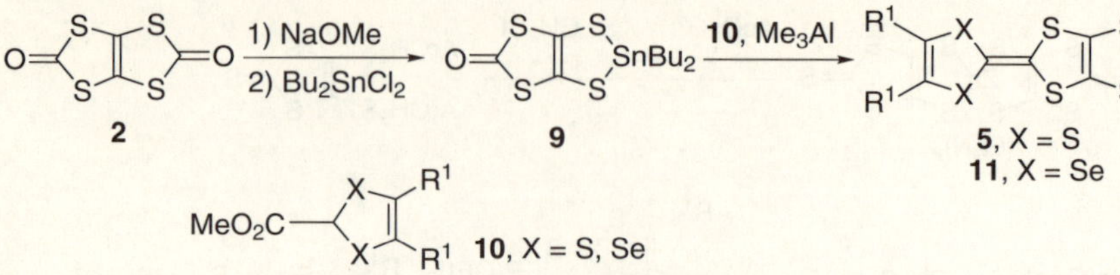

Scheme 10.3

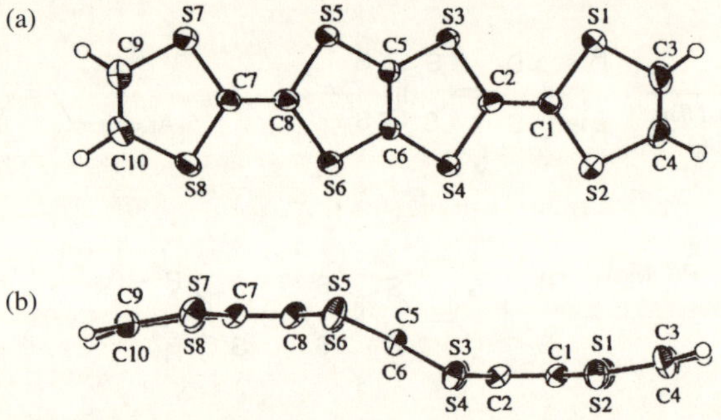

Fig. 10.2   Top (a) and side (b) views of the molecular structure of BDT-TTP.
(Reprinted with permission from ref. 6)

spite of an apparent extension of π-conjugation. Instead, the E_1 value of BDT-TTP is comparable to those of 2-isopropylidene-1,3-dithiolo[4,5-d]-TTF (**19**, 0.45 V)[30)] and 4,5-bis(methylthio)-TTF (**20**, 0.44 V) measured under identical conditions. These results indicate that the positive charge in (BDT-TTP)$^{+\bullet}$ is distributed mainly on one TTF moiety, and the other one acts as a sulfur-based substituent. This interpretation is consistent with the result that the tetrakis(methylthio) derivative (TTM-TTP, **13**) has an E_1 value (0.53 V) similar to the corresponding TTF derivative, TTM-TTF (**21**, 0.51 V). On the other hand, the ΔE (E_2 - E_1) value of BDT-TTP (0.18 V) is smaller by 0.24 V than that of TTF (0.42 V), suggesting a significantly decreased on-site Coulomb repulsion in the dicationic state.

Table 10.1   Redox Potentials of BDT-TTP Derivatives (V vs. SCE)^a

	E_1	E_2	E_3	E_4	ΔE
BDT-TTP	0.44	0.62	1.05	1.13^b	0.18
12	0.49	0.71	0.99	1.13^b	0.23
TTM-TTP, **13**	0.53	0.72	0.99	1.11	0.19
14	0.58	0.81	1.12	1.31^b	0.23
15	0.51	0.72	0.97	1.12	0.21
CPTM-TTP, **16**	0.46	0.69	0.98	1.17	0.25
17	0.56	0.73	1.00	1.20	0.17
ST-TTP, **18**	0.49	0.69	1.14^b		0.20
TTF	0.35	0.77			0.42
19	0.45	0.76	1.14^b		0.31
20	0.44	0.77			0.33
21	0.51	0.78			0.27

^{a}Measured in PhCN containing 0.1 M Bu$_4$NClO$_4$ at 25 °C with scan rate of 50 mV/s.
^{b}Irreversible step (an anodic peak potential).

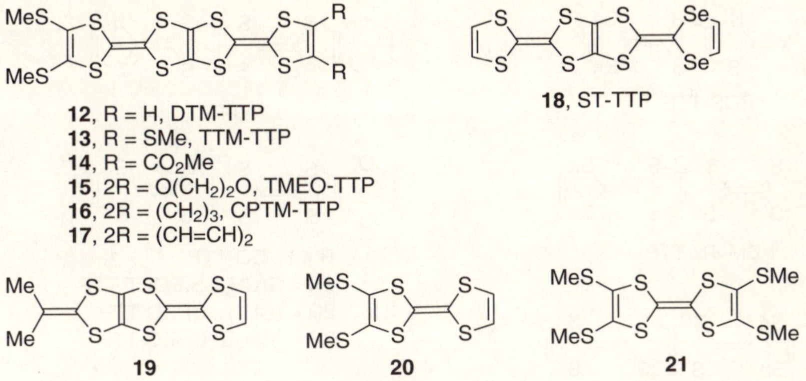

MeS
MeS
R
R
S
Se
Se
12, R = H, DTM-TTP
13, R = SMe, TTM-TTP
14, R = CO2Me
15, 2R = O(CH2)2O, TMEO-TTP
16, 2R = (CH2)3, CPTM-TTP
17, 2R = (CH=CH)2
18, ST-TTP
Me
Me
19
20
21
SMe

Chart 10.2

10.3 Structures and Physical Properties of Bis-fused TTF Conductors

A large number of radical cation salts and charge-transfer(CT) complexes based on BDT-TTP and its derivatives (Chart 10.3, p. 232) have been prepared.[9,11,14-26,31-51] Table 10.2 (p. 233) summarizes the donor packing patterns and conducting properties of TTP-type conductors whose crystal structures have been determined. In this section, representative materials based on BDT-TTP derivatives are discussed.

10.3.1 (BDT-TTP)A_x and $(ST\text{-}TTP)_2A$

Most radical cation salts based on BDT-TTP and its selenium analogue ST-TTP show high electrical conductivities of $\sigma_{rt} = 10^0$–10^3 S cm^{-1}, and metal-like conductive behavior down to 0.6–4.2 K regardless of the size and shape of the counteranions (Fig. 10.3).[9,32-34] The crystal

1.5
1.0
0.5
0.0
0
100
200
300
Normalized Resistivity
T / K
GaCl4
SbF6
Au(CN)2
AsF6
I3

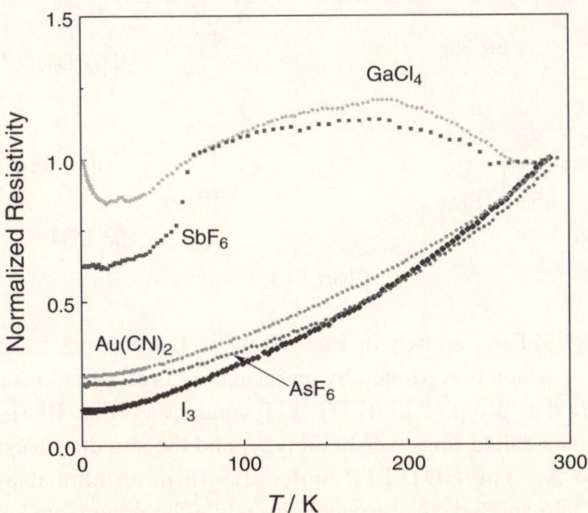

Fig. 10.3 Conducting behavior of BDT-TTP salts.

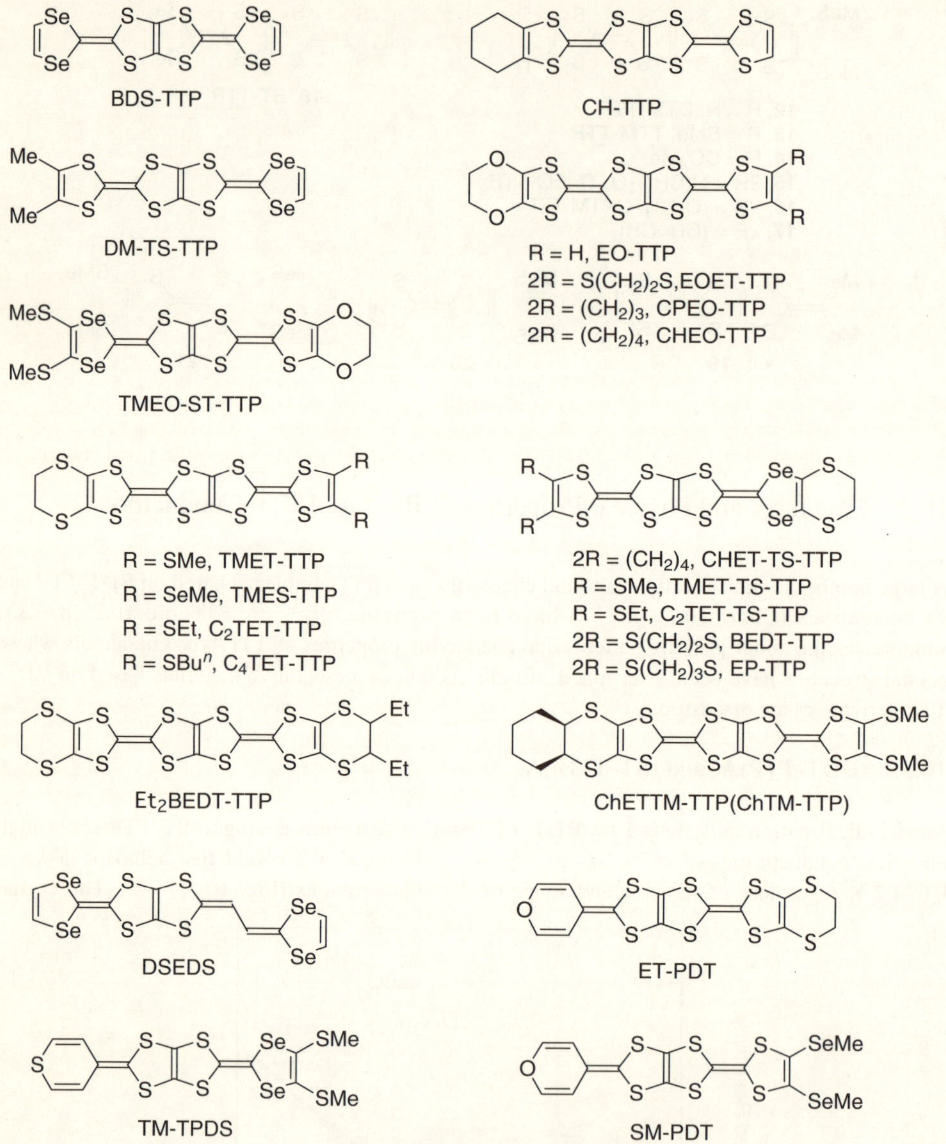

BDS-TTP

CH-TTP

DM-TS-TTP

R = H, EO-TTP
2R = S(CH$_2$)$_2$S, EOET-TTP
2R = (CH$_2$)$_3$, CPEO-TTP
2R = (CH$_2$)$_4$, CHEO-TTP

TMEO-ST-TTP

R = SMe, TMET-TTP
R = SeMe, TMES-TTP
R = SEt, C$_2$TET-TTP
R = SBu^n, C$_4$TET-TTP

2R = (CH$_2$)$_4$, CHET-TS-TTP
R = SMe, TMET-TS-TTP
R = SEt, C$_2$TET-TS-TTP
2R = S(CH$_2$)$_2$S, BEDT-TTP
2R = S(CH$_2$)$_3$S, EP-TTP

Et$_2$BEDT-TTP

ChETTM-TTP(ChTM-TTP)

DSEDS

ET-PDT

TM-TPDS

SM-PDT

Chart 10.3

structure of (BDT-TTP)$_2$SbF$_6$ is shown in Fig. 10.4.[32] The donors form two-dimensional conducting sheets, each of which is separated by an insulating layer of anions as is expected. The array of donors is classified as β-type[52] in BEDT-TTF conductors (Fig. 10.5). The overlap mode of donor molecules is the so-called ring-over-bond type, and the slip distance along the molecular axis in the stack is 1.6 Å. The BDT-TTP molecules form an almost uniform stack with interplanar distances of 3.46 and 3.47 Å, respectively, while the donors are strongly dimerized in β-BEDT-TTF salts.[52] As a result, the calculated overlap integrals in the stack are quite similar to

Table 10.2   Molecular Packing Pattern and Electrical Properties of Conducting Salts Based on BDT-TTP and Its Derivatives

Material	Donor packing	$\sigma_{rt}/S\ cm^{-1}$	Conducting behavior [b]	Ref(s).
(BDT-TTP)A_x	Uniform β-type	160–400	M'	31
(A = ClO$_4$, ReO$_4$, BF$_4$)				
(BDT-TTP)$_2$SbF$_6$	β-type	48	M	32
(BDT-TTP)$_6$A(TCE)$_2$ [a]	β-type	83–285	M	33
(A = Mo$_6$Cl$_{14}$, Re$_6$S$_6$Cl$_8$)				
(ST-TTP)$_2$AsF$_6$	β-type	200	M'	9
(BDS-TTP)$_2$AsF$_6$	β-type	70	M	34
(CH-TTP)(I$_3$)$_{0.31}$	κ-type	38	M	35
(CHEO-TTP)(ReO$_4$)$_{0.38}$	κ-type	11	M	22
(CHET-TS-TTP)(SbF$_6$)$_{0.36}$	κ-type	-	-	23
(DM-TS-TTP)$_2$PF$_6$	β-type	55	M	20
(EO-TTP)$_2$AsF$_6$	β-type	600	M	36
(CPEO-TTP)(SbF$_6$)$_{0.40}$	κ-type	60	M	19
(EOET-TTP)$_3$AsF$_6$	κ-type	600	M	11
(TMEO-TTP)$_2$Au(CN)$_2$	β"-type	200	M'	37
(TMEO-TTP)$_3$SbF$_6$	β'-type	9	I	38
(TMEO-TTP)AuBr$_2$(THF)	Dimerized 1D stack	0.12	I	39
(TMEO-ST-TTP)Au(CN)$_2$	Dimerized 1D stack	3.8×10^{-3}	I	40
(TMEO-ST-TTP)$_2$A	β-type	5–75	M	40
(A = PF$_6$, AsF$_6$, TaF$_6$)				
(TMEO-ST-TTP)$_2$ClO$_4$(DCE)[a]	β-type	0.07	I	21
(TMEO-ST-TTP)(TCNQ)(PhCl)	D-A stack	5.3×10^{-5}	I	21
(CPTM-TTP)$_4$A	λ-type	7–110	M' (A = PF$_6$, AsF$_6$)	14
(A = PF$_6$, AsF$_6$, SbF$_6$)			T_{MI} = 50 K (A = SbF$_6$)	
(ChETTM-TTP)$_2$A	β-type	20–80	T_{MI} = 100–150 K	15,41
(A = AuBr$_2$, Au(CN)$_2$, GaCl$_4$)				
(ChETTM-TTP)ReO$_4$	β-type	8×10^{-3}	I	41
(TMET-TTP)A_x	θ-type	9–45	T_{MI} = 200 K	42
(A = AuI$_2$, PF$_6$, ClO$_4$, ReO$_4$, IO$_4$, x = 0.13-0.34)				
(TMES-TTP)$_4$I$_3$	θ-type	13	I	16
(TMET-TS-TTP)$_2$TCNQ	θ-type	1–10	T_{MI} = 120 K	17
(C$_2$TET-TTP)$_2$ClO$_4$	Uniform β-type	12–55	M	24
(C$_2$TET-TS-TTP)$_2$A	Uniform β-type	900	M	25
(A = PF$_6$, ClO$_4$, BF$_4$)				
(C$_4$TET-TTP)I$_3$	Dimerized (non-columnar structure)	2.5×10^{-2}	I	24
(BEDT-TTP)$_2$I$_3$	β-type	900	M	43
(EP-TTP)$_2$Au(CN)$_2$	β-type	400	M	44
(TTM-TTP)$_2$I$_3$	Trimerized stack	0.03	I	45
(TTM-TTP)I$_3$	Uniform β-type	700	T_{MI}= 160 K	45
(TTM-TTP)(I$_3$)$_{5/3}$	Uniform 1D stack	200	T_{MI}= 20 K	46
(TTM-TTP)AuX$_2$ (X = Br, I)	Dimerized 1D stack	10–40	I	47
(TTM-TTP)A(PhCl)$_{0.5}$	Dimerized 1D stack	0.01–0.07	I	48,49
(A = FeCl$_4$, FeBr$_4$, GaCl$_4$, GaBr$_4$)				
(TTM-TTP)FeBr$_{1.8}$Cl$_{2.2}$	Uniform 1D stack	1000	T_{MI}= 160 K	48
(TTM-TTP)(PF$_6$)$_{0.267}$(THF)$_{0.6}$	Pentamerized stack	3×10^{-3}	I	50
(TSM-TTP)(I$_3$)$_{5/3}$	Uniform 1D stack	200	T_{MI} = 20 K	18
(TSM-TTP)$_3$(I$_3$)$_2$	Trimerized stack	3×10^{-2}	I	18
(DTM-TTP)(TCNQ)(TCE)	Uniform 1D stack	200	T_{MI} = 100 K	51
(Et$_2$BEDT-TTP)$_2$HgI$_3$	Pseudo λ-type	2	I	26
(DTEDT)$_3$Au(CN)$_2$	Uniform β-type	15	SC, T_c = 4 K	59
(DSEDS)$_3$TaF$_6$	Uniform β-type	9	M	70
(ET-PDT)$_4$PF$_6$(cn) [a]	β-type	50	M'	63
(TM-TPDS)$_2$AsF$_6$	Windmill type	240	T_{MI} = 100 K	74
(SM-PDT)PF$_6$(PhCl)$_x$	Pseudo β"-type	4.8	I	75
(SM-PDT)TCNQ(PhCl)	D-A stack	3.9×10^{-3}	I	76

[a]TCE = 1,1,2-trichloroethane, DCE = 1,2-dichloroethane, cn = 1-chloronaphthalene. [b]M: Metallic down to low temperature (≤ 4.2 K), M': Increase of resistivity at low temperature, but no metal-to-insulator transition, I: Semiconductor, T_{MI}: Metal-to-insulator transition temperature, SC: Superconductor, T_C: Superconducting transition temperature.

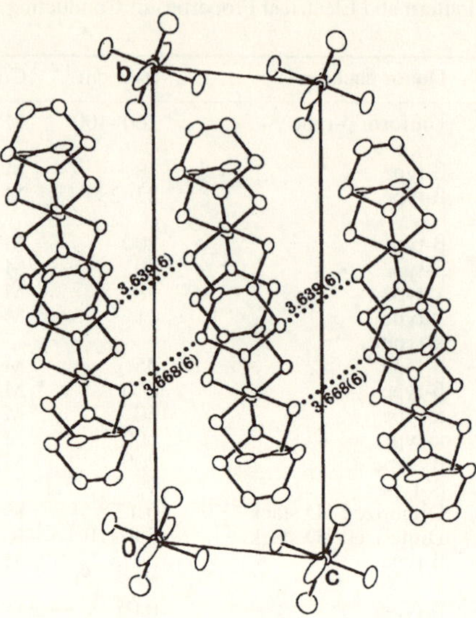

Fig. 10.4   Crystal structure of $(BDT\text{-}TTP)_2SbF_6$ showing intermolecular
S···S contacts (≤ 3.70 Å) with dotted line.
(Reprinted with permission from ref. 32)

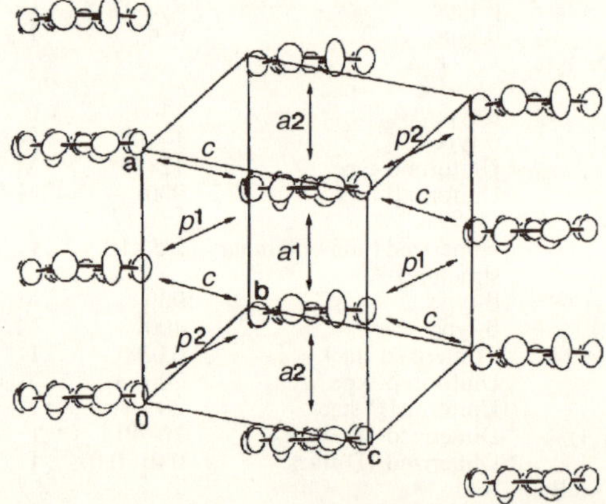

Fig. 10.5   Donor sheet structure of $(BDT\text{-}TTP)_2SbF_6$. The intermolecular
overlap integrals are $a1 = 25.1$, $a2 = 25.3$, $c = 0.8$, $p1 = 7.9$ and
$p2 = 8.6 \times 10^{-3}$. (Reprinted with permission from ref. 32)

each other ($a1 = 25.1$, $a2 = 25.3 \times 10^{-3}$). On the other hand, side-by-side interactions are about one third those along the stack. Due to such relatively strong interstack interactions, the calculated Fermi surface is closed in the conducting layer (Fig. 10.6). The tetrahedral anions (ClO_4^-, BF_4^- and ReO_4^-) also afford radical cation salts with similar donor packing.[31] In this case, the donors are arranged uniformly due to crystal symmetry. Similar donor packing has also been found in most radical cation salts based on BDT-TTP and its selenium analogues ST-TTP and BDS-TTP with various anions (PF_6^-, AsF_6^-, $Re_6S_6Cl_8^{2-}$, $Mo_6Cl_{14}^{2-}$, etc.).[9,32,34] Therefore, BDT-TTP is considered to have the strong self-aggregating property of forming uniform β-type packing.

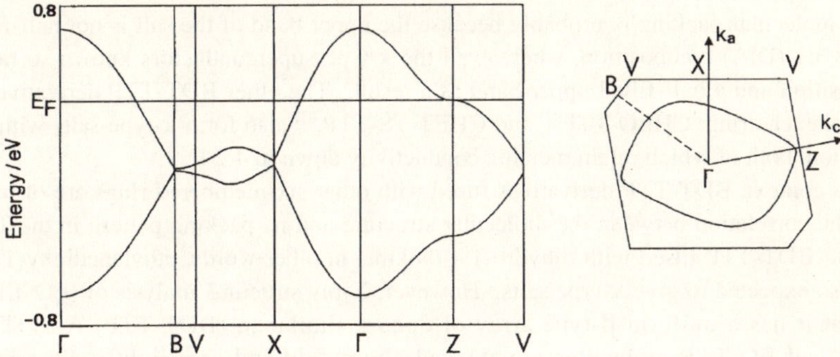

Fig. 10.6   Energy band and Fermi surface of $(BDT\text{-}TTP)_2SbF_6$. (Reprinted with permission from ref. 32)

10.3.2 κ-Type Salts Based on Bis-fused TTFs

Molecular packing other than the β-type is expected if an appropriate substituent is introduced to the BDT-TTP framework. In particular, construction of the κ-type structure has received considerable attention because many κ-type radical cation salts based on TTF derivatives are known to show superconducting transition.[53] For TTF systems, the introduction of capped alkylenedichalcogeno groups is indispensable to construct κ-type metals using TTF derivatives. The roles of the capped alkylenedichalcogeno group are regarded to be as follows: i) steric hindrance to avoid formation of stacking structure and to form dimers as a result, ii) realization of a 2D electronic structure through chalcogen-chalcogen side-by-side interaction. In fact, all the TTF donors affording κ-type salts have one or two ethylenedithio, methylendithio or ethylenedioxy groups. In contrast, the "capped" alkylenedichalcogeno group is not needed any more for the BDT-TTP system because it has sufficient sulfur atoms in the π-electron framework to realize side-by-side interaction. Due to such ability to form a 2D conducting layer, the introduction of a non-planar substituent such as a fused cyclohexene ring can be used to construct the κ-type structure. In fact, CH-TTP has afforded a κ-type salt with the I_3^- anion (Fig. 10.7).[35] This salt exhibits high room-temperature conductivity (σ_{rt}) of 38 S cm^{-1}, with metal-like

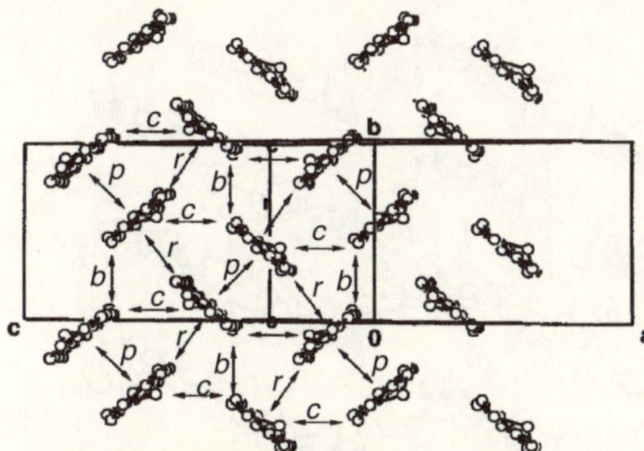

Fig. 10.7   Donor sheet structure of $(CH\text{-}TTP)(I_3)_{0.31}$ showing intermolecular S···S contacts (≤ 3.70 Å) with dotted line. The intermolecular overlap integrals are $c = -4.32$, $p = 22.2$, $r = 7.61$ and $b = 12.0$ x 10^{-3}. (Reprinted with permission from ref. 35)

temperature dependence down to 1.4 K. The fact that no superconductivity is observed in spite of the κ-type molecular packing is probably because the upper band of the salt is not half-filled due to a nearly 3:1 (D:A) composition, whereas all the κ-type superconductors known so far have a 2:1 composition and a half-filled upper band as a result. The other BDT-TTP derivatives with a fused cyclohexene ring, CHEO-TTP[22] and CHET-TS-TTP,[23] also form κ-type salts with a ReO_4^- or SbF_6^- anion, both of which retain metallic conductivity down to 4.2 K.

In this context, BDT-TTP derivatives fused with other six-membered rings are of interest to elucidate the correlation between the molecular structure and its packing pattern in the TTP-type conductors. BDT-TTP fused with dihydro-1,4-dioxine, in other words, ethylenedioxy-TTP (EO-TTP), is also expected to give κ-type salts. However, X-ray structure analysis of $(EO-TTP)_2AsF_6$ reveals that it has a uniform β-type array of donors similar to $(BDT-TTP)_2A$.[36] Thus, the unsymmetrical EO-TTP molecules are stacked alternately and are slightly dimerized with interplanar distances of 3.47 and 3.52 Å, respectively. The torsion angle around the ethylene bridge is 37°, which is about two thirds that in $(CH-TTP)(I_3)_{0.31}$ (57°). Such a smaller torsion angle as well as fewer hydrogen atoms in the dihydrodioxine ring compared with the cyclohexene ring may favor the β-type stacking structure rather than the strongly dimerized κ-type one. A tight-binding band calculation indicates that this salt has a closed Fermi surface similar to $(BDT-TTP)_2SbF_6$, although its 1D character is still strong due to large overlaps along the stacking direction. EO-TTP salts with tetrahedral and octahedral anions exhibit very high conductivities of σ_{rt} = 380–1300 S cm^{-1}, all of which exhibit simple metallic temperature dependence down to 1.5–4.2 K. On the other hand, CPEO-TTP[19] and EOET-TTP,[11] which have an additional substituent of an annelated cyclopentene ring or an ethylenedithio group, have yielded κ-type salts with octahedral anions, indicating that steric hindrance is indeed important for κ-type molecular packing.

10.3.3 TMEO-TTP and TMEO-ST-TTP Salts

TMEO-TTP affords many metallic radical cation salts with octahedral and linear anions, while tetrahedral anions give semiconductors. Of the metallic salts, the crystal structure of $(TMEO-TTP)_2Au(CN)_2$ has been investigated.[37] This salt has the so-called β''-type array of donors similar to $(BEDT-TTF)_2ClO_4(1,1,2-trichloroethane)$[3] (Fig. 10.8) and shows a resisitivity maximum around 180 K, while the other metallic TMEO-TTP salts exhibit simple metallic temperature

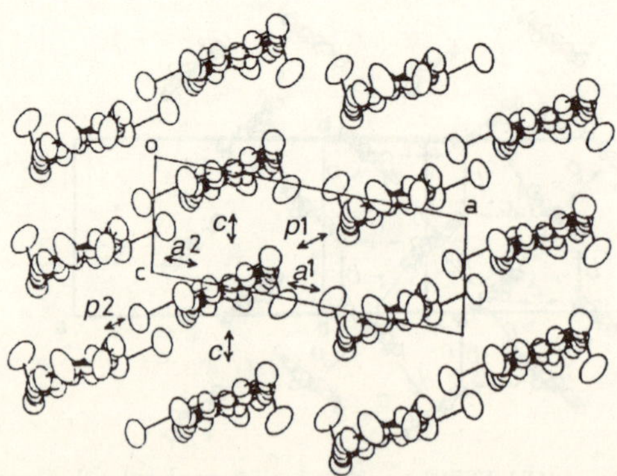

Fig. 10.8   Donor sheet structure of $(TMEO-TTP)_2 Au(CN)_2$. The intermolecular overlap integrals are c =4.2, $a1$ = 18.3, $a2$ = 7.3, $p1$ = -11.5 and $p2$ = -9.7 x 10^{-3}. (Reprinted with permission from ref. 37)

dependence. Thermoelectric power and ESR spectrum of this salt also exhibit anomalies around this temperature, indicating metal-to-metal transition.

In contrast to TMEO-TTP, a selenium analogue, TMEO-ST-TTP, forms a 1:1 salt with $Au(CN)_2^-$ anion.[40] The $Au(CN)_2^-$ anion does not form a layered structure and is located in a cavity surrounded by four donor molecules in the ac plane (Fig. 10.9). TMEO-ST-TTP molecules seem to form a 2D network in the ac plane at first glance. However, there is no significant side-by-side interaction along both the a and c axes. The donors are strongly dimerized in the stack, because one of the overlaps makes a large slip along both the donor long and short axes (Fig. 10.10). Therefore, this salt may be regarded as a 1D band insulator.

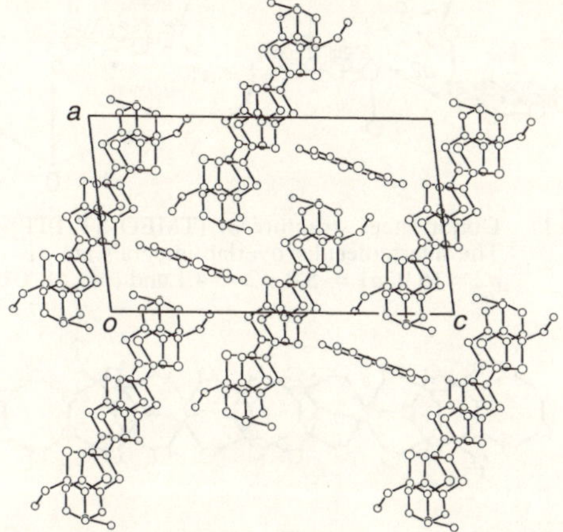

Fig. 10.9   Crystal structure of (TMEO-ST-TTP)Au(CN)₂. (Reprinted with permission from ref. 40)

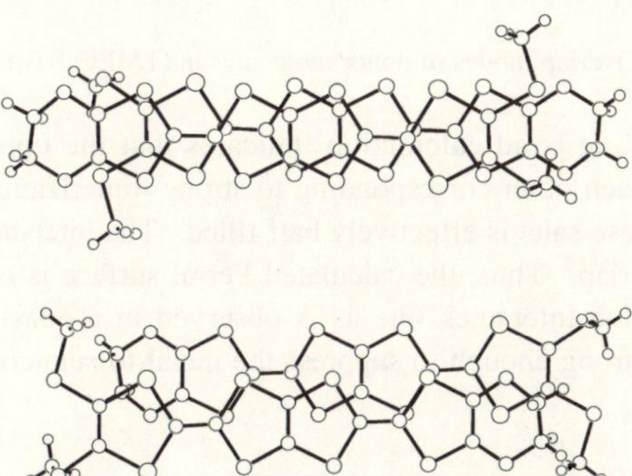

Fig. 10.10   Overlap modes of donor molecules in (TMEO-ST-TTP)Au(CN)₂.

On the other hand, TMEO-ST-TTP has yielded metallic salts down to 4.2 K with octahedral anions (PF_6^-, AsF_6^- and TaF_6^-).[40] X-Ray structure analyses reveal that they have β-type donor packing (Fig. 10.11). The overlap mode in the stack is of the ring-over-bond type (Fig. 10.12), and the slip distances along the molecular long axis are about half the 1,3-dithiole ring for $p1$ (1.7 Å), and one and a half for $p2$ (5.0 Å), respectively. As a result, the donors are electronically dimerized along the stacking direction, although interplanar distances are almost the same with each other (3.50 and 3.52 Å for the PF_6 salt). The ratio of the calculated overlap integrals ($p1/p2$)

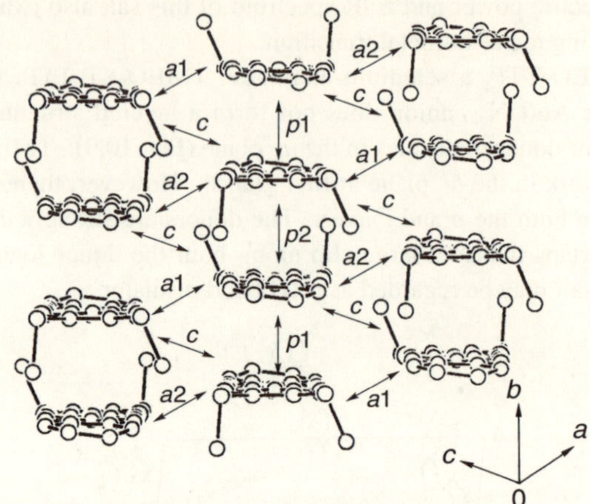

Fig. 10.11   Donor sheet structure of $(TMEO\text{-}ST\text{-}TTP)_2AsF_6$.
The intermolecular overlap integrals are $p1 = 38.1$,
$p2 = 16.9$, $a1 = -5.0$, $a2 = -4.1$ and $c = -7.8 \times 10^{-3}$.

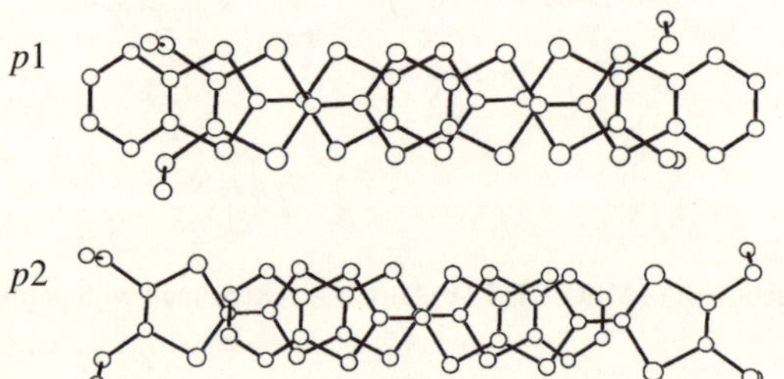

Fig. 10.12   Overlap modes of donor molecules in $(TMEO\text{-}ST\text{-}TTP)_2AsF_6$.

is 1.9–2.3. The tight-binding band calculation indicates that the upper and lower bands are completely separate from each other corresponding to strong dimerization along the stack. As a result, the upper band of these salts is effectively half filled. The interstack overlaps are 10–20% of the larger intrastack overlap. Thus, the calculated Fermi surface is closed along the stacking direction, but open along the interstack one as is observed in a quasi 1D system. However, interstack interactions are strong enough to suppress the metal-to-semiconductor transition at low temperatures.

10.3.4 Conducting Materials Based on TMET-TTP and Its Analogues

TMET-TTP and its selenium analogues TMES-TTP and TMET-TS-TTP have yielded θ-type salts with various anions even with TCNQ (Fig. 10.13).[16,17,42] The donors make pseudo stacks, and the dihedral angles between the molecular planes of the donors in adjacent stacks are as large as $118°–128°$. The θ-type salts based on TMET-TTP[42] and $(TMET\text{-}TS\text{-}TTP)_2TCNQ$[17] have uniform stacks, while the donors form a dimerized stack in $(TMES\text{-}TTP)_4I_3$.[16] The calculated Fermi surface of this I_3 salt is 2D similar to θ-type BEDT-TTF salts. These θ-type salts show metal-to-insulator transition at low temperatures, or exhibit semiconductive behavior from room temperature. Destabilization of the metallic state in TMET-TTP conductors is probably due to the

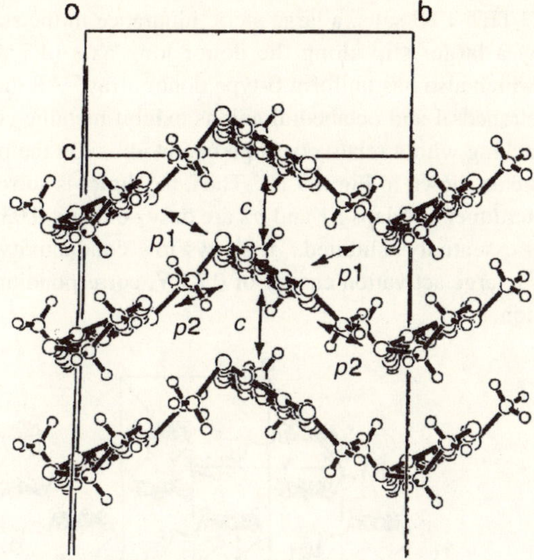

Fig. 10.13   Donor sheet structure of (TMET-TS-TTP)$_2$TCNQ. The intermolecular overlap integrals
are c = 4.9, $p1$ = 5.1 and $p2$ = 3.2 x 10^{-3}. (Reprinted with permission from ref. 17)

narrow bandwidth and resultant strong electron correlation derived from the large dihedral angles
(118°–128°) as noted in θ-BEDT-TTF salts by H. Mori and coworkers.[54]

T. Mori and coworkers have investigated BDT-TTP derivatives with long alkyl chains in
order to change the ratio of donor to anion in TMET-TTP salts, most of which have the
composition of (TMET-TTP)$_4$A. Among them, the crystal structures of (C$_2$TET-TTP)$_2$ClO$_4$,[24]
(C$_2$TET-TS-TTP)$_2$A (A = PF$_6$, ClO$_4$ and BF$_4$)[25] and (C$_4$TET-TTP)I$_3$[24] have been determined. The
radical cation salts based on C$_2$TET-TTP and C$_2$TET-TS-TTP are essentially isostructural with
each other. They have β-type donor packing contrary to the initial expectation of a θ-type
arrangement, although the donor-to-anion ratio indeed changed from 4:1 to 2:1 (Fig. 10.14). The
donors surprisingly form uniform stacks despite the presence of sterically hindered ethylthio
groups, while most donors are dimerized to avoid steric hindrance as is observed in CH-TTP (κ-
type) and TMEO-ST-TTP (dimerized β-type) salts. TMET-TTP exceptionally forms pseudo

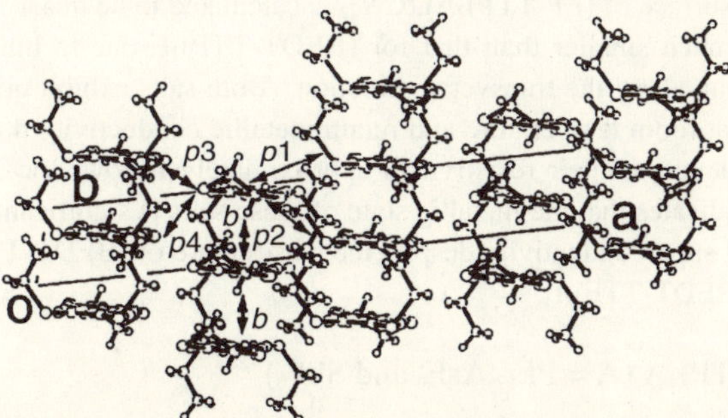

Fig. 10.14   Donor sheet structure of (C$_2$TET-TTP)$_2$ClO$_4$. The intermolecular overlap integrals are b = -20.2,
$p1$ = 7.0, $p2$ = 2.1, $p3$ = -0.9 and $p4$ = 9.6 x 10^{-3}. (Reprinted with permission from ref. 24)

stacks of the θ-type so that steric hindrance is avoidable by slip along the donor short axis. On the contrary, in the C_2TET-TTP salts, a large steric hindrance in the stack derived from the ethyl chains is reduced by a larger slip along the donor long axis (4.7 Å) compared with (BDT-TTP)$_2$ClO$_4$ (1.6 Å), which also has uniform β-type donor array. All the C_2TET-TTP and C_2TET-TS-TTP salts with tetrahedral and octahedral anions exhibit metallic conductivity down to 1.5 K thanks to uniform stacking with a relatively large bandwidth. On the other hand, (C$_4$TET-TTP)I$_3$ has a dimeric structure as shown in Fig. 10.15. Thus, the dimer is surrounded by four I$_3$ anions in the *ac* plane. The interdimer overlaps $p1$ and $p2$ are only –0.6 and 0.03 x 10^{-3}, with the result that formation of band is essentially inhibited. It shows low conductivity of 0.025 S cm^{-1} at room temperature, and has a large activation energy of 0.23 V, corresponding to its small bandwidth as well as 1:1 composition.

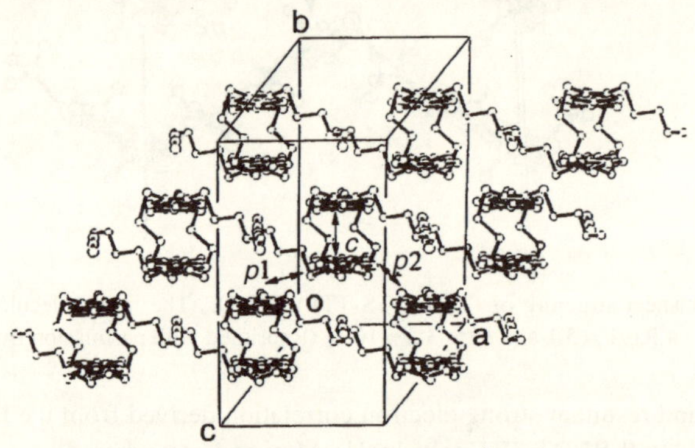

Fig. 10.15   Donor sheet structure of (C$_4$TET-TTP)I$_3$. The intermolecular overlap integrals are c = -16.6, $p1$ = -0.6 and $p2$ = 0.03 x 10^{-3}. (Reprinted with permission from ref. 24)

10.3.5 (BEDT-TTP)$_2$I$_3$ and (EP-TTP)$_2$Au(CN)$_2$

The packing pattern of donors in both salts is close to that of β-(BEDT-TTF)$_2$I$_3$.[43,44] However, the degree of dimerization in the stack is considerably smaller than in the BEDT-TTF salts, because the longer π-electron framework of BDT-TTP compared with that of TTF is less sensitive to overlap integral. The tight-binding band calculation of (BEDT-TTP)$_2$I$_3$ suggests the existence of a 2D Fermi surface,[43] which is also supported by angle- dependent magnetoresistance.[55] In contrast, the Fermi surface of (EP-TTP)$_2$Au(CN)$_2$ is calculated to be quasi 1D,[44] because side-by-side interaction is much smaller than that for (BEDT-TTP)$_2$I$_3$ due to hindrance by a projected propylenedithio group along the transverse direction. Both salts exhibit quite high conductivities of 600–900 S cm^{-1} at room temperature and retain metallic conductivity down to 0.5–0.6 K. The temperature dependence of their resistivities approximately follows the T^2 law. Such simple metallic behavior indicates that the metallic state of these salts is significantly stabilized, resulting in no observation of superconductivity despite the isostructure of (BEDT-TTP)$_2$I$_3$ with an ambient superconductor β-(BEDT-TTF)$_2$I$_3$.

10.3.6 (CPTM-TTP)$_4$A (A = PF$_6$, AsF$_6$ and SbF$_6$)

These salts have conducting sheets similar to the β-type BEDT-TTF salts.[14] Two crystallographically independent donors A and B are stacked with a four-folded period as AABB.

Thus, there are three overlap modes of donor molecules, rendering the molecular packing of those salts more closed to λ-(BETS)$_2$GaCl$_4$,[56] an ambient superconductor with $T_c = 8$ K, than the β-type BEDT-TTF salts. As shown in Fig. 10.16, the donor molecules are stacked in a head-to-tail manner in A-A ($p1$) and B-B ($p3$), while they are stacked in a head-to-head manner in A-B ($p2$). The interplanar distances in (CPTM-TTP)$_4$AsF$_6$ are 3.58, 3.61 and 3.40 Å for $p1$, $p2$ and $p3$, respectively. All of the overlap modes are of the ring-over-bond type; however, the slip distance along the molecular long axis (D) of $p3$ (4.73 Å) is longer than those of $p1$ (1.55 Å) and $p2$ (1.68 Å). Because BDT-TTP is about twice as long as TTF, the large D value of $p3$ does not affect the overlap integral so much. Thus, the overlap of $p3$ (18.1 x 10^{-3}) is smaller by only $ca.$ 20% than those of the other intrastack interactions (22.0 and 22.6 x 10^{-3}). On the other hand, there are relatively large interstack interactions (6.6 and 7.4 x 10^{-3}). As a result, the Fermi surface calculated by the tight-binding method is a typical ellipse characteristic of 2D metals. Among these salts, the PF$_6$ and AsF$_6$ salts are essentially metallic down to 4.2 K, although the maximum of resisitivity was observed around 80 K. In contrast, the SbF$_6$ salts exhibit metal-to-insulator transition around 50 K.

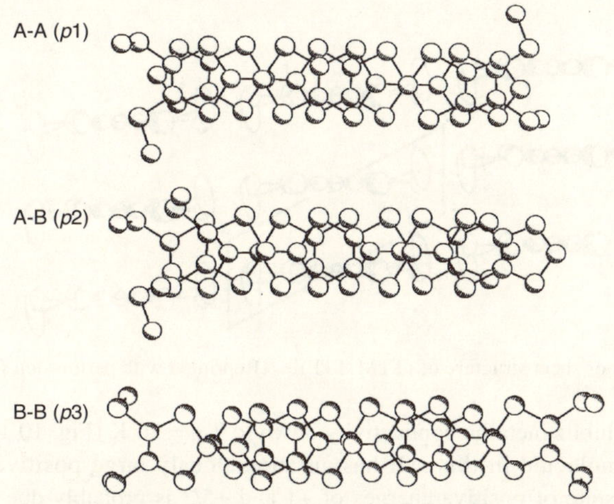

Fig. 10.16   Overlap modes of donor molecules in (CPTM-TTP)$_4$AsF$_6$.

10.3.7 (TTM-TTP)(I$_3$)$_x$

TTM-TTP has yielded three kinds of I$_3^-$ salts, (TTM-TTP)$_2$I$_3$, (TTM-TTP)I$_3$ and (TTM-TTP)(I$_3$)$_{5/3}$. (TTM-TTP)$_2$I$_3$ is a semiconductor with a trimerized column, and a neutral donor molecule is included in the anion layer.[45] (TTM-TTP)I$_3$ exhibits a very high conductivity of $\sigma_{rt} = 700$ S cm^{-1} in spite of its 1:1 composition.[5,45] It shows metal-like temperature dependence, and metal-to-insulator transition occurs around 160 K (Fig. 10.17). This is the first example of a metallic salt with a 1:1 composition.[57] The array of donors may be classified as β-type (Fig. 10.18). However, effective side-by-side interaction is inhibited due to methylthio groups jutting out from the molecular long axis. The donor molecules form uniform stacks with a relatively short interplanar distance of 3.45 Å. Tight-binding band calculation suggests a large bandwidth of $4t = 1$ eV. The nature of the insulating state of (TTM-TTP)I$_3$ has been actively investigated, but is not clear yet.[58] On the other hand, (TTM-TTP)(I$_3$)$_{5/3}$ also has a uniform donor column similar to

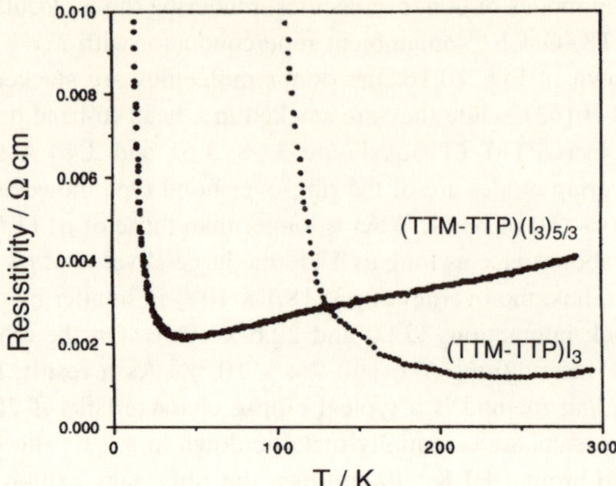

Fig. 10.17   Conducting behavior of (TTM-TTP)I₃ and (TTM-TTP)(I₃)₅/₃.

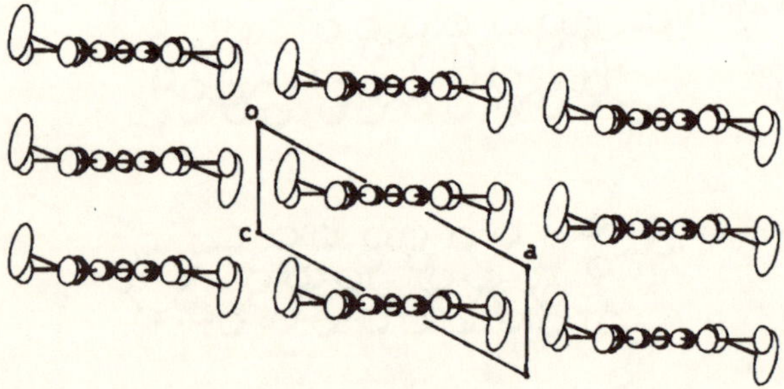

Fig. 10.18   Donor sheet structure of (TTM-TTP)I₃. (Reprinted with permission from ref. 45)

(TTM-TTP)I₃, and exhibits metallic conductivity down to T_{MI} = 20 K (Fig. 10.17).[45,46] It is noted that the TTM-TTP molecule in this salt has an anomalously large positive charge of +5/3. Metallic behavior in spite of positive charges of +1 and +5/3 is probably due to a small on-site Coulomb energy as well as a large bandwidth. The AuI_2 and $MX_4(PhCl)_{0.5}$ (M = Fe and Ga; X = Cl and Br) anions also afford 1:1 salts; however, such anions shorter than I_3^- form dimerized stacks.[47-49] As a result they show semiconductive behavior characteristic of band insulators. On the other hand, the alloy system (TTM-TTP)FeBr₁.₈Cl₂.₂ exhibits metallic behavior similar to that of (TTM-TTP)I₃, because it occasionally has a uniform donor column.[48]

10.4   Analogues of Bis-fused TTFs

π-Electron systems modifying the BDT-TTP skeleton are of significant interest from the viewpoint of the search for novel conducting materials. Various bis-fused donors possessing at least one non-TTF π-system have been synthesized.[59-64] Among them, the synthetic route of a vinylogous TTP (DTEDT) derivative (**22**) is outlined in Scheme 10.4.[60,64] Construction of a π-electron framework has been achieved by the successive introduction of DT units to the central tetrathiapentalene moiety using Wittig-Horner reaction followed by phosphite-mediated cross-coupling. Namely, an important precursor **23**,[64,65] which has reactive points for both phosphite

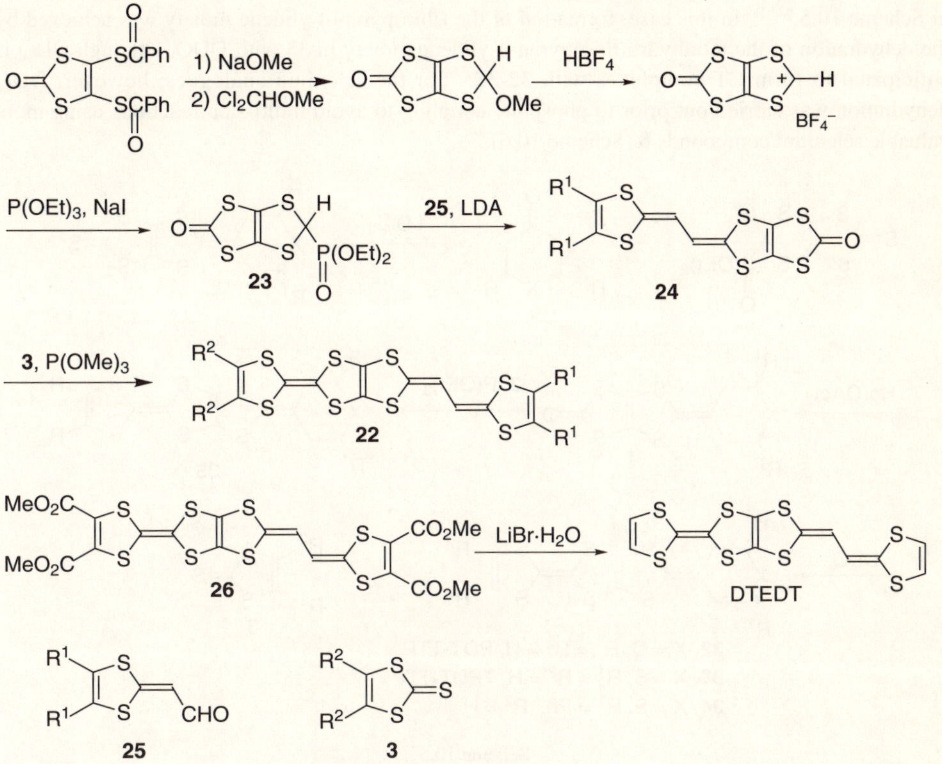

Scheme 10.4

Chart 10.4

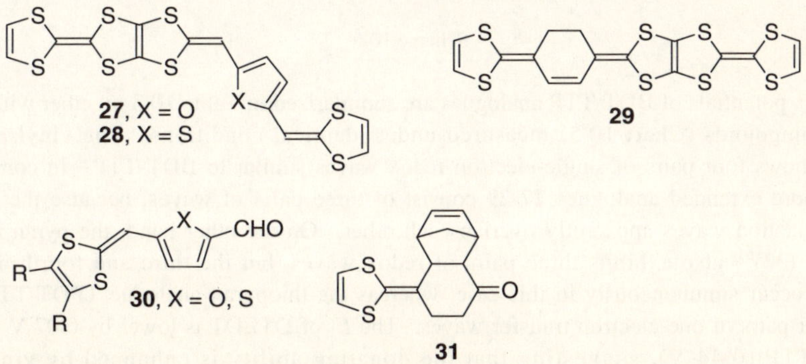

coupling and Wittig-Horner reaction, was converted to 1,3-dithiol-2-one fused with vinylogous TTFs (**24**) by Wittig-Horner reaction with aldehyde **25**. Then, **24** and appropriate 1,3-dithiole-2-thiones **3** were cross-coupled by using P(OMe)$_3$ in refluxing toluene to give **22**. The unsubstituted derivative DTEDT was prepared by heating tetraester (**26**) with an excess of LiBr·H$_2$O in HMPA. The related donors **27-29** (Chart 10.4)[60,61)] have also been synthesized by methods similar to those used for DTEDT, and aldehydes **30** or ketone **31** have been used instead of **25**. The pyran and thiopyran analogues **32-34** have also been synthesized similarly as outlined in Scheme 10.5.[62,63] In this case, formation of the (thio)pyran-4-ylidene moiety was achieved by the dehydration of the tetrahydro(thio)pyran-4-ylidene moiety in **35** with DDQ, although DDQ is anticipated to form CT complexes with **32-35**. For the selenium analogues, however, DDQ dehydration was carried out prior to phosphite coupling to avoid multi-step reactions using more valuable selenium compounds **8** (Scheme 10.6).[66]

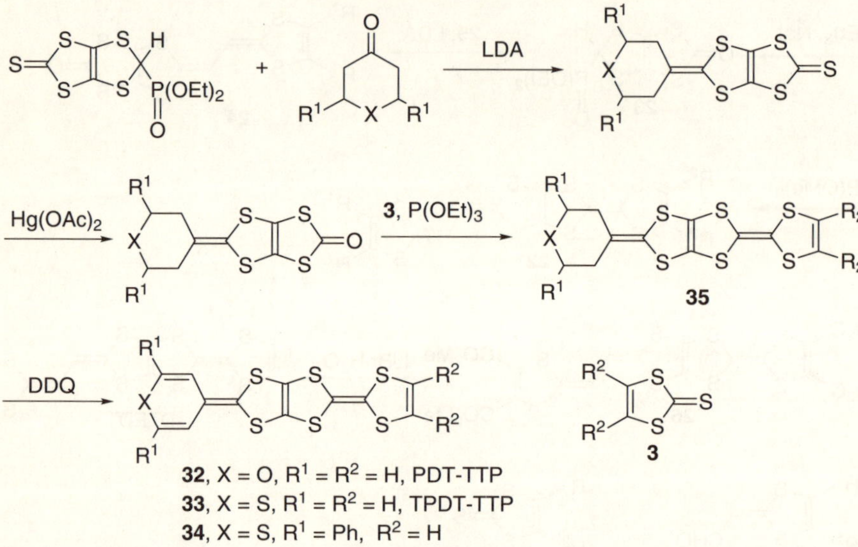

32, X = O, R^1 = R^2 = H, PDT-TTP
33, X = S, R^1 = R^2 = H, TPDT-TTP
34, X = S, R^1 = Ph, R^2 = H

Scheme 10.5

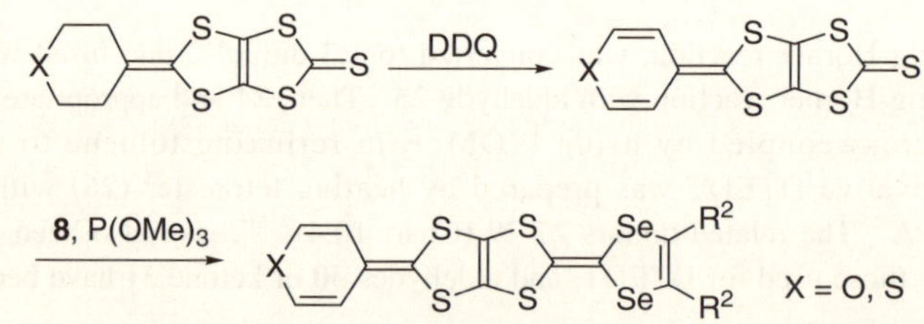

X = O, S

Scheme 10.6

Redox potentials of BDT-TTP analogues are summarized in Table 10.3 together with several related compounds (Chart 10.5) measured under identical conditions. The vinyl analogue DTEDT shows four pairs of single-electron redox waves similar to BDT-TTP. In contrast, the CVs of more extended analogues **27-29** consist of three pairs of waves, because the first and second oxidation waves apparently overlap each other. On the other hand, the pyran analogue PDT-TTP (**32**)[63] also exhibits three pairs of redox waves, but the third and fourth oxidation processes occur simultaneously in this case, whereas the thiopyran analogue TPDT-TTP (**33**)[62] shows four pairs of one-electron transfer waves. The E_1 of DTEDT is lower by 0.07 V than that of BDT-TTP (0.44 V), suggesting that the donating ability is enhanced by vinylogous modification. The negative shift of the E_1 value is also observed in the furan- (**27**)[60] or

Table 10.3   Redox Potentials of BDT-TTP Analogues (V *vs.* SCE)[a]

	E_1	E_{m1}	E_2	E_3	E_{m2}	E_4	ΔE
DTEDT	0.37		0.50	0.81		1.05[b]	0.13
36	0.38		0.52	0.86		1.07[b]	0.14
37	0.43		0.54	0.83		1.00	0.11
27		0.36		0.75		1.01	-
28		0.47		0.79		1.01	-
29		0.39		0.79		1.22	-
PDT-TTP, **32**	0.42		0.63		1.07[b]		0.21
TPDT-TTP, **33**	0.37		0.60	0.94		1.11	0.23
34	0.35		0.58	0.89		1.15	0.23
38	0.38		0.64	1.07		1.30	0.26
39	0.36		0.61	0.89		1.12	0.25
BDT-TTP	0.44		0.64	1.05[b]		1.13[b]	0.18
TTF	0.35		0.77				0.42
40	0.29		0.49				0.20
41	0.37		0.52				0.15
42		0.29					-
43		0.39					-
44		0.30					-
45	0.29		0.73				0.44
46	0.28		0.69				0.41
47	0.27		0.63				0.36
48	0.50		0.80				0.30
49	0.34		0.63				0.29

[a] Measured in PhCN containing 0.1 M Bu_4NClO_4 at 25 °C with scan rate of 50 mV/s.
[b] Irreversible step (an anodic peak potential).

cyclohexene-inserted TTPs (**29**)[61] and the thiopyran analogue (**33**), while **28**[60] and **32** show no significant enhancement of donating ability. However, the E_1 values of these BDT-TTP analogues are higher by 0.02–0.12 V and 0.07–0.13 V compared with those of each donor unit of TTF (0.35 V) and analogous TTF **40**[67] and **42-44**[67,68] (0.29–0.39 V). For example, the E_1 of

Chart 10.5

DTEDT is almost equal to that of 4,5-bis(methylthio)-**40** (**41**). These results indicate that the donating ability becomes somewhat weaker in spite of an apparent extension of π-conjugation by the fusion of two donor units as is observed in BDT-TTP. The E_1 values of DTEDT derivatives are susceptible to substitution of functional groups on the 1,3-dithiole rings in their vinylogous TTF moieties. For instance, increase of the E_1 value due to bis(methylthio) substitution is 0.06 V, which is almost equal to the difference in the E_1 values between **40** and **41** (0.08 V). In contrast, they are little affected by the introduction of substituents on the 1,3-dithiole rings in their TTF moieties. Thus, bis(methylthio) substitution raises the E_1 value only by 0.01 V. On the other hand, **38**, which has two strong electron-withdrawing methoxycarbonyl groups, shows the E_1 value higher by only 0.03 V compared with the corresponding unsubstituted derivative (**34**). As a result, the donating ability of **38** is still equal to that of the parent TPDT-TTP (**33**), and is superior to BEDT-TTF (E_1 = 0.51 V). However, a half unit of **38** (*i.e.*, **48**)[69] exhibits a large substituent effect of 0.22 V by the introduction of two methoxycarbonyl groups. Based on these results, it is concluded that the positive charge in the radical cation of BDT-TTP analogues lies mainly in the analogous TTF unit, possessing more powerful donating ability than TTF, and that the TTF part plays a role as a sulfur-based substituent rather than the π-conjugation system participating in the first one-electron redox. On the other hand, the ΔE of DTEDT (0.13 V) is smaller by 0.05–0.29 V than those of BDT-TTP (0.18 V), TTF (0.42 V) and **40** (0.20 V). This suggests a decrease in the on-site Coulomb repulsion in the dication by the delocalization of two positive charges on the entire molecule. The (thio)pyran analogues **32** and **33** exhibit the ΔE values of 0.21 and 0.23 V, respectively, being smaller by 0.18–0.23 V compared with TTF, **45** and **46** (0.41–0.44 V).

10.5 Radical Cation Salts Based on DTEDT and DSEDS

Vinylogous TTP donor DTEDT has a crooked structure and π-orbitals of sulfur atoms at the terminal of the vinylogous 1,3-dithiole ring are out of phase with those of the other sulfur atoms in HOMO (Fig. 10.19). Therefore, it is anticipated to cause both sterically and electronically less effective side-by-side interaction in the CT salts compared with strip-like BDT-TTP in which all sulfur atoms have the same phase. In spite of the above anticipation, similar to BDT-TTP, DTEDT affords many metallic radical cation salts regardless of the size and shape of counteranions.[59] Furthermore, (DTEDT)$_3$Au(CN)$_2$ exhibits superconducting transition at 4 K under ambient pressure (Fig. 10.20).[59] The crystal structure of (DTEDT)$_3$Au(CN)$_2$ is shown in Fig. 10.21. The donors form conducting sheets parallel to the *ac* plane, and the packing pattern of donors is classified as the β-type similar to BDT-TTP salts (Fig. 10.22). In contrast to most β-type salts based on unsymmetrical donors, DTEDT molecules are arranged so the stacking and transverse directions are parallel. Because DTEDT has a crooked structure and the phase of sulfur atoms in HOMO is reversed at the terminal vinylogous 1,3-dithiole, such a parallel array results in the most effective intermolecular interaction both sterically and electronically.

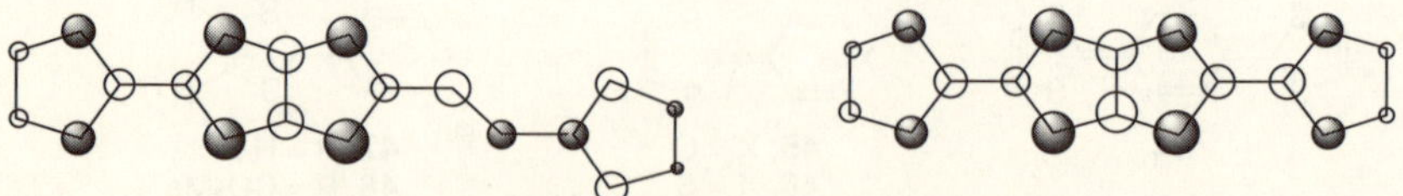

Fig. 10.19   HOMOs of DTEDT (left) and BDT-TTP (right).

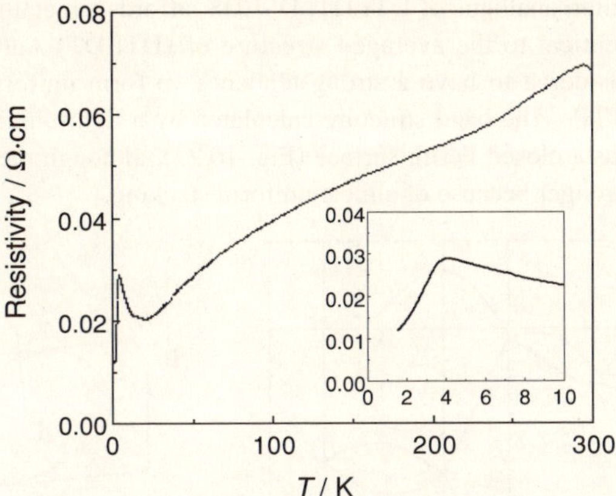

Fig. 10.20   Conducting behavior of (DTEDT)₃Au(CN)₂. (Reprinted with permission from ref. 59)

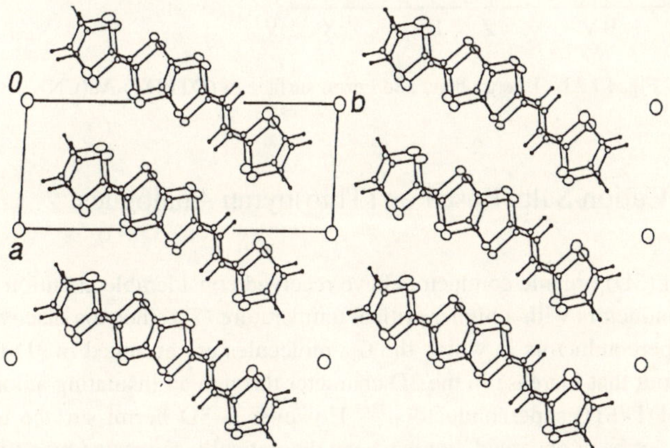

Fig. 10.21   Crystal structure of (DTEDT)₃Au(CN)₂. (Reprinted with permission from ref. 59)

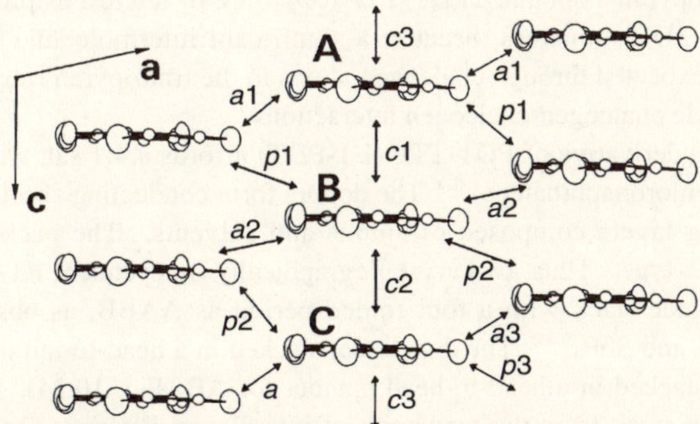

Fig. 10.22   Donor sheet structure of (DTEDT)₃Au(CN)₂. The intermolecular overlap integrals are $c1 = -19.2$, $c2 = -19.9$, $c3 = -10.8$, $a1 = -1.5$, $a2 = -1.3$, $a3 = -0.7$, $p1 = -3.8$, $p2 = -3.4$ and $p3 = -6.9 \times 10^{-3}$. (Reprinted with permission from ref. 59)

On the other hand, a selenium analogue of DTEDT, DSEDS, affords a metallic TaF_6 salt in which the array of donors is identical to the averaged structure of $(DTEDT)_3Au(CN)_2$.[70] Therefore, DTEDT may also be considered to have a strong tendency to form uniform β-type molecular packing similar to BDT-TTP. The band structure calculated by a tight-binding method suggests that $(DTEDT)_3Au(CN)_2$ has a closed Fermi surface (Fig. 10.23), although the 1D character along the stacking direction is stronger because of almost uniform stacking.

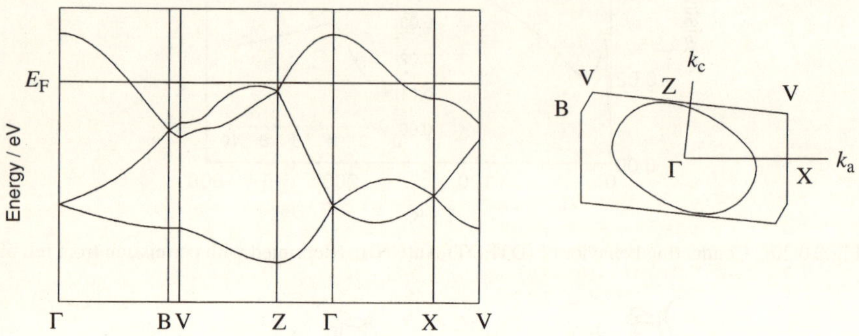

Fig. 10.23   Energy band and Fermi surface of $(DTEDT)_3Au(CN)_2$.

10.6   Radical Cation Salts Based on (Thio)pyran Analogues

Three-dimensional (3D) organic conductors have received considerable attention in the search for molecular superconductors with a higher critical temperature (T_c) since the discovery of M_3C_{60} (M = alkali metal) superconductors in which the C_{60} molecules are arranged in 3D fashion.[71] It has also been pointed out that increase in the 3D character through an insulating anion layer results in a higher T_c in BEDT-TTF superconductors.[72] However, a 3D Fermi surface based on organic molecules has not yet been observed, but has been theoretically suggested by a tight-binding band calculation in a metal dithiolene system, $Me_xH_{4-x}N[Pd(dmise)_2]_2$ (x = 2, 3),[73] where dmise is 4,5-dimercapto-1,3-dithiole-2-selone. A TTP-type donor containing a pyran-4-ylidene moiety, PDT-TTP (**32**),[63] and its thiopyran analogue TPDT-TTP (**33**)[62] are of interest as promising candidates for the preparation of 3D conductors, because a significant intermolecular overlap along the molecular long axis is expected through chalcogen atoms in the (thio)pyran ring in addition to π-π stacking and side-by-side chalcogen-chalcogen interactions.

The ethylenedithio derivative of PDT-TTP (ET-PDT) affords a 4:1 salt with PF_6^- anion (ET-PDT)$_4$PF$_6$(cn) (cn = 1-chloronaphthalene).[63] The donors form conducting sheets, each of which is separated by insulating layers composed of anions and solvents. The packing pattern of the donors is classified as λ-type. Thus, two crystallographically independent ET-PDT molecules A and B form a face-to-face stack with a four-folded period as AABB, as observed in (CPTM-TTP)$_4$A (A = PF$_6$, AsF$_6$ and SbF$_6$).[14] The donors are stacked in a head-to-tail manner for AA and BB, whereas they are stacked in a head-to-head manner for AB (Fig. 10.24). This salt seems to have an almost uniform stack from the viewpoint of interplanar distances (3.60–3.62 Å) at first glance. However, the head-to-tail overlap between the unsymmetrical π-electron frameworks of ET-PDT prevents an effective intrastack overlap, because the phases of the molecular coefficients between the outer 1,3-dithiole ring and the pyran ring are not consistent. In fact, the calculated

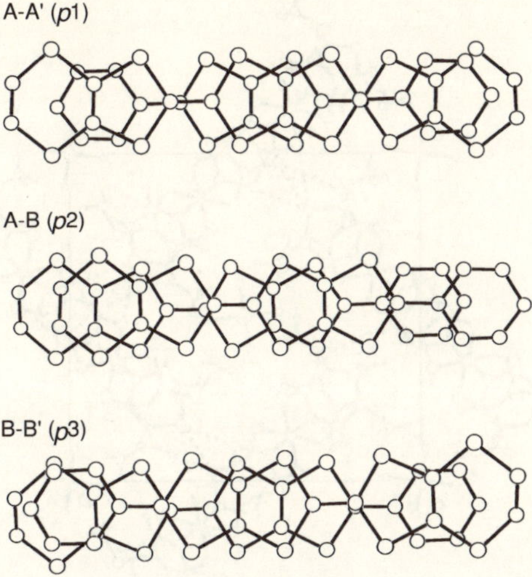

Fig. 10.24   Overlap modes of the donor molecules in (ET-PDT)$_4$PF$_6$(cn).

overlap integrals of the head-to-tail stack, $p1$ (8.6 x 10^{-3}) and $p3$ (4.9 x 10^{-3}), are only about one half or one quarter the head-to-head overlap $p2$ (-16.4 x 10^{-3}). Therefore, (ET-PDT)$_4$PF$_6$(cn) may be considered to have a strongly tetramerized electronic structure along the stack direction. Such difference in the manner of overlap does not affect the overlap integrals in (CPTM-TTP)$_4$A (19.7–21.3 x 10^{-3} for A = PF$_6$) having the donor stacks with the symmetrical π-electron frameworks of BDT-TTP. ET-PDT forms a highly conducting TCNQ complex and radical cation salts with octahedral and linear anions (σ_{rt} = 2–80 S cm^{-1}), all of which exhibit metallic temperature dependence down to 1.5–4.2 K.

In contrast, the radical cation salt of a methylthio derivative TM-TPDS adopts quite different donor packing from (ET-PDT)$_4$PF$_6$(cn). Fig. 10.25 shows the crystal structure of (TM-TPDS)$_2$AsF$_6$.[74] This salt does not adopt a 2D conducting sheet in contrast to the usual TTP conductors. Instead, the donors are arranged in a "windmill" manner in the bc plane and are stacked along the a axis. The anions are located in a cavity formed in the center of the windmill, and are surrounded by hydrogen atoms in the methylthio groups and thiopyran rings in the donors. The unsymmetrical TM-TPDS molecules are stacked alternately along the a axis, and are slightly dimerized. There are many sulfur···sulfur contacts shorter than the sum of van der Waals radii (3.60 Å) between the central tetrathiapentalene moiety and the thiopyran ring or methylthio groups. Due to the relatively large molecular orbital coefficient of the sulfur atom in the thiopyran ring as well as short sulfur···sulfur contacts, the calculated overlap integrals between columns are as large as approximately 10% of the intracolumn overlaps. However, the calculated Fermi surface is unfortunately closed only along the k_a direction because intrastack interactions are still dominant in the present salt. (TM-TPDS)$_2$AsF$_6$ showed a high conductivity of σ_{rt} = 240 S cm^{-1} and exhibited metal-like temperature dependence down to 100 K. Below this temperature, the resisitivity gradually increased. This metal-to-semiconductor transition should not be a Peierls transition derived from a highly 1D character because no obvious decrease of the magnetic susceptibility was observed below T_{MI}.

The above results indicate exchange of sulfur (or oxygen) atoms with seleniums on the tetrathiapentalene, methylthio and (thio)pyran-4-ylidene moieties, raising much expectation for

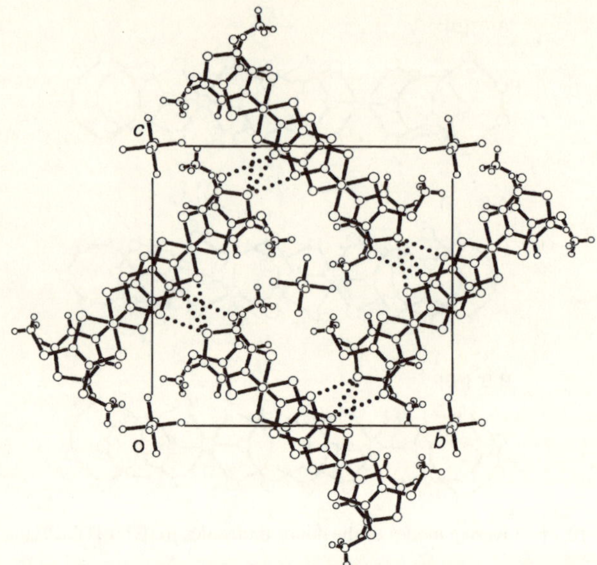

Fig. 10.25   Crystal structure of (TM-TPDS)$_2$AsF$_6$ onto the bc plane showing intermolecular S···S contacts (≤ 3.60 Å) with dotted lines. (Reprinted with permission from ref. 74)

the formation of a windmill-type structure with higher 3D character. Thus, bis(methylseleno) derivatives have been investigated.[75,76] However, SM-PDT forms a 1:1 salt with octahedral anion (SM-PDT)PF$_6$(PhCl)$_x$,[75] in which the packing pattern of the donors is quite different from that of (TM-TPDS)$_2$AsF$_6$. The donor columns are completely divided by counteranions and disordered solvents along the a axis (Fig. 10.26). Thus, no side-by-side interaction is observed for this salt. Instead, short selenium···selenium contacts within the sum of the van der Waals radii are observed between methylseleno moieties. The donor columns are bridged by these Se···Se interactions to form a 2D sheet-like structure. The Se···Se contacts between methylseleno groups in this salt may be more preferable than the edge-to-side interactions through the tetrathiapentalene moiety as is seen in (TM-TPDS)$_2$AsF$_6$. The donor molecules are dimerized in the stack with slip distances along the donor long axis of 1.4 and 4.7 Å. Interstack overlap integral through methylseleno

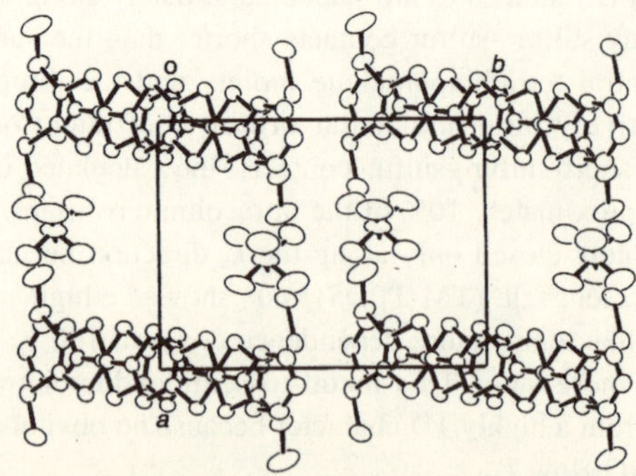

Fig. 10.26   Crystal structure of (SM-PDT)PF$_6$(PhCl)$_x$ viewed along the c axis. Disordered solvents are located on the cavity.

groups is approximately one fourth as large as the intrastack ones (Fig. 10.27) and is significantly as large as the longitudinal interaction. The band calculation suggests that this salt is a band insulator because of its 1:1 composition and dimerized column. However, it shows a relatively high conductivity of σ_{rt} = 4.8 S cm^{-1}. Considering such high conductivity in spite of its being a 1:1 salt as well as the relatively large edge-to-edge interaction, the 2:1 salts with the same donor array are expected to exhibit metallic conductivity down to low temperatures.

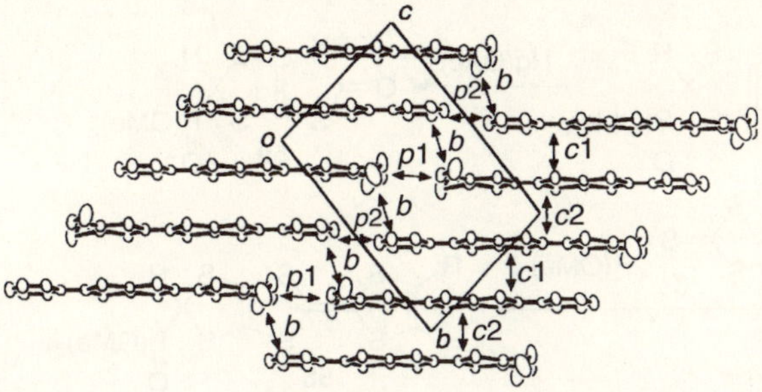

Fig. 10.27   Donor column structure viewed along the a axis and definition of the overlap integrals of (SM-PDT)PF$_6$(PhCl)$_x$. The intermolecular overlap integrals are $c1$ = 6.3, $c2$ = 8.1, $p1$ = -2.3, $p2$ = 0.00 and b = -0.8 ($\times 10^{-3}$). (Reprinted with permission from ref. 75)

10.7   Tris-fused TTFs

The higher homologues of BDT-TTP are also of interest not only as promising components for molecular metals but also as candidates for molecular electronics. The tris-fused TTFs (**50**, Chart 10.6) have been prepared by the P(OEt)$_3$-mediated coupling reaction of 1,3-dithiol-2-one fused with TTFs (**5**).[77] On the other hand, derivatives of the thiopyran analogue **51** have been obtained by the cross-coupling reaction between **5** and **52** with P(OMe)$_3$.[78] The donors **51** have also been synthesized *via* the phosphite-mediated cross-coupling reaction between **5** and **53** followed

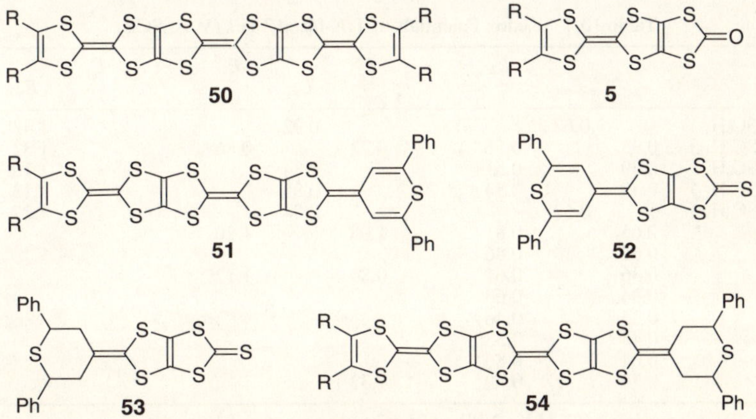

Chart 10.6

by dehydration of the resulting **54** with DDQ.[78] On the other hand, synthesis of extended-TTF analogues **55-57** in which an aromatic ring is inserted in the central TTF moiety have been achieved using a different procedure as outlined in Scheme 10.7.[79] Thus, MDT-TTF phosphonate esters **58**, which have been prepared by the cross-coupling reaction of ketone **59** with 4,5-bis(alkylthio)-1,3-dithiole-2-thiones **3**, can be used as important precursors. The target molecules have been obtained in 53–75% yields by Wittig-Horner reaction of dicarboxaldehyde **60** and **58** with LDA.

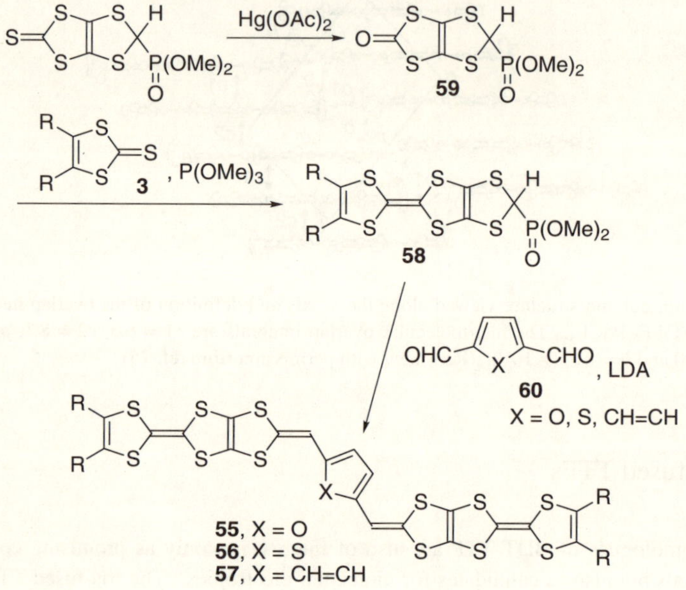

Scheme 10.7

Redox potentials of bis- or tetrakis(hexylthio) derivatives of tris-fused donors measured by CV are summarized in Table 10.4. The CV of **50** is composed of three pairs of redox waves, each

Table 10.4   Redox Potentials of Tris-fused TTFs (V *vs.* SCE)[a]

	E_1	E_{m1}	E_2	E_3	E_{m2}	E_4	E_5	E_{m3}	E_6
50 (R = SC$_6$H$_{13}$)[b]		0.67			0.92			1.43[c]	
51 (R = SC$_6$H$_{13}$)	0.35		0.56	0.72		0.86[c]		1.32	
55 (R = SC$_6$H$_{13}$)	0.39		0.51		0.84			1.10	
56 (R = SC$_6$H$_{13}$)	0.45		0.54		0.85			1.13	
57 (R = SC$_6$H$_{13}$)		0.52			0.84			1.27	
61	0.65		0.81	1.08		1.20			
62	0.62		0.86						
39	0.36		0.61	0.89		1.12[b]			
49	0.34		0.63						
63	0.35		0.46						
64	0.45		0.53						
65	0.64[b]		0.86[b]						
66	0.54		0.89	1.32					

[a]Measured in PhCN containing 0.1 M Bu$_4$NPF$_6$ at 25 °C with scan rate of 100 mV/s.
[b]Measured in CS$_2$-PhCN (1:1, v/v) containing 0.1 M Bu$_4$NBF$_4$ at 25 °C with scan rate of 50 mV/s.
[c]Irreversible step (an anodic peak potential).

of which corresponds to a two-electron transfer. On the contrary, the thiopyran analogue **51** shows five pairs of redox waves, and only the highest voltage wave corresponds to a two-electron transfer. Among the aromatic inserted donors, the furan and thiophene derivatives **55** and **56** showed two one-electron oxidation waves and successively two stages of two-electron redox waves, while the benzene derivative **57** exhibited three pairs of two-electron redox waves. The order of the first redox potentials (**55** < **56** < **57**) is consistent with that of aromaticity in the central aromatic rings. On the other hand, the ΔE values of **55** and **56** are comparable to those of **63** and **64** (Chart 10.7), indicating that two positive charges in $\mathbf{55}^{2+}$ and $\mathbf{56}^{2+}$ distribute mainly on the central extended-TTF moiety as is often observed in extended-TTP donors. In contrast, the redox potentials of the benzene derivative **57** are quite similar to those of the phenyl-substituted DT-TTF **66**. The above results strongly indicate that the two positive charges in $\mathbf{57}^{2+}$ are located on the outer TTF units so as to preserve the aromaticity of the central benzene ring and to avoid steric hindrance between the *peri*-hydrogen atoms on the benzene ring and the 1,3-dithiole rings as a result of the formation of a planar quinoid structure. Therefore, it may be concluded that the benzene derivative **57** should be regarded as a DT-TTF dimer rather than a tris-fused TTF system.

All the tris-fused TTF derivatives **50** afford highly conducting I_3^- salts (σ_{rt} = 0.7–16 S cm^{-1} on compressed pellets).[77] Several of them have very small activation energies of 0.017–0.033 eV, indicating metallic conductivity on a single crystal. The conducting materials based on the other tris-fused donors **51** and **55-57** also exhibit relatively high conductivities of σ_{rt} = 10^{-1}–10^{0} S cm^{-1} on compressed pellets.[78,79]

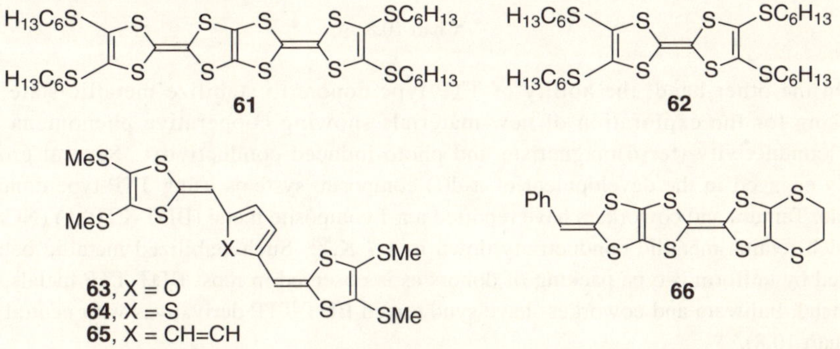

Chart 10.7

10.8   Summary and Outlook

In summary, BDT-TTP derivatives and their analogues have yielded a great number of radical cation salts and CT complexes, retaining metallic conductivity down to low temperatures ($\leq$ 4.2 K). Furthermore, the TTP vinylogue DTEDT has yielded an organic superconductor (DTEDT)$_3$Au(CN)$_2$, although its half unit of vinylogous TTF (**40**, Chart 10.5) and its derivatives have never yielded any metallic salts down to low temperature. Stabilization of the metallic state in TTP conductors is mainly achieved by increase in dimensionality due to side-by-side interaction through chalcogen atoms in the tetrathiapentalene moiety as well as those in the outer DT units. Furthermore, because of a lower Coulombic repulsion than TTF or **45**(Chart 10.5), **32**(Scheme 10.5) is also preferable for avoiding the Mott insulating ground state and retaining metallic conductivity. Over 130 TTP-type materials exhibit metallic behavior down to liquid helium temperatures. However, only one superconductor has been obtained so far, suggesting that strong stabilization of the metallic state is disadvantageous to inducing superconducting transition. Recently, it has been pointed out by Kanoda that the superconducting state lies between the metal and antiferromagnetic Mott insulating state.[80] Therefore, for the construction of a strong electron-correlated system it is desirable to enhance the on-site Coulomb repulsion or to narrow the bandwidth against stable metals down to low temperatures. For TTP systems, the introduction of strong dimerization and/or a head-to-tail stacking of the unsymmetrical π-electron framework are promising for the construction of narrow bands. On the other hand, reduction of the π-electron system or 2D molecular packing with a completely oxidized state (+1) is important to increase the on-site Coulomb repulsion. In fact, Yamada and coworkers have recently found that a BDT-TTP analogue composed of a reduced π-electron system BDA-TTP (Chart 10.8) has yielded superconductors (BDA-TTP)$_2$A (A = PF$_6$, AsF$_6$, SbF$_6$),[81] in which the donors form strongly dimerized β-type conducting sheets. Therefore, more organic superconductors may be prepared from TTP-type donors by appropriate modification of BDT-TTP.

BDA-TTP   67

R = H, 2R = -(CH$_2$)$_3$-

Chart 10.8

On the other hand, the ability of TTP-type donors to stabilize metallic state may be promising for the exploration of new materials showing cooperative phenomena such as (super)conductivity-(ferro)magnetism and photo-induced conductivity. Several groups are actually engaged in the development of π-d(f) composite systems using TTP-type donors. For example, Tamura and coworkers have reported a π-f composite metal (BDT-TTP)$_x$[Y(NCS)$_6$] (x $\approx$ 8), which retains metallic conductivity down to 1.7 K.[82] Such stabilized metallic behavior is achieved by uniform β-type packing of donors as is observed in most BDT-TTP metals. On the other hand, Fujiwara and coworkers have synthesized BDT-TTP derivatives with neutral radicals (**67**, Chart 10.8).[83]

The results mentioned here give rise to much expectation that the fusion of two 1,3-dithiole donors, *i.e.*, the insertion of a tetrathiapentalene moiety into a π-electron donor, is a reliable guiding principle for stabilization of the metallic state in conducting organic solids. In the future, modification or utilization of bis-fused donors will lead to the discovery of novel materials exhibiting exotic properties such as high T_c superconductors, ferromagnetic metals, magnetic superconductors, photo-induced (super)conductors and conducting parts for molecular electronics.

References

1. J. M. Williams, J. R. Ferraro, R. J. Thorn, K. D. Carlson, U. Geiser, H. H. Wang, A. M. Kini and M.-H. Whangbo, *Organic Superconductors*, Prentice Hall, New Jersey (1992); T. Ishiguro, K. Yamaji and G. Saito, *Organic Superconductors*, 2nd ed., Springer Ser. Solid-State Sci., Vol. 88, Springer, Berlin (1998).
2. M. Adam and K. Müllen, *Adv. Mater.*, **6**, 439 (1994); T. Otsubo, Y. Aso and K. Takimiya, *Adv. Mater.*, **8**, 203 (1996); J. Becher, J. Lan and P. Mørk, *Electronic Materials: The Oligomer Approach*, Wiley-VCH, Weinheim (1998), p. 198; J. L. Segura and N. Martin, *Angew. Chem. Int. Ed. Engl.*, **40**, 1409 (2001).
3. G. Saito, T. Enoki, K. Toriumi and H. Inokuchi, *Solid State Commun.*, **42**, 557 (1982).

4. R. R. Schumaker and E. M. Engler, *J. Am. Chem. Soc.*, **102**, 6652 (1980).
5. Y. Misaki, H. Nishikawa, K. Kawakami, S. Koyanagi, T. Yamabe and M. Shiro, *Chem. Lett.*, 2321 (1992).
6. Y. Misaki, T. Matsui, K. Kawakami, H. Nishikawa, T. Yamabe and M. Shiro, *Chem. Lett.*, 1337 (1993).
7. K.B. Simonsen, N. Svensrup, J. Lau, O. Simonsen, P. Mork, G. J. Kristensen and J. Becher, *Synthesis*, 407 (1996).
8. M. Aragaki, T. Mori, Y. Misaki, K. Tanaka and T. Yamabe, *Synth. Met.*, **102**, 1601 (1999).
9. Y. Misaki, T. Kochi, T. Yamabe and T. Mori, *Adv. Mater.*, **10**, 588 (1998).
10. Y. Misaki, H. Nishikawa, T. Yamabe, T. Mori, H. Inokuchi, H. Mori and S. Tanaka, *Chem. Lett.*, 729 (1993).
11. Y. Misaki, H. Nishikawa, K. Kawakami, T. Yamabe, T. Mori, H. Inokuchi, H. Mori and S. Tanaka, *Chem. Lett.*, 2073 (1993).
12. Y. Misaki, K. Kawakami, T. Matsui, T. Yamabe and M. Shiro, *J. Chem. Soc., Chem. Commun.*, 459 (1994).
13. T. Mori, Y. Misaki, K. Kawakami, T. Yamabe, H. Mori and S. Tanaka, *Synth. Met.*, **70**, 875 (1995).
14. Y. Misaki, K. Kawakami, H. Fujiwara, T. Miura, T. Kochi, M. Taniguchi, T. Yamabe, T. Mori, H. Mori and S. Tanaka, *Mol. Cryst. Liq. Cryst.*, **296**, 77 (1997).
15. M. Ashizawa, M. Aragaki, T. Mori, Y. Misaki and T. Yamabe, *Chem. Lett.*, 649 (1997).
16. T. Mori, M. Ashizawa, M. Aragaki, K. Murata, Y. Misaki and K. Tanaka, *Chem. Lett.*, 253 (1998).
17. M. Aragaki, H. Hoshino, T. Mori , Y. Misaki, K. Tanaka, H. Mori and S. Tanaka, *Adv. Mater.*, **12**, 983 (2000).
18. T. Mori, T. Kawamoto, Y. Misaki, K. Tanaka, H. Mori and S. Tanaka, *Bull. Chem. Soc. Jpn.*, **71**, 1321 (1998).
19. Y. Misaki, K. Tanaka, M. Taniguchi, T. Yamabe and T. Mori, *Adv. Mater.*, **11**, 25 (1999).
20. M. Taniguchi, Y. Misaki, K. Tanaka, T. Yamabe and T. Mori, *Synth. Met.*, **102**, 1721 (1999).
21. Y. Misaki, T. Kaibuki, T. Monobe, K. Tanaka, M. Taniguchi, K. Tanaka, T. Yamabe, K. Takimiya, A. Morikami, T. Otsubo and T. Mori, *Synth. Met.*, **102**, 1781 (1999).
22. M. Taniguchi, T. Miura, Y. Misaki, K. Tanaka, T. Yamabe and T. Mori, *Mol. Cryst. Liq. Cryst.*, **380**, 151 (2002).
23. K. Takahashi, K. Tanaka, M. Taniguchi, Y. Misaki, K. Tanaka and T. Mori, *Synth. Met.*, **120**, 955 (2001).
24. S. Kimura, H. Kurai, T. Mori, H. Mori and S. Tanaka, *Bull. Chem. Soc. Jpn.*, **74**, 59 (2001).
25. M. Aragaki, S. Kimura, M. Katsuhara, H. Kurai and T. Mori, *Bull. Chem. Soc. Jpn.*, **74**, 833 (2001).
26. S. Kimura, S. Hanazato, H. Kurai, T. Mori, Y. Misaki and K. Tanaka, *Tetrahedron Lett.*, **42**, 5729 (2001).
27. K. Tanaka, M. Taniguchi, Y. Misaki, T. Yamabe and T. Mori, *Synth. Met.*, **102**, 1720 (1999).
28. G. C. Papavassilliou, Y. Misaki, K. Takahashi, J. Yamada, G. A. Mousdis, T. Shirahata and T. Ise, *Z. Naturforsch.*, **56b**, 297 (2001).
29. J. Yamada, S. Satoki, S. Mishima, N. Araki, K. Takahashi, N. Masuda, Y. Nishimoto, S. Takasaki and H. Anzai, *J. Org. Chem.*, **61**, 3987 (1996).
30. Y. Misaki, H. Nishikawa, H. Fujiwara, K. Kawakami, T. Yamabe, H. Yamochi and G. Saito, *J. Chem. Soc., Chem. Commun.*, 1408 (1992).
31. T. Mori, Y. Misaki, H. Fujiwara and T. Yamabe, *Bull. Chem. Soc. Jpn.*, **67**, 2685 (1994).
32. Y. Misaki, H. Fujiwara, T. Yamabe, T. Mori, H. Mori and S. Tanaka, *Chem. Lett.*, 1653 (1994).
33. A. Deluzet, P. Batail, Y. Misaki, P. Auban-Senzier and E. Canadell, *Adv. Mater.*, **12**, 436 (2000).
34. Y. Misaki, T. Kochi, M. Taniguchi, T. Yamabe, K. Tanaka, K. Takimiya, A. Morikami, T. Otsubo and T. Mori, *Synth. Met.*, **102**, 1675 (1999).
35. Y. Misaki, T. Miura, M. Taniguchi, H. Fujiwara, T. Yamabe, T. Mori, H. Mori and S. Tanaka, *Adv. Mater.*, **9**, 714 (1997).
36. Y. Misaki, K. Tanaka, M. Taniguchi, T. Yamabe, T. Kawamoto and T. Mori, *Chem. Lett.*, 1249 (1999).
37. T. Mori, H. Inokuchi, Y. Misaki, H. Nishikawa, T. Yamabe, H. Mori and S. Tanaka, *Chem. Lett.*, 2085 (1993).
38. T. Mori, Y. Misaki, H. Nishikawa, T. Yamabe, H. Mori and S. Tanaka, *Synth. Met.*, **70**, 873 (1995).
39. Y. Misaki, H. Nishikawa, T. Yamabe, T. Mori and H. Inokuchi, *Bull. Chem. Soc. Jpn.*, **67**, 2368 (1994).
40. Y. Misaki, M. Taniguchi, K. Tanaka, K. Takimiya, A. Morikami, T. Otsubo and T. Mori, *J. Solid State Chem.*, **168**, 608 (2002).
41. T. Kawamoto, M. Ashizawa, T. Mori, J. Yamaura, R. Kato, Y. Misaki and K. Tanaka, *Bull. Chem. Soc. Jpn.*, **75**, 435 (2002).
42. T. Mori, H. Inokuchi, Y. Misaki, H. Nishikawa, T. Yamabe, H. Mori and S. Tanaka, *Chem. Lett.*, 733 (1993).

43. T. Mori, Y. Misaki, T. Yamabe, H. Mori and S. Tanaka, *Chem. Lett.*, 549 (1995).

44. T. Mori, Y. Misaki and T. Yamabe, *Bull. Chem. Soc. Jpn.*, **67**, 3187 (1994).

45. T. Mori, H. Inokuchi, Y. Misaki, T. Yamabe, H. Mori and S. Tanaka, *Bull. Chem. Soc. Jpn.*, **67**, 661 (1994).

46. T. Mori, Y. Misaki and T. Yamabe, *Bull. Chem. Soc. Jpn.*, **70**, 1809 (1997).

47. T. Kawamoto, M. Aragaki, T. Mori, Y. Misaki and T. Yamabe, *J. Mater. Chem.*, **8**, 285 (1998).

48. M. Katsuhara, M. Aragaki, T. Mori, Y. Misaki and K. Tanaka, *Chem. Mater.*, **12**, 3186 (2000).

49. M. Katsuhara, M. Aragaki, S. Kimura, T. Mori, Y. Misaki and K. Tanaka, *J. Mater. Chem.*, **11**, 2125 (2001).

50. Y. Misaki, H. Nishikawa, T. Yamabe, T. Mori, H. Mori and S. Tanaka, *Synth. Met.*, **70**, 1153 (1995).

51. T. Mori, K. Murata, Y. Misaki and K. Tanaka, *Chem. Lett.*, 435 (1998).

52. V. F. Kaminskii, T. G. Prokhoroea, R. P. Shibaeva and E. B. Yagubskii, *Pis'ma Zh. Eksp. Teor. Fiz.*, **39**, 15 (1984); T. Mori, A. Kobayashi, Y. Sasaki, H. Kobayashi, G. Saito and H. Inokuchi, *Chem. Lett.*, 957 (1984).

53. R. Kato, H. Kobayashi, A. Kobayashi, S. Moriyama, Y. Nishio, K. Kajita and W. Sasaki, *Chem. Lett.*, 459 (1987); H. Urayama, H. Yamochi, G. Saito, K. Nozawa, T. Sugano, M. Kinoshita, S. Sato, K. Oshima, A. Kawamoto and J. Tanaka, *Chem. Lett.*, 55 (1988); A. M. Kini, U. Geiser, H. H. Wang, K. D. Carlson, J. M. Williams, W.-K. Kwok, K. G. Vandervoort, J. E. Thompson, D. L. Stupka, D. Jung and M.-H. Whangbo, *Inorg. Chem.*, **29**, 2555 (1990); G. C. Papavassiliou, G. A. Mousdis and J. S. Zambounis, *Synth. Met.*, **27**, B379 (1988); K. Kikuchi, Y. Honda, Y. Ishikawa, K. Saito, I. Ikemoto, K. Murata, H. Anzai, T. Ishiguro and K. Kobayashi, *Solid State Commun.*, **66**, 405 (1988).

54. H. Mori, S. Tanaka and T. Mori, *Phys. Rev., B* **57**, 12023 (1998).

55. T. Mori, S. Ono, Y. Misaki, T. Yamabe, H. Mori and S. Tanaka, *Synth. Met.*, **85**, 1571 (1995).

56. H. Kobayashi, T. Udagawa, H. Tomita, K. Bun, T. Naito and A. Kobayashi, *Chem. Lett.*, 1559 (1993).

57. For another 1:1 metallic salt, see K. Takimiya, A. Ohnishi, Y. Aso, F. Ogura, K. Kawabata, K. Tanaka and M. Mizutani, *Bull. Chem. Soc. Jpn.*, **67**, 766 (1994); K. Kawabata, M. Yanai, T. Sambongi, Y. Aso, K. Takimiya and T. Otsubo, *Mol. Cryst. Liq. Cryst.*, **296**, 197 (1997).

58. T. Mori, T. Kawamoto, J. Yamaura, T. Enoki, Y. Misaki, T. Yamabe, H. Mori and S. Tanaka, *Phys. Rev. Lett.*, **79**, 1705 (1997); M. Maesato, Y. Saso, S. Kagoshima, T. Mori, T. Kawamoto, Y. Misaki and T. Yamabe, *Synth. Met.*, **103**, 2109 (1999); N. Fujimura, A. Namba, T. Kambe, Y. Nogami, K. Oshima, T. Mori, T. Kawamoto, Y. Misaki and T. Yamabe, *Synth. Met.*, **103**, 2111 (1999); M. Onuki, K. Hiraki, T. Takahashi, D. Jinno, T. Kawamoto, T. Mori, K. Tanaka and Y. Misaki, *J. Phys. Chem. Solid*, **62**, 405 (2001).

59. Y. Misaki, N. Higuchi, H. Fujiwara, T. Yamabe, T. Mori, H. Mori and S. Tanaka, *Angew. Chem. Int. Ed. Engl.*, **34**, 1222 (1995); Y. Misaki, N. Higuchi, T. Ohta, H. Fujiwara, T. Yamabe, T. Mori, H. Mori and S. Tanaka, *Mol. Cryst. Liq. Cryst.*, **284**, 27 (1996).

60. Y. Misaki, T. Sasaki, T. Ohta, H. Fujiwara and T. Yamabe, *Adv. Mater.*, **8**, 804 (1996).

61. Y. Misaki, T. Kaibuki, K. Takahashi, H. Tanioka and K. Tanaka, *Mol. Cryst. Liq. Cryst.*, **380**, 151 (2002).

62. Y. Misaki, H. Fujiwara and T. Yamabe, *J. Org. Chem.*, **61**, 3650 (1996).

63. Y. Misaki, H. Fujiwara, T. Maruyama, M. Taniguchi, T. Yamabe, T. Mori, H. Mori and S. Tanaka, *Chem. Mater.*, **11**, 2360 (1999).

64. Y. Misaki, T. Ohta, N. Higuchi, H. Fujiwara, T. Yamabe, T. Mori, H. Mori and S. Tanaka, *J. Mater. Chem.*, **5**, 1571 (1995).

65. Y. Misaki, H. Nishikawa, K. Kawakami, T. Uehara and T. Yamabe, *Tetrahedron Lett.*, **33**, 4321 (1992).

66. T. Kaibuki, Y. Misaki, K. Tanaka, K. Takimiya, A. Morikami and T. Otsubo, *Synth. Met.*, **102**, 1621 (1999).

67. T. Sugimoto, H. Awaji, I. Sugimoto, Y. Misaki, T. Kawase, S. Yoneda, Z. Yoshida, T. Kobayashi and H. Anzai, *Chem. Mater.*, **1**, 535 (1989); A. J. Moore and M. R. Bryce, *Tetrahedron Lett.*, **33**, 1373 (1992).

68. T. K. Hansen, M. V. Lakshmikantham, M. P. Cava, R. E. Niziurski-Mann, F. Jensen and J. Becher, *J. Am. Chem. Soc.*, **114**, 5035 (1992); A. Benahmed-Gasmi, P. Frére, B. Garrigues, A. Gorgues, M. Jubault, R. Carlier and F. Texier, *Tetrahedron Lett.*, **33**, 6457 (1992); K. Takahashi, T. Nihira, M. Yoshifuji and K. Tomitani, *Bull. Chem. Soc. Jpn.*, **66**, 2330 (1993).

69. D. J. Sandman, A. P. Fisher III, T. J. Holmes and A. J. Epstein, *J. Chem. Soc., Chem. Commun.*, 687 (1977); T. Otsubo, Y. Shiomi, M. Imamura, R. Kittaka, A. Ohnishi, H. Tagawa, Y. Aso and F. Ogura, *J. Chem. Soc., Perkin Trans. II*, 1815 (1993).

70. H. Fujiwara, Y. Misaki, T. Yamabe, T. Mori, H. Mori and S. Tanaka, *J. Mater. Chem.*, **10**, 1565 (2000).

71. R. C. Hadden, *Acc. Chem. Res.*, **25**, 127 (1992).

72. H. Yamochi, T. Komatsu, N. Matsukawa, G. Saito, T. Mori, M. Kusunoki and K. Sakaguchi, *J. Am.*

Chem. Soc., **115**, 11319 (1993); T. Nakamura, T. Komatsu, G. Saito, T. Osada, S. Kagoshima, N. Miura, K. Kato, Y. Maruyama and K. Oshima, *J. Phys. Soc. Jpn.*, **62**, 4373 (1993).
73. A. Sato, H. Kobayashi, T. Naito, F. Sakai and A. Kobayashi, *Inorg. Chem.*, **36**, 5262 (1997).
74. Y. Misaki, T. Kaibuki, M. Taniguchi, K. Tanaka, T. Kawamoto, T. Mori and T. Nakamura, *Chem. Lett.*, 1274 (2000).
75. K. Takahashi, T. Nakayashiki, M. Taniguchi, Y. Misaki and K. Tanaka, *Chem. Lett.*, 126 (2001).
76. K. Takahashi, T. Nakayashiki, Y. Misaki and K. Tanaka, *Mol. Cryst. Liq. Cryst.*, **376**, 107 (2002).
77. H. Nishikawa, S. Kawauchi, Y. Misaki and T. Yamabe, *Chem. Lett.*, 43 (1996).
78. H. Fujiwara, T. Nishikawa, Y. Misaki and T. Yamabe, *Synth. Met.*, **102**, 1737 (1999).
79. K. Takahashi, H. Tanioka, H. Fueno, Y. Misaki and K. Tanaka, *Chem. Lett.*, 1002 (2002).
80. K. Kanoda, *Hyperfine Interact.*, **104**, 235 (1997).
81. J. Yamada, M. Watanabe, H. Akutsu, S. Nakatsuji, H. Nishikawa, I. Ikemoto and K. Kikuchi, *J. Am. Chem. Soc.*, **123**, 4174 (2001).
82. M. Tamura, K. Yamanaka, Y. Mori, Y. Nishio, K. Kajita, H. Mori, S. Tanaka, J.-I. Yamaura, T. Imakubo, R. Kato, Y. Misaki and K. Tanaka, *Synth. Met.*, **120**, 1041 (2001).
83. H. Fujiwara, E. Fujiwara and H. Kobayashi, *Chem. Lett.*, 1048 (2002).

III 1,3-Dithiol-2-ylidene Donors

11

Dihydro-TTFs and Bis-fused 1,3-Dithiol-2-ylidene Donors

11.1 Introduction

1,3-Dithiol-2-ylidene (DT) derivatives, including tetrathiafulvalene (TTF **1**, Fig. 11.1) and its analogues, serve as important donor components for developing organic charge-transfer (CT) materials with remarkable electrical properties such as metallic conductivity and superconductivity.[1] The most general feature of DT derivatives is their ability to easily release one electron since they are oxidized to generate a cation-radical species to which the 1,3-dithiolium cation stabilized by the heteroaromatic 6π electron system makes a significant contribution (Scheme 11.1). Additionally, the planarity of the resulting cation-radical species with a contribution from the 1,3-dithiolium cation allows the formation of ordered stacks or sheets with intermolecular S···S contacts. For example, it is well known that TTFs containing two DT units are powerful electron donors which combine with a variety of organic or inorganic acceptors to produce quasi one-dimensional (1D) and quasi two-dimensional (2D) conductors; the metallic (TTF)(TCNQ) (TCNQ = 7,7,8,8-tetracyanoquinodimethane **2**, Fig. 11.1) complex is a case of the former,[2] and the superconductor β-(BEDT-TTF)$_2$I$_3$ [BEDT-TTF = bis(ethylenedithio) tetrathiafulvalene **3**, Fig. 11.1] is a case of the latter.[3,4] Therefore, TTFs have so far been central to the development of new organic metals and superconductors.[5,6] Meanwhile, the challenging question in this field of whether the production of organic metals and superconductors from a π-electron donor system, which contains only one DT unit or two units in a different pattern from

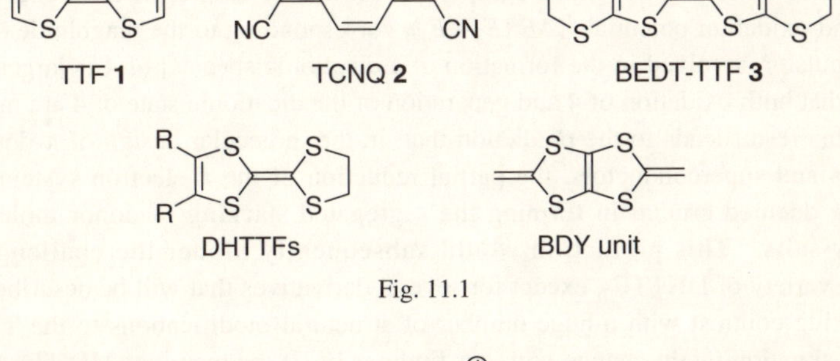

TTF **1**   TCNQ **2**   BEDT-TTF **3**

DHTTFs   BDY unit

Fig. 11.1

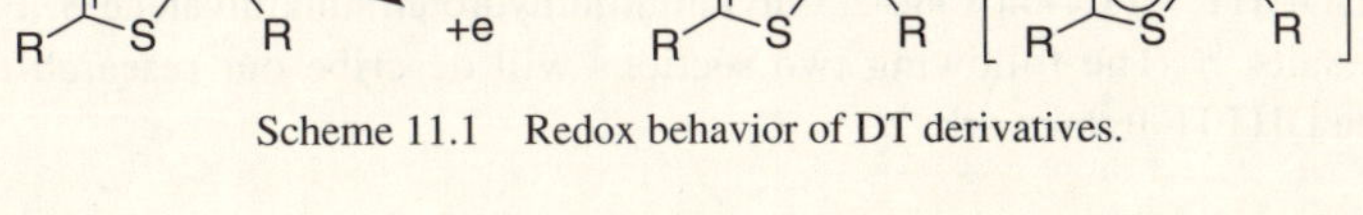

Scheme 11.1   Redox behavior of DT derivatives.

those in the TTF molecule, is feasible remains unsolved. The settlement of this issue would change the way of thinking of molecular designers for the design of organic metals and superconductors. The emphasis of this chapter is thus mainly on the recent development of organic metals and superconductors based on dihydro-TTFs (DHTTFs, Fig. 11.1), as π-donor systems with one DT unit, and new π-donors containing the bis-fused DT (BDY) unit (Fig. 11.1) as a π-system (hereafter designated as BDY donors).

11.2   DHTTFs

In 1971, Coffen and coworkers reported the synthesis and electrochemical properties of the parent DHTTF **4** (Scheme 11.2).[7] Formation of **4**, albeit as a minor product, was achieved by 2,3-dichloro-5,6-dicyano-1,4-benzoquinone (DDQ) oxidation of 2,2'-bi(1,3-dithiolanyl) **5** in boiling toluene. The major product in the reaction was 2,2'-bis(1,3-dithiolanylidene) **6**. An alternative synthetic method for the exclusive formation of **4** has been developed by Mori *et al.* *via* the reaction of bis(dimethylaluminum) *cis*-1,2-ethylenedithiolate with 2-ethoxycarbonyl-1,3-dithiolane.[8]

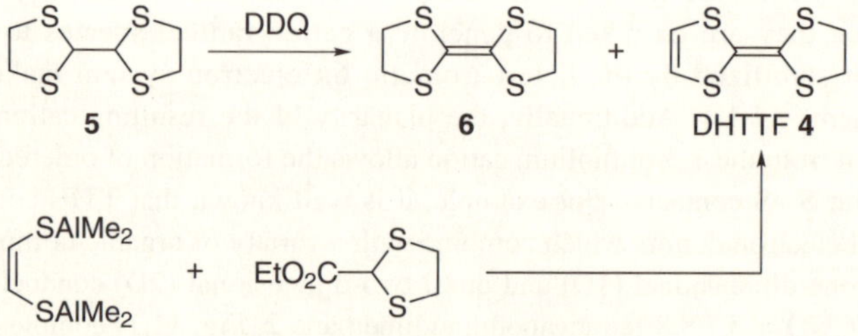

Scheme 11.2   Synthesis of DHTTF.

DHTTF **4**, like TTF **1**, is found to show two pairs of reversible redox waves by cyclic voltammetry (CV).[7] On comparing the respective values of oxidation potentials for **4** [$E_1 = +0.405$, $E_2 = +0.89$, $\Delta E\,(E_2 - E_1) = 0.485$ V (V *vs.* SCE)] with those for **1** ($E_1 = +0.33$, $E_2 = +0.70$, $\Delta E\,(E_2 - E_1) = 0.37$ V), the first oxidation potential (E_1, representing the electron-donating ability) of **4** is shifted to a more positive value compared with that of **1**, and the difference between the first and second oxidation potentials [$\Delta E\,(E_2 - E_1)$, corresponding to the magnitude of the on-site Coulombic repulsion involved in the formation of a dicationic species] of **4** is larger than that of **1**, suggesting that both oxidation of **4** and generation of the dicationic state of **4** are more difficult. Presumably, this result leads to the prediction that, in the molecular design of π-donors for new organic metals and superconductors, the partial reduction of the π-electron system of the TTF core would be deemed crucial in forming the segregated stacking of donor molecules in CT complexes or salts. This prediction would subsequently hinder the challenging task of synthesizing a variety of DHTTFs, except for several derivatives that will be described in Section 11.2.3, in striking contrast with a huge number of structural modifications to the TTF skeleton. However, this situation might change with our findings of (i) the metallic (MDHT)$_2$AuI$_2$ [MDHT = methylenedithio(dihydro)tetrathiafulvalene **7**, Fig. 11.2] salt[9] and (ii) the superconducting (DODHT)$_2$X [DODHT = (1,4-dioxane-2,3-diyldithio)dihydrotetrathiafulvalene **8**, Fig. 11.2, X = AsF$_6$ and PF$_6$] salts.[10] The following two sections will describe our research effort toward variations on the DHTTF framework.

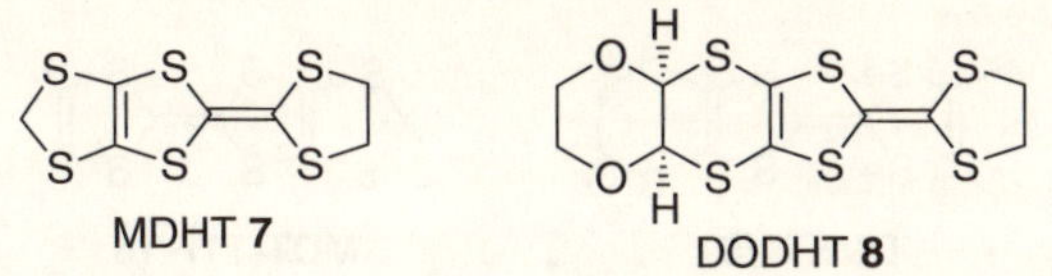

MDHT **7**

DODHT **8**

Fig. 11.2

11.2.1 Heterocycle-annulated DHTTFs and Their Seleno-analogues

The donor BEDT-TTF **3** (Fig. 11.1), needless to say, produces the largest number of superconducting salts, and the additional sulfur atoms, accompanied by fusion of a 1,4-dithiane ring onto both sides of the prototype TTF molecule, play an important role in the formation of 2D S···S sheets in the BEDT-TTF-based superconductors.[1] From this fact, it is expected that the additional chalcogen atoms by annulation of heterocycles onto the DHTTF skeleton lead to an enhancement in the intermolecular chalcogen···chalcogen interaction in the oxidation state, and, consequently, the formation of 1D stacks or 2D sheets conducive to metallic electronic structures becomes feasible. Thus, we investigated a systematic synthesis of heterocycle-annulated DHTTFs, including their selenium-containing analogues, and the preparation of their CT materials.

Synthesis of our target compounds **9–14** (Fig. 11.3), including MDHT **7** (Fig. 11.2), was attained by the Me_3Al-promoted coupling reaction (Scheme 11.3) of the corresponding tin-masked dithiolates or diselenolates with 2-ethoxycarbonyl-1,3-dithiolane in the following yields: **7**, 26%; **9**, 57%; **10**, 36%; **11**, 67%; **12**, 83%; **13**, 33%; **14**, 42%.[11–13] It is noteworthy that this Me_3Al-promoted coupling reaction is also applicable to the synthesis of diselenadithiafulvalenes (DSDTFs) and unsymmetrical TTFs (Scheme 11.3), such as dimethyl(ethylenedithio) diselenadithiafulvalene (DMET **15**, Fig. 11.4) and methylenedithiotetrathiafulvalene (MDT-TTF **16**, Fig. 11.4), which have given rise to superconducting salts.[14,15]

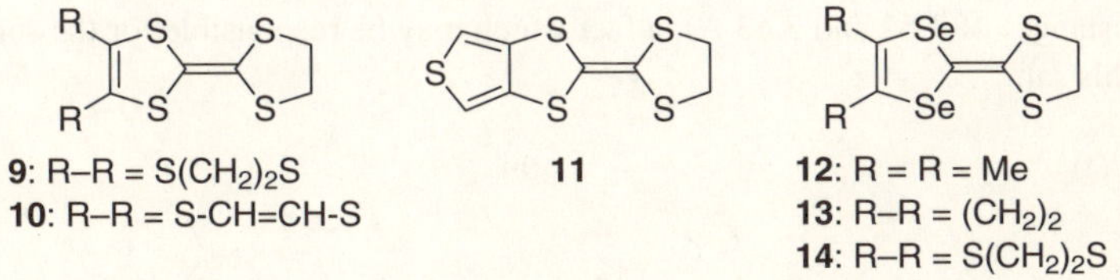

9: R–R = $S(CH_2)_2S$
10: R–R = S-CH=CH-S

11

12: R = R = Me
13: R–R = $(CH_2)_2$
14: R–R = $S(CH_2)_2S$

Fig. 11.3

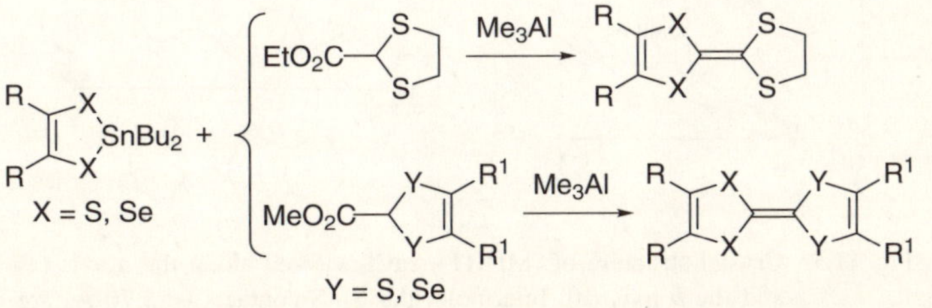

$X = S, Se$

$Y = S, Se$

Scheme 11.3 Me_3Al-promoted coupling reaction.

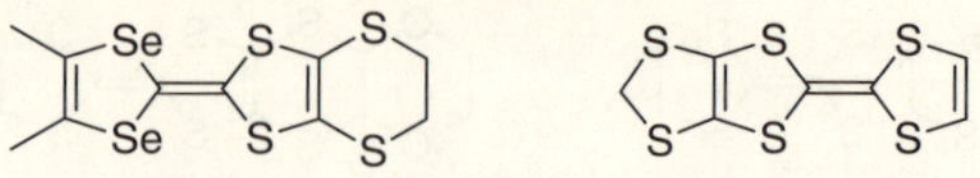

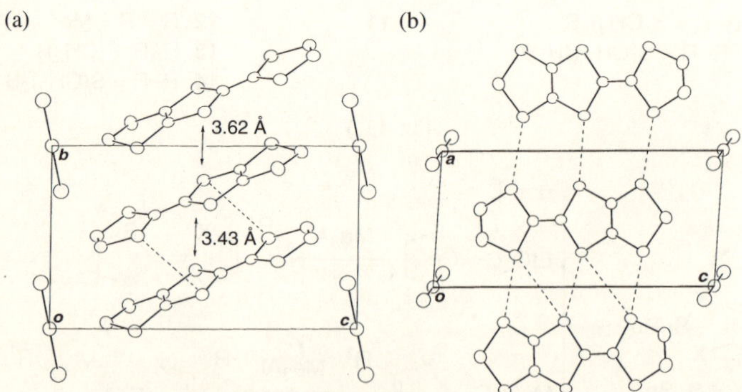

DMET **15**   MDT-TTF **16**

Fig. 11.4

As expected, the donor MDHT **7** showed a higher E_1 value [+0.60 V (*vs.* SCE)] and a larger ΔE ($E_2 - E_1$) value (0.40 V) compared to the corresponding values of its TTF analogue MDT-TTF **16** [E_1 = +0.49, ΔE ($E_2 - E_1$) = 0.29 V].   Nevertheless, DHTTF donors **7** and **9–11**, as well as the seleno-containing derivatives **12** and **13**, reacted with TCNQ to give the CT complexes, indicating that their donating ability would be sufficient to form organic conductors.   However, the room-temperature conductivities of the resulting TCNQ complexes were all very low (σ_{rt} < 10^{-6} S cm^{-1}).   We then attempted to prepare the cation-radical salts of **7**, **9**, **10** and **12–14** by the controlled-current electrocrystallization method.[16]   Among the salts so far prepared, the AuX$_2$ (X = CN, Cl, Br and I) salts of MDHT exhibited electrical conductivities of the order of 10^0–10^2 S cm^{-1} at room temperature for single crystals or compressed pellets.[13]   In particular, the (MDHT)$_2$AuI$_2$ salt (σ_{rt} = 60 S cm^{-1}) remained metallic all the way down to a very low temperature (1.4 K),[9] although the π-electron system of MDHT is less extended than that of MDT-TTF.   The other AuX$_2$ (X = CN, Cl and Br) salts were semiconductors with small activation energies (E_a = 6–20 meV).

Although our attempt to refine X-ray diffraction data of the metallic (MDHT)$_2$AuI$_2$ salt has not yet been successful, it appears that this salt has the θ-type donor arrangement.[17]   On the other hand, the crystal structure of a small gap semiconductor (MDHT)$_2$AuCl$_2$ (σ_{rt} = 250 S cm^{-1}, E_a = 6 meV) was determined by X-ray diffraction analysis.[13,17]   As illustrated in Fig. 11.5, the MDHT molecules in this salt are stacked along the *b* axis to form a column.   While there exist intermolecular S···S contacts shorter than the sum of the van der Waals radii (3.70 Å) between the columns, the dimerization of donor molecules within the column is fairly large (average interplanar distances of 3.62 and 3.43 Å), a fact which may be responsible for the semiconductive behavior of this salt.

Fig. 11.5   Crystal structure of (MDHT)$_2$AuCl$_2$ viewed along the *a* axis (a) and the *b* axis (b). Intermolecular S···S contacts (< 3.70 Å) are indicated by broken lines. (Reprinted with permission from ref. 13)

11.2.2 Superconducting DODHT Salts

The above-mentioned $(MDHT)_2AuI_2$ salt was found to be metallic to 1.4 K, but it failed to develop superconductivity. Therefore, further structural modification to the DHTTF donors examined was required to achieve superconductivity. We thus designed a derivative of the ethylenedithio-substituted DHTTF donor (**9**, Fig. 11.3) with the appended dioxane ring by *cis* fusion, *viz.*, DODHT **8** (Fig. 11.2),[10] because we found that the *cis*-fused dioxane ring, though bulky, on TTF derivatives does not necessarily break metallic electronic structures, as can be found in $(DOET)_2BF_4$ [DOET = (1,4-dioxane-2,3-diyldithio)ethylenedithiotetrathiafulvalene][18] and $(DOES)_2(AuI_2)_{0.75}$ [DOES = (1,4-dioxane-2,3-diyldithio)ethylenediselenotetrathiafulvalene].[19] Such a bulky donor containing only the DT unit as a π-system does not appear to hew to the conventional line of the molecular design for π-donors capable of providing superconductors; hence, the successful production of DODHT-based superconductors will lead to a new approach in the design of organic superconductors.

Our synthetic sequence for constructing DODHT is outlined in Scheme 11.4. Conversion of the dioxane-fused ketone **17**[20] into tin dithiolate **18**, and transmetalation of **18** with *n*-BuLi (2 equiv.) followed by treatment with methyl dichloroacetate gave ester **19** in 74% overall yield from **17**. Reaction of **19** with bis(dimethylaluminum) 1,2-ethanedithiolate furnished DODHT in 46% yield. The CV of DODHT showed two reversible oxidation waves at +0.64 (E_1) and +1.08 (E_2) V (*vs.* SCE). The E_1 value is higher than that of its TTF analogue (1,4-dioxane-2,3-diyldithio)tetrathiafulvalene (DOT **20**, Fig. 11.6, +0.51 V)[20] measured under the same conditions, and the ΔE ($E_2 - E_1$) value (0.44 V) is larger than that of DOT **20** (0.33 V). Fig. 11.6 shows the highest occupied molecular orbital (HOMO) of the DODHT molecule by the *ab initio* calculation (RHF/6-31G**), together with that of the DOT molecule. In the HOMO of DODHT, the atomic orbital coefficients of two sulfur atoms on the 1,3-dithiole ring are larger than those on the 1,3-dithiolane ring, whereas almost the same coefficients are observed for four sulfur atoms on the TTF backbone in the HOMO of DOT.

Single crystals of the AsF_6 and PF_6 salts of DODHT were prepared by electrocrystallization at a constant current (1.0 μA), X-ray analyses of which revealed that both salts crystallize isostructurally and have the β''-type (or β''_{211}-type[21]) donor arrangement. Fig. 11.7a shows the crystal structure of $(DODHT)_2AsF_6$, in which the donor molecules are stacked head-to-tail along

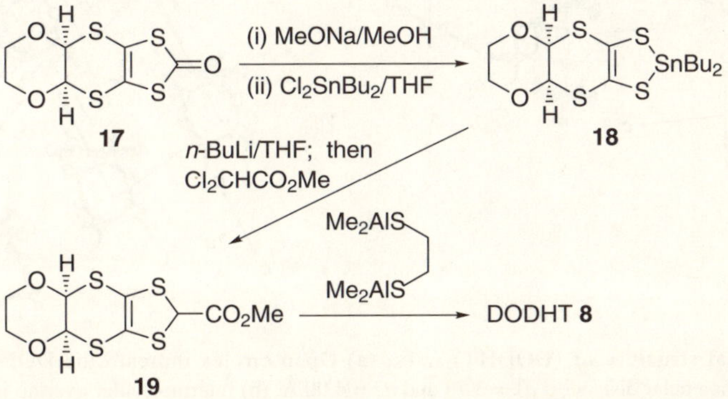

Scheme 11.4   Synthesis of DODHT.

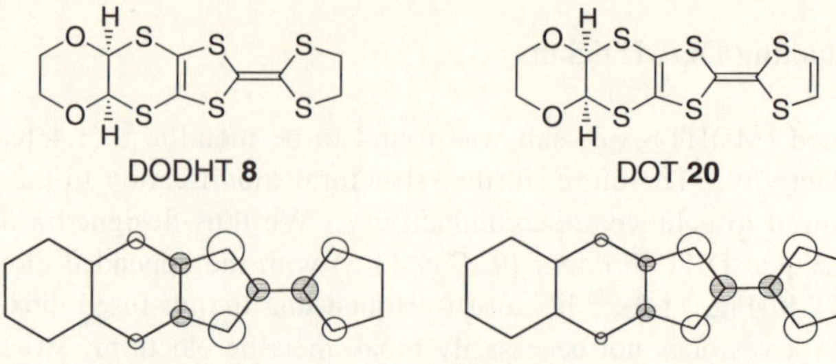

Fig. 11.6   The HOMOs of DODHT (a) and DOT (b).

the b axis. The interplanar distances (3.73 and 4.00 Å) within the stack are much longer than those in the metallic $(DOET)_2BF_4$ (3.50 and 3.77 Å) and $(DOES)_2(AuI_2)_{0.75}$ (3.47 and 3.78 Å) salts. There are several S···S contacts shorter than the van der Waals distance (3.70 Å) between stacks, whereas no S···S contact is observed within the stack. This S···S contact pattern reflects the anisotropy of the intermolecular overlap integrals (Fig. 11.7b), suggesting that the transverse interaction is superior to the interaction within the stack. Moreover, the two donor molecules with an interplanar distance of 4.00 Å are mutually slipped along both the molecular short and long axes, which causes the overlap integral $b2$ to be almost zero. These structural characteristics also hold for the isostructural $(DODHT)_2PF_6$.

The resistivities of $(DODHT)_2AsF_6$ ($\sigma_{300K} = 1.2$ S cm^{-1}) as a function of temperature under various pressures are shown in Fig. 11.8a. At ambient pressure, the resistivity of this salt was semiconductive with a broad shoulder around 120 K, and with increasing pressure, pronounced humps appeared at lower temperatures. At 16.5 kbar, the resistivity exhibited metallic behavior

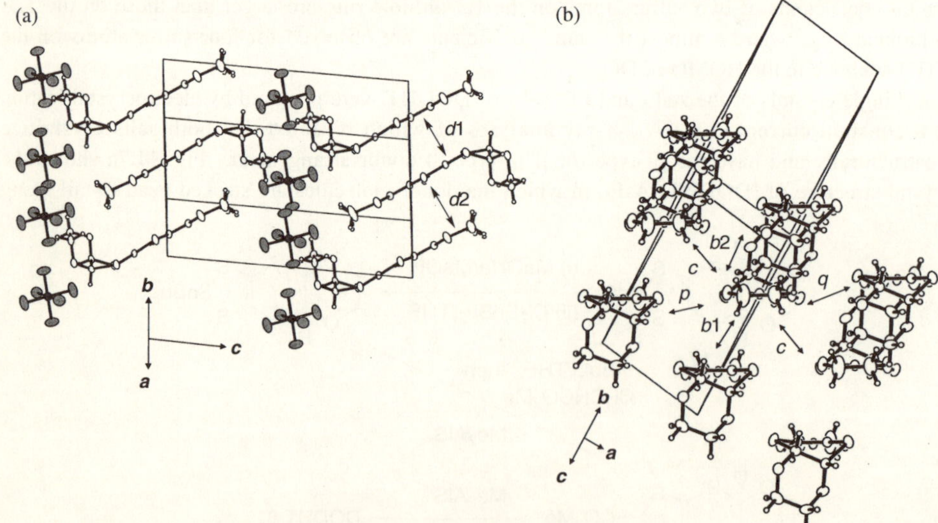

Fig. 11.7   Crystal structure of $(DODHT)_2AsF_6$. (a) Open circles indicate the DODHT molecules. Intermolecular distances: $d1 = 3.73$ and $d2 = 4.00$ Å. (b) Intermolecular overlap integrals ($\times\ 10^{-3}$) $b1$, $b2$, c, p and q are −2.81, −0.09, −10.95, −10.08 and −6.28, respectively.

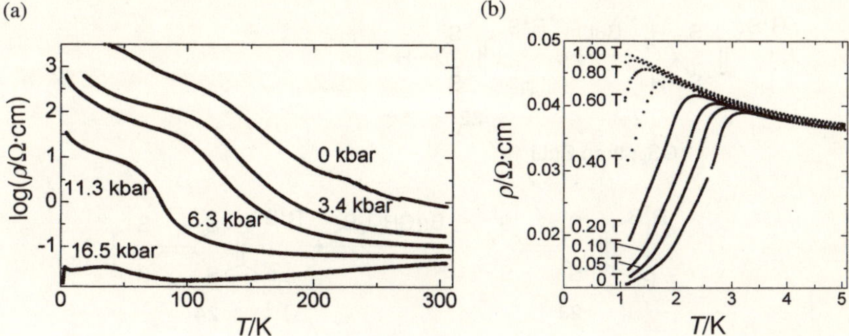

Fig. 11.8   (a) Temperature dependence of the resistivity for $(DODHT)_2AsF_6$ at 0 kbar and under pressure up to 16.5 kbar. (b) Magnetic field dependence of the superconducting transition of $(DODHT)_2AsF_6$ at 16.5 kbar. (Reprinted with permission from ref. 10)

below room temperature and an abrupt drop attributed to a superconducting transition with an onset of 3.3 K. Superconductivity in this salt was confirmed by the magnetic field dependence of the drop of the resistivity measured at 16.5 kbar. With increasing magnetic field, the superconducting transition of this salt shifted to a lower temperature, and was suppressed under 1 T (Fig. 11.8b). Similar to the $(DODHT)_2AsF_6$ salt, $(DODHT)_2PF_6$ ($\sigma_{300K} = 9.2 \times 10^{-1}$ S cm^{-1}) exhibited semiconducting behavior at ambient pressure and metallic behavior at 11.3 kbar, then underwent a superconducting transition with an onset of 3.1 K at 16.5 kbar, which disappeared under an applied magnetic field of 1 T.

11.2.3 Miscellaneous DHTTF-containing Compounds

In addition to our reported DHTTF-based donors, where a heterocycle or bis-fused heterocycle is condensed on the 1,3-dithiole ring of the DHTTF molecule, the benzene-fused DHTTF **21** (Scheme 11.5) has been synthesized by (i) coupling of bis(dimethylaluminum) 1,2-ethanedithiolate or bis(dimethylaluminum) 1,2-benzenedithiolate with the corresponding ester[8] and (ii) reaction of dilithium 1,2-benzenedithiolate with 2-dihalomethylene-1,3-dithiolane.[22] Also, DHTTFs with two different substituents attached to the 1,3-dithiole ring have been

Scheme 11.5   Synthesis of benzo-DHTTF.

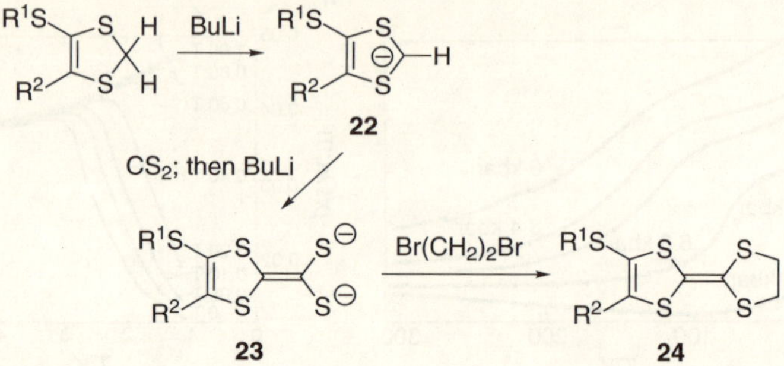

Scheme 11.6   Synthesis of DHTTFs bearing two different substituents.

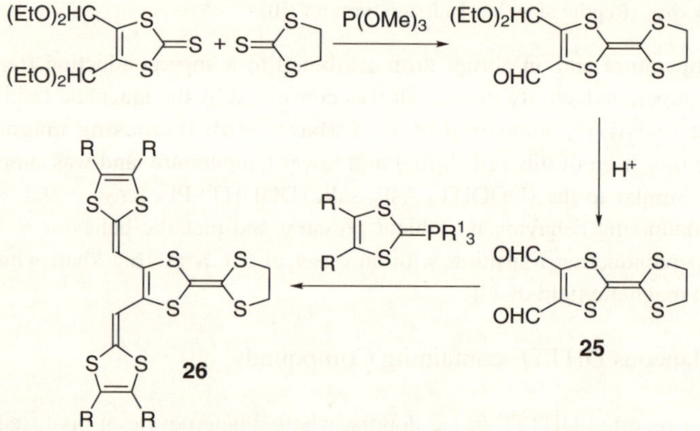

Scheme 11.7   Bis(formyl)-substituted DHTTF and its synthetic application.

reported.[23] As shown in Scheme 11.6, the synthesis of compound **24** proceeded *via* generation of *gem*-dithiolate **23** from 1,3-dithiole anion **22**, followed by treatment with 1,2-dibromoethane. One attractive precursor for constructing functionalized DHTTFs is the diformyl-attached DHTTF **25** (Scheme 11.7), the synthetic potential of which was exploited to synthesize the bis(1,3-dithiafulvenyl)-substituted DHTTF **26** by a twofold Wittig reaction with the corresponding phosphorus ylid.[24]

Another structural modification of the DHTTF framework is the attachment of substituents to the 1,3-dithiolane ring of the DHTTF molecule. The norbornane-fused DHTTF with the *exo* configuration, *i.e.*, compound **29** (Scheme 11.8), has been synthesized in a one-pot procedure involving (i) formation of a reactive phosphorane intermediate **28** by cycloaddition of the tributylphosphine-carbon disulfide complex **27** to norbornene, (ii) nucleophilic attack of **28** to carbon disulfide and (iii) cycloaddition with electron-deficient acetylenic esters.[25] The DHTTF-norbornane-DHTTF fused system **30** (Scheme 11.8) has also been obtainable by an analogous synthetic sequence using norbornadiene. As an intramolecular donor-acceptor molecule[26] based on DHTTF, the bis(1,3-indanedion-2-ylidene)-substituted DHTTF **31** has been studied.[27] The synthesis of **31** is shown in Scheme 11.9. A solution of **31** in benzene exhibited strong electronic absorption (λ_{max} = 850 nm), reaching a near-infrared region (> 800 nm), but showed almost no intramolecular CT interaction.

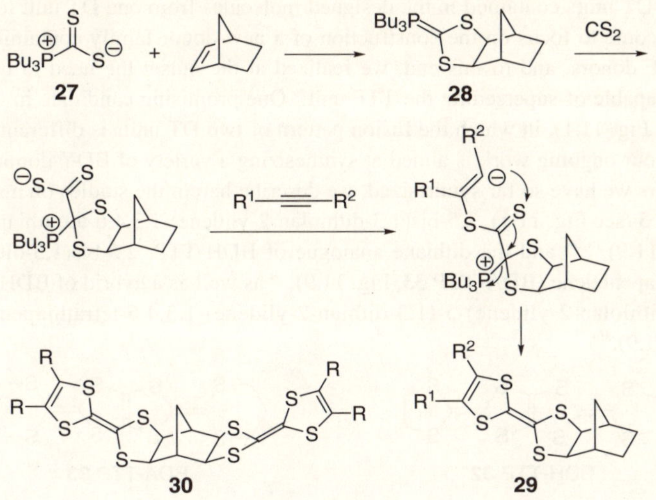

Scheme 11.8   Norbornane-fused DHTTFs and their bis(DHTTF) analogues.

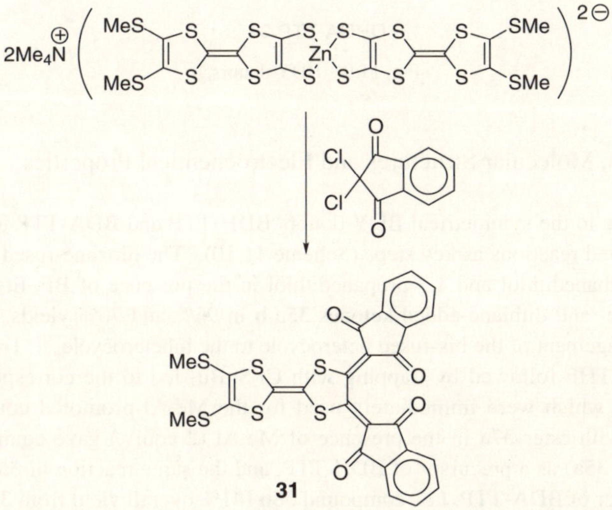

Scheme 11.9   Synthesis of bis(1,3-indanedion-2-ylidene)-substituted DHTTF.

11.3 BDY Donors

Our studies on the development of organic metals and superconductors composed of the DHTTF-based donors are directed toward the post-TTF era, and the results demonstrate that the DT unit alone is a sufficient π-electron system to form CT salts exhibiting metallic conductivity and superconductivity. Therefore, it can be anticipated that, in the molecular design of new π-donors, there is a good possibility of obtaining organic metals and superconductors by increasing the number of the DT units contained in the designed molecules from one DT unit to two units. Our attention thus come to focus on the construction of a new donor family containing two DT units other than TTF donors, and to this end, we realized at the outset the need to design a basic π-electron unit capable of superseding the TTF unit. One promising candidate in this regard is the BDY unit (see Fig. 11.1), in which the fusion pattern of two DT units is different from that in the TTF unit, and our ongoing work is aimed at synthesizing a variety of BDY donors. From among the BDY donors we have so far synthesized, we describe herein the studies on a structural isomer of BEDT-TTF **3** (see Fig. 11.1), 2,5-bis(1,3-dithiolan-2-ylidene)-1,3,4,6-tetrathiapentalene (BDH-TTP **32**, Fig. 11.9),[13,28] and the dithiane analogue of BDH-TTP, 2,5-bis(1,3-dithian-2-ylidene)-1,3,4,6-tetrathiapentalene (BDA-TTP **33**, Fig. 11.9),[29] as well as a hybrid of BDH-TTP and BDA-TTP, 2-(1,3-dithiolan-2-ylidene)-5-(1,3-dithian-2-ylidene)-1,3,4,6-tetrathiapentalene (DHDA-TTP **34**, Fig. 11.9).[30]

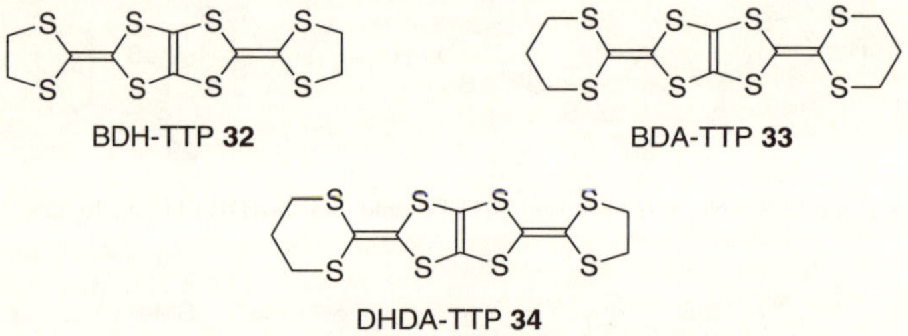

BDH-TTP **32**   BDA-TTP **33**

DHDA-TTP **34**

Fig. 11.9   BDY donors.

11.3.1 Synthesis, Molecular Structures and Electrochemical Properties

Our synthetic route to the symmetrical BDY donors BDH-TTP and BDA-TTP involved the BF_3- and Me_3Al-promoted reactions as key steps (Scheme 11.10). The dioxane-fused ketone **17**[20] was treated with 1,2-ethanedithiol and 1,3-propanedithiol in the presence of $BF_3 \cdot Et_2O$ (10 equiv.) to give the dithiolane- and dithiane-added ketones **35a,b** in 95% and 70% yields, respectively, *via* the skeletal rearrangement of the bis-fused heterocycle to the biheterocycle.[31] Treatment of **35a,b** with MeMgBr in THF followed by trapping with Cl_2SnBu_2 led to the corresponding labile tin dithiolates **36a,b**, which were immediately used for the Me_3Al-promoted coupling reactions. Reaction of **36a** with ester **37a** in the presence of Me_3Al (2 equiv.) gave compound **38a** (52% overall yield from **35a**) as a precursor of BDH-TTP, and the same reaction of **36b** with ester **37b** afforded a precursor of BDA-TTP, *i.e.*, compound **38b** (41% overall yield from **35b**). Subsequent DDQ oxidation of **38a,b** in refluxing toluene furnished BDH-TTP and BDA-TTP in 89% and 80% yields, respectively. A useful application of this synthetic route to BDH-TTP and BDA-TTP can be found in the synthesis of the unsymmetrical BDY donor DHDA-TTP. The Me_3Al-promoted reaction of tin dithiolate **36b** with ester **37a** gave compound **38c**, which upon oxidation with DDQ in refluxing toluene led to DHDA-TTP in 85% yield.

The molecular structures of BDH-TTP and BDA-TTP were determined by X-ray diffraction analyses. As depicted in Fig. 11.10, the common structural feature of each molecule is that the three tetrathioethylene units are found in an almost common plane. While the two ethylene end groups of BDH-TTP are slightly distorted from the common plane (Fig. 11.10a), the BDH-TTP molecule has a flatter structure compared to BEDT-TTF.[32] On the other hand, the two trimethylene end groups of BDA-TTP are far out of the common plane in opposite directions (Fig.

11.10b), and the dihedral angle around the intramolecular sulfur-to-sulfur axis in each dithiane ring is 53.6°. Accordingly, the two dithiane rings with an equivalent chair conformation

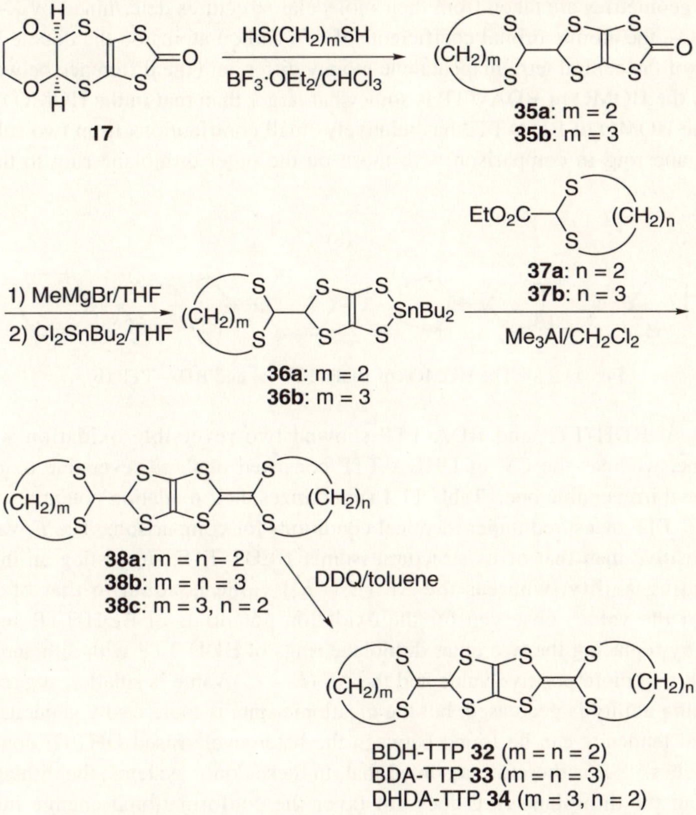

35a: m = 2
35b: m = 3

37a: n = 2
37b: n = 3

36a: m = 2
36b: m = 3

38a: m = n = 2
38b: m = n = 3
38c: m = 3, n = 2

BDH-TTP **32** (m = n = 2)
BDA-TTP **33** (m = n = 3)
DHDA-TTP **34** (m = 3, n = 2)

Scheme 11.10   Synthesis of BDH-TTP, BDA-TTP and DHDA-TTP.

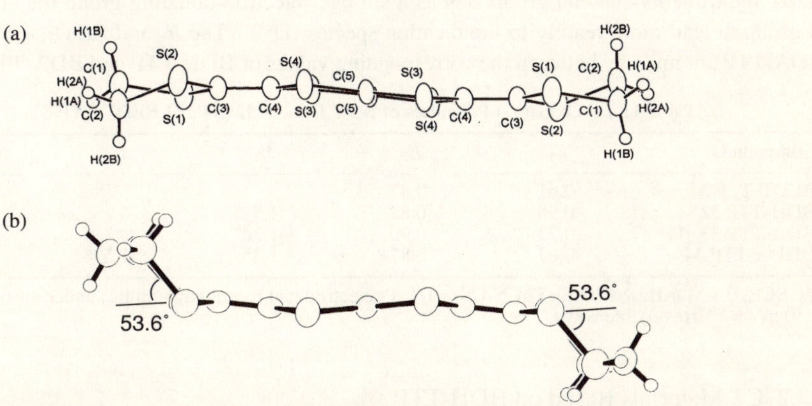

Fig. 11.10   Side views of the molecular structures of BDH-TTP(a) and BDA-TTP(b). (Reprinted with permission from refs. 28 and 29, respectively)

cause the whole molecular structure of BDA-TTP to be less planar than that of BDH-TTP. Fig. 11.11 shows the HOMOs of BDH-TTP and BDA-TTP calculated at the *ab initio* level (RHF/6-31G**). The geometries are taken from their molecular structures determined by X-ray analyses. In both HOMOs, the atomic orbital coefficients of two carbon atoms on the double bond situated on the outside of the central tetrathiapentalene moiety differ, and the difference between these two coefficients in the HOMO of BDA-TTP is somewhat larger than that in the HOMO of BDH-TTP. In addition, the HOMO of BDA-TTP has relatively small contributions from two sulfur atoms on the outer dithiane ring in comparison with those on the outer dithiolane ring to the HOMO of BDH-TTP.

(a)   (b)

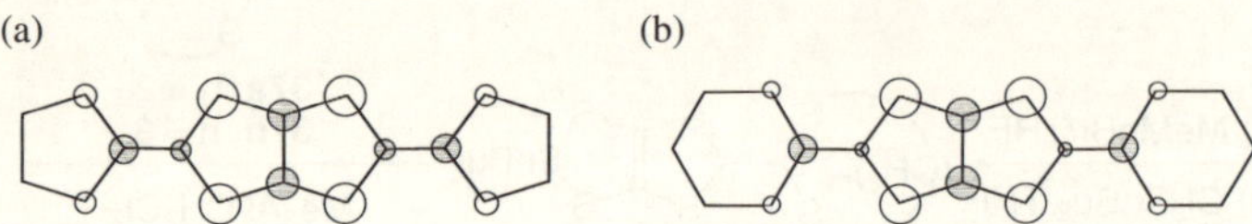

Fig. 11.11   The HOMOs of BDH-TTP (a) and BDA-TTP (b).

The CVs of BDH-TTP and BDA-TTP showed two reversible oxidation waves and an irreversible one, whereas the CV of DHDA-TTP consisted of three reversible oxidation waves and an additional irreversible one. Table 11.1 summarizes their oxidation potentials together with those of BEDT-TTF measured under identical conditions for comparison. The E_1 value of BDH-TTP is less positive than that of its structural isomer BEDT-TTF, suggesting an increase in the electron-donating ability, whereas the ΔE ($E_2 - E_1$) value is equal to that of BEDT-TTF. Comparison of the values observed for the oxidation potentials of BDH-TTP and BDA-TTP revealed that, by replacing the two outer dithiolane rings of BDH-TTP with dithiane rings, the E_1 value is shifted to a more positive value and the ΔE ($E_2 - E_1$) value is smaller, suggesting that the electron-donating ability is decreased, but the dicationic state is more easily generated. A similar electrochemical tendency can be found between the heterocycle-fused DHTTF donors and their dithiane analogues.[12] It is therefore assumed that, in these donor systems, the dithiane ring being less planar than the dithiolane ring does not favor the conformational change into the flatter arrangements required to generate the cation-radical species (D$^{\bullet +}$), but once these species are formed, the trimethylene end group acts as a stronger electron-donating group than the ethylene end group to lead more readily to the dication species (D^{2+}). The E_1 and ΔE ($E_2 - E_1$) values of DHDA-TTP are midway between the corresponding values of BDH-TTP and BDA-TTP.

Table 11.1   Oxidation Potentials of BDY Donors **32–34** and BEDT-TTF **3**[a]

Compound	E_1	E_2	E_3	E_4	ΔE (E_2–E_1)
BEDT-TTF **3**	0.61	0.87			0.26
BDH-TTP **32**	0.56	0.82	1.52[b]		0.26
BDA-TTP **33**	0.72	0.90	1.26[b]		0.18
DHDA-TTP **34**	0.67	0.87	1.35	1.53[b]	0.20

[a]V *vs.* SCE; 0.1 M *n*-Bu$_4$NClO$_4$ in PhCN/CS$_2$ =1/1; Pt electrode; at room temperature; under nitrogen; scan rate 50 mV s^{-1}. [b]Irreversible wave.

11.3.2 CT Materials Based on BDH-TTP

The conducting behavior of the CT complexes and salts based on BDH-TTP is summarized in Table 11.2. As expected from its E_1 value, BDH-TTP reacted with TCNQ and its tetrafluoro

Table 11.2   Conducting Behavior of the BDH-TTP-based CT Materials

Acceptor	D:A[a]	σ_{rt}/S cm^{-1} [b]
TCNQ	1:1	2.0×10^{-7} ($E_a = 430$ meV)
TCNQF$_4$	3:2	5.6×10^{-5} ($E_a = 420$ meV)[c]
I$_3^-$	2:1	230 (metallic > 2.0 K)
AuI$_2^-$	2:1	49 (metallic > 2.0 K)
BF$_4^-$	2:1	33 (metallic > 2.0 K)
ClO$_4^-$	—[d]	106 (metallic > 2.2 K)
FeCl$_4^-$	2:1[e]	39 (metallic > 1.5 K)
PF$_6^-$	2:1	102 (metallic > 2.1 K)
AsF$_6^-$	2:1	49 (metallic > 2.0 K)

[a]Determined by elemental analysis unless otherwise noted. [b]Room-temperature conductivity measured by a four-probe technique on a single crystal unless otherwise noted. [c]Measured on a compressed pellet. [d]Not determined because this salt may explode during analysis. [e]Determined by X-ray analysis.

analogue (TCNQF$_4$) at room temperature to form CT complexes, but their conductivities at room temperature were fairly low. In contrast, the BDH-TTP salts prepared by the controlled-current electrocrystallization method remained metallic down to very low temperatures. Moreover, with decreasing temperature, their similar resistive behavior could be observed independent of the structure and volume of the counteranions used. On the other hand, the temperature dependence of the magnetic susceptibility of the BDH-TTP salt containing the paramagnetic FeCl$_4^-$ anion can be well fitted to the Curie-Weiss law, and the values of the Curie and Weiss constants (C and θ) are 4.25 emu K mol^{-1} and 0.041 K, respectively.[33] The fitted C is close to the value of 4.38 emu K mol^{-1} expected for a high-spin Fe^{+3} ion ($S = 5/2$, $g = 2.0$), so that the Fe atom in the anion clearly dominates the measured magnetization.

By X-ray diffraction analysis, the structure of the metallic (BDH-TTP)$_2$PF$_6$ salt was ascertained to consist of κ-type sheets of BDH-TTP donor molecules and sheets of PF$_6^-$ anions (Fig. 11.12).[13,28] The interplanar distance within a pair of donor molecules is 3.53 Å and the dihedral angle of the molecular planes between pairs is 82°. In the donor sheet, each molecule is linked by several S···S contacts shorter than the sum of the van der Waals radii (3.70 Å), as illustrated in Fig. 11.13. The large intermolecular overlap integrals are also calculated not only within a pair of donor molecules but also between pairs, which form 2D interaction in the bc plane. This 2D electronic structure is responsible for the metallic behavior down to low temperatures.

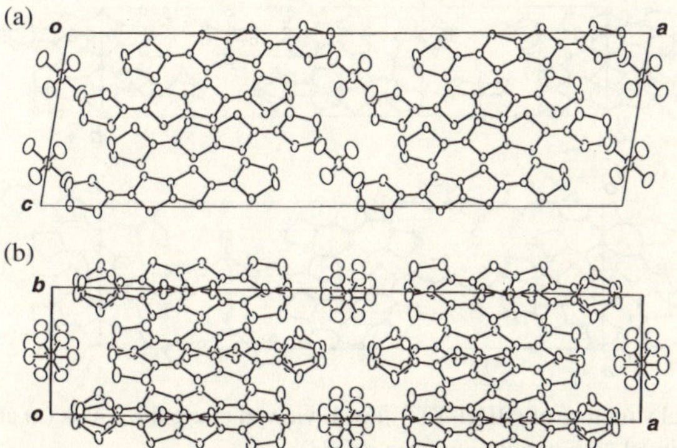

Fig. 11.12   Crystal structure of (BDH-TTP)$_2$PF$_6$ viewed along the b axis (a) and the c axis (b). (Reprinted with permission from refs. 13 and 28)

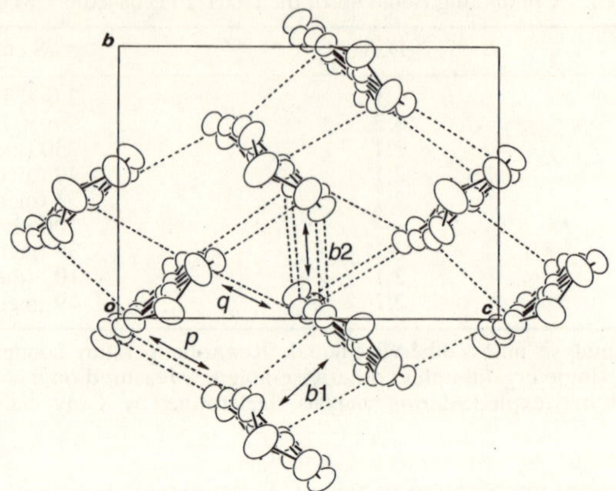

Fig. 11.13   Donor arrangement of $(BDH\text{-}TTP)_2PF_6$. Intermolecular S···S contacts (< 3.70 Å) are indicated by dotted lines. The values of intermolecular overlap integrals are $b1 = 20.7$, $b2 = 19.6$, $p = 5.9$ and $q = -7.5$ ($\times 10^{-3}$). (Reprinted with permission from refs. 13 and 28)

X-ray diffraction analysis of the metallic $(BDH\text{-}TTP)_2FeCl_4$ salt revealed that this structure is also composed of κ-type sheets of BDH-TTP donor molecules and sheets of $FeCl_4^-$ anions, as depicted in Fig. 11.14.[33] The interplanar distance within a pair of donor molecules is 3.59 Å and the dihedral angle of the molecular planes between pairs is 83.3°. Compared to κ-(BDH-TTP)$_2$PF$_6$, the donor packing in this salt is loose, resulting in no S···S contact shorter than the van der Walls distance (3.70 Å) within a pair of donor molecules (Fig. 11.15). On the other hand, there are several short S···S contacts between neighboring donor molecules located out of a pair and the large intermolecular overlap integrals are also calculated between pairs, which form 2D interaction in the *ac* plane. The $FeCl_4^-$ anions are separated by the donor sheets along the *b* axis. Within the anion sheet, consecutive anions form pairs with the Fe···Fe distance of 5.559(2) Å and the shortest Fe···Fe distance between pairs is 8.169(2) Å. This shortest distance seems to be too long to lead to direct interaction between the Fe moments, the phenomenon which is probably responsible for the Curie-Weiss paramagnetic behavior of this salt.

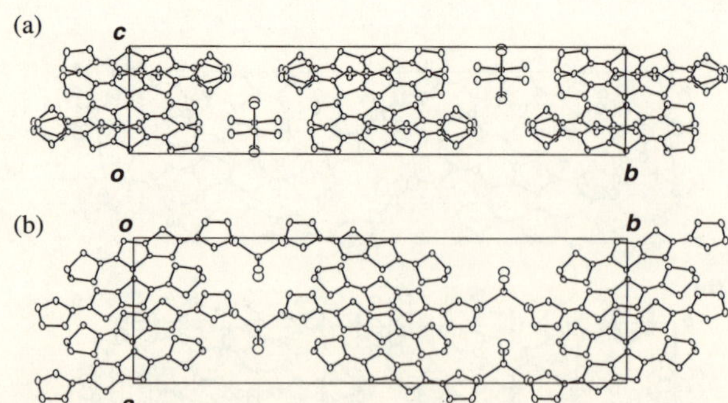

Fig. 11.14   Crystal structure of $(BDH\text{-}TTP)_2FeCl_4$ viewed along the *a* axis (a) and the *c* axis (b). (Reprinted with permission from ref. 33)

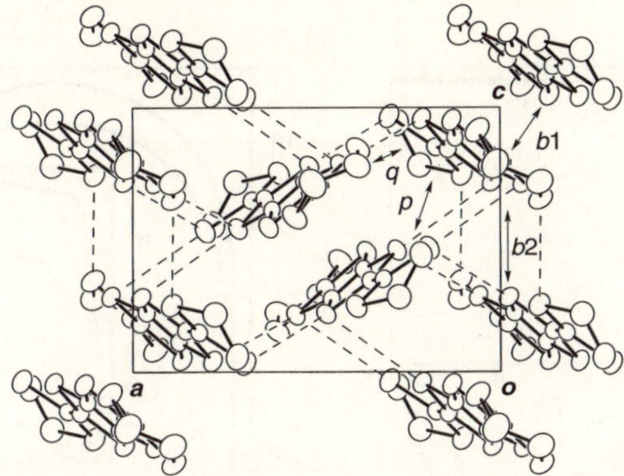

Fig. 11.15   Donor arrangement of (BDH-TTP)$_2$FeCl$_4$. Intermolecular S···S contacts (< 3.70 Å) are indicated by broken lines. The values of intermolecular overlap integrals ($\times$ 10^{-3}) $b1$, $b2$, p and q are 19.3, 15.7, 6.39 and –6.79, respectively. (Reprinted with permission from ref. 33)

11.3.3 Superconducting BDA-TTP Salts

Before our finding of the superconductivity in (DODHT)$_2$X (X = AsF$_6$ and PF$_6$),[10] we had discovered a series of ambient-pressure superconductors β-(BDA-TTP)$_2$X (X = SbF$_6$, AsF$_6$ and PF$_6$), which are the first organic superconductors containing no tetrachalcogenafulvalene (TCF) molecule in a donor component capable of contributing to superconductivity.[29] The zero-field-cooled (ZFC) temperature dependence of the dc magnetization for these superconducting salts, prepared by the controlled-current electrocrystallization method, revealed that the onset of diamagnetic transition occurs at a temperature of 6.9 K for the SbF$_6$ salt and of 5.9 K for both the AsF$_6$ and PF$_6$ salts (Fig. 11.16a). From their ZFC susceptibility curves at 2 K, we estimated that the diamagnetic shielding effects of the SbF$_6$, AsF$_6$ and PF$_6$ salts are *ca.* 15%, 20% and 15% of the perfect diamagnetism (–1/4π emu cm^{-3}), respectively. Superconductivity in the SbF$_6$ and AsF$_6$ salts was also recorded by the temperature dependence of their electrical resistivities under ambient pressure, as shown in Fig. 11.16b. On the resistive behavior of the SbF$_6$ salt (σ_{rt} = 1.5 S cm^{-1}), the resistivity increased like that in a semiconductor on cooling from room temperature to near 150 K, then exhibited metallic behavior down to 7.5 K. Below this temperature, the resistivity displayed an abrupt drop and became almost zero below 6.5 K. The onset of the superconducting transition observed in the subsequent heating process (the inset in Fig. 11.16b) occurred at 7.5 K, which is slightly higher than that in the magnetization measurement. Similar resistive maximum and superconductive behavior were found in the temperature dependence of the resistivity of the AsF$_6$ salt (σ_{rt} = 2.9 S cm^{-1}). Below 160 K, the resistivity of this salt decreased with decreasing temperature down to a superconducting transition with onset at 5.8 K, which is comparable to that of the magnetic transition, and zero resistance was observed below 4.4 K. Although we carried out resistive measurements on numerous single crystals of the PF$_6$ salt (σ_{rt} = 3.8 S cm^{-1}), all the crystal specimens were found to disintegrate at temperatures in the range of 110–140 K. In this temperature region, because no anomaly in the magnetic susceptibility of this salt was observed, it appears that there is no marked change in the electronic state. The decomposition of these crystal specimens could therefore be attributed to the stresses of the resistance-measuring probes or some other artifact of the measurements. Owing to the

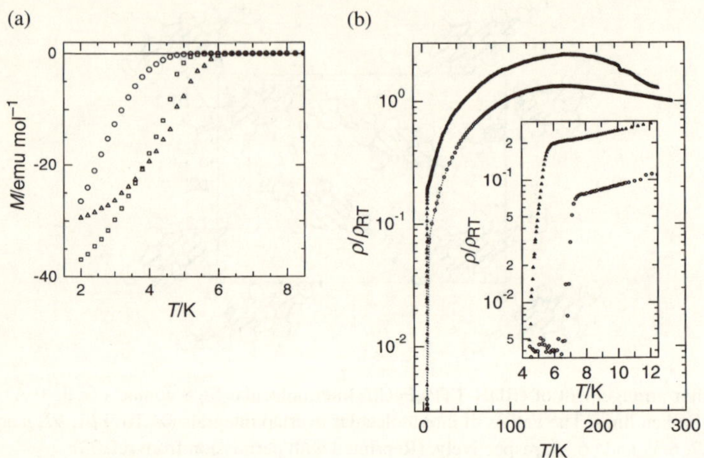

Fig. 11.16   (a) Temperature dependence of the ZFC dc magnetization for the SbF_6 (triangles), AsF_6 (squares) and PF_6 (circles) salts of BDA-TTP under an applied magnetic field of 1 Oe. (b) Temperature dependence of the relative resistivities for the SbF_6 (circles) and AsF_6 (triangles) salts of BDA-TTP. The dotted line is a guide for the eye. The inset shows the relative resistivities in the low-temperature region. (Reprinted with permission from ref. 29)

fragile nature of the crystals, the superconducting transition of this salt could not be determined by resistive measurements. Further investigation of the superconductivity in β-(BDA-TTP)$_2$SbF$_6$ has been reported by Ishiguro *et al.*[34]

All these superconductors crystallized in the triclinic space group *P*1 and are isostructual. Fig. 11.17 shows the crystal structure of (BDA-TTP)$_2$SbF$_6$, in which the molecular packing mode is very similar to that in β-(BEDT-TTF)$_2$I$_3$.[4]   As shown in Fig. 11.18, the whole molecular structure of BDA-TTP in this salt is somewhat flatter than that in the neutral state (see Fig. 11.10b). Although the chair conformations of two outer dithiane rings remain unchanged, they are nonequivalent and their dihedral angles around the intramolecular sulfur-to-sulfur axis are 47.9° and 33.2°, respectively. The BDA-TTP molecules are stacked along the [101] direction and are somewhat dimerized: the donor molecules alternate at the average interplanar distances of 3.52 and 3.80 Å (Fig. 11.17a). The two donor molecules being 3.52 Å apart mutually shifted, while one pair with interplanar spacing of 3.80 Å has a nearly eclipsed arrangement in which there are two intermolecular S···S contacts slightly shorter than the van der Waals distance (3.70 Å). As shown in Fig. 11.17b, three intermolecular S···S distances close to 3.70 Å and two S···S contacts relatively shorter than 3.70 Å are observed between stacks. These structural characteristics of β-(BDA-TTP)$_2$SbF$_6$ also hold for the isostructural β-(BDA-TTP)$_2$AsF$_6$ and β-(BDA-TTP)$_2$PF$_6$.

The S···S contact pattern found in intra- and interstacks of each BDA-TTP superconductor reflects the small anisotropy of the intermolecular overlap integrals in the *ac* plane (Fig. 11.19). A tight-binding band calculation based on these overlap integrals also leads to the 2D band dispersion relation and nearly isotropic closed Fermi surface, which are peculiar to the β-structure, for each BDA-TTP superconductor (Fig. 11.20). Furthermore, it is noteworthy that the largest overlap integrals in all three BDA-TTP superconductors (14.7, 14.8 and 13.9 × 10^{-3} for the SbF$_6$, AsF$_6$ and PF$_6$ salts, respectively) are roughly half those in the known organic

(a)

(b)

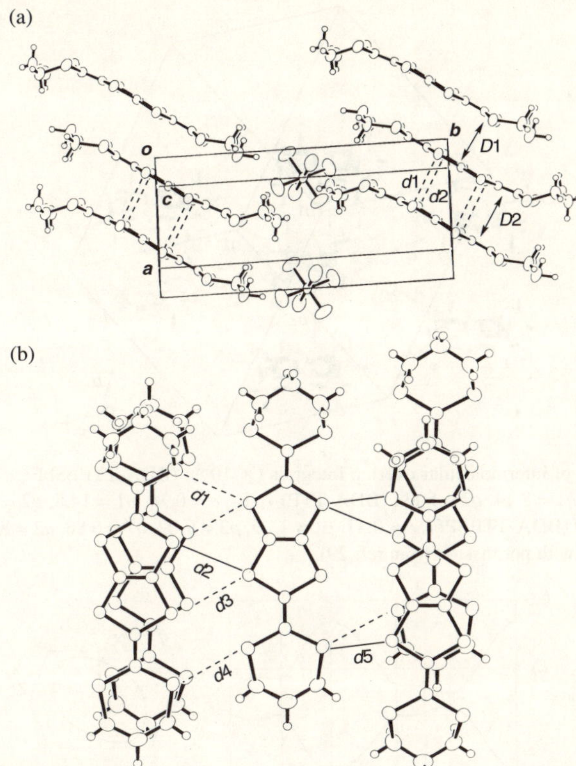

Fig. 11.17   Crystal structure of (BDA-TTP)$_2$SbF$_6$. (a) Intermolecular S⋯S contacts (< 3.70 Å) are indicated
by dotted lines: $d1$ = 3.694(1), $d2$ = 3.686(1) Å. Interplanar distances of the BDA-TTP column
are 3.52 ($D1$) and 3.80 ($D2$) Å. (b) Intermolecular S⋯S distances close to 3.70 Å are indicated
by dotted lines: $d1$ = 3.722(1), $d3$ = 3.694(1), $d4$ = 3.697(1) Å. Short intermolecular S⋯S contacts are indicated by thin lines: $d2$ = 3.645(1), $d5$ = 3.649(1) Å.
(Reprinted with permission from ref. 29)

Fig. 11.18   Molecular structure of BDA-TTP in (BDA-TTP)$_2$SbF$_6$. (Reprinted with permission from ref. 29)

superconductors such as β-(BEDT-TTF)$_2$I$_3$ (24.5 × 10^{-3})[35] and κ-(BEDT-TTF)$_2$Cu(NCS)$_2$ (25.7
× 10^{-3}),[36] indicating that they are characterized as a superconductor system with loose donor
packing.

11.3.4 Tetrachloroferrate (III) Salts of BDA-TTP

Recent studies on the development of molecular magnetic conductors by combining organic π-

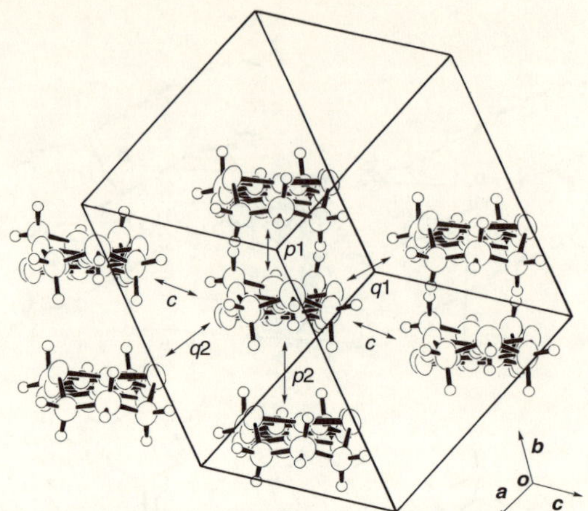

Fig. 11.19   The values of intermolecular overlap integrals ($\times$ 10^{-3}): (BDA-TTP)$_2$SbF$_6$, $c = -0.44$, $p1 = 14.7$, $p2 = 6.26$, $q1 = 8.14$, $q2 = 8.89$; (BDA-TTP)$_2$AsF$_6$, $c = 0.34$, $p1 = 14.8$, $p2 = 5.31$, $q1 = 7.59$, $q2 = 9.00$; and (BDA-TTP)$_2$PF$_6$, $c = 0.41$, $p1 = 13.9$, $p2 = 9.02$, $q1 = 6.88$, $q2 = 8.63$. (Reprinted with permission from ref. 29)

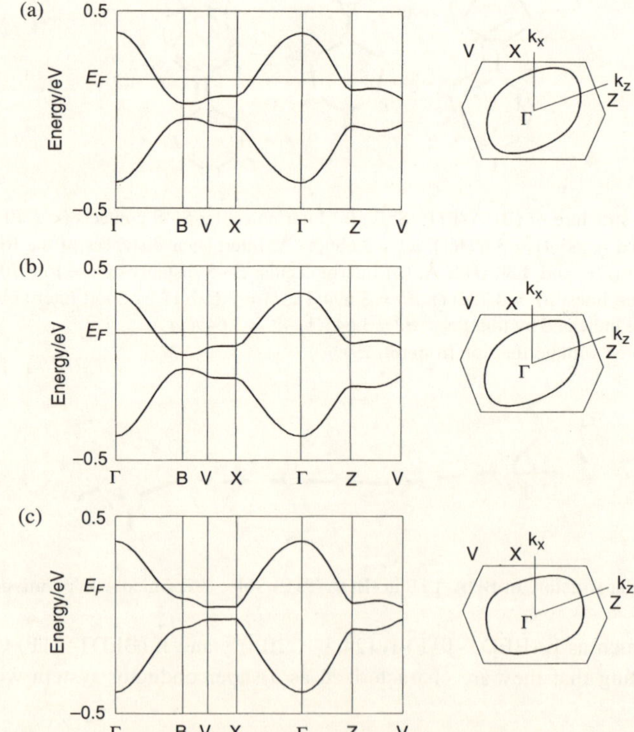

Fig. 11.20   Energy band structures and Fermi surfaces of (BDA-TTP)$_2$SbF$_6$ (a), (BDA-TTP)$_2$AsF$_6$ (b) and (BDA-TTP)$_2$PF$_6$ (c). (Reprinted with permission from ref. 29)

electron donors and inorganic paramagnetic anions have resulted in a new class of CT salts, such as an antiferromagnetic organic metal,[37] ferromagnetic organic metals[38] and antiferromagnetic organic metals exhibiting superconductivity,[39] in which magnetic order and metallic conductivity coexist. However, these CT salts are derived from TCF π-donors. Therefore, the successful formation of this class of multiproperty molecular materials consisting of non-TCF donors and paramagnetic anions has obvious appeal, but it has proven to be a challenge. For example, the $CuBr_2$ complexes based on new bis(1,3-dithiol-2-ylidene) donors containing no TTF unit have been studied, but these complexes failed to exhibit metallic conducting behavior.[40] In addition, as previously described, while κ-$(BDH\text{-}TTP)_2FeCl_4$ remained metallic down to a very low temperature (see Table 11.2), this salt did not exhibit magnetic order. Thus, we attempted to prepare the $FeCl_4$ salt of BDA-TTP.[33,41]

Two types of BDA-TTP salts containing the $FeCl_4^-$ anion, β-$(BDA\text{-}TTP)_2FeCl_4$ and the solvated $(BDA\text{-}TTP)_3FeCl_4\cdot PhCl$ salt, were obtained simultaneously during electrocrystallization in 5% EtOH–PhCl by the controlled-current method.

The crystal structure of β-$(BDA\text{-}TTP)_2FeCl_4$ was determined both at room temperature and at 95 K. Fig. 11.21a shows the crystal structure of this salt at room temperature, consisting of one $FeCl_4^-$ anion and two crystallographically independent BDA-TTP molecules. In this salt, the BDA-TTP molecules and the $FeCl_4^-$ anions are arranged in alternating layers along the b axis, so that the Fe$\cdots$Fe distance between the most adjacent anions along the b axis [19.535(6) Å] is considerably longer than those along the a and c axes [6.227(3) and 7.731(4) Å]. At 95 K, the Fe$\cdots$Fe distances along the a and c axes become shortened to 6.192(2) and 7.580(2) Å, respectively. Two independent BDA-TTP molecules have similar molecular structures, in both of

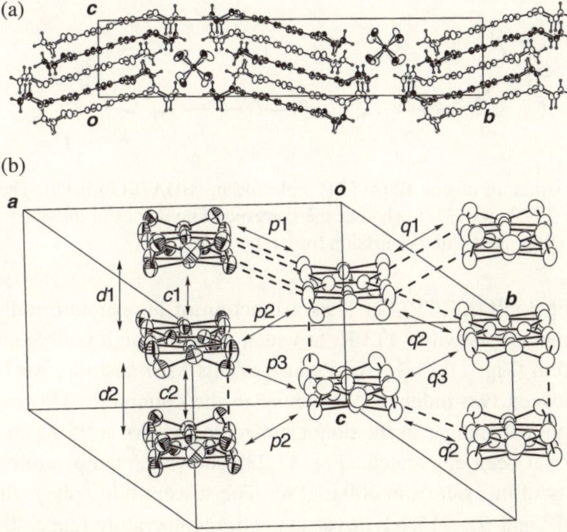

Fig. 11. 21   (a) Crystal structure of $(BDA\text{-}TTP)_2FeCl_4$ viewed down the a axis; open circles indicate the back molecules. (b) Donor arrangement of $(BDA\text{-}TTP)_2FeCl_4$. The interplanar distances of the BDA-TTP column are 3.56 ($d1$) and 3.89 ($d2$) Å. Intermolecular S$\cdots$S contacts (< 3.70 Å) are drawn by broken lines. The values of intermolecular overlap integrals ($\times 10^{-3}$) $c1$, $c2$, $p1$, $p2$, $p3$, $q1$, $q2$ and $q3$ are 14.4, 13.5, 1.88, –7.34, 3.44, 6.99, –5.62 and 5.49, respectively, whereas the corresponding values at 95 K are 15.4, 12.5, 1.49, –7.55, 3.19, 7.11, –5.65 and 5.39, respectively. (Reprinted with permission from refs. 33 and 41)

which the two outer dithiane rings adopt nonequivalent chair conformations: their respective dihedral angles around the intramolecular sulfur-to-sulfur axes in one BDA-TTP molecule are 38.6° and 13.0°, and the corresponding angles in the other are 39.1° and 13.0° (Fig. 11.22). Thus one dithiane ring of each independent BDA-TTP molecule is much flatter than that of each BDA-TTP molecule in the superconductors β-(BDA-TTP)$_2$X (X = SbF$_6$, AsF$_6$ and PF$_6$) (see *e.g.*, Fig. 11.18). Additionally, the two trimethylene end groups of each independent BDA-TTP molecule are found on the same side of the molecular plane; this is in contrast with the fact that those of the BDA-TTP molecules in both the neutral state and the superconducting salts appear with opposite orientation with respect to the molecular plane (see Figs. 11.10b and 11.18). The BDA-TTP molecules form slipped stacks along the c axis with some dimerization (average interplanar distances of 3.56 and 3.89 Å), so as to avoid the steric hindrance of the less flat dithiane ring. There is only one S···S contact shorter than the van der Waals distance (3.70 Å) within the stack, whereas several short intermolecular S···S contacts exist between stacks (Fig. 11.21b). At 95 K, although the interplanar distances in the stack are shortened to 3.47 and 3.84 Å, respectively, the number of short S···S contacts does not increase. The S···S contact pattern observed at room temperature, however, does not always reflect the magnitude of the intermolecular overlap integrals, calculated on the donor layer in the ac plane.[21] It is noteworthy that the largest overlap integral (14.4 × 10^{-3}) is almost equal to those found in the BDA-TTP superconductors, suggesting that the BDA-TTP molecules in this salt are also packed loosely. This loose donor packing motif remains almost unchanged at 95 K, as can be seen from the values of the intermolecular overlap integrals given in the caption to Fig. 11.21b. Moreover, even at 95 K, the conformations of two independent BDA-TTP molecules are analogous to those found at room temperature.

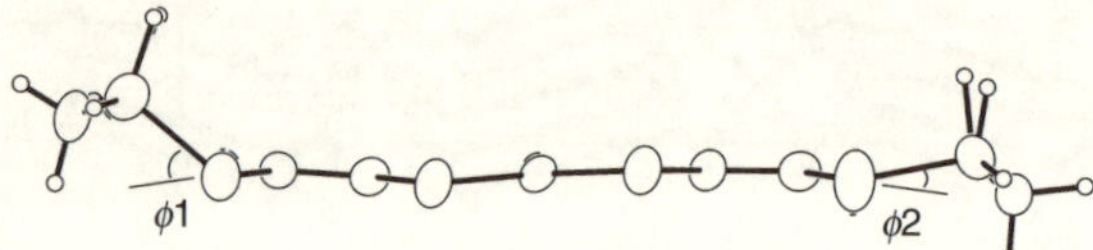

Fig. 11.22   Molecular structure of one BDA-TTP molecule in (BDA-TTP)$_2$FeCl$_4$. The dihedral angles ϕ1 and ϕ2 are 38.6° and 13.0°, whereas the corresponding angles in the other BDA-TTP are 39.1° and 13.0°. (Reprinted with permission from refs. 33 and 41)

The resistivity of β-(BDA-TTP)$_2$FeCl$_4$ as a function of temperature indicated that this salt (σ_{rt} = 9.4 S cm^{-1}) is metallic down to 113 K, at which temperature it undergoes a sharp metal-to-insulator (MI) transition (Fig. 11.23a). As a cause of this MI transition, we first anticipated the charge separation between two independent donors in the column.[41] However, in this salt, no appreciable conformational change in the donor molecules occurs at 95 K, so the reason for this MI transition has not yet been elucidated. Fig. 11.23b shows the temperature dependence of the magnetic susceptibility of this salt from 300 to 2 K. The susceptibility obeys the Curie-Weiss law (C = 4.48 emu K mol^{-1} and θ = −15.1 K) over the entire temperature range 40–300 K. From the fitted C close to the value (4.38 emu K mol^{-1}) predicted for a high-spin Fe^{3+} ion, the magnetic properties of this salt can be expected to be dominated by the anions. Below 40 K, the susceptibility increased to a maximum near 8.5 K, after which it decreased rapidly. As shown in Fig. 11.23c, the susceptibilities, which were measured under the magnetic fields applied along the directions approximately parallel to the crystallographic a, b and c axes, were anisotropic below near 8.5 K, indicating antiferromagnetic ordering with the Néel temperature (T_N) of near 8.5 K. Below T_N, the easy spin axis seems to lie close to the intrastacking c-direction of the donor

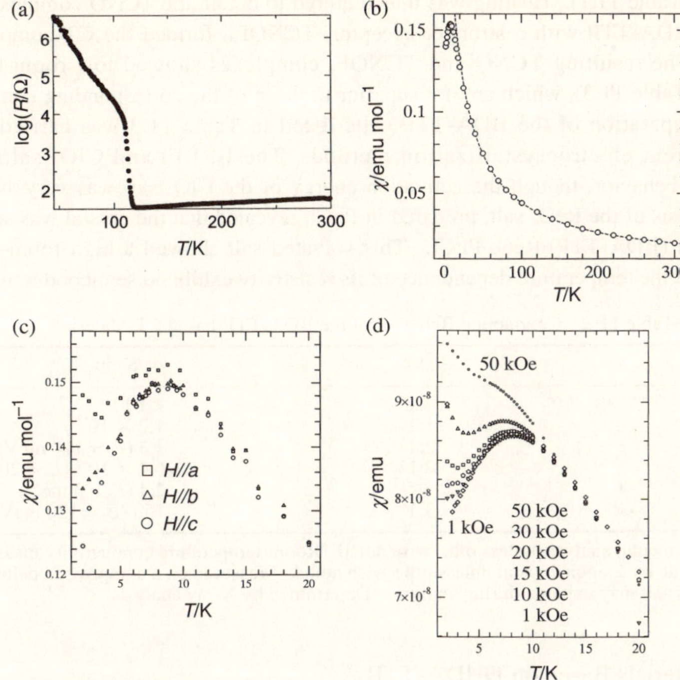

Fig. 11.23   (a) Temperature dependence of the resistance of $(BDA\text{-}TTP)_2FeCl_4$. (Reprinted with permission from refs. 33 and 41) (b) Temperature dependence of the susceptibility of $(BDA\text{-}TTP)_2FeCl_4$. The dashed line is a Curie-Weiss fit (see text). (Reprinted with permission from refs. 33 and 41) (c) Magnetic anisotropies in magnetic fields approximately parallel to the a, b and c axes ($H//a$, $H//b$ and $H//c$). (Reprinted with permission from refs. 33 and 41) (d) Magnetic field dependence of the susceptibilities of $(BDA\text{-}TTP)_2FeCl_4$. (Reprinted with permission from ref. 33)

molecules. So, considering that the shortest Fe···Fe distance along the c axis is longer than 6 Å, the donor molecules could mediate the observed antiferromagnetic order between the Fe^{3+} ions.[42] In addition, the magnetic field dependence of the susceptibilities using randomly orientated multiple crystals indicated that the antiferromagnetic transition begins to disappear near 30 kOe (Fig. 11.23d). These results demonstrate that we succeeded in finding the first non-TCF-based salt $\beta\text{-}(BDA\text{-}TTP)_2FeCl_4$ in which metallic conductivity and antiferromagnetism can coexist.

On the other hand, the solvated $(BDA\text{-}TTP)_3FeCl_4\cdot PhCl$ salt exhibited semiconducting behavior with an activation energy of 0.11 eV ($\sigma_{rt} = 2.0 \times 10^{-2}$ S cm^{-1}) and obeys the Curie-Weiss law with fitted $C = 4.42$ emu K mol^{-1} and $\theta = -0.35$ K from 300 to 2 K. This small negative Weiss constant signifies very weak antiferromagnetic interaction between the Fe moments.

11.3.5 Other CT Materials Based on BDA-TTP

Although the preparation of the TCNQ complex with BDH-TTP could be carried out at room temperature, our attempt to prepare the crystalline TCNQ complex with BDA-TTP at room temperature was unsuccessful probably due to the higher E_1 value of BDA-TTP relative to that of

BDH-TTP (see Table 11.1). Heating was thus required to obtain the TCNQ complex with BDA-TTP, whereas BDA-TTP with a stronger acceptor, TCNQF$_4$, formed the CT complex at room temperature. The resulting TCNQ and TCNQF$_4$ complexes showed low room-temperature conductivities (Table 11.3), which are not superior to those of the corresponding complexes with BDH-TTP. Preparation of the BDA-TTP salts listed in Table 11.3 was carried out by the controlled-current electrocrystallization method. The I$_3$, BF$_4$ and ClO$_4$ salts exhibited semiconductive behavior, though the activation energy of the ClO$_4$ salt was very small. X-ray diffraction analysis of the ReO$_4$ salt, prepared in PhCl, revealed that the crystal was solvated with the composition (BDA-TTP)$_3$ReO$_4\cdot$PhCl. This solvated salt showed a high room-temperature conductivity, but the temperature dependence of its resistivity exhibited semiconductive behavior.

Table 11.3   Conducting Behavior of the BDA-TTP-based CT Materials

Acceptor	D:A[a]	σ_{rt}/S cm^{-1}[b]
TCNQ	2:1	$< 10^{-6}$
TCNQF$_4$	1:1	1.2×10^{-5}[c]
I$_3^-$	2:1	1.3 (E_a = 110 meV)
BF$_4^-$	2:1	6.4×10^{-3} (E_a = 200 meV)
ClO$_4^-$	—[d]	2.4 (E_a = 3 meV)
ReO$_4^-$	3:1[e]	150 (E_a = 120 meV)

[a]Determined by elemental analysis unless otherwise noted. [b]Room-temperature conductivity measured by a four-probe technique on a single crystal unless otherwise noted. [c]Measured on a compressed pellet. [d]Not determined because this salt may explode during analysis. [e]Determined by X-ray analysis.

11.3.6 CT Materials Based on DHDA-TTP

Similar to the BDH-TTP donor, DHDA-TTP reacted with TCNQ and TCNQF$_4$ at room temperature to give CT complexes whose room-temperature conductivities were, however, very low (Table 11.4). The conducting behavior of the DHDA-TTP salts, prepared by the controlled-current electrocrystallization method, is also listed in Table 11.4. The DHDA-TTP salts with linear anions and a relatively small tetrahedral anion, such as I$_3^-$, AuI$_2^-$ and BF$_4^-$, remained metallic down to 2 K. Among these salts, the AuI$_2$ salt was found to have the κ-type structure with a 4:1 stoichiometry by X-ray analysis. Also, the larger tetrahedral ClO$_4^-$ anion and the octahedral PF$_6^-$ and AsF$_6^-$ anions led to metallic DHDA-TTP salts, which underwent MI transitions near 80, 30 and 60 K, respectively. X-ray analyses of these salts revealed that they were all β-type salts and isostructural. On the other hand, the larger octahedral SbF$_6^-$ anion with

Table 11.4   Conducting Behavior of the DHDA-TTP-based CT Materials

Acceptor	D:A[a]	σ_{rt}/S cm^{-1}[b]
TCNQ	3:1	$< 10^{-6}$[c]
TCNQF$_4$	3:1	$< 10^{-6}$[c]
I$_3^-$	3:1	53 (metallic > 2K)
AuI$_2^-$	4:1[d]	7.4 (metallic > 2 K)
BF$_4^-$	2:1	2.3 (metallic > 2 K)
ClO$_4^-$	2:1[d]	9.1 (T_{MI}[e] = $ca.$ 80 K)
PF$_6^-$	2:1	8.1 (T_{MI}[e] = $ca.$ 30 K)
AsF$_6^-$	2:1	39 (T_{MI}[e] = $ca.$ 60 K)
SbF$_6^-$	2:1[d]	2.6 (E_a = 16 meV)

[a]Determined by elemental analysis unless otherwise noted. [b]Room-temperature conductivity measured by a four-probe technique on a single crystal unless otherwise noted. [c]Measured on a compressed pellet. [d]Determined by X-ray analysis. [e]Temperature of MI transition.

DHDA-TTP formed a small gap semiconductor with an activation energy of 16 meV, and preliminary X-ray analysis suggested a κ-type donor arrangement.

Figure 11.24a shows the crystal structure of the metallic β-(DHDA-TTP)$_2$PF$_6$ salt with an MI transition in which the molecular packing motif is similar to those in the three isostructural superconductors β-(BDA-TTP)$_2$X (X = SbF$_6$, AsF$_6$ and PF$_6$). In this salt, the ethylene end group of the DHDA-TTP molecule is found essentially in the molecular plane, whereas the trimethylene end group is far out of the molecular plane. The outer dithiane ring adopts a chair conformation, and the dihedral angle around the intramolecular sulfur-to-sulfur axis is 40.6°, which is almost equal to the average value of the two corresponding angles of the BDA-TTP molecule in each BDA-TTP superconductor. The DHDA-TTP molecules are stacked along the [101] direction and slightly dimerized: the donor molecules alternate at the average interplanar distances of 3.66 and 3.70 Å. One pair with interplanar spacing of 3.66 Å has a slipped arrangement while the other with interplanar spacing of 3.70 Å has a nearly eclipsed arrangement. There are several S···S contacts shorter than the van der Waals distance (3.70 Å) between stacks (Fig. 11.24b), but no short S···S contact can be observed within the stack; this is in contrast to the S···S contact patterns found in the three BDA-TTP superconductors. The tight-binding band calculation, performed on the donor layer of this salt, results in the 2D band dispersion relation and closed Fermi surface shown in Fig. 11.25. The Fermi surface of this salt is less round than those of the three BDA-TTP superconductors, suggesting the occurrence of nesting responsible for the MI transition near 30 K.

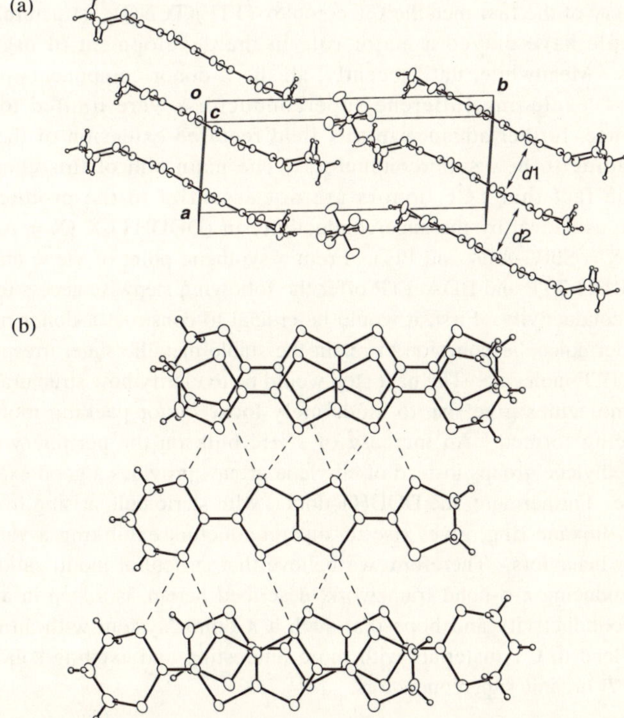

Fig. 11.24   Crystal structure of (DHDA-TTP)$_2$PF$_6$; (a) interplanar distances: $d1$ = 3.70, $d2$ = 3.66 Å; (b) intermolecular S···S contacts (< 3.70 Å) are indicated by broken lines.
(Reprinted with permission from ref. 30)

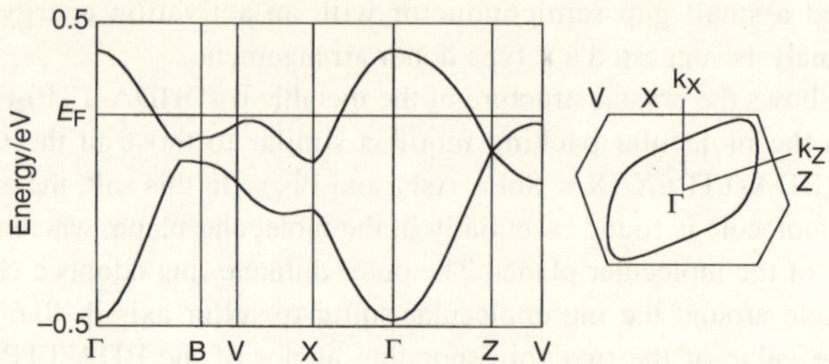

Fig. 11.25   Energy band structure and Fermi surface of $(DHDA\text{-}TTP)_2PF_6$.
(Reprinted with permission from ref. 30)

Eventually, the DHDA-TTP donor inherits to some extent the ability to form stable metallic CT salts from BDH-TTP. On the other hand, the octahedral PF_6^- and AsF_6^- anions, providing the isostructural superconducting BDA-TTP salts, also give the isostructural metallic DHDA-TTP salts, which exhibit MI transitions instead of superconductivity at ambient pressure.

11.4   Summary and Outlook

Since the discovery of the first metallic CT complex $(TTF)(TCNQ)$,[2] structural modifications of the TTF molecule have played a major role in the development of organic metals and superconductors. Meanwhile, until recently, all the π-donor components providing organic superconductors, excluding fullerene superconductors, were limited to the TCF-based donors;[1,43,44] hence, further advance in this field required extension of the scope of donor components leading to new superconductors. The main aim of this chapter has been to demonstrate the fact that TCF donors are not essential to the production of organic superconductors, as shown by the superconductivity in $(DODHT)_2X$ ($X = AsF_6$ and PF_6) and $(BDA\text{-}TTP)_2X$ ($X = SbF_6$, AsF_6 and PF_6). From a synthetic point of view, our results obtained from studies of BDH-TTP and BDA-TTP offer the following stepwise access to the achievement of organic superconductivity. First, it would be crucial to construct π-donor molecules that can generate tight intermolecular cohesion to retain the stable metallic state, irrespective of whether they contain the TCF molecule. The next step would be to clarify how structural modifications to those π-donor molecules give rise to moderately loose donor packing motifs when the CT materials are being formed. An increase of steric bulk on the periphery of BDH-TTP by introducing trimethylene groups instead of ethylene groups provides a good example of a way to address this issue. Furthermore, the DODHT donor, with steric bulk arising from the attachment of the *cis*-fused dioxane ring, gives rise to superconductors exhibiting a variety of pressure-induced resistive behaviors. Therefore, we believe that structural modification of the π-donor molecule by introducing a σ-bond framework, described herein, is a step in a new direction in achieving superconductivity and hope that such a π-donor system with an extended σ-bond framework will lead to CT materials with more interesting and exciting solid-state properties, including higher T_c organic superconductors.

References

1. J. M. Williams, J. R. Ferraro, R. J. Thorn, K. D. Carlson, U. Geiser, H. H. Wang, A. M. Kini and M.-H. Whangbo, *Organic Superconductors (Including Fullerenes): Synthesis, Structure, Properties, and*

Theory, Prentice Hall, New Jersey (1992); T. Ishiguro, K. Yamaji and G. Saito, *Organic Superconductors*, 2nd ed., Springer Ser. Solid-State Sci., Vol. 88, Springer, Berlin (1998).

2. J. Ferraris, D. O. Cowan, V. Walatka and J. H. Perlstein, *J. Am. Chem. Soc.*, **95**, 948 (1973); T. J. Kistenmacher, T. E. Phillips and D. O. Cowan, *Acta Cryst.*, B**30**, 763 (1974).

3. E. B. Yagubskii, I. F. Shchegolev, V. N. Laukhin, P. A. Kononovich, M. V. Kartsovnik, A. V. Zvarykina and L. I. Buravov, *Zh. Eksp. Teor. Fiz., Pis'ma Red.*, **39**, 12 (1984); V. F. Kaminskii, T. G. Prokhorova, R. P. Shibaeva and E. B. Yagubskii, *Zh. Eksp. Teor. Fiz., Pis'ma Red.*, **39**, 15 (1984).

4. J. M. Williams, T. J. Emge, H. H. Wang, M. A. Beno, P. T. Copps, L. N. Hall, K. D. Carlson and G. W. Crabtree, *Inorg. Chem.*, **23**, 2560 (1984).

5. G. C. Papavassiliou, A. Terzis and P. Delhaes, *Handbook of Organic Conductive Molecules and Polymers*, J. Wiley & Sons, Chichester (1997), Vol. 1, p 151; J. Becher, J. Lau and P. Mørk, *Electronic Materials: The Oligomer Approach*, Wily-VCH, Weinheim (1998), p. 198; J. L. Segura and N. Martín, *Angew. Chem. Int. Ed.*, **40**, 1372 (2001) and references therein.

6. Proceedings of the International Conference on Science and Technology of Synthetic Metals (ICSM) '98, Montpellier, *Synth. Met.*, **101–103** (1999); Proceedings of ICSM 2000, Gastein, *Synth. Met.*, **119–121** (2001) ; Proceedings of ICSM 2002, Shanghai, *Synth. Met.*, **135–137** (2003).

7. D. L. Coffen, J. Q. Chambers, D. R. Williams, P. E. Garrett and N. D. Canfield, *J. Am. Chem. Soc.*, **93**, 2258 (1971).

8. T. Mori and H. Inokuchi, *Chem. Lett.*, 1873 (1992).

9. J. Yamada, S. Takasaki, M. Kobayashi, H. Anzai, N. Tajima, M. Tamura, Y. Nishio and K. Kajita, *Chem. Lett.*, 1069 (1995).

10. H. Nishikawa, T. Morimoto, T. Kodama, I. Ikemoto, K. Kikuchi, J. Yamada, H. Yoshino and K. Murata, *J. Am. Chem. Soc.*, **124**, 730 (2002).

11. J. Yamada, Y. Amano, S. Takasaki, R. Nakanishi, K. Matsumoto, S. Satoki and H. Anzai, *J. Am. Chem. Soc.*, **117**, 1149 (1995); J. Yamada, S. Satoki, S. Mishima, N. Akashi, K. Takahashi, N. Masuda, Y. Nishimoto, S. Takasaki and H. Anzai, *J. Org. Chem.*, **61**, 3987 (1996).

12. J. Yamada, *Recent Res. Devel. in Organic Chem.*, Transworld Research Network, Trivandrum (1998), Vol. 2, p. 151.

13. J. Yamada, H. Nishikawa and K. Kikuchi, *J. Mater. Chem.*, **9**, 617 (1999).

14. K. Kikuchi, M. Kikuchi, T. Namiki, K. Saito, I. Ikemoto, K. Murata, T. Ishiguro and K. Kobayashi, *Chem. Lett.*, 931 (1987).

15. G. C. Papavassiliou, G. A. Mousdis, J. S. Zameounis, A. Terzis, A. Hountas, B. Hilti, C. W. Mayer and J. Pfeiffer, *Synth. Met.*, **27**, B379 (1988).

16. H. Anzai, J. M. Delrieu, S. Takasaki, S. Nakatsuji and J. Yamada, *J. Cryst. Growth*, **154**, 145 (1995); H. Nishikawa, T. Sato, T. Kodama, I. Ikemoto, K. Kikuchi, H. Anzai and J. Yamada, *J. Mater. Chem.*, **9**, 693 (1999).

17. K. Kikuchi, H. Nishikawa, T. Sato, T. Isaka, T. Kodama, I. Ikemoto, H. Anzai and J. Yamada, *Synth. Met.*, **102**, 1624 (1999).

18. J. Yamada, S. Tanaka, H. Anzai, T. Sato, H. Nishikawa, I. Ikemoto and K. Kikuchi, *J. Mater. Chem.*, **7**, 1311 (1997).

19. H. Nishikawa, H. Ishikawa, T. Sato, T. Kodama, I. Ikemoto, K. Kikuchi, S. Tanaka, H. Anzai and J. Yamada, *J. Mater. Chem.*, **8**, 1321 (1998).

20. J. Yamada, S. Tanaka, J. Segawa, M. Hamasaki, K. Hagiya, H. Anzai, H. Nishikawa, I. Ikemoto and K. Kikuchi, *J. Org. Chem.*, **63**, 3952 (1998).

21. T. Mori, *Bull. Chem. Soc. Jpn.*, **71**, 2509 (1998).

22. Y. Gimbert, A. Moradpour, G. Dive, D. Dehareng and K. Lahlil, *J. Org. Chem.*, **58**, 4685 (1993).

23. D. Lorcy, M.-P. L. Paillard and A. Robert, *Tetrahedron Lett.*, **34**, 5289 (1993).

24. M. Sallé, A. Gorgues, M. Jubault, K. Boubekeur, P. Batail and R. Carlier, *Bull. Soc. Chim. Fr.*, **133**, 417 (1996).

25. R. A. Aitken, L. Hill and P. Lightfoot, *Tetrahedron Lett.*, **39**, 7927 (1997).

26. M. R. Bryce, *Adv. Mater.*, **11**, 11 (1999).

27. K. Ueda, M. Iwamatsu, T. Sugimoto and H. Fujita, *Chem. Lett.*, 1197 (1997).

28. J. Yamada, M. Watanabe, H. Anzai, H. Nishikawa, I. Ikemoto and K. Kikuchi, *Angew. Chem. Int. Ed.*, **38**, 810 (1999).

29. J. Yamada, M. Watanabe, H. Akutsu, S. Nakatsuji, H. Nishikawa, I. Ikemoto and K. Kikuchi, *J. Am. Chem. Soc.*, **123**, 4174 (2001).

30. J. Yamada, M. Watanabe, T. Toita, H. Akutsu, S. Nakatsuji, H. Nishikawa, I. Ikemoto and K. Kikuchi, *Chem. Commun.*, 1118 (2002).

31. J. Yamada, R. Oka, T. Mangetsu, H. Akutsu, S. Nakatsuji, H. Nishikawa, I. Ikemoto and K. Kikuchi, *Chem. Mater.*, **13**, 1770 (2001).

32. H. Kobayashi, A. Kobayashi, Y. Sasaki, G. Saito and H. Inokuchi, *Bull. Chem. Soc. Jpn.*, **59**, 301 (1986).
33. K. Kikuchi, H. Nishikawa, I. Ikemoto, T. Toita, H. Akutsu, S. Nakatsuji and J. Yamada, *J. Solid State Chem.*, **168**, 503 (2002).
34. Y. Shimojo, T. Ishiguro, T. Toita and J. Yamada, *J. Phys. Soc. Jpn.*, **71**, 717 (2002).
35. T. Mori, A. Kobayashi, Y. Sasaki, H. Kobayashi, G. Saito and H. Inokuchi, *Chem. Lett.*, 957 (1984).
36. K. Oshima, T. Mori, H. Inokuchi, H. Urayama, H. Yamachi and G. Saito, *Phys. Rev.*, B **38**, 938 (1988).
37. T. Enoki, T. Umeyama, A. Miyazaki, H. Nishikawa, I. Ikemoto and K. Kikuchi, *Phy. Rev. Lett.*, **81**, 3719 (1998).
38. E. Coronado, J. R. Galán-Mascarós, C. J. Gómez-García and V. Laukhin, *Nature*, **408**, 447 (2000); J. Nishijo, E. Ogura, J. Yamaura, A. Miyazaki, T. Enoki, T. Tanaka, Y. Kuwatani and M. Iyoda, *Solid State Commun.*, **116**, 661 (2000).
39. H. Kobayashi, A. Kobayashi and P. Cassoux, *Chem. Soc. Rev.*, **29**, 325 (2000); H. Fujiwara, E. Fujiwara, Y. Nakazawa, B. Z. Narymbetov, K. Kato, H. Kobayashi, A. Kobayashi, M. Tokumoto and P. Cassoux, *J. Am. Chem. Soc.*, **123**, 306 (2001); T. Otsuka, A. Kobayashi, Y. Miyamoto, J. Kiuchi, S. Nakamura, N. Wada, E. Fujiwara, H. Fujiwara and H. Kobayashi, *J. Solid State Chem.*, **159**, 407 (2001).
40. M. Iwamatsu, T. Kominami, K. Ueda, T. Sugimoto, T. Adachi, H. Fujita, H. Yoshino, Y. Mizuno, K. Murata and M. Shiro, *Inorg. Chem.*, **39**, 3810 (2000).
41. J. Yamada, T. Toita, H. Akutsu, S. Nakatsuji, H. Nishikawa, I. Ikemoto and K. Kikuchi, *Chem. Commun.*, 2538 (2001).
42. H. Tanaka, T. Adachi, E. Ojima, H. Fujiwara, K. Kato, H. Kobayashi, A. Kobayashi and P. Cassoux, *J. Am. Chem. Soc.*, **121**, 11243 (1999).
43. K. Takimiya, Y. Kataoka, Y. Aso, T. Otsubo, H. Fukuoka and S. Yamanaka, *Angew. Chem. Int. Ed.*, **40**, 1122 (2001).
44. T. Imakubo, N. Tajima, M. Tamura, R. Kato, Y. Nishio and K. Kajita, *J. Mater. Chem.*, **12**, 159 (2002).

12

Bis(1,3-dithiol-2-ylidene) Donors with a Conjugated Spacer Group

12.1 Introduction

Insertion of a π-conjugated group into the tetrathiafulvalene (TTF) skeleton can modify remarkably the electronic structure as well as the molecular and crystal structures. Various kinds of compounds are obtained by changing the spacer groups. The most advantageous feature of this modification is that the Coulomb repulsion in the dication states is decreased due to the extended π-conjugation. Coulomb repulsion is one of the largest obstacles to electron transfer between molecules and its decrease is considered to lead to high conductivities. Small Coulomb repulsion is also favorable for π-stacking of the cationic species. Another advantage is that multidimensional π-π interactions are possible in the π-conjugated molecules. Increase in dimensionality is important to stabilize the metallic state, and intermolecular interactions through heteroatom contacts such as S$\cdots$S are usually used for this purpose. In addition, functional groups can be introduced into the π-conjugated units for exhibiting interesting physical properties. In this chapter TTF analogues containing various π-conjugated groups as spacers are described.

12.2 TTF Vinylogues

The parent compound of TTF vinylogue **1a** inserting one double bond into the TTF skeleton was first reported by Yoshida *et al.*[1] The electron-donating properties and Coulomb repulsion can be estimated from the cyclic voltammograms (CVs). The first oxidation potential is an estimate of the electron donating ability. The stronger electron donors exhibit lower oxidation potentials. On the other hand, the difference between the first and second oxidation potentials (ΔE) provides an estimate of the on-site Coulomb repulsion. The molecule with smaller Coulomb repulsion exhibits a smaller value of ΔE. Table 12.1 shows that the TTF vinylogue **1a** has a lower first oxidation potential and a smaller ΔE value than TTF, indicating that it is a stronger electron donor than TTF and has decreased Coulomb repulsion. The stronger electron-donating ability is attributed to the higher HOMO level as well as stabilization of the resulting cation radical by delocalization. The reduced Coulomb repulsion is related to the stabilization of the dication state.

Similar to the TTF series, several derivatives of the TTF vinylogues such as bis(ethylenedithio) derivative **1b** have been prepared.[2,3] Introduction of the substituent shifts the oxidation potentials to positive sides, indicating that the donor ability is weakened by the substitution. This trend is consistent with the substituent effect observed in TTF. However, the oxidation potential of **1b** is lower than that of BEDT-TTF, as shown in Table 12.1. The ethylenedithio groups have been replaced by other groups to give the derivatives including an

Table 12.1   Oxidation Potentials of TTF Vinylogues

Donor	E_1/V		E_2/V	$E_2 - E_1$/V	Ref.
1a	0.20		0.36	0.16	1
1b	0.48		0.71	0.23	3
2	0.33		0.47	0.14	4
4a		0.61 (2e)		0	8
4b	0.53		0.75	0.22	8
4c		0.67 (2e)		0	8
4d	0.66		0.85	0.19	8
4e	0.44		0.64	0.20	12
4f		0.49 (2e)		0	8
4g	0.38		0.60	0.22	8
4h		0.52 (2e)		0	8
4i	0.51		0.70	0.19	8
4j	0.31		0.51	0.20	12
17b	0.18		0.29	0.11	23
17c	0.45		0.53	0.08	23
17d	0.36		0.42	0.06	23
TTF	0.35		0.79	0.44	8
BEDT-TTF	0.50		0.85	0.35	8

unsymmetrical molecule **1c**.[3] X-ray structure analysis of **1c** reveals the transoid conformation which is expected for steric reasons. As novel substituents, hydroxymethyl groups are introduced to give **1d,e**.[4] Although hydroxy groups are of interest for the construction of supramolecular structures by hydrogen bonding, single crystals based on **1d,e** have not been obtained. The selenium analogue **2** has also been prepared.[5] It has higher oxidation potentials compared with the corresponding sulfur compounds. This is consistent with the fact that the oxidation potentials of tetraselenafulvalene (TSF) are higher than those of TTF.

1a R = R' = H
1b R,R = SCH$_2$CH$_2$S, R',R' = SCH$_2$CH$_2$S
1c R = Me, R',R' = SCH$_2$CH$_2$S
1d R = CH$_2$OH, R',R' = SCH$_2$CH$_2$S
1e R,R = CH$_2$OH

Although **1** and the simple derivatives have not afforded complexes showing metallic conductivities, Misaki *et al.* have found that the TTF fused compound **3** affords a cation radical salt exhibiting superconductivity.[6] This fact indicates that the TTF vinylogues are promising candidates as component molecules for superconductors.

The derivatives of the TTF vinylogues **4** containing substituents at the vinyl positions have been synthesized by oxidative coupling reaction of 1,4-dithiafulvenes **5** which are easily obtained by the Wittig reaction of a phosphonium salt **6** or the Wittig-Horner reaction of a phosphonate ester **7** with the corresponding benzaldehydes.[7,8] The reaction mechanism involves intermediates **8** followed by deprotonation to give **4**. Introduction of bulky groups at the vinyl positions usually causes steric interactions to result in nonplanar molecular structures which are unfavorable for intermolecular π-π interactions.[7] However, the derivatives containing phenyl substituents can take the planar geometry of the TTF vinylogue skeleton by twisting the phenyl groups.[8] The oxidation potentials of the TTF vinylogues **4** listed in Table 12.1 indicate that they are strong electron donors even though electron-withdrawing substituents are introduced on the phenyl

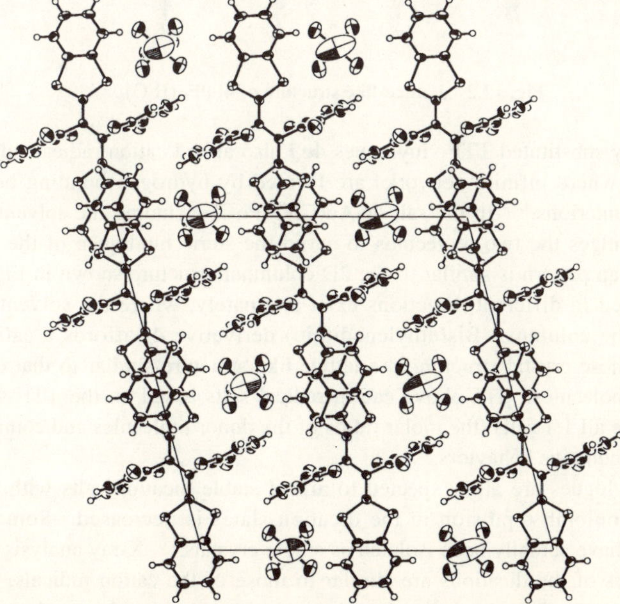

4a R,R = benzo, R' = C$_6$H$_5$
4b R,R = benzo, R' = o-ClC$_6$H$_4$
4c R,R = benzo, R' = p-ClC$_6$H$_4$
4d R,R = benzo, R' = 2,6-F$_2$C$_6$H$_3$
4e R,R = benzo, R' = o-HOC$_6$H$_4$
4f R,R = SCH$_2$CH$_2$S, R' = C$_6$H$_5$
4g R,R = SCH$_2$CH$_2$S, R' = o-ClC$_6$H$_4$
4h R,R = SCH$_2$CH$_2$S, R' = p-ClC$_6$H$_4$
4i R,R = SCH$_2$CH$_2$S, R' = 2,6-F$_2$C$_6$H$_3$
4j R,R = SCH$_2$CH$_2$S, R' = o-HOC$_6$H$_4$

groups. The redox behaviors are strongly affected by the substituents. Thus, the *p*-substituted derivatives as well as the non-substituted derivatives undergo one-stage two-electron oxidation, while the *o*-substituted derivatives show two reversible one-electron oxidation waves. This result suggests that the cation radicals of *o*-substituted phenyl derivatives are thermodynamically stable. This can be explained by considering that the *o*-substituents increase the steric interactions and make the phenyl groups twist. The cation radicals can be stabilized by delocalization in the planar TTF skeleton. Detailed studies on the redox behavior of the TTF vinylogues have been carried out to clarify the substituent effects and conformational change.[9]

Some cation radical salts have actually been obtained as single crystals from the *o*-substituted phenyl derivatives. For example, the cation radical salt **4d**·PF$_6$ has a two-dimensional (2D) columnar structure as shown in Fig. 12.1, where one donor molecule bridges two others to

Fig. 12.1   Two-dimensional crystal structure of **4d**·PF$_6$.

avoid steric interactions.[8] The 2D structures are tuned by changing the counteranion to bigger $FeCl_4$ and ReO_4.[10] In the $FeCl_4$ salt, the degree of overlap of the 1,3-benzodithiole parts decreases compared to the full-overlapping found in the $4d \cdot PF_6$ salt to involve the bigger size of the anion. In the ReO_4 salt, a zigzag 2D structure with an angle of nearly 90° is observed. The 2D structures have been observed only in the dibenzo derivatives, suggesting that π-π interactions are necessary for the formation of the unusual structures. On the other hand, the derivative **4i** containing bis(ethylenedithio) groups affords the cation radical salt $4i \cdot PF_6 \cdot (H_2O)_8$ with an unusual crystal structure.[10] The donor molecules are stacked in a one-dimensional (1D) manner, where the distance between the molecules is large (7.5 Å) due to the steric interactions of the phenyl groups. Therefore, a lattice-like structure with cavities is constructed as shown in Fig. 12.2, where water molecules are incorporated in the cavities. In the $4g \cdot Au(CN)_2$ crystal, 1D stacking of the donor molecules bearing bis(ethylenedithio) groups also occurs and the counteranions are sandwiched between the donor molecules.[11] The similar structure is formed when the chloro groups are replaced by the bromo ones.

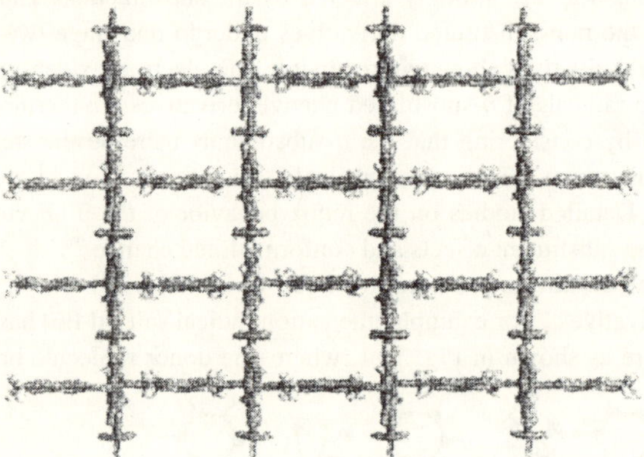

Fig. 12.2   Lattice-like structure of $4i \cdot PF_6 \cdot (H_2O)_8$.

Hydroxypheny substituted TTF vinylogues **4e,j** also afford cation radical salts with unusual crystal structures, where infinite networks are formed by hydrogen bonding between the OH groups and the counterions.[12] In the salt $4e \cdot Au(CN)_2 \cdot PhCl$ including the solvent molecule, one donor molecule bridges the two molecules to avoid the steric hindrance of the phenyl groups. Although the overlap pattern is similar to the 2D columnar structure shown in Fig. 12.1, the two 2D columns stacked in different directions exist alternately, where the solvent molecules are located between the columns. Bis(ethylenedithio) derivative **4j** affords a cation radical salt $4j \cdot ReO_4 \cdot (H_2O)_8$, whose crystal structure is a lattice-like structure similar to that of the salt of **4i** containing water molecules. The above cation radical salts based on the TTF vinylogues with diphenyl groups are all 1:1 salts (the molar ratios of the donor molecules and counterions are 1:1) and exhibit semiconductive behaviors.

The TTF vinylogues are also expected to afford stable dication salts with unusual crystal structures since Coulomb repulsion in the dication states is decreased. Some dication salts including Cu salts have actually been isolated as single crystals.[13] X-ray analysis reveals that the molecular structures of the dications are similar to those of the cation radicals, where the TTF vinylogue skeletons are planar and the phenyl groups are nearly orthogonal to them. Various unusual crystal structures have been observed, depending on the substituents and counteranions.

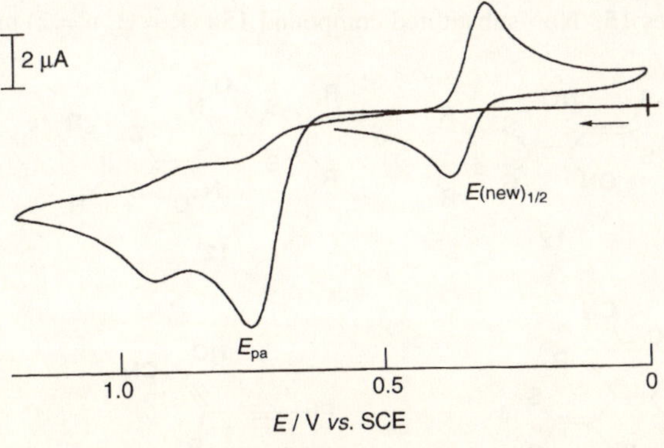

a R₁, R₂ = H
b R₁ = H, R₂ = SMe
c R₁ = H, R₂ = benzo
d R₁ = Me, R₂ = H
e R₁ = Me, R₂ = SMe
f R₁ = Me, R₂ = benzo

Instead of the phenyl groups, thiophene[14] and furan[15] have been substituted. The ferrocene-containing compounds have also been prepared by Bryce *et al.*[16] They exhibit four reversible one-oxidation waves in their CVs. The neutral molecular structure of one of the derivatives has a deformed geometry owing to steric interactions. The cation radical salts have not been isolated from the ferrocene derivatives.

The oxidative cyclization can be applied for an intramolecular reaction to give new bis(1,3-dithiole) electron donors. 2,2'-Bis(1,4-dithiafulven-6-yl)-3,3'-bithienyls **9** have been prepared by using either the Wittig or Wittig-Horner reaction from the corresponding dialdehydes.[17] The CVs of **9** reveal that they are irreversibly oxidized to give new redox active products which show reversible waves at lower potentials. The CV of **9b** is shown in Fig. 12.3. Chemical oxidation of **9** with tri(4-bromophenyl)aminium hexachloroantimonate affords cyclization products **10b-f** as dication salts in 86–95% yields. The formation of the dications **10** can be explained by oxidative coupling of **9** followed by dehydrogenation. Chemical reduction of **10** with zinc affords new bis(1,3-dithiole) compounds **11** whose oxidation potentials are consistent with those of the new peaks found in the CVs of **9** and are fairly low (*e.g.*, **11c**: 0.18 eV *vs.* SCE). They undergo two-electron oxidation at one stage. **11d** affords a conductive charge-transfer (CT) complex with TCNQ. X-ray structure analyses of the neutral and dication states of **11e** reveal that there is a large conformational change between them as shown Fig. 12.4. Thus, in the neutral state, the two 1,3-dithiole rings deviate largely from the plane of the tricyclic system to avoid the steric interaction between the S atoms of the 1,3-dithiole rings. Short intramolecular S···S contacts of 3.10 Å are observed as shown in Fig. 12.4a. In contrast to the neutral molecule, in the dications

Fig. 12.3   CV of **9b**.

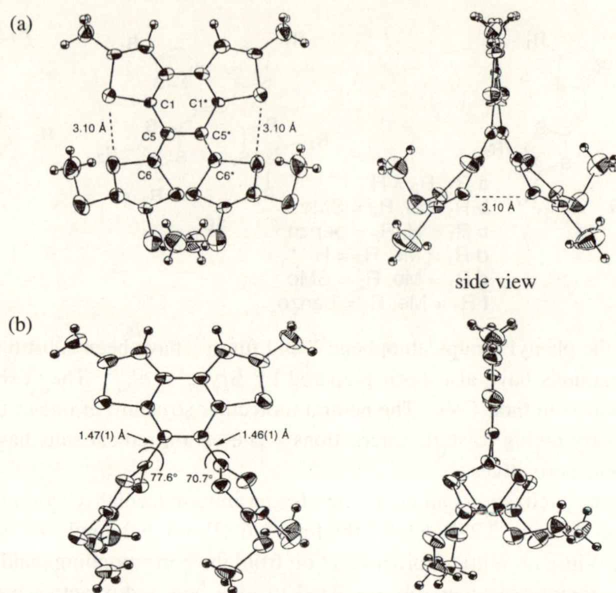

Fig. 12.4   Molecular structure of **11e** : (a) neutral state, (b) dication state.

the two 1,3-dithiole rings are almost orthogonal to the completely planar tricyclic system (Fig. 12.4b). The dihedral angles formed by the dithiole rings and the benzodithiophene moiety are 77.6° and 70.7°. It appears that the steric interactions between the S atoms found in the neutral molecule are reduced by the rotation of the 1,3-dithiole rings.

As unusual derivatives containing substituents at the vinyl positions of TTF vinylogues, **12** and **13** were reported. Although nitroso compounds are usually unstable, bis(nitroso) compounds **12** are stable due to the strong contribution of the resonance structure **12'**.[18] Compounds **13** containing a dicyanomethylene group have been prepared by an unusual metathesis reaction of TTF with acetylenes **14**. Bis(1,3-dithiole) compounds **13** are a kind of donor-acceptor compounds and the longest absorption maximum of **13** (R = CN) is observed at 511 nm.[19]

The number of spacer double bonds of the vinylogous TTF has been increased to give more extended TTF analogues **15**. Non-substituted compound **15a** (R = H, n = 2) undergoes one-stage

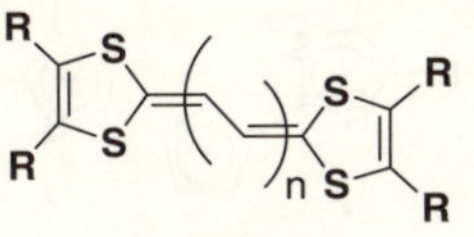

15

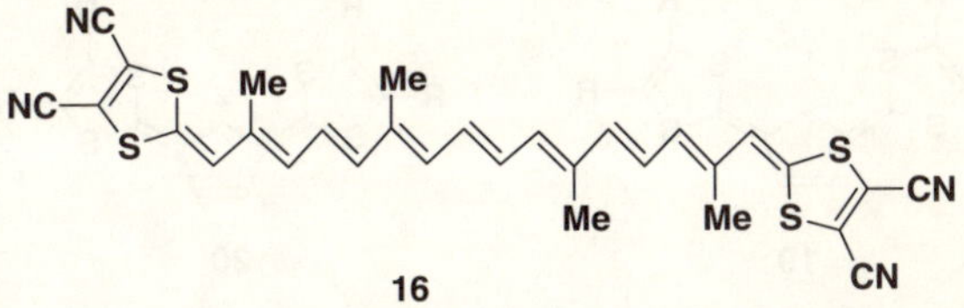

16

two-electron oxidation in contrast to **1a** due to the more reduced Coulomb repulsion.[20] A series of compounds (n = 0~6) containing dicyano groups have been synthesized.[21] The electron-withdrawing groups are introduced to stabilize the conjugated system. The π-extended compounds **15** (R = CN) with n = 2 and more exhibit a one-stage two-electron oxidation wave and the oxidation potentials gradually decrease with increasing conjugation. Interestingly, the longest absorption maximum shows a constant bathochromic shift of 16~20 nm with an increase of n number. Compound **16** containing methyl groups at the vinyl positions is the most extended one which shows the longest absorption maximum at 518 nm, which is 34 nm longer than that of **15** (R = CN, n = 6), indicating that no saturation of the absorption shift with increase of conjugation occurs until n = 8. X-ray crystal structure analysis of **15** (R = CN, n = 4) reveals that the rod-like planar molecule with transoid geometry forms a 1D columnar structure.[21]

When a cyclic olefin is inserted into the TTF skeleton, the double bond is fixed to a *cis* form in contrast to a *trans* form of **15**. Compound **17a** containing a cyclohexene unit between the TTF skeleton has been prepared by Sugimoto *et al.*,[22] whose CV shows two-electron oxidation at one stage similar to that of **15a** (R = H, n = 2). The derivatives **17b-d** have been prepared by the elimination reaction of cyclopentadiene from 1,3-dithiole compounds **18** obtained by using the Wittig-Horner reaction.[23] The derivatives **17b-d** undergo stepwise one-electron oxidation in contrast to **17a**, although the differences between the first and second oxidation potentials are small, as shown in Table 12.1. The donor **17b** affords the cation radical salts [(**17b**)$_2$·PF$_6$ and (**17b**)$_2$·AsF$_6$] as single crystals whose conductivities are almost constant to *ca.* 270 K. The crystal structures are similar to those of superconducting TMTSF salts, where the molecules are observed in a completely disordered manner and the distances between the molecules are long (3.66 and 3.67 Å) due to the steric interactions caused by the methylene protons.

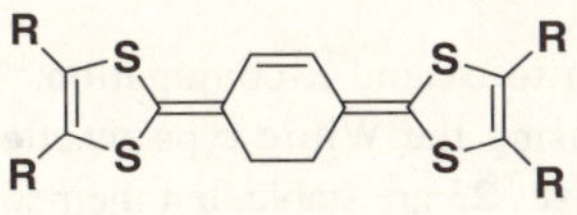

17 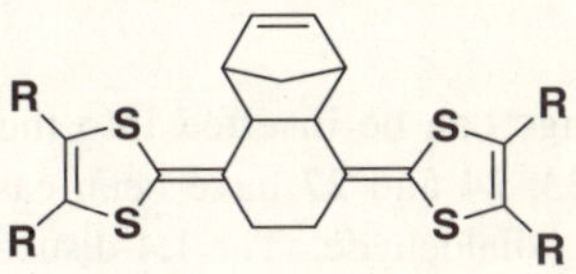   **18**

a R = H
b R = Me
c R,R = SCH$_2$CH$_2$S
d R,R = benzo

Compounds **19** containing five-membered rings have also been prepared using the Wittig-Horner reaction from cyclopentene-1,3-dione.[24] The CVs show irreversible waves and their cathodic peaks appear at much lower potentials than the anodic peaks. This result is explained by the fact that deprotonation reaction to give cations **20** occurs during the CV measurement. Actually, interesting deeply colored cations **20** have been isolated by chemical oxidation. They are stable cations due to the strong contribution of resonance form **20'**. The absorption wavelengths are strongly dependent on the substituents of the 1,3-dithiole ring and the alkylthio

19   **20**   **20'**

groups shift the absorption in the longer wavelength region. The absorption maximum of the bis(ethylenedithio) derivative is observed at 668 nm. The tetramethyl derivative shows strong fluorescence emission.

Bis(1,3-dithole) compounds containing acetylene units as spacer have been reported. Mono-acetylene compounds **21** have been obtained using the Wittig-Horner reaction from acetylenedicarboxyaldehyde or its acetal.[25] They show poorly reversible oxidation potentials in their CVs, suggesting that they are not good donors for affording conducting complexes. On the other hand, diacetylene compound **22b** possessing acetylenic substituents shows reversible two-stage one-electron oxidation waves with a potential difference of 0.12 V.[26] X-ray structure analysis of **22a** reveals that the two fulvene double bonds adopt the *trans* conformation to the diacetylene moiety. Compounds bearing diacetylene groups as substituent R of **22** were also synthesized. These donor molecules have not afforded conducting complexes.[26]

21

22a R = H
22b R = ———≡—Si(*i*-Pr)$_3$

12.3   TTF Analogues Containing Aromatic Rings as Spacer

Aromatic rings can be inserted into the TTF skeleton to extend π-conjugation.[27] Benzene derivatives **23**, **24** and **27** have been easily prepared using the Wittig type reaction from the corresponding dialdehyde. The 1,4-disubstituted derivatives **23** are stable, but their solubilities in common organic solvents are very low. The oxidation potentials are higher than that of the corresponding TTF.[28] It is general that compounds containing aromatic spacer units show lower electron donating abilities since the neutral states are stabilized by the aromatic ring. The 1,2-disubstituted derivatives **24** are labile to acid to rearrange to isomers **25**.[29] New heterocycles **26** were obtained by the reaction of **24** with bromine, where **25** are intermediates of the reaction. On the other hand, the 1,3-disubstituted derivatives **27** undergo an oxidative coupling reaction to give oligomers **28** containing TTF vinylogue units.[30] This reaction is considered to be caused by the fact that the 1,3-dithiole groups are not conjugated to each other due to the *meta* linkage.

Naphthalene derivatives **29** have also been prepared using the Wittig-Horner reaction.[31] Their CVs show irreversible oxidation peaks on the first scan and after successive scans new reversible peaks appear at lower potentials. This can be attributed to the formation of oligomers **30** by oxidative oligomerization as found in **27**. The weak interactions between the 1,4-dithiafulvene

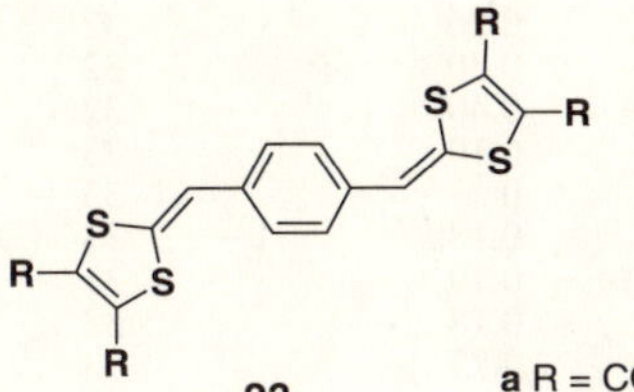

23   **a** R = CO$_2$Me   **b** R,R = benzo

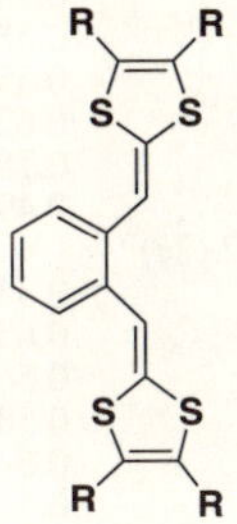

24 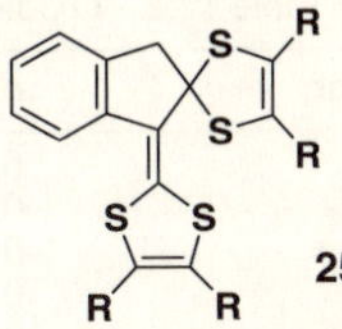   **25** 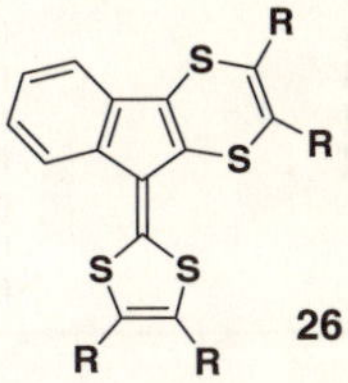   **26**

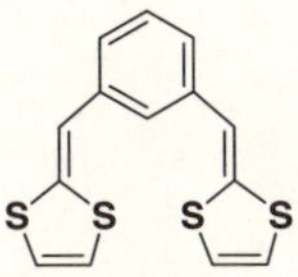

27 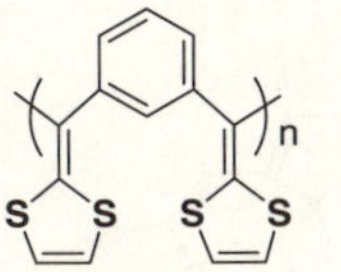   **28** 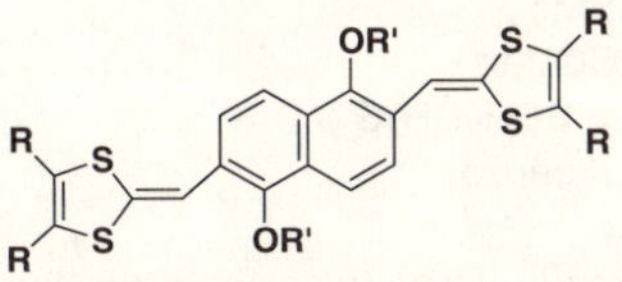   **29** 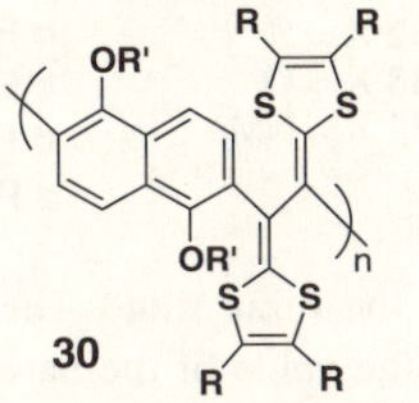   **30** 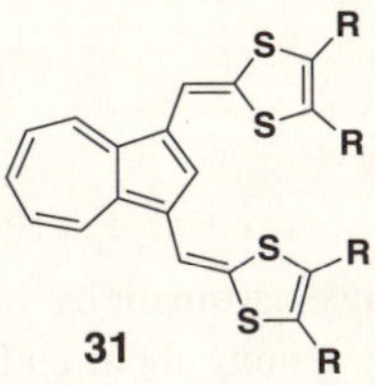   **31**

a R = H
b R = SMe
c R = SCH$_2$CH$_2$S
d R = benzo

units in the naphthalene derivatives seem to cause such an oligomerization reaction.

Novel bis(1,3-dithiole) donors **31** containing an azulene spacer unit have been prepared from 1,3-diformylazulene using the Wittig-Horner reaction.[32] The cyclic voltammetry reveals that they are stronger electron donors than TTF, as shown in Table 12.2. Even the dibenzo derivative **31d** shows lower oxidation potentials than TTF. Interestingly, they show reduction potentials (-1.42~ -1.49 V), resulting in an amphoteric character. X-ray structure analysis of **31d** reveals that one 1,4-dithiafulvenyl group and the azulene moiety are in an almost coplanar plane, while the other twists with a torsion angle of 20.8°. This is ascribed to the crystal packing force. The color of **31** in dichloromethane is greenish-yellow and the longest absorption maxima are observed at 733–762 nm.

The TTF analogues incorporating five-membered heterocycles such as thiophene, furan and pyrrole are expected to be better electron donors than the corresponding benzene derivatives since the aromaticity of these heterocycles are smaller than that of benzene. These compounds **32-34** have been synthesized by several groups.[33-36] As expected, their oxidation potentials are lower than those of the benzene derivatives, as shown in Table 12.2, indicating that the electron donating properties are increased compared to the corresponding benzene derivatives. The first oxidation potentials in the same substituent series decrease in the order thiophene > furan > pyrrole. The furan derivatives show lower oxidation potentials than the corresponding thiophene derivatives although the electronegativity of oxygen is larger than that of sulfur. This fact is

Table 12.2   Oxidation Potentials of TTF Analogues Containing Aromatic Rings as Spacer

Donor	E_1/V	E_2/V	$E_2 - E_1$/V	Ref.
31a	0.21	0.49	0.28	32
31b	0.32	0.58	0.26	32
31c	0.29	0.55	0.26	32
31d	0.32	0.62	0.30	32
TTF	0.38	0.78	0.40	32
32a	0.39	0.49	0.10	33
33a		0.32 (2e)	0	33
34a	0.20	0.38	0.18	33
32d	0.46	0.65	0.19	33
33d	0.38	0.56	0.18	33
34d	0.31	0.58	0.27	33
35c	0.55	0.64	0.09	36
37	0.38	0.46	0.08	39
38	0.38	0.56	0.18	39

32 X = S
33 X = O
34 X = NMe

a R = H
b R = SMe
c R = CO$_2$Me
d R,R = SCH$_2$CH$_2$S
e R,R = benzo

attributed to the lower aromaticity of the furan ring. The pyrrole derivatives show the lowest oxidation potentials among them and the value of the parent compound **34a** is even lower than that of TTF.[33] This finding is ascribed to the electron donating property of the pyrrole ring stabilizing the cationic species. The differences between the first and second oxidation potentials are reduced compared with that of TTF due to the extended π-conjugation. X-ray analysis reveals that **32a** has a *syn* conformation, where intramolecular short S···S contacts between the S atoms of the thiophene and 1,3-dithiole rings are observed.[35] The heteroatom contacts seem to favor the *syn* conformation. The cation radical salts of **32a** and **33a** have been isolated as 1:1 salts and the molecular and crystal structures are revealed by X-ray analyses. The donor molecules of both crystals have a *syn* conformation similar to the neutral structure of **32a**, where intramolecular S···S and S···O short contacts favorable for the *syn* conformation are observed.[37] Bond length analysis reveals that the positive charge is delocalized over the whole molecule and the degree is larger in the furan derivative **33a** than in the thiophene derivative **32a**. This is considered to be due to the difference in aromaticity between furan and thiophene. In **32a**·ClO$_4$ the cation radicals are dimerized, while a uniform stacking is observed in **33a**·BF$_4$. The conductivity of **33a**·BF$_4$ (10^{-2} S cm^{-1}) is higher than that of **32a**·ClO$_4$ (7×10^{-5} S cm^{-1}) owing to the enhanced delocalization.[37] TCNQ complexes are obtained from **33a** and **34c**. The conductivity of (**34c**)$_2$·TCNQ (10^{-2} S cm^{-1}) is better than that of (**33a**)$_2$·TCNQ$_3$ (10^{-5} S cm^{-1}).[33]

Bi- and terthiophene units have also been inserted into the TTF skeleton to give TTF analogues **35** and **36**.[36,38] In the case of the bithiophene spacer **35** (n = 1), the first and second oxidation potentials are coalesced as often observed in the π-extended TTF series. In the case of the terthiophene spacer **36** (n = 2), however, the difference between the first and second oxidation

35 (n = 1)
36 (n = 2)

a R = CO$_2$Me, **b** R,R = benzo, **c** R = SMe

37

38

potentials becomes larger. On the other hand, the bathochromic shifts in the electronic absorption spectra with increase of the conjugation length are not prominent in contrast to the steady shifts found in thiophene oligomers. These unusual properties can be explained by the fact that π-conjugation through the thiophene units is reduced in the case of bi- and terthiophenes due to rotation around the single bond linking the two thiophene rings although the 1,3-dithiole and thiophene parts are held planar by S$\cdots$S intramolecular contacts.

Such rotation can be suppressed by bridging the thiophene rings. The TTF analogue **37** containing such a rigid bithiophene spacer shows a much lower oxidation potential than the corresponding bithiophene derivative **35c**.[39] Even the oxidation potential of **38** containing an electron-accepting cyclopentadienone unit is lower than that of **35c**. Interestingly, **37** and **38** exhibit the first reduction potentials in a high potential region, resulting in a small difference between the first oxidation potential and the first reduction potential (0.90 and 0.94 V). Since this difference corresponds to the HOMO-LUMO energy gap, the redox data show that **37** and **38** have a small HOMO-LUMO energy gap. Their optical bandgaps estimated from the electronic absorption spectra are 1.3–1.4 and 1.2–1.3 eV for **37** and **38**, respectively. This result suggests that rigidification is important for reducing the HOMO-LUMO energy gap.

Poly(thienylvinylene)s exhibit smaller bandgaps than polythiophenes due to the decrease in aromatic character as well as the more planar structures. The thienylvinylene groups have been introduced as spacers into the TTF skeleton to give TTF analogues **39** and **40**.[40] Since the solubility decreases with increase in conjugation, alkyl groups have been introduced to enhance the solubility.[41] As expected, the absorption spectra show steady bathochromic shifts with increase in conjugation due to the constant decrease of HOMO-LUMO gaps. The derivatives **41** with the longest conjugation show the absorption maxima in the range of 540–563 nm, where the saturation of the conjugation does not occur yet. This optical gap is close to the bandgap of the parent polymer poly(thienylenevinylene). The first oxidation peaks in the CVs, whose potentials are almost independent of the length of the conjugation, involve a two-electron process. This suggests that the dications formed are localized at the 1,3-dithiole parts. The second and third peaks, which involve a one-electron oxidation process, are related to the formation of the trication

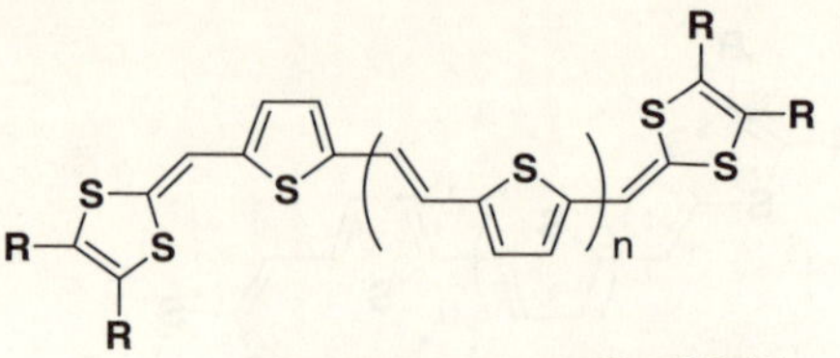

39a-d (n = 1)   **a** R = CO$_2$Me
40b,c (n = 2)   **b** R = SMe
                     c R = Pr
                     d R = H

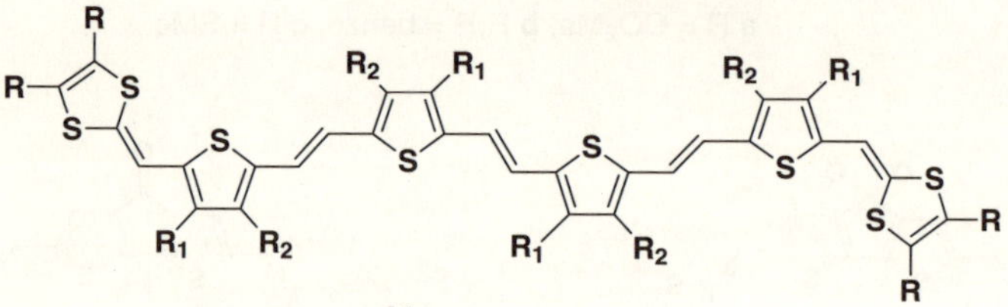

41

a R$_1$ = H, R$_2$ = n-Octyl, **b** R$_1$,R$_2$ = Butyl
R = SMe, n-Pr

radical and the tetracation. These potentials are little dependent on the substituents of the 1,3-dithiole, suggesting that the trication radical and the tetracation are formed by oxidation of the spacer part.

Furan-containing molecules **42** and **43** have also been prepared.[42] Since the aromatic character of furan is less than that of thiophene, more electron delocalization is expected in **42** and **43** than in **35** and **36**. Actually their oxidation potentials are lower than those of the corresponding thiophene analogues. Similar to the thiophenevinylene derivatives, the first oxidation peaks in their CVs involve a two-electron process. X-ray structure analysis of **42d** (R = H) reveals that the molecule is almost planar and short intramolecular S···O contacts between the 1,3-dithiole and furan units are observed, leading to a *syn* conformation. The methylthio derivative **42b** has a structure similar to that of **42d**. The cation radical salts of **42b** have been isolated as 1:1 salts.[43] The ClO$_4$ salt has a conductivity of 10^{-4} S cm^{-1}. The donor molecules in the crystal adopt a *syn* conformation similar to the neutral state. The intramolecular S···O distances are shorter than those in the neutral molecule, indicating a strengthening of the intramolecular S···O interactions in the oxidized state.

The π-conjugation in **39** and **42** has been further extended by introducing two or three double bonds into the heterocyclic rings to give **44-47**.[44] The first oxidation peaks in the CVs of these

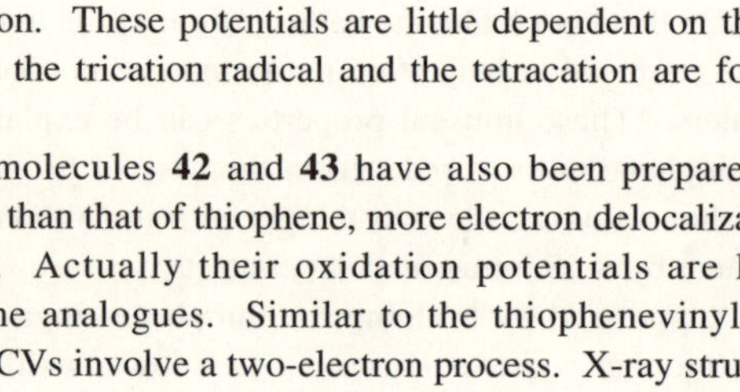

42 (n = 1)   **a** R = CO$_2$Me
43 (n = 2)   **b** R = SMe
                  c R = Pr
                  d R = H

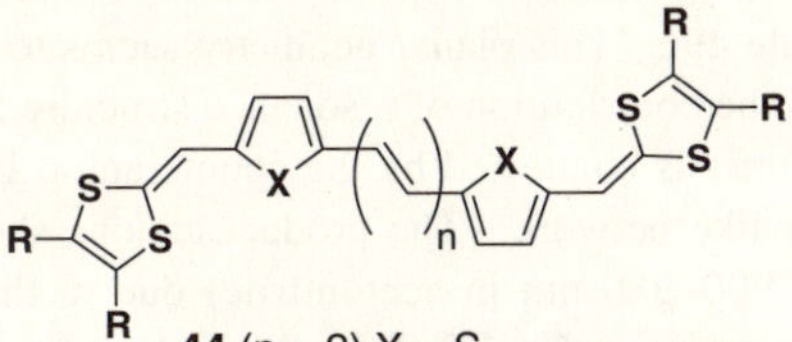

44 (n = 2) X = S
45 (n = 3) X = S
46 (n = 2) X = O
47 (n = 3) X = O

a R = CO$_2$Me
b R = SMe
c R = Pr

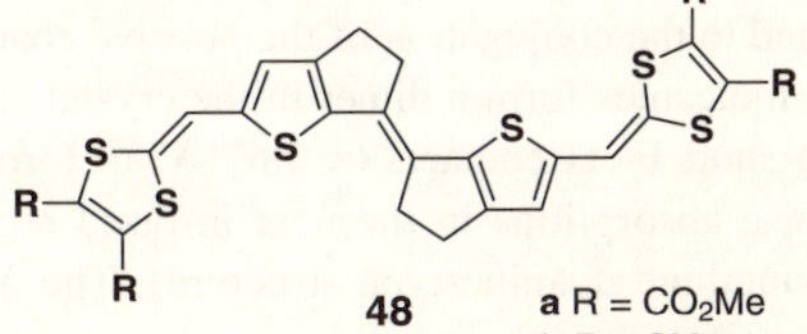

48

a R = CO$_2$Me
b R = SMe
c R = Pr

molecules involve a two-electron oxidation process to give their dications. The oxidation potentials become slightly lower with increasing the conjugation length. Furan-containing compounds show lower oxidation potentials than the corresponding thiophene analogues due to the lower aromaticity of furan. The difuryl compound **47c** with the longest conjugation shows the first oxidation potential at 0.16 eV *vs.* SCE, indicating that it is a much stronger donor than TTF. On the other hand, the absorption maxima of **45c** and **47c** are observed at a similar wavelength (512 nm), indicating that both compounds have a similar HOMO-LUMO gap.

TTF analogues **48** containing a rigid dithienylethylene spacer have been prepared by using the Wittig-Horner reaction of the corresponding dialdehyde.[45] They show two successive one-electron oxidation waves in their CVs in contrast to one-stage two-electron oxidation of the unbridged analogues. This is attributed to the decrease in the potentials for the formation of cation radicals in the bridged system, a fact that is explained by the enhanced delocalization of electron by rigidification. The oxidation potentials shift in more negative regions compared to those of the unbridged compounds. However, the extent of the shifts is smaller than that found in **37**, suggesting that the unbridged compounds **39** are already planar. X-ray analysis of **48c** reveals a planar geometry with intramolecular S⋯S interactions and a significant reduction of the bond length alternation by rigidification.

Bis(1,3-dithiole) compounds **49** and **50** have been prepared by the Wittig-Horner reaction of the corresponding diacetyl and dicaroboxyaldehydes, respectively.[46] X-ray structure analysis of **49c** reveals that the two thiophene rings are orthogonal, while the thienyldithiafulvenyl subunits are nearly planar. The interatomic distances between the S atoms of the thiophene and dithiole rings (3.01 and 3.02 Å) are shorter than the sum of the van der Waals radii, suggesting the presence of attractive interactions between them. The CVs of **49** and **50** show irreversible oxidation waves, but in the reverse scan new reduction peaks appear due to the formation of new products during the CV measurements. The products are novel cations **51** and **52**, which are formed by the deprotonation of the methylene protons in **49** and **50**. The cations **51** are prepared by oxidation of **49** with nitrosyl tetrafluoroborate (NOBF$_4$) or by hydride abstraction with trityl tetrafluoroborate. The successful preparation of **52b·BF$_4$** is achieved by using the latter method.

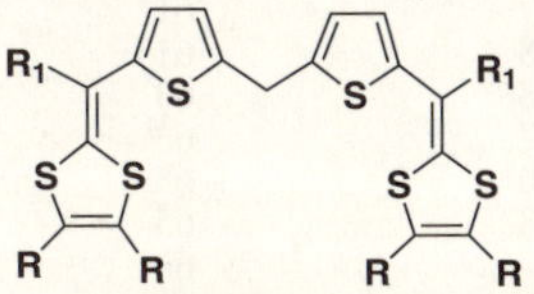

49 R$_1$ = Me
50 R$_1$ = H

a R = H
b R = SMe
c R,R = SCH$_2$CH$_2$S
d R,R = benzo

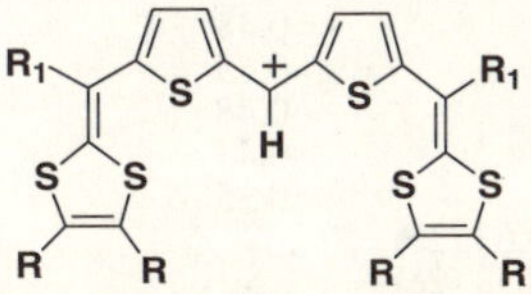

51 R$_1$ = Me
52 R$_1$ = H

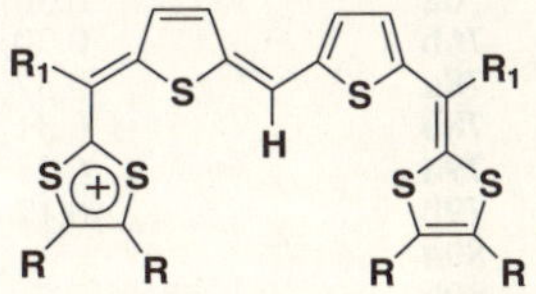

53

X-ray structure analysis of **52b**·BF$_4$ reveals that the thienylvinylene parts are almost planar in contrast to the twisted geometry of the neutral molecule **49c**. This planar geometry seems to be related to the conjugation of the positive charge due to the contribution of resonance structure **53**. The molecules form a dimer in the crystal and each dimer is connected by the counteranion BF$_4$ with short F···H contacts (< 2.67 Å) to form a ladder-like network. The product cations show intense absorptions in the near-infrared region (*ca.* 900–980 nm in acetonitrile) due to their polymethine cyanine-type structure. The alkylthio groups on the 1,3-dithiole rings make the absorptions strongly red-shift.

12.4   *p*-Quinodimethane Analogues of TTF

Insertion of a quinoid structure into the TTF skeleton enhances the donor ability since a new aromatic ring is formed upon oxidation, stabilizing the cationic species. Non-substituted 2,2'-*p*-quinobis(1,3-dithiole) **54a** has been synthesized using the retro-Diels-Alder reaction of **55a**, which is obtained using the Wittig-Horner reaction.[47] The oxidation potentials of **54a** are extremely low as shown in Table 12.3 and therefore unstable in solution. However, **54a** is stable in solid and can be isolated by carrying out the retro-Diels-Alder reaction of **55a** in the solid state. The derivatives **54b,c** possessing substituents on the 1,3-dithiole rings and **56** bearing substituents on the quinoid ring have been similarly prepared.[48] Introduction of substituents increase the oxidation potentials, as shown in Table 12.3, enhancing the stability.

This synthetic method including protection with cyclopentadiene has been applied for the synthesis of benzene and naphthalene fused compounds **57** and **58**.[47] Elimination reaction of cylopentadiene from **59** proceeds at around 200 °C and interestingly, this reaction easily occurs upon oxidation. Thus, during the CV measurement of **59**, new peaks assigned to those of the quinoid compounds appear. The CVs of **57** and **58** show that their oxidation potentials are still very low and one-stage two-electron oxidation occurs in contrast to **54**. This result indicates that the cation radical states in **57** and **58** are thermodynamically unstable probably due to the steric interaction between the peri-hydrogen atoms and the sulfur atoms in the 1,3-dithiole rings.

There exist larger steric interactions in dibenzo derivatives **60** caused by the peri-hydrogens. The dibenzo derivative **60d** has been prepared using the Wittig-Horner reaction of a phosphonate

Table 12.3   Oxidation Potentials of *p*-Quinodimethane Analogues of TTF

Donor	E_1/V	E_2/V	$E_2 - E_1$/V	Ref.
54a	-0.11	-0.04	0.07	47
56 (R = SMe)	0.26	0.47	0.21	48
57a		0.00 (2e)	0	47
58b		0.24 (2e)	0	47
60a		0.25 (2e, irrev.)	0	47
76a	0.36	0.53	0.17	61
76b	0.30	0.48	0.18	61
78a	0.37	0.55	0.18	63
78b	0.31	0.48	0.17	63
79a	0.23	0.43	0.20	65
79b	0.17	0.35	0.18	65
80a		0.63 (2e)	0	66
80b		0.58 (2e)	0	66
80d		0.80 (2e)	0	66
81a		0.73 (2e)	0	67
81b		0.66 (2e)	0	67
81c	0.76	0.85	0.09	67

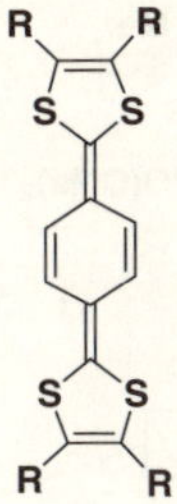

54
a R = H
b R = SMe
c R,R = benzo

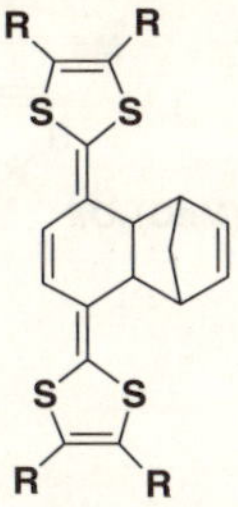

55

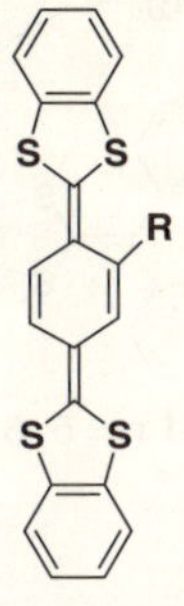

56
R = SMe, SC$_5$H$_{11}$
SC$_7$H$_{15}$, SC$_{10}$H$_{21}$,
Cl

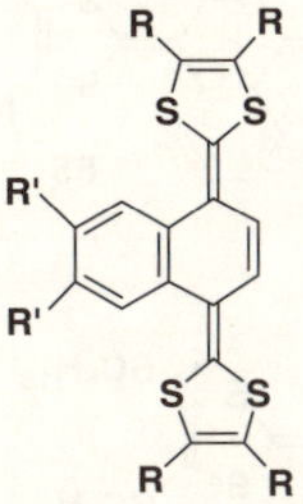

57 R' = H
58 R',R' = benzo

59
a R = H
b R,R = benzo

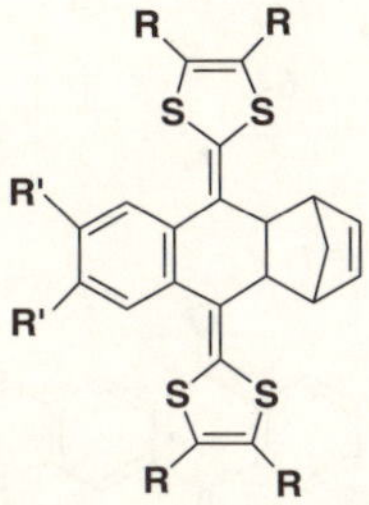

60
a R = H
b R = Me
c R,R = SCH$_2$CH$_2$S
d R,R = benzo

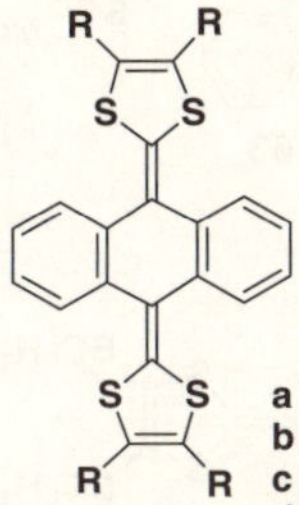

ester with anthaquinone by Akiba *et al.*[49] This was the first report of the use of the Wittig-Horner reaction to prepare bis(1,3-dithiole) electron donors. Later, the parent compound **60a**[47] and its derivatives **60b,c** have been synthesized by the Wittig-Horner reaction of the corresponding phosphonate esters with anthraquinone.[50] The corresponding 1,3-selenathiole compounds have also been prepared.[51] The CVs of **60** show one-stage two-electron oxidation peaks and the differences between the anodic and cathodic peak potentials are very large, suggesting that a conformation change takes place upon oxidation. X-ray analysis of the derivative actually reveals that the neutral molecule takes a nonplanar butterfly structure,[52] while the dication has a planar anthracene skeleton with almost orthogonal 1,3-dithiole groups.[53] The stable property of the dication is attributed to the aromatic 1,3-dithiolium ion as well as to the formation of an aromatic anthracene ring. The tetramethyl derivative **60b** forms a highly conductive 1:4 complex with TCNQ, which is paramagnetic, where the donor exists as a dication and a conducting path is formed along the TCNQ stacks.[53]

Modification of substituents on the 1,3-dithiole units leads to unusual bis(1,3-dithiole) compounds. Cyclophanes **61** have been obtained by the Wittig-Horner reaction of bis(1,3-dithiolyl)diphosphonate reagents **62** with anthraquinone.[54] X-ray structure analysis of **61** (n = 5) reveals the deformed molecular structure with a cavity, where the folding of the anthracene moiety and the 1,3-dithiole rings is enhanced by the pentamethylene bridge. The cyclic voltammetry data show that the rigidity of **61** restricts the conformational change upon oxidation. Diol derivatives **63** react with dicarbonyl chloride derivatives of benzene, thiophene, ferrocene and diphenyl ether **64** to give a series of cyclophanes **65**. These molecules show two-electron, quasi-reversible oxidation waves to afford the dications. The unusual structures involving a saddle-shaped conformation of the bis(1,3-dithiol-2-ylidene)-9,10-dihydroanthracene moiety are revealed by X-ray structure analysis.[55]

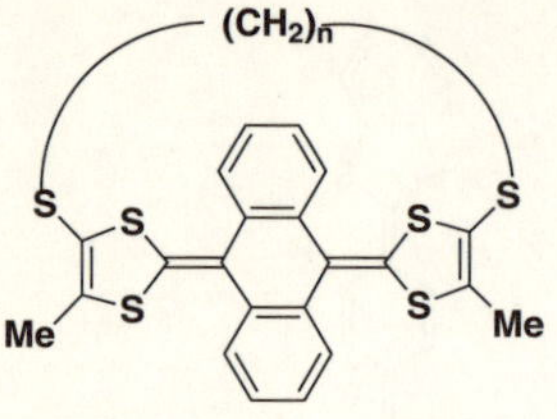

61 n = 5, 6

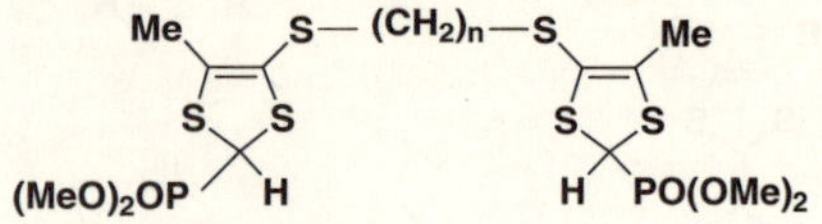

62 n = 5, 6

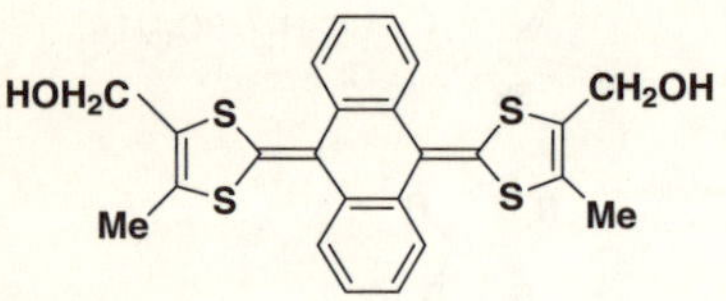

63

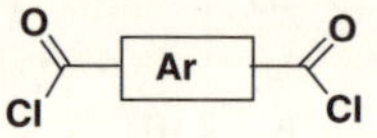

64

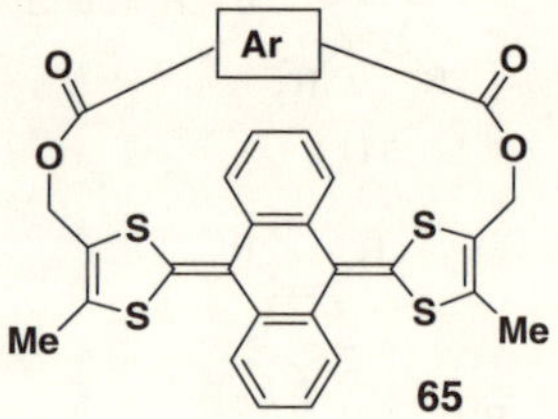

65

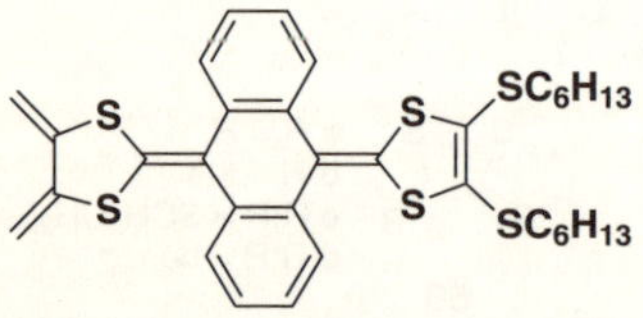

66

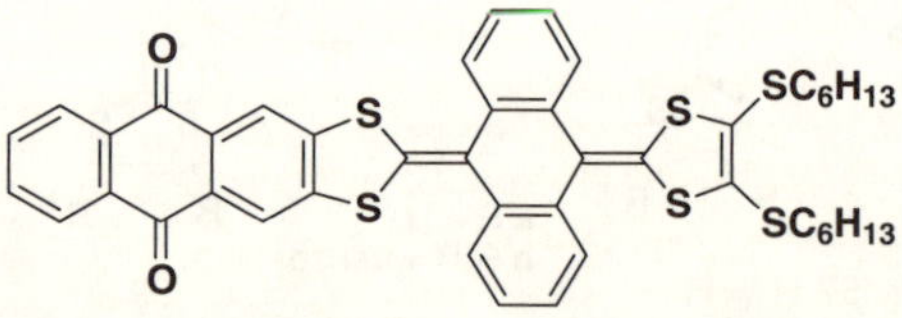

67

Bryce *et al.* have also prepared compound **66** containing a butadiene moiety, which can be used as a diene for the Diels-Alder reaction. The compound **66** reacts with naphthoquinone to give **67**, which is an interesting donor-π-acceptor compound. It has a saddle-like deformed structure and both the 1,3-dithiole rings are folded. A quasi-reversible reduction wave of the anthraquinone moiety is observed in addition to a quasi-reversible two-electron oxidation wave at +0.64 V *vs.* Ag/AgCl. The longest absorption maximum is observed at 428 nm, indicating that no effective intramolecular CT occurs.[56]

The fused benzene rings of **60** are also easily modified. For example, one of the benzene rings has been replaced by naphthalene and anthracene to give **68**, **69** and **70**.[57] The CVs show two-electron oxidation waves at one stage to give dications. The theoretical calculations show that the neutral and cation states adopt butterfly-shaped structures due to the steric hindrance

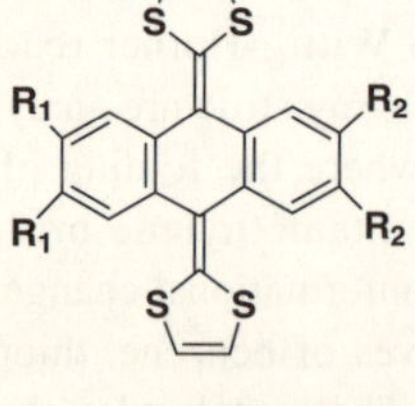

68 R$_1$= H, R$_2$,R$_2$ = benzo
69 R$_1$,R$_1$ = benzo, R$_2$,R$_2$ = benzo
70 R$_1$ = H, R$_2$,R$_2$ = naphtho

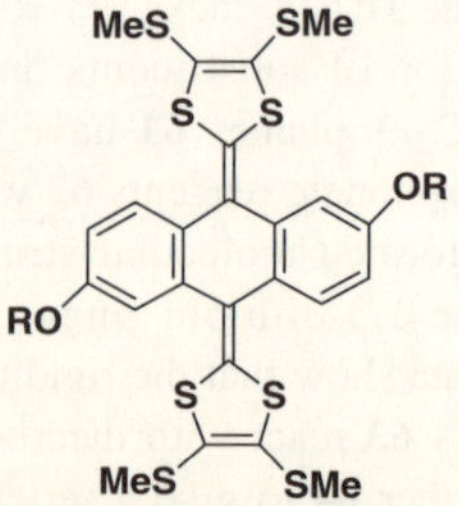

71 R = SiPh$_2$^{t}Bu
72a R = CO-ferrocene
72b R = CO-TTF

introduced by the benzoannelation, while the dications have a planar polyacene moiety with two orthogonal charged 1,3-dithiole rings. They afford CT complexes with a stronger electron acceptor tetrafluoro-TCNQ (TCNQF$_4$) although the complexes with TCNQ have not been obtained. The absorption maxima of **70** bearing an anthracene unit are red-shifted compared to those of **68** and **69**, suggesting that a CT from the 1,3-dithiol-2-ylidene moieties to the fused anthacene occurs.

The benzene ring of **60** can be substituted with functional groups. Bis(silyloxy) derivative **71**, which is obtained from 2,6-dihydroxyanthraquinone, has been used as an intermediate to give derivatives **72** containing ferrocene and TTF units which are a novel type of donor-σ-donor-σ-donor compounds.[58] The cyclic voltammetry data show that there are some inter- or intramolecular interactions between the different redox moieties in the system. Diol derivatives **73** containing the skeleton of bis(1,3-dithiole) compounds **60** have been similarly prepared using the Wittig-Horner reaction.[59] The diols **73** afford cyclophanes **74** by reaction with 1,4-benzenedicarbonyl chloride. The saddle-like deformed structure is revealed by X-ray analysis. The compound **74** (n = 0) forms a dication salt **74**$^{2+}$·(I$_3^-$)·(I$_2$)$_{0.5}$ whose structure is different from that of the neutral molecule, confirming that a drastic conformational change occurs upon oxidation.

Dimeric compounds **75** have been reported by Martín *et al.*[60] These dimers behave electrochemically as two independent monomers although the potentials are slightly negatively shifted due to the electronegative oxygen atom. The dimeric donors **75** do not form CT complexes with TCNQ due to their weak electron-donating properties. However, they afford CT complexes with stronger electron acceptors such as TCNQF$_4$ and DDQ. These complexes with a composition of 1:3 (donor:acceptor) are semiconductors.

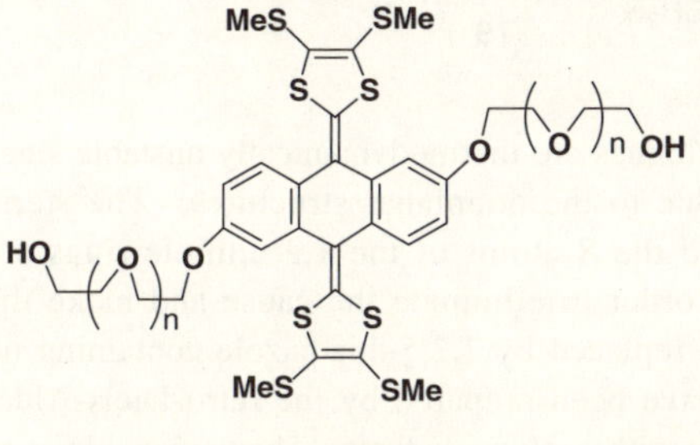

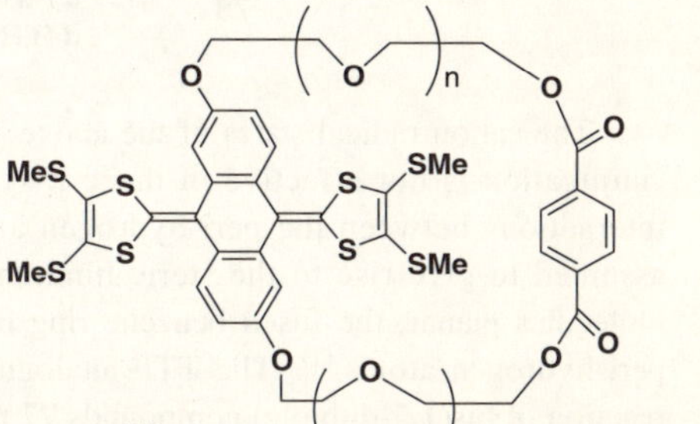

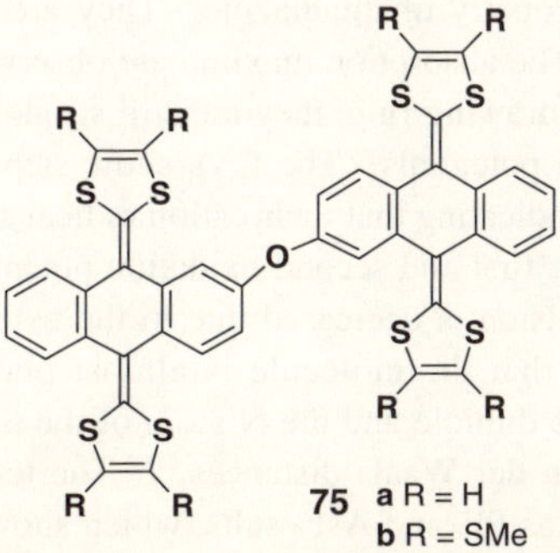

76

77

a R = H
b R = Me
c R,R = SCH$_2$CH$_2$S
d R,R = benzo

78

79

a R = H
b R = Me
c R,R = SCH$_2$CH$_2$S
d R,R = benzo

The cation radical states of the above TTF analogues are thermodynamically unstable since conjugation is not effective in those π-systems due to the nonplanar structures. The steric interactions between the peri-hydrogen atoms and the S atoms of the 1,3-dithiole rings are assumed to give rise to the steric hindrance. In order to eliminate the cause and make the molecules planar, the fused benzene ring has been replaced by 1,2,5-thiadiazole containing no peri-hydrogen atoms.[61] The TTF analogues **76** have been prepared by the retro-Diels-Alder reaction of bis(1,3-dithiole) compounds **77** protected with cyclopentadiene. They are stable even in solution in contrast to the parent compounds **54** because the oxidation potentials are increased due to the electron-withdrawing property of thiadiazole. They are a kind of donor-π-acceptor compounds and highly polarized. The absorption maxima are observed around 500 nm. Despite of the presence of the electron-withdrawing ring, they are still stronger electron donors than TTF as revealed by the lower oxidation potentials. The CVs show stepwise one-electron oxidation potentials as shown in Table 12.3, indicating that their cation radical states are thermodynamically stable. The differences between the first and second oxidation potentials are smaller than that of TTF, indicating that Coulomb repulsion is decreased due to the extended π-conjugation. X-ray structure analysis of **76d** reveals that the molecule is almost planar and the intramolecular distances between the S atom of the dithiole and the N atom of the thiadiazole (2.81 and 2.83 Å) are shorter than the sum of the van der Waals distances.[61] The tetramethyl derivative affords single crystals of the cation radical as PF$_6$ and AsF$_6$ salts, which show metallic behavior down to 100 K. Their crystal structures are similar to those of the superconducting TMTSF salts.[62] Similarly, bis(1,3-dithiole) compounds containing a selenadiazole ring instead of the thiadiazole ring have been prepared to give **78**.[63] The oxidation potentials of **78** are almost the same as those of **76** as shown in Table 12.3. The tetramethyl derivative affords a PF_6 salt of the cation radical, which shows metallic behavior down to 100 K. Pyrazine rings containing no peri-hydrogen atoms have also been introduced to give **79**.[64,65] They are stronger electron donors than the corresponding thiadiazole derivatives **76** due to the larger contribution of the quinoid structure in **79**. X-ray structure analysis of **79d** reveals that the planar molecules are unusually stacked in two directions.[65] Although the tetramethyl derivative **79b** affords a single crystal of a ClO_4 salt of the cation radical (3:2 salt), it shows semiconducting behavior.

Benzo-fused analogues **80** have been prepared by using the Wittig-type reaction from the corresponding ketones.[66] These molecules are expected to be nonplanar owing to the presence of the peri-hydrogen atoms. The butterfly-shaped structure of **80d** is actually revealed by X-ray analysis. Interestingly, the molecules are uniformly stacked along the *a* axis. The good overlap between the molecules results in a conductivity of 2.3 x 10^{-7} S cm^{-1} as a single component. Although the CV shows one-stage two-electron oxidation, the single crystals of the cation radical of **80d** have been isolated as PF_6 and ClO_4 salts upon electrochemical oxidation. They exhibit high conductivities despite the 1:1 composition. X-ray analysis of the PF_6 salt reveals that the donor molecule is nonplanar, similar to the neutral molecule, and the deformed donor molecules are uniformly stacked in a 1D manner as shown in Fig. 12.5. The good π-overlap between molecules results in the high conductivity as a 1:1 salt.[66]

Naphthalene fused derivatives **81** have also been synthesized using the Wittig-Horner reaction.[67] The bis(ethylenedithio) derivative **81c** shows a stepwise oxidation wave with a small difference between the first and second oxidation potentials in contrast to other derivatives **81a,b,d** undergoing one-stage two-electron oxidation. The donor **81c** affords cation radical salts as single crystals by electrochemical oxidation in tetrahydrofuran (THF). The crystals include the solvent molecule and the composition is 2:1:1 (donor:anion:solvent). The PF_6 and AsF_6 salts show metallic behavior down to 3 K.[68] X-ray analysis reveals that the butterfly-shaped donor molecules are uniformly stacked as shown in Fig. 12.6, where good π overlap between molecules is observed.[67] In addition, short intermolecular S⋯S contacts exist. The calculated overlap integral along the intercolumnar direction is about half that of the stacking direction, indicating that the salt is 2D in nature. Single crystals containing dihydrofuran (DHF) and dioxorane (DO) instead of THF are obtained and are found to have isomorphous structures similar to that of the THF-containing salt. Interestingly, the physical properties are strongly dependent on the solvent molecules. Thus, the stability of the metallic state decreases in the order: THF-containing salt < DO-containing salt < DHF-containing salt. This order is consistent with the ease of the ordering

80   **a** R = H   **81**   **82**

b R = Me

c R,R = SCH$_2$CH$_2$S

d R,R = benzo

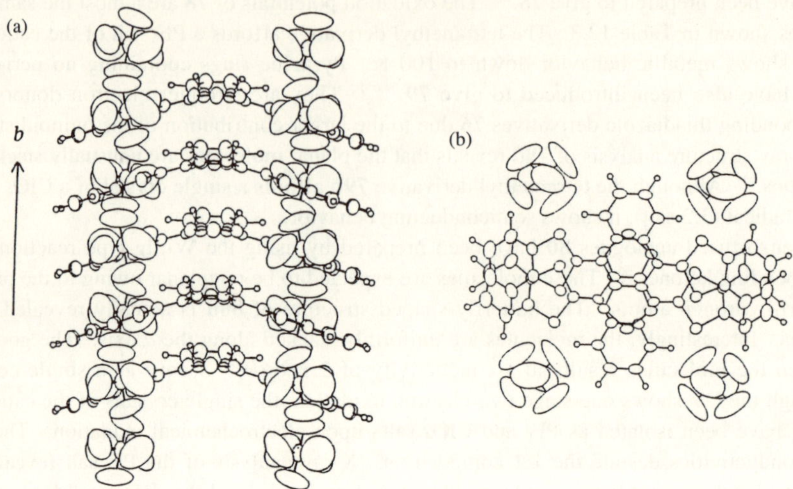

Fig. 12.5   Crystal structure of **80d**· PF$_6$: (a) stacking along the *b* axis, (b) overlap mode.

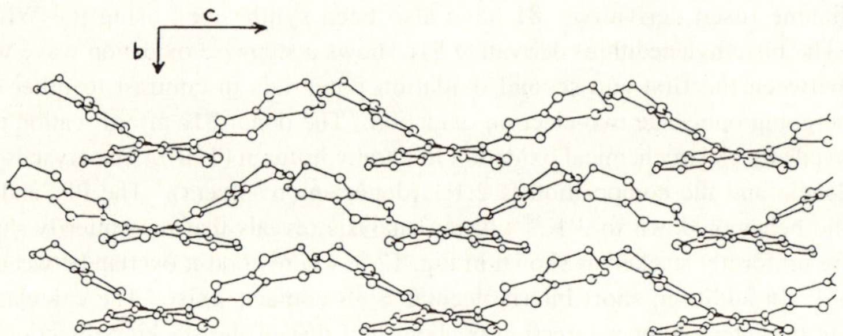

Fig. 12.6   Stacking mode of donor molecules in (**81c**)$_2$·PF$_6$·THF.

of solvent molecules at low temperatures, suggesting that the solvent effect is related to the ordering of the included molecules. Similar crystals and solvent effects are observed in the dimethyl derivatives **82**.[69] The size of the solvent molecules is found to be important for being included in the crystal.

In contrast to the above compounds containing fused-benzene or naphthalene rings, bis(thiadiazolo) compound **83a** bis [1,2,5]thiadiazolo-*p*-quinobis(1,3-dithiole) (BTQBT) is a planar molecule due to the absence of peri-hydrogen atoms.[70] The compound **83a** and its derivatives **83b-d** have been prepared using the Wittig-Horner reaction. BTQBT shows a good conductivity of 8.3 x 10^{-4} S cm^{-1} as a single component and a small anisotropy of conductivity (the ratio of the conductivity along the stacking direction to that along the intercolumnar direction is *ca.* 2). X-ray structure analysis reveals that the planar molecule forms a sheet-like network by short S···S contacts (3.26 Å) as shown in Fig. 12.7. The small anisotropy of conductivity is attributed to the electron conduction along the heteroatom contacts. The conductivities of the derivatives **83b-d** are poorer than that of the parent BTQBT, suggesting that the unique structure plays an important role in the good conductivity. Tetramethyl derivative **83b** has been found to have a different crystal structure from that of BTQBT. The high conductivity of BTQBT is

83   **a** R = H (BTQBT)
     b R = Me
     c R,R = SCH$_2$CH$_2$S
     d R,R = benzo

84 X = S (STQBT)
85 X = Se (BSQBT)

attributed to the strong intermolecular interactions leading to a wide bandwidth since the bandgap is not so small. In addition, BTQBT shows the Hall effect which is unusual in organic semiconductors.[71] The carrier is a hole and the mobility is high (4 cm^2 s^{-1} V^{-1} at room temperature). BTQBT affords a good quality film. Angle-resolved ultraviolet photoelectron spectra using synchrotron radiation have been measured for oriented thin films of BTQBT on a graphite surface.[72] The molecules in the thin film are estimated to lie flat on the surface. The film shows p-type semiconducting behavior.[73] The hole mobility and on/off ratio of the BTQBT film under high vacuum conditions reaches to 0.2 cm^2 s^{-1} V^{-1} and 10^8. These values are in the same order as those of the pentacene film, suggesting that BTQBT has great potential for use as the active layer in organic electronic devices.

The thiadiazole rings of BTQBT have been replaced by selenadiazole rings to give **84** (STQBT) and **85** (BSQBT).[63] These compounds have high melting points (> 400 °C) and are sublimed around 400 °C/0.1 torr. They scarcely dissolve in any solvents. The conductivity of STQBT as a single molecule is slightly higher than that of BTQBT. X-ray structure analysis of **84** shows that it crystallizes isomorphously with BTQBT, and the S and Se atoms in **84** are disordered in the crystal. The molecules form a sheet-like network by short S···S contacts (3.26

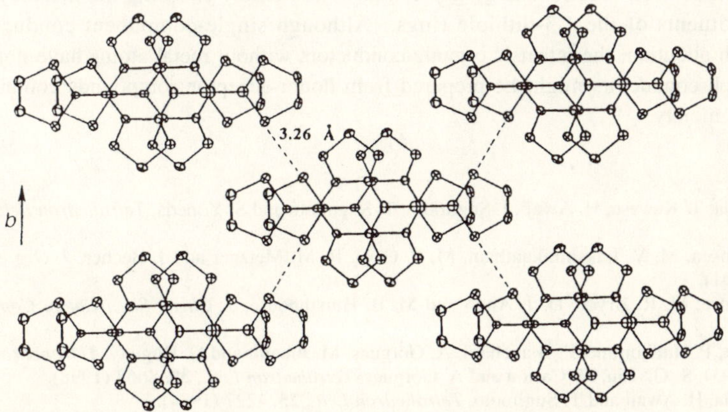

Fig.12.7   Crystal structure of BTQBT **83a**.

Å) and the sheet is well overlapped with an intermolecular distance of 3.48 Å. The superior conductivity of **84** over BTQBT is attributed to the selenium atom, which can increase the intermolecular interaction along the stack.

12.5   Summary and Outlook

TTF analogues containing π-conjugated spacer groups have been highlighted in this chapter. Many novel compounds with various π-conjugation units have been reported. The most remarkable feature found in these molecules is that Coulomb repulsion is decreased in the dication states due to extended π-conjugation.

TTF vinylogues with double bonds inserted in the TTF skeleton exhibit higher electron-donating properties as well as reduced on-site Coulomb repulsion. A TTF-fused derivative **3** afforded a cation radical salt showing superconductivity. This fact suggests that the TTF vinylogues are promising electron donors to afford organic superconductors. The introduction of substituents at the vinyl positions leads to the formation of 2D crystal structures to avoid steric interactions. This result indicates that TTF vinylogues can be used to construct multidimensional structures, which are important to stabilize the metallic state.

Aromatic rings have also been inserted into the TTF skeleton to extend π-conjugation. Various compounds have been prepared by changing the kinds and numbers of the aromatic rings, where the Wittig or Wittig-Horner reaction is conveniently used to synthesize them. The structures and properties depend on the aromatic spacer part. Among them, the derivatives containing heterocyclic oligomers as spacer units are particularly interesting since they have characteristics of both TTF derivatives and oligomers. Highly electron-donating linear π-conjugated systems are promising electron donors for conducting materials.

The use of a quinoid structure as a spacer is effective to raise the HOMO, enhancing the electron-donating property. The parent compound, 2,2'-*p*-quinobis(1,3-dithiole), **54a** is the strongest electron donor in the TTF analogues. The introduction of electron-withdrawing heterocyclic rings in the quinoid part leads to novel donor-acceptor compounds, some of which exhibit unusual physical properties such as semiconducting properties and non-linear optical properties. The HOMO-LUMO energy gap can be controlled by choosing the heterocyclic rings and the substituents of the 1,3-dithiole rings. Although single-component conductors have attracted much attention, the intrinsic organic conductors without metal atoms have not yet been obtained. Such conductors might be prepared from donor-acceptor compounds containing a π-extended TTF moiety.

References

1. Z. Yoshida, T. Kawase, H. Awaji, I. Sugimoto, T. Sugimoto and S. Yoneda, *Tetrahedron Lett.*, **24**, 3469 (1983).
2. T. K. Hansen, M. V. Lakshmikantham, M. P. Cava, R. M. Metzger and J. Becher, *J. Org. Chem.*, **56**, 2720 (1991).
3. A. J. Moore, M. R. Bryce, D. J. Ando and M. B. Hursthhouse, *J. Chem. Soc., Chem. Commun.*, 320 (1991).
4. C. Guillot, P. Hudhomme, P. Blanchard, A. Gorgues, M. Jubault and G. Duguay, *Tetrahedron Lett.*, **36**, 1645 (1995); S.-G. Liu, M. Cariou and A. Gorgues, *Tetrahedron Lett.*, **39**, 8663 (1998).
5. Z. Yoshida, H. Awaji and T. Sugimoto, *Tetrahedron Lett.*, **25**, 4227 (1984).
6. Y. Misaki, N. Higuchi, H. Fujiwara, T. Yamabe, T. Mori, H. Mori and S. Tanaka, *Angew. Chem., Int. Ed. Engl.*, **34**, 1222 (1995).
7. D. Lorcy, R. Carlier, A. Robert, A. Tallec, P. Le Maguerès and L. Quahab, *J. Org. Chem.*, **60**, 2443

(1995).

8. Y. Yamashita, M. Tomura, M. B. Zaman and K. Imaeda, *Chem. Commun.*, 1657 (1998).

9. N. Bellec, K. Boubekeur, R. Carlier, P. Hapiot, D. Lorcy and A. Tallec, *J. Phys. Chem. A*, **104**, 9750 (2000).

10. M. Tomura and Y. Yamashita, *Cryst. Eng. Comm.*, 14 (2000).

11. Y. Yamashita, M. Tomura, S. Tanaka and K. Imaeda, *Synth. Met.*, **102**, 1730 (1999).

12. Y. Yamashita, M. Tomura and K. Imaeda, *Tetrahedron Lett.*, **42**, 4191 (2001).

13. Y. Yamashita and M. Tomura, *J. Solid State Chem.*, **168**, 427 (2002).

14. A. Benahamed-Gasmi, P. Frère, J. Roncali, E. Elandaloussei, J. Orduna, J. Garin, M. Jubault and A. Gorugues, *Tetrahedron Lett.*, **36**, 2983 (1995).

15. U. Schöberl, J. Salbeck and J. Daub, *Adv. Mater.*, **4**, 41 (1992).

16. A. J. Moore, M. R. Bryce, P. J. Skabara, A. S. Batsanov, L. M. Goldenberg and J. A. K. Howard, *J. Chem. Soc., Perkin Trans. I*, 3443 (1997).

17. A. Ohta and Y. Yamashita, *J. Chem. Soc., Chem. Commun.*, 1761 (1995); A. Ohta and Y. Yamashita, *Mol. Cryst. Liq. Cryst.*, **296**, 1 (1997).

18. Y. A. Jackson, J. P. Parakka, M. V. Lakshmikantham and M. P. Cava, *J. Org. Chem.*, **62**, 2616 (1997).

19. H. Hopf, M. Kreutzer and P. G. Jones, *Angew. Chem., Int. Ed. Engl.*, **30**, 1127 (1991).

20. Z. Yoshida, T. Kawase, H. Awaji and S. Yoneda, *Tetrahedron Lett.*, **24**, 3473 (1983).

21. G. Märkl, A. Pöll, N. G. Aschenbrenner, C. Schmaus, T. Troll, P. Kreitmeier, H. Nöth and M. Schmidt, *Helv. Chim. Acta*, **79**, 1497 (1996).

22. T. Sugimoto, H. Awaji, I. Sugimoto, Y. Misaki, T. Kawase, S. Yoneda and Z. Yoshida, *Chem. Mater.*, **1**, 535 (1989).

23. Y. Yamashita, M. Tomura, M. Uruichi and K. Yakushi, *Mol. Cryst. Liq. Cryst.*, **376**, 19 (2002).

24. Y. Yamashita, S. Tanaka and M. Tomura, *J. Chem. Soc., Chem. Commun.*, 652 (1993).

25. A. Khanous, A. Gorgues and F. Texier, *Tetrahedron Lett.*, **31**, 7307 (1990); A. Khanous, A. Gorgues and M. Jubault, *Tetrahedron Lett.*, **31**, 7311 (1990).

26. M. B. Nielsen, N. N. P. Moonen, C. Boudon, J.-P. Giesselbrecht, P. Seiler, M. Gross and F. Diederich, *Chem. Commun.*, 1848 (2001).

27. J. Roncali, *J. Mater. Chem.*, **7**, 2307 (1997).

28. M. Salle, A. Belyasmine, A. Gorgues, M. Jubault and N. Soyer, *Tetrahedron Lett.*, **32**, 2897 (1991).

29. P. Frère, A. Gorgues, M. Jubault, A. Riou, Y. Gouriou and J. Roncali, *Tetrahedron Lett.*, **35**, 1991 (1994).

30. P. Frère, A. Benahamed-Gasumi, J. Roncali, M. Jubault and A. Gorgues, *J. Chim. Phys.*, **92**, 863 (1995).

31. S. González, N. Martín, L. Sánchez, J. L. Segura, C. Seoane, I. Fonseca, F. H. Cano, J. Sedó, J. Vidal-Gancedo and C. Rovira, *J. Org. Chem.*, **64**, 3498 (1999).

32. A. Ohta, K. Yamaguchi, N. Fujisawa, Y. Yamashita and K. Fujimori, *Heterocycles*, **54**, 377 (2001).

33. T. K. Hansen, M. V. Lakshmikantham, M. P. Cava, R. E. Niziurski-Mann, F. Jensen and J. Becher, *J. Am. Chem. Soc.*, **114**, 5035 (1992).

34. K. Takahashi, T. Nihira, M. Yoshifuji and K. Tomitani, *Bull. Chem. Soc. Jpn.*, **66**, 2330 (1993).

35. A. Benahamed-Gasmi, P. Frère, M. Jubault, A. Gorgues, J. Cousseau and B. Garrigues, *Synth. Met.*, **55-57**, 1751 (1993).

36. J. Roncali, L. Rasumussen, C. Thobie-Gautier, P. Frère, H. Brisset, M. Sallé, J. Becher, O. Simonsen, T. K. Hansen, A. Benahmed-Gasumi, J. Orduna, J. Garin, M. Jubault and A. Gorgues, *Adv. Mater.*, **6**, 841 (1994).

37. J.-F. Favard, P. Frère, A. Riou, A. Benahamed-Gasmi, M. Jubault, A. Gorgues and J. Roncali, *J. Mater. Chem.*, **8**, 363 (1998).

38. J. Roncali, M. Giffard, P. Frère, M. Jubault and A. Gorgues, *J. Chem. Soc., Chem. Commun.*, 689 (1993).

39. H. Brisset, C. Thobie-Gautier, M. Jubault, A. Gorgues and J. Roncali, *J. Chem. Soc., Chem. Commun.*, 1765 (1994).

40. E. Elandaloussi, P. Frère, J. Roncali, P. Richomme, M. Jubault and A. Gorgues, *Adv. Mater.*, **7**, 390 (1995).

41. E. Elandaloussi, P. Frère and J. Roncali, *Tetrahedron Lett.*, **37**, 6121 (1996); E. Elandaloussi, P. Frère and J. Roncali, *Chem. Commun.*, 301 (1997).

42. E. Elandaloussi, P. Frère, A. Benahamed-Gasmi, A. Riou, A. Gorgues and J. Roncali, *J. Mater. Chem.*, **6**, 1859 (1996).

43. E. Elandaloussi, P. Frère, A. Riou and J. Roncali, *New. J. Chem.*, 1051 (1998).

44. A. Benahamed-Gasmi, P. Frère, E. Elandaloussi, J. Roncali, J. Orduna, J. Garin, M. Jubault, A. Riou

and A. Gorgues, *Chem. Mater.*, **8**, 2291 (1996).

45. H. Brisset, S. Le Moustarder, P. Blanchard, B. Illien, A. Riou, J. Orduna, J. Garin and J. Roncali, *J. Mater. Chem.*, **7**, 2027 (1997).

46. A. Ohta and Y. Yamashita, *J. Chem. Soc., Chem. Commun.*, 557 (1995); A. Ohta and Y. Yamashita, *Heterocycles*, **44**, 263 (1997).

47. Y. Yamashita, Y. Kobayashi and T. Miyashi, *Angew. Chem., Int. Ed. Engl.*, **28**, 1052 (1989).

48. K. Saito, C. Sugiura, E. Tanimoto, K. Saito and Y. Yamashita, *Heterocycles*, **38**, 2153 (1994).

49. K. Akiba, K. Ishikawa and N. Inamoto, *Bull. Chem. Soc. Jpn.*, **51**, 2674 (1978).

50. M. R. Bryce and A. J. Moore, *Synth. Met.*, **27**, B557 (1988); A. J. Moore and M. R. Bryce, *Synthesis*, 26 (1991).

51. A. J. Moore and M. R. Bryce, *J. Chem. Soc., Perkin Trans. 1*, 157 (1991).

52. M. R. Bryce, M. A. Coffin, M. B. Hursthouse, A. I. Karaulov, K. Müllen and H. Scheich, *Tetrahedron Lett.*, **42**, 6029 (1991); M. R. Bryce, T. Finn, A. J. Moore, A. S. Batsanov and J. A. K. Howard, *Eur. J. Org. Chem.*, 51 (2000).

53. M. R. Bryce, A. J. Moore, M. Hasan, G. J. Ashwell, A. T. Fraser, W. Clegg, M. B. Hursthouse and A. I. Karaulov, *Angew. Chem., Int. Ed. Engl.*, **29**, 1450 (1990).

54. T. Finn, M. R. Bryce, A. S. Batsanov and J. A. K. Howard, *Chem. Commun.*, 1835 (1999).

55. N. Godbert, A. S. Batsanov, M. R. Bryce and J. A. K. Howard, *J. Org. Chem.*, **66**, 713 (2001).

56. C. A. Christensen, M. R. Bryce, A. S. Batsanov, J. A. K. Howard, J. O. Jeppesen and J. Becher, *Chem. Commun.*, 2433 (1999).

57. N. Martín, L. Sánchez, C. Seoane and C. Fernández, *Synth. Met.*, **78**, 137 (1996); N. Martín, L. Sánchez, C. Seoane, E. Ortí, P. M. Viruela and R. Viruela, *J. Org. Chem.*, **63**, 1268 (1998).

58. G. J. Marshallsay and M. R. Bryce, *J. Org. Chem.*, **59**, 6847 (1994).

59. C. A. Christensen, A. S. Batsanov, M. R. Bryce and J. A. K. Howard, *J. Org. Chem.*, **66**, 3313 (2001).

60. N. Martín, I. Pérez, L. Sánchez and C. Seoane, *J. Org. Chem.*, **62**, 870 (1997).

61   Y. Yamashita, S. Tanaka, K. Imaeda, H. Inokuchi and M. Sano, *Chem. Lett.*, 419 (1992).

62. Y. Yamashita and S. Tanaka, *Chem. Lett.*, 73 (1993).

63. Y. Yamashita, S. Tanaka and M. Tomura, *Phosphorus, Sulfur, and Silicon*, **67**, 327 (1992).

64. Y. Yamashita, T. Suzuki and T. Miyashi, *Chem. Lett.*, 1607(1989).

65. Y. Yamashita, S. Tanaka, K. Imaeda, H. Inokuchi and M. Sano, *J. Chem. Soc., Chem. Commun.*, 1132 (1991).

66. Y. Yamashita, K. Ono, S. Tanaka, K. Imaeda and H. Inokuchi, *Adv. Mater.*, **6**, 295 (1994).

67. Y. Yamashita, S. Tanaka and K. Imaeda, *Synth. Met.*, **71**, 1965 (1995).

68. K. Imaeda, Y. Yamashita, S. Tanaka and H. Inokuchi, *Synth. Met.*, **73**, 107 (1995).

69. Y. Yamashita, M. Tomura and K. Imaeda, *Chem. Commun.*, 2021 (1996).

70. Y. Yamashita, S. Tanaka, K. Imaeda and H. Inokuchi, *Chem. Lett.*, 1213 (1991); Y. Yamashita, S. Tanaka, K. Imaeda, H. Inokuchi and M. Sano, *J. Org. Chem.*, **57**, 5517 (1992).

71. K. Imaeda, Y. Yamashita, Y. Li, T. Mori, H. Inokuchi and M. Sano, *J. Mater. Chem.*, **2**, 115 (1992); K. Imaeda, Y. Li, Y. Yamashita, H. Inokuchi and M. Sano, *J. Mater. Chem.*, **5**, 861 (1995).

72. S. Hasegawa, S. Tanaka, Y. Yamashita, H. Inokuchi, H. Fujimoto, K. Kamiya, K. Seki and N. Ueno, *Phys. Rev., B* **48**, 2596 (1993).

73. M. Takada, H. Graaf, Y. Yamashita and H. Tada, *Jpn. J. Applied Phys.*, **41**, L4 (2002).

13

Bis(1,3-dithiol-2-ylidene) Donors with a Heteroquinonoid-spacer Group

13.1   Introduction

Conjugation-extended tetrathiafulvalene (TTF) analogues containing a linking π-spacer group between the 1,3-dithiole rings of TTF are expected to act as promising electron donors for organic conductors because the π-elongation of the TTF type donor is a very important molecular design strategy to decrease the on-site Coulombic repulsive energy[1] and then to increase the transfer integral by stabilizing the dication state of the donor relative to TTF. This molecular design strategy is much more important for the bis(ethylenedithio)tetrathiafulvalene (BEDT-TTF) type donors with chalcogen atoms at the peripheral part affording two-dimensional metals or superconductors, because it has recently been pointed out that the superconducting transition temperatures (T_c's) of superconductors are roughly proportional to the volume of the most effective space (V_{mes}) in which the carrier can delocalize effectively and the V_{mes} can be increased with increasing thickness of the effective conducting donor layer.[2] Therefore, conjugation-extended BEDT-TTF analogues containing a linking π-spacer group between the 4,5-ethylenedithio-1,3-dithiole rings are currently receiving attention in the search for new high T_c organic superconductors. However, the extended donors with a linking π-bridge consisting of a more than two sp^2 carbon chain such as an 1,4-butenediylidene group[3] or those extended by a benzoquinonoid bridge[4,5] are rather unstable in the neutral state and tend to give unstable cation radicals for which no single electron-donating nature has been observed in the redox reactions. Thus, the discussion has been focused on the point of how to prepare stable π-extended donors and how to thermodynamically stabilize the cation radicals of such π-extended donor systems, since the high conductivity is essentially associated with the stable cation radical state or stable partial electron donated state of the donor columns in the conducting salts.

We have already designed and synthesized various kinds of conjugation-extended quinonoid compounds with heteroquinonoid-spacer group(s)[6] and in these efforts we originally found that the heteroquinonoid-spacer group such as dihydrothiophenediylidene (thienoquinonoid), dihydrofurandiylidene (furoquinonoid) and dihydroselenophenediylidene (selenienoquinonoid) moieties are very effective to stabilize not only the π-extended quinons but also their anion radicals as well as their cation radicals.[7,8] This finding prompted us to design and synthesize new types of conjugation-extended donors with a heteroquinonoid-spacer group, namely, heteroquinonoid-extended TTF, heteroquinonoid-extended BEDT-TTF and heteroquinonoid-extended bis(ethylenedithio)tetraselenafulvalene (BETS) type donors. These heteroquinonoid-extended donor systems maintain rigid and coplanar conformations throughout the redox process (Scheme 13.1) in which cation radicals as well as dications are thermodynamically stable. We have succeeded in obtaining various kinds of single crystalline cation radical salts of these

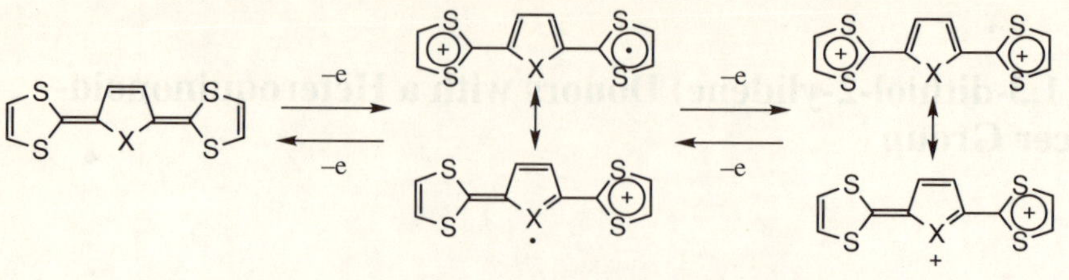

Scheme 13.1   Redox scheme of BDTT (X=S) and BDTS (X=Se).

π-extended donors with a heteroquinonoid-spacer group and clarified the crystal structures and electrical properties of these salts. The following discussions are focused on these aspects.

13.2   Bis(1,3-dithiol-2-ylidene) Donors with a Thienoquinonoid- or Selenienoquinonoid-spacer Group: BDTT and BDTS Type Donors

13.2.1 Synthesis

The first examples of bis(1,3-dithiol-2-ylidene) donors with a thienoquinonoid-spacer group, namely, bis(1,3-dithiol-2-ylidene)-2,5-dihydrothiophene (BDTT: **1**)-type donors were synthesized by our group in 1989 in preparing dibenzo derivatives DB-BDTTs (**2a-c**), according to the routes shown in Scheme 13.2.[9] Tetramethyl and diphenyl derivatives of BDTT were also prepared as the dication forms **3a,b**, according to the routes shown in Scheme 13.3.[10]

However, such synthetic methodologies were very limited since only these dibenzo derivatives and dication salts of tetramethyl and diphenyl derivatives were synthesized. After many fruitless efforts, we have found advantageous synthetic methods applicable widely for the systematic synthesis of various derivatives of BDTT-type donors, together with unsubstituted BDTT (**1**) itself.[11] The new synthetic routes are outlined in Scheme 13.4. Succinic thioanhydride

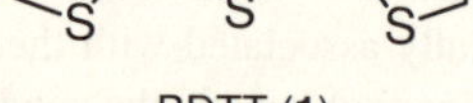

BDTT (**1**)

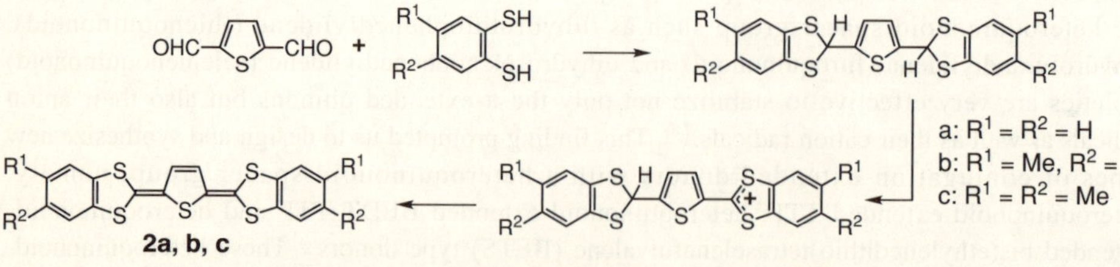

a: R^1 = R^2 = H
b: R^1 = Me, R^2 = H
c: R^1 = R^2 = Me

2a, b, c

Scheme 13.2

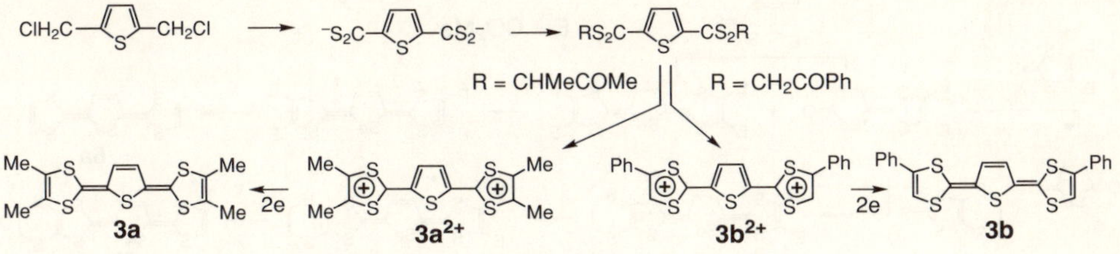

Scheme 13.3

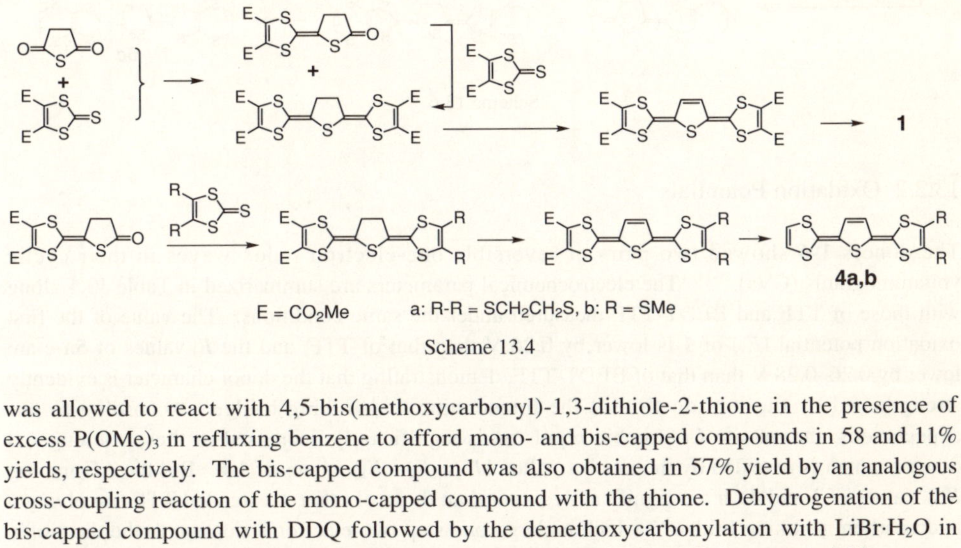

Scheme 13.4

E = CO$_2$Me   a: R-R = SCH$_2$CH$_2$S, b: R = SMe

was allowed to react with 4,5-bis(methoxycarbonyl)-1,3-dithiole-2-thione in the presence of excess P(OMe)$_3$ in refluxing benzene to afford mono- and bis-capped compounds in 58 and 11% yields, respectively. The bis-capped compound was also obtained in 57% yield by an analogous cross-coupling reaction of the mono-capped compound with the thione. Dehydrogenation of the bis-capped compound with DDQ followed by the demethoxycarbonylation with LiBr·H$_2$O in hexamethylphosphoric triamide (HMPA) to give unsubstituted BDTT (**1**) in a good yield.

The unsymmetrical donors **4a,b** were also prepared by the P(OMe)$_3$ cross-coupling reaction of the mono-capped compound with 4,5-ethylenedithio- and with 4,5-bis(methylthio)-1,3-dithiole-2-thione, respectively, followed by dehydrogenation and demethoxycarbonylation. Moreover, bis(alkylenedithio-1,3-dithiol-2-ylidene) donors with a thienoquinonoid-spacer group **5a-c** were synthesized according to the routes shown in Scheme 13.5.[12] When succinic selenoanhydride was allowed to react with 1,3-dithiole-2-thione derivatives under similar reaction conditions, bis(1,3-dithiol-2-ylidene) donors with a selenienoquinonoid-spacer group **6a-c** were synthesized according to the routes shown in Scheme 13.6.[13,14]

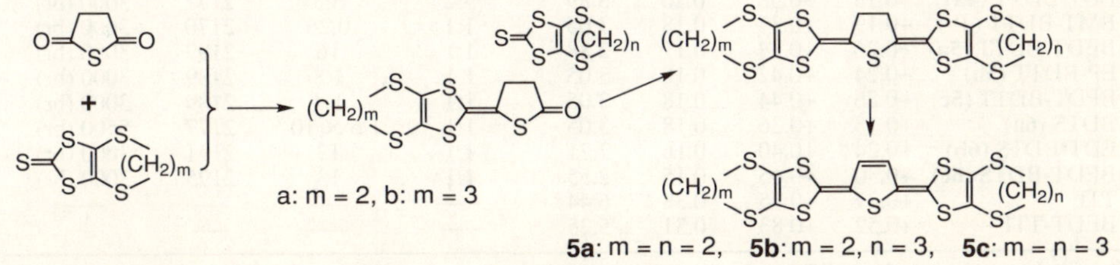

a: m = 2, b: m = 3

5a: m = n = 2,   **5b**: m = 2, n = 3,   **5c**: m = n = 3

Scheme 13.5

E = CO$_2$Me

Scheme 13.6

13.2.2 Oxidation Potentials

The donors **1-6** showed two pairs of reversible one-electron redox waves in their cyclic voltammograms (CVs).[11-14] The electrochemical parameters are summarized in Table 13.1 along with those of TTF and BEDT-TTF measured under the same conditions. The value of the first oxidation potential (E_1) of **1** is lower by 0.26 V than that of TTF, and the E_1 values of **5a-c** are lower by 0.26–0.28 V than that of BEDT-TTF, demonstrating that the donor character is evidently strengthened by the introduction of the thienoquinonoid linking bridge. According to our expectations, the on-site Coulombic repulsion is significantly decreased in these π-elongated donors, which is revealed by the much smaller ΔE (= $E_2 - E_1$) values of **1-6**. But the ΔE and log K_{sem} values of **1** and **4a,b** are significantly larger than those of other π-extended TTFs involving a hydrocarbon π-linking bridge[3-5] which undergo one-step two-electron oxidation to the dicationic state. The thermodynamic stabilization of the cation radicals of our donors is ascribed to the

Table 13.1   Electrochemical Data of **1-6**, TTF and BEDT-TTF, and Electrical Conductivities of Their TCNQ Complexes

Donor	Electrochemical property[a]				Conductivity[b] and IR band[c] of the TCNQ complex			
	E_1	E_2	ΔE	log K_{sem}	D:A	σ_{rt} /S cm^{-1}	ν_{CN} /cm^{-1}	ν_{CT} /cm^{-1}
BDTT (**1**)	+0.11	+0.34	0.23	3.90	1:1	6.9 × 10^{-3}	2152	5500 (br)
2b	+0.30	+0.54	0.24	4.14	1:1	0.02		
3a	+0.05	+0.31	0.26	4.50	1:2	0.013		
EDT-BDTT (**4a**)	+0.18	+0.38	0.20	3.39	3:2	3.3	2177	3000 (br)
BMT-BDTT (**4b**)	+0.19	+0.37	0.18	3.05	1:1	0.26	2170	3000 (br)
BEDT-BDTT (**5a**)	+0.26	+0.34	0.17	2.88	1:1	16	2197	3000 (br)
EP-BDTT (**5b**)	+0.24	+0.42	0.18	3.05	1:1	1.8	2189	3000 (br)
BPDT-BDTT (**5c**)	+0.26	+0.44	0.18	3.05	1:1	1.4	2189	3000 (br)
BDTS (**6a**)	+0.18	+0.36	0.18	3.05	1:1	1.6 × 10^{-2}	2177	5500 (br)
EDT-BDTS (**6b**)	+0.24	+0.40	0.16	2.71	1:1	13	2191	3000 (br)
BEDT-BDTS (**6c**)	+0.30	+0.45	0.15	2.55	1:1	13	2195	3000 (br)
TTF	+0.37	+0.75	0.38	6.44	——	——	——	——
BEDT-TTF	+0.52	+0.83	0.31	5.25	——	——	——	——

[a]Potentials are given in V *vs*. SCE and were determined by cyclic voltammetry/1.0 mM solutions in PhCN with 0.1 M TBAP: 50 mV s^{-1}.
[b]Four-probe method on a compressed pellet. [c]ν_{CN} of the neutral TCNQ: 2224 cm^{-1}.

distribution of an unpaired electron all over the three rings, and in consequence, the conformational change associated with the rotation of the 1,3-dithiole rings about the intercyclic bonds is difficult since the intercyclic bonds have a double bond character even in the cation radical and dication states, as shown in Scheme 13.1. The E_1 values of **5a-c** are higher by 0.13–0.15 V than that of BDTT (**1**) having no outer chalcogen atom. The selenienoquinonoid-extended BDTS system exhibits a slightly lower electron-donating ability than the thienoquinonoid-extended BDTT system, because the E_1 of **6a,b** are more positive by 0.07 and 0.06 V than those of **1** and **4a**, respectively. The cation radical is formed more easily in the BDTT system than in the BDTS system since the central heteroquinonoid ring can achieve hetero-aromaticity at the cation radical stage as shown in Scheme 13.1 and the thiophene moiety has a higher aromaticity than the selenophene moiety. The ΔE values as well as the log K_{sem} values of **6a,b** are smaller than the values of **1** and **4a**, indicating that the cation radicals of **6a,b** are slightly more unstable thermodynamically than those of **1** and **4a**, respectively. These phenomena can be attributed to the fact that the conjugative interaction between the two terminal dithiol rings through the C–Se–C bond of the central ring is relatively weak compared with the conjugation through the corresponding C–S–C bond.

From X-ray crystal analysis of BEDT-BDTS (**6c**),[14] it is revealed that there are some intermolecular S···S contacts (3.70–3.77 Å) along the stacking a-axis. However, there are many more short S···S contacts (3.48–3.64 Å) along the molecular short c-axis, where the molecules are arranged so as to direct the central selenium atom to the same side of the molecule. Thus BEDT-BDTS may have a strong tendency to form conducting donor layers of two-dimensional character in its cation radical salts.

The donors **1** and **3a** appear to be too much electron-donating to form highly conducting charge-transfer (CT) complexes with tetracyanoquinodimethane (TCNQ). Indeed, the room temperature conductivities of the TCNQ complexes of **5a-c** are 10^4–10^3 times higher than that of the TCNQ complex of BDTT (**1**), as shown in Table 13.1. Moreover, the 1:1 TCNQ complex of BEDT-BDTT (**5a**) showed metallic temperature dependence of the conductivity down to 248 K and exhibited semiconducting behavior with a very low activation energy (E_a) of 0.023 eV in the temperature range of 240–80 K on the compressed pellet.[12] The solid state IR spectra of the TCNQ salts of **5a-c** in KBr showed a very broad intrastack CT absorption band (ν_{CT}) at around 3000 cm^{-1} (Table 13.1) indicating that these salts possess a segregated stacking mode in a mixed valence state. The I_3 salt of **6c**, namely, (BEDT-BDTS)$_3$I$_3$, prepared by electrochemical oxidation, showed semiconducting character in the temperature region of 300–100 K, with a small activation energy of 0.03 eV in the temperature region of 250–130 K on a compressed pellet.[14]

13.3 Bis(1,3-dithiol-2-ylidene) Donors with a Fused Heteroquinonoid-spacer Group: BDTBF, BDTBT and BDTBS Type Donors

Although BDTT and BDTS type donors discussed above are stable in the solid state and gave a fairly stable radical cation in solution, they are powerful donors and in consequence slightly air-sensitive. Therefore, a new type of bis(1,3-dithiol-2-ylidene) donor with a heteroquinonoid-spacer group having a slightly weaker electron-donating ability are desirable for the isolation of air-stable single crystalline cation radical salts by electrochemical crystallization. The electron-donating abilities of BDTT and BDTS type donors may be weakened by condensation of a benzene ring on the central heterocyclic quinonoid ring, because the fused spacer group takes an unstable *ortho*-quinodimethane structure (**A**) in the cation radical stage as well as in the dicationic

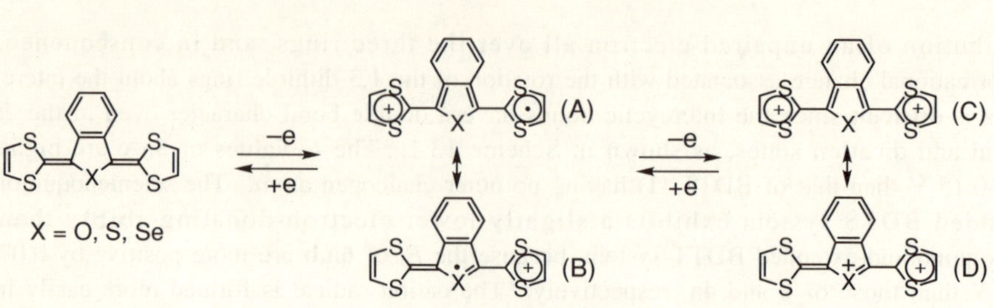

X = O, S, Se

Scheme 13.7

stage (**C**), as shown in Scheme 13.7. From this point of view, we have designed and successfully synthesized bis(1,3-dithiol-2-ylidene) donors with a benzo[c]heteroquinonoid-spacer group.

13.3.1 Synthesis

The synthetic methods are very convenient including short-step reactions starting from commercially available or easily accessible acid anhydrides. As shown in Scheme 13.8,[15] mono-capped compounds (X = O, S and Se) were synthesized in 60–75% yields by the cross-coupling reactions of the acid anhydrides with 4,5-bis(methoxycarbonyl)-1,3-dithiole-2-thione in the presence of an excess of P(OMe)$_3$ in refluxing benzene or toluene. The bis-capped tetraesters (X = O, S and Se) were synthesized in 41–52% yields by the reactions of the mono-capped compounds with the thione in P(OMe)$_3$ at 90–100 °C. The new donors **7-9** were obtained in 78–87% yields by heating the tetraester derivatives with an excess of LiBr·H$_2$O in HMPA. The reactions of the mono-capped compounds (X = S and Se) with 4,5-alkylenedithio-1,3-dithiole-2-thiones in an excess of P(OMe)$_3$ afforded unsymmetrically substituted compounds which were demethoxycarbonylated to give the bis(1,3-dithiol-2-ylidene) donors with a benzo[c] thienoquinonoid-spacer group **10-12** and donors with a benzo[c]selenienoquinonoid-spacer group **13-15**, respectively.[16,17] Bis(ethylenedithio) substituted donors with a fused heteroquinonoid-spacer group, BEDT-BDTBF (**16**),[18] BEDT-BDTBT (**17**)[19] and BEDT-BDTBS (**18**)[20] were also prepared in fairly good overall yields of *ca.* 40–60% in a manner similar to that shown in Scheme 13.9. Donors **7-18** are stable both in solid state and in solution. The 1,3-dithiole rings and the central benzo[c]heteroquinonoid ring of these donors were proved to be in essentially coplanar conformations since a good planar geometry of the dimethoxycarbonyl derivative of **14** was found by X-ray crystallographic analysis.[17]

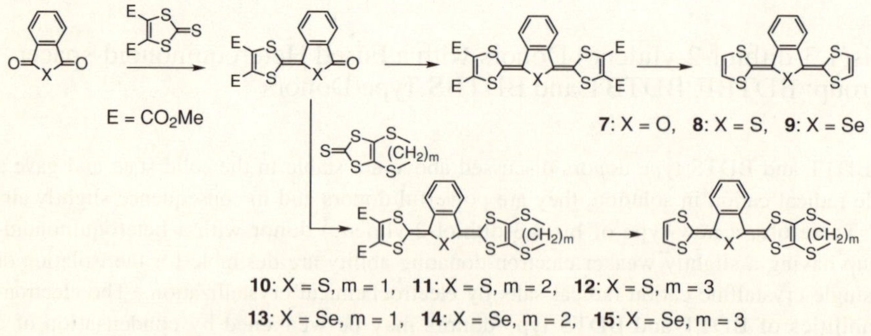

E = CO$_2$Me

7: X = O,   **8**: X = S,   **9**: X = Se

10: X = S, m = 1,   **11**: X = S, m = 2,   **12**: X = S, m = 3
13: X = Se, m = 1,   **14**: X = Se, m = 2,   **15**: X = Se, m = 3

Scheme 13.8

BEDT-BDTBF (**16**): X = O
BEDT-BDTBT (**17**): X = S
BEDT-BDTBS (**18**): X = Se

Scheme 13.9

13.3.2 Oxidation Potentials

The CVs of **7-18** showed two well-defined reversible one-electron oxidation waves. The E_1 values of **8** and **9** are higher by 0.20 and 0.19 V than those of the non-fused donors, **1** and **6a**, respectively (Table 13.2).[15-20] This can be ascribed mainly to the contribution of the non-aromatic *ortho*-quinodimethane structure (**A**) in the cation radicals of **8** and **9** (Scheme 13.7). The E_1 values of **10-12** are higher by 0.18–0.19 V than that of the non-fused donor **4a**. The E_1 values of **13-15** are higher by 0.16–0.19 V than that of non-fused donor **6b**. The higher E_1 value of BDTBS (**9**) by 0.06 V than that of BDTBT (**8**) can be ascribed to the lower aromaticity of the selenophene ring than the thiophene ring in **A** (Scheme 13.7), *i.e.*, the energetic benefit of gaining aromaticity in the central heterocyclic ring on affording the cation radical is less in **9** than in **8**, since selenophene is less aromatic than thiophene. The same trend was observed in the E_1 values of **1** and **6a**. Interestingly, the E_1 values of the donors with a benzo[c]furoquinonoid-spacer group, **7** and **16**, are lower by 0.07 and 0.06 V than those of the donors with a benzo[c]thienoquinonoid-spacer group, **8** and **17**, respectively. The low E_1 values of **7** and **16** mean that the cation radicals of **7** and **16** should be more stable than those of **8** and **17**, respectively. This fact can not be explained by the structure **A**, because the furan ring should have much less aromaticity than the thiophene ring. Therefore, we must consider another resonance structure **B** (Scheme 13.7), which contributes more significantly in the cation radical of **7** and **16** than those of **8** and **9**, since **B** has

Table 13.2   Electrochemical Data of Fused Donors **7-19**, and Conductivities and IR Data of Their TCNQ Complexes

Donor	Electrochemical property[a]				Conductivity[b] and IR band[c] of the TCNQ complex				
	E_1	E_2	ΔE	$\log K_{sem}$	D:A	σ_{rt} /S cm^{-1}	ν_{CN} /cm^{-1}	ν_{CT} /cm^{-1}	
BDTBF (**7**)	+0.24	+0.55	0.31	5.25	1:1	1.9	2179, 2156	3000 (br)	
BDTBT (**8**)	+0.31	+0.55	0.24	4.07	1:1	5.8	2177	3000 (br)	
BDTBS (**9**)	+0.37	+0.59	0.22	3.73	1:1	4.4	2175	2000 (br)	
MDT-BDTBT (**10**)	+0.36	+0.55	0.19	3.22	1:1	15	2177	3000 (br)	
EDT-BDTBT (**11**)	+0.37	+0.58	0.21	3.56	1:1	4.8	2177	3000 (br)	
PDT-BDTBT (**12**)	+0.37	+0.58	0.21	3.56	2:1	4.2×10^{-2}	2177, 2148	3000 (br)	
MDT-BDTBS (**13**)	+0.42	+0.61	0.19	3.22	2:1	1.2	2177	3000 (br)	
EDT-BDTBS (**14**)	+0.43	+0.63	0.20	3.39	2:1	2.4	2177, 2148	3000 (br)	
PDT-BDTBS (**15**)	+0.40	+0.62	0.22	3.73	3:2	0.32	2177	3000 (br)	
BEDT-BDTBF (**16**)	+0.37	+0.53	0.16	2.71	1:1	4.8	2191	3000 (br)	
BEDT-BDTBT (**17**)	+0.43	+0.65	0.22	3.73	2:3	6.7	2198	3000 (br)	
BEDT-BDTBS (**18**)	+0.47	+0.68	0.21	3.56	2:3	4.9	2198	2800 (br)	
BDTFP (**19**)	+0.36	+0.62	0.26	4.50	1:1	20	2195	3000 (br)	

[a]Potentials are given in V *vs.* SCE and were determined by cyclic voltammetry/1.0 mM solutions in PhCN with 0.1 M TBAP: 50 mV s^{-1}.
[b]Four-probe method on a compressed pellet. [c]ν_{CN} of the neutral TCNQ: 2224 cm^{-1}.

no non-aromatic *ortho*-quinodimethane structure, so that **B** can acquire extra stabilization energy by the aromatization of the fused benzene ring instead of missing the small aromatic stabilization energy of the furan ring. For the same reason, the resonance structure **D** may contribute more significantly to the dications of **7** and **16** than resonance structure **C**.

13.3.3 TCNQ Complexes

The room temperature conductivities measured on compressed pellets and characteristic IR bands of TCNQ complexes of the fused donors are summarized in Table 13.2.[15-20] The conductivities of the 1:1 TCNQ complexes of **8** and **9**, *i.e.*, BDTBT·TCNQ and BDTBS·TCNQ, are 10^3 and 10^2 times higher than those of the 1:1 TCNQ complexes of **1** and **6a**, *i.e.*, BDTT·TCNQ and BDTS·TCNQ, respectively. However, the conductivities of the TCNQ complexes of the fused donors with peripheral chalcogen atoms **11**, **14**, **17** and **18** do not differ significantly from those of the non-fused donors **4a**, **6b**, **5b** and **6c**, respectively. All the complexes of the fused donors showed an extremely broad and characteristic v_{CT} at a very low energy region of 2000–3000 cm^{-1}, revealing a segregated stacking mode in a mixed valence state.[21] Interestingly, the conductivities of the 1:1 TCNQ complexes of the donors with a benzo[c]furoquinonoid-spacer group **7** and **16**, *i.e.*, BDTBF·TCNQ and BEDT-BDTBF·TCNQ, are almost identical with those of the TCNQ complexes of the corresponding donors with a benzo[c]thienoquinonoid- or benzo[c] selenienoquinonoid-spacer group **8**, **9**, **17** and **18**, although **7** and **16** do not contain the heavy chalcogen atom at the central spacer group. The existence of strong intermolecular contacts is probable in the donor columns of BDTBF·TCNQ and BEDT-BDTBF·TCNQ owing to the rigid and tight donor stacking in the conducting columns.

13.3.4 Crystal Structures and Electrical Properties of (BEDT-BDTBT)$_3$(ReO$_4$)$_2$, BEDT-BDTBF·BF$_4$ and (BEDT-BDTBS)$_2$(X)$_3$ (X = ClO$_4$, BF$_4$ and ReO$_4$)

In contrast with TTF and BEDT-TTF, bis(1,3-dithiol-2-ylidene) donors with a heteroquinonoid-spacer group are not linear molecules. These donors are bent along the donor long axes due to the 2,5-bent bonds of the central five-membered heterocycles as shown in Scheme 13.10. The bent angles (α) of the donor molecules determined from the molecular structures of their cation radical salts are *ca.* 121° for the BEDT-BDTBF molecule, *ca.* 142° for the BEDT-BDTBT molecule and *ca.* 155 ° for the BEDT-BDTBS molecule. This is quite reasonable considering the short C–O bond length and the long C–Se bond length. Thus BEDT-BDTBS (**18**) is more linear than BEDT-BDTBT (**17**), and BEDT-BDTBT is more linear than BEDT-BDTBF (**16**). We have succeeded in obtaining single-crystalline cation radical salts of **16-18** with ClO$_4$, BF$_4$, ReO$_4$ and PF$_6$ by conventional electrochemical oxidation.[18-20] We have found that the donor-to-anion ratio and the

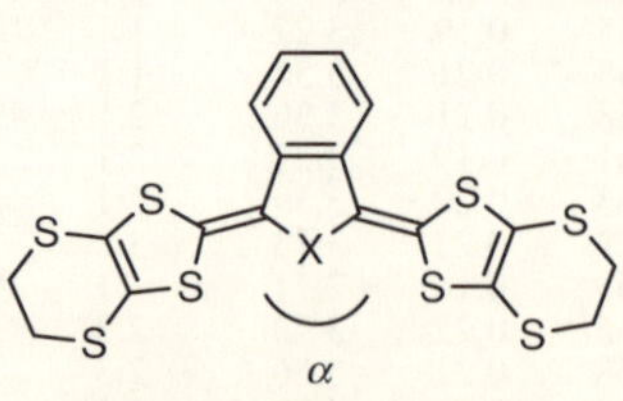

Scheme 13.10   The bent angle (α) of bis (1,3-dithiol-2-ylidene) donors with
a fused heteroquinonoid-spacer group.

formal charge on a donor are closely related to the bent angle (α) of the donor molecule. Namely, the donor with a large bent angle (α) tends to give salts with an anion-rich stoichiometry.

A single crystalline ReO_4 salt of BEDT-BDTBT (**17**), $(BEDT\text{-}BDTBT)_3(ReO_4)_2$, showing a room temperature conductivity of 0.10 S cm^{-1}, exhibited semiconducting temperature dependence with an activation energy of 0.165 eV down to 180 K.[19] X-Ray crystal structural analysis showed that the exact donor-to-anion ratio is a donor-rich stoichiometry of 3:2. Thus the formal charge on the BEDT-BDTBT molecule in this salt is +2/3. The BEDT-BDTBT molecules are stacked with the protruding benzene rings directed to the same side in the donor layer. The protruding benzene rings in one column can meet the dent of the neighboring column in the side-by-side direction, along the donor short a–c axis. In consequence, there is no space for the anion to enter along the donor short a–c axis, and there exist favorable S$\cdots$S contacts along the intermolecular side-by-side direction. Moreover, in this salt, the protruding benzene rings in the neighboring columns along the donor long b-axis are directed to the opposite side, providing smallish cavities between the neighboring donor layers along the donor long axis, as shown in Fig. 13.1. Thus only a limited number of anions can enter the small space along the donor long axis, resulting in a donor-rich stoichiometry of 3:2.

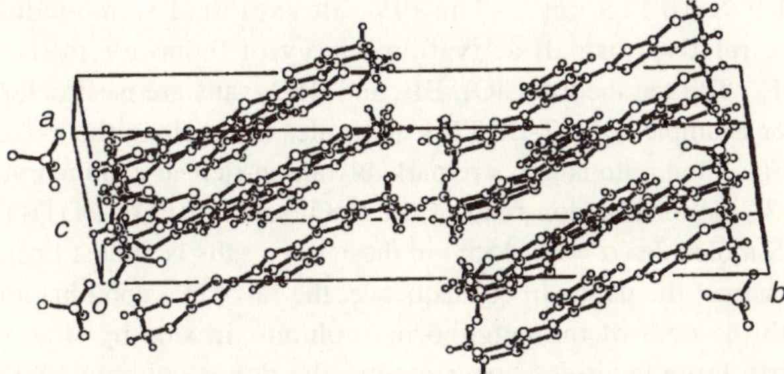

Fig. 13.1   Crystal structure of $(BEDT\text{-}BDTBT)_3(ReO_4)_2$. (Reprinted with permission from ref. 19)

Single crystalline BF_4 and ClO_4 salts of BEDT-BDTBF (**16**), *i.e.*, BEDT-BDTBF·BF_4 and BEDT-BDTBF·ClO_4, showing room temperature conductivities of 2.9 and 1.3 × 10^{-3} S cm^{-1}, respectively, exhibited semiconducting temperature dependence with activation energies of 0.205 eV for the BF_4 salt and 0.322 eV for the ClO_4 salt in the temperature range from room temperature to 180 K.[22] These salts are isostructural with each other in the crystal structures. The donors are stacked uniformly along the c-axis and the donor-to-anion ratio is 1:1, then the formal charge on a donor is +1.0. The calculated band structures of these salts are half-filled. The low conductivities of these salts can be ascribed to the Mott type transition at around room temperature. As shown in Fig. 13.2, the donor molecules are stacked directing the protruding benzene rings to the same side in the donor column. Moreover, the protruding benzene rings in the neighboring columns are directed toward the same direction along the donor short b-axis, so the protruding benzene rings in one column just meet the dent of the neighboring column in the side-by-side direction, resulting in no space in the donor short b-axis to let the anion come in. Moreover, the protruding benzene rings in the neighboring columns along the donor long a-axis are also directed to the same side, resulting in medium-size cavities permitting the anions to come in. Thus the stoichiometries of these salts are more anion-rich than $(BEDT\text{-}BDTBT)_3(ReO_4)_2$, giving a donor-to-anion ratio of 1:1.

We have obtained single-crystalline ClO_4, BF_4, ReO_4 and PF_6 salts of BEDT-BDTBS (**18**), namely, $(BEDT\text{-}BDTBS)_2X_3$ (X = ClO_4, BF_4, ReO_4 and PF_6).[20] Of these, the BF_4 and PF_6 salts

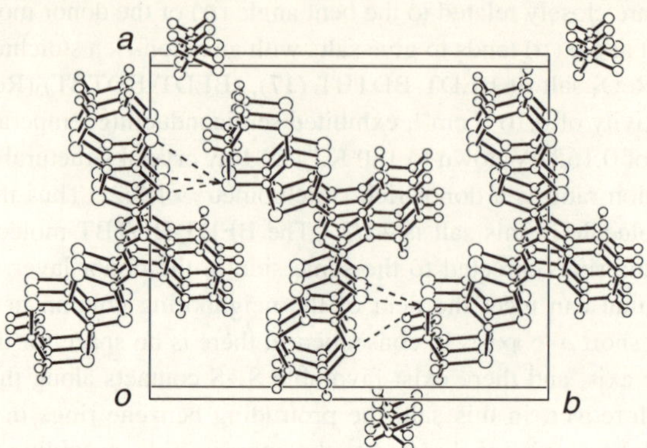

Fig. 13.2   Crystal structure of BEDT-BDTBF·BF$_4$. The dotted lines indicate the short S···S contacts.

exhibited fairly high conductivities of 1.0 and 2.8 S cm^{-1}, respectively, while the conductivity of the ClO$_4$ salt is 1.9 × 10^{-4} S cm^{-1}. The PF$_6$ salt exhibited semiconducting temperature dependence with a relatively small activation energy of 0.066 eV in the range from room temperature to 80 K. The tetrahedral ClO$_4$, BF$_4$ and ReO$_4$ salts are isostructural with each other. In the ClO$_4$ salt, for example, BEDT-BDTBS molecules stack along the c-axis as shown in Fig. 13.3 and the donor-to-anion ratio is 2:3, a remarkably anion-rich stoichiometry, which is distinctly different from the 3:2 donor-to-anion ratio of the ReO$_4$ salt of BEDT-BDTBT mentioned above. The BEDT-BDTBS molecule ($\alpha = ca.$ 155°) in these three salts is almost linear. Thus, the donor columns do not construct the dent. In consequence, the favorable combination of the protruding benzene rings with the dent of the neighboring columns in side-by-side directions becomes impossible, so fairly large cavities appear among the donor columns along the side-by-side direction as shown in Fig. 13.4. Thus, anions enter not only the cavities in the donor long axis, but also the cavities in the donor short axis, giving an extremely anion-rich composition ratio of 2:3. The donor columns of the ClO$_4$ salt are strongly dimerized, while the dimerization of the BF$_4$ salt is very weak. This is the main reason why the BF$_4$ salt exhibits 10^4 times higher conductivity than the ClO$_4$ salt, since these salts have one-dimensional band structures.

The schematic presentation of donor and anion packing patterns is shown in Fig. 13.5. The most interesting finding from these investigations is that the bent angle governs roughly the donor packing motif, especially the donor-to-anion ratio of the cation radical salts with tetrahedral anions. In the field of organic conductors, control of band-filling is one of the most important

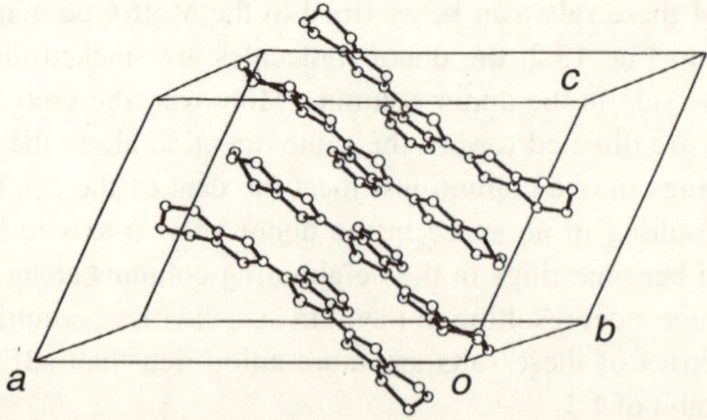

Fig. 13.3   Crystal structure of (BEDT-BDTBS)$_2$(ClO$_4$)$_3$. (Reprinted with permission from ref. 20)

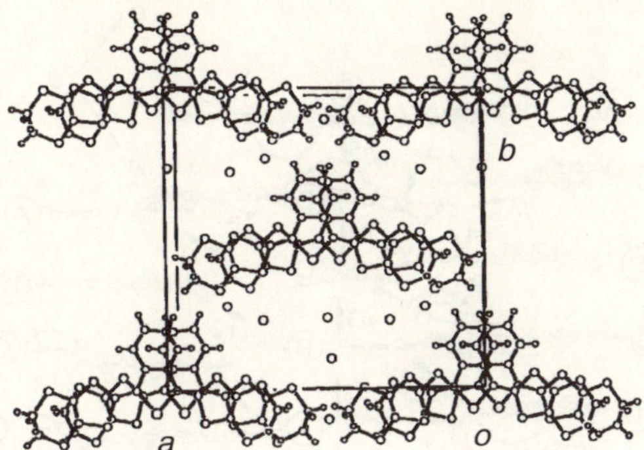

Fig. 13.4   Crystal structure of $(BEDT\text{-}BDTBS)_2(ClO_4)_3$ viewed along the c-axis.

Fig. 13.5   Schematic presentation of donor and anion packing patterns in
cation radical salts of bis(ethylenedithio) substituted donors with a
benzo[c]heteroquinonoid-spacer group with tetrahedral anions.

aims. This observation presents an idea for solving the band-filling problem in organic conductors by controlling the molecular shape of the component molecule.

13.3.5 Crystal Structures and Electrical Properties of $(BEDT\text{-}BDTBF)_2X$ (X = PF_6 and AsF_6) and $(BEDT\text{-}BDTBF)_2SbF_6(PhCl)_{0.5}$

Single crystalline octahedral PF_6 and AsF_6 anion salts of BEDT-BDTBF (**16**) with a 2:1 stoichiometry, $(BEDT\text{-}BDTBF)_2X$ (X = PF_6 and AsF_6),[18] showed semiconducting temperature dependence with E_a = 0.191 and 0.190 eV in the temperature range from room temperature to 160 K, respectively. These two salts are isostructural with each other. The donors stack along the $-a+c$ axis and the donor columns are strongly tetramerized, as shown in Fig. 13.6a. The PF_6 or AsF_6 anion exists in the opening between the conducting ac-planes. Quite interestingly, there exists a fairly large calculated inter-column overlap integral, $r3$, along the c-axis, which plays an important role for the conduction, because the $r3$ interaction combines all the tetramers belonging to all the neighboring columns along the c-axis, as shown in Fig. 13.6b. The overlap integral $r3$ of the PF_6 salt (1.3×10^{-3}) is much larger than that of the AsF_6 salt (0.2×10^{-3}). Indeed, the room temperature conductivity of the PF_6 salt (σ_{rt} = 7.3 S cm^{-1}) is higher by 26 times than that of the AsF_6 salt (σ_{rt} = 0.28 S cm^{-1}). Hence, the band structures of $(BEDT\text{-}BDTBF)_2X$ (X = PF_6 and AsF_6) are one-dimensional along the c-axis, although the donor stacks along the $-a+c$ axis.

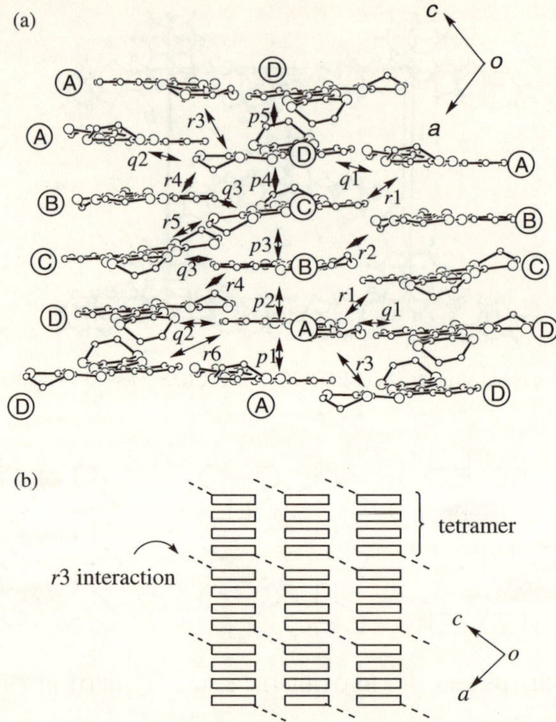

Fig. 13.6   (a) Donor arrangement in (BEDT-BDTBF)$_2$PF$_6$, viewed along donor long axis. (b) Schematic drawing of the donor arrangement in (BEDT-BDTBF)$_2$PF$_6$. The dotted lines indicate $r3$ interactions. (Reprinted with permission from ref. 18)

We have also succeeded in obtaining a SbF$_6$ salt of BEDT-BDTBF (**16**), (BEDT-BDTBF)$_2$SbF$_6$(PhCl)$_{0.5}$.[23] Although SbF$_6$ is the same octahedral anion as PF$_6$ and AsF$_6$, the donor-packing mode and the band structure of the SbF$_6$ salt are significantly different from those of (BEDT-BDTBF)$_2$X (X = PF$_6$ and AsF$_6$), probably due to the large size of the SbF$_6$ anion.

The introduction of a fused-benzene ring at the linking five-membered heteroquinonoid ring resulted in a considerable advantage by weakening the electron-donating ability and giving single crystals of various kinds of cation radical salts. However, there remain several problems to be solved. One is to make uniform stacking crystal structures and the other is to increase intermolecular interactions. Although uniform stacking structures are actually observed in BEDT-BDTBF·X (X = BF$_4$ and ClO$_4$), other salts of the fused donors are composed of dimerized, trimerized or tetramerized donor columns. The protruding benzene ring tends to avoid intermolecular interactions. In order to overcome these disadvantages, we have designed a new donor with a dihydrofuropyrazinediylidene-spacer group, which will be discussed in Section 13.4.

13.4   Bis(1,3-dithiol-2-ylidene) Donor with a 5,7-Dihydrofuro[3,4-b]pyrazine-5,7-diylidene-spacer Group

Another effective strategy to decrease the donor abilities of BDTT and BDTS would be the introduction of an electron-deficient pyrazine ring on the linking five-membered heteroquinonoid-spacer group. With this in mind, we have designed and synthesized 5,7-bis(1,3-dithiol-2-ylidene)-5,7-dihydrofuro[3,4-b]pyrazine (BDTFP) (**19**).

13.4.1 Synthesis, Physical Properties and Oxidation Potentials

BDTFP (**19**) was synthesized in a fairly good yield by the route outlined in Scheme 13.11, starting from pyrazine-2,3-dicarboxylic anhydride,[24] conveniently feasible by the dehydration of pyrazine-2,3-dicarboxylic acid.

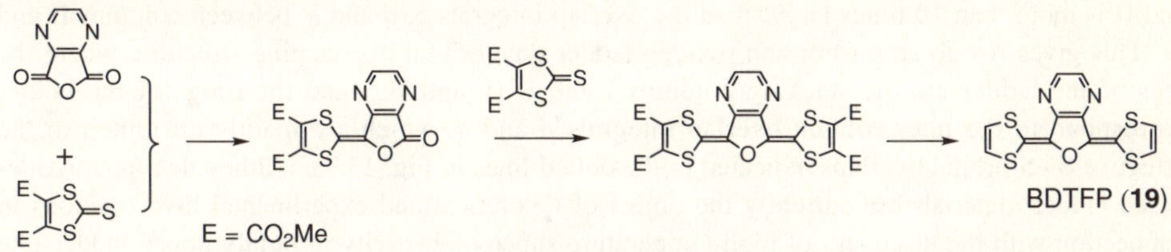

$$E = CO_2Me$$

Scheme 13.11

BDTFP showed a new broad absorption band at around 580 nm in the UV spectrum which is attributed to an intramolecular π-π^* transition in the polarized structure (**19B**) with intramolecular N$\cdots$S contacts as shown in Scheme 13.12, since this new absorption band is not observed in the donor with a benzo[c]furoquinonoid-spacer group, BDTBF (**7**). Indeed, the intramolecular N$\cdots$S distances (3.21–3.15 Å) in the BDTFP molecules incorporated in the PF_6 and AsF_6 salts of BDTFP, described in the following section, are far shorter than the van der Waals distance (3.35 Å).

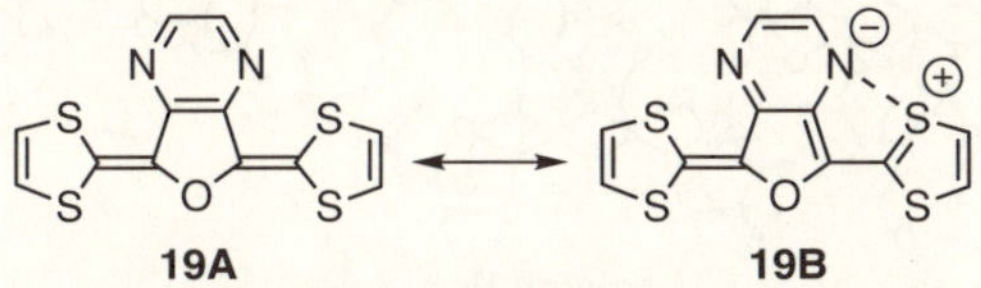

19A   **19B**

Scheme 13.12   Resonance structures of BDTFP (**19**).

Electrochemical data of BDTFP are listed in Table 13.2. The E_1 of BDTFP (+0.36 V) is more positive by 0.25 V and by 0.12 V than those of BDTT (**1**) (+0.11 V) and BDTBF (**7**) (+0.24 V), respectively, reflecting the low energy level of the HOMO of BDTFP (**19**) by the annulation of the electron-deficient pyrazine ring.[24]

The 1:1 TCNQ complex of BDTFP, BDTFP·TCNQ, exhibiting a room temperature conductivity of 20 S cm^{-1} is considered to be metallic from room temperature down to 150 K since the Pauli-like paramagnetic susceptibility ($\chi_\text{n} = 2 \times 10^{-4}$ emu mol^{-1}) is almost invariable down to around 150 K.

13.4.2 Crystal Structures and Electrical Properties of $(BDTFP)_2X(PhCl)_{0.5}$ (X = PF_6 and AsF_6)

Good quality black needle-like single crystals of $(BDTFP)_2PF_6(PhCl)_{0.5}$ and $(BDTFP)_2$-$AsF_6(PhCl)_{0.5}$ were grown by conventional electrochemical oxidation.[24] X-Ray crystallographic analyses disclosed that these salts are isostructural with each other and the exact donor-to-anion ratio is D:A:Solvent = 2:1:0.5, so the formal charge on a donor is +0.5. Two crystallographically independent donor molecules (A and B) stack along the *c*-axis in a A-B-A-B manner in two-fold period repetition directing the pyrazine ring to the same side of the molecules as shown in Fig. 13.7b,c. Looking for inter-column interactions, the pyrazine rings in the neighboring columns are directed opposite from each other (Fig. 13.7a). There are some S···S contacts (3.23–3.54 Å) less than the van der Waals distance (3.6 Å) between columns I and II (and between columns I' and II') in which the 1,3-dithiole rings in neighboring columns come close to each other. On the other hand, there is no side-by-side inter-column contact between columns II and I', in which the pyrazine rings come close to each other. The inter-column overlap integral *p* between columns I and II is more than 10 times larger than the overlap integrals *b*, *q* and *q'* between columns II and I'. This gives rise to an uncommon two-leg ladder type orbital overlapping structure, where the legs of the ladder are the stacking columns I and II (I' and II') and the rungs of the ladder correspond to the inter-column overlap integrals *p* and *a*, which are mainly attributed to the effective S···S orbital overlaps indicated in the dotted lines in Fig. 13.7a. Although copper oxide-based ladder materials are currently the object of theoretical and experimental investigations in connection with the discovery of high-temperature superconductivity in lightly doped ladder-like antiferromagnets,[25] very few examples of molecular spin ladder systems have been reported so far.[26]

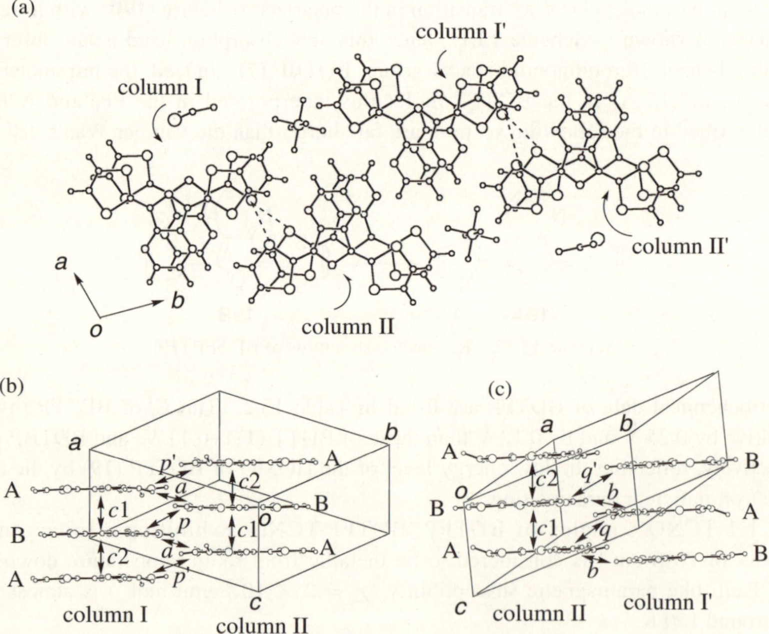

Fig. 13.7   Crystal structure of (BDTFP)$_2$PF$_6$(PhCl)$_{0.5}$: (a) the projection onto the *ab*-plane showing the ladder type structure; (b) the overlap mode between columns I and II; (c) the overlap mode between columns II and I'. (Reprinted with permission from ref. 24)

The PF_6 and AsF_6 salts showed room temperature conductivities of 40 and 4.0 S cm^{-1} with very small activation energies of 0.076 (at 300–180 K) and 0.078 eV (at 300–230 K), respectively.[24]   However, the calculated band structure is metallic with half-filled upper bands (Fig. 13.8), due to the dimerization of the donor columns.   The calculated Fermi surface consisting of two pairs of linear planes is quasi-one-dimensional.   Although the crystal structures and the electronic states of these two salts are very similar at around room temperature, the electronic states of these two salts are quite different from each other at low temperatures.   Quite interestingly, phase transitions were observed at 175 K (Fig. 13.9a) and 220 K for the PF_6 and AsF_6 salts, respectively.   As shown in Fig. 13.9b, the magnetic susceptibility of the PF_6 salt, determined by the SQUID method, is almost invariable from 300 K down to 175 K and decreased drastically at 175 K with the transition to a diamagnetic phase.[24,27]   EPR measurement provided the same result as the static susceptibility measurement.[27]   These facts indicate that the PF_6 salt undergoes a spin-singlet transition at around 170 K, which may originate from a structural phase transition.   On the other hand, the AsF_6 salt is magnetic down to low temperature.   From measurement of the temperature dependence of the EPR linewidth, it is clear that the AsF_6 salt

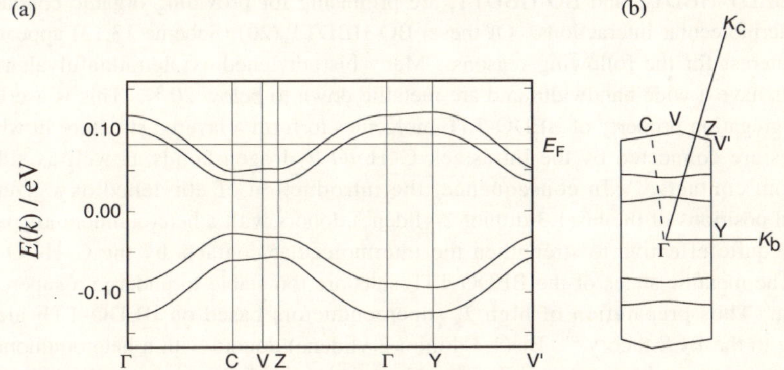

Fig. 13.8   (a) Band structure and (b) Fermi surface of $(BDTFP)_2PF_6(PhCl)_{0.5}$, calculated by the tight binding approximation method. (Reprinted with permission from ref. 24)

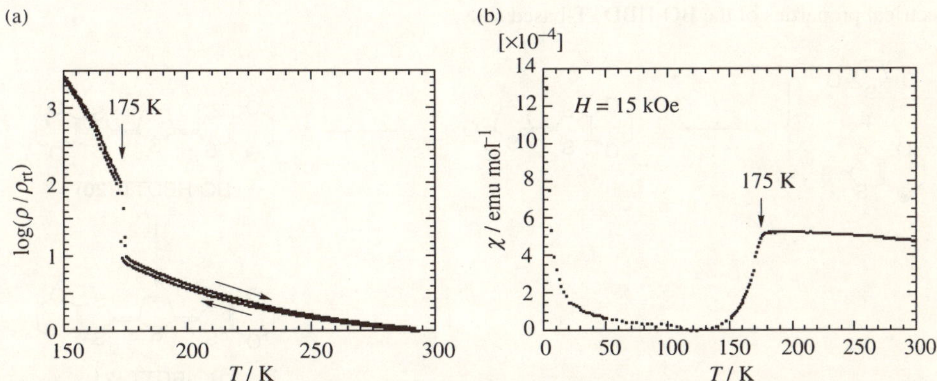

Fig. 13.9   (a) Temperature dependence of the normalized resistivity and (b) temperature dependence of the static magnetic susceptibility of $(BDTFP)_2PF_6(PhCl)_{0.5}$.
(Reprinted with permission from ref. 24)

undergoes a huge transition of the first order at around 230 K, associated with an abrupt jump in spin susceptibility. The χ_{spin} of the AsF$_6$ salt shows a Curie-like enhancement down to *ca.* 50 K. The AsF$_6$ salt shows spin-gap behavior below 50 K and undergoes an antiferromagnetic transition at 14 K. Further detailed investigations are in progress.

13.5   Bis(4,5-ethylenedioxy-1,3-dithiol-2-ylidene) Donor with a Dihydrothienoquinonoid-spacer Group

The electron-donating abilities of BDTT and BDTS type donors should be weakened by hydrogenation of the C=C double bond of the central heteroquinonoid-spacer group, since the energy level of the HOMO of the BDTT or BDTS type donor molecule is significantly stabilized by the reduction of two π-electrons. This molecular design strategy may be superior to the condensation of the benzene ring at the central spacer group because the steric hindrance of the protruding benzene ring is rejected. Thus, donors with a dihydroheteroquinonoid-spacer group, such as BEDT-HBDTT and BO-HBDTT, are promising for providing organic conductors with strong intermolecular interactions. Of these, BO-HBDTT (**20**) (Scheme 13.13) appears to be of special interest for the following reasons. Many bis(ethylenedioxy)tetrathiafulvalene (BEDO-TTF) salts have a wide bandwidth and are metallic down to below 20 K. This is ascribed to the strong aggregation property of BEDO-TTF molecules to form a layered structure in which donor molecules are connected by the intrastack C–H···O hydrogen bonds as well as side-by-side heteroatom contacts.[28] In consequence, the introduction of ethylenedioxy groups in the peripheral positions of the bis(1,3-dithiol-2-ylidene) donors with a heteroquinonoid-spacer group would be quite effective to strengthen the intermolecular contacts by the C–H···O hydrogen bonds. The metallic states of the BEDO-TTF salts are too stable to undergo a superconducting transition. Thus preparation of high T_c superconductors based on BEDO-TTF are difficult according to the BCS theory.[29] Bis(1,3-dithiol-2-ylidene) donors with a heteroquinonoid-spacer group prepared so far have a narrow bandwidth compared with the corresponding bis(1,3-dithiol-2-ylidene) donors without a spacer group. Thus, BO-HBDTT-based salts will have a relatively narrow bandwidth and their metallic state will be relatively unstable compared with the BEDO-TTF-based salts. With this in mind, we have synthesized BO-HBDTT and investigated the electrical properties of the BO-HBDTT-based salts.

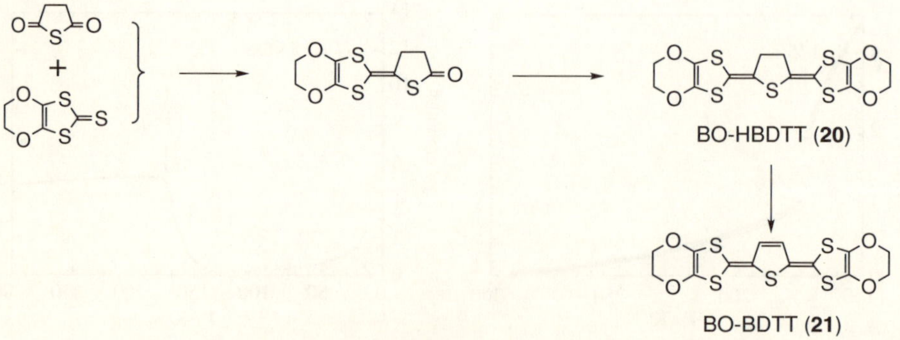

BO-HBDTT (**20**)

BO-BDTT (**21**)

Scheme 13.13

13.5.1 Synthesis, Physical Properties and Oxidation Potentials of BO-HBDTT and BO-BDTT

The synthetic routes for BO-HBDTT (**20**) and BO-BDTT (**21**) are outlined in Scheme 13.13.[30] When succinic thioanhydride was allowed to react with 2 equiv. of 4,5-ethylenedioxy-1,3-dithiole-2-thione in the presence of excess of P(OMe)$_3$, the mono-capped intermediate was mainly obtained. When succinic thioanhydride was allowed to react with 3 equiv. of the thione, **20** was obtained in 68% yield. Treatment of the mono-capped compound with the thione gave **20** in 75% yield. Dehydrogenation of **20** with chloranil afforded the unsaturated donor **21** in 51% yield.

Although the C=C bond of the linking heteroquinonoid ring is saturated, BO-HBDTT (**20**) showed a reversible one-electron two-step redox reaction in the CV, suggesting that the cation radical state of **20** is stable.[30] This reveals that the conjugation between the two terminal 1,3-dithiole rings in **20** is significant and that the sulfur-bonding orbital interaction[31] plays a very important role in the conjugation of the two terminal 1,3-dithiole rings. Indeed, the π-HOMO atomic orbital coefficient of the central sulfur atom is significantly large, as shown in Scheme 13.14. The E_1 and E_2 values of BO-HBDTT (**20**) (+0.53 and +0.75 V, respectively) are higher than those of BEDO-TTF (+0.42 and +0.74 V) in PhCN at 25 °C. The E_1 and E_2 values of BO-BDTT (**21**) (+0.19 and +0.37 V, respectively, in PhCN) are much lower than those of **20**. Reflecting the low E_1 value, **21** is too unstable to prepare cation radical salts by electrochemical crystallization.

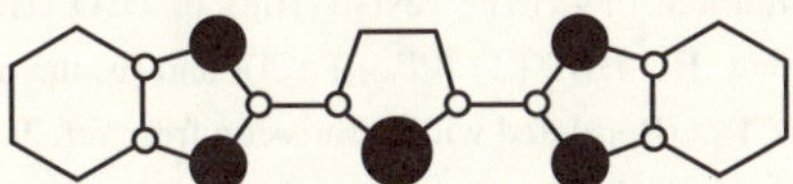

Scheme 13.14 π-AO coefficients in the HOMO of BO-HBDTT (**20**) calculated by PM3 method.

BO-HBDTT and BO-BDTT formed 1:1 and 3:1 TCNQ complexes exhibiting room temperature conductivities of 1.3 and 1.2 S cm^{-1}, respectively, measured on compressed pellets. The ν_{CN} value of 2181 cm^{-1} and the broad ν_{CT}'s at 2500–3500 cm^{-1} indicate that these salts exist in partial CT conditions adopting segregated stacking modes of mixed valence states.

13.5.2 Crystal Structures and Electrical Properties of BO-HBDTT Salts

Black prism-like single crystalline cation radical salts of BO-HBDTT, *i.e.*, (BO-HBDTT)$_5$-(ClO$_4$)$_2$(PhCl)$_2$, (BO-HBDTT)$_2$BF$_4$·PhCl and (BO-HBDTT)$_5$(PF$_6$)$_2$(PhCl)$_2$, showing room temperature conductivities of 22, 1.1, and 0.68 S cm^{-1}, were grown by conventional electrochemical crystallization in chlorobenzene.[30] The ClO$_4$, BF$_4$ and PF$_6$ salts are metallic from room temperature down to 30, 10 and 25 K, respectively, as shown in Fig. 13.10a. The temperature dependence of the magnetic susceptibility of the ClO$_4$ salt is almost invariable from room temperature down to 5 K; this is considered to be Pauli paramagnetism, as shown in Fig. 13.10b. Thus, the ClO$_4$ salt is metallic down to 5 K. X-Ray crystal analysis was performed for the ClO$_4$ salt, but the structure analysis of the BF$_4$ and PF$_6$ salts was unsuccessful. X-Ray structural analysis revealed that the metallic property of the ClO$_4$ salt is ascribed to the uniform stacking structure of the donor columns constructed from the intrastack C–H···O hydrogen bonds. Unfortunately, the conventional $R(Rw)$ factor [$R(Rw)$ = 0.300(0.337)] of the X-ray structural analysis is larger than the allowable values at this stage due to the spontaneous loss of the PhCl

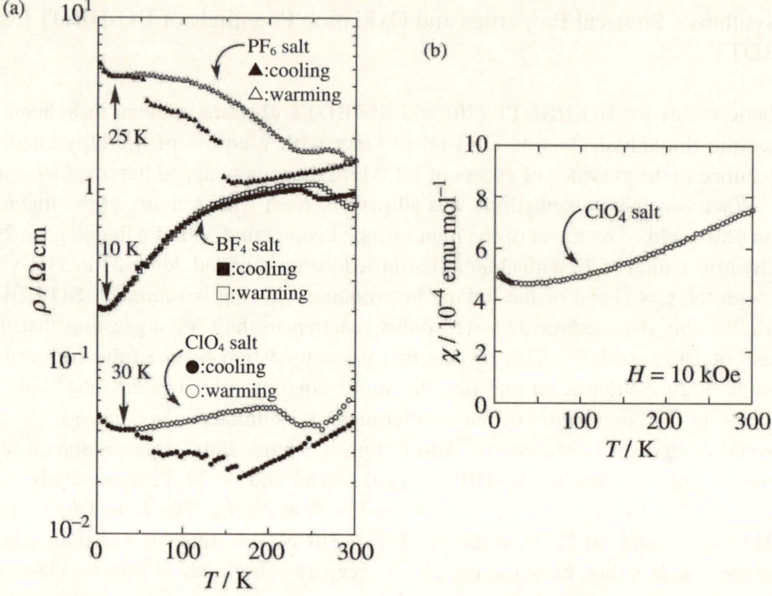

Fig. 13.10   Temperature dependence of (a) the resistivities of $(BO\text{-}HBDTT)_5(ClO_4)_2(PhCl)_2$, $(BO\text{-}HBDTT)_2BF_4\cdot PhCl$ and $(BO\text{-}HBDTT)_5(PF_6)_2(PhCl)_2$ and (b) the magnetic susceptibility of $(BO\text{-}HBDTT)_5(ClO_4)_2(PhCl)_2$. (Reprinted with permission from ref. 30)

molecules from the crystal during X-ray data collection. However, the choice of the space group is unambiguous. The donor molecules in the ClO_4 salt stack uniformly along the a-axis directing the central sulfur atom to the same side, as shown in Fig. 13.11a. Interestingly, two short intrastack C–H$\cdots$O hydrogen bonds exist on both sides of the donor molecule with four ethylenedioxy oxygen atoms and terminal ethylene protons of the neighboring donor molecules participating, as shown in Fig. 13.11b. These intrastack C–H$\cdots$O contacts play an important role in the formation of a uniform stack. Moreover, the ClO_4 salt has two-dimensional conducting character, since there are S$\cdots$S and S$\cdots$O contacts along the side-by-side ($a+c$ axis) and oblique directions (the overlap integrals, s, q and p are 3.7, –7.9 and 6.8 × 10^{-3}, respectively), constructing rigid and tight two-dimensional layered intermolecular networks along the ac-plane, as shown in

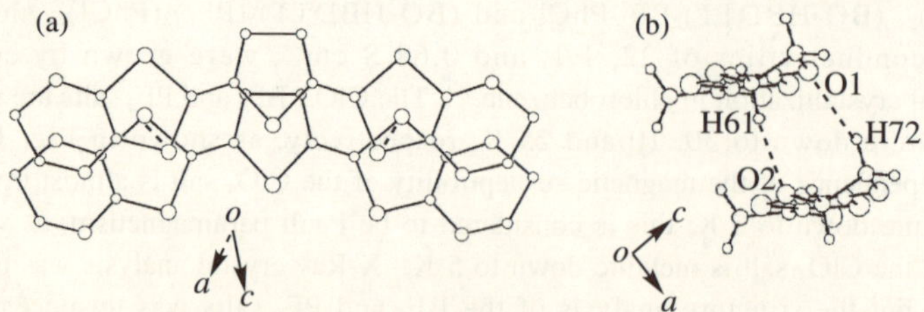

Fig. 13.11   Intrastack intermolecular contacts in $(BO\text{-}HBDTT)_5(ClO_4)_2(PhCl)_2$:
(a) overlapping mode of donor molecules viewed along the a–c axis;
(b) intrastack intermolecular hydrogen bonds viewed along the
donor long axis (b-axis). (Reprinted with permission from ref. 30)

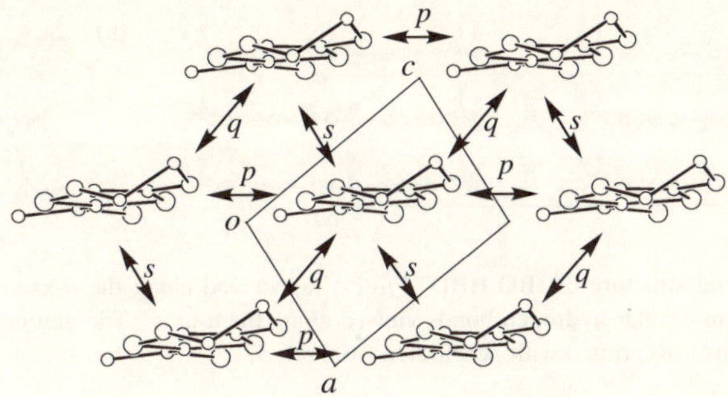

Fig. 13.12   Donor arrangement in (BO-HBDTT)$_5$(ClO$_4$)$_2$(PhCl)$_2$ viewed along the donor long axis (*b*-axis). (Reprinted with permission from ref. 30)

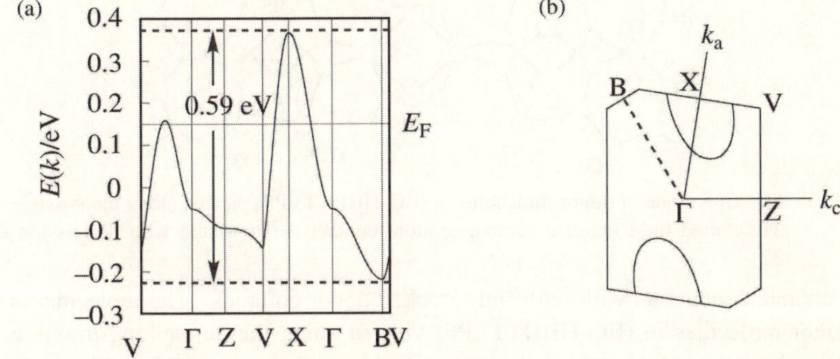

Fig. 13.13   (a) Band structure and (b) Fermi surface of (BO-HBDTT)$_5$(ClO$_4$)$_2$(PhCl)$_2$. (Reprinted with permission from ref. 30)

Fig. 13.12. The 4/5 filled metallic band structure and Fermi surface with two-dimensional nature are calculated for the ClO$_4$ salt as shown in Fig. 13.13a,b. The bandwidth of the ClO$_4$ salt (0.59 eV) is about half that of (BEDO-TTF)$_{2.4}$I$_3$ (1.13 eV).[28] The two-dimensional character is relatively enhanced in the ClO$_4$ salt (*p*/*s* = 1.8, *q*/*s* = 2.1) compared with the I$_3$ salt (*p*/*s* = 2.76, *q*/*s* = 2.36). These facts indicate that BO-HBDTT-based salts with a relatively narrow bandwidth and having two-dimensional character are promising in their possibilities for the creation of high T_c organic superconductors.

On the other hand, a black needle-like single crystalline salt, (BO-HBDTT)$_4$PF$_6$ was obtained by electrochemical crystallization in THF.[32] (BO-HBDTT)$_4$PF$_6$ showed a fairly high room temperature conductivity of 16 S cm^{-1} and metallic temperature dependence of the conductivity down to 200 K. The donor molecules stack uniformly along the *c*-axis (Fig. 13.14a) directing the central sulfur atom to the opposite side without slipping (Fig. 13.15). Similar to (BO-HBDTT)$_5$(ClO$_4$)$_2$(PhCl)$_2$, there exist short intrastack C–H···O hydrogen bonds on both sides of the donor molecule with ethylenedioxy oxygen atoms and terminal ethylene protons of the neighboring donor molecules as shown in Fig. 13.14b. There are also short interstack C–H···O hydrogen bonds along the oblique direction as shown in the dotted lines in Fig. 13.16. The intrastack C–H···O contacts also play an important role in the uniform stacking of the donor molecules in (BO-HBDTT)$_4$PF$_6$. Thus the incorporation of ethylenedioxy moieties in conjugation-elongated donor molecules is a very useful molecular design strategy to produce

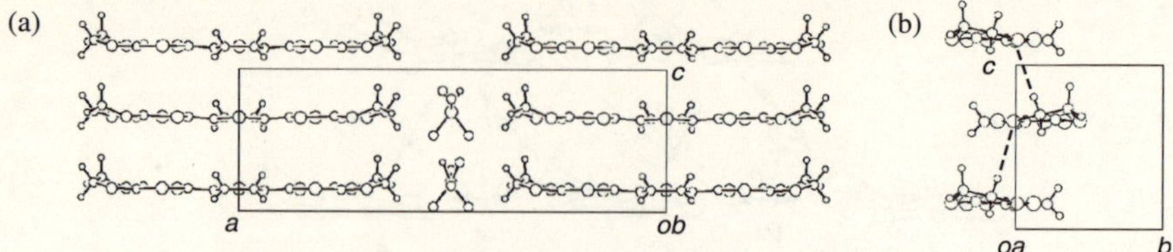

Fig. 13.14   Crystal structure of (BO-HBDTT)$_4$PF$_6$: (a) viewed along the *b*-axis; (b) intrastack intermolecular hydrogen bonds viewed along the *a*-axis. The unit cell contains 0.5 anions. (Reprinted with permission from ref. 32)

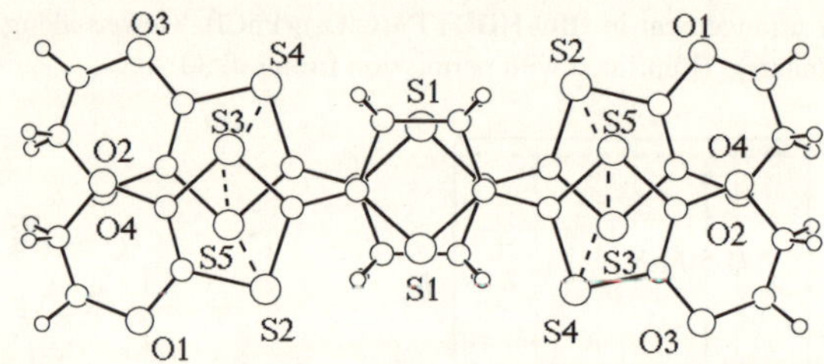

Fig. 13.15   Stacking mode of donor molecules in (BO-HBDTT)$_4$PF$_6$ viewed along the *c*-axis. The dotted lines indicate chalcogen atom contacts. (Reprinted with permission from ref. 32)

metallic organic conductors with uniformly stacked donor columns. The molecular arrangement of the donor molecules in (BO-HBDTT)$_4$PF$_6$ viewed along the donor long *a*-axis is shown in Fig. 13.16, where the calculated overlap integrals *s*, *p* and *q* are -10.26, 2.93 and 2.32×10^{-3}, respectively. As shown in Fig. 13.16, the layered intermolecular π-electron overlapping networks are constructed along the *bc*-plane. The intrastack overlap integral *s* is about three times larger than the inter-column overlap integrals *p* and *q*. Therefore, this salt has a quasi-one-dimensional character along the stacking direction. Metallic band structure and quasi-one-dimensional Fermi surface were calculated by the tight binding approximation method as shown in Fig. 13.17a, b. The calculated bandwidth of this salt is 0.57 eV.

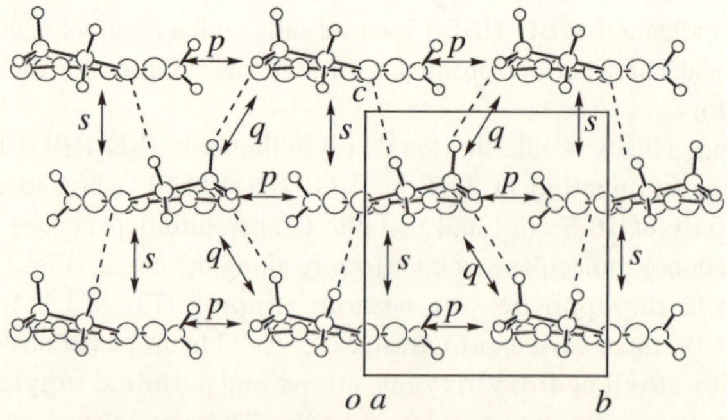

Fig. 13.16   Donor arrangement in (BO-HBDTT)$_4$PF$_6$ viewed along the donor long axis (*a*-axis). The dotted lines indicate intermolecular C–H⋯O contacts. (Reprinted with permission from ref. 32)

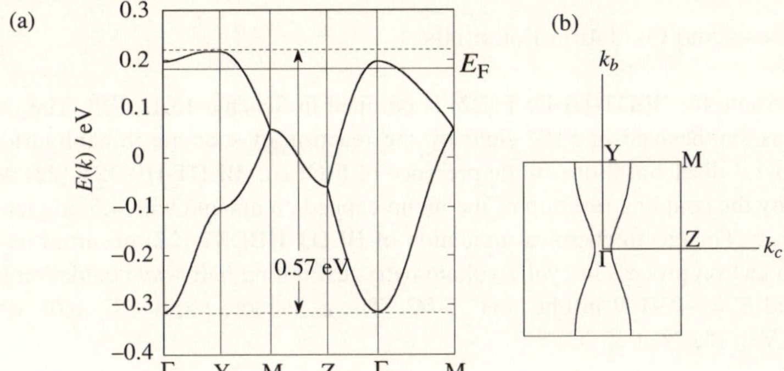

Fig. 13.17   (a) Band structure and (b) Fermi surface of $(BO\text{-}HBDTT)_4PF_6$,
calculated by the tight binding approximation method.
(Reprinted with permission from ref. 32)

The donor packing pattern of $(BO\text{-}HBDTT)_4PF_6$ is similar to those of α'-type (BEDT-TTF)$_2$X (X = AuBr$_2$, Au(CN)$_2$ and Ag(CN)$_2$).[33,34] These α'-type BEDT-TTF salts, without C–H···O hydrogen bonds, are insulators having strongly dimerized donor columns. This indicates the importance of the existence of the intrastack or oblique C–H···O hydrogen bonds to provide metallic organic conductors with uniform stacking structures.

13.6   Bis(4,5-ethylenedithio-1,3-diselenol-2-ylidene) Donor with a Dihydrothienoquinonoid-spacer Group

One of the most exciting topics in recent studies on molecular conductors are π-d interactions between π-conduction electrons in the donor layer and localized d-spins in the magnetic anion layer, which is very interesting in the study of the interplay of conduction and magnetism. For example, λ-(BETS)$_2$FeCl$_4$ undergoes a sharp metal-insulator transition at 8.5 K, where the insulating phase is strongly π-d coupled and antiferromagnetic.[35,36] For the strong π-d coupling in λ-(BETS)$_2$FeCl$_4$, the close Se···Cl contacts in the crystal structure play an important role. More recently weak π-d interactions were also found in κ-(BETS)$_2$FeBr$_4$ and κ-(BETS)$_2$FeCl$_4$.[35,37-39] Therefore, a novel conjugation-elongated BETS type donor, 2,5-bis(4,5-ethylenedithio-1,3-diselenol-2-ylidene)-2,3,4,5-tetrahydrothiophene (BEDT-HBDST, **22**), incorporating four inner Se atoms, appears to be a promising candidate for the donor component of the magnetic anion salts for developing π-d interactions. Moreover, the four inner Se atoms with large 4pπ orbitals would be quite suitable to induce good transverse intermolecular overlaps producing wide bandwidths. We have also designed and synthesized a new donor, BEDT-HBDST (**22**), and clarified its crystal structure as well as the electrical and magnetic properties of the magnetic anion salts of BEDT-HBDST (**22**), *i.e.*, (BEDT-HBDST)$_2$FeCl$_4$ and (BEDT-HBDST)$_2$FeBr$_4$, revealing that antiferromagnetic interactions of the localized 3d-spins of the Fe^{3+} ion may occur *via* the π-electrons of the donor molecules at low temperature regions in these two salts. Moreover, metallic ClO$_4$, ReO$_4$, PF$_6$, AsF$_6$ and SbF$_6$ salts of BEDT-HBDST were obtained and their electrical properties clarified.

13.6.1 Synthesis and Oxidation Potentials

The synthetic route for BEDT-HBDST (**22**) is outlined in Scheme 13.15.[40,41] The mono-capped compound was synthesized in 51% yield by the reaction of succinic thioanhydride with 4,5-ethylenedithio-1,3-diselenol-2-one in the presence of P(OEt)$_3$. BEDT-HBDST (**22**) was obtained in 42% yield by the coupling reaction of the mono-capped compound with 4,5-ethylenedithio-1,3-diselenol-2-one. The electrochemical oxidation of BEDT-HBDST (**22**) occurred in a reversible two-step one-electron process in cyclic voltammetry, exhibiting half-wave oxidation potentials of $E_1 = +0.70$ and $E_2 = +0.91$ V in PhCN at 25 °C. The E_1 is more positive by 0.03 V than that of BETS (+0.67 V in PhCN at 25 °C).[41]

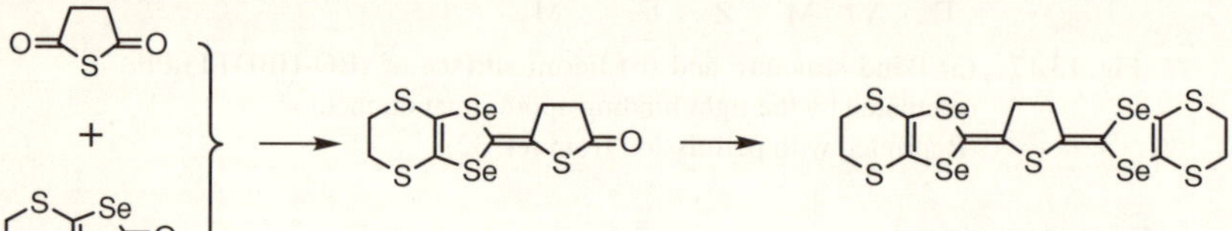

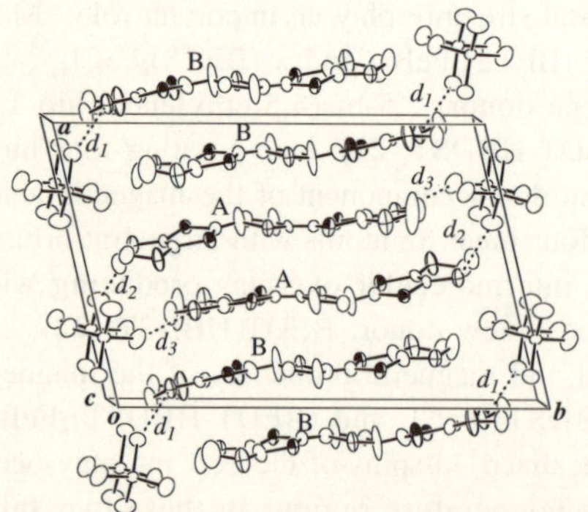

BEDT-HBDST (**22**)

Scheme 13.15

13.6.2 Crystal Structures and π-d Interactions in δ'-(BEDT-HBDST)$_2$FeX$_4$ (X = Cl and Br)

Black plate-like δ'-(BEDT-HBDST)$_2$FeCl$_4$ and δ'-(BEDT-HBDST)$_2$FeBr$_4$ showing room temperature conductivities of 0.89 and 2.2 S cm^{-1}, respectively, were obtained by electrochemical oxidation in chlorobenzene.[40] Both FeCl$_4$ and FeBr$_4$ salts exhibited semiconducting behavior with activation energies of 0.064 and 0.059 eV, respectively. The crystal structures of these two salts are shown in Figs. 13.18 and 13.19. The FeCl$_4$ and FeBr$_4$ anions are located in the space between the donor layers. The unit cell of these two salts contains four donor molecules and two anions, giving a donor-to-anion ratio of 2:1, from which the formal charge on a BEDT-HBDST molecule is determined to be +0.5. In the FeCl$_4$ salt, two crystallographically independent donors

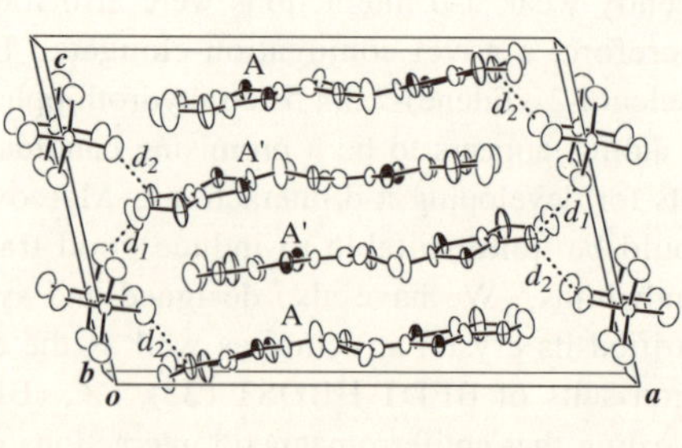

Fig. 13.18 Crystal structure of (BEDT-HBDST)$_2$-FeCl$_4$. The dotted lines indicate S···Cl contacts: $d_1 = 3.634$ and $d_2 = d_3 = 3.705$ Å.

Fig. 13.19 Crystal structure of (BEDT-HBDST)$_2$-FeBr$_4$. The dotted lines indicate S···Br contacts: $d_1 = 3.681$ and $d_2 = 3.734$ Å.

A and B stack in a B-A-A-B manner along the a-axis forming dimers. In the $FeBr_4$ salt, one crystallographically independent donor A (A') stacks in an A-A'-A'-A manner along the c-axis. Thus, the donor arrangements of these two salts are essentially the same, since the donor molecules stack in the manner of four-fold repetition. The inter-dimer overlap integrals are about 1/9 to 1/4 of the intra-dimer overlap integrals, so the dimerization is fairly strong in these salts. In contrast, the donor molecules are coplanarly and uniformly linked along the transverse direction, directing the central sulfur atom to the same side, resulting in considerably large overlap integrals in the transverse direction. Thus, band structures of these two salts are quasi-one-dimensional along the transverse direction. Both salts adopt the δ'-type donor arrangement similar to those of δ'-(BEDT-TTF)$_2$X (X = FeCl$_4$ and GaCl$_4$).[42]

Temperature dependence of the magnetic susceptibility of δ'-(BEDT-HBDST)$_2$FeCl$_4$ was measured down to 2 K using a SQUID magnetometer.[40] As shown in Fig. 13.20a, the χT -T and $1/\chi$ -T plots follow the Curie-Weiss law from around 100 to 2 K with the Curie constant of 4.6 emu K mol^{-1} and the extrapolation of the $1/\chi$ to zero gives the Weiss temperature (θ) of -1.2 K. Moreover, the magnetic field dependence of the magnetization was also measured by the SQUID method. As shown in Fig. 13.20b, the experimentally measured magnetization is well fitted to the solid line calculated using the Brillouin function with the parameters of S (J) = 5/2 and $\theta = -1.2$ K for $T^* = T-\theta$. By the same method, the Curie constant of 4.8 emu K mol^{-1} and the Weiss temperature of -4.9 K were obtained for δ'-(BEDT-HBDST)$_2$FeBr$_4$. The absolute value of this Weiss temperature is much larger than that of κ-(BETS)$_2$FeCl$_4$.[43] From the negative values of the Weiss temperatures, it is revealed that fluctuation of the short range antiferromagnetic ordering of the localized 3d-spins of the Fe^{3+} ion in δ'-(BEDT-HBDST)$_2$FeX$_4$ (X = Cl and Br) may occur at the low temperature region, although the ordering ability is inferior.

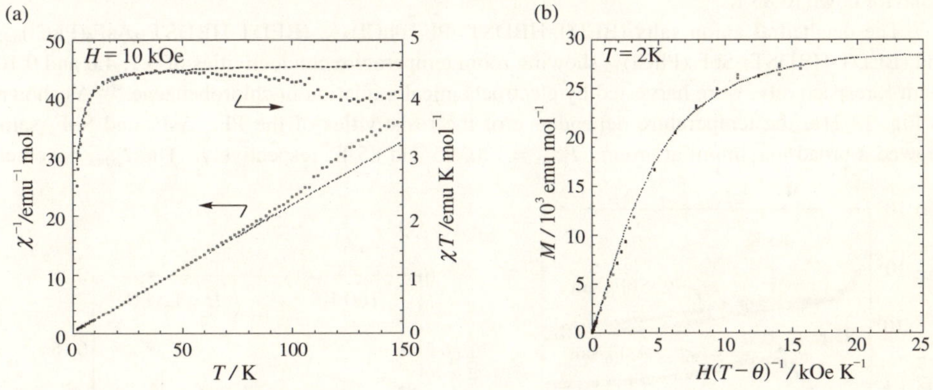

Fig. 13.20   (a) χ^{-1}–T and χT–T plots and (b) magnetization as a function of $H/(T$-$\theta)$ for (BEDT-HBDST)$_2$FeCl$_4$.

Although there is no short Se···Cl or Se···Br contact, there are sort S···Cl and S···Br contacts in δ'-(BEDT-HBDST)$_2$FeCl$_4$ and δ'-(BEDT-HBDST)$_2$FeBr$_4$ as shown in Table 13.3 and in Figs. 13.18 and 13.19. Direct anion-to-anion interactions appear to be difficult to occur, since the shortest Cl···Cl and Br···Br distances of these salts are far longer than the van der Waals distances. Moreover, the shortest Fe^{3+}···Fe^{3+} distances of these salts are longer by more than 0.78 Å than that of κ-(BETS)$_2$FeCl$_4$ (5.879 Å).[35] Based on the π-d and d-d transfer integrals obtained by the extended Hückel MO calculations, magnetic interactions $J_{\pi d}$ and J_{dd} are evaluated using the

Table 13.3   Intermolecular Atom-atom Distance/Å and θ Values/K of FeCl$_4$ and FeBr$_4$ Salts[a]

| | Donor-Anion | Anion-Anion | | |
	S···Cl (S···Br)	Fe···Fe	Cl(Br)···Cl(Br)	θ
δ'-(BEDT-HBDST)$_2$FeCl$_4$	3.634	6.658	4.810	−1.2
	3.705	7.563	4.819	
δ'-(BEDT-HBDST)$_2$FeBr$_4$	3.681	6.672	4.744	−4.9
	3.734	7.599	4.768	
κ-(BETS)$_2$FeCl$_4$	3.547	5.879	4.12	−0.8
	3.592			

[a]The atom-atom distance and Weiss temperature of κ-(BETS)$_2$FeCl$_4$ are taken from refs. 35 and 43, respectively.

equation $J = -2t^2/U$ (t: transfer integral and U: Coulomb integral).[44] The largest $J_{\pi d}$ (1.69 K) is about 85 times that of the largest J_{dd} (0.020 K). Therefore, it may be possible to assume that the d-spins of the Fe^{3+} ion of these salts antiferromagnetically interact indirectly *via* the π-electrons of the donor layer in the low temperature insulating state, although further investigation should be continued.

13.6.3 Electrical Properties of Tetrahedral and Octahedral Anion Salts of BEDT-HBDST

The tetrahedral anion salts (BEDT-HBDST)$_2$ClO$_4$(PhCl)$_{0.5}$ and (BEDT-HBDST)$_2$ReO$_4$(PhCl)$_{0.2}$ showing room temperature conductivities of 20 and 10 S cm^{-1}, respectively, were grown by conventional electrochemical oxidation in chlorobenzene. Both of these salts showed metallic behavior down to 85 K.[41]

The octahedral anion salts (BEDT-HBDST)$_2$PF$_6$(PhCl)$_{0.75}$, (BEDT-HBDST)$_2$AsF$_6$(PhCl)$_{0.75}$ and (BEDT-HBDST)$_2$SbF$_6$(PhCl)$_{0.5}$ showing room temperature conductivities of 3.1, 4.0 and 0.10 S cm^{-1}, respectively, were harvested by electrochemical oxidation in chlorobenzene.[41] As shown in Fig. 13.21a, the temperature dependence of the resistivities of the PF$_6$, AsF$_6$ and SbF$_6$ salts showed a broad maximum at around $T_{\rho max}$ = 120, 95 and 85 K, respectively. The $T_{\rho max}$ becomes

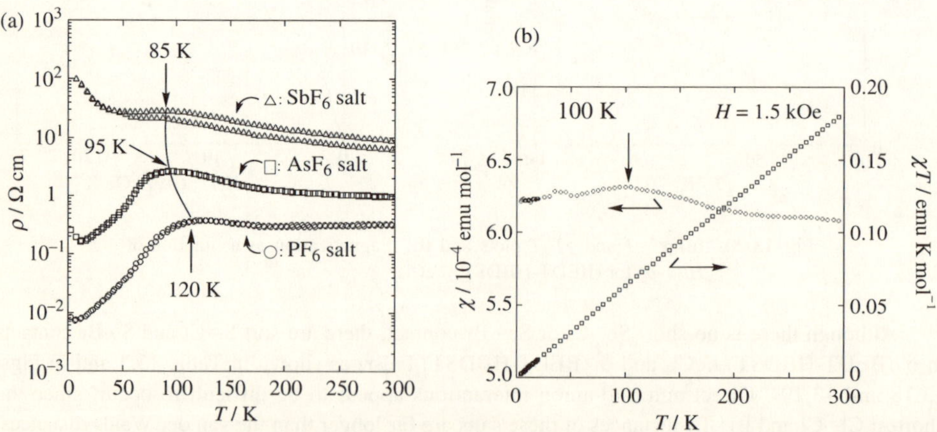

Fig. 13.21   (a) Temperature dependence of the resistivities of (BEDT-HBDST)$_2$PF$_6$(PhCl)$_{0.75}$, (BEDT-HBDST)$_2$AsF$_6$(PhCl)$_{0.75}$ and (BEDT-HBDST)$_2$SbF$_6$(PhCl)$_{0.5}$ and (b) temperature dependence of the magnetic susceptibility of (BEDT-HBDST)$_2$PF$_6$(PhCl)$_{0.75}$.

higher with decrease in the anion size. Thus it is suggested that semiconducting to metal transition temperature rises with increase in the chemical pressure. These conducting properties closely resemble high-temperature conducting properties of high T_c organic superconductors such as κ-phase BEDT-TTF and λ-phase BETS salts since these salts also show a broad maximum in temperature dependence of the resistivity plots and the $T_{\rho max}$ shifts to high temperature on application of physical or chemical pressure.[45] However, the present PF_6, AsF_6 and SbF_6 salts did not show superconductivity down to 2 K. As shown in Fig. 13.21b, the magnetic susceptibility of the PF_6 salt is almost invariable from room temperature down to 2 K, a phenomenon which can be explained by a Pauli paramagnetism. The relatively large χ value ($\chi_{rt} = 6.1 \times 10^{-4}$ emu mol^{-1}) can be ascribed to the relatively narrow bandwidth of this salt compared with the usual paramagnetic conductors, indicating that this salt exists in a fairly strong π-electron correlation state. Thus, the conducting and magnetic behaviors shown in Fig. 13.21a,b are consistent with each other, strongly suggesting that the electronic state of this salt is located near the boundary between the nonmetallic and metallic phases. X-Ray crystal analyses of these salts are now in progress.

13.7   Summary and Outlook

A series of bis(1,3-dithiol-2-ylidene) donors with a heteroquinonoid-spacer group has been successfully synthesized. The synthetic methodologies involve a convenient method and short step reactions which can be manipulated quite easily. Moreover, the synthetic methodology is widely applicable to the synthesis of a variety of modified donors with a heteroquinonoid-, fused heteroquinonoid- or dihydroheteroquinonoid-spacer group having different oxidation potentials and having a different bent angle. The introduction of not only ethylenedithio groups but also ethylenedioxy groups in the peripheral part of the donor molecules is easily accomplished.

BDTT and BDTS type donors have fairly high electron-donating abilities. Therefore, the formation of good quality single crystalline cation radical salts by electrochemical crystallization was difficult. However, their high donor abilities are weakened by the condensation of a benzene ring on the linking heteroquinonoid ring providing BDTBF, BDTBT, and BDTBS type donors. From these fused donors, various kinds of single crystalline cation radical salts were obtained. In these fused donor-based salts, the bent angle of the donor molecule governs roughly the donor packing motif, especially the donor-to-anion ratio.

The incorporation of the electron-deficient pyrazine ring in the linking heteroquinonoid-spacer group is very effective in decreasing electron-donating ability. Consequently, stability of BDTFP is highly increased. $(BDTFP)_2PF_6(PhCl)_{0.5}$ and $(BDTFP)_2AsF_6(PhCl)_{0.5}$ exhibit conductivities of 40 and 4.0 S cm^{-1}, respectively. EPR measurement reveals that the PF_6 salt undergoes a spin-singlet transition at around 170 K, while the AsF_6 salt undergoes a huge transition of the first order at around 230 K, associated with an abrupt jump in the spin susceptibility. The AsF_6 salt is magnetic down to low temperature. X-Ray crystal structural analysis has clarified that the uncommon two-leg ladder type overlapping mode is attained in the PF_6 and AsF_6 salts. Detailed investigations on the mechanisms of the remarkable changes in the magnetic and electrical properties observed in these salts are now in progress.

As for the third approach for decreasing the electron-donating abilities of BDTT type donors, donors with a dihydrothienoquinonoid-spacer group such as BO-HBDTT and BEDT-HBDST have been developed. BO-HBDTT afforded many metallic tetrahedral anion salts in which uniform stacking structures of the donor columns constructed from the intrastack C–H···O hydrogen bonds have been revealed. Of these, $(BO\text{-}HBDTT)_5(ClO_4)_2(PhCl)_2$ showed two-dimensional metallic character with a relatively narrow bandwidth of 0.59 eV, which is promising as a high T_c organic superconductor. To obtain metallic organic conductors with uniform stacking structure, the introduction of ethylenedioxy moieties to the peripheral part of the donor molecule is especially important. BEDT-HBDST also afforded many metallic salts. The magnetic susceptibilities of δ'-(BEDT-HBDST)$_2$FeCl$_4$ and δ'-(BEDT-HBDST)$_2$FeBr$_4$ follow the Curie-Weiss law with Curie constants of 4.6 and 4.8 emu K mol^{-1} ($S = 5/2$) and negative Weiss temperatures of -1.2 and -4.9 K, respectively, revealing that weak antiferromagnetic interactions of the 3d-spins of the Fe^{3+} ion *via* the π-electrons of the donor molecules might be possible.

We have also designed and successfully synthesized conceptionally novel electron-acceptors incorporating three electron-accepting groups and exhibiting fairly high electron-accepting abilities (Scheme 13.16). Me$_4$X(CPDT)$_2$ (X = N, P and As) are extremely one-dimensional along the acceptor stacking *b*-axis. These salts are the first examples of heterophene-TCNQ type anion radicals showing metallic behavior.[46] The [3]radialene-type acceptors have four bulky *tert*-butyl groups and only one dicyanomethylene group. Nevertheless, the single crystalline TTF and tetrathiotetracene (TTT) complexes of these acceptors exhibited surprisingly high conductivities reaching 420 to 450 S cm^{-1} and are metallic down to 182–80 K, which is completely unexpected judging from the common knowledge so far accumulated in the field of organic conductors.[47]

CPDT

X = S, Se and Te

Scheme 13.16

Thus the concept of the incorporation of heteroquinonoid-spacer group can be widely applicable to the molecular design and synthesis of other related systems, providing many possibilities for the creation of various new types of molecular systems with novel electrical and magnetic functions.

References

1. A. F. Garito and A. J. Heeger, *Acc. Chem. Res.*, **7**, 232 (1974); J. B. Torrance, *Acc. Chem. Res.*, **12**, 79 (1979); F. Wudl, *Acc. Chem. Res.*, **7**, 227 (1984).
2. H. Yamochi, T. Komatsu, N. Matsukawa, G. Saito, T. Mori, M. Kusunoki and K. Sakaguchi, *J. Am. Chem. Soc.*, **115**, 11319 (1993).
3. Z. Yoshida, T. Kawase, H. Awaji and S. Yoneda *Tetrahedron Lett.*, **24**, 3473 (1983).
4. Y. Yamashita, Y. Kobayashi and T. Miyashi, *Angew. Chem. Int. Ed. Engl.*, **28**, 1052 (1989); Y. Yamashita, S. Tanaka and M. Tomura, *J. Chem. Soc., Chem. Commun.*, 652 (1993).
5. A. J. Moore and M. R. Bryce, *J. Chem. Soc., Perkin Trans. 1*, 157 (1991); M. Bryce, *J. Chem. Soc., Perkin Trans. 1*, 1675 (1985).
6. K. Takahashi and T. Suzuki, *J. Am. Chem. Soc.*, **111**, 5483 (1989); K. Takahashi, T. Suzuki, K. Akiyama, Y. Ikegami and Y. Fukazawa, *J. Am. Chem. Soc.*, **113**, 4576 (1991); K. Takahashi and M. Ogiyama, *J. Chem. Soc., Chem. Commun.*, 1196 (1990).
7. K. Takahashi, A. Gunji and K. Akiyama, *Chem. Lett.*, 863 (1994).
8. K. Takahashi, A. Gunji, K. Yanagi and M. Miki, *J. Org. Chem.*, **61**, 4784 (1996).
9. K. Takahashi, T. Nihira, K. Takase and K. Shibata, *Tetrahedron Lett.*, **30**, 2091 (1989).
10. K. Takahashi and T. Nihira, *Tetrahedron Lett.*, **30**, 5903 (1989).

11. K. Takahashi, T. Nihira and K. Tomitani, *J. Chem. Soc., Chem. Commun.*, 1617 (1993).
12. K. Takahashi and K. Tomitani, *J. Chem. Soc., Chem. Commun.*, 821 (1995).
13. K. Takahashi, K. Tomitani, T. Ise and T. Shirahata, *Chem. Lett.*, 619 (1995).
14. K. Takahashi, T. Shirahata and K. Tomitani, *J. Mater. Chem.*, **7**, 2375 (1997).
15. K. Takahashi and T. Ise, *Heterocycles*, **45**, 1051 (1997).
16. K. Takahashi and T. Ise, *Chem. Lett.*, 77 (1995).
17. K. Takahashi and T. Ise, *Phosphorus, Sulfur, and Silicon*, **120-121**, 415 (1997).
18. K. Takahashi, T. Ise, T. Mori, H. Mori and S. Tanaka, *Chem. Lett.*, 1147 (1998).
19. T. Ise, T. Mori and K. Takahashi, *Chem. Lett.*, 1013 (1997).
20. K. Takahashi, T. Ise, T. Mori, H. Mori and S. Tanaka, *Chem. Lett.*, 1001 (1996).
21. J. B. Torrance, B. A. Scott and F. B. Kaufman, *Solid State Commun.*, **17**, 1369 (1975); J. Tanaka, M. Tanaka, T. Kawai, T. Takabe and O. Maki, *Bull. Chem. Soc. Jpn.*, **49**, 2358 (1976).
22. K. Takahashi, T. Ise and T. Mori, to be published.
23. T. Ise and K. Takahashi, *J. Cryst. Growth*, **229**, 591 (2001).
24. T. Ise, T. Mori and K. Takahashi, *J. Mater. Chem.*, **11**, 264 (2001).
25. M. Uehara, T. Nagata, J. Akimitsu, H. Takahashi, N. Mori and K. Kinoshita, *J. Phys. Soc. Jpn.*, **65**, 2764 (1996).
26. C. Rovira, *Chem. Eur. J.*, **6**, 1723 (2000).
27. T. Nakamura, K. Takahashi, T. Shirahata, M. Uruichi, K. Yakushi and T. Mori, *J. Phys. Soc. Jpn.*, **71**, 2022 (2002).
28. S. Horiuchi, H. Yamochi, G. Saito, K. Sakaguchi and M. Kusunoki, *J. Am. Chem. Soc.*, **118**, 8604 (1996).
29. J. M. Williams, J. R. Ferraro, R. J. Thorn, K. D. Carlson, U. Geiser, H. H. Wang, A. M. Kini and M.-H. Whangbo, *Organic Superconductors*, Prentice Hall, NJ (1992), p. 115.
30. T. Shirahata and K. Takahashi, *Mol. Cryst. Liq. Cryst.*, **376**, 1 (2002).
31. M. M. Urberg and E. T. Kaiser, *Radical Ions*, Wiley, New York (1968), p. 312.
32. K. Takahashi and T. Shirahata, *Chem. Lett.*, 514 (2001).
33. M. A. Beno, M. A. Firestone, P. C. W. Leung, L. M. Sowa, H. H. Wang, J. M. Williams and M.-H. Whangbo, *Solid State Commun.*, **57**, 735 (1986).
34. E. Amberger, H. Fuchs and K. Polborn, *Angew. Chem. Int. Ed. Engl.*, **25**, 729 (1986).
35. H. Kobayashi, H. Tomita, T. Naito, A. Kobayashi, F. Sasaki, T. Watanabe and P. Cassoux, *J. Am. Chem. Soc.*, **118**, 368 (1996).
36. H. Kobayashi, A. Kobayashi and P. Cassoux, *Chem. Soc. Rev.*, **29**, 325 (2000).
37. E. Ojima, H. Fujiwara, K. Kato and H. Kobayashi, *J. Am. Chem. Soc.*, **121**, 5581 (1999).
38. T. Otsuka, A. Kobayashi, Y. Miyamoto, J. Kiuchi, N. Wada, E. Ojima, H. Fujiwara and H. Kobayashi, *Chem. Lett.*, 732 (2000).
39. H. Fujiwara, E. Fujiwara, Y. Nakazawa, B. Z. Narymbetov, K. Kato, H. Kobayashi, A. Kobayashi, M. Tokumoto and P. Cassoux, *J. Am. Chem. Soc.*, **123**, 306 (2001).
40. T. Shirahata, T. Mori and K. Takahashi, *J. solid state chem.*, in press.
41. T. Shirahata, T. Mori, R. Kato and K. Takahashi, *Synth. Met.*, **133-134**, 321 (2003).
42. T. Mallah, C. Hollis, S. Bott, M. Kurmoo, P. Day, M. Allan and R. H. Friend, *J. Chem. Soc., Dalton Trans.*, **42**, 2127 (1991).
43. H. Uozaki, K. Okamoto, S. Endo, H. Matsui, K. Ueda, T. Sugimoto and N. Toyota, *Synth. Met.*, **103**, 1984 (1999).
44. T. Mori and M. Katsuhara, *J. Phys. Soc. Jpn.*, **71**, 826 (2002).
45. H. Tanaka, A. Kobayashi, A. Sato, H. Akutsu and H. Kobayashi, *J. Am. Chem. Soc.*, **121**, 760 (1999); L. K. Montgomery, T. Burgin, J. C. Huffman, K. D. Carlson, J. D. Dudek, G. A. Yaconi, L. A. Megna, P. R. Mobley, W. K. Kwok, J. M. Williams, J. E. Schirber, D. L. Overmyer, J. Rew, C. Rovira and M.-H. Whangbo, *Synth. Met.*, **55-57**, 2090 (1993).
46. K. Takahashi and S. Tarutani, *J. Chem. Soc., Chem. Commun.*, 1233 (1998); J. Yamaura, R. Kato, S. Tarutani and K. Takahashi, *Synth. Met.*, **103**, 2212 (1999); J. Yamaura, M. Fujiwara, R. Kato, T. Chonan and K. Takahashi, *Synth. Met.*, **120**, 913 (2001); S. Tarutani, J. Yamaura, K. Takahashi and R. Kato, *Solid State Commun.*, **123**, 251 (2002).
47. K. Takahashi and S. Tarutani, *J. Chem. Soc., Chem. Commun.*, 519 (1994); K. Takahashi and S. Tarutani, *Synth. Met.*, **70**, 1165 (1995); K. Takahashi and S. Tarutani, *Adv. Mater.*, 639 (1995); K. Takahashi and S. Tarutani, *Mol. Cryst. Liq. Cryst.*, **296**, 145 (1997); S. Tarutani, T. Mori, H. Mori, S. Tanaka and K. Takahashi, *Chem. Lett.*, 627 (1997).

14

Donor Systems with Multi-1,3-dithiol-2-ylidene Units

14.1   Introduction

As discussed in the introductory section of Chapter 11, 1,3-dithiol-2-ylidene (DT) compounds display enough electron-donating ability to be oxidized to stable cation-radical species. This feature leads to the prediction that these compounds can be utilized as (i) donor components of charge-transfer (CT) materials exhibiting electrical conductivity, especially metallic conductivity and superconductivity and (ii) a spin source for achieving ferromagnetic interactions in organic materials. In fact, extensive studies on the synthesis of multi-DT-containing compounds have been made to develop new organic conductors and materials with interesting optical and magnetic properties. This chapter deals mainly with (i) the donor systems comprising more than three DT units, the frameworks of which are based on TTF, vinylogous TTF, radialene, dendralene and other cyclic compounds, and (ii) trimethylenemethane and tetramethylenethane systems with one or two appended DT units. In addition, their unique redox and physical properties are described.

14.2   TTFs with Multi-DT Units

Structural modifications of tetrathiafulvalene (TTF **1**, Fig. 14.1) continue to play an important role in the development of new donor components capable of producing highly conducting CT materials such as organic metals and superconductors.[1-3] TTF itself is composed of two DT units, each of which is oxidized to generate the 1,3-dithiolium cation stabilized by the heteroaromatic 6π electron system (see Scheme 11.1 in Chapter 11). Consequently, oxidation of TTF occurs in two successive one-electron steps with formation of the cation-radical and the dication (Scheme 14.1). In addition, the planarity of the oxidized TTF molecule allows the formation of ordered stacks [quasi one-dimensional (1D) character], as can be found in the metallic (TTF)(TCNQ) (TCNQ = tetracyanoquinodimetane **2**, Fig. 14.1) complex,[4] and the additional sulfur atoms by attachment of sulfur-containing substituents to the TTF molecule increase intermolecular S···S interactions resulting in formation of ordered sheets [quasi two-

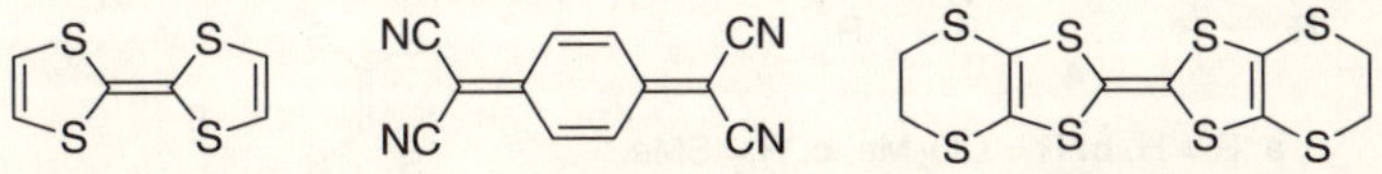

TTF **1**   TCNQ **2**   BEDT-TTF **3**

Fig. 14.1

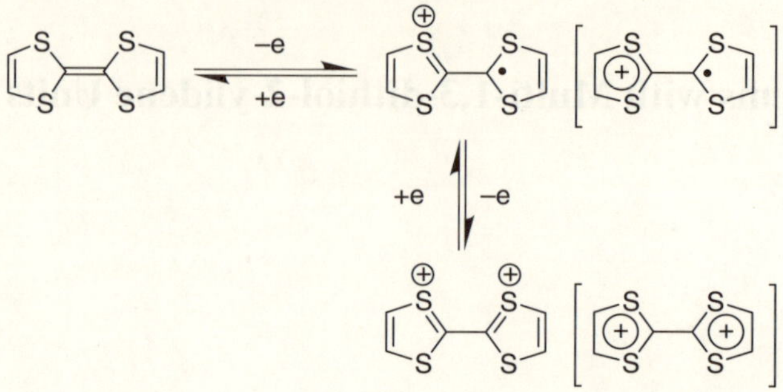

Scheme 14.1

dimensional (2D) character], as can be found in the superconducting BEDT-TTF [bis(ethyleneditio)tetrathiafulvalene **3**, Fig. 14.1] salts.[2]   Based on these facts, the design of introducing redox-active DT units into the periphery of the TTF core has evolved mainly from (i) the realization of increased dimensionality of the conduction pathway in CT materials, (ii) the expectation of a decrease in the on-site Coulombic repulsion involved in the generation of a dicationic species [an increase in delocalization of the generating positive charge(s)] and (iii) the construction of new organic redox systems.

Gorgues and coworkers have studied the tetrakis- and bis-(1,3-dithiafulvenyl)-substituted TTFs **4** and **5** (Fig. 14.2), as well as the bis(1,3-dithiafulvenyl)-substituted dihydro-TTF (DHTTF) system **6** (Fig. 14.2).[5–8]   Scheme 14.2 outlines the synthetic routes to **4** and **5**, which involve construction of the tetra- and di-formyl-attached TTFs **7** and **8** followed by fourfold and twofold Wittig olefinations, respectively.   In a manner similar to the synthesis of **5**, a twofold Wittig reaction of the diformyl derivative of DHTTF (see Scheme 11.7 in Chapter 11) with the corresponding phosphorus ylid gave **6**.   The electrochemical behavior of **4a–d**, **5a** and **6a,b** was investigated by cyclic voltammetry (CV).[8]   The CVs of **5a** and **6a,b** show two, reversible, one-electron oxidation waves, whereas the CVs of **4a–d** consist of three reversible oxidation waves; the two-electron transfer in the third oxidation wave was confirmed by thin-layer CV.   Table 14.1 summarizes the oxidation potentials of these compounds, together with those of TTF for comparison.   It should be noted that the attachment of four or two 1,3-dithiafulvenyl substituents

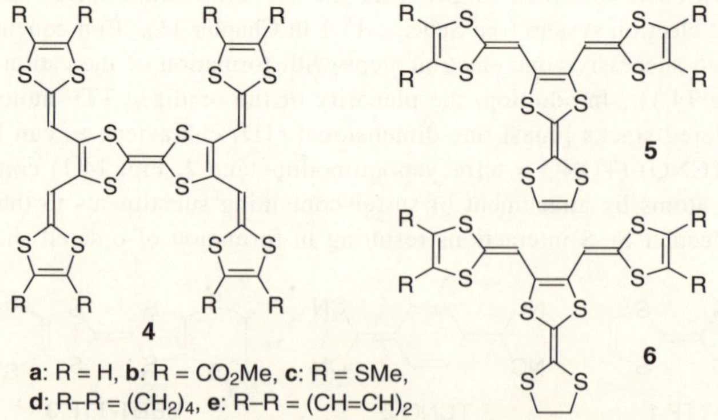

a: R = H, **b**: R = CO$_2$Me, **c**: R = SMe,
d: R–R = (CH$_2$)$_4$, **e**: R–R = (CH=CH)$_2$

Fig. 14.2   Multi-DT-containing TTF and DHTTF systems.

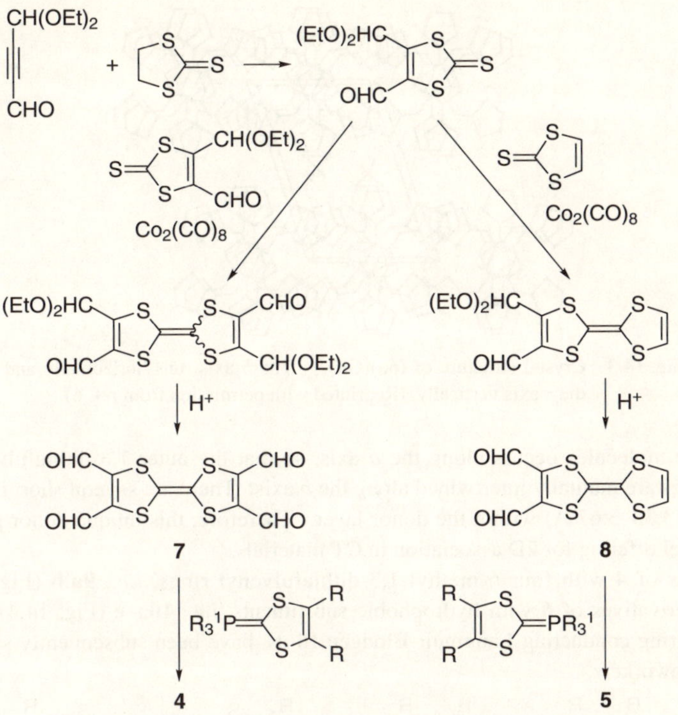

Scheme 14.2

Table 14.1   Anodic Peak Potentials (V *vs.* SCE) of **4–6** and TTF[8]

Compound	Solvent	E_{pa1}	E_{pa2}	E_{pa3}	$\Delta E_{pa}\,(E_{pa2} - E_{pa1})$
4a	DMF[a]	0.19	0.34	0.50	0.15
4b	TCE[b]	0.34	0.65	0.95	0.31
4c	TCE	0.09	0.42	0.57	0.33
4d	TCE	0.08	0.17	0.51	0.09
5a	DMF	0.22	0.30		0.08
6a	TCE	0.26	0.39		0.13
6b	TCE	0.51	0.74		0.23
TTF	DMF	0.42	0.72		0.30
	TCE	0.40	0.82		0.42

[a]Dimethylformamide. [b]1,1,2-Trichloroethane.

to TTF lowers the first oxidation potential (E_{pa1}) and reduces the difference between the first and second oxidation potentials [$\Delta E_{pa}\,(E_{pa2} - E_{pa1})$], even though compound **4b** has eight electron-withdrawing methoxycarbonyl groups. These provide evidence for both an increase in the electron-donating ability and a decrease in the on-site Coulombic repulsion.

DHTTF derivative **6a** was found to form a 2:3 complex with TCNQ, (**6a**)$_2$(TCNQ)$_3$, a compressed pellet of which showed a room-temperature conductivity (σ_{rt}) of 1 S cm^{-1}. In addition, a semiconductive 1:1 salt, (**6a**)ClO$_4$, with a thermal activation energy of 170 meV ($\sigma_{rt} = 0.38$ S cm^{-1}), has been prepared by electrocrystallization. Fig. 14.3 shows the crystal structure of this salt,[6] in which the donor molecules and the ClO$_4^-$ anions are arranged in alternating sheets along the c axis. Within a donor stack, head-to-tail overlap of the unsymmetrical central DHTTF

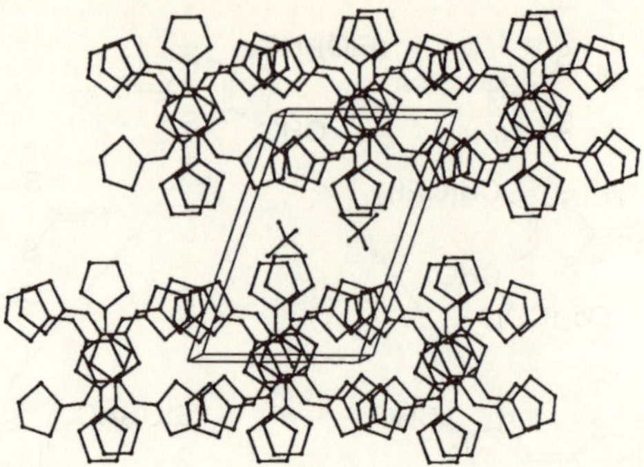

Fig. 14.3   Crystal structure of (**6a**)(ClO₄). The *b* axis runs horizontally and
the *c* axis vertically. (Reprinted with permission from ref. 6)

cores of donor molecules occurs along the *a* axis, so that the outer 1,3-dithiafulvenyl rings of donor molecules are mutually intertwined along the *b* axis. There are several short intermolecular S···S contacts (3.30–3.67 Å) within the donor layer. Therefore, this unique donor packing motif provides a novel offering for 2D association in CT materials.

Analogues of **4** with four 6-methyl-1,3-dithiafulvenyl rings, *i.e.*, **9a,b** (Fig. 14.4),[9] and amphiphilic derivatives of **6** with hydrophobic substituents, *i.e.*, **10a–c** (Fig. 14.4),[10] which are aimed at preparing conducting Langmuir-Blodgett films, have been subsequently synthesized by Gorgues and coworkers.

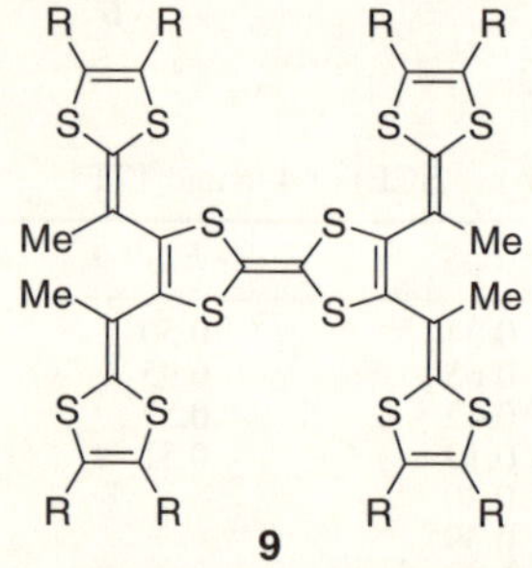

9

a: R = SMe, **b**: R–R = S(CH₂)₂S

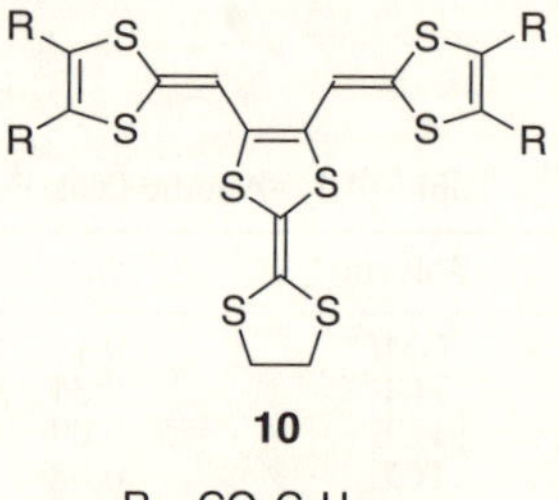

10

R = CO₂CₙH₂ₙ₊₁

a: n = 2, **b**: n = 4, **c**: n = 8

Fig. 14.4

14.3   TTF Vinylogues with Multi-DT Units

The parent TTF vinylogues **11a** and **12** (Fig. 14.5), in which conjugated π-systems are incorporated between the two 1,3-dithiole rings of TTF, were first reported by Yoshida and coworkers.[11] Gorgues *et al.* developed a synthetic method for introducing 1,3-dithiafulvenyl and 6-methyl-1,3-dithiafulvenyl substituents into **11a** and **12** in order to obtain vinylogous TTF derivatives with a highly extended π-conjugation.[7,9,12] The synthesis of **13–16** (Scheme 14.3) proceeded *via* Wittig-Horner reactions with phosphonate anions. A common feature of these compounds is that they all exhibit strong π-donor properties with very small on-site Coulombic

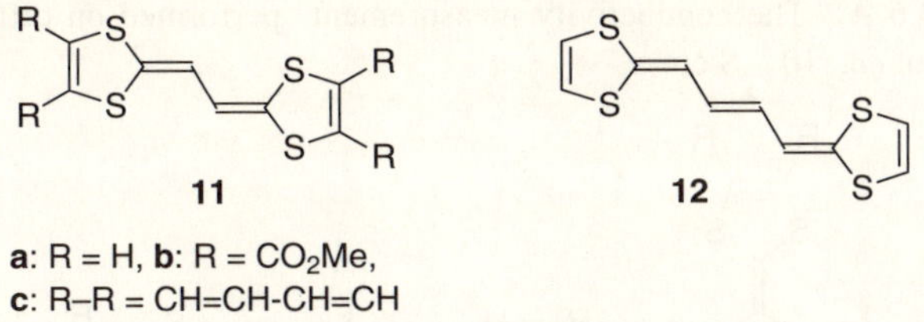

11   **12**

a: R = H, **b**: R = CO$_2$Me,
c: R–R = CH=CH-CH=CH

Fig. 14.5   Vinylogous TTFs.

repulsion energies. For example, the CV of the unsubstituted **15** (R = R^2 = H) consisted of two, two-electron, reversible oxidation waves at 0.12 (E_{pa1} = E_{pa2}) and 0.47 (E_{pa3} = E_{pa4}) V (*vs.* SCE), in which direct oxidation to the dication takes place at a single potential and the first oxidation potential (E_{pa1}) is considerably lower than that of TTF (0.38 V).[12] However, it is pointed out that, as shown in Scheme 14.4, compounds such as **13** and **15** may undergo intramolecular cyclization under acidic or oxidative conditions.[9] Nevertheless, **17a,b** and **18** (Fig. 14.6) have been found to give small dark crystals by electrocrystallization under galvanostatic conditions in the presence of Bu$_4$NX (X = PF$_6$ and ClO$_4$) in CH$_2$Cl$_2$.[13] The 1:1 stoichiometry and crystal structure of (**18**)(PF$_6$) was determined by X-ray diffraction analysis. As shown in Fig. 14.7, the structure of this salt is built up from sheets of donor molecules stacked along the *a* axis with an inclination of 22°. Within a donor sheet, each donor molecule is characterized by the plane formed by the four DT units, though the *p*-MeC$_6$H$_4$ group is located in a roughly perpendicular plane. The overlap mode of donor molecules minimizes the steric hindrance between bulky aryl groups and allows several close contacts between the neighboring donors. The distance between the planes in the donor

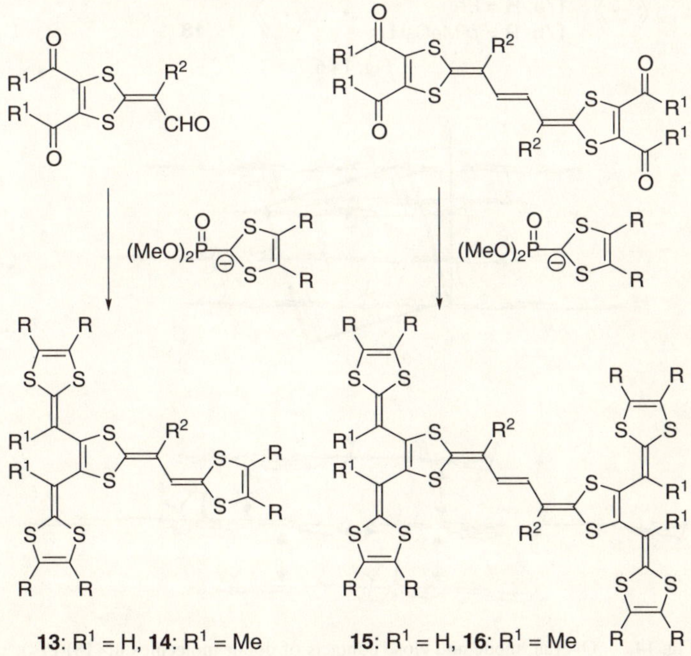

13: R^1 = H, **14**: R^1 = Me   **15**: R^1 = H, **16**: R^1 = Me

Scheme 14.3

stacking is about 3.6 Å. The conductivity measurements performed on different compressed pellets gave values of *ca.* 10^{-1} S cm^{-1}.

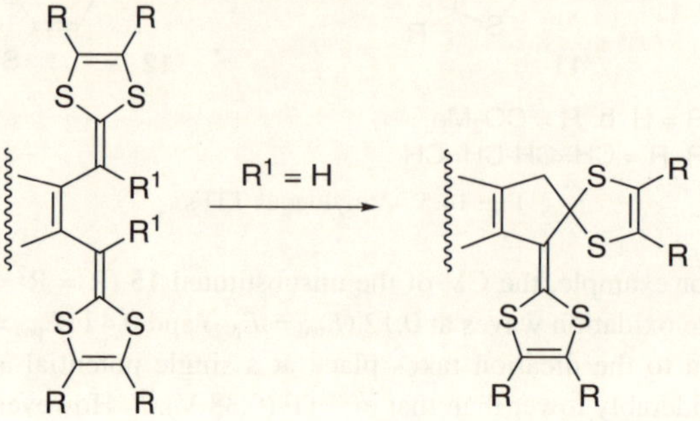

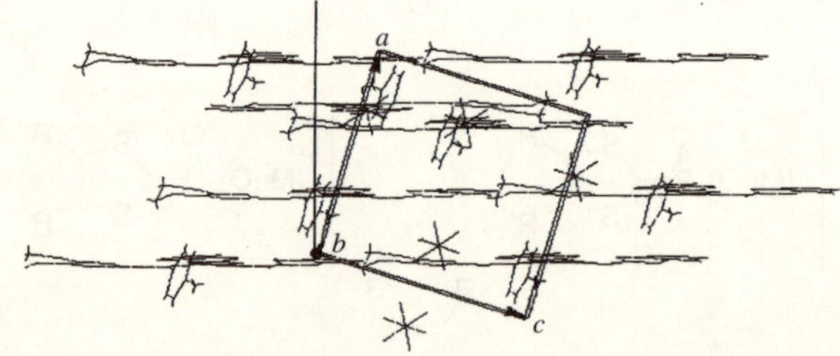

Scheme 14.4

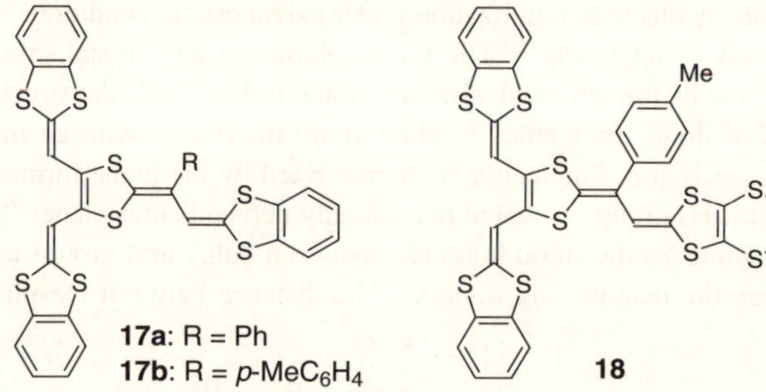

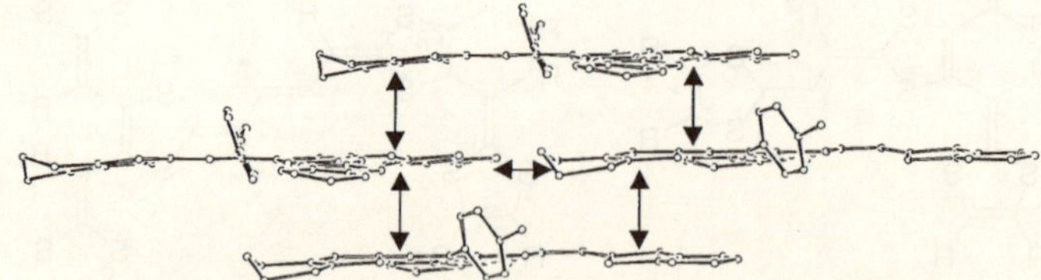

17a: R = Ph
17b: R = *p*-MeC$_6$H$_4$

18

Fig. 14.6

Fig. 14.7   Overlap mode and close contacts of donor molecules in (**18**)(PF$_6$).
(Reprinted with permission from ref. 13)

Bryce *et al.* synthesized an interesting series of donors containing covalently linked TTF and its vinylogue **11a**, *i.e.*, **19a–c**, *via* the route shown in Scheme 14.5.[14]   Although the CV data of **19a** have not been obtained due to its easy decomposition under electrochemical conditions, the CV measurements of **19b,c** demonstrated their strong π-donor properties [E_{pa1} = 0.18 V (**19b**), 0.19 V (**19c**) (V *vs.* SCE), *cf.* TTF, E_{pa1} = 0.34 V (*vs.* Ag/AgCl)] with ΔE_{pa} ($E_{pa2} - E_{pa1}$) values in the same range (0.20 V) as the parent vinylogous TTF unit.

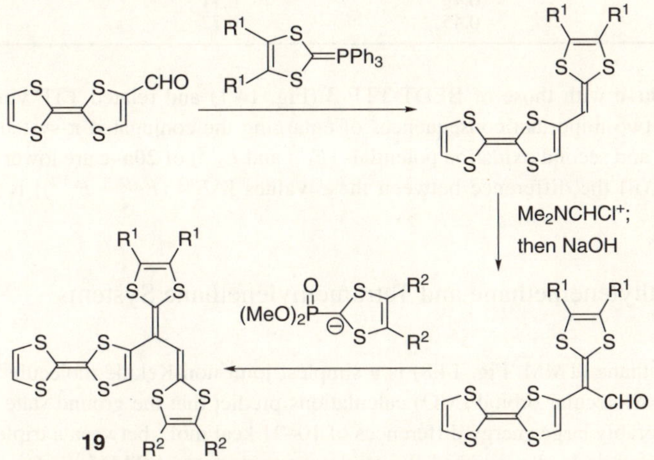

Scheme 14.5

19

a: R^1–R^1 = R^2–R^2 = S(CH$_2$)$_2$S

b: R^1 = SMe, R^2–R^2 = S(CH$_2$)$_2$S

c: R^1–R^1 = S(CH$_2$)$_2$S, R^2 = SMe

Wittig-Horner reaction using a phosphonate anion has been most widely applied to the synthesis of vinylogous TTFs with additional DT units.   An alternative synthetic approach to the mono(1,3-dithiafulvenyl)-substituted TTF vinylogues **20a–c** was reported by Zhu *et al.* (Scheme 14.6).[15]   The key step in the synthesis of **20a–c** is the base catalyzed decomposition of 1,2,3-thiadiazole derivatives **21a–c**.   From Table 14.2, which compares the values of oxidation

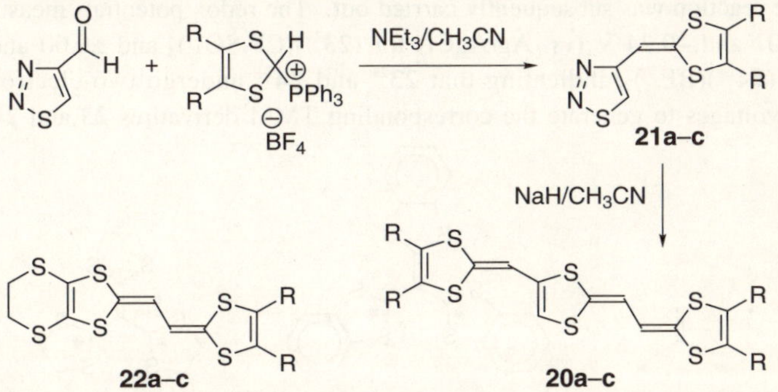

21a–c

22a–c   **20a–c**

a: R–R = S(CH$_2$)$_2$S, b: R–R = S(CH$_2$)$_3$S, c: R = SMe

Scheme 14.6

Table 14.2   CV Data of **20**, **3** and **22** in CH_2Cl_2 (V *vs.* SCE) [15]

Compound	$E_1^{1/2}$	$E_2^{1/2}$	$\Delta E^{1/2}$ $(E_2^{1/2} - E_1^{1/2})$
20a	0.41	0.62	0.21
20b	0.29	0.60	0.31
20c	0.30	0.61	0.31
BEDT-TTF **3**	0.46	0.96	0.50
22a	0.42	0.77	0.35
22b	0.49	0.71	0.22
22c	0.45	0.72	0.27

potentials of **20a–c** with those of BEDT-TTF **3** (Fig. 14.1) and related TTF vinylogues **22a–c** (Scheme 14.6), two important consequences of enlarging the conjugated π-system are observed: (i) both the first and second oxidation potentials ($E_1^{1/2}$ and $E_2^{1/2}$) of **20a–c** are lower than those of **3** and **22a–c** and (ii) the difference between these values [$\Delta E^{1/2}$ ($E_2^{1/2} - E_1^{1/2}$)] is reduced in the vinylogues.

14.4   Trimethylenemethane and Tetramethylenethane Systems

Trimethylenemethane (TMM, Fig. 14.8) is a simplest joint non-Kekulé molecule[16] and possesses two π-radicals. Molecular orbital (MO) calculations predict that the ground state is a triplet and there are considerably large energy differences of 10–21 kcal mol^{-1} between a triplet and a singlet. ESR measurements indeed supported the triplet ground state of TMM.[17] It can therefore be expected that TMM may serve as a promising component for the preparation of organic ferromagnets based on the guiding principle of McConnell's model, which suggests that ferromagnetic interaction between donor and acceptor molecules in an alternating stacking can be stabilized by admixing the lowest excited CT triplet state with the ground state.[18] However, TMM and its cation-radical (TMM$^{+\cdot}$) and dication (TMM^{2+}) species are too reactive to be used as a donor or an acceptor.

TMM^{2+} species substituted with three 1,2-benzenedithio or ethylenedithio groups (**23^{2+}** or **24^{2+}**, Fig. 14.8) was prepared as a stable $CF_3SO_3^-$ or BF_4^- salt *via* the reaction of potassium tricyanomethanide with 1,2-benzenedithiol or ethanedithiol in the presence of CF_3SO_3H in sulfolane at room temperature[19] (Scheme 14.7). For isolation of the **24^{2+}** salt, the $CF_3SO_3^-/BF_4^-$ anion exchange reaction was subsequently carried out. The redox potentials measured in DMF at 0 °C were –0.01 and –0.24 V (*vs.* Ag/AgCl) for (**23^{2+}**)($CF_3SO_3^-$)$_2$ and ±0.00 and –0.55 V (*vs.* Ag/AgCl) for (**24^{2+}**)(BF_4^-)$_2$, indicating that **23^{2+}** and **24^{2+}** undergo two-electron reduction at relatively low voltages to generate the corresponding TMM derivatires **23** and **24**. A degassed

Fig. 14.8   TMM and its derivatives.

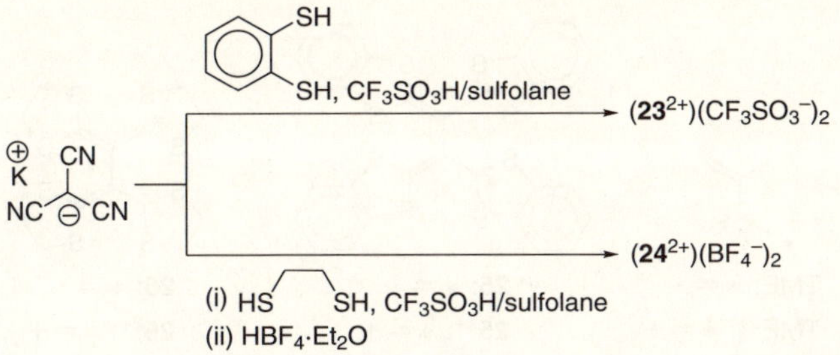

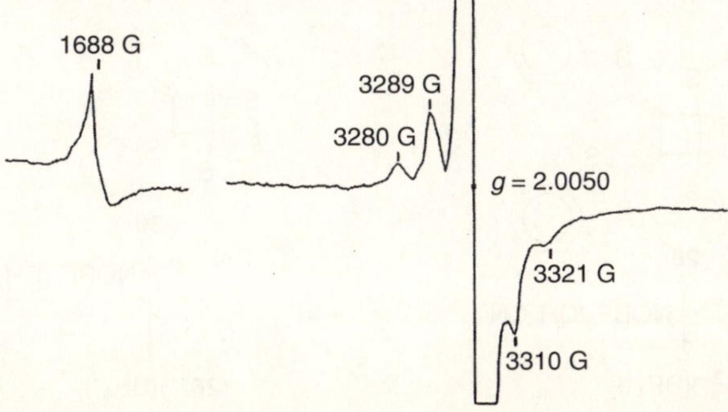

Scheme 14.7

CH$_3$CN solution containing $(\mathbf{23^{2+}})(CF_3SO_3^-)_2$ and a supporting electrolyte n-Bu$_4$NClO$_4$ was electrochemically reduced at -40 °C by gradually applying voltage. The ESR signal began to appear at *ca.* -1.6 V (not corrected by using a reference electrode) and the intensity reached maximum at -2.2 V. The solution was immediately frozen and the ESR spectrum was measured. As can be seen from the ESR spectrum shown in Fig. 14.9, four signals, two of which occur on each side of the central signal at $g = 2.0050$, were characteristic of nonoriented triplet species of **23**, the zero-field splitting parameters D and E of which are 0.0019 and $\sim$0 cm^{-1}, respectively. This smaller D value compared to that (0.025 cm^{-1}) in TMM suggests delocalization of the two interacting unpaired electrons over the 1,2-benzenedithio groups as well as the TMM moiety. In addition, a half-field resonance signal was observed at 1688 G. The temperature dependence of the signal intensity obeyed the Curie law in temperatures ranging from -50 to -196 °C, indicating that the triplet is present in the ground state. In a similar manner, $(\mathbf{24^{2+}})(BF_4^-)_2$ was electrochemically reduced in a degassed C$_2$H$_5$CN solution at -90 °C and, immediately, frozen at -120 °C, but the triplet ESR spectrum of **24** could not be detected because the reactivity of the diradical was so high as to permit rapid ring closure to methylenecyclopropane.

On the other hand, tetramethylenethane (TME, Fig. 14.10) is one of the simplest disjoint non-Kekulé molecules.[16] According to the prediction suggested by the MO theory, a singlet is preferable to a triplet for the ground state of TME. However, in contradiction to the theory, several experiments demonstrated that the ground state is a triplet irrespective of whether TME adopts the planar (D_{2h})[20] or orthogonal (D_{2d}) conformation.[21] If the ground states of the 1,2-benzenedithio- and ethylenedithio-substituted TMEs (**25** and **26**, Fig. 14.10) as well as TME are triplets, these TMEs could be used as components for preparing CT-based organic ferromagnets.

Fig. 14.9   ESR spectrum of **23** in CH$_3$CN at -50 °C. (Reprinted with permission from ref. 19)

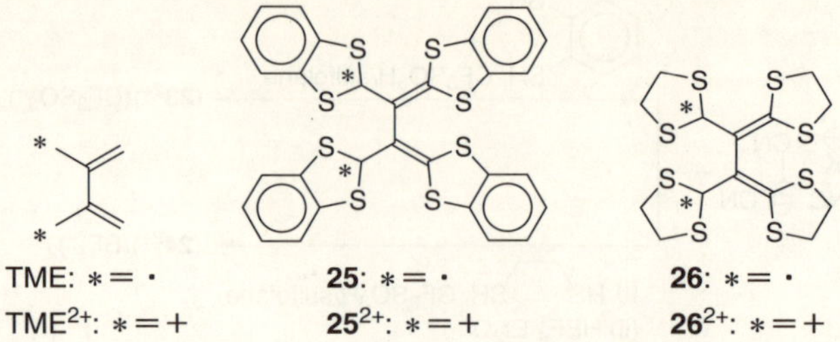

TME: * = ·
TME²⁺: * = +

25: * = ·
25²⁺: * = +

26: * = ·
26²⁺: * = +

Fig. 14.10   TME and its derivatives.

Synthesis of the BF_4^- salts of the dications **25** and **26** (**25²⁺** and **26²⁺**) was examined by the reaction sequence shown in Scheme 14.8.[22] Thus, the reaction of 1,1,3,3,3-pentachloropropene with the disodium salt of 1,2-benzenedithiolate in DME at room temperature followed by treatment with aqueous HBF_4 gave the BF_4^- salt **27**, which subsequently reacted with an excess of NaH in dry DMF at −20 °C to form the corresponding dimer **28**. Similar procedures using the disodium salt of ethylenedithiolate allowed the successive formation of the BF_4^- salt **29** and the dimer **30**. The dimer **28** was quantitatively converted into $(25^{2+})(BF_4^-)_2$ by oxidation with $NOBF_4$ in CH_3CN, whereas oxidation of the dimer **30** under similar reaction conditions was unsuccessful.

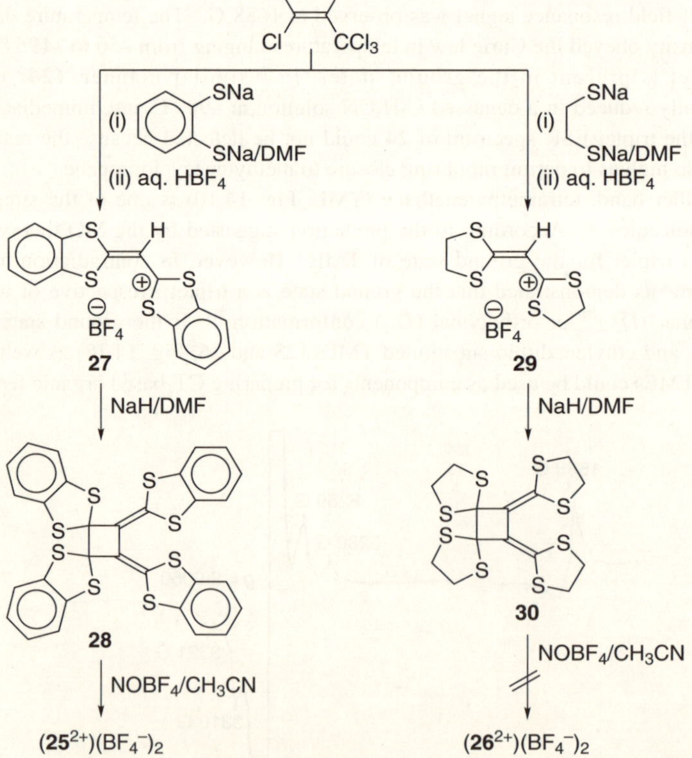

(i) SNa / SNa/DMF
(ii) aq. HBF_4

27   BF_4

29   BF_4

NaH/DMF

28

30

NaH/DMF

$NOBF_4/CH_3CN$

$NOBF_4/CH_3CN$

$(25^{2+})(BF_4^-)_2$

$(26^{2+})(BF_4^-)_2$

Scheme 14.8

In the CV of $(\mathbf{25}^{2+})(BF_4^-)_2$ measured in DMF at –60 °C, two pairs of slightly irreversible redox signals appeared at –0.22 and –0.38 V (*vs.* Ag/AgCl). Electrochemical reduction of $(\mathbf{25}^{2+})(BF_4^-)_2$ was performed at –60 °C in DMF and monitored by ESR. The solution that exhibited a change at a reduction voltage of –2.3 V (not corrected by using a reference electrode) was immediately frozen to –102 °C, and its ESR spectrum was measured. As shown in Fig. 14.11, one pair of weak signals presumably due to the fine structure of **25** appeared at 3282 and 3327 G together with a strong two-overlapped signal at $g = 2.00214$. From this result it is assumed that even at this low temperature the two unpaired electrons in **25** are not stabilized enough to inhibit intramolecular radical coupling, as evidenced by isolation of the original dimer **28** (Scheme 14.8) from the solution after the ESR measurement. Unfortunately, the ground state of **25** or **26** has not yet been determined.

$g = 2.00214$

3282 G   3327 G

Fig. 14.11   ESR spectrum of **25** in DMF at –102 °C. (Reprinted with permission from ref. 22)

The $(\mathbf{23}^{2+})(CF_3SO_3^-)_2$ and $(\mathbf{24}^{2+})(BF_4^-)_2$ salts were used to form the $\mathbf{23}^{2+}$- and $\mathbf{24}^{2+}$-based CT complexes with the dianions hexacyanotrimethylenecyclopropanide $(HCTMC^{2-})^{23)}$ and tetrafluorotetracyanoquinodimethanide $(TCNQF_4^{2-})$ (Fig. 14.12). The CT complexes, *i.e.*, $(\mathbf{23}^{2+})(HCTMC^{2-})$, $(\mathbf{23}^{2+})(TCNQF_4^{2-})$ and $(\mathbf{24}^{2+})(HCTMC^{2-})$, were obtained in good yields.[24] On comparing the first oxidation potentials of $HCTMC^{2-}$ and $TCNQF_4^{2-}$ (0.34 and 0.06 V *vs.* Ag/AgCl, respectively) with the first reduction potentials of $\mathbf{23}^{2+}$ (–0.01 V) and $\mathbf{24}^{2+}$ (±0.00 V), the CT from $HCTMC^{2-}$ or $TCNQF_4^{2-}$ to $\mathbf{23}^{2+}$ or $\mathbf{24}^{2+}$ is energetically unfavorable so that the degree of CT is expected to be very small. In fact, the CN stretching frequencies of the HCTMC and $TCNQF_4$ moieties are 2166 and 2187 cm^{-1} for $(\mathbf{23}^{2+})(HCTMC^{2-})$, 2129 and 2162 cm^{-1} for $(\mathbf{23}^{2+})(TCNQF_4^{2-})$ and 2170 and 2185 cm^{-1} for $(\mathbf{24}^{2+})(HCTMC^{2-})$, which are very close to those of $HCTMC^{2-}$ (2164 and 2181 cm^{-1}) and $TCNQF_4^{2-}$ (2131 and 2162 cm^{-1}), closer than those of the corresponding anion-radicals $HCTMC^{-\bullet}$ (2195 and 2209 cm^{-1}) and $TCNQF_4^{-\bullet}$ (2173 and 2195 cm^{-1}). The CT degree of each CT complex was accurately estimated from the spin amount measured by ESR (Fig. 14.13). For $(\mathbf{23}^{2+})(HCTMC^{2-})$ and $(\mathbf{24}^{2+})(HCTMC^{2-})$, three pairs of weak signals arising from the fine structures of their triplet species were observed together with doublet-like signals at $g = 2.00290$ and 2.00297, respectively. The zero-field splitting parameters D and E are

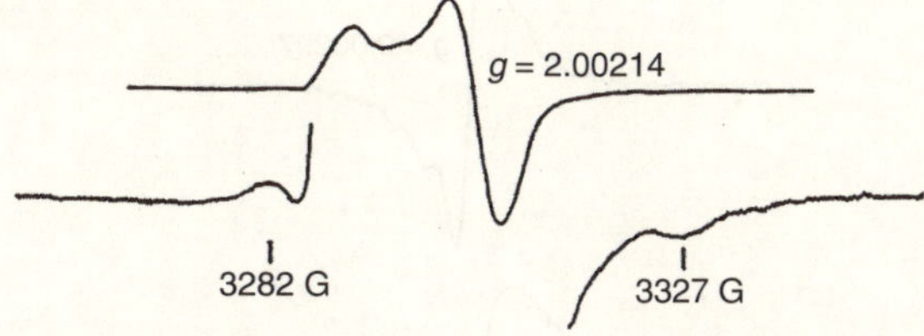

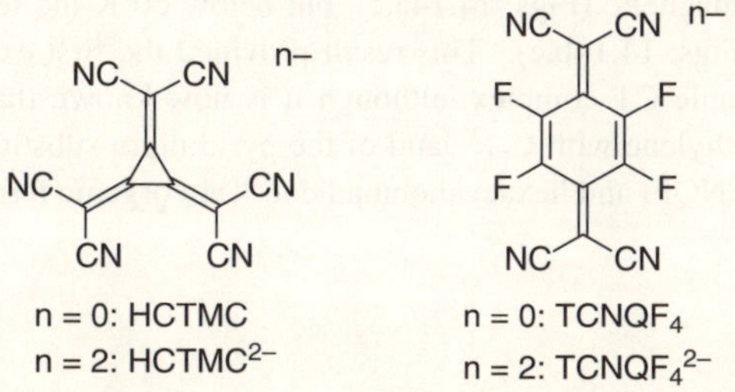

Fig. 14.12

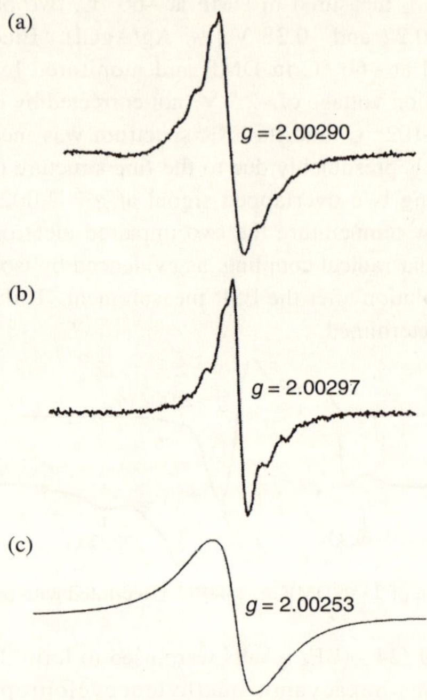

Fig. 14.13   ESR spectra of (a) $(\mathbf{23}^{2+})(HCTMC^{2-})$, (b) $(\mathbf{24}^{2+})(HCTMC^{2-})$ and (c) $(\mathbf{23}^{2+})(TCNQF_4^{2-})$ at 293 K. (Reprinted with permission from ref. 24)

12.5 and 1.8 G for $(\mathbf{23}^{2+})(HCTMC^{2-})$ and 11.8 and 1.7 G for $(\mathbf{24}^{2+})(HCTMC^{2-})$. On the other hand, $(\mathbf{23}^{2+})(TCNQF_4^{2-})$ exhibited only a broad, comparatively strong doublet signal. The spin amounts of these CT complexes at 293 K were calculated from the signal intensities calibrated with a reference 2,2,6,6-tetramethyl-4-piperidiol-1-oxyl: 1.64×10^{21}, 3.72×10^{20} and 3.55×10^{22} spins mol^{-1} for $(\mathbf{23}^{2+})(HCTMC^{2-})$, $(\mathbf{24}^{2+})(HCTMC^{2-})$ and $(\mathbf{23}^{2+})(TCNQF_4^{2-})$, respectively, which correspond to the spin contents of 0.27, 0.0062 and 5.89%, respectively. It is therefore assumed that there is almost no CT from $HCTMC^{2-}$ to $\mathbf{23}^{2+}$ or $\mathbf{24}^{2+}$, whereas a small degree of CT from $TCNQF_4^{2-}$ to $\mathbf{23}^{2+}$ occurs. The temperature dependence of the signal intensity (I) for $(\mathbf{23}^{2+})(TCNQF_4^{2-})$ was investigated in order to elucidate the interaction between the spins of $\mathbf{23}$ and the $TCNQF_4$ moieties. Fig. 14.14 shows the $1/I–T$ and $IT–T$ plots. Above $ca.$ 60 K the spin interaction was antiferromagnetic (Figs. 14.14a,c), but below 60 K the ferromagnetic interaction occurred preferentially (Figs. 14.14b,c). This result provided the first example of ferromagnetic interaction in a pure organic CT complex, although it is now known that the CT complexes of tetrakis(dimethylamino)ethylene with C_{60}[25] and of the pyridinium-substituted imidazolin-1-oxyls with anion-radicals of $TCNQF_4$ and hexacyanobutadiene[26] are organic ferromagnets.

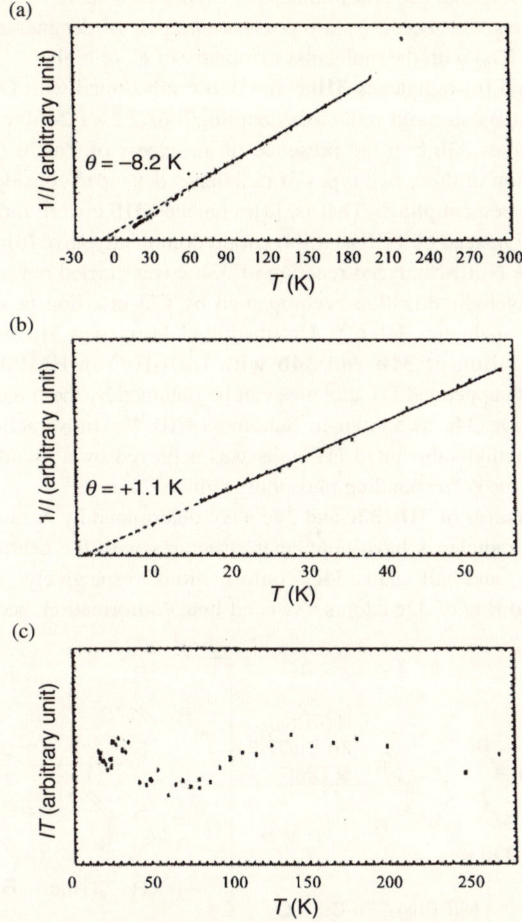

Fig. 14.14   Temperature dependence of the signal intensity (I) for $(\mathbf{23}^{2+})(\text{TCNQF}_4{}^{2-})$; (a) and (b) show the $1/I$–T plots and (c) is the IT–T plot. (Reprinted with permission from ref. 24)

14.5   [n]Radialenes (n = 4–6) with Multi-DT Units

[n]Radialene (polymethylenecycloalkane) is a kind of cyclic π-conjugated hydrocarbon composed of the n-membered ring and n pieces of exocyclic double bonds. Because of its unique structure with high symmetry of D_{nh}, much attention has been focused on the synthesis and electronic structures of various radialene derivatives in past years.[27] In particular, [n]radialenes with a strong electron-donating 1,3-dithiole ring attached to the end of each exocyclic double bond are attractive as new multi-step redox systems comprising n pieces of redox-active DT units. Such multi-DT-containing systems are also promising π-electron donors for the development of new organic conductors, because the on-site Coulombic repulsion in the dicationic state is considerably reduced due to delocalization of two positive charges over the whole multi-DT unit.

Furthermore, it is suggested that the odd-numbered [n]radialenes serve as precursors for the CT-type organic ferromagnets,[18, 28] because they possess one pair of degenerated highest occupied molecular orbitals (HOMOs) with the molecular symmetry of C_n or higher.

Synthesis of [4]- and [6]-radialenes **31b,c** and **32b,c** substituted with DT units[29,30] has been accomplished by the Ni(0)-catalyzed reductive coupling[31] of 2,2'-(1,2-dibromoethanediylidene)-bis(1,3-dithiole) derivatives **33b,c** in the presence of an excess of Zn-Cu (Scheme 14.9). The selectivity in the formation of these two types of radialenes depended considerably on the solvent used for the Ni(0)-catalyzed coupling. That is, [4]radialenes **31b,c** were formed mainly in THF, whereas the use of DMF instead of THF as a solvent predominantly gave [6]radialenes **32b,c**. On the other hand, when the Ni(0)-catalyzed reaction of **33b,c** was carried out in a carbon monoxide atmosphere, reductive cyclodimerization accompanied by CO insertion proceeded efficiently to give the cyclopentanone analogues **34b,c**.[32] Unsubstituted derivatives **31a** and **34a** were obtained by demethoxycarbonylation of **31b** and **34b** with LiBr·H₂O in HMPA, respectively.[11,33] [5]Radialenes **35a–c** with appended DT moieties can be obtained by the three-step synthesis from cyclopentanone derivative **34c** as shown in Scheme 14.10.[34] Introduction of the benzene-, bis(methytiho)- and dimethyl-substituted DT units was achieved by a pseudo Wittig reaction of cationic species **36** with the corresponding phosphorus ylids **37a–c**.

The molecular structures of **31b**, **32c** and **34c** were determined by X-ray diffraction analyses (Fig. 14.15). All these radialenes have nonplanar structures with the central rings of puckered (**31b**), twisted-boat (**32c**) and half-chair (**34c**) conformations, respectively. It is noteworthy that the central six-membered ring of **32c** adopts a twisted-boat conformation, because such a twisted-

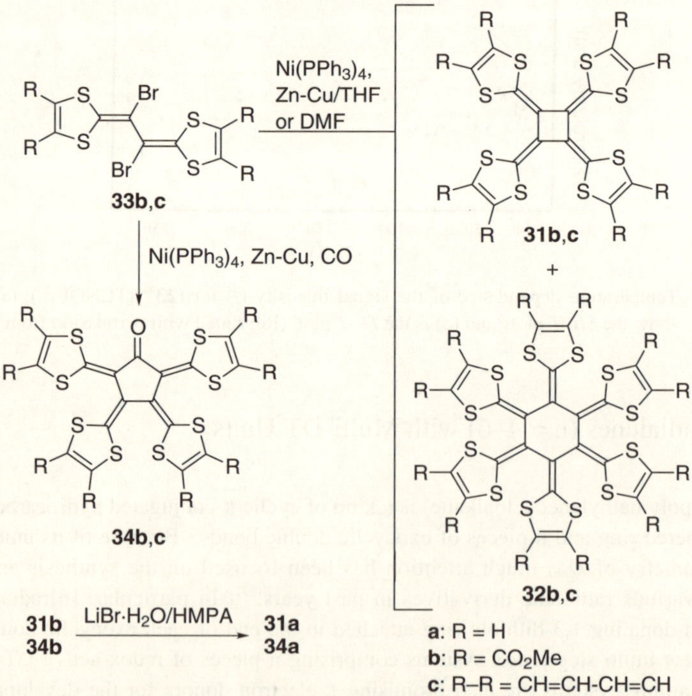

33b,c

Ni(PPh₃)₄,
Zn-Cu/THF
or DMF

31b,c

+

32b,c

Ni(PPh₃)₄, Zn-Cu, CO

34b,c

31b LiBr·H₂O/HMPA **31a**
34b ⟶ **34a**

a: R = H
b: R = CO₂Me
c: R–R = CH=CH-CH=CH

Scheme 14.9

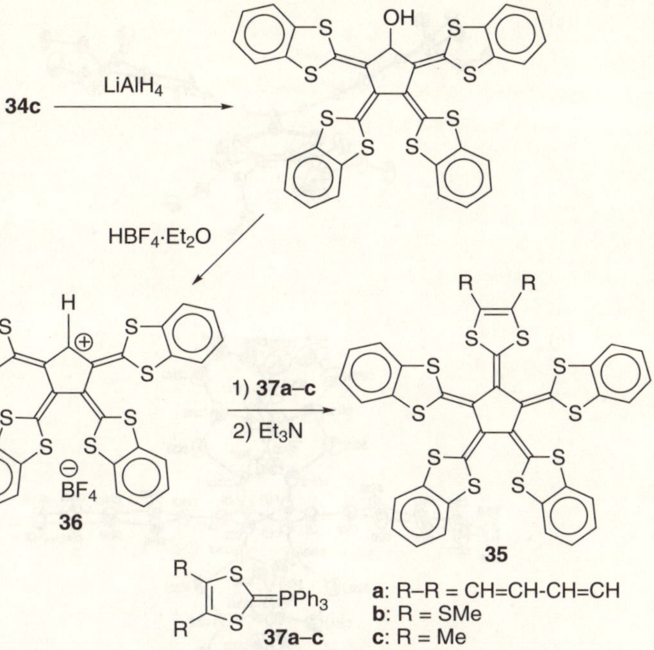

Scheme 14.10

37a–c   **a**: R–R = CH=CH-CH=CH
             b: R = SMe
             c: R = Me

boat conformation in the six-membered ring rarely occurs in crystallographic determination. The structure of **32c** is also different from that of hexakis(ethylidene)cyclohexane with a chair conformation of the six-membered ring.[35] The observed conformation in **32c** would seem to be a metastable structure kinetically formed by the Ni(0)-catalyzed coupling, because heating **32c** in solution induced a conformational change to give a more stable, perhaps chair conformer (**32'c**), the structure of which was confirmed by elemental analysis and ^{13}C-NMR and MS spectra.[30] A similar conformational conversion of **32b** into **32'b** took place by heating.

The redox potentials of new radialenes measured by CV are summarized in Table 14.3. As representative examples, Fig. 14.16 shows the cyclic voltammograms of the benzo-DT-containing derivatives **35a**, **31c** and **34c**, together with that of the bis(benzene)-fused TTF vinylogue **11c** (Fig. 14.5) for comparison. The CVs of [4]Radialenes **31a–c** consist of three oxidation waves in PhCN. A similar three-step oxidation was observed in the CVs of **31a,b** by changing the solvent used for the CV measurement from PhCN to CH_2Cl_2 [0.19, 0.98 and 1.30 V for **31a**; 0.70, 1.29 and 1.79 V for **31b** (V vs. Ag/AgCl)]. In each of these two CVs, the variation of peak current at the first step with respect to sample concentration revealed that this step involves a two-electron transfer, which is divided into two one-electron steps corresponding to E_1 and E_2 by the Myers-Shain method.[36] Consequently, the ΔE ($E_2 - E_1$) values are estimated to be 0.04 V for **31a** and 0.05 V for **31b**. On the other hand, vinylogous TTF **11a** and its tetrakis(methoxycarbonyl) analogue **11b** (Fig. 14.5), which have structures equivalent to halves of **31a,b**, have been reported to undergo two successive one-electron transfer processes with ΔE ($E_2 - E_1$) = 0.19–0.21 V.[11] These results indicate a significant decrease in the on-site Coulombic repulsion in **31a**$^{2+}$ or **31b**$^{2+}$ by the effective delocalization of two positive charges over the whole molecule: the same can be said for **31c** on comparing its E_{m1} value [0.46 ($E_1 = E_2$) V] with the ΔE value (0.18 V) of **11c**.

(a)

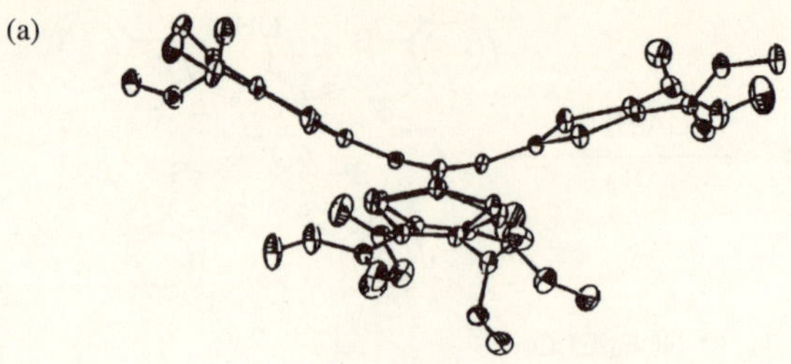

(b)

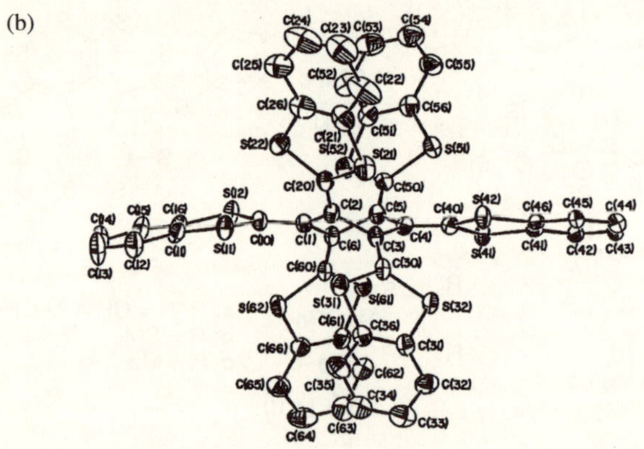

(c)

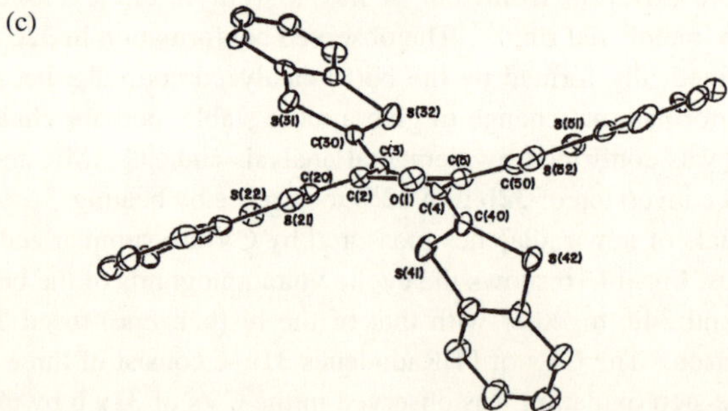

Fig. 14.15   ORTEP drawings of the molecular structures of (a) **31b**, (b) **32c** and (c) **34c**. The hydrogen atoms have been omitted for clarity. (Reprinted with permission from refs. 29, 30 and 32)

The cyclopentanone analogues **34a–c** showed two pairs of two-electron redox waves. The first redox potentials (E_{m1}) of **34a–c** are lower than those of the corresponding [4]radialenes **31a–c** in spite of the insertion of a strong electron-withdrawing carbonyl group into the central ring. Also, the second redox potentials (E_{m3}) are lower by *ca.* 0.3–0.4 V than the third redox potentials (E_4) of the corresponding **31a–c**. Considering the fact that antiaromatic cyclopentadienone and cyclobutadiene structures contribute to the tetracations **34**$^{4+}$ and **31**$^{4+}$ (Fig. 14.17), respectively, this

Table 14.3   Redox Potentials of Multi-DT-substituted [4]-, [5]- and [6]-Radialenes and Vinylogous TTFs (V *vs.* Ag/AgCl in PhCN)

Compound	E_1	$E_{m1}{}^a$	E_2	$E_{m2}{}^b$	E_3	$E_{m3}{}^c$	E_4
31a		0.23			1.00^d		1.43^d
31b		0.69			1.29^d		1.78^d
31c		0.46			1.09^d		1.59^d
32b	0.73		0.89^d				
32c	0.43		0.65^d				
32'b		1.14^d					
32'c		0.82^d					
34a		0.17				1.14	
34b		0.64				1.35	
34c		0.38				1.22	
35a^e				0.39			
35a^f		0.394				0.386	
35b^e				0.37			
35b^f		0.374				0.358	
35c^e		0.28				0.36	
11a	0.23		0.42				
11b	0.61		0.82				
11c	0.46		0.64				

[a]Redox potential involving a two-electron transfer (corresponding to E_1 and E_2) process. [b]Redox potential involving a four-electron transfer (corresponding to E_1–E_4) process. [c]Redox potential involving a two-electron transfer (corresponding to E_3 and E_4) process. [d]Irreversible step (anodic peak potential). [e]V *vs.* SCE. [f]Redox potential calculated by digital simulation of the two-step two-electron transfer mechanism.

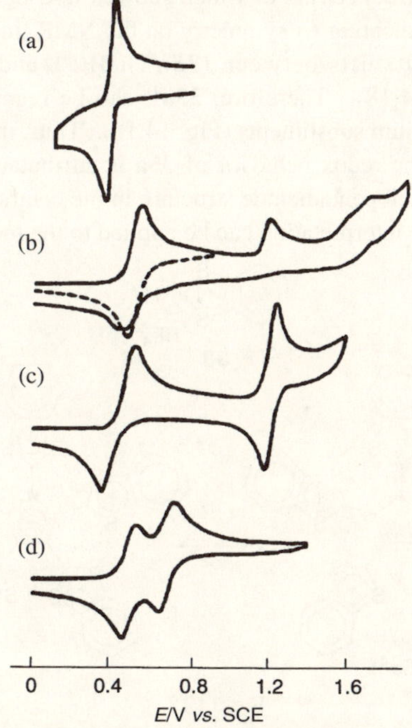

Fig. 14.16   Cyclic voltammograms of (a) **35a**, (b) **31c**, (c) **34c** and (d) **11c** in PhCN containing 0.1 M Et₄NClO₄ at 25 °C.

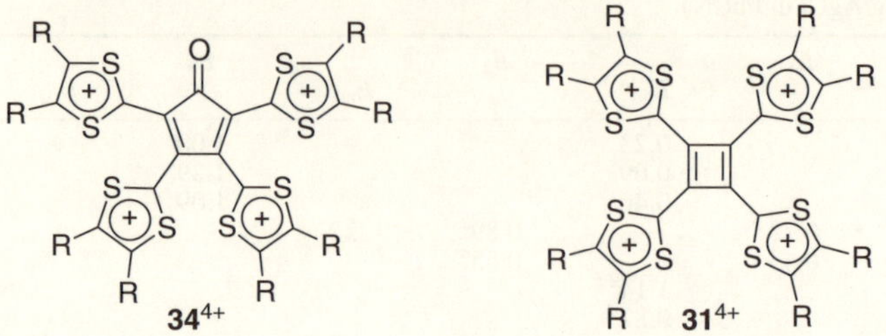

Fig. 14.17

result can be attributed to the lower antiaromaticity of cyclopentadienone than that of cyclobutadiene.[37]

In contrast to **31a–c** and **34a–c**, [5]radialene derivatives **35a,b** exhibited notable redox behavior in which a four-electron transfer occurred simultaneously. On the other hand, **35c** bearing two electron-donating methyl groups showed two pairs of two-electron waves similar to **34a–c**. Digital simulation analysis assuming Nernstian response[38] was undertaken to elucidate the detailed redox process of **35a**.[34] It was consequently found that **35a** is subject to a two-step two-electron transfer and sequentially converted into the dication **35a**$^{2+}$ and into the tetracation **35a**$^{4+}$. The estimated redox potentials for the **35a/35a**$^{2+}$ and **35a**$^{2+}$/**35a**$^{4+}$ couples are 0.394 and 0.386 V (*vs.* SCE), respectively. Chemical oxidation of **35a** with NOBF$_4$ afforded the tetracation salt (**35a**$^{4+}$)(BF$_4^-$)$_4$, the ^{13}C-NMR spectrum of which showed five signals at δ 123.7, 127.6, 132.3, 146.0 and 187.8 in CD$_3$CN, indicating C_5 symmetry on the NMR time scale. In addition, there is little difference in chemical shifts between (**35a**$^{4+}$)(BF$_4^-$)$_4$ and 4,5-benzo-1,3-dithiolium tetrafluoroborate (**38**) (Fig. 14.18). Therefore, **35a**$^{4+}$ can be regarded as a cyclopentadienide derivative with five 1,3-dithiolium substituents (Fig. 14.18). Thus, the simultaneous four-electron transfer process observed in the redox behavior of **35a** is attributable to a large stabilization of **31a**$^{4+}$ to which the aromatic cyclopentadienide structure in the central ring makes a strong (nearly 100%) contribution. The same interpretation can be applied to the redox behavior of **35b**.

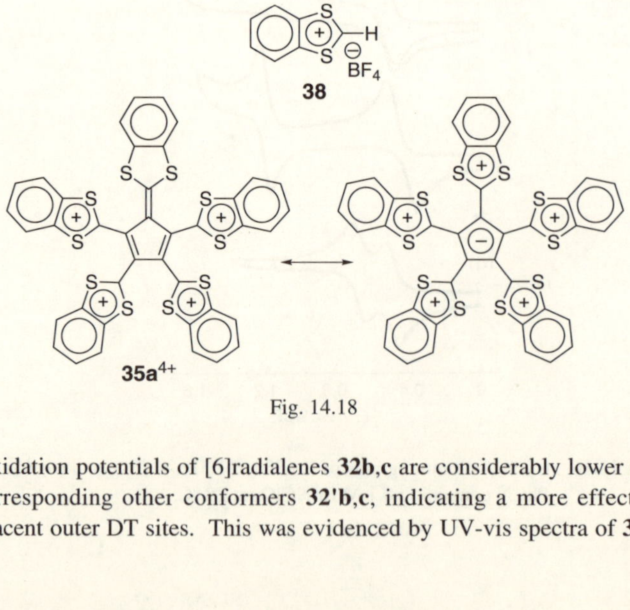

Fig. 14.18

The first oxidation potentials of [6]radialenes **32b,c** are considerably lower (*ca.* 0.4 V) than those of the corresponding other conformers **32'b,c**, indicating a more effective conjugation between the adjacent outer DT sites. This was evidenced by UV-vis spectra of **32b,c** and **32'b,c**, in which both **32'b,c** showed the absorption maxima at a shorter wavelength region compared to those of **32b,c** (354 nm for **32'b,c**; 386 nm for **32b**; 393 nm for **32c**). An interesting phenomenon was observed in the CV measurements of **32b,c**. The multi-scanning voltammetry indicated that conformational conversion of **32b,c** into **32'b,c** occurs in the dicationic state even at room temperature. From these results, it appears that the central six-membered ring in a thermodynamically more stable structure of **32'b** or **32'c** favors a chair conformation, which causes a less effective conjugation between two neighboring DT units.

The above radialenes produced conducting CT complexes or salts with organic or inorganic acceptors. For example, **31a** with TCNQ formed $(31a)(TCNQ)_2$, and **31c** reacted with TCNQF$_4$ to give $(31c)(TCNQF_4)$. While these complexes exhibited conductivities of the order of 10^{-2} S cm^{-1} on compressed pellets at room temperature,[29,39] the temperature dependence of their resistivities revealed semiconducting behavior with small activation energies (0.04–0.06 eV). Room-temperature conductivities of compressed pellets of the 2:1 and 1:1 salts, $(34a)_2I_3$ and $(34c)I_3$, were also of the order of 10^{-2} S cm^{-1}, whereas the 1:2 salts $(34a)(I_3)_2$ and $(34c)(I_3)_2$ showed low conductivities of 10^{-8}–10^{-5} S cm^{-1} at room temperature.

[5]Radialene derivative **35a** formed 1:2 (donor:acceptor) complexes with HCTMC (Fig. 14. 12) and dichlorodicyanobenzoquinone (DDQ), respectively, and a 1:4 (donor:acceptor) complex with TCNQF$_4$.[40] The CN stretching bands of these complexes in IR spectra indicated that they exist in the closed forms $(35a^{4+})(HCTMC^{2-})_2$, $(35a^{2+})(DDQ^{-\bullet})_2$ and $(35a^{4+})(TCNQF_4^{-\bullet})_4$, respectively. The paramagnetic susceptibility of $(35a^{2+})(DDQ^{-\bullet})_2$ at 298 K was 3.3×10^{-4} emu mol^{-1}, which corresponds to a few percent of that calculated by using a spin system in which there is no interaction between $35a^{2+}$ ($S = 1$, if the ground state is a triplet) and DDQ$^{-\bullet}$ ($S = 1/2$). This result suggests that only a small amount of triplet and doublet species survives in this complex probably due to a strong antiferromagnetic interaction between a pair of DDQ$^{-\bullet}$ anion-radicals as well as a Jahn-Teller distortion of $35a^{2+}$. Nevertheless, the ESR spectrum of $(35a^{2+})(DDQ^{-\bullet})_2$ showed a half-field resonance signal due to a triplet species, derived from a transition with $\Delta m_s = \pm 2$, at 163.9 mT together with a strong doublet signal at 327.6 mT ($g = 2.00523$) (Fig. 14.19). The temperature dependence of the spin intensity for the signal at 163.9 mT suggested that this contaminant triplet species is present in the ground state. No fine structure was observed around the signal at 327.6 mT probably due to the very small zero-field splitting parameters of $35a^{2+}$.

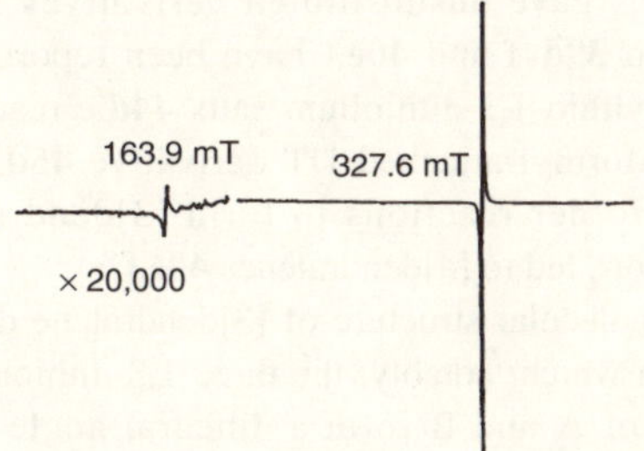

Fig. 14.19   ESR spectrum of $(35a^{2+})(DDQ^{-\bullet})_2$ at 298 K. (Reprinted with permission from ref. 40)

14.6   [n]Dendralenes (n = 3, 4) with Multi-DT Units

Multi-DT-containing [n]dendalenes are also of considerable interest as novel multi-step redox

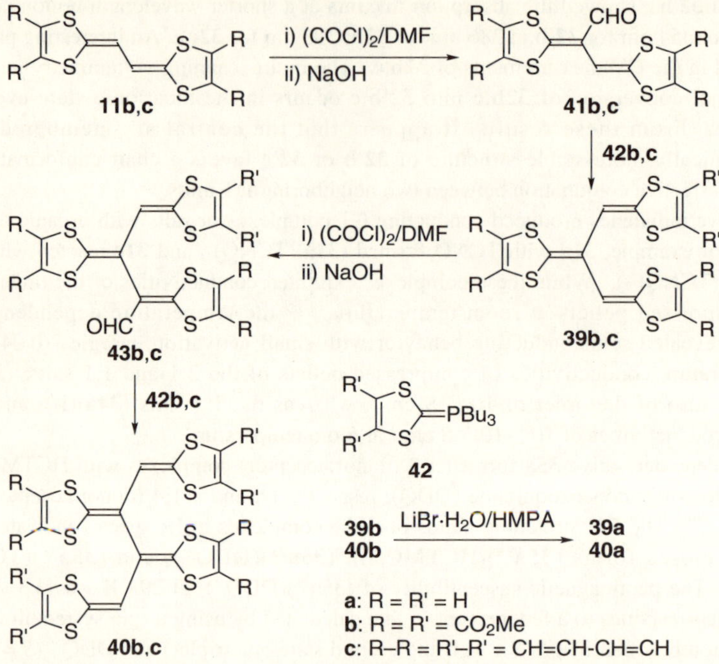

Scheme 14.11

systems based on unique cross-conjugated π-systems[41] as well as acyclic analogues of [n]radialenes. Synthesis of [3]- and [4]-dendralenes **39a–c** and **40a–c** substituted with DT units is shown in Scheme 14.11.[42] Beginning with vinylogous TTFs **11b,c** (Fig. 14.5), the reported sequence involved repetition of the Vilsmeier reaction (**11b,c** → **41b,c** and **39b,c** → **43b,c**) followed by the Wittig reaction (**41b,c** → **39b,c** and **43b,c** → **40b,c**) to allow successive construction of the [3]- and [4]-dendralene frameworks. Demethoxycarbonylation of **39b** and **40b** with LiBr·H$_2$O in HMPA gave unsubstituted derivatives **39a** and **40a**, respectively. Alternative synthetic routes to **39d–f** and **40e,f** have been reported by Bryce and coworkers (Scheme 14.12).[43] The methylthio-1,3-dithiolium salts **44d,e** reacted with the sodium salt of malonaldehyde to give the diformyl-attached DT derivative **45d,e**, respectively, which then underwent twofold Wittig-Horner reactions to form [3]dendralenes **39d–f**. Subsequent Vilsmeier/Wittig-Horner reactions led to [4]dendralenes **40e,f**.

Figure 14.20 shows the molecular structure of [3]dendralene derivative **39d** determined by X-ray diffraction analysis,[43] in which, notably, the three 1,3-dithiole rings (A, B and C) are not coplanar. The mean planes of A and B form a dihedral angle of 8.8°, while C is almost orthogonal, forming dihedral angles with A and B of 80.0° and 81.7°, respectively. In addition, there is a remarkable difference between the lengths of the central single bonds C(1)–C(7) and C(7)–C(13); the former is 1.409 Å, being strongly conjugated and significantly shorter than the single bond in conjugated butadienes (average length 1.455 Å), whereas the latter remains single (1.499 Å) due to non-conjugation (*cf.* the C(13)=C(14) has a normal C=C double bond length).

Table 14.4 summarizes the redox potentials of [3]- and [4]-dendralenes **39a–f** and **40a–c,e,f** measured by CV. All [3]dendralenes showed three pairs of one-electron transfer waves arising from three redox-active DT units. The E_1 value (0.26 V) of **39a** is close to that (0.23 V) of vinylogous **11a** (Table 14.3) despite the extended π-conjugation by attachment of the additional 1,3-dithiafulvenyl moiety. In addition, considerably high E_3 values are observed for **39a–f**. These results suggest the existence of a non-conjugated DT unit, so that it is likely that the conformations of oxidized species of **39a–f**, *i.e.*, their cation-radicals, dications and trication-radicals, in solution are similar to that of the neutral **39d** in the solid state.

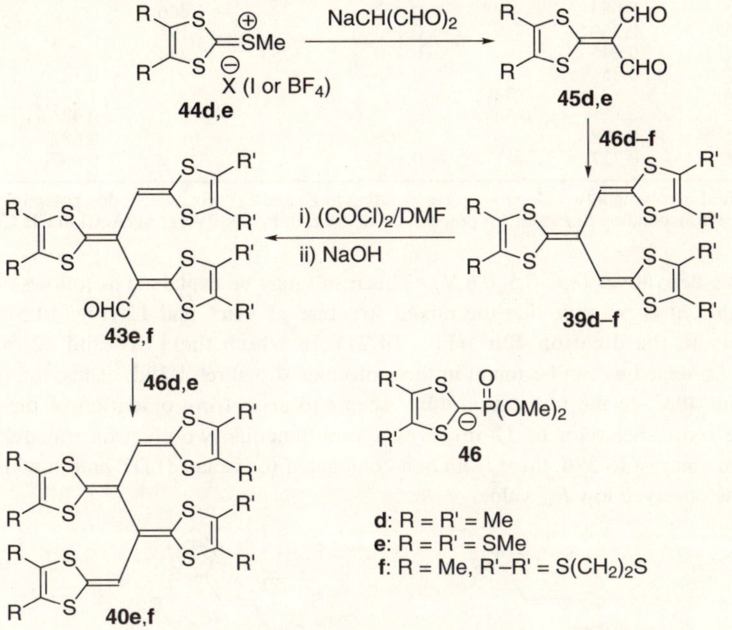

Scheme 14.12

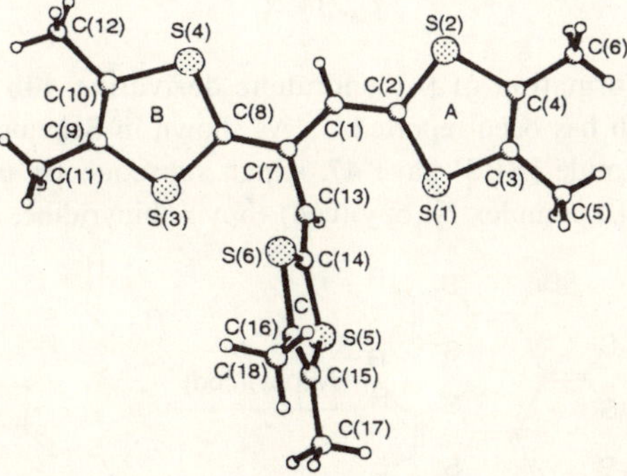

Fig. 14.20   Molecular structure of **39d**. (Reprinted with permission from ref. 43)

[4]Rendralenes **40a,e,f** also exhibited three pairs of redox waves. Since they have four redox-active DT units, the third wave is thought to be a two-electron transfer by the simultaneous occurrence of E_3 and E_4. On the other hand, **40b,c** showed two pairs of two-electron redox waves. Accordingly, in all the CVs of **40a–c,e,f**, each potential of E_{m2} involves a two-electron transfer, the value of which is drastically decreased compared with the E_3 value of the corresponding

Table 14.4   Redox Potentials of Multi-DT-substituted [3]- and [4]-Dendralenes

Compound	E_1	$E_{m1}{}^a$	E_2	E_3	$E_{m2}{}^b$	E_4
39a[c]	0.26		0.40	1.32		
39b[c]	0.70		0.79	1.75		
39c[c]	0.50		0.62	1.53		
39d[d]	0.084		0.327	1.246		
39e[d]	0.360		0.490	1.195		
39f[d]	0.245		0.470	1.244		
40a[c]	0.25		0.37		0.72	
40b[c]		0.73			1.14	
40c[c]		0.54			0.97	
40e[d]	0.258		0.366		0.687	
40f[d]	0.227		0.393		0.807	

[a] Redox potential involving a two-electron (corresponding to E_1 and E_2) process. [b] Redox potential involving a two-electron (corresponding to E_3 and E_4) process. [c] V *vs.* SCE in PhCN. [d] V *vs.* Ag/AgCl in CH_2Cl_2.

[3]dendralene derivatives (*ca.* 0.5–0.6 V). This result may be explained as follows. Taking **40a** as an example, it is possible that the mixed structure of **11a**$^{2+}$ and **12** (Fig. 14.5) contributes considerably to the dication **40a**$^{2+}$ (Fig. 14.21), in which the **11a**$^{2+}$ and **12** moieties are orthogonally oriented as can be found in the molecular structure of **39d**. Thus, the two-electron transfer from **40a**$^{2+}$ to the tetracation **40a**$^{4+}$ seems to arise from oxidation of the **12** moiety, because the redox behavior of **12** involves a simultaneous two-electron transfer process.[11] Therefore, in contrast to **39d**, there is no non-conjugated (or isolated) DT unit in **40a**$^{2+}$, probably leading to the observed low E_{m2} value.

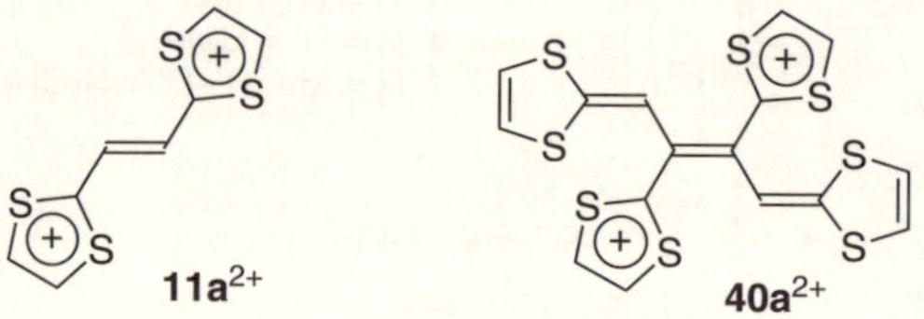

Fig. 14.21

Additionally, transformation of [4]dendralene derivative **40b** into the corresponding [4]radialene derivative **31b** has been reported.[42] As shown in Scheme 14.13, dibromination of **40b** with *N*-bromosuccinimide (NBS) gave **47**, which subsequently underwent intramolecular reductive coupling by Ni(0) complex Ni(bpy)(cod) (bpy = bipyridine, cod = cyclooctadiene) to

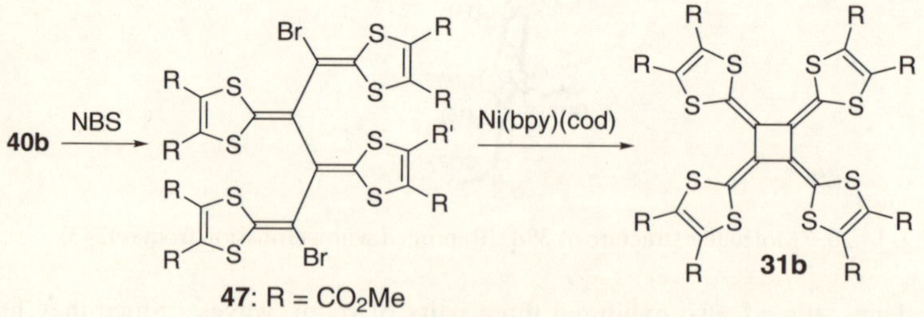

Scheme 14.13

produce [4]radialene derivative **31b**, but the yield was very low (3%). On the other hand, an attempt to convert [3]dendralenes **39b,c** into the corresponding [3]radialenes by a similar reaction sequence has been unsuccessful, although it is reported that intramolecular cyclization of the aryl-substituted [3]dendralene dibromides takes place in the presence of hexamethylditin and a catalytic amount of Pd(PPh$_3$)$_4$ to give the corresponding [3]radialenes.[44]

14.7   Miscellaneous Donors Containing Multi-DT Units

In addition to the multi-DT-containing donor molecules described above, several other related compounds **48–54** (Fig. 14.22) have been reported. The basic cyclic ring systems of **48–54** are classified as follows: (i) 1,3,5-cyclohexanetrione, leading to **48**[45] and **49**,[46] (ii) 1,3,5-trithiane, leading to **50**,[47] (iii) benzene, leading to **51–53**[48,49] and (iv) indan, leading to **54**.[50] Among these

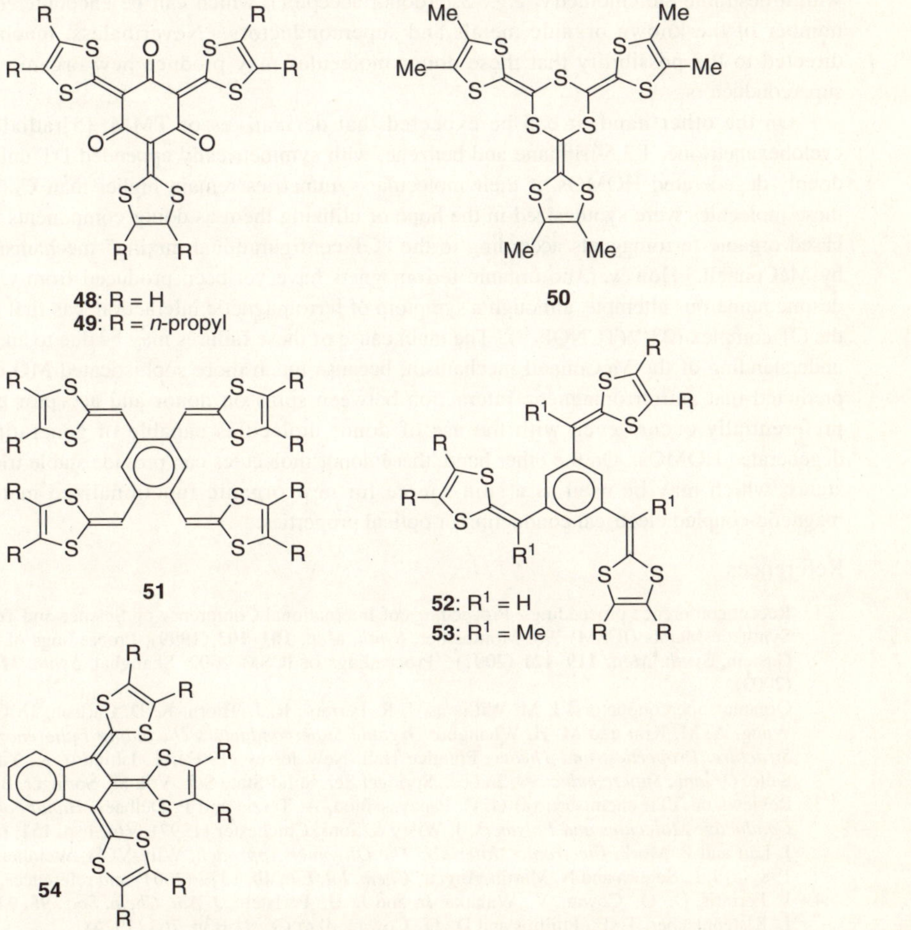

48: R = H
49: R = *n*-propyl

50

51

52: R^1 = H
53: R^1 = Me

54

Fig. 14.22   Miscellaneous multi-DT-containing compounds.

compounds, it is noteworthy that compound **49** exhibited four-stage, amphoteric, redox behavior, *i.e.*, the ability to undergo both anodic oxidation and cathodic reduction. Another point worth noting is that compounds **49**, **50**, **52** and **53**, with threefold symmetry, were designed to achieve ferromagnetic interactions since the generation of the dication present in a stable triplet ground state may be feasible owing to the degeneracy of the frontier orbitals.[51,52)]

14.8   Summary and Outlook

A fruitful result regarding the development of new organic metals and superconductors has not yet been obtained from the studies on the multi-DT-containing TTFs and their vinylogues. These donors have the advantage of a superior donating ability compared to that of the parent molecules, but both an increase in size and a decrease in planarity cause difficulty in forming CT materials with a desirable stoichiometry, *e.g.*, 2:1 (donor:acceptor), which can be encountered in a large number of the known organic metals and superconductors. Nevertheless, much interest is directed to the possibility that these donor molecules may produce new organic metals and superconductors.

On the other hand, it can be expected that derivatives of TMM, [5]radialene, 1,3,5-cyclohexanetrione, 1,3,5-trithiane and benzene, with symmetrically appended DT units, generate doubly-degenerated HOMOs, if their molecular symmetries remain higher than C_3. Therefore, these molecules were synthesized in the hope of utilizing them as donor components for the CT-based organic ferromagnets according to the "CT configurational mixing" mechanism proposed by McConnell. However, no organic ferromagnets have yet been produced from CT materials despite numerous attempts, although a symptom of ferromagnetic interaction was first observed in the CT complex $(\mathbf{23}^{2+})(TCNQF_4{}^{2-})$. The main cause of these failures may be due to an insufficient understanding of the McConnell mechanism, because much more sophisticated MO calculations predicted that antiferromagnetic interaction between spins on donor and acceptor components preferentially occurs even with the use of donor molecules capable of generating doubly-degenerated HOMOs. On the other hand, these donor molecules can provide stable triplet ground states, which may be used as a spin source for new organic functionalized materials with magnetic-coupled electrical conducting or optical properties.

References

1. Recent conference proceedings: Proceedings of International Conference on Science and Technology of Synthetic Metals (ICSM) '98, Montpellier, *Synth. Met.*, **101–103** (1999); Proceedings of ICSM 2000, Gastein, *Synth. Met.*, **119–121** (2001) ; Proceedings of ICSM 2002, Shanghai, *Synth. Met.*, **135–137** (2003).
2. Organic superconductors: J. M. Williams, J. R. Ferraro, R. J. Thorn, K. D. Carlson, U. Geiser, H. H. Wang, A. M. Kini and M.-H. Whangbo, *Organic Superconductors (Including Fullerenes): Synthesis, Structure, Properties, and Theory*, Prentice Hall, New Jersey (1992); T. Ishiguro, K. Yamaji and G. Saito, *Organic Superconductors*, 2nd ed., Springer Ser. Solid-State Sci., Vol. 88, Springer, Berlin (1998).
3. Reviews on TTF chemistry: (a) G. C. Papavassiliou, A. Terzis and P. Delhaes, *Handbook of Organic Conductive Molecules and Polymers*, J. Wiley & Sons, Chichester (1997), Vol. 1, p. 151; (b) J. Becher, J. Lau and P. Mørk, *Electronic Materials: The Oligomer Approach*, Wily-VCH, Weinheim (1998), p. 198; (c) J. L. Segura and N. Martín, *Angew. Chem. Int. Ed.*, **40**, 1372 (2001) and references therein.
4. J. Ferraris, D. O. Cowan, V. Walatka, Jr. and J. H. Perlstein, *J. Am. Chem. Soc.*, **95**, 948 (1973); T. J. Kistenmacher, T. E. Phillips and D. O. Cowan, *Acta Cryst.*, B**30**, 763 (1974).
5. M. Salle, A. Gorgues, M. Jubault and Y. Gouriou, *Synth. Met.*, **41–43**, 2575 (1991).
6. M. Sallé, M. Jubault, A. Gorgues, K. Boubekeur, M. Fourmigué, P. Batail and E. Canadell, *Chem. Mater.*, **5**, 1196 (1993).

7. A. Gorgues, M. Jubault, A. Belyasmine, M. Salle, P. Frere, V. Morisson and Y. Gouriou, *Phosphorus, Sulfur, and Silicon*, **95–96**, 235 (1994).

8. M. Sallé, A. Gorgues, M. Jubault, K. Boubekeur, P. Batail and R. Carlier, *Bull. Soc. Chim. Fr.*, **133**, 417 (1996).

9. P. Leriche, A. Belyasmine, M. Sallé, A. Gorgues, M. Jubault, J. Garín and J. Orduna, *Tetrahedron Lett.*, **38**, 1399 (1997).

10. V. Morisson, C. Mingotaud, B. Agricole, M. Sallé, A. Gorgues, C. Garrigou-Lagrange and P. Delhaès, *J. Mater. Chem.*, **5**, 1617 (1995).

11. Z. Yoshida, T. Kawase, H. Awaji, I. Sugimoto, T. Sugimoto and S. Yoneda, *Tetrahedron Lett.*, **24**, 3469 (1983); Z. Yoshida, T. Kawase, H. Awaji and S. Yoneda, *Tetrahedron Lett.*, **24**, 3473 (1983); T. Sugimoto, H. Awaji, I. Sugimoto, Y. Misaki, T. Kawase, S. Yoneda, Z. Yoshida, T. Kobayashi and H. Anzai, *Chem. Mater.*, **1**, 535 (1989).

12. A. Belyasmine, P. Frère, A. Gorgues, M. Jubault, G. Duguay and P. Hudhomme, *Tetrahedron Lett.*, **34**, 4005 (1993).

13. P. Frére, K. Boubekeur, M. Jubault, P. Batail and A. Gorgues, *Eur. J. Org. Chem.*, 3741 (2001).

14. M. Sallé, A. J. Moore, M. R. Bryce and M. Jubault, *Tetrahedron Lett.*, **34**, 7475 (1993).

15. L. Yu and D. Zhu, *Chem. Commun.*, 787 (1997).

16. W. T. Borden, *Diradicals*, John Wiley and Sons, New York (1982).

17. P. Dowd, *Acc. Chem. Res.*, **5**, 242 (1972).

18. H. M. McConnell, *Proc. R. A. Welch Found. Conf.*, **11**, 144 (1967).

19. T. Sugimoto, K. Ikeda and Y. Yamauchi, *Chem. Lett.*, 29 (1991).

20. W. R. Roth and G. Erker, *Angew. Chem.*, **85**, 510 (1973).

21. P. Dowd, *J. Am. Chem. Soc.*, **92**, 1066 (1970).

22. T. Sugimoto, A. Kawachi, K. Ikeda and J. Yamauchi, *Chem. Lett.*, 571 (1991).

23. T. Fukunaga, *J. Am. Chem. Soc.*, **98**, 610 (1976).

24. T. Sugimoto, E. Murahashi, K. Ikeda, Z. Yoshida, H. Nakatsuji, J. Yamauchi, Y. Kai and N. Kasai, *Mat. Res. Soc. Symp. Proc.*, **247**, 417 (1992).

25. P.-M. Allemand, K. C. Khemani, A. Koch, F. Wudl, K. Holczer, S. Donovan, G. Gruner and J. D. Thompson, *Science*, **253**, 301 (1991).

26. T. Sugimoto, M. Tsujii, T. Suga, H. Matsuura, N. Hosoito, N. Takeda, M. Ishikawa and M. Shiro, *Mol. Cryst. Liq. Cryst.*, **272**, 183 (1995).

27. E. Weltin, F. Gerson, J. N. Murrell and E. Heilbronner, *Helv. Chim. Acta*, **44**, 1400 (1961); T. Inauge, V. Gramlich, W. Amrein, F. Diedrich, M. Gross, C. Bouden and J.-P. Gisselbrech, *Angew. Chem. Int. Ed. Engl.*, **34**, 805 (1995); T. Enomoto, T. Kawase, H. Kurata and M. Oda, *Tetrahedron Lett.*, **38**, 2693 (1997).

28. R. Breslow, *Pure Appl. Chem.*, **54**, 927 (1982).

29. T. Sugimoto, H. Awaji, Y. Misaki, Z. Yoshida, Y. Kai, H. Nakagawa and N. Kasai, *J. Am. Chem. Soc.*, **107**, 5792 (1985).

30. T. Sugimoto, Y. Misaki, T. Kajita, Z. Yoshida, Y. Kai and N. Kasai, *J. Am. Chem. Soc.*, **109**, 4106 (1987).

31. M. Iyoda, S. Tanaka, M. Nose and M. Oda, *J. Chem. Soc., Chem. Commun.*, 1058 (1983).

32. T. Sugimoto, Y. Misaki, Y. Arai, Y. Yamamoto, Z. Yoshida, Y. Kai and N. Kasai, *J. Am. Chem. Soc.*, **110**, 628 (1988).

33. M. V. Lakshmikantham and M. P. Cava, *J. Org. Chem.*, **41**, 882 (1976); S. Yoneda, T. Kawase, Y. Yasuda and Z. Yoshida, *J. Org. Chem.*, **44**, 1728 (1979).

34. K. Kano, T. Sugimoto, Y. Misaki, T. Enoki, H. Hatakeyama, H. Oka, Y. Hosotani and Z. Yoshida, *J. Phys. Chem.*, **98**, 252 (1994).

35. W. Marsh and J. D. Dunitz, *Helv. Chim. Acta*, **58**, 707 (1975).

36. R. L. Myers and I. Shain, *Anal. Chem.*, **41**, 980 (1964).

37. R. Breslow, *Pure Appl. Chem.*, **28**, 111 (1971); R. Breslow, M. Oda and T. Sugimoto, *J. Am. Chem. Soc.*, **96**, 1974 (1974).

38. S. Feldberg, *Electroanalytical Chemistry*, Marcel Dekker, New York (1969), Vol. 3, p. 200; W. F. Sokol, D. H. Evans, K. Niki and T. Yagi, *J. Electroanal. Chem.*, **108**, 107 (1980)

39. T. Sugimoto, H. Awaji, I. Sugimoto, Y. Misaki and Z. Yoshida, *Synth. Met.*, **19**, 569 (1987).

40. T. Sugimoto, Y. Misaki, Z. Yoshida and J. Yamauchi, *Mol. Cryst. Liq. Cryst.*, **176**, 259 (1989).

41. R. Hopf, *Angew. Chem. Int. Ed. Engl.*, **23**, 948 (1984).

42. Y. Misaki, Y. Matsumura, T. Sugimoto and Z. Yoshida, *Tetrahedron Lett.*, **30**, 5289 (1989).

43. M. A. Coffin, M. R. Bryce, A. S. Batsanov and J. K. Howard, *J. Chem. Soc., Chem. Commun*, 552 (1993).

44. M. Iyoda, N. Nakamura, M. Todaka, S. Ohtsu, K. Hara, Y. Kuwatani, M. Yoshida, H. Matsuyama, M. Sugita, H. Tachibana and H. Inoue, *Tetrahedron Lett.*, **41**, 7059 (2000).

45. M. Kimura, W. H. Watson and J. Nakayama, *J. Org. Chem.*, **45**, 3719 (1980).
46. M. A. Coffin, M. R. Bryce and W. Clegg, *J. Chem. Soc., Chem. Commun.*, 401 (1992).
47. R. R. Schumaker, S. Rajeswari, M. V. Joshi, M. P. Cava, M. A. Takassi and R. M. Metzger, *J. Am. Chem. Soc.*, **111**, 308 (1989).
48. M. Salle, A. Belyasmine, A. Gorgues, M. Jubault and N. Soyer, *Tetrahedron Lett.*, **32**, 2897 (1991).
49. M. Fourmigué, I. Johannsen, K. Boubekeur, C. Nelson and P. Batail, *J. Am. Chem. Soc.*, **115**, 3752 (1993).
50. M. R. Bryce, M. A. Coffin, A. J. Moore, A. S. Batsanov and J. A. K. Howard, *Tetrahedron,* **55**, 9915 (1999).
51. Z. Yoshida and T. Sugimoto, *Angew. Chem. Int. Ed. Engl.*, **27**, 1573 (1988) and references therein.
52. L. Y. Chiang and H. Thomann, *J. Chem. Soc., Chem. Commun.*, 172 (1989).

IV Applications of TTFs

15

Macrocyclic, Molecular and Supramolecular TTF Systems

15.1   Introduction

Among sulfur-containing heterocycles, tetrathiafulvalene (TTF, **1**)[1] and its derivatives have been intensively studied during the past two decades on account of their unique π-electron donor properties. They were originally prepared for the development of electrically conducting materials and have, as such, been synonymous with the development of molecular organic metals. However, during the past few years, the utility of TTF derivatives as building blocks in macrocyclic and supramolecular chemistry has revealed that the TTF moiety is useful beyond the field of materials chemistry.[1b–d] Progress in synthetic TTF chemistry has enabled the preparation of elaborate molecular architectures and, in recent years, TTF has been incorporated into a number of molecular and supramolecular systems, such as cyclophanes, catenanes, rotaxanes, dendrimers, polymers, *etc.* Systems based on host-guest interactions may act as, for example, sensors or molecular switches. Among these systems mechanically interlocked architectures such as catenanes and rotaxanes are now prime candidates for the construction of artificial molecular machines[2] and the fabrication of molecular electronic devices.[2b,3]

This chapter will highlight recent developments in the field of macrocyclic, molecular and supramolecular TTF chemistry.

15.2   Key Properties of TTF

Although TTF is a planar 14π-electron system, it is non-aromatic according to the Hückel definition because the 14π-electrons lack cyclic conjugation.[4] Its oxidation to the radical cation **2** and dication **3** occurs (Fig. 15.1) sequentially and reversibly at low potentials. In contrast to the neutral TTF molecule, both the radical cation **2** and the dication **3** are aromatic in the Hückel sense as a result of the 6π-electron heteroaromaticity of the 1,3-dithiolium cation.

The key properties of TTF that make it an interesting building block in materials, macrocyclic and supramolecular chemistry are:

(i) TTF is a strong π-electron donor.

(ii) Oxidation of the TTF^0 ring system to the radical cation $TTF^{\bullet+}$ and dication TTF^{2+} occurs

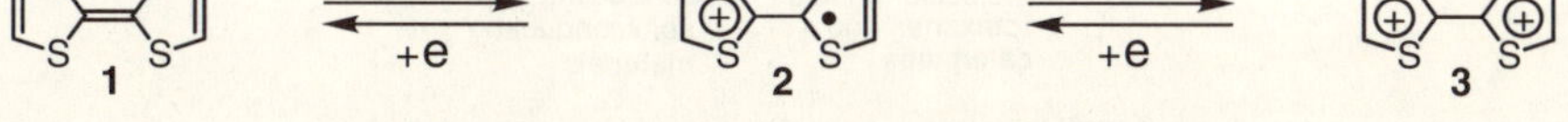

Fig. 15.1   Sequential and reversible oxidation of TTF (**1**).

sequentially and reversibly (see Fig. 15.1).

(iii) The oxidation potentials can be finely tuned by the attachment of electron-donating or electron-withdrawing substituents.

(iv) The TTF radical cation **2** and dication **3** are thermodynamically stable species.

(v) The UV–vis absorption spectra of TTF^0, $TTF^{\cdot+}$ and TTF^{2+} are decisively different from one another.

(vi) TTF derivatives readily form dimers, highly ordered stacks or two-dimensional sheets, which are stabilized by intermolecular π–π interactions and non-bonded sulfur–sulfur interactions.

(vii) TTF is stable to many synthetic transformations, although it is important to avoid strongly acidic conditions and strong oxidizing agents.

(viii) The contribution from 6π-electron heteroaromaticity in the 1,3-dithiolium cations explains the relative low oxidation potentials for the parent TTF ($E_{1/2}^{1} = 0.34$ V, $E_{1/2}^{2} = 0.73$ V $vs.$ Ag/AgCl in MeCN).

15.3   Applications of TTF Building Blocks

In recent years, different chemical modifications have been carried out on the TTF unit to prepare building blocks for macrocyclic and supramolecular chemistry at the molecular level. New TTF building blocks have been exploited in many areas of materials, macrocyclic and supramolecular chemistry. Some examples where TTFs have been used are shown in Fig. 15.2.[5] The multitude of interesting properties of the TTF unit has prompted much research into the synthesis of new TTF derivatives and excellent reviews have been published.[1,6] The comprehensive review by Segura and Martín[1d] deserves special mention as it covers the use of TTFs in molecular, supramolecular and materials chemistry up to the end of 1999. In the present chapter, we will therefore mainly describe recent examples of macrocyclic, molecular and supramolecular systems based on TTF, especially those based on pyrrolo-annulated TTFs.

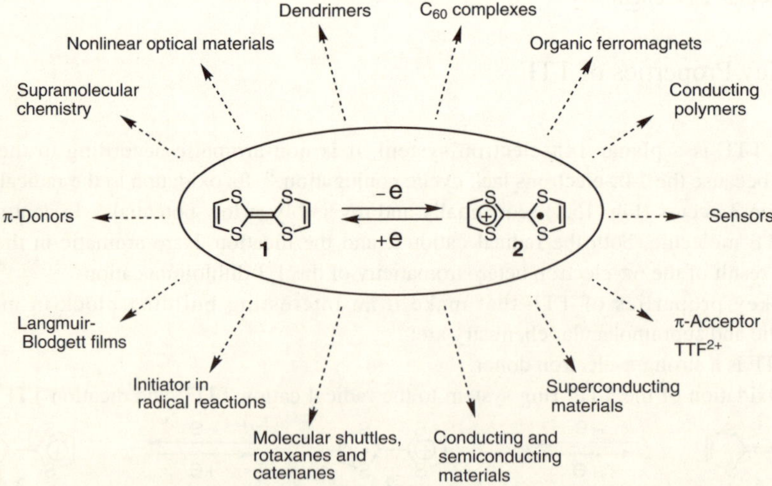

Fig. 15.2   Some uses of the redox-active TTF unit in molecular, supramolecular and materials chemistry.

15.4   TTF Systems Based on TTF Tetrathiolate

Versatile TTF building blocks have been developed in recent years, and many problems concerning the selective functionalization of TTF have been solved.[1,4,7] Tetrakis(2-cyanoethylthio)TTF[7b] **4** (Fig. 15.3) is a useful tetrafunctionalized TTF building block. The four cyanoethyl thiolate protecting groups in **4** are readily deprotected when treated with 4.5 equivalents of cesium hydroxide monohydrate, and the resulting TTF tetrathiolate **5** (Fig. 15.3) can subsequently be realkylated. In addition, it has been shown that selective and stepwise deprotection of the cyanoethyl thiolate protecting groups is possible by treatment with 1 equivalent of cesium hydroxide monohydrate.[4,5] When **4** is treated with 2 equivalents of cesium hydroxide monohydrate, symmetric/pseudo-symmetric deprotection takes place exclusively to produce a *cis/trans* mixture of bis-thiolate **8**, *i.e.*, no asymmetric bis-thiolate **7** is formed (Fig. 15.3).[4,5]

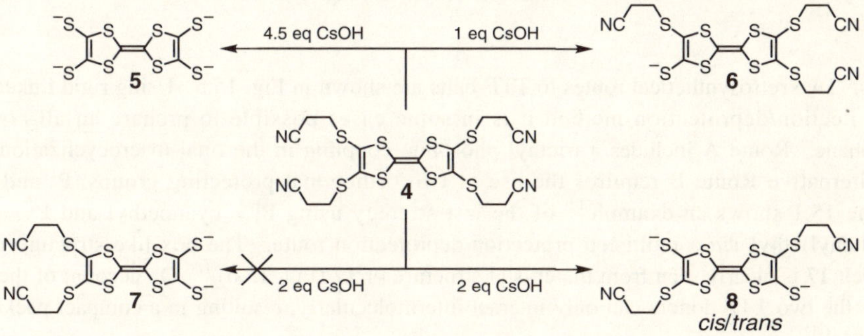

Fig. 15.3   The mono-, bis- and tetrathiolates which can be obtained by deprotection of the tetrakis(cyanoethylthio)TTF **4**.

15.4.1 TTF-macrocycles, -belts and -cages

Cyclophanes are fundamentally important compounds in many aspects of macrocyclic and supramolecular chemistry, and research in this field has expanded rapidly in recent years.[8,9] On account of their rigid framework, primarily defined by aromatic units, these molecules with very large cavities are ready to accommodate charged or neutral guest molecules.[10] Incorporation of the redox-active TTF unit into such cyclophanes may serve a dual purpose, namely increasing the host-guest interaction with a complementary electroactive guest and at the same time electrochemically signal the complexation event, making such molecules attractive components, for example, for sensor technology.[11]

Using the stepwise cyanoethyl protection/deprotection methods outlined in Fig. 15.3, it has been possible to prepare a variety of macrocyclic-, dendritic- and cage-TTF systems, and some examples are illustrated in Fig. 15.4. In the following, we will describe some recent approaches which have been used to obtain TTF-macrocycles, -belts and -cages. The TTF-belt[12] **10** and the criss-cross cyclophane[13] **11** in Fig. 15.4 were isolated as the *cis* isomers. However, in most cases the crude reaction mixture will also contain the corresponding *trans* isomers depending on the

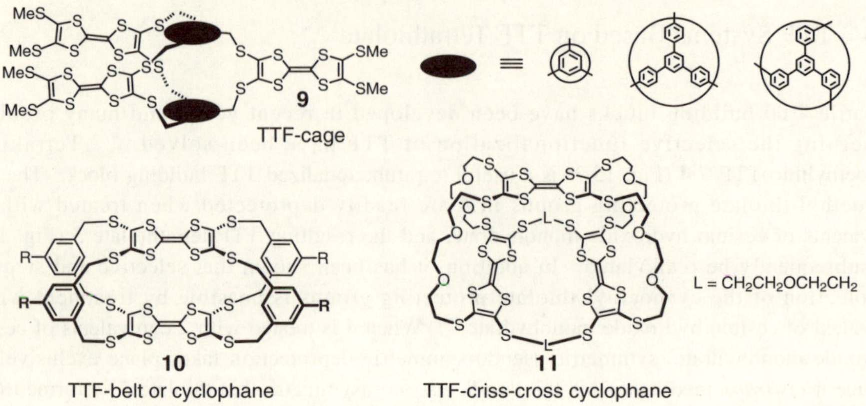

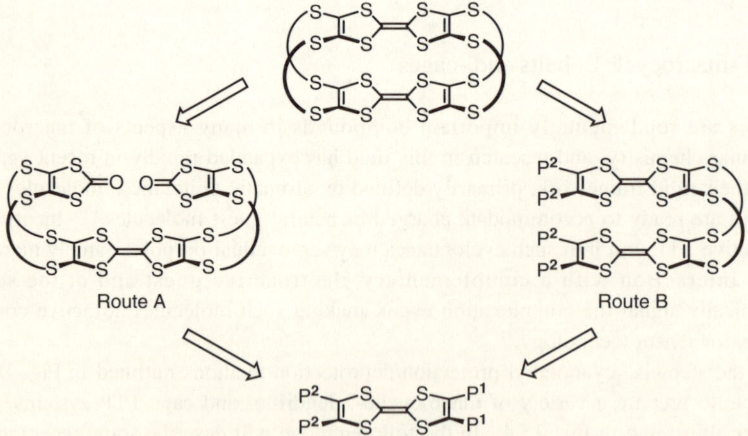

Fig. 15.4   Examples of belt- and cage-type TTFs.

linkers. Two retrosynthetical routes to TTF-belts are shown in Fig. 15.5. Using rigid linkers and the protection/deprotection method it is in some cases possible to prepare an all-*cis*-TTF cyclophane. Route A includes a triethyl phosphite coupling in the final macrocyclization step. The alternative Route B requires the use of two orthogonal protecting groups, P^1 and P^2.[14] Scheme 15.1 shows an example[12] of the last strategy using P^1 = cyanoethyl and P^2 = 2-(*p*-nitrophenyl)ethyl *via* a multistep protection/deprotection route. The box-like structure of the TTF-belt **17** is clearly seen from the crystal structure of **17** (Fig. 15.6).[12] On account of the rigid linkers the two TTF donors can only interact intermolecularly, resulting in a compact packing in the crystal.[12]

As a result of the effective S-alkylations used in the synthesis of such macrocycles, it is even possible to prepare TTF-cyclophanes of this type in a one-pot reaction, as shown in Scheme 15.2.

Fig. 15.5   Retrosynthetic approaches for the preparation of TTF-belt molecules. P^1 = cyanoethyl and P^2 = 2-(*p*-nitrophenyl)ethyl.

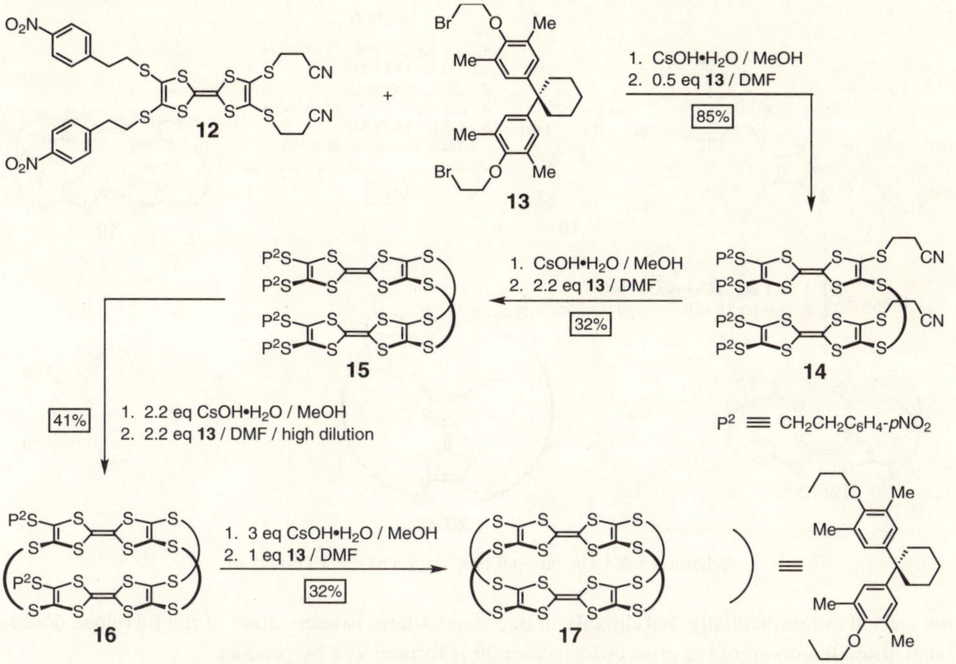

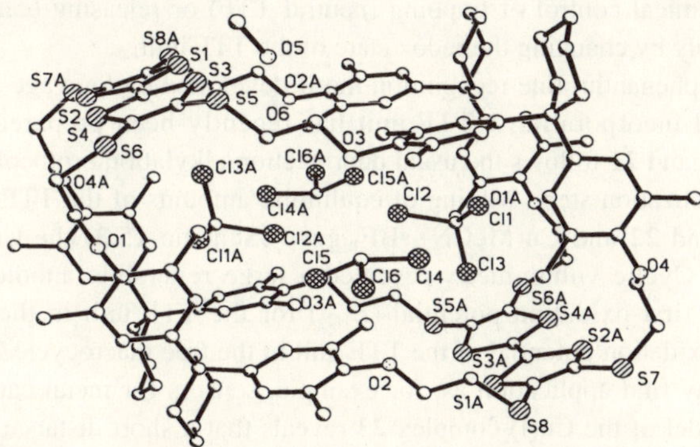

Scheme 15.1   Sequential synthesis of the rigid TTF-belt **17**.

By employing the rigid linker **18**, it has been demonstrated[15] that it is possible to prepare TTF-belts in up to 45% yield when a sequential one-pot reaction is used. Note that Step 3 in Route A involves an isomerization of the initial mixture of *cis* and *trans* isomers with a catalytic amount of acid. The rigidity of the linker **18** force the intermediate to adopt a correct *cis/cis* conformation before its final ring closure to complete the synthesis of the belt-molecule **19**. This reaction procedure only works when rigid linkers, such as dibromide **18**, are used – flexible and longer linkers result in the formation of complex reaction mixtures. Note also that, if the reactions are

Fig. 15.6   Crystal structure of the TTF-belt **17**. The bis-TTF cage includes two CHCl$_3$ guest molecules in its cavity. Hydrogen atoms are omitted for clarity.

Route A

1. 2.2 eq CsOH•H$_2$O / MeOH
2. 1.0 eq **18** / DMF
3. CF$_3$COOH (cat)
4. 2.2 eq CsOH•H$_2$O / MeOH
5. 1.0 eq **18** / DMF
   high dilution in all steps

45%

Route B

1. 4.1 eq CsOH•H$_2$O / MeOH
2. 2.0 eq **18** / DMF / high dilution

19: 35%

20: 5%

Scheme 15.2   The one-pot preparation of the TTF-belt **19**.

not carried out sequentially, but directly in one step without isomerization of the fulvalene double bond, a small amount of the criss-cross isomer **20** is formed as a by-product.

15.4.2   TTF-based Ligands

Attachment of ligands such as crown ethers to redox-active units in many cases result in ligands which can be tuned electrochemically. Control of the interactions in host-guest binding is relevant for the development of sensors and molecular devices. In the context of redox-responsive ligands, the TTF unit is an ideal redox-active unit in view of its unique π-electron donating properties. Its oxidation to the radical cation (TTF$^{•+}$) and dication (TTF^{2+}) occurs sequentially and reversibly at low potentials and such a reversibility of its redox processes can allow the electrochemical control of trapping (neutral TTF) or releasing (cationic TTF) a given metallic cation, simply by changing the redox state of the TTF unit.

Employing the phenanthroline recognition motif developed by Sauvage and coworkers,[16] an electroactive ligand incorporating a TTF unit has recently been prepared (Scheme 15.3).[17] Preparation of the ligand **21** follows the usual deprotection/alkylation protocol using high dilution conditions in the cyclization step. Mixing of equimolar amounts of the TTF-macrocycle **21**, the phenanthroline thread **22** and Cu(MeCN)$_4$•BF$_4$ gave (Scheme 15.3) the Cu(I) complex **23** in quantitative yield. Cyclic voltammetry revealed a large reversible anodic shift (< 100 mV) associated with the first oxidation potential ($E_{1/2}^1$) for the TTF unit in the Cu(I) complex **23** relative to the first oxidation potential of the TTF unit in the free macrocycle **21**. Such "Sauvage-type" complexes may find application as, for example, sensors for metal cations. Inspection of the space-filling model of the Cu(I) complex **23** reveals that a short distance between the redox-active TTF unit and the metal cation center can be established because of the flexibility of the glycol linkers. In the isolated TTF unit the largest HOMO coefficients are located on the central S$_2$C=CS$_2$ fulvalene fragment. Therefore, it can be expected that the introduction of a suitable long crown-ether unit (or another ligand) along the long axis (*i.e.*, at the 2,7-positions) of the redox-active

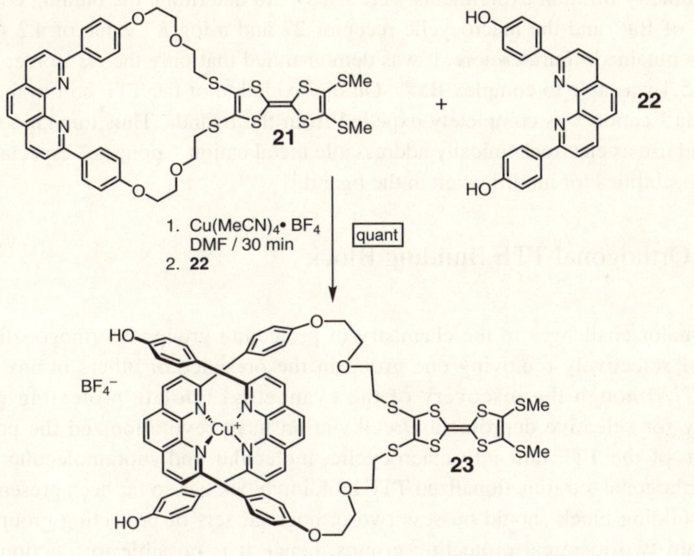

Scheme 15.3 Complexation of the phenanthroline thread **22** by the TTF-macrocycle **21**.

TTF unit will give a stronger electrochemical response upon complexation of an appropriate cation. Indeed, it has been demonstrated[18] that the anodic shift in $E_{1/2}^{1}$, upon complexation of a metal cation, is much stronger for 2,7-TTF crowns, such as **24** (Fig. 15.7) than for 2,3-TTF-crowns, such as **25**. The preparation of such 2,7-TTF crowns can be accomplished in a one-step reaction; thus deprotection of the TTF building block **26** and subsequent alkylation under high dilution conditions gave (Scheme 15.4) a 55% yield of the 2,7-TTF-crown **27**.[18e] ^{1}H NMR and

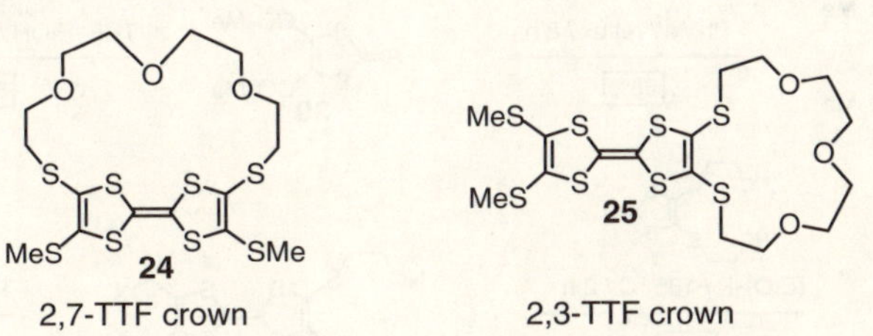

24

2,7-TTF crown

25

2,3-TTF crown

Fig. 15.7   2,7- and 2,3-TTF crown type ligands.

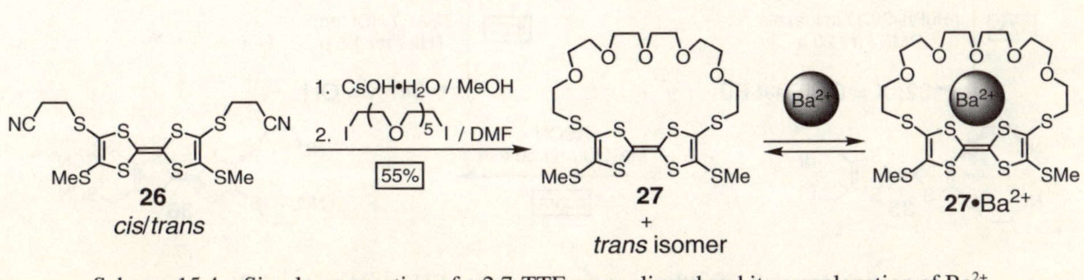

26
cis/trans

27
+
trans isomer

27•Ba^{2+}

Scheme 15.4   Simple preparation of a 2,7-TTF-crown ligand and its complexation of Ba^{2+}.

cyclic voltammetry titration experiments were used[18e] to determine the binding constant ($K°$) for the complex of Ba^{2+} and the macrocyclic receptor **27** and a log $K°$ value of 4.2 for the **27**•Ba^{2+} complex was obtained. Furthermore, it was demonstrated that only the *cis* isomer of **27** depicted in Scheme 15.4 was able to complex Ba^{2+}. On the oxidation of the TTF unit to the dication (*i.e.*, TTF^{2+}) the Ba^{2+} cation was completely expelled from the ligand. Thus this kind of crown-ether TTF may find use as electrochemically addressable metal cation "sponges," especially as there are numerous possibilities for modification in the ligand.[1b]

15.5   An Orthogonal TTF Building Block

One of the major challenges in the chemistry of protecting groups is orthogonality, that is, the possibility of selectively removing one group in the presence of others in any chronological sequence.[19] Although the discovery of the cyanoethyl thiolate protecting group and the methodology for selective deprotection/realkylation have revolutionized the possibilities for incorporation of the TTF unit into macrocyclic, molecular and supramolecular structures, no preformed orthogonal tetrafunctionalized TTF building block has so far been presented. A perfect orthogonal building block should possess two orthogonal sets of protecting groups.[14] Each set should contain two identical protecting groups, hence it is possible to functionalize the TTF moiety regioselectively in the 6- and 7-positions in one synthetic step, followed by a functionalization of the TTF moiety in the 2- and 3-positions in the next step. An orthogonal TTF building block possessing alternative functionalization possibilities has recently been reported.[20] The synthesis of this versatile tetrafunctionalized TTF building block, 2,3-bis(2-cyanoethylthio)-6,7-bis(thiocyanatomethyl)TTF **36** depicted in Scheme 15.5 has two thiolate groups in the 2- and 3-positions protected by the conventional cyanoethyl group[4,7b–c] whereas the TTF unit can be functionalized through two thiocyanatomethyl groups in the 6- and 7-positions. The two thiocyanatomethyl groups can be deprotected selectively with excess sodium borohydride in a

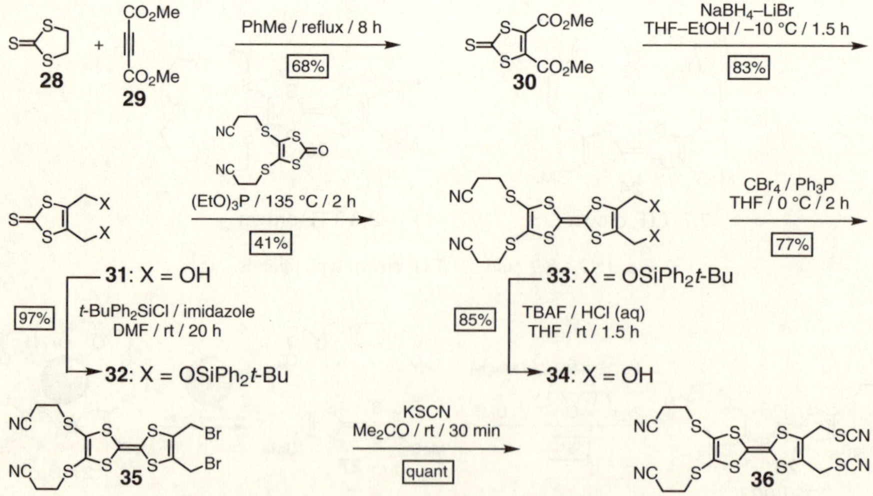

Scheme 15.5   Synthesis of the orthogonal TTF building block **36**.

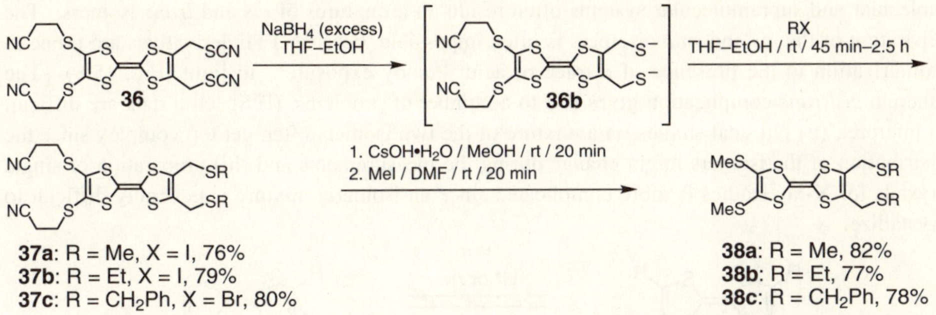

Scheme 15.6   Deprotection and realkylation of thiocyanate and cyanoethyl protecting groups.

37a: R = Me, X = I, 76%
37b: R = Et, X = I, 79%
37c: R = CH$_2$Ph, X = Br, 80%

38a: R = Me, 82%
38b: R = Et, 77%
38c: R = CH$_2$Ph, 78%

mixture of THF and EtOH to generate (Scheme 15.6) the reactive 2,3-bis(2-cyanoethylthio)TTF-6,7-bis(methylthiolate) **36b**, which can be trapped *in situ* with different electrophiles to give (Scheme 15.6) the *S*-alkylated TTFs **37a–c** in 76–80% yields. Subsequently, the remaining two cyanoethyl thiolate protecting groups in **37a–c** can be deprotected using 2.2 equivalents of cesium hydroxide monohydrate followed by the addition of iodomethane affording (Scheme 15.6) the corresponding 2,3-bis(methylthio)TTFs **38a–c** in 77–82% yields.

The potential of **36** as a building block for cyclophane-type compounds is demonstrated in Scheme 15.7. Thus under high dilution conditions, reaction of 2,6,2",6"-tetrakis(bromomethyl)-1,1':3',1"-terphenyl (**39**) with 2 equivalents of **36** gave the tetrafunctionalized bis(TTF)-cuppedophane **40** in 31% yield as an orange crystalline solid.[21] Subsequently, the four cyanoethyl thiolate protecting groups in **40** can be deprotected using 4.5 equivalents of cesium hydroxide monohydrate followed by the addition of iodomethane affording (Scheme 15.7) the corresponding tetrakis(methylthio)-bis(TTF)cuppedophane **41** in 57% yield after column chromatography.

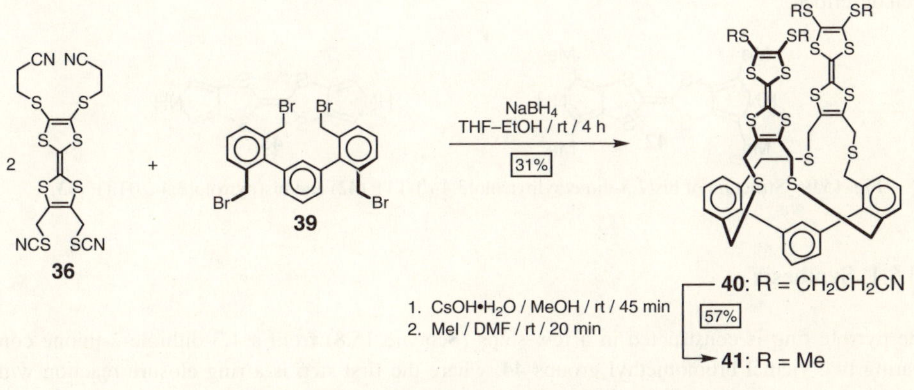

40: R = CH$_2$CH$_2$CN

41: R = Me

Scheme 15.7   Preparation of cuppedophane type cyclophanes.

15.6   Pyrrolo-annulated TTFs

As mentioned in Section 15.4, the regioselective functionalization of TTF is a challenge, since TTF has four identical potential attachment sites. Hence, incorporation of TTF into macrocyclic, molecular and supramolecular systems often results in a mixtures of *cis* and *trans* isomers. The separation of the *cis* and *trans* isomers is often impossible, because TTF derivatives are prone to isomerization in the presence of a trace of acid[22] or by exposure[23] to light (Fig. 15.8). The inherent *cis/trans* complication gives rise to a number of problems. (i) Spectral data are difficult to interpret, (ii) physical studies on a mixture of the two isomers often get too complex since the distribution of the isomers might change during the measurements and (iii) preparation of single crystals for X-ray studies is more complicated since an isomeric mixture is generally difficult to crystallize.

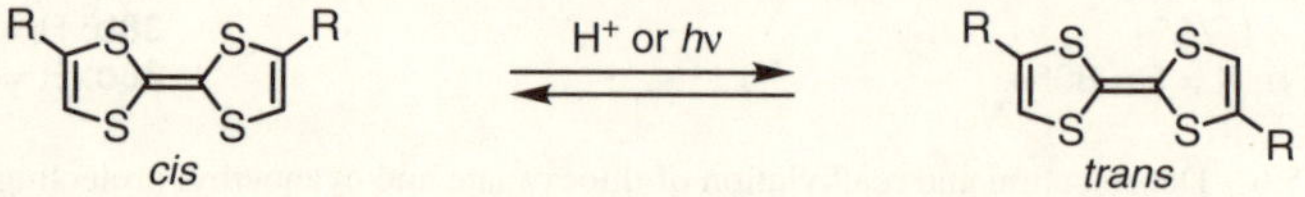

Fig. 15.8   Isomerization of TTF derivatives.

Cava and coworkers some years ago presented a detailed study of bis(2,5-dimethyl-pyrrolo[3,4-*d*])TTF (**42**) and their *N*-alkylated derivatives (Fig. 15.9).[24] The synthesis of **42** involves classical pyrrole chemistry, building up the 1,3-dithiole-2-thione moiety from a 2,5-dimethylpyrrole core. Annulation of TTF to two electron-rich 2,5-dimethylpyrrole rings produces a donor system with a lower oxidation potential than that of the parent TTF (**1**). **42** has been used as a building block in macrocyclic, molecular and supramolecular chemistry and produces systems devoid of *cis/trans* isomerism.[25] However, the four methyl groups in **42**, acting as protecting groups during the synthesis, block the α-positions in the pyrrole units, excluding the possibility for further *C*-functionalization in addition to *N*-alkylations. Furthermore, steric hindrance has been observed from the methyl groups when **42** is incorporated in macrocyclic, molecular and supramolecular systems. In this context the parent bis(pyrrolo[3,4-*d*])TTF (**43**) appeared to be interesting (Fig. 15.9), and a straightforward route to **43** and the related monopyrrolo annulated TTFs, using a simple and nonclassical pyrrole synthesis,[26] has recently been described.[27]

Fig. 15.9   Structure of bis(2,5-dimethylpyrrolo[3,4-*d*])TTF (**42**) and bis(pyrrolo[3,4-*d*])TTF (**43**).

15.6.1 Synthesis

The pyrrole ring is constructed in a few steps (Scheme 15.8) from a 1,3-dithiole-2-thione core bearing two vicinal bromomethyl groups **44**, where the first step is a ring closure reaction with sodium tosylamide followed by oxidation of the annulated dihydropyrrole ring to give the *N*-tosyl protected pyrrole **47**.

Self-coupling of the ketone **47** to form the TTF **48** using triethyl phosphite proceeded (Scheme 15.9) in 84% yield.[27] The tosyl protection groups of **48** were cleanly removed by refluxing a suspension of **48** and sodium methoxide in a 1:1 mixture of THF–MeOH affording (Scheme 15.9) the novel bis(pyrrolo[3,4-*d*])TTF (**43**) in quantitative yield. Finally, **43** was dialkylated to give (Scheme 15.9) the *N,N'*-dialkyl derivatives **49a,b** in 82–83% yields.

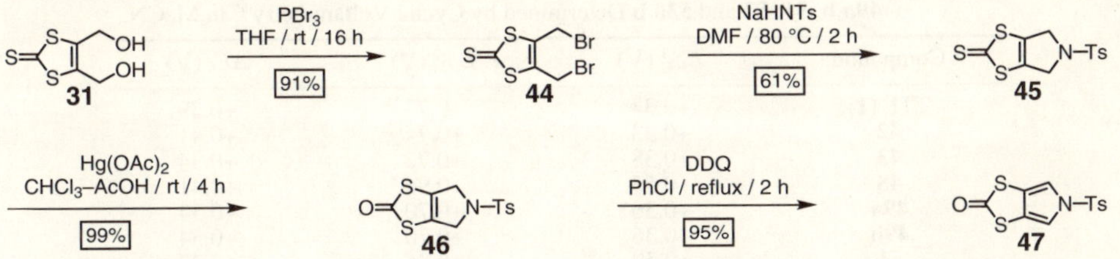

Scheme 15.8   Preparation of the parent [*c*]fused (1,3-dithiole)pyrrole ring system.

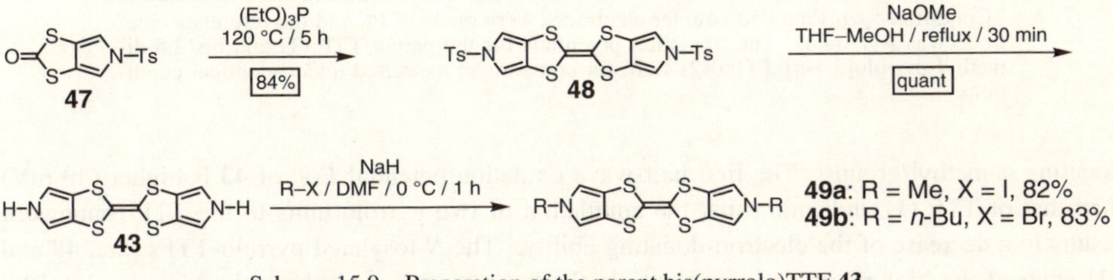

Scheme 15.9   Preparation of the parent bis(pyrrolo)TTF **43**.

The asymmetrical monopyrrolo-TTF **52** (Scheme 15.10) was obtained in optimum yield[27] by cross-coupling of **47** with 4,5-bis(methylthio)-1,3-dithiole-2-thione (**50**) in neat triethyl phosphite. Subsequent detosylation of **51** using sodium methoxide followed by *N*-alkylation proceeded smoothly to give the *N*-alkylated derivatives **53**.

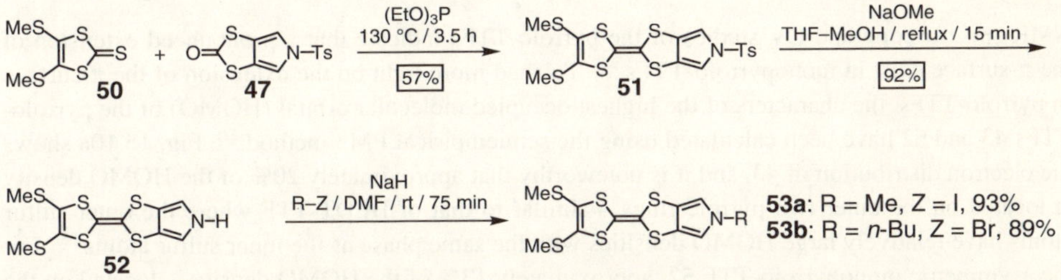

Scheme 15.10   Preparation of the parent monopyrrolo-TTF **52**.

Note that the tosyl group performs a triple function during the syntheses. First, it activates the nitrogen in the ring closure reaction. Second, it works as an excellent protecting group for the pyrrole nitrogen during the harsh triethyl phosphite coupling reaction. Finally, the tosyl group can be removed almost quantitatively from the aromatic pyrrole ring under mild conditions.

15.6.2 Electrochemistry

Solution oxidation potentials obtained from cyclic voltammograms (CVs) of pyrrolo-TTF π donors (D) are summarized in Table 15.1.[27] The CVs of all compounds showed two pairs of reversible redox waves, indicating good stability of the corresponding radical cation (D$^{\cdot+}$) and dication (D^{2+}). Compound **43** showed a higher first half-wave oxidation potential $E_{1/2}^1$ (50 mV) than its $\alpha,\alpha',\alpha'',\alpha'''$-tetramethylated analogue **42** on account of the absence of the electron-

Table 15.1   Oxidation Potentials $E_{1/2}^1$ and $E_{1/2}^2$ of Pyrrolo-TTF Derivatives **43**, **48**, **49a,b**, **51**, **52** and **53a,b** Determined by Cyclic Voltammetry[a,b] in MeCN

Compound	$E_{1/2}^1$ (V)	$E_{1/2}^2$ (V)	ΔE_p (V)
TTF (**1**)	+0.34	+0.73	+0.39
42	+0.33	+0.74	+0.41
43	+0.38	+0.72	+0.34
48	+0.55	+0.96	+0.41
49a	+0.36	+0.70	+0.34
49b	+0.36	+0.70	+0.34
51	+0.59	+0.86	+0.27
52	+0.44	+0.75	+0.31
53a	+0.42	+0.74	+0.32
53b	+0.42	+0.73	+0.31

[a] Conditions: working and counter electrodes were made of Pt, and the reference electrode was Ag/AgCl. [b] The oxidation potentials for the parent TTF (**1**) and bis(2,5-dimethyl-pyrrolo[3,4-*d*])TTF (**42**) were for comparison measured under identical conditions.

donating α-methyl groups. The first half-wave oxidation potential $E_{1/2}^1$ of **43** is higher (40 mV) than that of TTF (**1**), indicating that the annulation of two pyrrolo units to the TTF framework results in a decrease of the electron-donating ability. The *N*-tosylated pyrrolo-TTFs (*i.e.*, **48** and **51**) showed the highest oxidation potentials (both $E_{1/2}^1$ and $E_{1/2}^2$) in both the bis(pyrrolo)-TTF series and the monopyrrolo-TTF series, whereas the *N*-alkylated pyrrolo-TTFs (*i.e.*, **49a,b** and **53a,b**) revealed the lowest oxidation potentials (both $E_{1/2}^1$ and $E_{1/2}^2$), due to the inductive effect exhibited from the tosyl groups and alkyl groups, respectively.

15.6.3 HOMO Orbitals

NMR spectroscopy and CV studies of the pyrrolo-TTFs indicate that a pronounced extension of the π-surface exist in monopyrrolo-TTFs.[27] To shed more light on the extension of the π-surface in pyrrolo-TTFs, the characters of the highest occupied molecular orbital (HOMO) of the pyrrolo-TTFs **43** and **52** have been calculated using the semiempirical PM3 method.[27] Fig. 15.10a shows the electron distribution of **43**, and it is noteworthy that approximately 20% of the HOMO density is located on the outer two pyrrole rings – similar to that of BEDT-TTF where the outer sulfur atoms have relatively large HOMO densities with the same phase as the inner sulfur atoms.[28] For the asymmetric monopyrrolo-TTF **52**, approximately 13% of the HOMO density is located on the outer pyrrole ring (Fig. 15.10b), clearly demonstrating an extension of the π-surface in pyrrolo-TTFs.

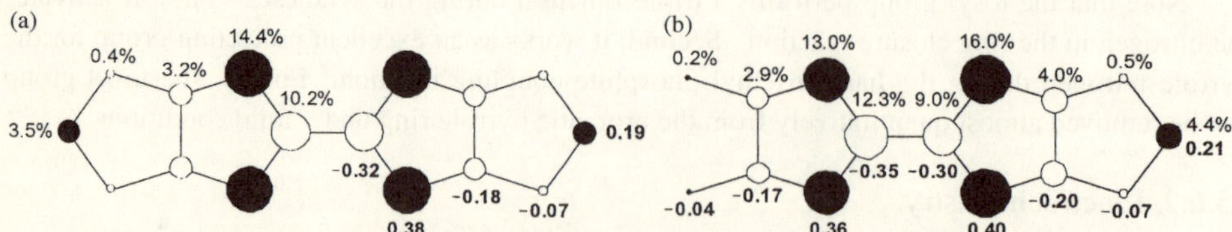

Fig. 15.10   HOMO orbitals of (a) **43** and (b) **52**; lightface numbers: HOMO electron density, boldface numbers: HOMO coefficients.

15.7 Applications of Pyrrolo-TTFs

In the following sections some recent applications of the new pyrrolo-TTF system in macrocyclic, molecular and supramolecular chemistry will be discussed.

15.7.1 Asymmetric Monopyrrolo-TTF Building Block

Combination of the new pyrrolo-TTF ring system with the conventional cyanoethyl thiolate protected TTF chemistry[7c] produces an asymmetric TTF building block, which can be functionalized regioselectively and is devoid of *cis/trans* isomerism. The asymmetric TTF building block **55** was synthesized as outlined in Scheme 15.11. Cross-coupling of the ketone **47** with 2 equivalents of the thione **54** in neat triethyl phosphite gave **55** (64%) in gram quantities after column chromatography.[29] **55** comprises two different protecting groups – a tosyl and a cyanoethyl protecting group – which can be deprotected stepwise allowing the TTF core to be functionalized regioselectively. Treatment of a THF solution of **55** with 1 equivalent of cesium hydroxide monohydrate generated the TTF-monothiolate selectively, without interfering with the tosyl protecting group. This TTF-monothiolate can subsequently be alkylated with a variety of alkylating reagents (R^1X) affording (Scheme 15.11) the alkylthio-TTF derivatives **56** in high yields.[3d,29] Removal of the tosyl protecting group can be carried out in near quantitative yield by refluxing **56** in a 1:1 mixture of THF–MeOH in the presence of excess sodium methoxide. Finally, *N*-alkylation of the pyrrole unit with different alkylating agents (R^2X) can be carried out in DMF containing sodium hydride affording (Scheme 15.11) the TTF derivatives **57**.[3d,29]

Examples in which the asymmetric TTF building block **55** is incorporated into supramolecular and molecular systems will be presented in Section 15.8.2.

Scheme 15.11   Synthesis and potential of asymmetric monopyrrolo-TTF building block **55**.

15.7.2 A Pyrrolo-TTF Belt

As mentioned in Section 15.4.1 dimeric TTF molecules[30] and TTF-belts[31] in which two TTF units are linked by one or more spacer groups have received particular attention due to the possibility of affecting the formation, structure and physical properties of their charge transfer (CT) complexes and ion radical salts.[30] Most synthetic strategies employed for the preparation of this kind of molecule have so far resulted in a mixture of *cis* and *trans* isomers. This problem can be circumvented using the bis-cyanoethyl thiolate protected monopyrrolo-TTF building block[27b] **58** (Scheme 15.12), which only possesses three attachment sites and is thereby devoid of *cis/trans*

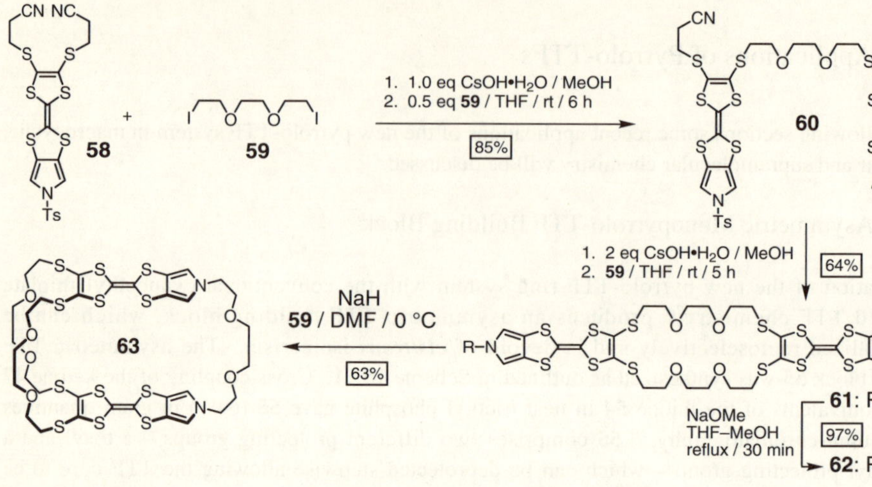

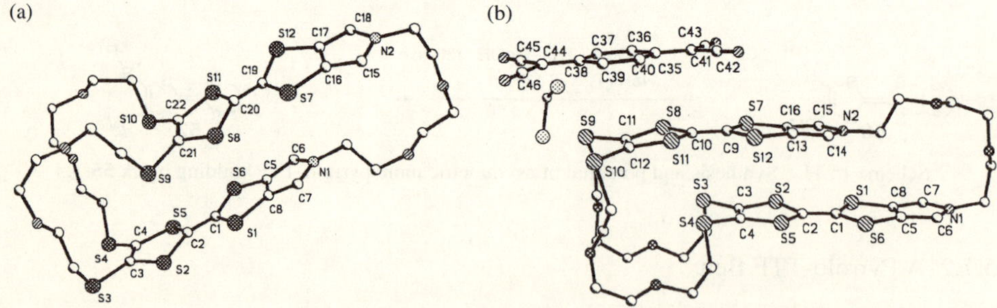

Scheme 15.12   Synthesis of a TTF-belt **63**.

problems. The monopyrrolo-TTF building block **58** was synthesized (not shown) by a triethyl phosphite mediated cross-coupling of the ketone **47** and 4,5-bis(2-cyanoethylthio)-1,3-dithiole-2-thione. By employing this building block a TTF-belt **63** devoid of *cis/trans* isomerism has been prepared in relatively few steps as illustrated in Scheme 15.12.[32] Fig. 15.11a shows the X-ray structure of the TTF-belt **63**.[32] The TTF-belt **63** was designed to allow complexation of the electron acceptor 7,7,8,8-tetracyano-*p*-quinodimethane (TCNQ) inside its cavity. However, a solid-state X-ray crystal structure analysis (Fig. 15.11b) of the CT complex **63•TCNQ** revealed that TCNQ is associated outside (alongside) one of the electron donors, reflecting that the complicated and subtle balance between all the individual noncovalent forces acting in cooperation are difficult to predict. The efficient synthetic methodology described here, however, allows, in a few steps, to make related TTF-belts in which the spacers can be varied in order to optimize the complexation properties of the macrocyclic host.

Fig. 15.11   (a) Crystal structure of the TTF-belt **63**. Hydrogen atoms are omitted for clarity.
(b) Crystal structure of **63•TCNQ**. Hydrogen atoms are omitted for clarity.

15.7.3 Tetrathiafulvaleno-annulated Porphyrins

Porphyrins are of fundamental importance in biological systems and are currently the focus of much attention for applications in supramolecular[33] and materials chemistry; therefore the porphyrin chromophore has been extensively modified to enhance the desired properties.[34] Although there have been some attempts to combine TTF chemistry with porphyrin chemistry, the direct combination of these two major fields has so far been unsuccessful, most likely due to the lack of an appropriate pyrrolo-TTF unit. With this building block at hand, the first examples of single molecules in which the intriguing optical and metal-ion binding abilities of the porphyrin molecule have been intimately coupled to the favorable redox properties of the TTF unit, have recently been reported.[35] The direct annulation of the porphyrin core with four TTF units situated at the periphery gives a porphyrin in which the annulene π-electron system is extended by direct conjugation with TTF units. By employing the new pyrrolo-TTF unit we have prepared the first tetrathiafulvaleno-annelated fully conjugated porphyrin **68** (**68** refer to the mixture of **68i** and **68ii**) using two different synthetic routes as illustrated in Scheme 15.13. Characterization[35] of the porphyrin **68** turned out to be a problem, since the ^{1}H NMR spectra (250 MHz) of compound **68** recorded in CDCl$_3$ at 298 K featured only very broadened peaks, which can be explained by the presence of radicals or slow tumbling resulting from aggregation in solution. Electron paramagnetic resonance (EPR) spectroscopy of **68** revealed a strong signal – arising from radicals – as a broad singlet without any detectable hyperfine structure at g = 2.0084, used either neat or in CH$_2$Cl$_2$ solution. The linewidth and g-value are consistent with the presence of a TTF radical cation.[36] Quantitative EPR measurements in CH$_2$Cl$_2$ showed that the isolated product **68** contained 19% unpaired spin, indicating that compound **68** consists of a mixture of the neutral porphyrin (*i.e.*, **68i**) and the radical cation porphyrin (*i.e.*, **68ii**) in a ratio of approximately 4:1

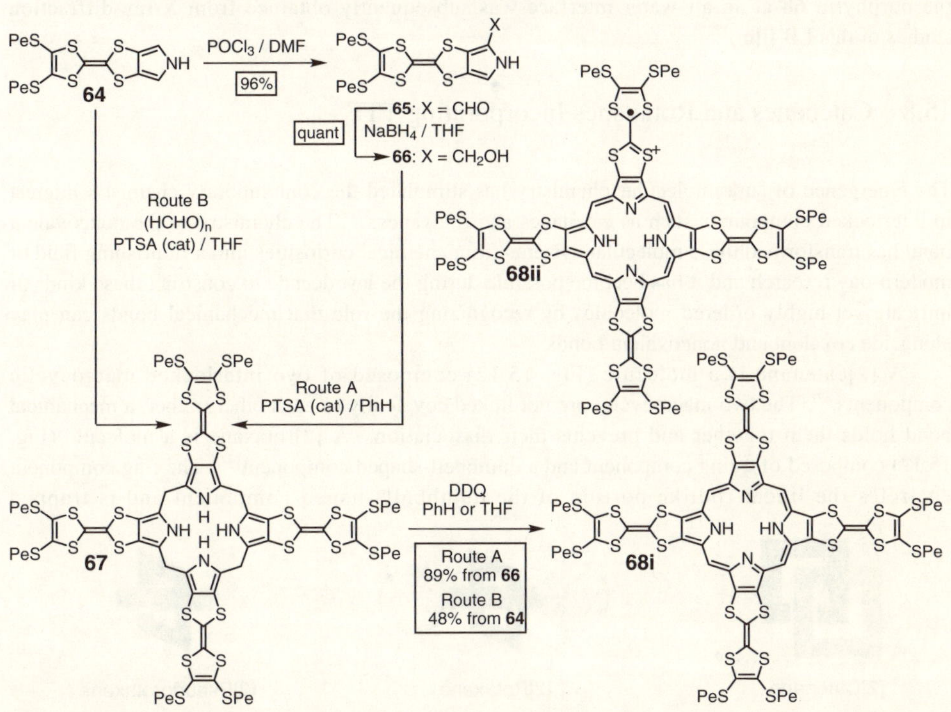

Scheme 15.13   Synthesis of the tetrathiafulvaleno-annelated porphyrin **68**.

(Scheme 15.13). All attempts to obtain the neutral porphyrin **68i** exclusively by chemical reduction of the paramagnetic mixture of the neutral porphyrin **68i** and the radical cation porphyrin **68ii** have so far been unsuccessful. A high resolution matrix-assisted laser-desorption/ionization mass spectrum (MALDI–MS) of **68** showed the exact mass, m/z 1830.076 corresponding to $C_{76}H_{94}N_4S_{24}^{\cdot+}$ (calculated mass $M^{\cdot+}$ = 1830.077) and together with the elemental analysis of the isolated product it proved to be the most unequivocal evidence[35] for the formation of **68**. CV and thin layer cyclic voltammetry (TLCV) of **68** revealed that some of the TTF units in **68** do not exhibit normal electrochemical characteristics. The CV (*vs.* Ag/AgCl) of **68** showed two pairs of reversible oxidation waves at $E_{1/2}^1$ = +0.63 V and $E_{1/2}^2$ = +1.10 V. TLCV and deconvoluted voltammograms of **68** revealed that the first oxidation waves actually correspond to two one-electron processes with $\Delta E < 100$ mV, whereas the second wave involves a one-step two-electron process. Thus, first a radical cation is formed, closely followed by the formation of a dication (probably as a biradical). Finally, a tetracation is formed. These results suggest the presence of an isolated electron-withdrawing 18π-electron porphyrin ring system in which the pyrrolo *c*-bonds in two of the TTF units are included in the 18π-electron ring system. Therefore, these two TTF units do not show the electrochemical characteristics of a normal TTF unit and only a total of four electrons are removed during oxidation of the neutral porphyrin. Furthermore, the deconvoluted voltammogram of compound **68** showed one reversible wave at $E_{1/2} = -1.3$ V (1 electron), which can be assigned to the first reduction of the porphyrin ring system. Finally, it has been demonstrated that compound **68** spreads readily from $CHCl_3$ solutions on H_2O resulting in the formation of a well-defined Langmuir-Blodgett (LB) film. Information about the packing of the porphyrin **68** at an air-water interface was subsequently obtained from X-ray diffraction studies of this LB film.[35]

15.8   Catenanes and Rotaxanes Incorporating TTF

The emergence of supramolecular chemistry has stimulated the contemporary chemist's interest in interlocked compounds such as catenanes and rotaxanes.[37] The chemistry of the noncovalent bond has transformed these molecular systems from chemical curiosities into a flourishing field of modern-day research and it has become possible during the last decade to construct these kinds of intricate yet highly ordered molecules by recognizing the role that mechanical bonds can play alongside covalent and noncovalent bonds.

A [2]catenane is a molecule (Fig. 15.12) composed of two interlocked macrocyclic components.[37] The two macrocycles are not linked covalently to each other; rather, a mechanical bond holds them together and prevents their dissociation. A [2]rotaxane is a molecule (Fig. 15.12) composed of a ring component and a dumbbell-shaped component.[37] The ring component encircles the linear rodlike portion of the dumbbell-shaped component and is trapped

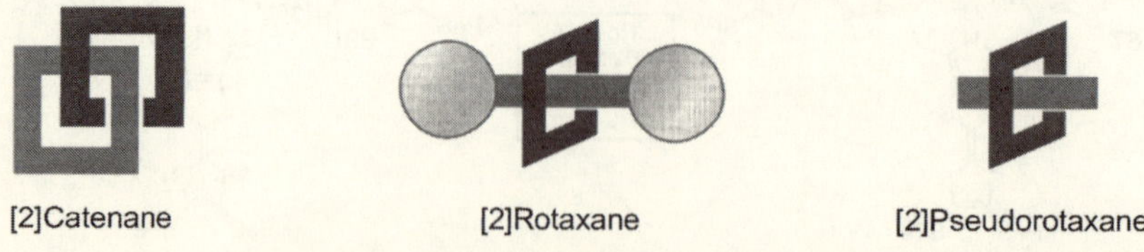

Fig. 15.12   Schematic representation of a [2]catenane, a [2]rotaxane and a [2]pseudorotaxane.

mechanically around it by two bulky stoppers. Thus, the two components cannot dissociate from one another, even although they are not linked covalently to each other. By contrast, in a [2]pseudorotaxane,[37] at least one of the stoppers on the dumbbell-shaped component is absent (Fig. 15.12) with the consequence that dissociation of the [2]pseudorotaxane into its components can occur, and the equilibrium between the species is controlled by the free energy of complexation, *i.e.*, a [2]pseudorotaxane is a supramolecular species.

Stoddart and coworkers[38] have developed a strategy for the formation of both catenanes and rotaxanes based on molecular recognition between π-electron donors and acceptors assisted by hydrogen bonding. The first catenane[38] synthesized in Stoddart's laboratories was the [2]catenane 74^{4+}, which was prepared in an extraordinary 70% yield from 1,4-bis(bromomethyl)benzene (**69**), the dication 70^{2+} and the macrocyclic polyether, bis-*p*-phenylene-34-crown-10 (**72**). The high yield can be explained from the mechanism of the catenane's formation, which is illustrated in Scheme 15.14. Most likely, one of the free nitrogen atoms in 70^{2+} first quaternizes upon reaction with the dibromide **69** to afford the π-electron-accepting tricationic intermediate 71^{3+}. This intermediate can thread its way through the cavity of the macrocyclic polyether **72** to form a [2]pseudorotaxane 73^{3+} (pre-catenane) which is stabilized by π–π stacking, CT interactions and electrostatic interactions between the 4,4'-bipyridinium moiety and the two hydroquinone units, and by hydrogen bonding between the α-H protons in the 4,4'-bipyridinium moiety and the oxygen atoms in the macrocyclic polyether **72**. The tricationic intermediate 73^{3+} is ideally disposed to form the [2]catenane 74^{4+} by an intramolecular nucleophilic attack of the residual free nitrogen atom on the remaining benzylic bromide carbon. This so-called clipping approach for the preparation of [2]catenanes is a good example of supramolecular assistance to covalent synthesis. This strategy can also be employed for the preparation of [2]rotaxanes and an illustrative example carried out by Becher and coworkers[39] is shown in Scheme 15.15. The dumbbell-shaped compound **75** – containing a bis(pyrrolo)TTF unit as the π-electron donor in its rod section – was used as the template for the production of the

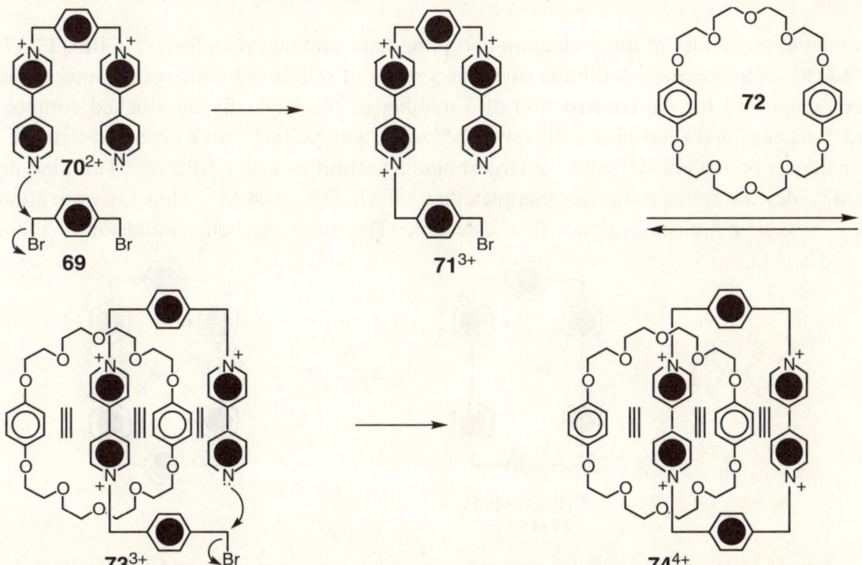

Scheme 15.14   Mechanism for the formation of the [2]catenane 74^{4+}.

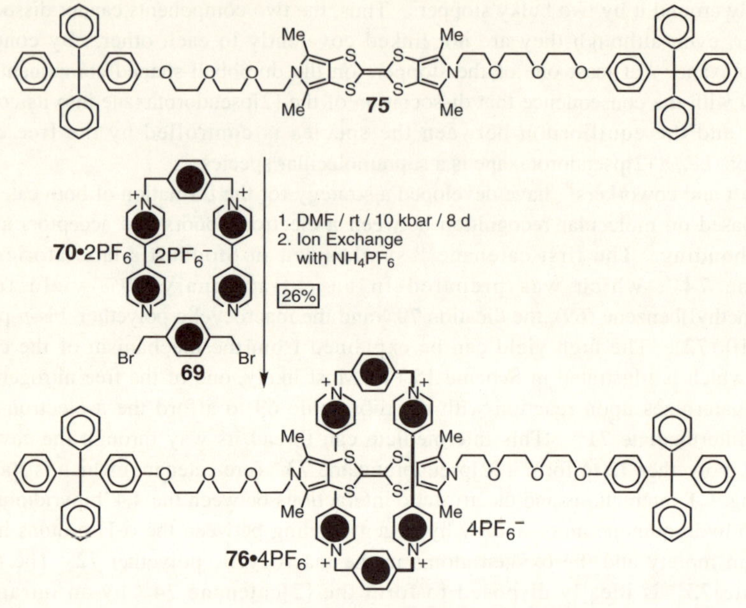

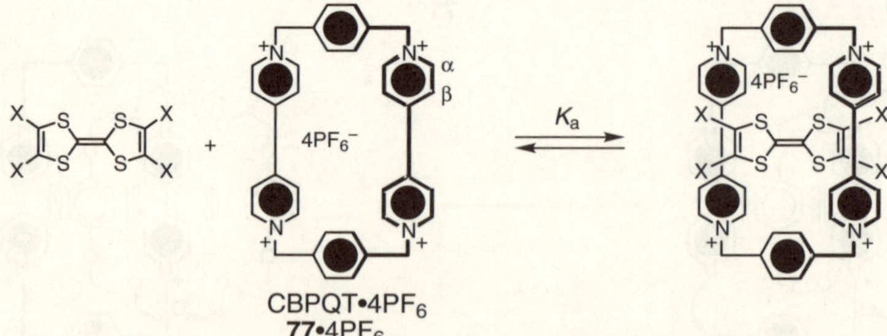

Scheme 15.15   Synthesis of the [2]rotaxane **76•4PF₆**.

cyclobis(paraquat-*p*-phenylene) tetracation (CBPQT⁴⁺) from the dicationic precursor **70•2PF₆** and the dibromide **69**, affording the [2]rotaxane **76•4PF₆** in 26% yield.

15.8.1 Binding Studies between TTF Derivatives and CBPQT⁴⁺

On account of the ability of the π-electron-accepting tetracationic cyclophane,[40] CBPQT⁴⁺ (**77⁴⁺**), to include π-electron donors within its cavity as a result of stabilizing noncovalent interactions, it has been employed for the construction of a number of mechanically interlocked compounds, such as catenanes and rotaxanes. However, Mirzoian and Kaifer[41] have observed that the best electron donors do not always exert the largest binding affinities with CBPQT⁴⁺. Thus, indole (K_a = 100 M⁻¹, Me₂CO) forms a stronger complex than catechol (K_a = 68 M⁻¹, Me₂CO), even although indole is a weaker π-electron donor than catechol. The relatively better inclusion of indole is

Scheme 15.16   Equilibrium for the formation of 1:1 complexes between CBPQT•4PF₆
(**77•4PF₆**) and TTF derivatives.

probably the consequence of its larger π-surface. Stoddart *et al.*, and later Bryce *et al.*, investigated[42] the green 1:1 complex formed between CBPQT[4+] and TTF (**1**) and found their association to be very strong indeed. A considerable range of catenanes, rotaxanes and pseudorotaxanes, incorporating different TTF units, and employing CBPQT[4+] as the encircling component, have been reported.[1,29] However, the binding affinities between differently substituted TTF derivatives and CBPQT[4+] have not been investigated in detail. Recently, some binding studies have been carried out to shed more light on the factors influencing the inclusion of different TTF derivatives inside the cavity of CBPQT[4+].[43]

The TTFs which were investigated are displayed in Fig. 15.13. All of these TTFs are devoid of polyether substituents capable of forming hydrogen bonds[44] with the α-bipyridinium hydrogen atoms on the tetracationic cyclophane CBPQT[4+]. Compounds **38a**, **42**, **78** and **79** were observed to be in slow exchange with their CBPQT[4+] complexes on the ^{1}H NMR timescale (250 MHz, 303 K), since both complexed and uncomplexed CBPQT[4+] resonances were present in the spectra. The cyclophane protons show significant shifts in their resonances upon complexation, making it possible to determine association constants (K_a) employing the ^{1}H NMR single-point method.[45] The K_a values and derived free energies for complexation ($-\Delta G°$) by CBPQT[4+]of the TTF derivatives **38a**, **42**, **78** and **79** are listed in Table 15.2. In contrast, TTF **1** and its derivative **43** were found to be in fast exchange with their CBPQT[4+] complexes on the ^{1}H NMR timescale (250 MHz, 303 K). Employing the UV–vis dilution method the association constants for the complexation of CBPQT[4+] with bis(pyrrolo)TTF **43** and its derivative **42** have been determined. The association constants (K_a) and derived $-\Delta G°$ values, obtained from the two UV–vis experiments, are recorded in Table 15.3. The K_a values listed in Table 15.2 reveal a significant trend. As the first redox potential ($E_{1/2}^1$) for the TTF derivatives decreases, the association constant (K_a) increases. Thus, the better the π-electron donor, the stronger the complex formed with CBPQT[4+]. However, the data recorded in Table 15.3 shows that the first redox potential ($E_{1/2}^1$) is not the only factor of importance. Note that, even though the bis(pyrrolo)TTF derivative **43** is a slightly weaker donor than TTF (**1**), it nevertheless exhibits a stronger association with

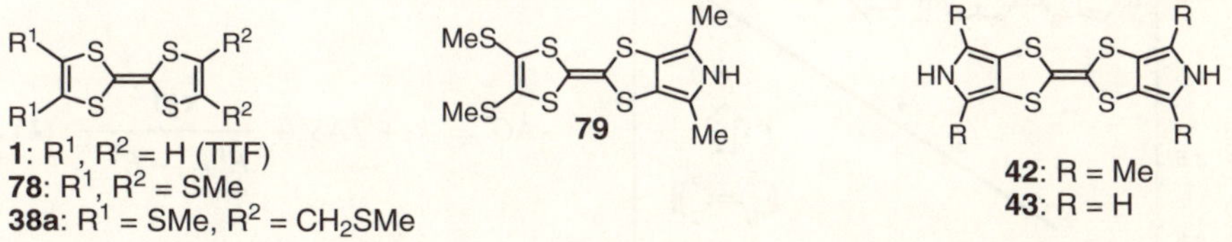

1: R^1, R^2 = H (TTF)
78: R^1, R^2 = SMe
38a: R^1 = SMe, R^2 = CH$_2$SMe

79

42: R = Me
43: R = H

Fig. 15.13   TTF (**1**) and the TTF derivatives **38a**, **42**, **43**, **78** and **79**.

Table 15.2   Comparison of Association Constants (K_a) and Derived Free Energies of Complexation ($-\Delta G°$) between Different TTFs and CBPQT[4+] Determined by the ^{1}H NMR Single-point Method with the First Redox Potential $E_{1/2}^1$ for Different TTFs. Temperature: 303 K unless Otherwise Stated

Compound	K_a (M^{-1})^a Me$_2$CO	$-\Delta G°$ (kcal mol^{-1}) Me$_2$CO	$E_{1/2}^1$ (V)^b MeCN
42	7900	5.4	+0.33
79	1000	4.2	+0.42
38a	490	3.7	+0.45
78	40	2.2	+0.51

^{a} Estimated error on K_a: ±10%. ^{b} Half-wave potentials $E_{1/2}^1$ (*vs.* Ag/AgCl) were obtained from CVs.

Table 15.3   Comparison of Association Constants (K_a) and Derived Free Energies of Complexation ($-\Delta G°$) between Different TTFs and CBPQT^{4+} Determined by UV–vis Spectroscopy with the First Redox Potential $E_{1/2}^1$ for Different TTFs. Temperature: 295 K unless Otherwise Stated

Compound	K_a (M^{-1})^a (Me$_2$CO)	$-\Delta G°$ (kcal mol^{-1}) (Me$_2$CO)	λ_{max} (nm) (Me$_2$CO)	$E_{1/2}^1$ (V)^b (MeCN)
42	18000	5.7	870	+0.33
43	12000	5.5	850	+0.38
1	2600^c	4.6	854	+0.34

^{a} Estimated error on K_a: ±15%. ^{b} Half-wave potentials $E_{1/2}^1$ (*vs.* Ag/AgCl) were obtained from CVs. ^{c} At 294 K, see ref. 42b.

CBPQT^{4+}. It seems that the extended π-surface of the bis(pyrrolo)TTF derivative **43** is exerting a stabilizing influence upon the complex as compared to the parent TTF (**1**). When four electron-donating methyl groups are attached to the pyrrole units, as in the TTF derivative **42**, the association constant increases even further. Thus, the thermodynamic data demonstrate clearly that both (i) the π-electron-donating properties (measured by the first redox potential $E_{1/2}^1$) and (ii) the area of the π-surface govern the strength of the complexation. It transpires that, for donor-acceptor interactions, the following relationship[46] is obeyed (*i.e.*, Equation 15.1, see Fig. 15.14), here k_1 and k_2 are constants and β is the overlap integral between donor and acceptor, while E_{don}^1 and E_{acc}^1 are the first redox potentials of the donor (TTF derivative) and acceptor (CBPQT^{4+}) entities. Fig. 15.14 shows a plot of $-\Delta G°$ in Me$_2$CO *vs.* the reciprocal difference $1/(E_{don}^1 - E_{acc}^1)$ in redox potentials for the TTFs **1**, **38a**, **42**, **43**, **78** and **79**, using for the cyclic CBPQT^{4+} acceptor[47] $E_{acc}^1 = -0.25$ V (*vs.* Ag/AgCl in MeCN). Deviations from a straight line originate mainly from different overlap integrals β. Note that the overlap is smaller for unsubstituted TTF (**1**), carrying neither sulfur-containing nor pyrrolo substituents, and larger for the bis(pyrrolo)TTF **43**.

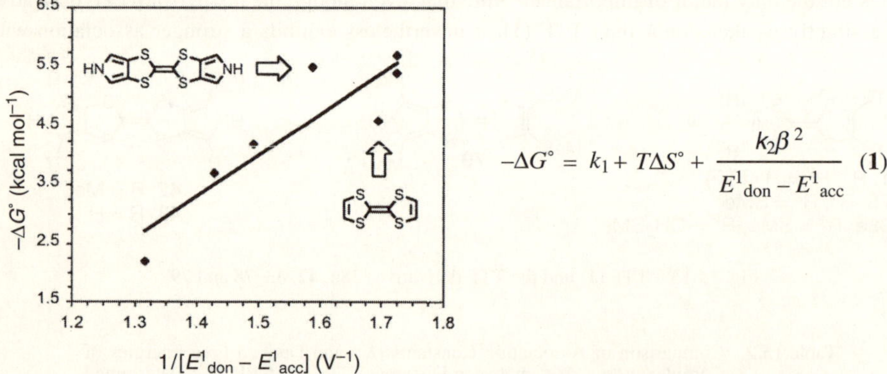

$$-\Delta G° = k_1 + T\Delta S° + \frac{k_2\beta^2}{E_{don}^1 - E_{acc}^1} \quad (1)$$

Fig. 15.14   Plot of $-\Delta G°$ in Me$_2$CO *vs.* the reciprocal difference in redox potentials between π-electron donor (TTF or its derivative) and π-electron acceptor (CBPQT^{4+}) units. The seven data points have been fitted by a best straight line.

Whether a TTF derivative and the cyclophane CBPQT^{4+} are in fast exchange or not with their complex on the ^{1}H NMR timescale is influenced largely by the bulkiness of the TTF derivative. Hence, when the TTF core is substituted with four bulky SMe groups, as in the case of **78** – or two SMe and two CH$_2$SMe groups, as in the case of **38a** – the kinetics for the complexation–decomplexation process is slow exchange on the 250 MHz timescale (303 K).

However, the bis(pyrrolo)TTF derivative **43** undergoes fast exchange with its $CBPQT^{4+}$ complex, but attachment of four Me substituents, as in the case of **42**, changes the kinetics into slow exchange on the 250 MHz timescale (303 K). Thus, the exchange rate is the result of a fine balance between electronic and steric factors.

In summary, three major conclusions can be drawn: (i) The stronger the donor, the higher the association constant (K_a). (ii) Extension of the π-system of the TTF derivative increases the extent of the association. (iii) The kinetics for the complexation–decomplexation are related to the bulkiness of the TTF derivative. These findings are of fundamental importance in designing (bistable) molecular switches[2,3c] based on catenanes, rotaxanes and pseudorotaxanes in which one of the π-electron-donating sites in one component is a derivatized TTF unit and the other recognition component is $CBPQT^{4+}$.

15.8.2 A Bistable [2]Rotaxane Incorporating a TTF Unit

Since the mechanically interlocked components of nondegenerate catenanes and rotaxanes can be induced to change their relative positions as a result of some well-chosen external stimulus, they are ideally suited for the construction of artificial molecular machines[2] and the fabrication of molecular electronic devices[2b,3] because large-amplitude motion can be envisaged within such architectures without the risk of damaging the chemical structure of the system. The relative movements of the interlocked components (Fig. 15.15) can be triggered by chemical, electrochemical and photochemical stimuli (input signal), forcing the molecule to switch between its two nondegenerate states (State 0 and State 1) that can be distinguished spectroscopically (output signal). Many catenanes and rotaxanes have been synthesized by template-directed methods and characterized in solution. However, it is the integration of these molecules into the device setting that has been receiving much attention over the last few years.[48] Among the desirable features for the redox-controllable amphiphilic [2]rotaxanes that have been employed to fabricate single-molecule-thick electrochemical junctions in electronic devices are (i) the siting of redox-active units along the rod section of the dumbbell component and (ii) the presence of both hydrophobic and hydrophilic groups as stoppers at the ends of the dumbbell component. TTF unique π-electron donor properties, together with its ability to form a strong green 1:1 complex with the tetracationic cyclophane $CBPQT^{4+}$ have been responsible for TTFs incorporation into a considerable range of catenanes and pseudorotaxanes. Although a TTF unit and a 1,5-

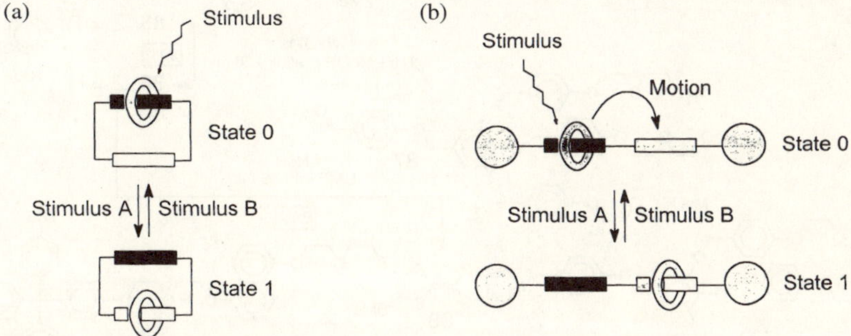

Fig. 15.15   Schematic representations of the mechanical movements relating to states (0 and 1) in nondegenerate (a) [2]catenanes and (b) [2]rotaxanes.

dioxynaphthalene (DNP) moiety have been incorporated[49] into the crown-ether ring component of a redox-switchable [2]catenane, which has already been employed in the fabrication of a solid-state electronically-reconfigurable switch,[3c] no rotaxanes employing these two recognition sites for a CBPQT[4+] component have been synthesized. In addition, rotaxanes incorporating TTF units in dumbbell components comprising two different stoppers were unknown, most likely because of the lack of an appropriate TTF building block. With the development of the new monopyrrolo-TTF building block **55** the synthesis of such an amphiphilic bistable [2]rotaxane has recently been accomplished.[50]

The huge effort to synthesize this kind of complicated molecule is well illustrated by the synthesis[50] of the amphiphilic bistable [2]rotaxane **89**•4PF$_6$, as shown in Schemes 15.17 and 15.18. A condensed version of the synthesis of the linear dumbbell component is outlined in Scheme 15.17. In order to complete the synthesis (Scheme 15.18) of the amphiphilic bistable [2]rotaxane **89**•4PF$_6$, the cyclophane CBPQT[4+] was introduced by a clipping reaction, which gave an analytically pure compound. The bistable [2]rotaxanes **89**•4PF$_6$ was isolated as a brown solid and ^{1}H NMR and UV–vis spectroscopies revealed the presence of both stable translational isomers, in an approximately ratio of 1:1 at ambient temperature. It turned out that the SMe group situated between the TTF and DNP recognition sites add a considerable activation barrier for the shuttling of the tetracationic cyclophane CBPQT[4+] between the two recognition sites. In fact, the steric hindrance exhibited from the SMe made it possible to isolate (Fig. 15.16a) the translational isomers **89**•4PF$_6$•RED and **89**•4PF$_6$•GREEN and to study the kinetics of the shuttling of CBPQT[4+] between the two recognition sites. The processes, which are accompanied by clearly

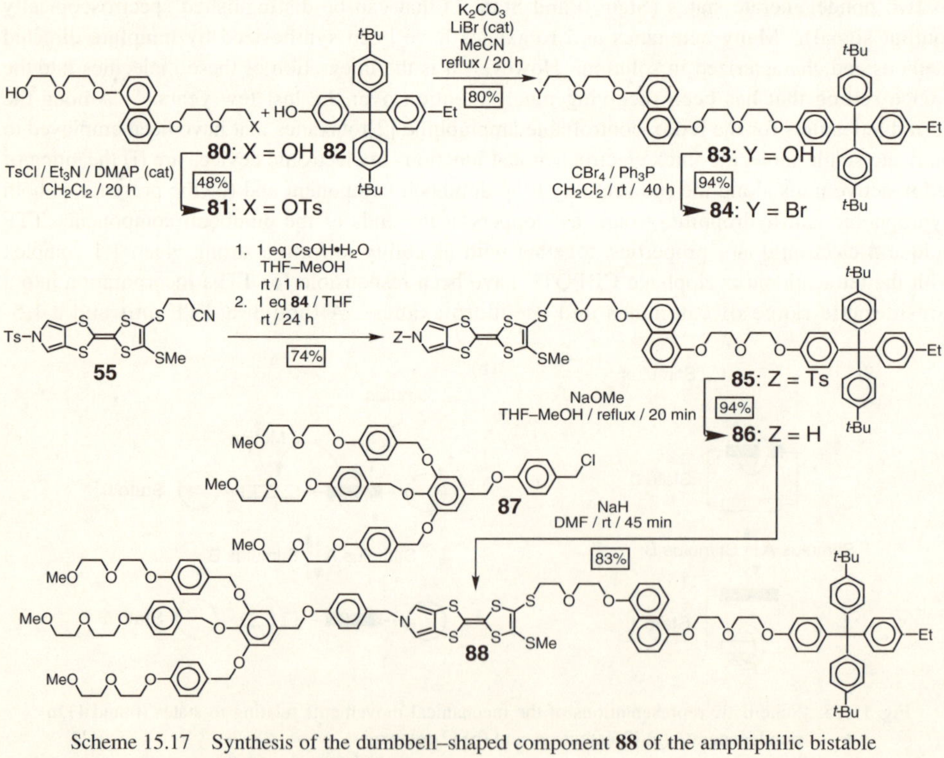

Scheme 15.17   Synthesis of the dumbbell–shaped component **88** of the amphiphilic bistable [2]rotaxane **89**•4PF$_6$.

detectable color changes, can be followed by ^{1}H NMR and UV–vis spectroscopies, allowing the rate constants (k_c) and the associated energies of activation ($\Delta G^\ddagger$) for both the shuttling

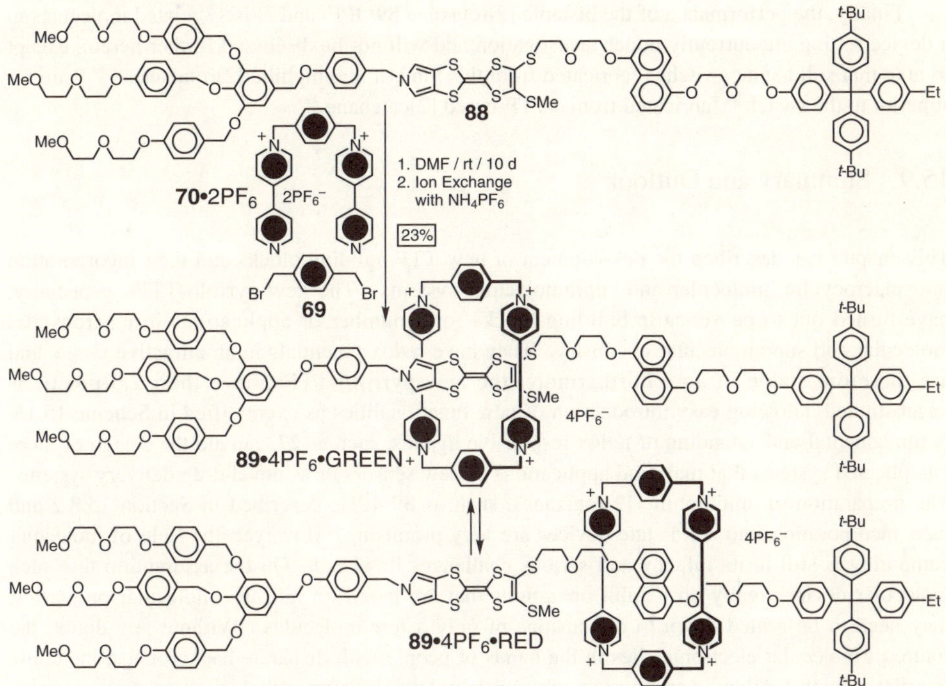

Scheme 15.18   The clipping reaction to give a mixture of translational isomers of the bistable [2]rotaxane **89•4PF₆**.

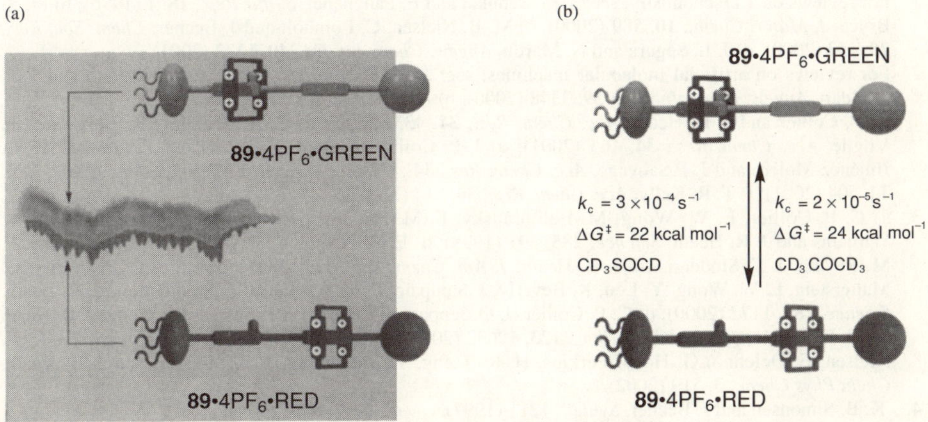

Fig. 15.16   (a) A preparative thin layer chromatogram, showing the separation of **89•4PF₆•RED** from **89•4PF₆•GREEN**. (b) Rate constants (k_c) and associated energies of activation ($\Delta G^\ddagger$) for the shuttling processes in the bistable [2]rotaxane **89•4PF₆**.

(Fig. 15.16b) of $CBPQT^{4+}$ from the DNP recognition site in **89•4PF$_6$•RED** to the TTF recognition site in **89•4PF$_6$•GREEN**, as well as the shuttling of $CBPQT^{4+}$ from the TTF recognition site in **89•4PF$_6$•GREEN** to the DNP recognition site in **89•4PF$_6$•RED**, to be determined.[50]

Finally, the performance of the bistable [2]rotaxane **89•4PF$_6$** and closely related molecules in a device setting are currently under investigation and will not be discussed further herein, except to note that solid-state switches fabricated from this kind of amphiphilic [2]rotaxanes[2b,3d-e] are far superior to the switch[3c] fabricated from a TTF-based [2]catenane.[49]

15.9   Summary and Outlook

This chapter has described the development of new TTF building blocks and their incorporation into macrocyclic, molecular and supramolecular systems. The new pyrrolo-TTFs, especially, have turned out to be versatile building blocks for a number of applications in macrocyclic, molecular and supramolecular chemistry. They have redox potentials in an attractive range, and are therefore stable in air. Furthermore, the new pyrrolo-TTFs, have their 2,5-positions unsubstituted, allowing easy introduction of new functionalities as exemplified in Scheme 15.13. A fundamental understanding of redox responsive ligands, such as **27**, can aid the design of more complicated systems that may find applications as new sensors or as novel drug delivery systems. The preparation of amphiphilic [2]rotaxanes, such as **89•4PF$_6$**, described in Section 15.8.2 and their incorporation into solid-state devices are very promising. However, the field of molecular computing is still in its infancy and serious challenges lie ahead. On the assumption that such molecular devices really do exhibit operations that are based on intrinsic molecular properties, they need to be scaled down to dimensions of only a few molecules. Without any doubt, the future of molecular electronics lies in the hands of people with disparate backgrounds, including chemists – both synthetic and physical, physicists, materials scientists, electronics engineers and, of course, computer scientists.

References

1. For reviews on TTF chemistry, see: a) G. Schukat and E. Fanghänel, *Sulfur Rep.*, **18**, 1 (1996); b) M. R. Bryce, *J. Mater. Chem.*, **10**, 589 (2000); c) M. B. Nielsen, C. Lomholt and J. Becher, *Chem. Soc. Rev.*, **29**, 153 (2000); d) J. L. Segura and N. Martín, *Angew. Chem. Int. Ed.*, **40**, 1372 (2001).
2. For reviews on artificial molecular machines, see: a) V. Balzani, A. Credi, F. M. Raymo and J. F. Stoddart, *Angew. Chem. Int. Ed.*, **39**, 3348 (2000); b) A. R. Pease, J. O. Jeppesen, J. F. Stoddart, Y. Luo, C. P. Collier and J. R. Heath, *Acc. Chem. Res.*, **34**, 433 (2001); c) C. A. Schalley, K. Beizai and F. Vögtle, *Acc. Chem. Res.*, **34**, 465 (2001); d) J.-P. Collin, C. Dietrich-Buchecker, P. Gavina, M. C. Jiménez-Molero and J.-P.Sauvage, *Acc. Chem. Res.*, **34**, 477 (2001); e) B. L. Feringa, *Acc. Chem. Res.*, **34**, 504 (2001); f) T. R. Kelly, *Acc. Chem. Res.*, **34**, 514 (2001).
3. a) C. P. Collier, E. W. Wong, M. Belohradsky, F. M. Raymo, J. F. Stoddart, P. J. Kuekes, R. S. Williams and J. R. Heath, *Science*, **285**, 391 (1999); b) E. W. Wong, C. P. Collier, M. Belohradsky, F. M. Raymo, J. F. Stoddart and J. R. Heath, *J. Am. Chem. Soc.*, **122**, 5831 (2000); c) C. P. Collier, G. Mattersteig, E. W. Wong, Y. Lou, K. Beverly, J. Sampaio, F. M. Raymo, J. F. Stoddart and J. R. Heath, *Science*, **289**, 1172 (2000); d) C. P. Collier, J. O. Jeppesen, Y. Luo, J. Perkins, E. W. Wong, J. R. Heath and J. F. Stoddart, *J. Am. Chem. Soc.*, **123**, 12632 (2001); e) Y. Luo, C. P. Collier, J. O. Jeppesen, K. A. Nielsen, E. Delonno, G. Ho, J. Perkins, H.-R. Tseng, T. Yamamoto, J. F. Stoddart and J. R. Heath, *Chem Phys Chem.*, **3**, 519 (2002).
4. K. B. Simonsen and J. Becher, *Synlett*, 1211 (1997).
5. J. Becher, Z.-T. Li, P. Blanchard, N. Svenstrup, J. Lau, M. B. Nielsen and P. Leriche, *Pure Appl. Chem.*, **69**, 465 (1997).
6. a) M. Narita and Jr. C. U. Pittman, *Synthesis*, 489 (1976); b) A. Krief, *Tetrahedron*, **42**, 1209 (1986).
7. a) J. Garin, J. Orduna, J. Uriel, A. J. Moore, M. R. Bryce, S. Wegener, D. S. Yufit and J. A. K. Howard, *Synthesis*, 489 (1994); b) N. Svenstrup, K. M. Rasmussen, T. K. Hansen and J. Becher, *Synthesis*, 809

(1994); c) K. B. Simonsen, N. Svenstrup, J. Lau, O. Simonsen, P. Mørk, G. J. Kristensen and J. Becher, *Synthesis*, 407 (1996).

8. a) *Cyclophanes I, Topics in Current Chemistry*, Springer, Berlin, (1983), Vol. 113; b) *Cyclophanes II, Topics in Current Chemistry*, Springer, Berlin (1983), Vol. 115; c) F. Vögtle, *Cyclophane Chemistry*, Wiley, Chichester (1993).

9. F. Diederich, *Cyclophanes, Monographs in Supramolecular Chemistry*, The Royal Society of Chemistry, London (1991).

10. C. Seel and F. Vögtle, *Angew. Chem. Int. Ed. Engl.*, **31**, 528 (1992).

11. P. L. Boulas, M. Gomez-Kaifer and L. Echegoyen, *Angew. Chem. Int. Ed.*, **37**, 216 (1998).

12. K. B. Simonsen, N. Svenstrup, J. Lau, N. Thorup and J. Becher, *Angew. Chem. Int. Ed.*, **38**, 1417(1999).

13. K. Takimiya, N. Thorup and J. Becher, *Chem. Eur. J.*, **6**, 1947 (2000).

14. a) G. Barany and R. B. Merrifield, *J. Am. Chem. Soc.*, **99**, 7363 (1977); b) G. Barany and F. J. Albericio, *J. Am. Chem. Soc.*, **107**, 4936 (1985).

15. H. Spanggaard, J. Prehn, M. B. Nielsen, E. Levillain, M. Allain and J. Becher, *J. Am. Chem. Soc.*, **122**, 9486 (2000).

16. J.-P. Sauvage and L. Raehm, *Struct. Bonding*, **99**, 55 (2001).

17. K. S. Bang, M. B. Nielsen, R. Zubarev and J. Becher, *Chem. Commun.*, 215 (2000).

18. a) F. Le Derf, M. Mazari, N. Mercier, E. Levillain, P. Richomme, J. Becher, J. Garín, J. Orduna, A. Gorgues and M. Sallé, *Inorg. Chem.*, **38**, 6096 (1999); b) H. L. Liu, S. Liu and L. Echegoyen, *Chem. Commun.*, 1493 (1999); c) M. R. Bryce, A. S. Batsanov, T. Finn, T. K. Hansen, J. A. K. Howard, M. Kamenjicki, I. K. Lednev and S. A. Asher, *Chem. Commun.*, 295 (2000); d) F. Le Derf, E. Levillain, G. Trippé, G. A. Gorgues, M. Sallé, R. M. Sebastian, A. M. Caminade and J. P. Majoral, *Angew. Chem. Int. Ed.*, **40**, 224 (2001); e) F. Le Derf, M. Mazari, N. Mercier, E. Levillain, G. Trippé, A. Riou, P. Richomme, J. Becher, J. Garín, J. Orduna, N. Gallego-Planas, A. Gorgues and M. Sallé, *Chem. Eur. J.*, **7**, 447 (2001).

19. T. W. Greene and P. G. M. Wuts, *Protective Groups in Organic Synthesis*, 3rd ed., Wiley, New York (1999).

20. J. O. Jeppesen, K. Takimiya, N. Thorup and J. Becher, *Synthesis*, 803 (1999).

21. J. O. Jeppesen, K. Takimiya and J. Becher, *Org. Lett.*, **2**, 2471 (2000).

22. A. Souizi, A. Robert, P. Batail and L. Ouahab, *J. Org. Chem.*, **52**, 1610 (1987).

23. Y. N. Kreitsberga, É. É. Liepin'sh, I. B. Mazheika and O. Y. Neilands, *Zh. Org. Khim.*, **22**, 416 (1986).

24. a) K. Zong, W. Chen, M. P. Cava and R. D. Rogers, *J. Org. Chem.*, **61**, 8117 (1996); b) K. Zong and M. P. Cava, *J. Org. Chem.*, **62**, 1903 (1997).

25. a) K. B. Simonsen, K. Zong, R. D. Rogers, M. P. Cava and J. Becher, *J. Org. Chem.*, **62**, 679 (1997); b) J. Lau, M. B. Nielsen, N. Thorup, M. P. Cava and J. Becher, *Eur. J. Org. Chem.*, 3335 (1999).

26. For a comprehensive review on pyrrole synthesis, see: C. W. Bird, *Comprehensive Heterocyclic Chemistry II*, Pergamon, Oxford (1996), Vol. 2, p. 1.

27. a) J. O. Jeppesen, K. Takimiya, F. Jensen and J. Becher, *Org. Lett.*, **1**, 1291 (1999); b) J. O. Jeppesen, K. Takimiya, F. Jensen, T. Brimert, K. Nielsen, N. Thorup and J. Becher, *J. Org. Chem.*, **65**, 5794 (2000).

28. T. Mori, A. Kobayashi, Y. Sasaki, G. Saito and H. Inokuchi, *Bull. Chem. Soc. Jpn.*, **57**, 627 (1984).

29. J. O. Jeppesen, J. Perkins, J. Becher and J. F. Stoddart, *Org. Lett.*, **2**, 3547 (2000).

30. T. Otsubo, Y. Aso and K. Takimiya, *Adv. Mater.*, **8**, 203 (1996).

31. In fact very few TTF-belts have been reported, see: a) M. Adam, V. Enkelmann, H.-J. Räder, J. Röhrich and K. Müllen, *Angew. Chem. Int. Ed. Engl.*, **31**, 309 (1992); b) K. Takimiya, K. Imamura, Y. Shibata, Y. Aso, F. Ogura and T. Otsubo, *J. Org. Chem.*, **62**, 5567 (1997); c) Ref.12 ; d) Ref.15.

32. K. Nielsen, J. O. Jeppesen, N. Thorup and J. Becher, *Org. Lett.*, **4**, 1327 (2002).

33. J. N. H. Reek, A. E. Rowan, R. de Gelder, P. T. Beurskens, M. J. Crossley, S. De Feuter, F. de Scryver and R. J. M. Nolte, *Angew. Chem. Int. Ed. Engl.*, **36**, 361 (1997).

34. *The Porphyrin Handbook*, Academic Press, San Diego (2000), Vol. 1.

35. J. Becher, T. Brimert, J. O. Jeppesen, J. Z. Pedersen, R. Zubarev, T. Bjørnholm, N. Reitzel, T. R. Jensen, K. Kjaer and E. Levillain, *Angew. Chem. Int. Ed.*, **41**, 2497 (2001).

36. F. Wudl, G. M. Smith and E. J. Hufnagel, *J. Chem. Soc. Chem. Commun.*, 1453 (1970).

37. a) *Molecular Catenanes, Rotaxanes and Knots*, VCH –Wiley, Weinheim (1999); b) L. Raehm, D. G. Hamilton and J. K. M. Sanders, *Synlett*, 1743 (2002).

38. P. R. Ashton, T. T. Goodnow, A. E. Kaifer, M. V. Reddington, A. M. Z. Slawin, N. Spencer, J. F. Stoddart, C. Vicent and D. J. Williams, *Angew. Chem. Int. Ed. Engl.*, **28**, 1396 (1989).

39. D. Jensen, M. B. Nielsen, J. Lau, K. B. Jensen, R. Zubarev, E. Levillain and J. Becher, *J. Mater. Chem.*, **10**, 2249 (2000).

40. M. Asakawa, W. Dehaen, G. L'abbé, S. Menzer, J. Nouwen, F. M. Raymo, J. F. Stoddart and D. J.

Williams, *J. Org. Chem.*, **61**, 9591 (1996).

41. A. Mirzoian and A. E. Kaifer, *J. Org. Chem.*, **60**, 8093 (1995).

42. a) D. Philp, A. M. Z. Slawin, N. Spencer, J. F. Stoddart and D. J. Williams, *J. Chem. Soc. Chem. Commun.*, 1584 (1991); b) W. Devonport, M. A. Blower, M. R. Bryce and L. M. Goldenberg, *J. Org. Chem.*, **62**, 885 (1997); c) P. R. Ashton, V. Balzani, J. Becher, A. Credi, M. C. T. Fyfe, G. Mattersteig, S. Menzer, M. B. Nielsen, F. M. Raymo, J. F. Stoddart, M. Venturi and D. J. Williams, *J. Am. Chem. Soc.*, **121**, 3951 (1999).

43. M. B. Nielsen, J. O. Jeppesen, J. Lau, C. Lomholt, D. Damgaard, J. P. Jacobsen, J. Becher and J. F. Stoddart, *J. Org. Chem.*, **66**, 3559 (2001).

44. K. N. Houk, S. Menzer, S. P. Newton, F. M. Raymo, J. F. Stoddart and D. J. Williams, *J. Am. Chem. Soc.*, **121**, 1479 (1999).

45. K. A. Connors, *Binding Constants*, Wiley Interscience, New York (1987).

46. *Donor-Acceptor Bond*, Koterpress Enterprises, Jerusalem, Israel (1975).

47. D. B. Amabilino, P.-L. Anelli, P. R. Ashton, G. R. Brown, E. Córdova, L. A. Godínez, W. Hayes, A. E. Kaifer, D. Philp, A. M. Z. Slawin, N. Spencer, J. F. Stoddart, M. S. Tolley and D. J. Williams, *J. Am. Chem. Soc.*, **117**, 11142 (1995).

48. a) R. F. Service, *Science*, **291**, 426 (2001); b) R. F. Service, *Science*, **294**, 2442 (2001); c) M. Jacoby, *Chem. Eng. News*, **80** (39), 38 (2002).

49. V. Balzani, A. Credi, G. Mattersteig, O. A. Matthews, F. M. Raymo, J. F. Stoddart, M. Venturi, A. J. P. White and D. J. Williams, *J. Org. Chem.*, **65**, 1924 (2000).

50. J. O. Jeppesen, J. Perkins, J. Becher and J. F. Stoddart, *Angew. Chem. Int. Ed.*, **40**, 1216 (2001).

16

TTF-acceptor Type Molecules

16.1 Introduction

The electronic interaction between electron donor (D) and acceptor (A) molecules is an important topic in chemistry which has been studied in different fields in the search for a variety of nonconventional properties. The mode in which two electroactive species interact strongly depends on the type of connectivity between them (Fig. 16.1). Thus, through-space interactions between segregated stacks of D and A molecules are the base for the generation of salts and charge transfer (CT) complexes exhibiting electrical properties in which a partial degree of electron transfer takes place from the D to the A unit.[1] The CT mechanism can be regarded as a kind of coordination bond between both D and A species. Part of the valence electrons flow from the highest occupied molecular orbital (HOMO) of the D molecule to the lowest unoccupied molecular orbital (LUMO) of the A molecule, resulting in an electron delocalization over the D-A pair. This mechanism lowers the total energy, which becomes the origin of the CT interaction. Thus, the CT force not only contributes to form the supramolecular assembly, but also defines the electronic properties of the resulting electric-conducting materials.[1] This important topic that has been studied during the last three decades is considered in Part I of the present volume where the latest results have been comprehensively presented.

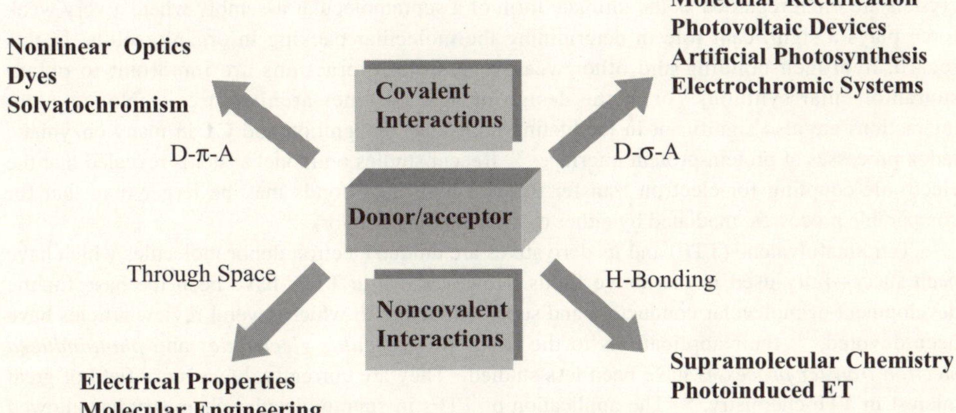

Fig. 16.1

The linkage between both D and A units can take place through a covalent saturated ("sigma") bridge, giving rise to a D-σ-A system. This type of molecules has received a great deal of attention in the chemical literatures since 1974, when Aviram and Ratner proposed that a single molecule with this constitution could act as a rectifier of electrical current.[2] Currently, the concept of *molecular electronics* is being pursued as one of the most outstanding and realistic applications of molecular materials in materials science.[3]

D-σ-A molecules are also of interest in other fields involving excited states in which a photoinduced electron transfer from the D to the A moiety occurs. These assemblies have been recently used in the preparation of artificial photosynthetic systems and photovoltaic devices since, under the appropriate conditions, they are able to mimic the photosynthetic solar energy transduction. In this regard, different approaches have been developed for the application of the basic physical and chemical principles of photosynthesis to the preparation of artificial systems.[4] Artificial molecular constructions are of interest for the experimental verification of electron transfer theories as well as for the understanding of the primary processes involved in the photosynthesis of plants and bacteria. These molecular structures also have potential uses in molecular-scale optoelectronics, photonics, sensor design and other aspects of nanotechnology.

In addition, both D and A moieties can be covalently attached by means of a π-conjugated bridge, giving rise to D-π-A organic materials exhibiting nonlinear optical (NLO) properties. D-A substituted organic molecules with large second-order NLO properties have been the subject of considerable research efforts due to their potential applications in areas such as optical modulation, molecular switching, optical memory and frequency doubling.[5] Although organic molecules are not as robust as inorganics, they have received a great deal of attention in the NLO field given that they offer some advantages over traditional inorganic crystals. Thus, i) organic materials present a high electronic susceptibility ($\chi^{(2)}$) through a high molecular hyperpolarizability (β) and fast response time; ii) they are cheaper to produce and iii) easier to fabricate; iv) they are compatible with existing semiconductor technology and v) their structures can be tailored in different ways allowing for the fine tuning of NLO properties for desired applications.

Finally, noncovalent interactions between D and A moieties are increasingly being used in the molecular self-assembly of well-defined supramolecular structures and materials.[6] Molecular crystals are often regarded as the ultimate form of a supramolecular assembly where a very weak force plays a significant role in determining the molecular packing in organic solids. In this regard, hydrogen bonding and other weak reversible interactions are important to obtain supramolecular synthons[7] or in the design of new polymer architectures.[8] Noncovalent interactions are also significant in facilitating molecular recognition and CT in many enzymatic redox processes at protein-protein interfaces.[9] Recent studies on model systems revealed that the electronic coupling for electron transfer through hydrogen bonds may be larger than that for comparable processes, mediated by either σ- or π-bonding networks.[10]

Tetrathiafulvalene (TTF) and its derivatives are unique electron donor molecules which have been successfully used in the above fields. However, while TTFs have been the base for the development of molecular conductors and superconductors to which several review articles have been devoted,[11] their applications to the fields of *molecular electronics* and *photoinduced electron transfer processes* have been less studied. They are currently, however, a field of great interest in TTF chemistry.[12] The application of TTFs in supramolecular chemistry has allowed the preparation of a number of molecular systems based on host-guest interactions being able to act as sensors, molecular switches or mediators for specific reactions. Special attention has been devoted to interlocked architectures such as rotaxanes and catenanes, which have been elegantly collected in a recent review by Becher[13] as well as discussed in Chapter 15 of this volume by the same author.

Therefore, the aim of this chapter is to update the most recent and relevant aspects of TTF-acceptor type molecules, focusing our efforts on those new applications of TTFs different from extensively studied electrical properties. Thus, this chapter concentrates on the use of TTFs in *molecular electronics* and *photoinduced electron transfer processes* directed to the study of optoelectronic properties as well as on the recent applications in the search for NLO properties.

16.2   Nonconventional Applications of TTF

TTF (**1**) is a nonaromatic 14 π-electron system in which the oxidation to the cation radical and dication occurs sequentially and reversibly (Scheme 16.1) at relatively low potentials ($E^1_{1/2} = 0.37$ V and $E^2_{1/2} = 0.67$ V in dichloromethane *vs.* SCE for the unsubstituted TTF). Because of the 6π-electron heteroaromaticity of the 1,3-dithiolium cation, and in contrast to the neutral TTF, both cation radical and dication species are aromatic in the Hückel sense and, therefore, while TTF^{+} and TTF $^{2+}$ have a planar D_{2h} symmetry, neutral TTF has a boatlike C_{2v} symmetry equilibrium structure.[14] However, because of the small difference between the planar and boatlike conformations, TTF is very flexible and can appear in various conformations depending on the D-D and D-A interactions in the crystals.

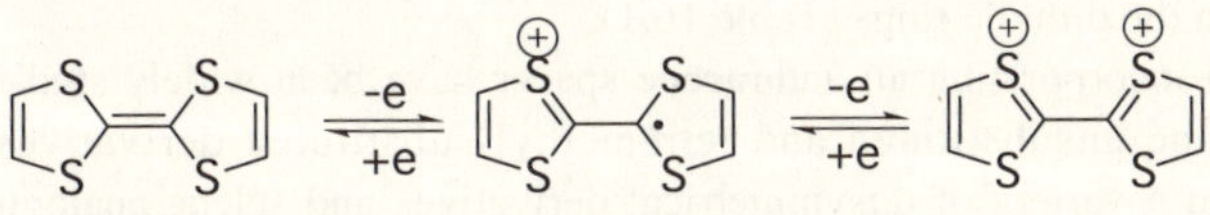

TTF, 1   -e / +e   -e / +e

Scheme 16.1

TTF, which was first reported by Wudl in 1970,[15] formed the first paradigmatic organic metal tetrathiafulvalene-tetracyano-*p*-quinodimethane (TTF-TCNQ).[16] The first molecule-based superconductor based on the Bechgaard salts (TMTSF)$_2$X (X = PF$_6^-$, AsF$_6^-$)[17] was prepared in 1979 by using a selenium-containing TTF analogue, namely tetramethyltetraselenafulvalene (TMTSF). Since then a huge amount of work was done to enhance the electron-donating power of TTF analogues in order to improve the conductivitiy of salts and CT complexes derived from them. However, just over the past few years, the utility of TTF derivatives (i) as building blocks in macromolecular and supramolecular structures, (ii) as molecule-based ferromagnetic compounds, (iii) as synthetic intermediates in organic chemistry, (iv) as a donor moiety in intramolecular D-A systems including NLO materials and the important association with the fullerene core and (v) in the preparation of liquid crystalline materials and Langmuir-Blodgett (LB) films, has made TTF one of the most extensively studied molecules.[12]

Recently, an appealing review article on TTFs for intramolecular CT (ICT) materials has been published by Bryce[11q] in which the different TTF-σ-acceptor systems have been organized according to the chemical structure of the acceptor moiety linked to the TTF unit.

In the following sections we will outline the main research lines towards the development of new TTF-acceptor systems in the search for new nonconventional properties. Aspects such as molecular rectification, chromophores for dyes and nonlinear optics and excited-state energy and electron transfer processes for application in artificial photosynthetic systems and photovoltaic devices will be discussed.

16.3   Polarized TTF Molecules

Polarized TTFs can be produced by the presence of polarizable substituents or electron-accepting units on the TTF molecule. The latter type of compounds form the TTF-acceptor systems, which will be discussed in detail below. Moreover, polarized TTFs present several advantages to be used in the preparation of organic conductors since the intermolecular interactions can be enhanced due to the presence of electrostatic interactions, thus increasing the dimensionality in the solid state. In addition, electron-withdrawing groups on the TTF molecule can stabilize unstable strong electron donors.

Recently, Yamashita[18] has reviewed this topic in an article devoted to highly polarized electron donors, acceptors and D-A compounds for the preparation of organic conductors. The reader is referred to this article for additional information.

We have reported that highly π-extended TTF analogues with a quinonoid structure are polarized molecules which exhibit an ICT from the 1,3-dithiol-2-ylidene moiety to the fused hydrocarbon skeleton.[19]

As an alternative to the TTF donors, a large number of p-quinodimethane analogues of TTF, in which the central ring spacer of TTF is substituted by a π-conjugated polycyclic structure (**2**), have been prepared. We focus on π-extended TTFs such as **3**, which introduce a polyacenic unit as a spacer between the dithiole rings (Table 16.1).

Compounds **3** incorporating an anthracene spacer have been widely studied. Bryce *et al.*[20] first synthesized the unsubstituted and tetramethyl-substituted derivatives **3a** and **3b** and afterwards prepared a variety of unsymmetrical derivatives and selena-analogues. Yamashita *et al.*[21] reported independently the synthesis of **3a** and some derivatives and analogues.

The solution electrochemistry reveals that compounds **3** undergo a two-electron oxidation, which is observed by cyclic voltammetry (CV) as a single, quasi-reversible redox wave.[20,21] The coalescence of the two one-electron processes characteristic of TTF ($E^1_{1/2}$ = +0.34 V; $E^2_{1/2}$ = +0.71 V; *vs.* Ag/AgCl in CH_3CN) under a single oxidation wave is taken as an indication of the decrease of the intramolecular Coulombic repulsion in the doubly ionized state due to the introduction of the anthracene spacer. Compounds **3** therefore yield dication species upon oxidation, the intermediate cation radical not being observed even at -70 °C.[22]

The π-radical cation as well as the dication species of a series of quinonoid π-extended TTF (**3-5a-c**) have recently been generated by means of time-resolved and steady-state radiation chemical techniques. The inaccessibility of the π-extended TTFs radical cation by, for example, chemical oxidation techniques or conventional CV prompted us to probe the selective one-electron oxidation by radiation chemical methods. The spectral characteristics of the π-radical cation revealed fingerprint absorptions, which are predominantly in the red region of the spectrum (570–690 nm). The maxima vary markedly depending upon the substitution pattern of the 1,3-dithiole rings and the different backbone structures (anthracene, tetracene and pentacene). This first experimental evidence confirms the existence of the π-radical cation of the extended TTF derivatives.[23]

The oxidation potentials measured for **3** (**3a**: E^1_{ap} = +0.40 V; **3b**: E^1_{ap} = +0.42 V; anodic peak values *vs.* Ag/AgCl in CH_3CN)[20e] are slightly higher than the first oxidation potential reported for TTF ($E^1_{1/2}$ = +0.34 V), indicating that **3** are slightly poorer donors than TTF. The π-extension of the quinonoid spacer in passing from the parent p-quinodimethane derivative **2a** to the anthracene derivatives **3** therefore implies the loss of the exceptional donor ability of **2a** ($E^1_{1/2}$ = -0.11 V; *vs.*

Table 16.1 Cyclic Voltammetry and UV-vis Data for π-Extended p-Quinodimethane Analogues of TTF

Compound	Formula	$E^1_{a.p.}{}^a$	$E^2_{a.p.}{}^a$	λ_{max} (nm)^b
TTF		0.37	0.67	317^c
2a		-0.11^d	-0.04^d	495^d
3a		0.44	—	415
4a: R = H		0.39	1.47	417
4b: R = SMe		0.55	1.47	422
4c: R,R = (SCH$_2$)$_2$		0.52	1.45	434
5a: R = H		0.50	1.48	404
5b: R = SMe		0.63	1.41	406
5c: R,R = (SCH$_2$)$_2$		0.65	1.41	420
6a: R = H		0.49	1.31	458
6b: R = SMe		0.59	—	460
6c: R,R = (SCH$_2$)$_2$		0.59	—	462

^{a} Experimental conditions: CH$_2$Cl$_2$, V $vs.$ SCE, glassy carbon electrode (GCE) as a working electrode, NBu$_4{}^+$ ClO$_4{}^-$ as a supporting electrolyte, scan rate: 200 mV s^{-1}. ^{b} in CH$_2$Cl$_2$. ^{c} D. L. Coffen, J. Q. Chambers, D. R. Williams, P. E. Garrett and N. D. Canfield, *J. Am. Chem. Soc.*, **93**, 2258 (1971). ^{d} Ref. 21.

SCE in CH$_3$CN).[21]

Bryce *et al.* have determined the X-ray crystal structure of compound **3b**[24] and of its 2,3-dipentyl derivative.[22] Both molecules exhibit a butterfly-shaped structure with the central quinonoid ring severely distorted into a boat form. Although the lack of planarity usually precludes the obtainment of organic conductors, compounds **3** form CT complexes with high conductivities. The tetramethyl derivative **3b** yields a 1:4 complex with TCNQ with a room temperature conductivity (σ_{rt}) of 60 S cm^{-1}, which varies only slightly upon cooling to 90 K and then decreases sharply.[24] As expected from the electrochemical properties, the donor is present as the dication in the complex $3b^{2+}(TCNQ)_4^{2-}$. The dication adopts a conformation completely different from that of the neutral species, with the two dithiole rings nearly orthogonal to the planar anthracene group. The pathway for conduction is provided by plane-to-plane stacks of TCNQ molecules with a uniform intermolecular spacing of 3.36 Å.[24]

For a deeper understanding of the electrochemical behavior and structural properties of compounds **3**, the unsubstituted derivative **3a** was theoretically investigated using different methodological approaches.[25] The minimum-energy conformation of **3a** corresponds to a butterfly-shaped nonplanar structure (C_{2v} symmetry). The planar conformation is strongly hindered by the very short contacts between the sulfur atoms and the hydrogen atoms in *peri*-positions, separated by only 2.02 Å. To avoid these interactions, the central ring folds into a boat conformation and the molecule adopts a butterfly structure where the lateral benzene rings point upward and the dithiole rings point downward. The resulting conformation corresponds exactly to that observed experimentally for **3b**.[24]

The loss of the quinonoid character together with the folding of the central ring have important consequences on the electronic properties. As can be seen in Fig. 16.2, the HOMO of **3a** (-6.89 eV) is stabilized with respect to the HOMO of **2a** (-6.05 eV) and lies slightly below the HOMO of TTF (-6.81 eV). These MO energies provide a first-approach rationalization of the relative donor abilities observed experimentally. Within a MO picture, oxidation implies the extraction of an electron from the HOMO, and more positive oxidation potentials are expected for compounds with lower energy HOMOs. The HOMO energies thus justify the drastic reduction of the donor ability in passing from **2a** ($E^1_{1/2}$ = -0.11 V) to the more extended **3a** (E^1_{ap} = +0.40 V) and the slightly poorer donor capacity of **3a** compared to TTF ($E^1_{1/2}$ = +0.34 V).

Much less is known about *p*-quinodimethane analogues of TTF incorporating more extended polyacenic units as π-conjugated spacers between the dithiole rings. Compounds **4-6** have been recently synthesized by Wittig-Horner reaction of carbanions of adequately substituted 2-dimethylphosphono-1,3-dithioles with the respective acenequinones[25] (Table 16.1). Compound **4a** and its tetramethyl derivative were previously prepared by Bryce *et al.*[20]

The UV-visible spectra of these largely π-extended donors present interesting features (see

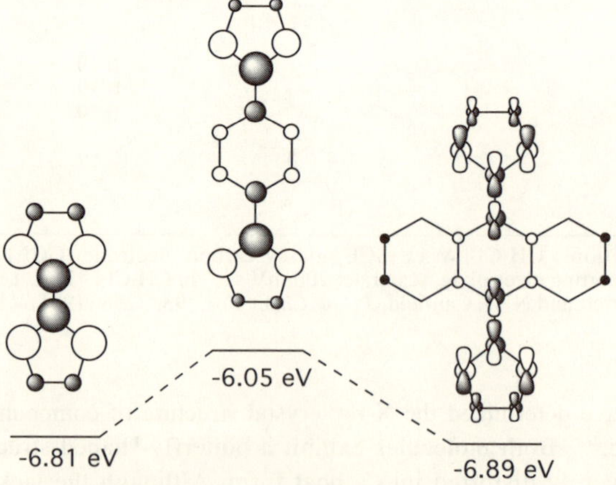

Fig. 16.2

Table 16.1). On the one hand, the lowest energy absorption band appearing above 400 nm is bathochromically shifted in comparison to TTF, but it is blue-shifted with respect to **2a**. These shifts indicate that part of the π-conjugation gained by **2** with respect to TTF is lost for compounds **4-6**. On the other hand, the position of λ_{max} depends on the number of benzene rings fused on each side of the *p*-quinodimethane system. λ_{max} for donors **6a-c** is red-shifted compared to compounds **3**, **4** and **5**.

The loss of π-conjugation is determined by the nonplanar structures of these compounds, which adopt a butterfly-shaped conformation identical to that found for **3**. It is important to note that, as discussed for **3a**, the laterally fused benzene, naphthalene and/or anthracene units preserve their structural identity. This means that, from a structural standpoint, compounds **4-6** must be visualized as formed by polyaromatic units (benzene, naphthalene and anthracene) linked together by 1,3-dithiol-2-ylidene units.

These structural findings have important consequences on the optical properties. The stabilization of the HOMO together with the destabilization of the LUMO determine that the HOMO-LUMO energy gap increases in passing from **2** to **3** and to **4-6**. This explains the blue shift for the first absorption in the UV-visible spectra (see Table 16.1). For the more extended systems **4-6** a new orbital located on the lateral naphthalene (**4**, **5**) or anthracene (**6**) units appears close in energy to the LUMO. This orbital becomes the LUMO for **4** and reduces the HOMO-LUMO energy gap to 2.59 eV (479 nm) compared with the gap of 3.00 eV (413 nm) obtained for **3**. This explains the red shift observed for the first absorption band of **4a** (458 nm), which must be assigned to an ICT band. Lateral benzoannulation therefore induces the appearance of a photoinduced intramolecular electron transfer in compounds **6** from the donor 1,3-dithiol-2-ylidene moieties (HOMO) to the anthracene fragment acting as the acceptor part of the molecule (LUMO). This band also exists for **4** and **5** but it cannot be clearly differentiated.

In conclusion, lateral benzoannulation of the parent *p*-quinodimethane analogue **2** determines the loss of planarity of the molecule. This precludes the formation of radical cations and very stable, highly aromatic dications are obtained upon oxidation. Largely extended systems present a photoinduced intramolecular electron transfer between the 1,3-dithiol-2-ylidene moieties, acting as donors, and the laterally fused polyacenic units, acting as acceptors.

16.4 TTF-σ-acceptor Molecules: Molecular Rectification

Aviram and Ratner proposed, in 1974, the first example of rectification of electrical current through a single molecule D-σ-A, where a strong donor moiety is covalently attached to a strong acceptor moiety through a covalent, saturated σ bridge.[2] As a prototype material, molecule **7** was chosen, given the excellent donor and acceptor properties of TTF and TCNQ, respectively. In order to achieve molecular rectification the molecule should exhibit anisotropic electrical properties. Thus, these materials should be aligned uniformly between two electrodes (M_1 and M_2) so they can allow electron flow from the cathode, M_2, to the acceptor end, from there to the donor through the σ bridge *via* electron tunneling, and then from the donor terminus to the anode, M_1. In Fig. 16.3 shows how such a device is asymmetric, because the HOMO of D is relatively low while the LUMO of A is relatively high. The device will work if the inelastic through-bond tunneling is more likely than the elastic through-space tunneling, which is unaffected by molecular orbitals.[26]

The groups of Panetta and Metzger have over the last two decades conducted a great deal of work in order to bring this theoretical project into reality.[27] However, to the best of our

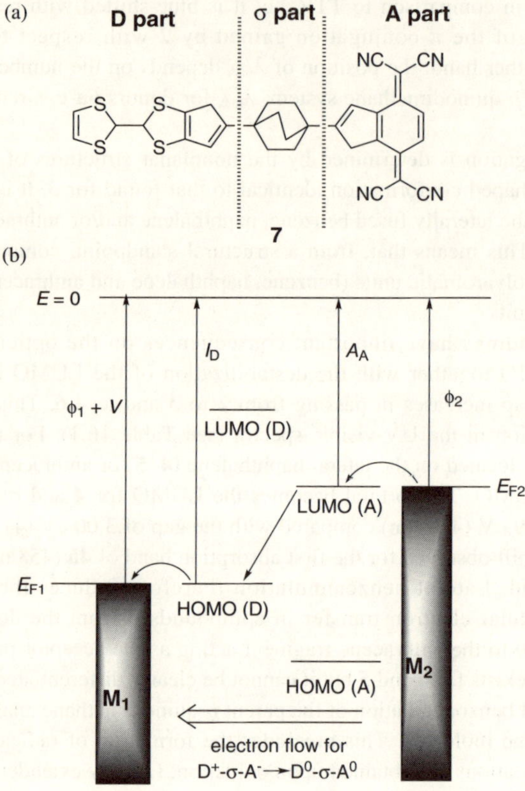

Fig. 16.3

knowledge, D-σ-A structure **7** has not yet been synthesized. They have developed a systematic synthetic plan aimed to the preparation of a series of model compounds in which different D and A moieties as well as σ bridges are incorporated. This effort has enabled the synthesis of different D-σ-A molecules and several ground state zwitterions, D^+-π-A^-, one of which became the first confirmed unimolecular rectifier.[28] Fig. 16.4 shows selected examples of TTF containing D-σ-A molecules proposed as molecular rectifiers.

The first two D-σ-A products prepared (**8, 9**) containing the strong donor TTF and the strong acceptor TCNQ were metastable materials; they were amorphous, difficult to purify and readily decomposed. The proton NMR studies of these products resulted in spectra in which many of the resonances were extremely broad, perhaps indicative of paramagnetic species or CT complexes.[29] A ground state biradical character of these compounds is also suggested by the intense broad signal observed in the ESR spectrum of a powdered sample of **9** carried out at 300 K. In order to avoid these problems, other weaker donor moieties (dialkylamine, pyrene, anthracene) were covalently attached through different bridges (ester, carbamate) to the TCNQ moiety, yielding stable D-σ-A systems some of them exhibiting rectification.[27] On the other hand, TTF has been incorporated in stable D-σ-A systems by replacing the strong TCNQ acceptor moiety by other

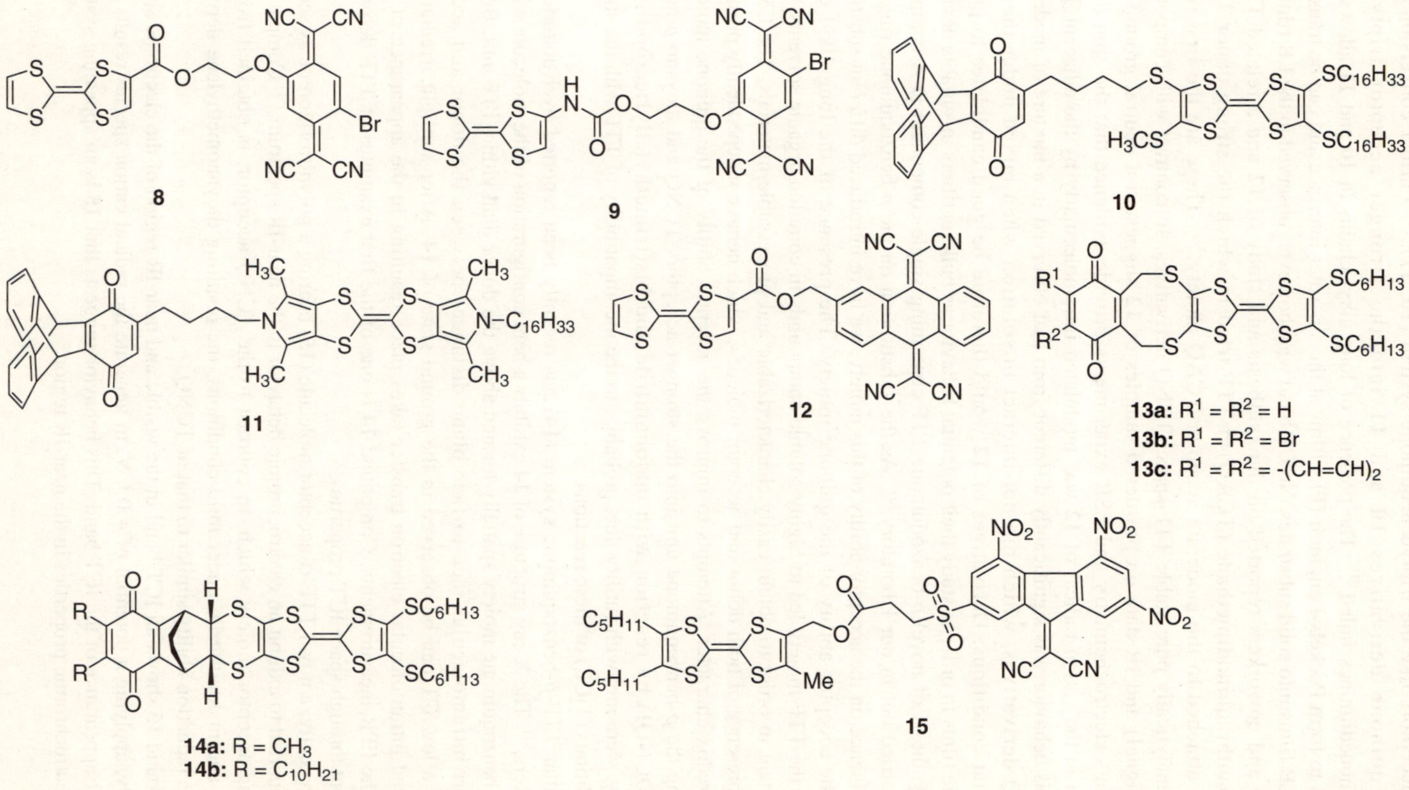

8
9
10
11
12
13a: R¹ = R² = H
13b: R¹ = R² = Br
13c: R¹ = R² = -(CH=CH)₂
14a: R = CH₃
14b: R = C₁₀H₂₁
15

Fig. 16.4

moderate acceptors like the triptycenequinone system (**10**, **11**).[30] Further conversion of the triptycenequinone derivatives **10** and **11** into the stronger acceptor triptycene-dicyanoquinonediimines failed.[31] The presence of long alkyl chains in **10** and **11** allows these compounds to form Pockels-Langmuir (PL) films at the air-water interface and can be transferred mostly as LB films into solid substrates. No CT band was, however, observed in the LB films.

Bryce and coworkers reported on the synthesis and study of **12** and a related TTF-σ-tetracyanoanthraquinodimethane (TCAQ)-σ-TTF triad in which the strong donor TTF is covalently attached to the moderate acceptor TCAQ moiety.[32] These are the first readily available, analytically pure stable TTF-spacer-TCNQ derivatives. In contrast with compound **9**, ESR (no signal) and IR data on powdered samples of **12** suggested a neutral ground state. Simultaneous electrochemistry and ESR experiments provide evidence that the spin density distribution in the cation radical of **12** was modulated intramolecularly by the adjacent TCAQ moiety. This behavior was significantly different from that observed in a mixture of model TTF and TCAQ derivatives, which did not interact in solution when mixed under the same experimental conditions. Derivatives of **12** would therefore be good candidates to prepare macroscopic films in order to study their potential behavior as bulk rectifiers in suitable devices.

The synthesis of novel *p*-benzoquinone-TTF containing single-component D Λ compounds (**13**) was carried out in our laboratory.[33] As the substitution on the *p*-benzoquinone ring has a striking influence on the acceptor ability of this moiety, we have introduced different substituents to adjust the acceptor ability of the quinone moiety. The presence of the long alkyl chains attached to the TTF moiety led to highly soluble compounds in common organic solvents; this is very important in order to obtain easily characterizable and processable materials. The CV data reveal the presence of both donor and acceptor moieties, which behave independently preserving their own redox character. Attempts to improve the acceptor ability of the quinone moiety by transforming the *p*-benzoquinone unit into the stronger acceptors TCNQ and dicyano-*p*-quinon-diimine (DCNQI) by reaction with malononitrile and bis(trimethylsilyl)carbodiimide, respectively, were met with futility due, probably, to the incompatibility of TTF with the strongly acidic conditions (TiCl$_4$) of these reactions.

A similar TTF-*p*-benzoquinone system (**14**) has recently been reported by Khodorkovsky and coworkers.[34] The X-ray structure of **14** exhibits a bent configuration of the molecule with the acceptor *p*-benzoquinone moiety spatially located above the donor dialkylthio-TTF unit. Because of the short intramolecular face-to-face plane distances between the donor and acceptor molecules, a low CT can be observed in the ground state of **14**. After sunlight irradiation, a photoinduced intramolecular electron transfer takes place resulting in the appearance of broad signals in the EPR measurements. Compound **14** is one of the first examples of TTF derivatives with inherent through space ICT properties.

Very recently, a novel TTF-σ-acceptor molecule (**15**) bearing a polynitrofluorene as acceptor has been reported to exhibit an electrochromic behavior in the near-IR spectrum.[35] Compound **15** has a strong acceptor moiety which, in contrast to the TCNQ acceptor, is obtained from the fluorenone and malononitrile under mild conditions, the resulting dicyanomethylene derivative presenting a reduction ability similar to that of TCNQ.

Compound **15** shows an ICT band in the visible and near-IR region of the electronic spectra. Interestingly, applying a potential of + 0.6 V, to generate the radical cation species results in the complete disappearance of the ICT band. This behavior suggests that **15** is an appealing system to study its electrochromic properties in the near-IR region.

16.5   TTF-σ-acceptor Molecules: Photovoltaic Applications

The solar energy received by our planet is of about 10^{22} kJ/year from which only 0.02–0.05 % is transformed into biological material.[36]   In contrast to green or purple bacteria, which have only one photosynthetic unit (PS II) to carry out the light-to-chemical product conversion, green plants use two systems (PS I and PS II). The photosynthetic unit is basically constituted by the light-harvesting (LH) system and the photosynthetic reaction center (RC). In the RC the light-induced electron transfer process forms the long-lived primary charge-separated (CS) state, thus initiating further consecutive chemical transformations.

The energy transfer from the LH system to the RC takes place on a picosecond (100 ps) time scale with a 100 % quantum yield.

The design and synthesis of bichromophoric organic molecules constituted by electroactive donor and acceptor fragments connected through a spacer displaying photoinduced charge separation has been carried out as an important task in chemistry. These molecules can be used as biomimetic models to transform sunlight into chemical energy.[4,37]

Fullerenes, and particularly [60]fullerene(C_{60}), have revealed promising features for the preparation of photovoltaic devices based on the unique photophysical properties they display.[38] Thus, one of the most outstanding properties of C_{60} in electron transfer processes is that it accelerates the photoinduced charge separation and retards the charge recombination in the absence of light. These findings suggest that C_{60} and its derivatives are promising candidates for the preparation of artificial photosynthetic systems and photovoltaic devices, making it one of the most realistic and outstanding applications of fullerenes.

During the last years, different C_{60}-donor systems have been prepared in the search of electron transfer properties. We have recently reviewed this topic and a wide variety of C_{60}-based intermolecular as well as intramolecular complexes have been synthesized.[39]

It is important to note that most of the donor molecules used in the preparation of C_{60}-σ-donor systems have an aromatic character in the ground state which is partially or totally lost upon oxidation to form the radical cation species. In contrast, TTF and its derivatives are nonaromatic molecules which upon oxidation form the 1,3-dithiolium cation, which presents an aromatic character (Scheme 16.1). This gain of aromaticity in forming the radical cation and dication species of TTF in the oxidation process is an important improvement to further increase the stabilization of the CS state (Fig. 16.5).

The first C_{60}-TTF dyad (**16**) was simultaneously and independently reported by Prato[40] and Martín[41] in 1996. Since then, other C_{60}-TTF derivatives such as **17**,[42] **18**,[43] **19**,[44] **20**,[45] **21**[46] and **22**[47] (Fig. 16.6) have been published in the search for photoinduced electron transfer processes.

C_{60}-TTF dyad (**19**) was prepared by Diels-Alder cycloaddition of the *o*-quinodimethane

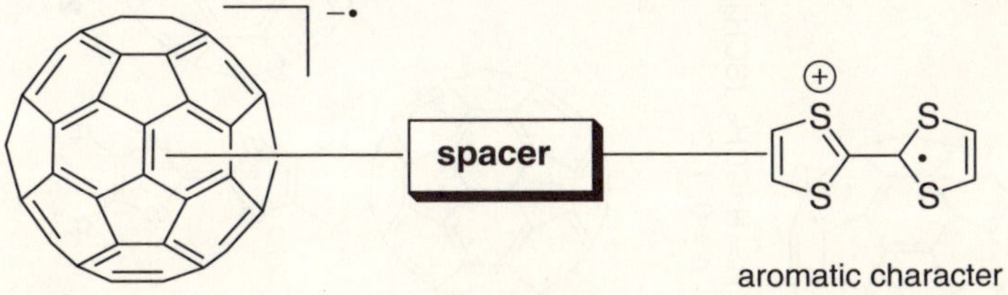

Fig. 16.5

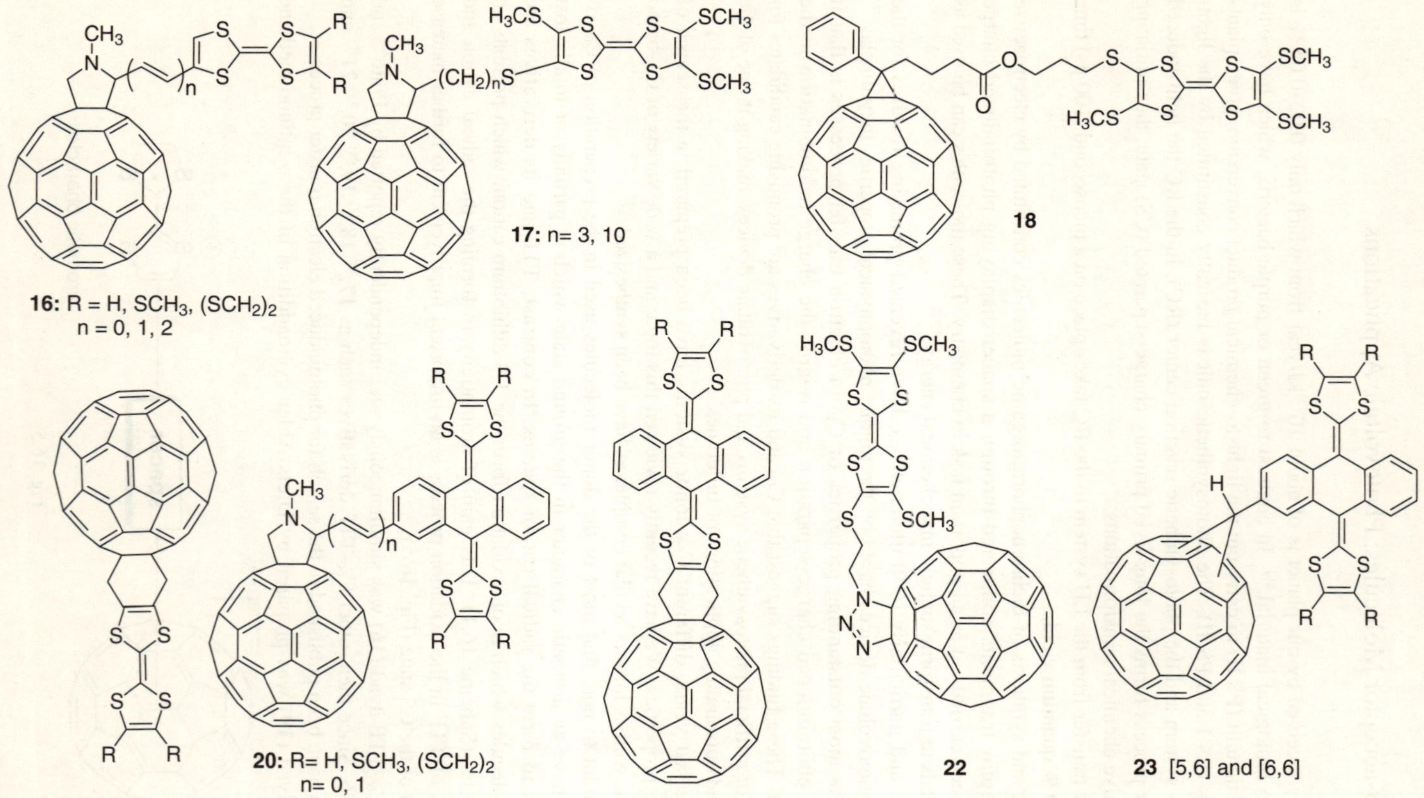

16: R = H, SCH$_3$, (SCH$_2$)$_2$
n = 0, 1, 2

17: n= 3, 10

18

19: R= H, CH$_3$,
SCH$_3$, CO$_2$CH$_3$

20: R= H, SCH$_3$, (SCH$_2$)$_2$
n= 0, 1

21: R= H, SCH$_3$

22

23 [5,6] and [6,6]

Fig. 16.6

analogue of TTF to C_{60}. In nanosecond-resolved flash photolysis, these systems undergo a rapid quenching of the triplet excited states, generating transient CS open-shell species with lifetimes typically around 75 μs.[48] These lifetime values are, however, far away of those measured for dyads such as **16** and **21-23**.

The nature and length of the spacer linking both electroactive moieties has a profound impact on the rate and efficiency of intramolecular electron transfer processes. Thus, we have prepared dyads (**16**) in which the TTF donor unit is attached to the C_{60} core through a single σ-bond and one (n = 1) or two (n = 2) vinyl spacers. We have recently reported the photophysics of the TTF-containing dyads **16**[49] and **22**.[47] Absorption of light in these dyads leads to the formation of the fullerene excited singlet state which undergoes an intramolecular electron transfer to form the CS state (Fig. 16.5). Evidence for the electron transfer process comes from steady-state and time-resolved photolysis. It is worth mentioning that an increase in the stabilization of the CS state takes place by gaining aromaticity upon oxidation of the TTF moiety.

Charge recombination proceeds mainly *via* formation of the fullerene triplet excited state due to the strong second-order vibronic spin-orbit coupling induced by the presence of the sulfur atoms of the TTF unit. This charge-recombination behavior forming a fullerene triplet excited state is different from most of the previously reported C_{60}-based dyads bearing donor moieties other than TTF [49,50] (Fig. 16.7).

The lifetimes of the CS states in all these C_{60}-TTF dyads are in the range of 1–2 ns, which is similar to that found in other C_{60}-based tetraphenylporphyrin dyads (~0.5 ns) endowed with similar spacers.[51]

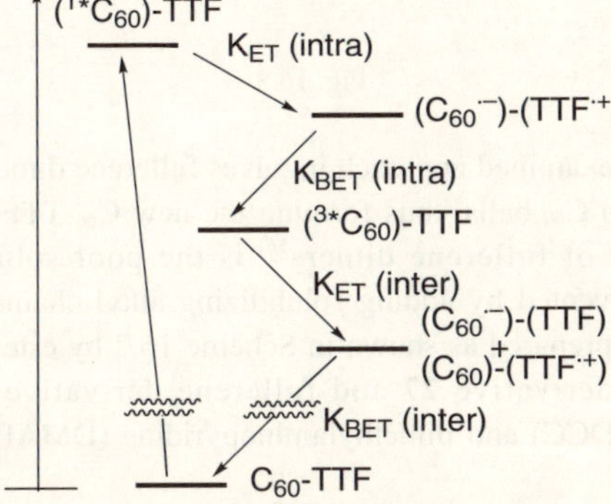

Fig. 16.7

More recently, we have found that C_{60}-based dyads bearing a π-extended TTF fragment (**23**), which upon oxidation form very stable cationic species, exhibit a highly remarkable stability of the CS state which is improved by around two orders of magnitude in comparison with other C_{60}-TTF dyads.[52]

Other more complex electroactive systems involving C_{60} and TTF moieties are shown in Fig. 16.8. These systems (**24**,[53] **25**[54]) have recently been prepared in the search for a gradient of redox centers which could stabilize the formed CS state. Work is currently in progress to determine this behavior.

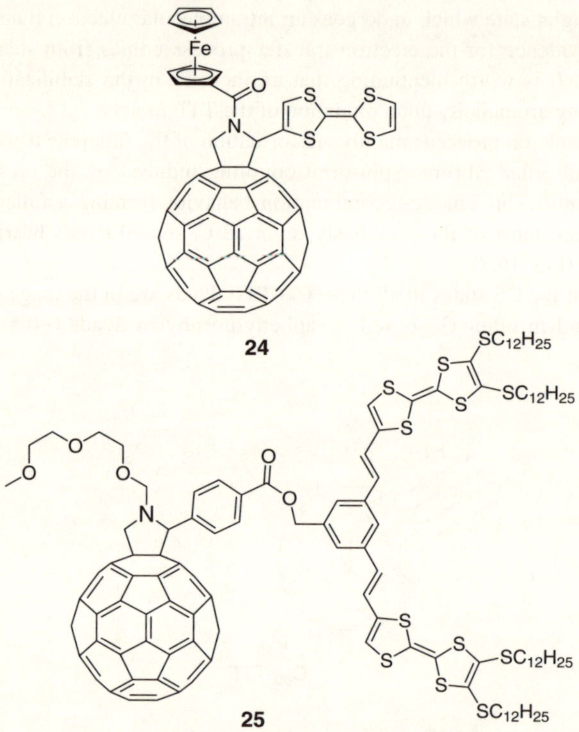

24

25

Fig. 16.8

An intriguing and less examined approach involves fullerene dimers in which the TTF unit is covalently connected to two C_{60} balls, thus forming the new C_{60}-TTF-C_{60} triad (**28**).[55] A major problem in the chemistry of fullerene dimers[56] is the poor solubility they exhibit. This shortcoming can be circumvented by adding solubilizing alkyl chains either on the TTF or C_{60} moiety. Compound **28** was prepared as shown in Scheme 16.2 by esterification reaction from the bis(hydroxymethyl)TTF derivative **27** and fullerene derivative **26**[57] in the presence of dicyclohexylcarbodiimide (DCC) and dimethylaminopyridine (DMAP) to yield triad **28** in 43% yield.

Considering the electroactive character of both TTF and C_{60} units, a charge separation was expected to dominate the deactivation of the fullerene singlet excited state. The steady-state

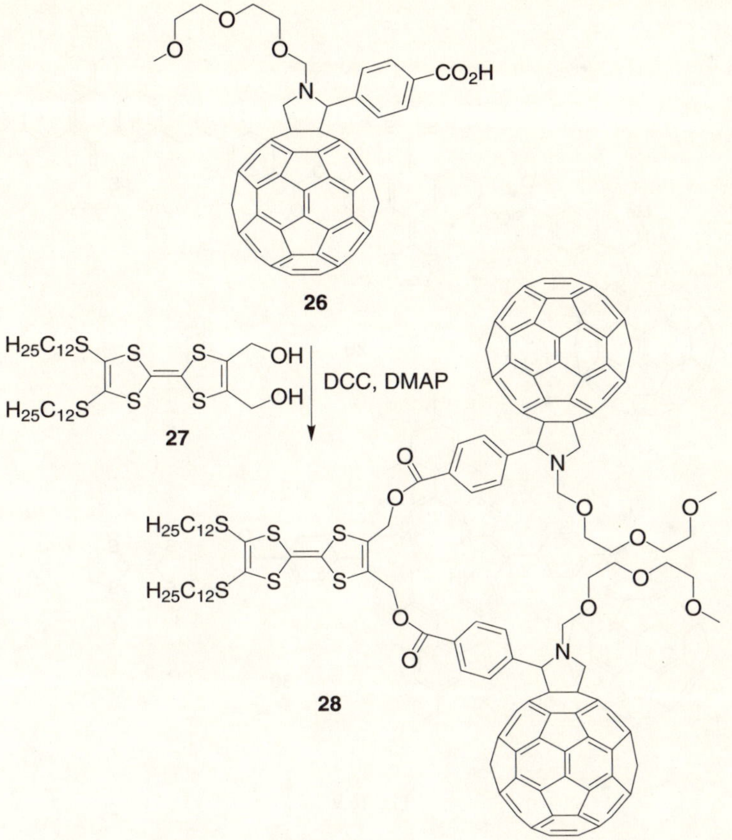

Scheme 16.2

fluorescence spectra of **28** in different solvents showed a clear correlation between the fluorescence quantum yields (Φ) and solvent polarity: $\Phi_{toluene} > \Phi_{o\text{-dichlorobenzene}} > \Phi_{benzonitrile} > \Phi_{DMF}$.

Time-resolved pico- and nanosecond transient absorption spectroscopy enabled the observation of a set of Vis and NIR maxima at 430 and 1000 nm for the TTF$^{\bullet+}$ and $C_{60}^{\bullet-}$, respectively, thus confirming the formation of the TTF$^{\bullet+}$-$C_{60}^{\bullet-}$ radical pair *via* a photoinduced electron transfer evolving from the TTF donor to the electron-accepting singlet excited state of the fullerene. The measured lifetime for the radical pair ranges from 54 ns in *o*-dichlorobenzene and 121 ns in DMF. These data reveal that C_{60} dimers are appealing ensembles and potentially useful in photovoltaic devices.

Very recently, TTF and its derivatives have also been used in the preparation of the first electroactive triads of type C_{60}-exTTF$_1$-exTTF$_2$ (**29**)[58] and C_{60}-Porphyrin-TTF (**30**)[59] (Fig. 16.9).

Excitation of the fullerene moiety in **29** yields the first excited singlet state $^1C_{60}$-exTTF$_1$-exTTF$_2$, which undergoes photoinduced electron transfer to give the CS state $C_{60}^{\bullet-}$-exTTF$_1^{\bullet+}$-exTTF$_2$. A second electron transfer from exTTF$_2$ competes with charge recombination to form the final $C_{60}^{\bullet-}$-exTTF$_1$-exTTF$_2^{\bullet+}$ state which eventually recombines to the ground state.[58] Remarkable lifetimes were measured for the final CS state of triad **29** with values ranging from 50 to 100 μs. These values are among the highest lifetimes found for electroactive C_{60}-based triads[58] and can be

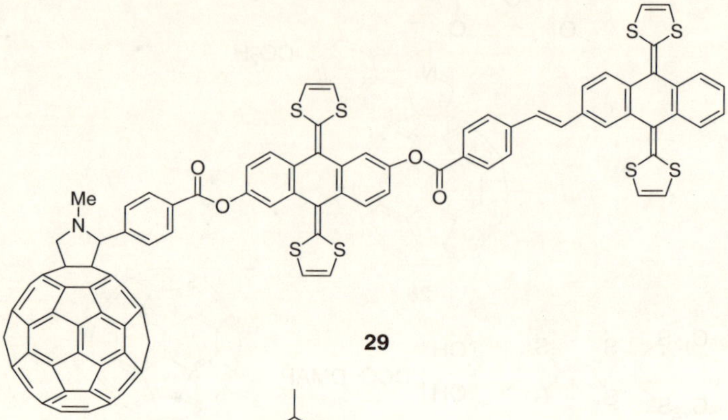

29

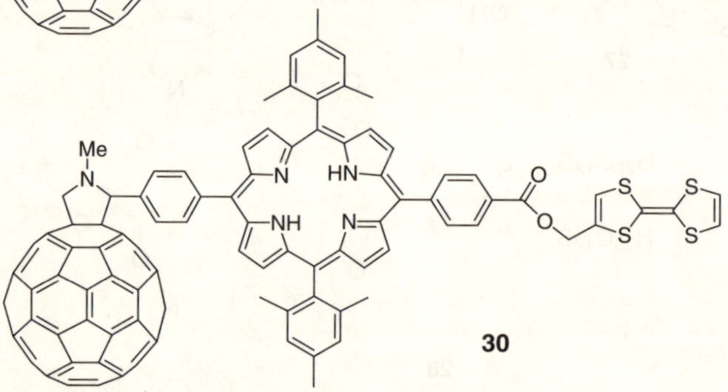

30

Fig. 16.9

accounted for by the additional stabilization provided by the gain of aromaticity and planarity of the π-extended TTF unit upon oxidation.

A similar behavior is observed for triad C_{60}-porphyrin-TTF (**30**) in which excitation takes place on the porphyrin moiety to form the singlet excited state C_{60}-^{1}porphyrin-TTF which undergoes two consecutive electron transfers through a well-defined structure to form the final state $C_{60}^{\bullet-}$-porphyrin-TTF$^{\bullet+}$. The final CS state for **30** has a lifetime of 660 ns for both free and Zn porphyrins.

In summary, biomimetic models of type C_{60}-TTF dyads and triads are promising candidates for further preparation of solar energy conversion devices. The search for novel photovoltaic devices and artificial photosynthetic systems is currently a major task in chemistry and the photophysical properties shown by dyads and triads involving TTF and π-extended TTFs are among the most outstanding found for these purposes.

16.6 TTF-π-acceptor Molecules: Nonlinear Optical Applications

As previously stated, D-π-A compounds are suitable candidates for dipolar chromophores endowed with second-order NLO properties. In contrast to 1,3-dithiole derivatives, which have received much attention as donor units in NLO-phores[60] since the pioneering works of Katz[61] and

Lehn,[62] TTFs have only recently emerged as promising donors in this type of structure, a field first developed in the present authors' laboratories.

Although the outer carbon atoms of TTF have a relatively small coefficient at the HOMO, the redox chemistry of this donor should allow a facile entry to redox-switchable NLO materials [63] (through a decrease of the first molecular hyperpolarizability, β, on oxidation) and multiproperty materials.[64]

The synthesis of NLO-phores where the TTF core acts as the electron-donating moiety invariably rests on the use of formyl-TTF [65] or vinylogues thereof[66] (Fig. 16.10). These aldehydes (**31 a-c**) are then subjected to Knoevenagel reactions with active methylene compounds, or Wittig reactions with functionalized phosphorus ylides.

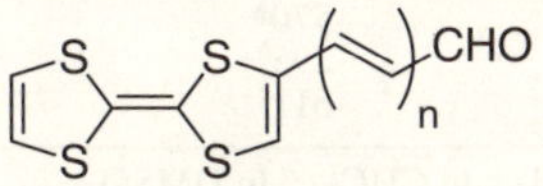

31 a-c (n = 0-2)

Fig. 16.10

Not unexpectedly, compounds **31** showed weak efficiency at second harmonic generation (SHG), as determined by the electric field induced second harmonic (EFISH) technique, with $\mu\beta(0)$ values (in 10^{-48} esu units) in the range 33 (for **31a**) to 80 (for **31c**).[67-69]

On the other hand, when the formyl group was replaced by a stronger acceptor, such as dicyanomethylene (**32 a-c**) (Fig. 16.11), a four- to fivefold increase in $\mu\beta(0)$ values was observed.[66,68,70]

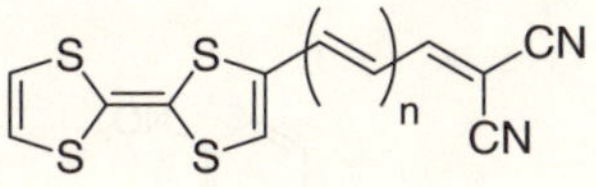

32 a-c (n = 0-2)

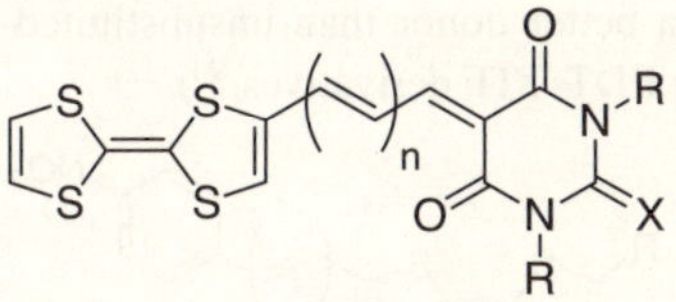

33 a-b (n = 0,1; R = H; X = O)
34 a-b (n = 0,1; R = Me; X = O)
35 a-c (n = 0-2; R = Et; X = S)

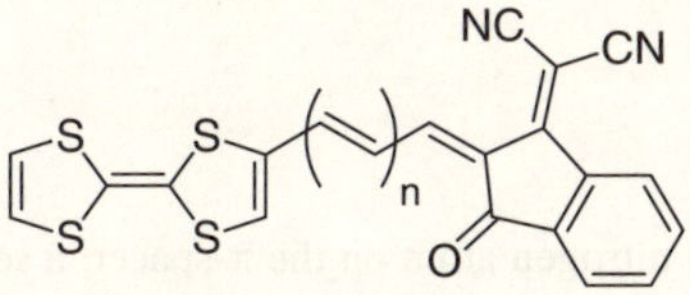

36 a-c (n = 0-2)

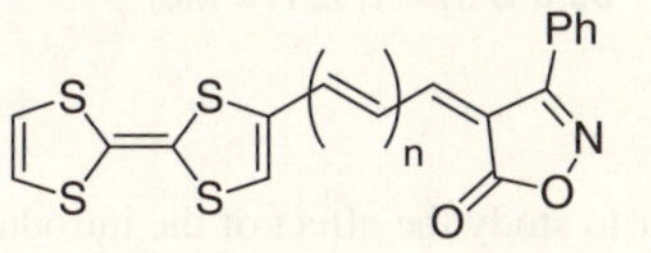

37 a-c (n = 0-2)

Fig. 16.11

In order to obtain more efficient NLO-phores, (thio)barbituric acid derivatives **33**, **34** and **35** (Fig. 16.11) were prepared from aldehydes **31**.[71] As expected, compounds containing the thiobarbituric moiety showed higher $\mu\beta$ values than the corresponding barbituric derivatives, and less negative reduction potentials, in agreement with the stronger electron-acceptor character of the thiobarbiturate end group. Moreover, all of these derivatives showed good thermal stability.

In a similar vein, Knoevenagel reactions of **31** with 3-dicyanomethylenindan-1-one and 3-phenyl-5-isoxazolone gave rise to compounds **36** and **37**, respectively[70,72] (Fig. 16.11). Derivatives **36c** and **37c** containing the longest polyenic spacer display $\mu\beta(0)$ values similar to

Table 16.2 $\mu\beta(0)$ and UV-vis Data for Selected TTF-containing NLO-phores

Compound	$\mu\beta(0)^a$	λ_{max} (nm)^b
31a	33^c	495^c
31b	46^c	510^c
31c	80^c	517^c
32a		624
32b	234^c	618
32c	320^c	608
35a	180^d	718
35b	347^d	672
35c	455^d	648
36a		798
36b	259^b	726
36c	570^b	690
37b	295^b	658
37c	611^b	630

^{a} In 10^{-48} esu. ^{b} In CH_2Cl_2. ^{c} In $CHCl_3$. ^{d} In DMSO.

that of the classical NLO dye Dispersed Red 1 (DR1) (*ca.* 600. 10^{-48} esu) (Table 16.2).

CV measurements indicate that in compounds **32**, **35**, **36** and **37** the oxidation potential values of the TTF moiety are shifted toward more positive values in comparison with TTF itself, which is consistent with an increase in the oxidation potential due to the presence of the acceptor moieties. This shift becomes larger as the length of the spacer decreases, indicating a stronger interaction between the donor and acceptor groups when they are closer.

Using Wittig reactions, chromophores **38** and **39** have been prepared (Fig. 16.12). They exhibit moderate second-order NLO properties, although it is noteworthy that the trimethyl-TTF moiety is a better donor than unsubstituted TTF.[69] (A similar observation has been reported concerning EDT-TTF derivatives.[68])

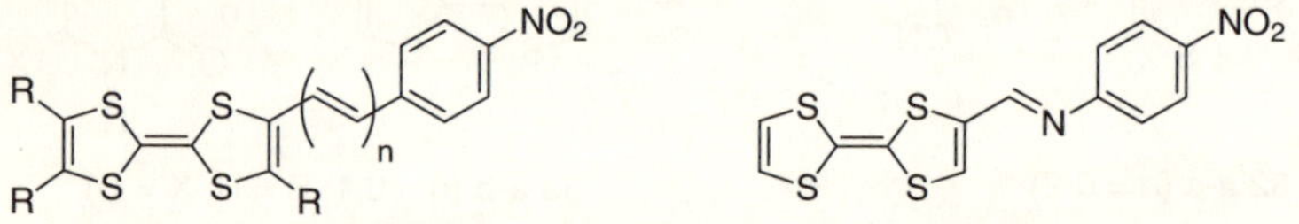

38 a-c (n = 1-3; R = H)

39 a-b (n = 1, 2; R = Me)

40

Fig. 16.12

In order to study the effect of the introduction of a nitrogen atom on the π-spacer, a series of imines, such as **40**, were prepared; the $\mu\beta(0)$ value determined for this compound (62.10^{-48} esu) was notably lower than that obtained for the vinyl analogue **38a** (150.10^{-48} esu). A similar decrease in SHG efficiency occurs in 1,3-dithiole NLO-phores when polyenic spacers are replaced by azine spacers, since the latter act as conjugation stoppers.[73]

Compounds dealt with in this section show the expected increase in $\mu\beta(0)$ values when the length of the polyenic spacer is increased. Nevertheless, a most interesting and unusual effect is observed in the UV-vis spectra (Table 16.2) of NLO-phores in which the TTF moiety is conjugated with a strong electron acceptor (compounds **32** to **37**), namely, the lengthening of the π-spacer is accompanied by a *hypsochromic* shift of the ICT band.[70] This phenomenon has been reported in only a few instances (vinylogous indigoid chromophores[74] and some push-pull stilbenes[75]) and is of interest in the context of NLO properties, with a view to overcoming the well-known transparency/efficiency trade-off. On the other hand, aldehydes **31**, in which the TTF core is conjugated to a weaker acceptor, show the expected bathochromic shift on passing from **31a** to **31c**, whereas compounds **38** and **39** show an intermediate behavior (their ICT bands are nearly independent on the length of the spacer). Semiempirical (ZINDO, CNDO/S) and Hartree-Fock methods (CIS) were unable to reproduce the experimental trends, but a TD-DFT study (6-31+G*//B3P86/6-31G*) of compounds **31** and **35** not only revealed the HOMO → LUMO character of the ICT band, but also accounted for the different bathochromic and hypsochromic shifts observed in these series.[76] Moreover, an IR and Raman spectroscopy study reveals that the degree of ICT decreases with increasing the number of vinylenic units in the π-spacer.[70]

A similar but less noticeable hypsochromic shift has been observed in chromophores **41** (Fig. 16.13), the only ones reported to date in which an extended-TTF moiety acts as a donor unit.[77]

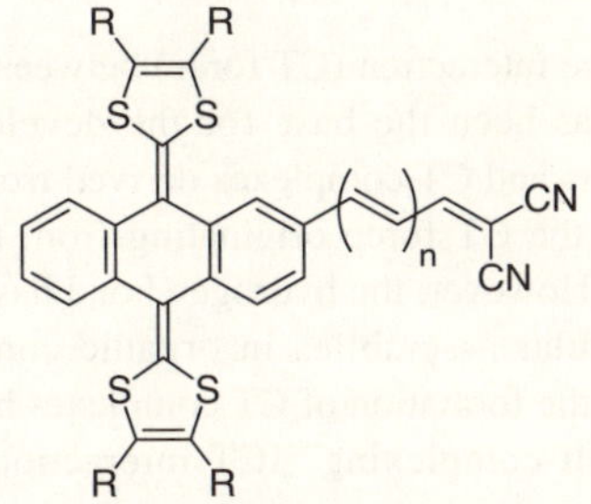

41 (n = 0, 1)

Fig. 16.13

Taking advantage of the good performance of semiempirical (FF-PM3) and *ab initio* (CPHF) methods at predicting the NLO properties of TTF-based chromophores,[70] compound **42a** (Fig. 16.14) seems to be a promising candidate, having a calculated high β value.

42a (X = I) was in fact prepared[78] but, unfortunately, it crystallized in a centrosymmetric space group, therefore exhibiting no efficiency in SHG. In order to ensure noncentrosymmetry, **42b** (X = L-hydrogentartrate) was synthesized,[78,79] but it did not exhibit any efficiency either (Kurtz-Perry powder test). This was attributed either to a quasi antiparallel arrangement of the chromophores or to the poor crystallinity of the sample. To gain better control over the arrangement of the NLO-phores, amphiphilic dye **42c** (X = I) was eventually prepared, and LB monolayers of this compound, with non-centrosymmetrically aligned molecules, showed SHG with an effective susceptibility of 70 pm V^{-1}.[80] It should be noted that the only previously

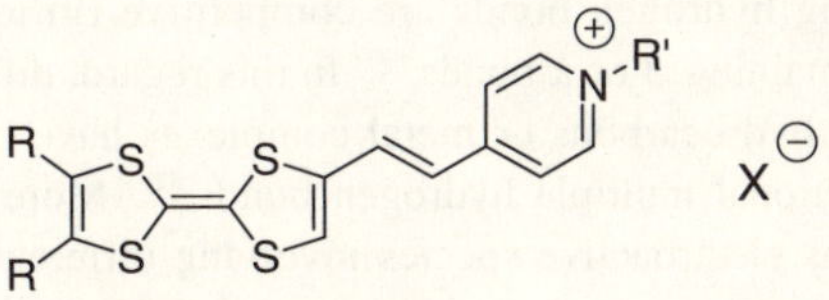

42 a (R = H; R' = Me)
42 b (R = H; R' = H)
42 c (R = SC$_{18}$H$_{37}$; R' = Me)

Fig. 16.14

reported NLO data for a TTF/pyridinium derivative showed a small NLO coefficient for the monolayer, which was attributed to an unfavorable molecular conformation, where the TTF moiety and the pyridinium acceptor are nearly orthogonal.[81]   Such misalignment is overcome in the case of **42c**, since both moieties should be coplanar.

Interestingly enough, a recent theoretical paper shows that dyads **16** (R = H, n = 0-6) should have good second-order NLO properties.[82]

Finally, it should be mentioned that third-order NLO properties of some TTF derivatives incorporating acceptor groups have been reported.[83]

16.7   Noncovalent TTF-acceptor Interactions: Supramolecular Applications

As stated above, the through-space interaction (CT force) between segregated stacks of donor and acceptor arrays of molecules has been the base for the development of the conducting and superconducting properties of salts and CT complexes derived from TTFs. The local nature of the hydrogen bond is in contrast to the CT force originating from the face-to-face or side-by-side intermolecular π-orbital overlap. However, the hydrogen bond has played a very important role in the construction of supramolecular assemblies in organic conductors, and the effect of the intermolecular hydrogen bond in the formation of CT complexes has been widely studied.[1]

On the other hand, the "self-complexing" ICT interactions of TTF derivatives forming catenanes, rotaxanes and pseudorotaxanes have been mainly studied by the groups of Becher[13] and Balzani and Stoddart[84] in the search of novel switches and devices. This important aspect of the TTF chemistry has been thoroughly revised in a different chapter of this volume, and we strongly recommend reading it for further information.

Here, we focus our attention on the much less studied supramolecular aspects involving hydrogen bonding formation between D and A species resulting in a noncovalently assembled heterosupramolecule.[85] A supramolecule is distinguished from a large molecule in the following aspects: i) the molecular components of a supramolecule are noncovalently linked, ii) the intrinsic properties of these molecular components largely persist and iii) the properties of a supramolecule are different from the constituent molecular components, that is, there exists a defined supramolecular function.[86] More recently, this concept has also been extended to covalently connected molecular components in which there exists a supramolecular function.[6]

It is well known that hydrogen bonding networks are also very important in biological electron transfer processes. Therefore, considerable effort is being currently devoted to the synthesis and study of noncovalently constructed D-A model systems in order to determine whether pathways containing hydrogen bonds are competitive (in terms of allowable electron transfer rates) with those containing σ or π bonds.[87]   In this regard, different electron donors such as porphyrins, polyaromatic hydrocarbons or metal complexes have been linked to the acceptor moiety through highly directional multiple hydrogen bonds.[87]   More recently, the formation of supramolecular C_{60} dimers as electroactive species involving different hydrogen bonding motifs has also been reported.[57]

To the best of our knowledge, the formation of TTF-based D-A systems connected through hydrogen bonds is an area that still remains unexplored in TTF chemistry. Goldenberg and Neilands have recently reported a pioneering system in which a TTF, fused to a 2,4-dioxopyrimidine ring, is connected to a complementary 2,6-di(*N*-acetamido)pyridine through a threefold hydrogen bond (**43**)[88] (Fig. 16.15).

The electronic spectrum of this TTF derivative shows a small red shift in the absorption

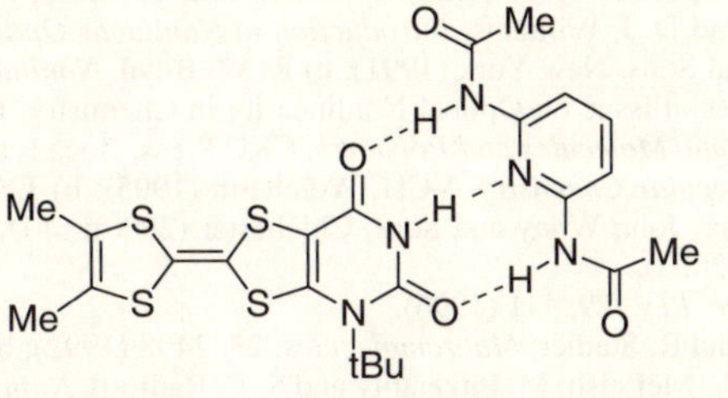

43

Fig. 16.15

maximum upon complexation, as well as a positive shift (30 mV) in the oxidation potential of the TTF unit, thus showing the electronic interaction between both molecules through the threefold hydrogen bonding.

In our laboratory we are currently working on the synthesis and study of the photoinduced electron transfer of novel TTF and C_{60} systems, electronically interacting through a highly directional hydrogen-bonding array.

16.8   Summary and Outlook

TTF is a mature molecule which has been successfully used as the electron donor in the formation of a wide variety of intermolecular salts and CT complexes exhibiting electrically conducting properties. However, due to its remarkable redox properties, TTF is also a highly versatile building block for the construction of more sophisticated structures of interest in supramolecular and materials chemistry.

Among the different TTF-based systems prepared, TTF-acceptor molecules are of particular interest since, depending upon the spacer connecting the donor and acceptor moieties, different properties are observed. The aim of the present chapter has been to emphasize these properties in connection with the nature of the chemical spacer.

At the dawn of this new century, TTF still maintains the interest of the chemical community and, considering the potential applications of TTF in fields such as molecular rectification, dyes, nonlinear optics, sensors or photoinduced energy and electron transfer processes, the chemistry of TTF and its derivatives is currently an active area of research which surpasses the frontiers of chemistry to lie in a multidisciplinary field.

References

1. a) M. R. Bryce, *Chem. Soc. Rev.*, **20**, 355 (1991); b) J. M. Williams, J. Ferraro, R. J. Thorn, K. D. Carlson, U. Geiser, H. H. Wang, A. M. Kini and M.-H. Whangbo, *Organic Superconductors (including Fullerenes)*, Prentice Hall, New Jersey (1992); c) *Organic Conductors, Fundamentals and Applications*, Marcel Dekker, New York (1994); d) *Handbook of Organic Conductive Molecules and Polymers*, John Wiley and Sons, New York (1997), Vol. 1; e) *Handbook of Advanced Electronic and Photonic Materials and Devices*, Academic Press, San Diego (2001), Vol. 3.
2. A. Aviram and M. A. Ratner, *Chem. Phys. Lett.*, **29**, 277 (1974).
3. a) R. M. Metzger, *Acc. Chem. Res.*, **32**, 950 (1999); b) *Introduction to Molecular Electronics*, Oxford University Press, New York (1995).
4. a) M. R. Wasielewski, *Chem. Rev.*, **92**, 435 (1992); b) D. Gust, T. A. Moore and A. L. Moore, *Acc. Chem. Res.*, **34**, 40(2001).

5. a) *Nonlinear Optical Properties of Organic Molecules and Crystals*, Academic Press, New York (1987); b) P. N. Prasad and D. J. Williams, *Introduction to Nonlinear Optical Effects in Molecules and Polymers*, John Wiley and Sons, New York (1991); c) R. W. Boyd, *Nonlinear Optics*, Academic Press, San Diego (1992); d) Special issue on Optical Nonlinearity in Chemistry, *Chem. Rev.*, **94**, 1 (1994); e) *Nonlinear Optics of Organic Molecules and Polymers*, CRC Press, Boca Raton (1997).

6. a) J. M. Lehn, *Supramolecular Chemistry*, VCH, Weinheim (1995); b) J. W. Steed and J. L. Atwood, *Supramolecular Chemistry*, John Wiley and Sons, Chichester (2000); c) D. M. Guldi and N. Martín, *J. Mater. Chem.*, **12**, 1978 (2002).

7. G. R. Desiraju, *Acc. Chem. Res.*, **29**, 441 (1996).

8. a) C. Hilger, M. Dräger and R. Stadler, *Macromolecules*, **25**, 2498 (1992); b) A. Aggeli, N. Boden, J. N. Keen, P. F. Knowles, T. C. McLeish, M. Pitkeathly and S. E. Radford, *Nature*, **386**, 259 (1997).

9. a) H. Pelletier and J. Kraut, *Science*, **258**, 1748 (1992); b) J. M. Nocek, J. S. Zhou, S. DeForest, S. Priyadarshy, D. N. Beratan, J. N. Onuchic and B. M. Hoffman, *Chem. Rev.*, **96**, 2459 (1996).

10. P. J. F. De Rege, S. A. Williams and M. J. Therien, *Science*, **269**, 1409 (1995).

11. a) V. Khodorkovsky and J. Y. Becker, *Organic Conductors*, Marcel Dekker, New York (1994), p. 5; b) M. R. Bryce, *J. Mater. Chem.*, **5**, 148 (1995); c) G. Schukat, A. M. Richter and E. Fanghänel, *Sulfur Rep.*, **7**, 155 (1987); d) G. Schukat and E. Fanghänel, *Sulfur Rep.*, **14**, 245 (1993); e) G. Schukat and E. Fanghänel, *Sulfur Rep.*, **18**, 1 (1996); f) J. Garín, *Adv. Heterocycl. Chem.* **62**, 249 (1995); g) K. B. Simonsen, N. Svenstrup, J. Lau, O. Simonsen, P. Mørk, G. J. Kristensen and J. Becher, *Synthesis* 407 (1996); h) T. Otsubo, Y. Aso and K. Takimiya, *Adv. Mater.*, **8**, 203 (1996); i) M. Adam and K. Müllen, *Adv. Mater.*, **6**, 439 (1994); j) J. Becher, J. Lau and P. Mørk, *Electronic Materials : The Oligomer Approach*, Wiley-VCH, Weinheim(1998), p. 198; k) M. R. Bryce, W. Davenport, L. M. Goldenberg and C. Wang, *Chem. Commun.*, 945 (1998); l) K. B. Simonsen and J. Becher, *Synlett*, 1211 (1997); m) M. B. Nielsen and J. Becher, *Liebigs Ann./Recueil*, 2177(1997); n) E. Coronado and C. J. Gómez-García, *Chem. Rev.*, **98**, 273 (1998); o) J. Roncali, *J. Mater. Chem.*, **7**, 2307 (1997); p) P. Day and M. Kurmoo, *J. Mater. Chem.*, **7**, 1291 (1997); q) M. R. Bryce, *Adv. Mater.*, **11**, 11 (1999); r) M. R. Bryce, *J. Mater. Chem.*, **10**, 589 (2000).

12. J. L. Segura and N. Martín, *Angew. Chem. Int. Ed.*, **40**, 1372 (2001).

13. a) T. Jorgensen, T. K. Hansen and J. Becher, *Chem. Soc. Rev.*, **23**, 41 (1994); b) B. B. Nielsen, C. Lomholt and J. Becher, *Chem. Soc. Rev.*, **29**, 153 (2000).

14. a) C. Katan, *J. Phys. Chem.*, **103**, 1407 (1999); b) I. Hargittai, J. Brunvoll, M. Kolonits and V. Khodorkovsky, *J. Mol. Struct.*, **317**, 273 (1994).

15. a) F. Wudl, G. M. Smith and E. J. Hufnagel, *J. Chem. Soc. Chem. Commun.*, 1453 (1970); b) S. Hünig, G. Kiesslich, D. Scheutzow, R. Zahradnik and P. Carsky, *Int. J. Sulfur Chem. Part C*, **6**, 109 (1971).

16. a) J. Ferraris, D. O. Cowan, V. V. Walatka and J. H. Perlstein, *J. Am. Chem. Soc.*, **95**, 948 (1973); b) L. B. Coleman, M. J. Cohen, D. J. Sandman, F. G. Yamagishi, A. F. Garito and A. J. Heeger, *Solid State Commun.*, **12**, 1125 (1973).

17. a) A. Andrieux, C. Duroure, D. Jérome and K. Bechgaard, *J. Phys. Lett.*, **40**, 381 (1979); b) D. Jérome, A. Mazaud, M. Ribault and K. Bechgaard, *J. Phys. Lett.*, **41**, L195 (1980).

18. Y. Yamashita and M. Tomura, *J. Mater. Chem.*, **8**, 1933 (1998).

19. N. Martín and E. Ortí, *Handbook of Advanced Electronic and Photonic Materials and Devices*, Academic Press, San Diego (2001), Vol. 3, p. 245.

20. a) M. R. Bryce and A. J. Moore, *Synth. Met.*, **27**, B557 (1988); b) M. R. Bryce and A. J. Moore, *Synth. Met.*, **27**, 203 (1988); c) M. R. Bryce, A. J. Moore, D. Lorcy, A. S. Dhindsa and A. Robert, *J. Chem. Soc., Chem. Commun.*, 470 (1990); d) M. R. Bryce and A. J. Moore, *Pure Appl. Chem.*, **62**, 473 (1990); e) A. J. Moore and M. R. Bryce, *J. Chem. Soc., Perkin Trans. 1*, 157 (1991).

21. Y. Yamashita, Y. Kobayashi and T. Miyashi, *Angew. Chem. Int. Ed. Engl.*, **28**, 1052 (1989).

22. M. R. Bryce, M. A. Coffin, M. B. Hursthouse, A. I. Karaulov, K. Müllen and H. Scheich, *Tetrahedron Lett.*, **32**, 6029 (1991).

23. a) D. M. Guldi, L. Sánchez and N. Martín, *J. Phys. Chem. B*, **105**, 7139 (2001); b) A. E. Jones, C. Christensen, D. F. Perepichka, A. S. Batsanov, A. Beeby, P. J. Low, M. R. Bryce and A. W. Parker, *Chem. Eur. J.*, **7**, 973 (2001).

24. M. R. Bryce, A. J. Moore, M. Hasan, G. J. Ashwell, A. T. Fraser, W. Clegg, M. B. Hursthouse and A. I. Karaulov, *Angew. Chem. Int. Ed. Engl.*, **29**, 1450 (1990).

25. N. Martín, L. Sánchez, C. Seoane, E. Ortí, P. M. Viruela and R. Viruela, *J. Org. Chem.*, **63**, 1268 (1998).

26. R. Hoffmann, *Acc. Chem. Res.*, **4**, 1 (1971).

27. R. M. Metzger, *J. Mater. Chem.*, **9**, 2027 (1999).

28. R. M. Metzger, B. Chen, U. Höpfner, M. V. Lakshmikantham, D. Villaume, T. Kawai, X. Wu, H. Tachibana, T. V. Hughes, H. Sakurai, J. W. Daldwin, C. Hosch, M. P. Cava, L. Brehmer and G. J.

Ashwell, *J. Am. Chem. Soc.*, **119**, 10455 (1997).

29. C. A. Panetta, N. E. Heimer, C. L. Hussey and R. M. Metzger, *Synlett*, 301 (1991).
30. R. M. Metzger, H. Tabibana, X. Wu, U. Höpfner, B. Chen, M. V. Lakshmikantham and M. P. Cava, *Synth. Met.*, **85**, 1359 (1997).
31. S. Scheib, M. P. Cava, J. W. Baldwin and R. M. Metzger, *J. Org. Chem.*, **63**, 1198 (1998).
32. P. De Miguel, M. R. Bryce, L. M. Goldenberg, A. Beeby, V. Khodorkovsky, L. Shapiro, A. Niemz, A. O. Cuello and V. Rotello, *J. Mater. Chem.*, **8**, 71 (1998).
33. a) J. L. Segura, N. Martín, C. Seoane and M. Hanack, *Tetrahedron Lett.*, **37**, 2503 (1996); b) M. González, B. Illescas, N. Martín, J. L. Segura, C. Seoane and M. Hanack, *Tetrahedron*, **54**, 2853 (1998).
34. E. Tsiperman, T. Regev, J. Y. Becker, J. Bernstein, A. Ellern, V. Khodorskovsky, A. Shames and L. Shapiro, *Chem. Commun.*, 1125 (1999).
35. D. F. Perepichka, M. R. Bryce, E. J. L. Mcinnes and J. P. Zhao, *Org. Lett.*, **3**, 1431 (2001).
36. J. Barber and B. Andersson, *Nature*, **370**, 31 (1994).
37. a) W. Retig, *Angew. Chem. Int. Ed. Engl.*, **25**, 971 (1986); b) T. J. Meyer, *Acc. Chem. Res.*, **22**, 163 (1989); c) H. Kurreck and M. Huber, *Angew. Chem. Int. Ed. Engl.*, **34**, 849 (1995).
38. *Fullerenes: From Synthesis to Optoelectronic Applications*, Kluwer Academic Publishers, Dordrecht (2002).
39. N. Martín, L. Sánchez, B. Illescas and I. Pérez, *Chem. Rev.*, **98**, 2527 (1998).
40. M. Prato, M. Maggini, C. Giacometti, G. Scorrano, G. Sandoná and G. Farnia, *Tetrahedron*, **52**, 5221 (1996).
41. N. Martín, L. Sánchez, C. Seoane, R. Andreu, J. Garín and J. Orduna, *Tetrahedron Lett.*, **37**, 5979 (1996).
42. K. B. Simonsen, V. V. Konovalov, T. A. Konovalova, T. Kawai, M. P. Cava, L. D. Kispert, R. M. Metzger and J. Becher, *J. Chem. Soc., Perkin Trans. 2*, 657 (1999).
43. S. Ravaine, P. Delhaès, P. Leriche and M. Sallé, *Synth. Met.*, **87**, 93 (1997).
44. a) P. Hudhomme, C. Boullé, J. M. Rabreau, M. Cariou, M. Jubault and A. Gorgues, *Synth. Met.*, **94**, 73 (1998); b) C. Boullé, J. M. Rabreau, P. Hudhomme, M. Cariou, M. Jubault, A. Gorgues, J. Orduna and J. Garín, *Tetrahedron Lett.*, **38**, 3909 (1997); c) J. Llacay, M. Mas, E. Molins, J. Veciana, D. Powell and C. Rovira, *Chem. Commun.*, 659 (1997).
45. a) N. Martín, L. Sánchez, I. Pérez and C. Seoane, *J. Org. Chem.*, **62**, 5690 (1997); b) M. A. Herranz, N. Martín, L. Sánchez, C. Seoane and D. M. Guldi, *J. Organomet. Chem.*, **599**, 2 (2000).
46. M. A. Herranz and N. Martín, *Org. Lett.*, **1**, 2005 (1999).
47. D. M. Guldi, S. González, N. Martín, A. Antón, J. Garín and J. Orduna, *J. Org. Chem.*, **65**, 1978 (2000).
48. J. Llacay, J. Veciana, J. Vidal-Gancedo, J. L. Bourdelande, R. González-Moreno and C. Rovira, *J. Org. Chem.*, **63**, 5201 (1998).
49. N. Martín, L. Sánchez, M. A. Herranz and D. M. Guldi *J. Phys. Chem. A*, **104**, 4648 (2000).
50. N. Martín, L. Sánchez, B. Illescas, S. González, M. A. Herranz and D. M. Guldi *Carbon*, **38**, 1577 (2000).
51. a) H. Imahori and Y. Sakata, *Adv. Mater.*, **9**, 537 (1997); b) H. Imahori and Y. Sakata *Eur. J. Org. Chem.*, 2445 (1999).
52. N. Martín, L. Sánchez and D. M. Guldi, *Chem. Commun.*, 113 (2000).
53. M. A. Herranz, B. Illescas, N. Martín, C. Luo and D. M. Guldi, *J. Org. Chem.*, **65**, 5728 (2000).
54. J. L. Segura, E. M. Priego and N. Martín, *Tetrahedron Lett.*, **41**, 7737 (2000).
55. J. L. Segura, E. M. Priego, N. Martín, C. Luo and D. M. Guldi, *Org. Lett.*, **2**, 4021 (2000).
56. J. L. Segura and N. Martín, *Chem. Soc. Rev.*, **29**, 13 (2000).
57. J. J. González, S. González, E. Priego, C. Luo, D. M. Guldi, J. De Mendoza and N. Martín, *Chem. Commun.*, 163 (2001).
58. L. Sánchez, I. Pérez, N. Martín and D. M. Guldi, *Chem. Eur. J.*, **9**, 2457 (2003).
59. P. A. Liddell, G. Kodis, L. De la Garza, J. L. Bahr, A. L. Moore, T. A. Moore and D. Gust, *Helv. Chim. Acta*, **84**, 2765 (2001).
60. J. Garín, J. Orduna and R. Andreu, *Recent Res. Devel. Organic Chem.*, **5**, 77 (2001).
61. H. E. Katz, K. D. Singer, J. E. Sohn, C. W. Dirk, L. A. King and H. M. Gordon, *J. Am. Chem. Soc.*, **109**, 6561 (1987).
62. M. Blanchard-Desce, I. Ledoux, J.-M. Lehn, J. Malthête and J. Zyss, *J. Chem. Soc., Chem. Commun.*, 737 (1988).
63. a) B. J. Coe, *Chem. Eur. J.*, **5**, 2464 (1999); b) M. Malaun, Z. R. Reeves, R. L. Paul, J. C. Jeffery, J. A. McCleverty, M. D. Ward, I. Asselberghs, K. Clays and A. Persoons, *Chem. Commun.*, 49 (2001).
64. a) R. Andreu, I. Malfant, P. G. Lacroix, H. Gornitzka and K. Nakatani, *Chem. Mater.*, **11**, 840 (1999); b) P. G. Lacroix, *Chem. Mater.*, **13**, 3495 (2001).
65. J. Garín, J. Orduna, S. Uriel, A. J. Moore, M. R. Bryce, S. Wegener, D. S. Yufit and J. A. K. Howard,

Synthesis, 489 (1994).
66. M. González, N. Martín, J. L. Segura, J. Garín and J. Orduna, *Tetrahedron Lett.*, **39**, 3269 (1998).
67. R. Andreu, A. I. de Lucas, J. Garín, N. Martín, J. Orduna, L. Sánchez and C. Seoane, *Synth. Met.*, **86**, 1817 (1997).
68. A. I. De Lucas, N. Martín, L. Sánchez, C. Seoane, R. Andreu, J. Garín, J. Orduna, R. Alcalá and B. Villacampa, *Tetrahedron*, **54**, 4655 (1998).
69. M. R. Bryce, A. Green, A. J. Moore, D. F. Perepichka, A. S. Batsanov, J. A. K. Howard, I. Ledoux-Rak, M. González, N. Martín, J. L. Segura, J. Garín, J. Orduna, R. Alcalá and B. Villacampa, *Eur. J. Org. Chem.*, 1927 (2001).
70. M. González, J. L. Segura, C. Seoane, N. Martín, J. Garín, J. Orduna, R. Alcalá, B. Villacampa, V. Hernández and J. T. López Navarrete, *J. Org. Chem.*, **66**, 8872 (2001).
71. J. Garín, J. Orduna, J. I. Rupérez, R. Alcalá, B. Villacampa, C. Sánchez, N. Martín, J. L. Segura and M. González, *Tetrahedron Lett.*, **39**, 3577 (1998).
72. M. González, N. Martín, J. L. Segura, C. Seoane, J. Garín, J. Orduna, R. Alcalá, C. Sáchez and B. Villacampa, *Tetrahedron Lett.*, **40**, 8599 (1999).
73. M. Moreno-Mañas, R. Pleixats, R. Andreu, J. Garín, J. Orduna, B. Villacampa, E. Levillain and M. Sallé, *J. Mater. Chem.*, **11**, 374 (2001).
74. J. Fabian and H. Hartmann, *Light Absorption of Organic Colorants*, Springer, Berlin (1980), p. 124.
75. H. Meier, R. Petermann and J. Gerold, *Chem. Commun.*, 977 (1999).
76. R. Andreu, J. Garín and J. Orduna, *Tetrahedron*, **57**, 7883 (2001).
77. a) M. A. Herranz, N. Martín, L. Sánchez, J. Garín, J. Orduna, R. Alcalá, B. Villacampa and C. Sánchez, *Tetrahedron*, **54**, 11651 (1998); b) M. Otero, M. A. Herranz, C. Seoane, N. Martín, J. Garín, J. Orduna, R. Alcalá and B. Villacampa, *Tetrahedron*, **58**, 7463 (2002).
78. R. Andreu, I. Malfant, P. G. Lacroix and P. Cassoux, *Eur. J. Org. Chem.*, 737 (2000).
79. R. Andreu, I. Malfant, P. G. Lacroix and P. Cassoux, *Synth. Met.*, **102**, 1575 (1999).
80. R. Andreu, J. Garín, J. Orduna, G. J. Ashwell, M. A. Amiri and R. Hamilton, *Thin Solid Films*, **408**, 236 (2002).
81. L. M. Goldenberg, J. Y. Becker, O. Paz-Tal Levi, V. Y. Khodorkovsky, L. M. Shapiro, M. R. Bryce, J. P. Cresswell and M. C. Petty, *J. Mater. Chem.*, **7**, 901 (1997).
82. W. Fu, H.-X. Zhang, J.-K. Feng and W. Li, *Can. J. Chem.*, **79**, 1366 (2001).
83. a) B. Sahraoui, R. Chevalier, G. Rivoire, J. Zaremba and M. Sallé, *Optics Commun.*, **135**, 109 (1997); b) N. Terkia-Dedra, R. Andreu, M. Sallé, E. Levillain, J. Orduna, J. Garín, E. Ortí, R. Viruela, R. Pou-Amérigo, B. Sahraoui, A. Gorgues, J.-F. Favard and A. Riou, *Chem. Eur. J.*, **6**, 1199 (2000).
84. M. Asakawa, P. R. Ashton, V. Balzani, A. Credi, C. Hamers, G. Mattersteig, M. Montalti, A. N. Shipway, N. Spencer, J. F. Soddart, M. S. Tolley, M. Venturi, A. J. P. White and D. J. Williams, *Angew. Chem. Int. Ed.*, **37**, 333 (1998) and references cited therein.
85. X. Marguerettaz, A. Merrins and D. Fitzmaurice, *J. Mater. Chem.*, **8**, 2157 (1998).
86. a) J.-M. Lehn, *Angew. Chem. Int. Ed. Engl.*, **27**, 89 (1988); b) D. J. Cram, *Angew. Chem. Int. Ed. Engl.*, **27**, 1009 (1988); c) C. J. Pedersen, *Angew. Chem. Int. Ed. Engl.*, **27**, 1021 (1988).
87. As representative recent examples, see: a) E. Prasad and K. R. Gopidas, *J. Am. Chem. Soc.*, **122**, 3191 (2000); b) T. H. Ghaddar, E. W. Castner and S. Isied, *J. Am. Chem. Soc.*, **122**, 1233 (2000); c) R. T. Hayes, M. R. Wasielewski and D. Gosztola, *J. Am. Chem. Soc.*, **122**, 5563 (2000).
88. a) L. M. Goldenberg and O. Neilands, *J. Electroanal. Chem.*, **463**, 212 (1999); b) A. S. F. Boyd, G. Cooke, F. M. A. Duclairoir and V. M. Rotello, *Tetrahedron Lett.*, **44**, 303 (2003).

17

TTF Polymers and Dendritic TTFs

17.1 General Aspects and Background

Prior to this review, work on tetrathiafulvalene (TTF) polymers has been summarized in the context of synthetic TTF chemistry[1] and as part of general reviews[2] and more specific articles on multi-TTFs[3] and dendritic macromolecules.[4] This chapter covers advances in TTF polymers and dendrimers published up to June 2002, with emphasis on recent results.

The combination of properties which make TTF a special building block for polymer systems are these: (i) oxidation of the TTF ring system to the radical cation and dication species occurs sequentially and reversibly within a very accessible potential window in a range of organic solvents (for unsubstituted TTF, $E_1^{1/2} = +0.34$ and $E_2^{1/2} = +0.78$ V, *vs.* Ag/AgCl in acetonitrile); (ii) the oxidation potentials can be finely tuned by the attachment of appropriate substituents; (iii) oxidized TTF species are thermodynamically stable (due to 6π-electron heteroaromaticity of the 1,3-dithiolium cation); (iv) TTF derivatives and their derived cation radicals readily form dimers, highly-ordered stacks or two-dimensional sheets, which are stabilized by intermolecular π-π interactions and non-bonded sulfur···sulfur interactions; and (v) TTF is stable to many synthetic transformations, although it is important to avoid strongly acidic conditions and oxidizing agents.

Charge-transfer (CT) complexes and radical anion salts which display outstanding electronic properties, including superconductivity,[5] tend to be brittle and unprocessable as single crystals or amorphous powders. From the early days of monomeric TTF chemistry, it has been recognized that this problem might be overcome by incorporation of TTF units into polymers that possess the beneficial properties of good processability and thin film formation. However, polymeric TTFs have been developed only sporadically, mainly because of the synthetic challenge of obtaining suitably functionalized TTF units in reasonable quantities. In the last decade this barrier has been overcome with major progress in three branches of synthetic TTF chemistry: (i) TTF can now be readily synthesized in 20 g batches[6]; (ii) methodology involving electrophilic substitution of lithiated-TTF has been optimized to afford a wide range of mono-substituted derivatives in high yield[7]; and (iii) selective protection/deprotection chemistry of the thiolate groups of the 1,3-dithiole-2-thione-4,5-dithiolate (dmit) system (and derived TTFs) has been developed, primarily by Becher *et al.*[8] to provide multi-gram quantities of versatile building blocks.

17.2 Side-chain Polymers

Early attempts gave materials which were generally quite intractable and hence poorly characterized. Kaplan *et al.* established that TTF is an inhibitor of anionic and free radical

3 spacer = -C$_{12}$H$_{24}$- or -(CH$_2$)$_2$O(CH$_2$)$_2$O(CH$_2$)$_2$-

2 n = 0.06, m = 0.94

polymerization. Polymer **1** was obtained in quantitative yield by thermolysis (130 °C, 3 days, sealed tube) of (4-vinylphenyl)TTF. Polymer **1** was brittle and insoluble in all the usual organic solvents. After bromine doping only low room-temperature conductivity (σ_{rt}) was observed (σ_{rt} = 5 x 10^{-7} S cm^{-1}) attributed to low tacticity preventing extended overlap of TTF units.[9]

Pittman *et al.* copolymerized TTF-phenylmethacrylate with methyl methacrylate (AIBN, 60 °C, DMF) to obtain copolymer **2** (21 % yield), which was thermally stable up to *ca.* 250 °C and soluble in a range of organic solvents. Complexes with bromine and DDQ were obtained, but no conductivity data were reported.[10]

Kaufman *et al.* sought to obtain TTF polymer-modified electrodes by spin coating. Films of a TTF substituted polystyrene copolymer physically adsorbed on a platinum electrode were stable to cycling between –0.2 and +0.7 V many thousands of times without any decrease in the amount of current passed. Coulometric data suggested that all the TTF units in the film could be oxidized regardless of film thickness.[11]

Müllen and coworkers studied a series of side-chain TTF polymers, *e.g.*, **3**, obtained from reaction of poly(hydroxystyrene) with the appropriate terminal iodo-substituted TTF derivative under basic conditions. Polymers **3** exhibited no liquid crystalline phase, according to differential scanning calorimetry (DSC) and polarization microscopy measurements, and the complexes obtained after doping in solution (with DDQ, TCNQF$_4$ and iodine) showed low conductivity (σ_{rt} = 10^{-9} S cm^{-1}).[12]

Shimizu and Yamamoto polymerized 2-ethynylTTF using Rh(μ-Cl)(η^4-nbd)$_2$ (nbd = norbornadiene) as the catalyst in the presence of NEt$_3$ at 20 °C in THF to obtain polymer **4** (47% yield) with M_n = 11,700. The polymer is soluble in DMSO and DMF. The π-π* absorption band of the monomer (λ_{max} 395 nm) was replaced by a broad band at λ_{max} 490 nm in the polymer **4** due to the expansion of the π-system. In the cyclic valtammetry (CV) of cast films of the polymer **4** the two TTF redox waves were clearly seen although they partly overlapped [$E_{pa}(1) - E_{pa}(2)$ = 240 mV]. Compressed pellets of a CT complex of polymer **4** had σ_{rt} = 2.1 x 10^{-3} S cm^{-1}.[13]

We synthesized a prototype TTF-σ-thiophene system **5** and proposed that long-range order in such hybrid poly(thiophene)s might be modulated by stacking of the pendant TTF units.[14] However, electropolymerization of the monomer did not proceed smoothly; the orange color of **5** suggested a short conjugation length. This idea has been successfully pursued by Roncali *et al.* who sought to meet the stringent electronic and steric requirements needed for electronic conductivity by two parallel charge-transport mechanisms: *viz.*, migration along mixed valence TTF stacks and polaron/bipolaron conduction in the poly(thiophene) backbone.[15] Replacement of thiophene by bithiophene enabled polymerization to proceed at much lower potentials, and

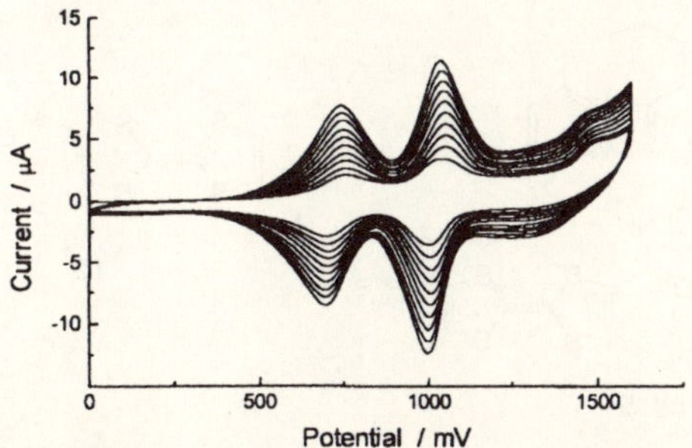

compounds **6** and **7** gave interesting results. The CVs of these polymers exhibit an additional anodic wave between the two TTF oxidation waves, and the redox and spectroelectrochemical features were explained by self-organization among the TTF units.

Skabara *et al.* obtained polymer **8** by electropolymerization of the terthienyl monomer.[16] Fig. 17.1 shows the electropolymerization of monomer **8**. The electronic absorption spectra of polymer **8** on indium tin oxide (ITO) glass indicated that the as-grown polymer remains in its oxidized state. The neutral polymer could be obtained by cycling of the film over the range –0.3 to 0 V. The experimentally determined optical bandgap of neutral polymer **8** is 1.69 eV. Monomer **9** did not polymerize electrochemically, suggesting that the 1,4-dithiin bridge in **8** serves to provide favourable spin density for polymerization to proceed.[16] Recently, polymer **10** was obtained in a high yield by chemical polymerization [Ni(0) catalyst, cyclooctatetraene, 2,2'-bipyridyl, DMF] of the corresponding 2,5-dibromothiophene monomer. This is the first directly fused TTF-thiophene polymer. Polymer **10** has poor solubility in common organic solvents and

Fig. 17.1   Electropolymerization of monomer **8** in dichloromethane-hexane (2:1) over 10 cycles. (Reprinted with permission from ref. 16)

molecular weight data could not be obtained for the bulk material. From the electronic absorption spectrum of a film on quartz the band gap was estimated to be 1.77 eV. Broad redox waves were observed for the polymer on a gold electrode consistent with both TTF (0.59 and 0.78 V) and polythiophene (1.07 V, *vs.* Ag/AgCl in MeCN) components. Films of polymer **10** doped with TCNQ had $\sigma_{rt} = 0.1$ S cm^{-1}.[17]

Rotello *et al.* have studied TTF-containing polymer **11** in a different context, *viz.*, redox-controlled molecular recognition.[18] Upon addition of a monomeric or polymeric anthracene derivative the first oxidation of **11** becomes easier (by -10 to -26 mV) whereas TTF dication formation becomes harder (by $+24$ mV). These changes are ascribed to interactions between (neutral) anthracene and TTF^0 and TTF$^{+\bullet}$, respectively. As noted by the authors, it is surprising that the first oxidation potential is lowered, as both TTF and anthracene are electron-rich moieties. The fluorescence of the anthracene derivatives in solution was quenched by the addition of polymer **11** (although the monomer TTF-C(O)NHC$_{12}$H$_{25}$ was a more efficient quencher). When polymer **11** was oxidized prior to the addition of the anthracene derivatives, a steeper initial decrease in fluorescence intensity was observed, which is indicative that the π-deficient TTF$^{+\bullet}$ units bind more strongly than TTF^0 units to the π-rich anthracene systems.

Polymer resin-supported TTF reagent **12** has been used by Murphy, Sherrington *et al.* in radical-polar crossover reactions of importance in organic synthesis, *e.g.*, the transformation **13** $\rightarrow$ **14** (Scheme 17.1).[19] Isolation of the product was easier than for analogous solution phase experiments, as all the TTF-related products were retained on the polymer support. The recovered polymer had the TTF units primarily in the radical cation state and could be reactivated by treatment with sodium borohydride and used in further cycles.

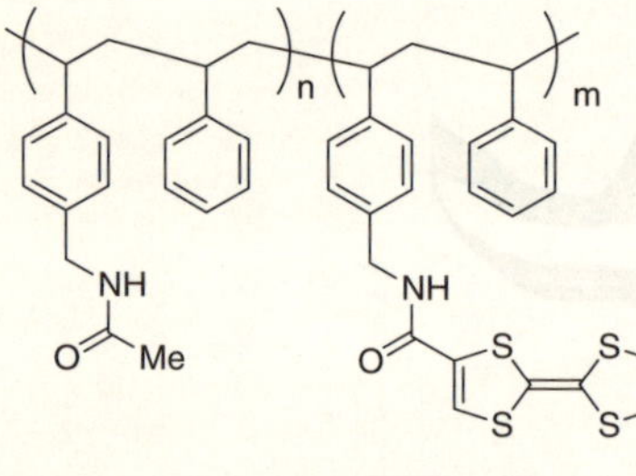

11 n = *ca.* 15; m = *ca.* 5

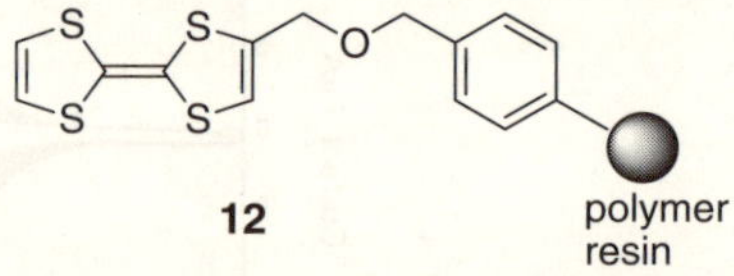

12

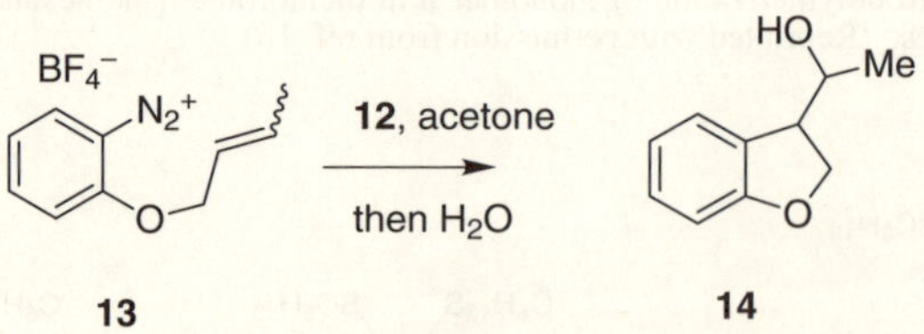

polymer resin

Scheme 17.1

17.3   Main-chain Polymers

In early work Hertler prepared an insoluble TTF-containing polyurethane by polycondensation of

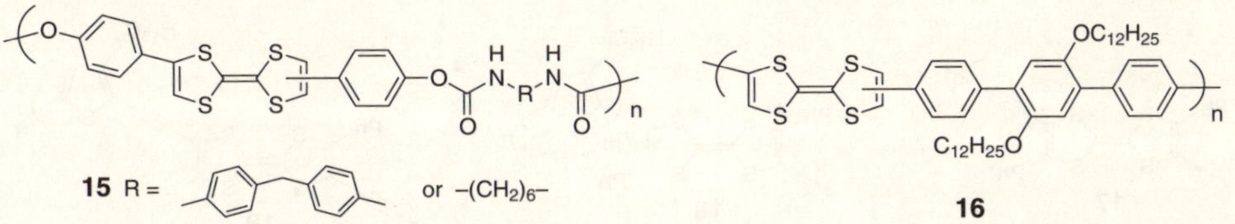

TTF(NCO)$_2$ and TTF(CH$_2$OH)$_2$: low conductivity was reported for an iodine complex.[20] Pittman *et al.* similarly obtained polyurethanes **15** and polysulfonates by polycondensation of 4,4'(5')-bis(4-hydroxyphenyl)TTF with diisocyanates and disulfonyl chlorides, respectively.[21] These polymers were soluble in CF$_3$CO$_2$H, DMF, DMSO and HMPA. Bromide salts were prepared.

More recently, Tamura *et al.* prepared the soluble polymer **16** (M_w = 36,700) from 4,4'(5')-bis(4-bromophenyl)TTF and 1,4-dialkoxybenzene-2,5-diboronic acid under Suzuki conditions.[22] Polymer films were obtained by spin coating from chloroform solution. CV data for the films on ITO glass showed the two characteristic TTF redox couples, with reversible color changes from yellow, blue and red-purple depending on the applied potential (Fig. 17.2). This electrochromic response was very rapid and was reproducible over 1000 cycles.

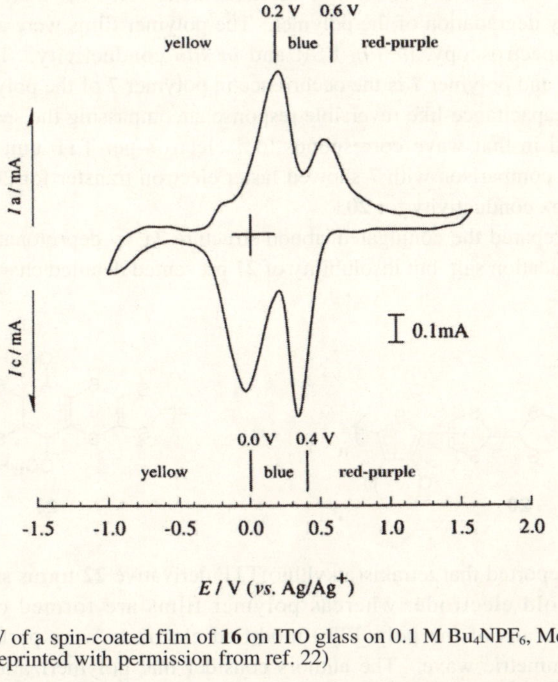

Fig. 17.2 CV of a spin-coated film of **16** on ITO glass on 0.1 M Bu$_4$NPF$_6$, MeCN solution. (Reprinted with permission from ref. 22)

Main-chain polymeric TTF derivative **17** and hybrid TTF-thiophenes **18** and **19** have been studied by Yamamoto and Shimizu.[23] Extended π-conjugation in **17** and **19** (compared to their precursor monomers and polymer **18**) is evident from a red shift in their UV-vis absorption spectra: the ethynyl spacer relieves steric repulsion and hence reduces twisting within the backbone. Limited solubility of these polymers was reported and electroactivity of the TTF units was observed for cast films. Iodine doping raised the conductivity of the polymers by *ca.* four orders of magnitude to *ca.* 10^{-3} S cm^{-1}.

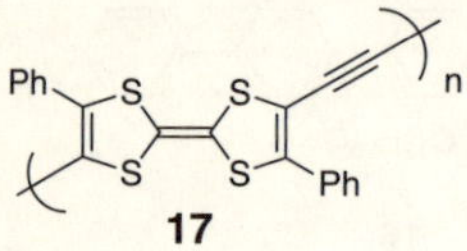

Zotti *et al.* have studied the new TTF–bis(EDOT) (EDOT = ethylenedioxythiophene) regular copolymer **20** prepared by anodic coupling.[24] It was recognized that the TTF spacer, which interrupts conjugation in its neutral state, could allow conjugation along the chain in its dication state. The ethylenedioxy moieties enhance stability due to capping of the thienyl β-positions and lowering of the oxidation potential. The first oxidation potentials of the separate TTF and bis(EDOT) subunits are very different (-0.03 and +0.51 V, respectively, *vs.* Ag/Ag$^+$). Initial TTF oxidation in polymer **20** causes an electron deficiency in the chain, with a concomitant positive shift of the intrachain bis(EDOT) oxidation beyond the limit of sulfur overoxidation to form the sulfone. Thus the response of the polymer up to +1.2 V (*vs.* Ag/Ag$^+$) beyond which it becomes unstable to potential cycling involves the TTF moieties alone. At +1.5 V the thiophene oxidation is seen to accompany degradation of the polymer. The polymer films were also characterized by FTIR and UV-vis spectroscopy, *in situ* ESR and *in situ* conductivity. The main difference between polymer **20** and polymer **7** is the occurrence in polymer **7** of the polythiophene oxidation which appears as a capacitance-like reversible response encompassing the second TTF oxidation. The charge involved in that wave corresponds to 1 electron per TTF unit + 0.3 electrons per thiophene unit.[24] A comparison with **7** showed faster electron transfer for **20** at both TTF redox steps and higher redox conductivity for **20**.

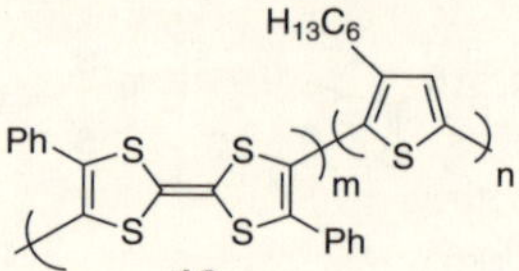

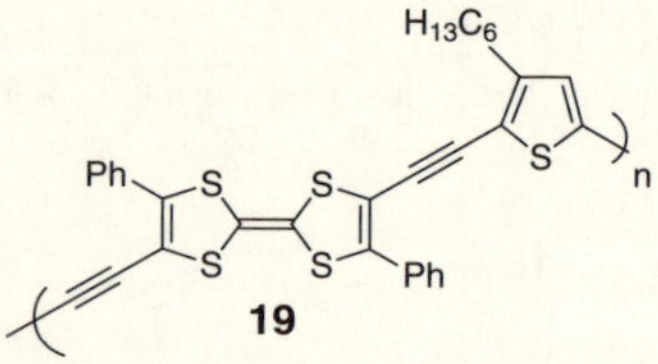

Müllen *et al.* prepared the conjugated ribbon structure **21** by deprotonation of the precursor bis(1,3-dithiolium) dication salt, but insolubility of **21** prevented detailed characterization.[12]

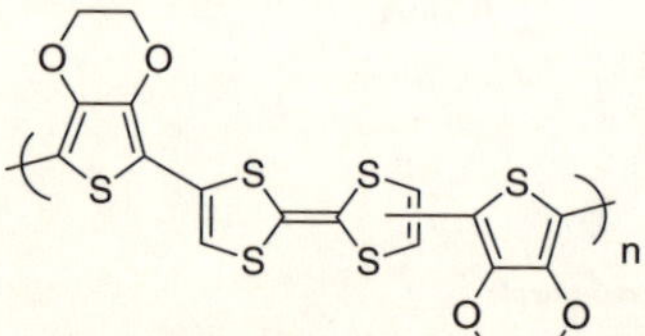

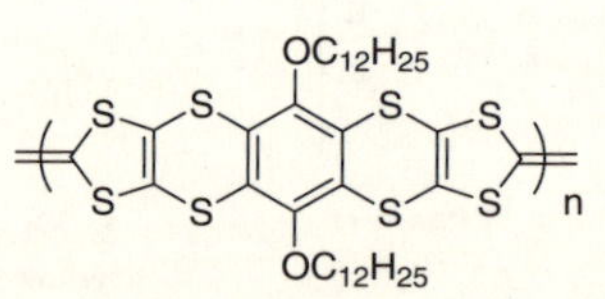

Fujihara *et al.* reported that tetrakis(alkylthio)TTF derivative **22** forms stable self-assembled monolayers on a gold electrode, whereas polymer films are formed on a glassy carbon electrode.[25] Many scans of the poly(TTF) modified electrode were performed without any change in the voltammetric wave. The authors consider that polymerization may proceed *via* formation of disulfide linkages upon oxidation of the thiol groups.

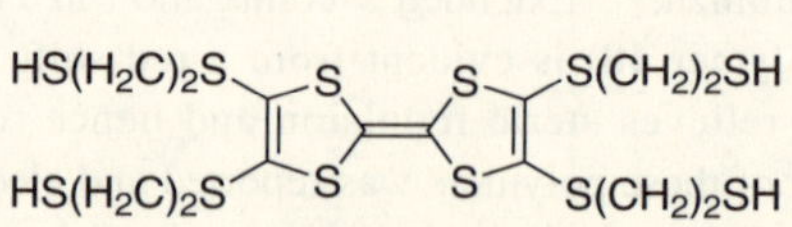

Polymers prepared by electrochemical oxidation of various linearly extended (vinylogous) TTF systems,[26-28] *e.g.*, **23**[26] and **24**,[28] have been reported. In neutral polymer **24**, studied by Lorcy *et al.*, the vinylogous TTF core will not be planar, whereas in its oxidized states this unit should become planar with the fluorene groups located orthogonal to it.[28] In the CV sweeps p-doping of the polyfluorene chain is observed together with formation of cation radicals and dications of the vinylogous TTF units. Polymer **24** is photoluminescent, as usual for fluorene-containing polymers, which may offer applications in optoelectronic devices.

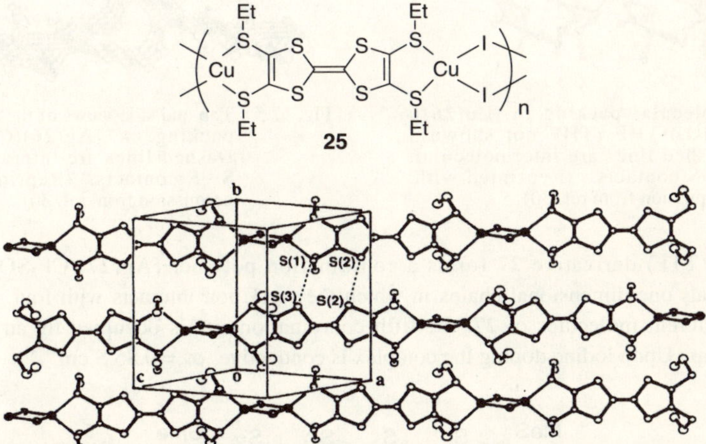

17.4   Metal Coordination Polymers

TTF derivatives, especially those containing thioalkyl substituents, have been coordinated to metal salts to obtain metallo-TTF polymers with interesting structural properties. Reaction of copper(I) iodide with tetrakis(ethylthio)TTF gave the polymer [(CuI)$_2$TTF(SEt)$_4$] **25** as a neutral 2:1 metal:ligand system.[29] X-Ray crystal analysis established that each copper atom is tetrahedrally coordinated by two bridging iodide ions and two sulfur atoms from SEt groups of the ligand. The crystal structure of **25** is shown in Fig. 17.3. The iodine-doped polymer is a semiconductor, $\sigma_{rt} = 2 \times 10^{-3}$ S cm^{-1}.

Fig. 17.3   Polymeric chain arrangement in (CuI)$_2$TTF(SEt)$_4$. Dashed lines are intermolecular S$\cdots$S contacts. (Reprinted with permission from ref. 29)

Munakata *et al.* subsequently reported a two-dimensional coordination polymer of **26** of formula [Cu(**26**)](ClO$_4$)•THF, formed by alternate four-coordinate copper ions and tridentate TTF molecules **26**.[30] The distorted tetrahedral coordination around Cu comprises two sulfur atoms of SMe groups and two nitrogen atoms of three separate molecules of **26** (Fig. 17.4). A silver coordination polymer of **26** [Ag(**26**)(CF$_3$SO$_3$)] has a different structure: it consists of an infinite linear chain arrangement with each **26** bridging two Ag centers (Fig. 17.5), rather than three Cu centers, as in the Cu complex.[30] Both these complexes of **26** and their oxidized species are poor electrical conductors.

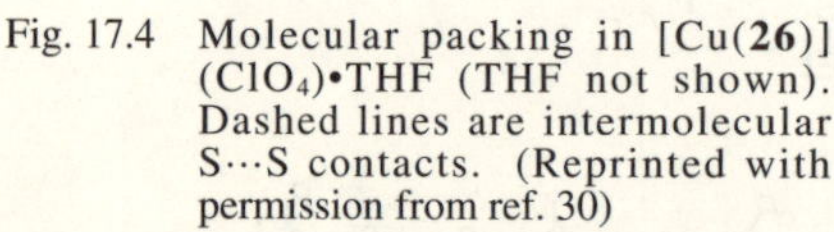

26

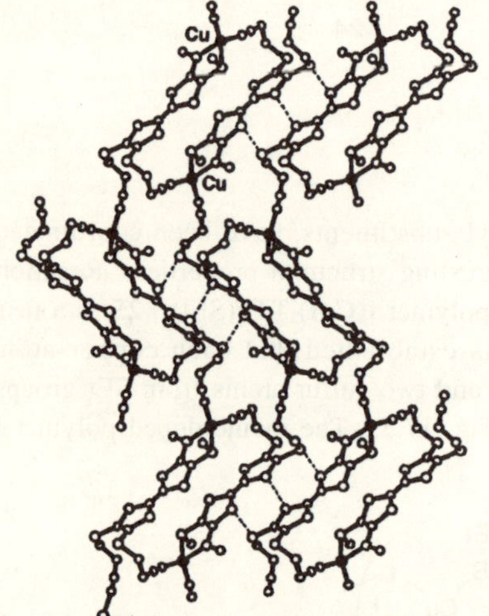

Fig. 17.4   Molecular packing in [Cu(**26**)] (ClO$_4$)•THF (THF not shown). Dashed lines are intermolecular S···S contacts. (Reprinted with permission from ref. 30)

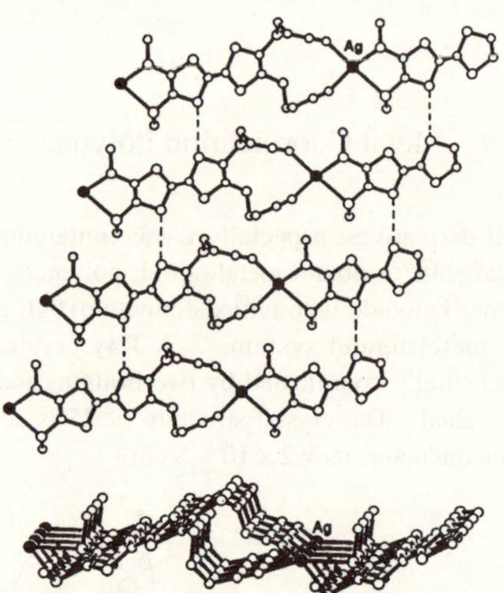

Fig. 17.5   Top and side views of the molecular packing in [Ag(**26**)(CF$_3$SO$_3$)]. Dashed lines are intermolecular S···S contacts. (Reprinted with permission from ref. 30)

The bis(TTF) derivative **27** forms a coordination polymer [Ag(**27**)(CF$_3$SO$_3$)]. X-Ray analysis reveals one-dimensional chains in which the Ag(I) ion interacts with four sulfur atoms from two different molecules of **27**: the fifth coordination site is occupied by an O from the CF$_3$SO$_3^-$ group. Upon iodine doping the complex is conductive, $\sigma_{rt} = 0.85$ S cm^{-1}.[31]

27

17.5 Dendritic TTFs

The study of dendrimers, cascade molecules and related hyper-branched systems is a rapidly expanding topic in macromolecular science.[32] They are structurally ordered monodisperse polymers with their size and architecture being precisely regulated in their synthesis. They provide unique molecular frameworks for placing a high density of functional groups in predetermined spatial arrangements. In the context of functional dendrimers, a variety of redox-active substituents have been built into the structures.[33] These materials are relevant to the development of: (i) new electron-transfer catalysts; (ii) studies on the dynamics of electron transport at surfaces and within restricted reaction spaces; (iii) new materials for charge storage and energy conversion; (iv) organic semiconductors; and (v) mimics of biological redox processes. The redox groups may behave independently in multi-electron processes (n identical non-interacting electroactive centers giving rise to a single n-electron wave) or they may interact intra- or intermolecularly, in which case overlapping or closely-spaced redox waves are observed at different potentials. Compared to analogous linear polymers, TTF dendrimers generally enjoy good solubility in a range of organic solvents, which has enabled their properties to be well characterized.

The first TTF dendrimers possessed peripheral TTF groups built upon arylester core and branch units and they were surface-functionalized with up to 12 peripheral TTF groups.[34] Within this series of macromolecules, stability decreased with increasing generation, so no attempts were made to assemble molecules of higher molecular weight than **28**. Compound **28** is an oil which is

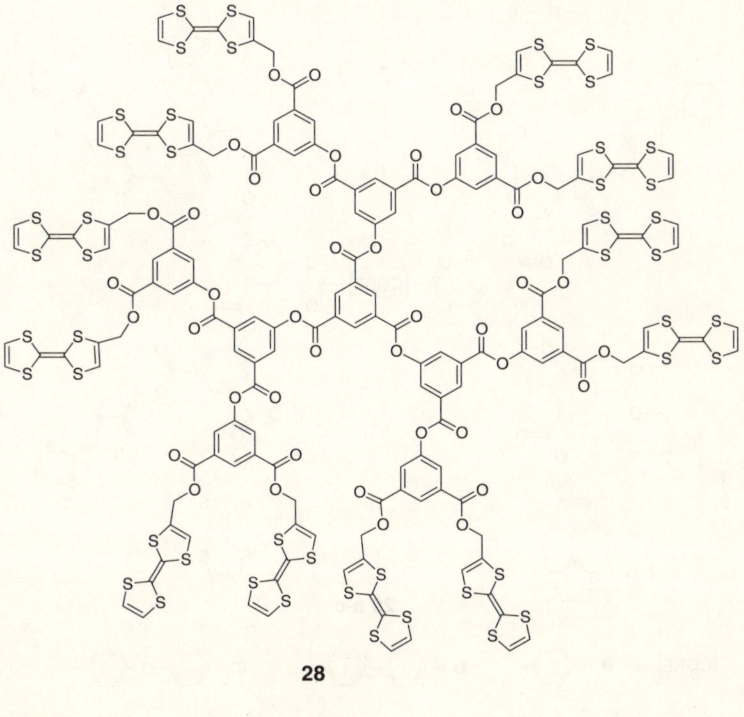

28

readily soluble in polar organic solvents and is reasonably stable to storage at < 0 °C; ^{1}H NMR spectroscopy and plasma desorption mass spectra (PDMS) were entirely consistent with the monodisperse structure **28**.

The stability of (TTF)$_x$ arylester dendrimers was improved by changing from trifunctional to bifunctional core units, which gave more "open" structures. Thus, the series of (TTF)$_8$ systems **29a-c** was constructed using terephthaloyl chloride, 4,4'-biphenyldiacid chloride and 4,4'-biphenyletherdiacid chloride as core reagents.[35] These three compounds were all stable upon storage at room temperature in air for at least one year. However, **29a,b** were only sparingly soluble in organic solvents, whereas analogue **29c**, with the more flexible biphenylether core unit, showed good solubility in polar organic solvents (*e.g.*, acetone and dichloromethane). We have also used these multi-TTF dendrons in the synthesis of the polyester "co-block" (TTF)$_4$(AQ)$_2$ **30** and (TTF)$_8$(AQ)$_4$ dendrimers (AQ = anthraquinone).[36]

The use of entirely different building blocks afforded another series of TTF dendrimers with *p*-xylyl spacers, *e.g.*, compound **31** containing thirteen TTF units.[37] Compound **31** was characterized unambiguously by a combination of elemental analysis, matrix-assisted laser desorption time-of-flight (MALDI TOF) mass spectrometry and ^{1}H NMR spectrometry. Thus for the first time TTF units were placed at all layers of the structural hierarchy of a dendrimer, including the core. The (TTF)$_{21}$ glycol system **37** has TTF units at the branch points and at the periphery.[38] Electrochemically controlled supramolecular interactions have been shown to occur between **37** and an electron-rich phenyleneglycol oligomer. Electrochemical oxidation of the dendrimer triggers the interactions which have been monitored by CV data and fluorescence spectroscopy.[38c] All the above systems **28-31** and **37** were obtained by convergent methodology.

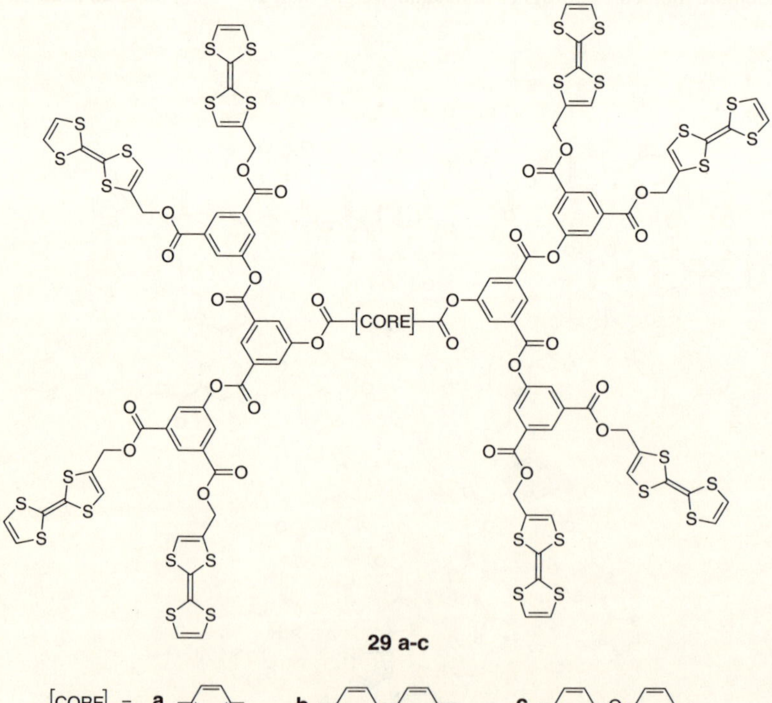

29 a-c

$[CORE]$ =   **a** ⟨benzene-1,4-diyl⟩   **b** ⟨biphenyl-4,4'-diyl⟩   **c** ⟨diphenylether-4,4'-diyl⟩

30

(In a convergent synthesis, dendrimer construction begins at what will become the surface of the molecule, and progresses inwards *via* a series of dendron wedges, which are attached to the core unit in the final step.) This is illustrated in the synthesis of **37** (Scheme 17.2). The synthesis of these dendrimers has become possible only in the light of the advances in the functionalization of TTF units, referred to in Section 17.1. 4-(Hydroxymethyl)-TTF[7] served as the starting material for dendrimers **28** and **29**, and Becher's selective protection/deprotection methodology for TTF-thiolates[8] was employed in the syntheses of **31** and **37**.

In collaboration with Becher we have used hexakis(bromomethylbenzene) as a six-directional core unit in the synthesis of **38**[39] and **39**[38] using reagents **33** (6 equiv.) and **34** (6 equiv.) respectively.

The solution redox properties of all these TTF dendrimers have been evaluated by a variety of electrochemical techniques. A general feature is that all the compounds exhibit two redox couples typical of the TTF. With the exception of higher generations in the arylester series, reversibility of the redox waves was observed for all the dendrimers. Thin layer cyclic voltammetry (TLCV) techniques have been most useful because, in contrast to conventional CV, in TLCV the current is not limited by the mass transport to the electrode. Integrating the voltammetric waves against the one-electron reduction peak of 2,3-dichloronaphthoquinone (DCNQ) as the internal standard provides convincing evidence that complete oxidation occurs for all the TTF units in these compounds. We have noted that sometimes the second TTF oxidation wave is slightly narrower than the first wave, which could be due to adsorption phenomena. The data for dendrimer **37** are representative and show the generation of the 42+ oxidation state of the dendrimer[38a] (Fig. 17.6). No evidence was obtained of raised oxidation potentials for the inner TTF units, which would be expected if they were shielded by oxidized peripheral units. A special feature of **30** is that CV reveals a clean combination of TTF and anthraquinone redox behavior with switching between the +8, +4, 0, –2 and –4 redox states. We are exploring these molecules in the fields of molecular batteries, electrooptical switches, electrochromic materials and redox-controlled molecular encapsulation.

31

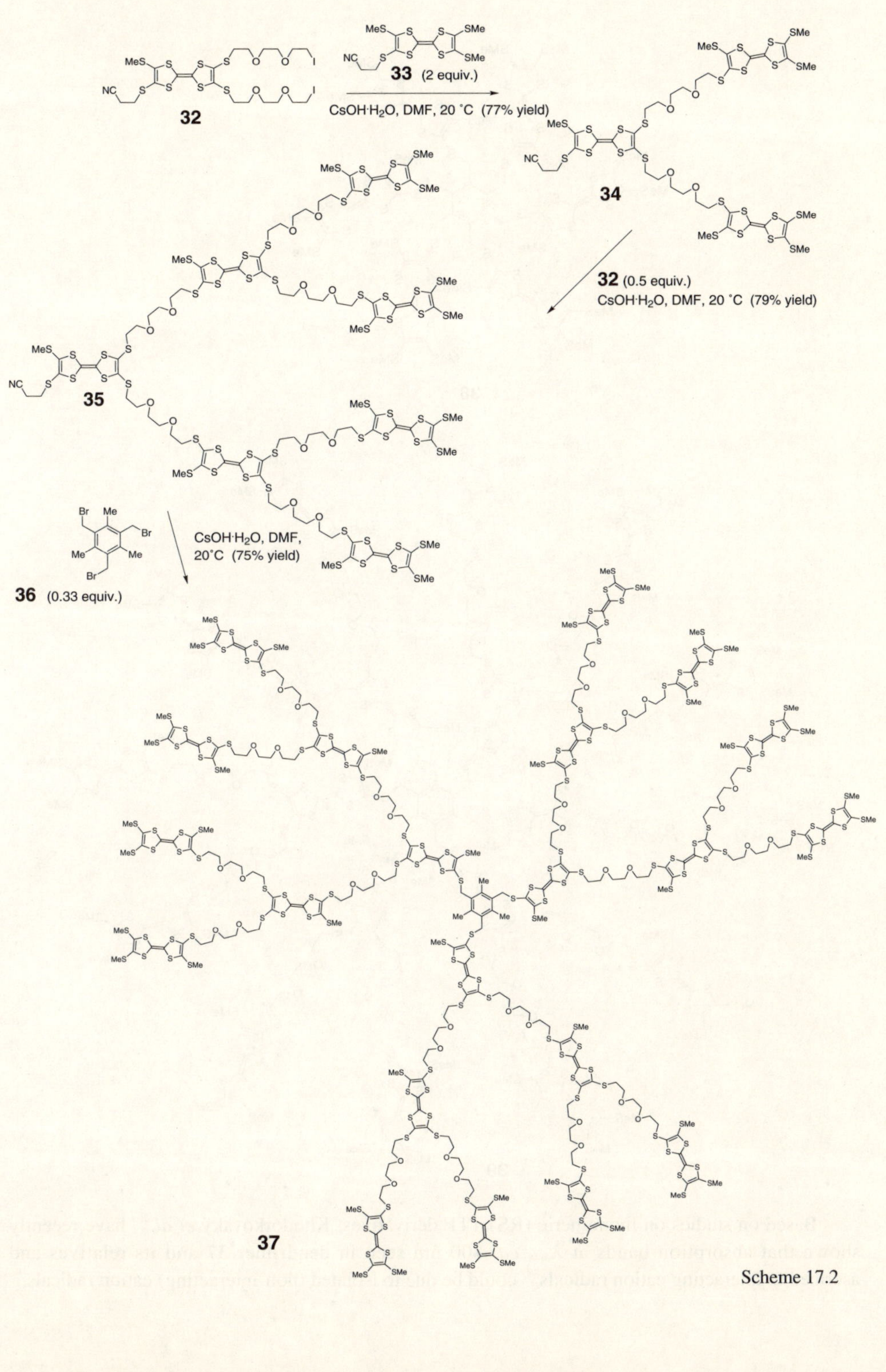

32

33 (2 equiv.)

CsOH·H₂O, DMF, 20 °C (77% yield)

34

32 (0.5 equiv.)
CsOH·H₂O, DMF, 20 °C (79% yield)

35

36 (0.33 equiv.)

CsOH·H₂O, DMF,
20°C (75% yield)

37

Scheme 17.2

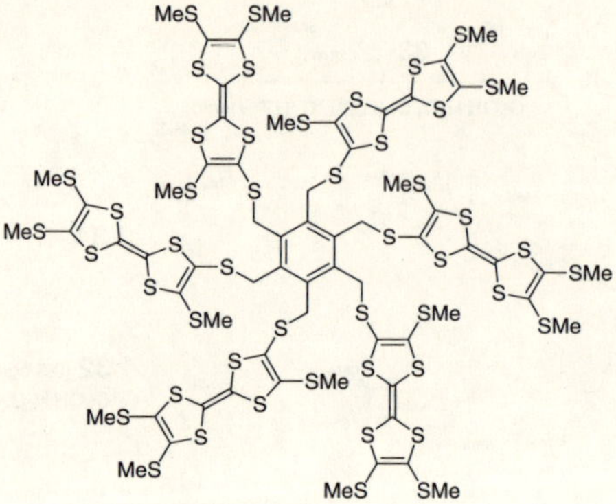

38

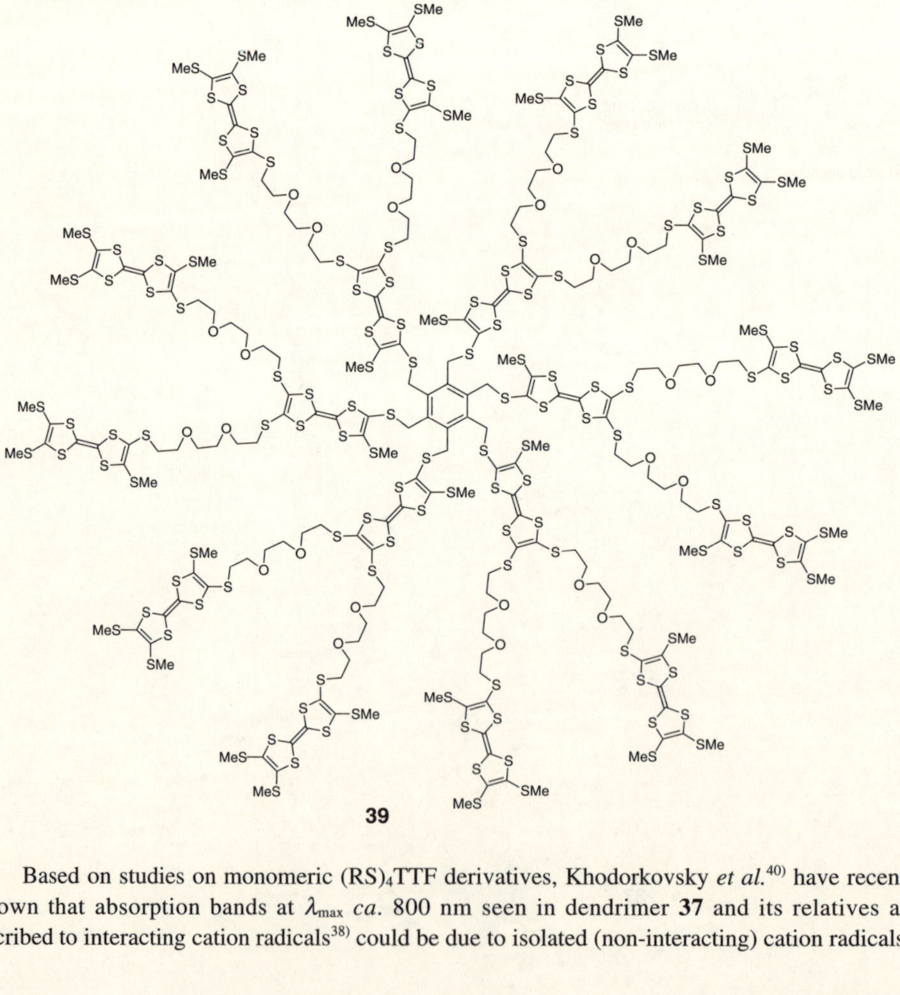

39

Based on studies on monomeric (RS)$_4$TTF derivatives, Khodorkovsky *et al.*[40)] have recently shown that absorption bands at λ_{max} *ca.* 800 nm seen in dendrimer **37** and its relatives and ascribed to interacting cation radicals[38)] could be due to isolated (non-interacting) cation radicals.

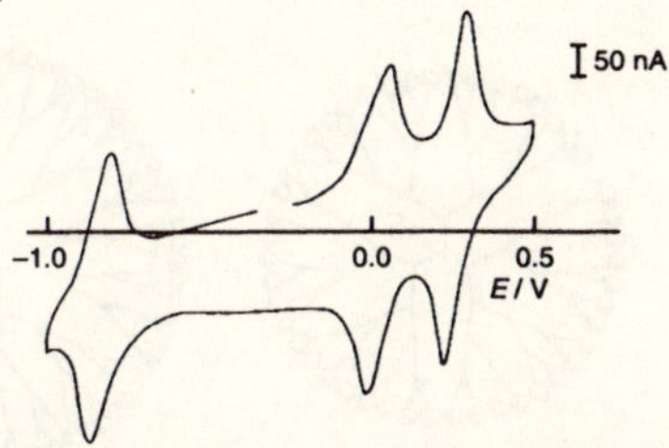

Fig. 17.6   TLCV of dendrimer **37** (0.5 x 10^{-4} M) and DCNQ (1.05 x 10^{-3} M) as internal standard (giving the cathodic wave at *ca.* –0.8 V) in Bu$_4$NClO$_4$ (1 M) CH$_2$Cl$_2$ solution, *vs.* Ag/Ag$^+$, scan rate 2 mV s^{-1}. (Reprinted with permission from ref. 38a)

These macromolecules provide some of the first all-organic dendritic poly(radical cations) and poly(dications), which can be considered as highly charged organic nanoparticles.[33b] Chemical and electrochemical doping cleanly affords TTF cation radical and dication species as judged by UV-vis spectroscopic data. A semiconducting TCNQ complex of **29c** (σ_{rt} = 2.2 x 10^{-3} S cm^{-1}) was isolated.[35]

TTF dendritic wedges, *e.g.*, **34**, have been used in our studies on the quenching of the fluorescence of phthalocyanines (pcs) for which we have prepared a series of silicon phthalocyanine bis-esters with axial ligands containing TTF units, including compound **40**.[41] It was possible to achieve selective and reversible oxidation of the TTF groups as shown by new absorption bands at λ_{max} *ca.* 450 and 850 nm arising from TTF$^{+\bullet}$ while the pc Q bands retained their shape and intensity. The intensity of the pc emission in **40** is reduced due to quenching by the electron donor TTF units by *ca.* 94% relative to a model silicon pc derivative lacking the TTF moieties. The potential for rapid electrochemical off-on switching of the pc fluorescence by oxidation of the TTF units was explored. However, upon oxidation of the TTF units the pc emission spectrum was unchanged. We tentatively explain this by a reversal of the electron transfer path, *i.e.*, from pc* to TTF$^{+\bullet}$ after oxidation with TTF$^{+\bullet}$ now acting as the electron acceptor moiety. Model compounds established that these are *intra*molecular quenching processes. This phenomenon is conceptually similar to the *inter*molecular quenching of

40

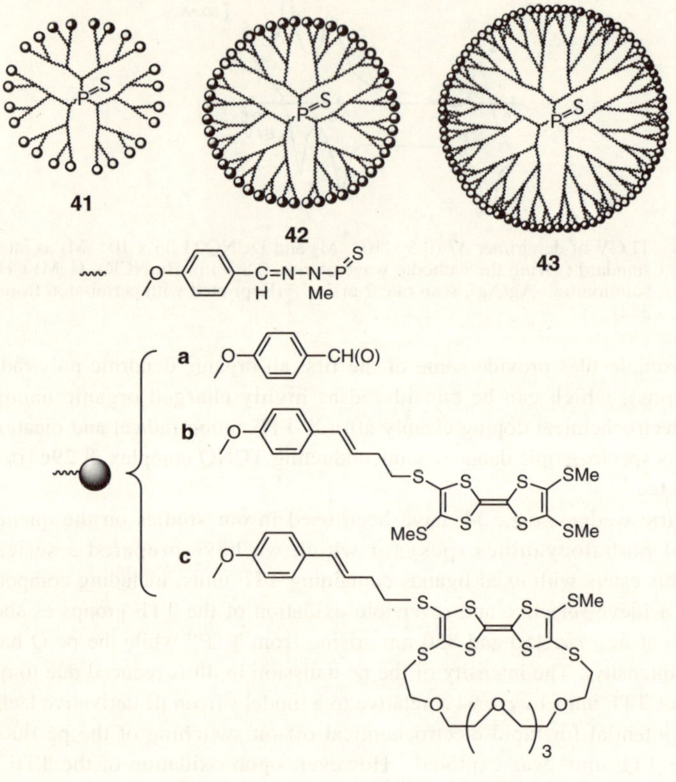

41

42

43

anthracene fluorescence by TTF/TTF$^{+\bullet}$ discussed above.[18]

LeDerf *et al.* have studied a series of dendrimers **41-43** with up to 96 TTF moieties on the periphery for fifth generation (G$_5$) systems **43**.[42] The narrow shape of the redox waves in the CVs showed a clear tendancy for these dendrimers to deposit on the electrode surface, especially from THF solution. The data were characteristic for surface-confined redox couples. Ionophoric crowns were incorporated into structures **41c**, **42c** and **43c**, and these dendrimer films displayed the same cation recognition properties as observed in homogeneous solution. For example, when an electrode modified with **41c** was exposed to a solution containing Ba^{2+} ions there was a positive shift in the first oxidation potential due to Ba^{2+} complexation [$\Delta(E_{pa1})$ = 85 mV] without affecting the reversibility of the redox wave. This is similar to solution behavior. However, in solution, E_{pa2} is unaffected by Ba^{2+}, but a small positive shift is observed for the films [$\Delta(E_{pa2})$ = 30 mV] attributed to cooperative effects between neighboring crown rings in the films (Fig. 17.7).

17.6 Summary and Outlook

Early hopes that the remarkable electrical and magnetic properties of TTF salts could be retained, or even enhanced, in processable polymeric derivatives have not been realized, and indeed the structural requirements for metallic and superconducting behavior in this class of molecular

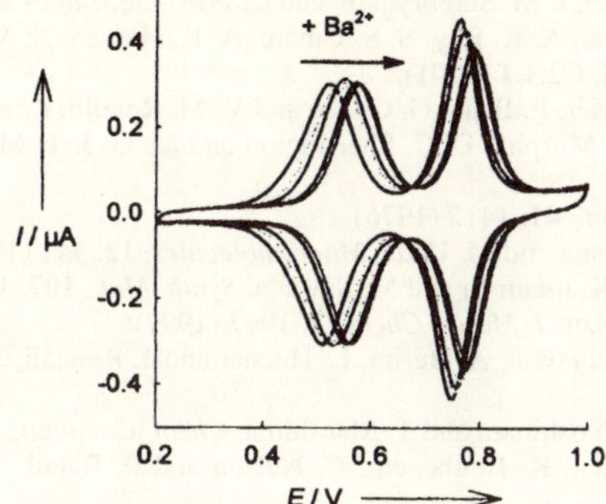

Fig. 17.7   Electrochemical response of a Pt electrode modified with dendrimer **41c**, CH_2Cl_2/CH_3CN, Bu_4NPF_6 (0.1 M), scan rate 0.1 V s^{-1}, *vs.* Ag/AgCl, in the presence of increasing concentrations of $Ba(ClO_4)_2$. (Reprinted with permission from ref. 42)

material are so stringent that this may remain an elusive goal in polymer systems. For main-chain and side-chain systems insolubility has frequently been a problem, although many semiconducting TTF containing polymers have been reported. Many soluble and well-characterized TTF dendrimers are now known. The synthetic chemistry has recently been optimized to afford practically useful quantities of the functionalized building blocks needed to incorporate TTF moieties into a wide range of polymer architectures, including dendrimers. The redox and optoelectronic properties of TTF polymers are now attracting as much attention as the electrical properties, and device applications can be envisaged, involving redox-switching, molecular recognition and charge storage and transport. There is no doubt that TTF polymers will continue to be appealing and challenging targets of relevance to several areas of organic chemistry and materials science.

References

1. G. Schukat and E. Fanghänel, *Sulfur Rep.*, **13**, 254 (1993).
2. M. R. Bryce, *J. Mater. Chem.*, **10**, 589 (2000); J. L. Segura and N. Martín, *Angew. Chem. Int. Ed.*, **40**, 1372 (2001).
3. M. Adam and K. Müllen, *Adv. Mater.*, **6**, 439 (1994).
4. M. R. Bryce, W. Devonport, L. M. Goldenberg and C. Wang, *Chem. Commun.*, 945 (1998).
5. T. Ishiguro, K. Yamaji and G. Saito, *Organic Superconductors*, 2nd ed., Springer Ser. Solid-State Sci., Vol. 88, Springer, Berlin (1998); J. Yamada, H. Nishikawa and K. Kikuchi, *J. Mater. Chem.*, **9**, 617 (1999).
6. A. J. Moore and M. R. Bryce, *Synthesis*, 407 (1997).
7. J. Garín, J. Orduna, S. Uriel, A. J. Moore, M. R. Bryce, S. Wegener, D. S. Yufit and J. A. K. Howard, *Synthesis*, 489 (1994); J. Garín, *Adv. Heterocyl. Chem.*, **62**, 249 (1995).
8. N. Svenstrup and J. Becher, *Synthesis*, 215 (1995); M. B. Nielsen and J. Becher, *Liebigs Ann. Recl.*, 2177 (1997).
9. M. L. Kaplan, R. C. Haddon, F. Wudl and E. D. Feit, *J. Org. Chem.*, **43**, 4643 (1978).
10. C. U. Pittman, Jr., M. Ueda and Y. F. Liang, *J. Org. Chem.*, **44**, 3639 (1979).
11. F. B. Kaufman, A. H. Schroeder, E. M. Engler, S. R. Kramer and J. Q. Chambers, *J. Am. Chem. Soc.*, **102**, 483 (1980).
12. S. Frenzel, S. Arndt, R. M. Gregorious and K. Müllen, *J. Mater. Chem.*, **5**, 1529 (1995).
13. T. Shimizu and T. Yamamoto, *Chem. Commun.*, 515 (1999).
14. M. R. Bryce, A. D. Chissel, J. Gopal, P. Kathirgamanathan and D. Parker, *Synth. Met.*, **39**, 397 (1991).
15. (a) C. Thobie-Gautier, A. Gorgues, M. Jubault and J. Roncali, *Macromolecules*, **6**, 4094 (1993); (b) L. Huchet, S. Akoudad and J. Roncali, *Adv. Mater.*, **10**, 541 (1998); (c) L. Huchet, S. Akoudad, E. Levillain, J. Roncali, A. Emge and P. Bäuerle, *J. Chem. Phys.*, **102**, 7776 (1998).

16. P. J. Skabara, D. M. Roberts, I. M. Serebryakov and C. Pozo-Gonzalo, *Chem. Commun.*, 1005 (2000).
17. P. J. Skabara, D. M. Roberts, A. K. Ray, S. S. Umare, A. K. Hassan, A. V. Nabok and K. Müllen, *Proc. Red. Soc. Symp. Proc.,* **665**, C2.1.1 (2001).
18. H. A. de Cremiers, G. Clavier, F. Ilhan, G. Cooke and V. M. Rotello, *Chem. Commun.*, 2232 (2001).
19. B. Patro, M. Merrett, J. A. Murphy, D. C. Sherrington and M. G. J. T. Morrison, *Tetrahedron Lett.*, **40**, 7857 (1999).
20. W. R. Hertler, *J. Org. Chem.*, **41**, 1412 (1976).
21. C. U. Pittman, Jr., Y.-F. Liang and M. Ueda, *Macromolecules*, **12**, 541 (1979).
22. H. Tamura, T. Watanabe, K. Imanishi and M. Sawada, *Synth. Met.*, **107**, 19 (1999).
23. T. Yamamoto and T. Shimizu, *J. Mater. Chem.*, **7**, 1967 (1997).
24. G. Zotti, S. Zecchin, G. Schiavon, A. Berlin, L. Huchet and J. Roncali, *J. Electroanal. Chem.*, **504**, 64 (2001).
25. H. Fujihara, H. Nakai, M. Yoshihara and T. Maeshima, *Chem. Commun.*, 737 (1999).
26. M. Fourmigué, I. Johannsen, K. Boubekeur, C. Nelson and P. Batail, *J. Am. Chem. Soc.*, **115**, 3752 (1993).
27. H. Müller, F. Salhi, B. Divisia-Blohorn, F. Genoud, T. Narayanan, M. Lorenzen and C. Ferrero, *Chem. Commun.*, 1407 (1999).
28. D. Lorcy, J. Rault-Berthelot and C. Poriel, *Electrochem. Commun.*, **2**, 382 (2000).
29. X. Gan, M. Munakata, T. Kuroda-Sowa and M. Maekawa, *Bull. Chem. Soc. Jpn.*, **67**, 3009 (1994).
30. L. P. Wu, J. Dai, M. Munakata, T. Kuroda-Sowa, M. Maekawa, Y. Suenaga and Y. Ohno, *J. Chem. Soc., Dalton Trans.*, 3255 (1998).
31. J. C. Zhong, Y. Miaski, M. Munakata, T. Kuroda-Sowa, M. Maekawa, Y. Suenaga and H. Konada, *Inorg. Chem.*, **40**, 7096 (2001).
32. G. R. Newkome, C. N. Moorefield and F. Vögtle, *Dendritic Macromolecules: Concepts, Synthesis, Perspectives,* VCH, Weinheim (1996); S. Hecht and J. M. J. Frécher, *Angew. Chem. Int. Ed.*, **40**, 74 (2001).
33. (a) M. R. Bryce and W. Devonport, *Advances in Dendritic Macromolecules*, JAI Press, London (1996), Vol. 3, p. 115; (b) N. Godbert and M. R. Bryce, *J. Mater. Chem.*, **12**, 27 (2002).
34. (a) M. R. Bryce, W. Devonport and A. J. Moore, *Angew. Chem. Int. Ed. Engl.*, **33**, 1761 (1994); (b) W. Devonport, M. R. Bryce, G. J. Marshallsay, A. J. Moore and L. M. Goldenberg, *J. Mater. Chem.*, **8**, 1361 (1998).
35. M. R. Bryce and W. Devonport, *Synth. Met.*, **76**, 305 (1996).
36. M. R. Bryce, P. de Miguel and W. Devonport, *Chem. Commun.*, 2565 (1998).
37. C. Wang, M. R. Bryce, A. S. Batsanov, L. M. Goldenberg and J. A. K. Howard, *J. Mater. Chem.*, **7**, 1189 (1997).
38. (a) C. A. Christensen, L. M. Goldenberg, M. R. Bryce and J. Becher, *Chem. Commun.*, 509 (1998); (b) C. A. Christensen, M. R. Bryce and J. Becher, *Synthesis*, 1695 (2000); (c) A. Beeby, M. R. Bryce, C. A. Christensen, G. Cooke, F. M. A. Duclairoir and V. Rotello, *Chem. Commun.*, 2950 (2002).
39. C. A. Christensen, M. R. Bryce, A. S. Batsanov and J. Becher, *Chem. Commun.*, 331 (2000).
40. V. Khodorkovsky, L. Shapiro, P. Krief, A. Shames, G. Mabon, A. Gorgues and M. Giffard, *Chem. Commun.*, 2736 (2001).
41. C. Farren, C. A. Christensen, S. FitzGerald, M. R. Bryce and A. Beeby, *J. Org. Chem.*, **67**, 9130 (2002).
42. F. LeDerf, E. Levillain, G. Trippé, A. Gorgues, M. Sallé, R.-M. Sebastían, A.-M. Caminade and J.-P. Marjoral, *Angew. Chem. Int. Ed.,* **40**, 224 (2001).

Subject Index

Compound Index